Enzyme Catalysis in Organic Synthesis
Volume III

Edited by K. Drauz and H. Waldmann

Second Edition

*Related Titles
from Wiley-VCH*

B. Cornils, W. A. Herrmann (Eds.)

Applied Homogeneous Catalysis with Organometallic Compounds

Second, Completely Revised and Enlarged Edition
Three Volumes

2000, ISBN 3-527-30434-7

B. Cornils, W. A. Herrmann, R. Schlögl, C.-H. Wong (Eds.)

Catalysis from A–Z

A Concise Encyclopedia

2000, ISBN 3-527-29855-X

D. E. De Vos, I. F. J. Vankelecom, P. A. Jacobs

Chiral Catalysts Immobilization and Recycling

2001, ISBN 3-527-19295-2

U. Th. Bornscheuer, R. J. Kazlauskas

Hydrolases in Organic Synthesis

Regio- and Stereoselective Biotransformations

1999, ISBN 3-527-30104-6

R. A. Sheldon, H. van Bekkum

Fine Chemicals through Heterogeneous Catalysis

2001, ISBN 3-527-29951-3

Enzyme Catalysis in Organic Synthesis

A Comprehensive Handbook

Volume III

Edited by
Karlheinz Drauz and Herbert Waldmann

Second, Completely Revised and Enlarged Edition

Editors:

Prof. Dr. Karlheinz Drauz
Degussa AG
1 ZN Wolfgang, Bereich FC-TRM
Rodenbacher Chaussee 4
63457 Hanau-Wolfgang
Germany

Prof. Dr. Herbert Waldmann
Max-Planck-Institut für Molekulare Physiologie
Otto-Hahn-Straße 11
44227 Dortmund
Germany

Library of Congress Card No.:
applied for

British Library Cataloguing-in-Publication Data
A catalogue record for this book is available from the British Library.

Die Deutsche Bibliothek – CIP Cataloguing-in-Publication-Data
A catalogue record for this publication is available from Die Deutsche Bibliothek

Printed on acid-free paper.

Printed in the Federal Republic of Germany.

Cover   Gesine Schulte, Max-Planck-Institut für Molekulare Physiologie, Dortmund
Composition   Typomedia, Ostfildern
Printing   Strauss Offsetdruck, Mörlenbach
Bookbinding   Buchbinderei Schaumann GmbH, Darmstadt

ISBN   3-527-29949-1

Foreword

That biological systems are masterful chemists is a fact long appreciated by those who study how living things build complexity from simple compounds in the environment. Enzymes catalyze the interconversion of vast numbers of chemical species, providing materials and energy to fuel cell survival and growth. Enzymes build the intricate natural products, which, for their potential utility in treating disease, pose almost unlimited new challenges for ambitious synthetic chemists. But, unlike most industrial chemical processes, Nature's catalysts generate few waste products and effect their transformations under mild conditions–in water, at room temperature and atmospheric pressure. Biocatalysts are models of energy-efficient, environmentally-conscious chemistry and will play a prominent role in the 21st century's chemicals industry.

The world of biocatalysis has undergone significant change in the eight years since the first edition of this handbook appeared. Most of the news is good, with enzymes showing up in many more organic syntheses and a number of important new industrial processes coming on line. Apart from continuing clever insights into how to integrate biocatalysis into synthetic chemistry, several forces are accelerating a move to biocatalytic processes. In the first place, the search for better, enantiomerically pure drugs has forced many chemists to turn to enzymes for assistance in their preparation. Ever increasing demands for environmentally acceptable processes push in the same direction. At the same time, rapidly-developing technologies for making better catalysts through genetic enginering and for discovering new catalysts are are offering new process opportunities which in the past were either not economical or not even conceivable. A plethora of new catalysts to choose from, as well as a high probability that a catalyst can be further improved during the process design and engineering phases, means that we can respond rapidly to new synthetic needs with biocatalytic solutions.

The organization of these volumes into specific technologies and transformations provides a comprehensive coverage of practical biocatalysis that no other single source provides. The work of experts in each of the fields, the individual chapters review vast relevant literature and synthesize it in order to present key concepts and many illustrative examples. This coverage should give organic chemists immediate access to the wealth of experience that has accumulated in the biocatalysis world and allow them to identify the most promising ways to use biocatalysts in their own syntheses. Biocatalysts should feature prominently in the repertoire of synthetic chemistry, and this handbook deserves a prominent place in the modern chemist's library.

Pasadena, January, 2002

Frances Arnold

Preface

Nearly eight years have passed since we the First Edition of „Enzyme Catalysis in Organic Synthesis" was issued but much of what we had written in its preface then still applies today. The application of biocatalysis in organic synthesis is a powerful technique. It has grown steadily and today this field is well-established in both academia and industry. With increasing application and acceptance the need for a comprehensive and up to date overview of the state of the art has grown. In addition numerous colleagues have approached us and asked for an update of "the Handbook".

In response to these demands and in recognition of the new and groundbreaking strides taken since the first half of the nineties the Second Edition which is now in the hand of the reader was prepared. In comparing it with the First edition one discovers that we have not changed the overall arrangement in the volumes. Therefore we continue to have a part that addresses general principles (Chapters 1–10) and another one which summarizes the application of enzymes in organic synthesis according to reaction type (Chapters 11–20). This arrangement was very well received by the readers before and we hope that it will be for the Second Edition as well.

However, the entire text was streamlined and in many cases regrouped to ensure for a better presentation. Also a few chapters which in the long run turned out to be less relevant to organic synthesis were not included again. In contrast other aspects were now integrated and attention was given to techniques of enzyme evolution, bioinformatics and enzymatic reactions in low-water media, areas that have developed with great pace and that we believe to be of major importance in the time to come.

We hope that the Second Edition of the "Handbook" will be a plentiful source of information just as valuable as the First Edition was eight years ago.

Dortmund and Hanau, February 2002　　　　　*Karlheinz Drauz, Herbert Waldmann*

Contents

14 Formation of C–C Bonds 931
Chi-Huey Wong

18 Introduction and Removal of Protecting Groups 1333

Dieter Kadereit, Reinhard Reents, Duraiswamy A. Feyaraj, and Herbert Waldmann

List of Contributors

Garabed Antranikian
Technische Universität Hamburg-
Harburg
Institut für Biotechnologie
Kasernestraße 12
21073 Hamburg
Germany

Frances H. Arnold
California Institute of Technology
Department of Chemical Engineering
MC 210–41
Pasadena CA 91125
USA

Haruyuki Atomi
Department of Synthetic Chemistry and
Biological Chemistry
Graduate School of Engineering
Kyoto University
Yoshida, Sakyo-ku
Kyoto, 606–8501
Japan

Constanzo Bertoldo
Technische Universität Hamburg-
Harburg
Institut für Biotechnologie
Kasernestraße 12
21073 Hamburg
Germany

Manfred Biselli
Fachhochschule Aachen
Abteilung Jülich
Labor für Zellkulturtechnik
Ginsterweg 1
52428 Jülich
Germany

Andreas S. Bommarius
School of Chemical Engineering
Georgia Institute of Technology
315 Ferst Drive
Atlanta, GA 30332–0363
USA

Bruno Bühler
Institut für Biotechnologie
ETH Hönggerberg, HPT
8093 Zürich
Switzerland

Nobuyoshi Esaki
Institute for Chemical Research
Kyoto University
Uji
Kyoto-fu 611
Japan

Kurt Faber
Department of Chemistry
Organic and Bioorganic Chemistry
University of Graz
Heinrichstrasse 28
8010 Graz
Austria

Duraiswamy A. Feyaraj
Max-Planck-Institut für Molekulare
Physiologie
Abteilung Chemische Biologie
Otto-Hahn-Straße 11
44227 Dortmund
Germany

Martin H. Fechter
Institut für Organische Chemie
Technische Universität Graz
Stremayrgasse 16
8010-Graz
Austria

Sabine Flitsch
Department of Chemistry
The University of Edinburgh
West Mains Road
The King's Building
Edinburgh EH9 3JJ
United Kingdom

Hans-Joachim Gais
Institut für Organische Chemie
RWTH Aachen
Professor-Pirlet-Straße 1
52056 Aachen
Germany

Herfried Griengl
Institut für Organische Chemie
Technische Universität Graz
Stremayrgasse 16
8010-Graz
Austria

Gideon Grogan
Department of Chemistry
The University of Edinburgh
West Mains Road
The King's Building
Edinburgh EH9 3JJ
United Kingdom

Peter Halling
Department of Chemistry
University of Strathclyde
Glasgow G1 1XL
United Kingdom

Yoshihiko Hirose
Amano Enzyme Inc.
Gifu R & D Center
4-179-35, Sue, Kakamighara
Gifu 509-0108
Japan

Kay Hofmann
Bioinformatics Group
MEMOREC Stoffel GmbH
Stöckheimer Weg 1
50829 Köln
Germany

Frank Hollmann
Institut für Biotechnologie
ETH Hönggerberg, HPT
CH-8093 Zürich
Switzerland

Tadayuki Imanaka
Department of Synthetic Chemistry and
Biological Chemistry
Graduate School of Engineering
Kyoto University
Yoshida-Honmachi, Sakyo-ku
Kyoto 606–8501
Japan

Hans-Dieter Jakubke
Fakultät für Biowissenschaften,
Psychologie und Pharmazie
Institut für Biochemie
Universität Leipzig
Talstraße 3
04103 Leipzig
Germany

Dieter Kadereit
Johann-Strauß-Straße 18a
65779 Kelkheim
Germany

Udo Kragl
Universität Rostock
Fachbereich Chemie
Buchbinderstraße 9
18051 Rostock
Germany

Maria-Regina Kula
Institut für Enzymtechnologie
Heinrich-Heine Universität Düsseldorf
im Forschungszentrum Jülich
Stetternicher Forst
52428 Jülich
Germany

Tatsuo Kurihara
Institute for Chemical Research
Kyoto University
Uji
Kyoto-Fu 611
Japan

James J. Lalonde
Altus Biologics Inc.
625 Putnam Avenue
Cambridge
MA 02139–4807
USA

Andreas Liese
Forschungszentrum Jülich GmbH
IBT
Leo-Brandt-Straße
D-52428 Jülich
Germany

Tomoko Matsuda
Department of Materials Chemistry
Faculty of Science and Technology
Ryukoku University
Otsu
Shiga 520–2194
Japan

Oliver May
Degussa-Hüls AG
Rodenbacher Chaussee 4
63457 Hanau
Germany

Kaoru Nakamura
Institute for Chemical Research
Kyoto University
Uji
Kyoto 611–0011
Japan

Claudia Neri
School of Chemistry
University of Bath
Claverton Down
Bath BA2 7AY
United Kingdom

Romano A. Orru
Division of Chemistry
Bio-organic Chemistry
Vrije University Amsterdam
De Boelelaan 1083
1081 HV Amsterdam
The Netherlands

Rebekka J. Parker
School of Chemistry
University of Bath
Claverton Down
Bath BA2 7AY
United Kingdom

Markus Pietzsch
Institute for Bioprocess Engineering
University of Stuttgart
Department of Microbial Physiology
Allmandring 31
70569 Stuttgart
Germany

Reinhard Reents
Max-Planck-Institut für Molekulare
Physiologie
Abteilung Chemische Biologie
Otto-Hahn-Straße 11
44227 Dortmund
Germany

Peter Rasor
Industrial Biochemicals Business
BB-PS, Roche Molecular Biochemicals
Roche Diagnostic GmbH
Nonnenwald 2
82372 Penzberg
Germany

J. David Rozzell
School of Chemical Engineering
Georgia Institute of Technology
315 Ferst Drive
Atlanta, GA 30332–0363
USA

Kenji Soda
Faculty of Engineering
Kansai University
Yamate-cho
Suita
Osaka-Fu 564
Japan

Andreas Schmid
Institut für Biotechnologie
ETH Hönggerberg, HPT
CH-8093 Zürich
Switzerland

Birgit Schulze
DSM Food Specialties
Nutritional Ingredients
P. O. Box 1
2600 MA Delft
The Netherlands

Christoph Syldatk
Institute for Bioprocess Engineering
University of Stuttgart
Department of Microbial Physiology
Allmandring 31
70569 Stuttgart
Germany

Fritz Theil
ASCA Angewandte Synthesechemie
Adlershof GmbH
Richard-Willstätter-Straße 12
12489 Berlin
Germany

Karl-Heinz van Pée
Institut für Biochemie
Technische Universität Dresden
Mommsenstraße 13
01062 Dresden
Germany

Christopher A. Voigt
California Institute of Technology
MC 210–41
Pasadena CA 91125
USA

Erik de Vroom
DSM Food Specialties
Nutritional Ingredients
P. O. Box 1
2600 MA Delft
The Netherlands

Herbert Waldmann
Max-Planck-Institut für Molekulare
Physiologie
Abteilung Chemische Biologie
Otto-Hahn-Straße 11
44227 Dortmund
Germany

Christian Wandrey
Forschungszentrum Jülich GmbH
Institut für Biotechnologie
52425 Jülich
Germany

George M. Whitesides
Department of Chemistry
Harvard University
12 Oxford Street
Cambridge, MA 02138–2902
USA

Jonathan M. J. Williams
School of Chemistry
University of Bath
Claverton Down
Bath BA2 7AY
United Kingdom

Chi-Huey Wong
Department of Chemistry
The Scripps Research Institute
10550 Torrey Pines Road
La Jolla, CA 92037
USA

Marcel Wubbolts
Manager Research & Development
DSM Biotech GmbH
Karl-Heinz-Beckurts-Straße 13
52428 Jülich
Germany

15
Reduction Reactions

15.1
Reduction of Ketones

Kaoru Nakamura and Tomoko Matsuda

15.1.1
Introduction

15.1.1.1
Enzyme Classfication and Reaction Mechanism

Research on the asymmetric reduction of ketones by biocatalysis is expanding, and its practical applications to organic chemistry have resulted in success in the enantioselective synthesis of pharmaceuticals, agrochemicals and natural products [1–4]. It is attracting increasing attention because of the following advantages:

– providing a green and sustainable process (natural catalysis),
– high enantio-, regio- and chemo-selectivity compared with most man-made reagents and catalysts,
– achiral ketones can be transformed into the corresponding alcohols with 100% yield and 100% ee theoretically, whereas kinetic resolution of racemic substrates by hydrolytic enzymes such as lipases yields only 50% of products to achieve 100% ee.
– the resulting alcohol functionality can be easily transformed, without racemization, into other useful functional groups such as halides, thiols, amines, azides, *etc.*

Dehydrogenases, classified under E.C.1.1., are enzymes that catalyze reduction and oxidation of carbonyl groups and alcohols, respectively [5]. The natural substrates of the enzymes are alcohols such as ethanol, lactate, glycerol, *etc.* and the corresponding carbonyl compounds, but unnatural ketones can also be reduced enantioselectively. To exhibit catalytic activities, the enzymes require a coenzyme; most of the dehydrogenases use NADH or NADPH, and a few use flavin, pyrroloquinoline quinone, *etc.* The reaction mechanism of the dehydrogenase reduction is as follows:

Step 1 A holoenzyme (an enzyme with its coenzyme) binds a ketone.

Step 2 A hydride on the coenzyme is transferred to the ketone to produce an alcohol. (With concurrent oxidation of the coenzyme)

Step 3 The enzyme releases the product alcohol.

Step 4 The oxidized coenzyme is transformed back into the reduced form. (With concurrent oxidation of an auxiliary substrate)

There are four stereochemical patterns in the transfer of the hydride from the coenzyme, NAD(P)H, to the substrate (Step 2) as shown in Fig. 15-1. With E1 and E2 enzymes, the hydride attacks the *si*-face of the carbonyl group, whereas with E3 and E4 enzymes, the hydride attacks the *re*-face, which results in the formation of *R* and *S* alcohols, respectively. On the other hand, E1 and E3 enzymes transfer the pro-*R* hydride of the coenzymes, and E2 and E4 enzymes use the pro-*S* hydride. Examples of the E1–E3 enzymes are as follows:

E1: *Pseudomonas sp.* alcohol dehydrogenase [6]
 Lactobacillus kefir alcohol dehydrogenase [7]

E2: *Geotrichum candidum* glycerol dehydrogenase [8–10]
 Mucor javanicus dihydroxyacetone reductase [11]

E3: Yeast alcohol dehydrogenase [12]
 Horse liver alcohol dehydrogenase [13–16]
 Moraxella sp. alcohol dehydrogenase [17]

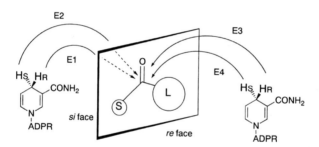

Figure 15-1. Stereochemistry of the hydride transfer from NAD(P)H to the carbonyl carbon on the substrate (S is a small group and L is a large group).

15.1.1.2

Coenzyme Regeneration

Reduction of the substrate accompanies the oxidation of the coenzyme (Step 2). Before the next cycle of the reduction of the main substrate can occur, the coenzyme has to be reduced (Step 4). Many methods for the regeneration of the reduced form of coenzyme [NAD(P)H] have been developed, so that only a catalytic amount of the coenzyme is required for the reaction. The coenzyme regeneration methods can be classified into two types:

– two-enzmye system: different enzymes reduce the substrate and NAD(P)$^+$,
– one-enzyme system: the substrate and NAD(P)$^+$ are both reduced by the same enzyme.

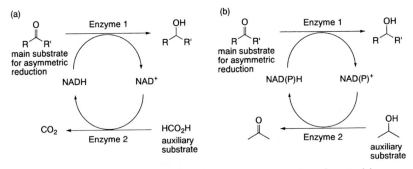

Figure 15-2. Regeneration of NAD(P)H: (a) Two-enzyme system using a formate dehydrogenase as an auxiliary enzyme and formic acid as an auxiliary substrate; Enzyme 1 = Enzyme for the reduction of the main substrate
Enzyme 2 = Formate dehydrogenase
(b) one-enzyme system using 2-propanol as an auxiliary substrate.
Enzyme 1 = Enzyme 2 For example alcohol dehydrogenase from *Thermoanaerobium brockii*[18, 19], *Pseudomonas sp.*[6], *Lactobacillus kefir*[7], and *Geotrichum candidum*[20, 21].

One of the examples of the two-enzyme system uses a formate dehydrogenase for the recycling of coenzyme [Fig. 15-2(a)][1, 3, 22–24]. It catalyzes oxidation of HCO_2H to CO_2 in order to drive the reduction of NAD^+ to NADH. The system is one of the most widely used due to the advantages such as: 1) the enzyme is commercially available, 2) CO_2 can be easily removed from the reaction, 3) formate is strongly reducing, therefore no back reaction occurs, and 4) CO_2 and HCO_2H are innocuous to enzymes. For example, the reduction of ethyl 4-chloro-3-oxobutanoate by a carbonyl reductase from *Rhodococcus eruthropolis* uses NAD^+/formate dehydrogenase as shown in Fig. 15-3[25]. As exemplified, the system is very useful for the recycling of NADH. However, it does not accept $NADP^+$, so it cannot be used for the direct reduction of $NADP^+$. For the reduction of $NADP^+$ using formate dehydrogenase, catalytic amounts of NAD^+ and $NAD(P)^+$ transhydrogenase are required. Changing the coenzyme specificity of a formate dehydrogenase using genetic methods is discussed in Sect. 15.1.3.7.

Two-enzyme systems using glucose dehydrogenase or glucose-6-phosphate dehydrogenase (commercially available enzymes) have also been widely employed[26–31].

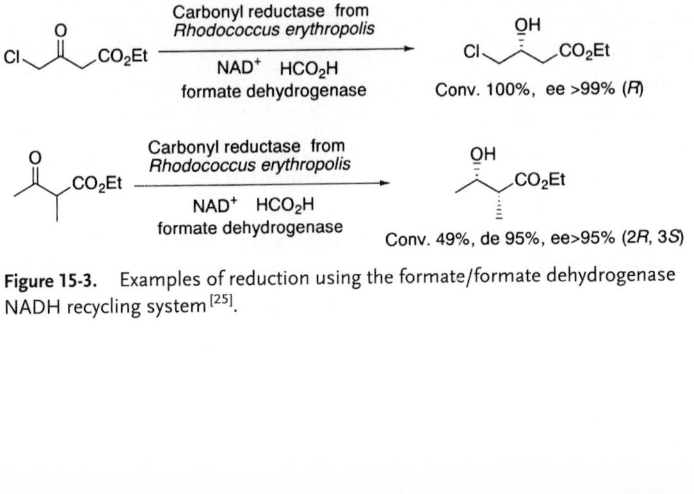

Carbonyl reductase from *Rhodococcus erythropolis*

NAD^+ HCO_2H
formate dehydrogenase

Conv. 100%, ee >99% (*R*)

Carbonyl reductase from *Rhodococcus erythropolis*

NAD^+ HCO_2H
formate dehydrogenase

Conv. 49%, de 95%, ee>95% (2*R*, 3*S*)

Figure 15-3. Examples of reduction using the formate/formate dehydrogenase NADH recycling system[25].

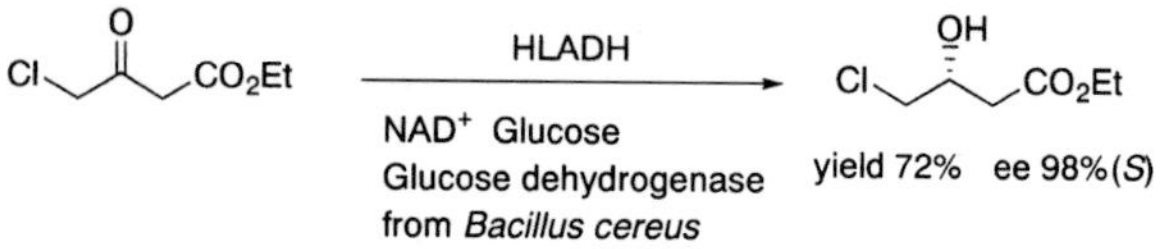

Figure 15-4. Example of reduction using glucose/glucose dehydrogenase NADH recycling system [26].

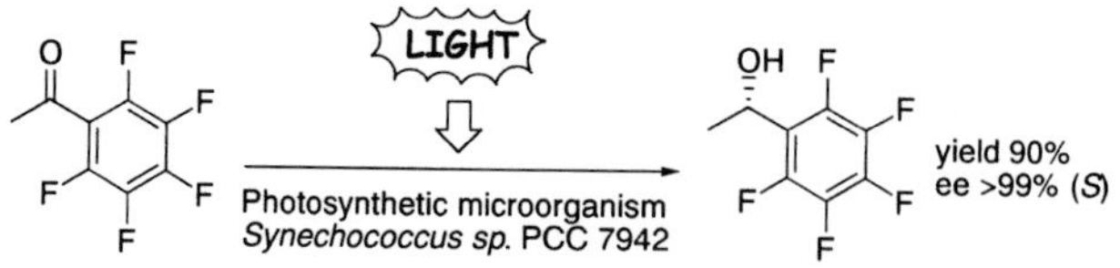

Figure 15-5. Utilization of light energy for an efficient reduction [32].

They oxidize glucose or glucose-6-phosphate to form gluconolactone or glucono-lactone-6-phosphate, respectively, which is spontaneously hydrolyzed to give gluconic acid. Both NAD^+ and $NADP^+$ act as substrates for these enzymes. For example, a thermostable glucose dehydrogenase form *Bacillus cereus* was used to recycle NADH in the asymmetric reduction of ethyl 4-chloro-3-oxobutanoate by horse liver alcohol dehydrogenase (HLADH) as shown in Fig. 15-4 [26].

Another example of a two-enzyme system involves molecular hydrogen and a hydrogenase [1]. Hydrogenases catalyze the reduction of NAD^+ or other redox dyes by dihydrogen. The system is attractive because dihydrogen is inexpensive, strongly reducing and innocuous to enzymes and NAD(H), and no by-product is formed. However, a drawback is the extreme sensitivity of the hydrogenase enzymes to inactivation by dioxygen, preventing this system from being widely used.

To provide an environmentally friendly system, photochemical methods have been developed, which utilize light energy for the regeneration of NAD(P)H [1, 32, 33]. Recently, the use of cyanobacterium, a photosynthetic biocatalyst, for the reduction was reported where the effective reduction occurred under illumination (Fig. 15-5) [32]. When a photosynthetic organisms is omitted, the addition of a photosensitizer is necessary. The methods utilize light energy to promote the transfer of an electron from a photosensitizer *via* an electron transport reagent to NAD(P)+[1].

One-enzyme recycling systems are also well developed. One of the most frequently utilized is the alcohol-alcohol dehydrogenase system as shown in Fig. 15-2(b). The system does not need an auxiliary enzyme, but an auxiliary substrate is necessary. Ethanol or 2-propanol is frequently used as an auxiliary substrate. For example, HLADH uses ethanol as shown in Fig. 15-6 [13–16] and *Thermoanaerobium brockii* [18, 19], *Pseudomonas sp.* [6], *Lactobacillus kefir* [7], and *Geotrichum candidum* [20, 21] alcohol dehydrogenases recycle NAD(P)H by employing an excess of 2-propanol. A detailed investigation of the type and amount of the auxiliary substrate needed by *G. candidum* revealed that it can use 2-alkanols from 2-propanol to 2-octanol (and cyclopentanol as well), and 15–20 equivalents of the supplementary alcohol are necessary to shift the equilibrium (between the oxidation and reduction) towards the reduction of the main substrate. Because a much higher concentration

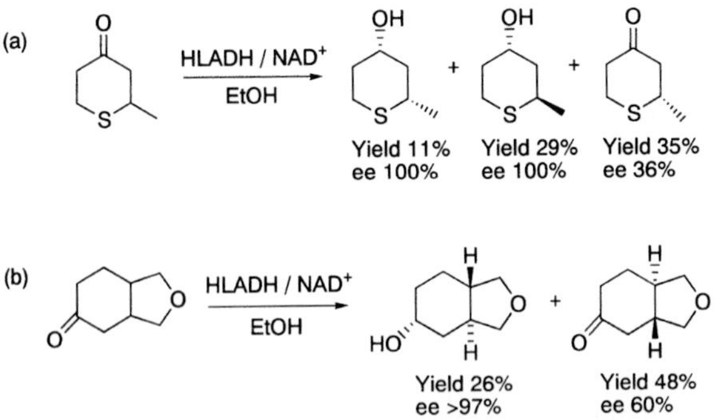

Figure 15-6. Reduction of heterocyclic ketones by HLADH using ethanol as an auxiliary substrate [15, 16].

of the auxiliary substrate to that of the main substrate is required, 2-propanol is deemed most suitable for synthetic purposes due to its high volatility.

Electrochemical regeneration of NAD(P)H represents another interesting method [34–36]. The system involves electron transfer from the electrode to the electron mediator such as methyl viologen or acetophenone *etc.*, then to the NAD(P)⁺ (which is catalyzed by an electrocatalyst such as ferredoxin-NADP⁺ reductase or alcohol dehydrogenase, *etc.*) [34]. Other methods involve the direct reduction of NAD⁺ on the electrode [35]. Both one-enzyme systems and two-enzyme systems have been reported.

15.1.1.3
Form of the Biocatalysts: Isolated Enzyme vs. Whole Cell

Enzymes in a pure form, in a partially purified form, and in the whole cell can be used for organic synthesis, and each has advantages and disadvantages [3]. The proper choice of the form of the biocatalyst is important because it affects the enantio-, regio- and chemo-selectivities, the requirement (or not) of a coenzyme and an auxiliary enzyme, the ease of catalyst preparation and work up procedures, *etc.* as shown in Table 15-1.

The most widely used whole cell biocatalyst is bakers' yeast. Since it has many different kinds of enzymes, many kinds of substrate can thus be reduced, and various types of the reactions are expected. For example, β-keto esters, aromatic, aliphatic, cyclic and acyclic ketones can be reduced with high yield [1, 37–39]. There-fore, it is a versatile "all-round" reagent. However, since bakers' yeast contains many kinds of dehydrogenases, some of them may be *S* selective, while others are *R* selective, so that the enantioselectivities can be low to high depending on the substrate structure. Further degradation of the product may also be a problem, again associated with the fact that there are many kinds of enzymes in the cell.

Not only the enzymes but also the cellular components such as coenzymes and carbohydrates are conserved in the cell, which makes the whole cell processes favorable. For example, the addition of an expensive coenzyme and an auxiliary enzyme for coenzyme regeneration is not necessary, which makes the system simple and economical when comparing with the equivalent isolated enzyme process.

Table 15-1. The form of biocatalyst: whole cell vs. isolated enzyme.

Parameter	Whole Cell	Isolated Enzyme
Kinds of enzymes	Many	One
Kinds of reactions	Many	One
Regio- and enantioselectivity	Low to high	High
Coenzyme	Unnecessary	Necessary
Catalyst preparation	Easy	Difficult
Work up	Difficult	Easy
Example	Bakers' yeast	Horse liver alcohol dehydrogenase

However, the product isolation may be complicated due to large amounts of biomass and metabolites.

On the other hand, isolated enzyme processes also have many advantages. The problem associated with the product isolation and overmetabolism can be avoided using an isolated enzyme. More importantly, chemo-, regio-, and enantioselectivities of isolated enzyme systems are usually higher than that of whole cell processes because two competing enzymes with different stereoselectivities are not present. One of the most widely used isolated enzymes is horse liver alcohol dehydrogenase (HLADH) which reduces, for example, S-heterocyclic ketones to give the corresponding tetrahydrothiopyran-4-ol with 100 % ee [Fig. 15-6(a)][15]. However, when the selectivity is so high, the substrate specificity is not wide; thus HLADH can reduce cyclic ketones with excellent enantioselectivity but cannot reduce acyclic ketones.

Another advantage of the isolated enzyme system is that the reaction pathway can be understood and predictions made. For example, for HLADH, the crystal structure[40–42] and the active site (diamond lattice) model[13, 14] are available to understand the reduction, whereas, in a whole cell process, even the catalytic species itself may not be clear.

In summary, whole cell and isolated enzyme biocatalysts both have various advantages and disadvantages. Using a recombinant yeast having the gene of a requisite enzyme is the way to access a single predominant enzyme in a microorganisms, a strategy which will be further discussed in Sect. 15.1.3.2.

15.1.1.4
Origin of Enzymes

Enzymes from various sources have been used for asymmetric reductions in organic synthesis. Microorganisms are the most important sources. There are a huge number of species (mostly in soil), containing a variety of enzymes. Commercially available microbial dehydrogenases are alcohol dehydrogenases from yeast, *Thermoanaerobium brockii* (TBADH), and the hydroxysteroid dehydrogenase from *Pseudomonas testosteroni*.

One of the most attractive kinds of microorganisms for organic synthesis is a thermophilic microorganism such as *Thermoanaerobium brockii*[18, 19], or *Thermoanaerobacter ethanolicus*, etc.[43–49]. The thermostability of the dehydrogenase enzymes from these microorganisms is very high; TBADH is stable even at 86 °C[18, 50] and an alcohol dehydrogenase from *Thermoanaerobacter ethanolicus* can be used at 50–60 °C[43, 47]. Since the enzymes with high thermostability usually have a high tolerance to organic solvent or substrates, the enzymes from thermophilic microorganisms are most suitable for organic synthesis.

Another interesting class of biocatalyst encompasses the photosynthetic microorganisms, the algae[32, 51]. Owing do the high growth rate, a large amount of the biomass for use as the biocatalyst is available. Importantly, such organisms can use light energy as power for coenzyme recycling as described in Sect. 15.1.1.2, so an environmentally friendly system can be constructed using them.

The second most widely studied source of enzymes are mammalian enzymes as exemplified by horse liver alcohol dehydrogenase (HLADH). Detailed investigations on this enzyme have been reviewed elsewhere[13, 14].

The third and least studies source is from plant cell cultures, which have only recently been used in biocatalysis[51–57]. Although the number of species available are much less than microorganisms, plants possess a much larger gene. More importantly since plants can effect photosynthesis, different types of enzymes exist in plants to those of microorganisms. Therefore, different enzymes which catalyze unique reactions with man-made substrates may be expected. Despite the strong possibility of the discovery of interesting enzymes, plant cell cultures have not been fully investigated for use in biocatalysis due to their relatively slow growth rate.

15.1.2
Stereochemical Control

15.1.2.1
Enantioselectivity of Reduction Reactions

The synthesis of enantiomerically pure compounds is becoming increasingly important for research and development in chemistry and biochemistry[58], especially in the pharmaceutical industry, as chiral drugs now represent close to one-third of all pharmaceutical sales world wide[59]. In most of the cases, one enantiomer is more effective as a drug than the other. The influence on the environment is also different between the enantiomers; different enantiomers of chiral pollutants in soils are preferentially degraded by microorganisms in various environments[60]. Therefore, synthetic methods exhibiting extremely high enantioselectivities are necessary.

The enzymatic reactions occurring in Nature involving natural substrates usually show very high enantioselectivities. On the other hand, with man-made substrates the enantioselectivity can also be high (> 99 % ee) but this is not always the case as shown in Fig. 15-7. Low enantioselectivity results when the catalyst is a low selectivity enzyme [Fig. 15-7 (C)] and/or when there are more than two competing enzymes with different enantioselectivities [Fig. 15-7 (D)]. In case (C), either an enzyme or substrate has to be changed. On the other hand, in case (D), a change in a microorganism or substrate as well as a change in reaction conditions may be effective in improving the enantioselectivity. In case (D), by choosing the proper

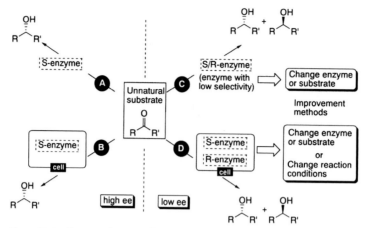

Figure 15-7. Enantioselectivity of the product and improvement methods.

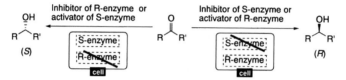

Figure 15-8. Synthesis of both enantiomers using one microorganism by choosing appropriate conditions.

conditions, both enantiomers can be synthesized by using only one microorganism; when a selective inhibitor for an *S*-directing enzyme or on *R*-directing enzyme is added to the reaction mixture, the *(R)*-alcohol or *(S)*-alcohol will be enantiose-lectively produced, respectively, as shown in Fig. 15-8.

15.1.2.2
Modification of the Substrate: Use of an "Enantiocontrolling" Group

The enantioselectivity of a biocatalytic reduction can be controlled by modifying the substrate because the enantioselectivity of the reduction reaction is profoundly affected by the structure of substrates. For example, in the reduction of 4-chloro-3-oxobutanoate by bakers' yeast, the ester moiety can be used to control the stereochemical course of the reduction[61–63]. When the ester moiety was smaller than a butyl group, then *(S)*-alcohols were obtained, and when it was larger than a pentyl group then *(R)*-alcohols were obtained as shown in Fig. 15-9.

After the reduction, the ester moiety can be exchanged easily without racemization, so both enantiomers of an equivalent synthetic building block are obtained using the same reaction system by changing an "enantiocontrolling" group, the ester moiety. The "enantiocontrolling" group can also be introduced into the keto esters at the α- or α'-positions. For example, sulfur functionalities such as methyl- and

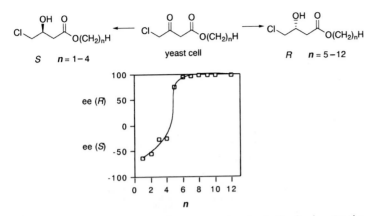

Figure 15-9. Stereochemical control on yeast-catalyzed reduction by changing the ester group [61–63].

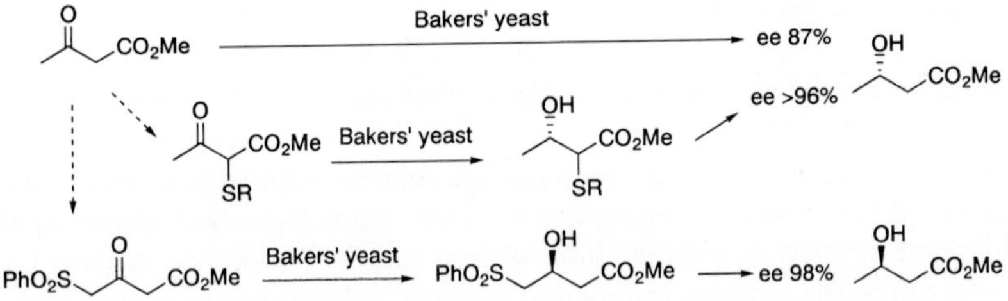

Figure 15-10. Stereochemical control on yeast-catalyzed reduction by introducing sulfur functionalities [64, 65].

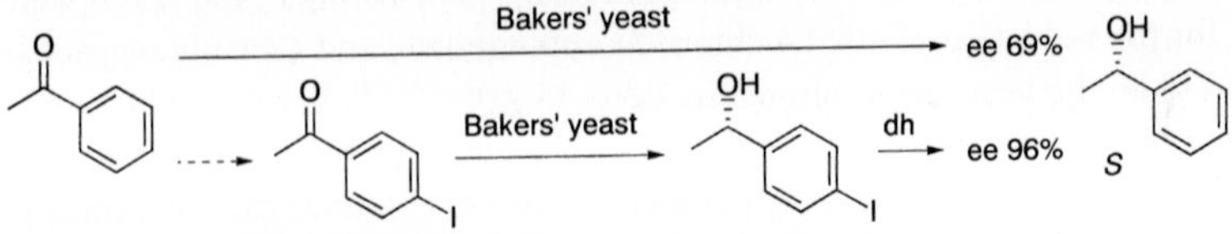

Figure 15-11. Improvement of enantioselectivity by substituting iodide at the *para* position; yeast reduction followed by dehalogenation (dh) [65].

phenylthio [64] and phenylsulfonyl [65] groups can be used to improve the enantioselectivities as shown in Fig. 15-10.

Other types of ketones can also be modified to improve the enantioselectivities, and various functionalities can be used to modify the substrate to produce the corresponding alcohol with higher enantioselectivities. For example, the reduction of acetophenone by yeast results in the formation of phenylethanol in 69 % ee, whereas the reduction of p-iodoacetophenone followed by the dehalogenation results in a product of 96 % ee (Fig. 15-11) [65].

As shown above, the substrate modification and "de"modification steps can be used to improve the enantioselectivity, although on the negative side the strategy may introduce extra steps into a synthetic route.

Table 15-2. Screening for the synthesis of important chiral building blocks.

Reactions	Microorganisms screened	Result	Reference
	400 yeasts	*Candida magnoliae* 90 g/L, 96.6% ee (99% ee after heat treatment)	67
	191 bacteria 59 actinomycetes 230 yeasts 81 molds 42 basidiomycetes	45 mg/mL stoichiometric yield *Rhodotorula minuta* IFO 0920: 86% ee *Candida parapsilosis* IFO 0708: 87% ee *Aspergillus niger* IFO 4415: 87% ee	68
	450 bacteria	*Klebsiella pneumoniae* IFO 3319 99% de, >99% ee, 99% yield (2 Kg in 200 L fermentor)	70

15.1.2.3
Screening of Microorganisms

Screening for a novel enzyme is a classical method and one of the most powerful tools available to find the system to convert a selected ketone into a desired alcohol[66–71]. It is possible to discover a suitable enzyme or microorganisms by the application of the newest screening and selection technologies that allows rapid identification of enzyme activities from diverse sources[66]. Enzyme sources for screening can be soil samples, commercial enzymes, culture sources, a clone bank, *etc.* From these sources, enzymes which are regularly expressed and enzymes which are not expressed in the original host can be tested to establish whether they are suitable for the transformation of certain substrates[66]. For example, 400 yeasts were screened for the reduction of ethyl 4-chloro-3-oxobutanoate, and *Candida magnoliae* was found to be the best one as shown in Table 15-2[67, 72, 73]. For the reduction of ketopantoyl lactone, various kinds of microorganisms were screened, and several microorganisms which produce D-(−)-pantoyl lactone stoichiometrically at a concentration of 45 mg mL^{-1} with high enantioselectivity were found[68]. For the reduction of ethyl 2-methyl-3-oxobutanoate, out of 450 bacteria, *Klebsiella pneumoniae* IFO 3319 and 4 other strains were found to give the corresponding (2R, 3S)-hydroxyesters with more than 98% de and > 99% ee[70].

Screening techniques have also been applied for the purpose of drug synthesis. For example, a key intermediate in the synthesis of the anti-asthma drug, Montelukast, was prepared from the ketone 1 by microbial transformation as shown in Fig. 15-12[71]. The biotransforming organism, *Microbacterium campoquemadoensis* (MB5614), was discovered as a result of an extensive screening programme.

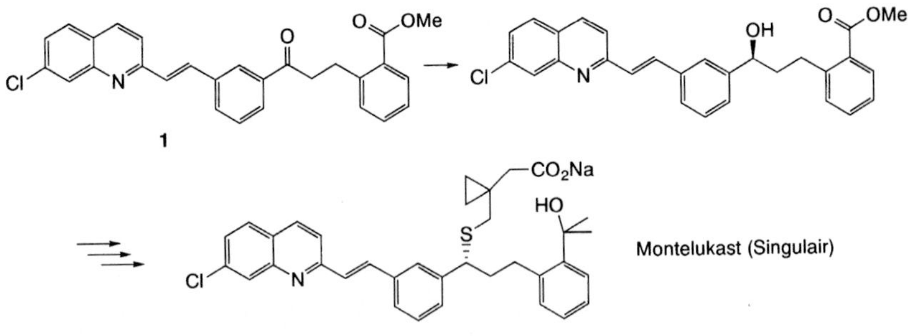

Figure 15-12. Reduction of a ketone by *Microbacterium campoquemadoensis* (MB5614) in a synthesis of the anti-asthma drug, Montelukast [71].

Table 15-3. Control on diastereoselectivity by heat treatment[74].

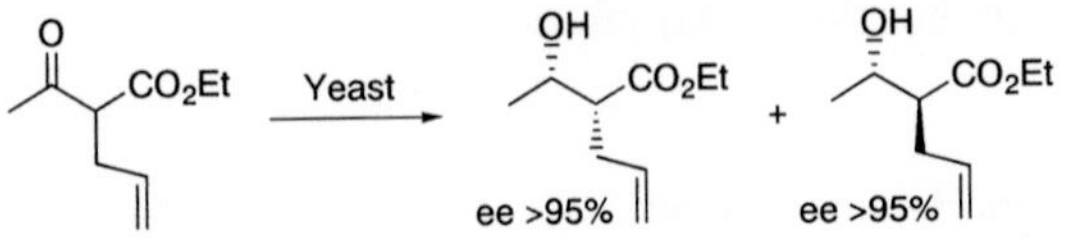

Yeast cell	Syn (%)	Anti (%)
No heat treatment	30	70
50 °C, 30 min	65	30
heat + inhibitor	96	4

15.1.2.4
Treatment of the Cell: Heat Treatment

Treatment of the cell before the reaction is sometimes an effective method of controlling the selectivity of some biocatalysts. When reducing with a whole cell and the selectivity is not as is desired due to the presence of plural enzymes with different selectivities, heat treatment of the cell to selectively deactivate one or more enzymes can change the selectivity of the reduction. For example, the diastereoselectivity in the yeast reduction of 2-allyl-3-oxobutanoate was changed from *anti*-selectivity to *syn*-selectivity by pre-treatment of the yeast before the reaction as shown in Table 15-3 [74]. In this case, the diastereoselectivity is further improved to 96 : 4 by using an enzyme inhibitor.

Another example is the use of heat treatment as a supplement to the screening process. The enantioselectivity of the reduction of ethyl 4-chloro-3-oxobutanoate by *Candida magnoliae* was improved from 96.6 % ee *(S)* using untreated cells to 99 % ee *(S)* with heat treated cells [67].

15.1.2.5
Treatment of the Cell: Aging

When a whole cell system is used for a reduction, the substrate is usually added to the cultivation medium after a certain growth period, or to the mixture of the medium and freshly harvested cells. However, when the mycelium of a local strain of *Geotrichum candidum* was not used immediately after growth, but filtered and preincubated by shaking in deionized water for 24 hours at 27 °C ("aged mycelium"), then used for the reduction of ethyl 3-oxobutanoate, the stereochemistry of the product alcohol was different from that obtained from the reduction using fresh mycelium [75–78]. When fresh mycelium was used, the enantioselectivity and the absolute configuration of the product shifted from *S* (26% ee) to *R* (58% ee) on raising the substrate concentration from 1 to 20 g L^{-1}. When aged mycelium was used, the absolute configuration was always *R* and showed constant enantioselectivity (ca. 50% ee) regardless of the substrate concentration, although the reduction proceeded at a slightly slower rate.

In the aging process, an *S*-forming activity, was lost, leaving unaffected either one low-specificity reducing enzyme with major *R*-forming activity, or several enzymes having opposite enantioselectivities but similar K_M values.

15.1.2.6
Treatment of the Cell: High Pressure Homogenization

High pressure homogenization is a new technology in food processing. It was found that the same technology can be applied to effect the microbial reduction of chemical compounds [79]. The cell culture with substrate (such as acetophenone, 5-hexen-2-one, *etc.*) was poured into the high pressure homogenizer, and then it was incubated for 48 h and the enantioselectivity of the product was evaluated. During the process the reaction mixture was forced under pressure through a narrow gap where it was subjected to rapid acceleration [1 (blank experiment), 500, 1000, 1500 bar] after which it undergoes an extreme drop in pressure. Various strains of *Saccharomyces cerevisiae* and *Yarrowia lipolytica* are utilized in the reduction processes and higher enantioselectivities were generally achieved albeit in lower yields than the standard process.

15.1.2.7
Treatment of the Cell: Acetone Dehydration

A dried cell mass is often used as a biocatalyst for a reduction, since it can be stored for a long time and can be used whenever needed, without cultivation. One of the useful methods to dry the cell mass is acetone dehydration [80]. For example, the cells of *Geotrichum candidum* IFO 4597 were mixed with cold acetone (–20 °C) and the cells were collected by filtration [20, 21]. The procedure was repeated five times and then the cells were dried under reduced pressure. The dried cells (acetone powder of *G. candidum* IFO 4597: APG4) were obtained; they can be stored for a long time in the freezer.

The drying of the cell not only aids the preservation of the cell but also contributes to the stereochemical control as shown in Table 15-4. The reduction of acetophenone catalyzed by *G. candidum* IFO 4597 resulted in poor enantioselectivity [28% ee(*R*)]. When the form of the catalyst was changed from wet whole-cell to dried powdered-

Table 15-4. Acetone treatment of *Geotrichum candidum* for the improvement of enantioselectivity[20, 21].

OH Untreated whole cell O Acetone dried cell (APG4) OH
|‾|‾Ph ←———————— |‾|‾Ph ————————————→ |‾|‾Ph
28% ee (R) NAD+ or NADP+ >99% ee (S)
  2-propanol or cyclopentanol

Catalyst	Coenzyme	Additive	Yield (%)	ee (%)
Untreated whole cell	none	none	52	28(R)
Acetone dried cell (APG4)	none	none	0	–
Acetone dried cell (APG4)	NAD+	2-propanol	89	>99(S)
Acetone dried cell (APG4)	NAD+	cyclopentanol	97	>99(S)
Acetone dried cell (APG4)	NADP+	cyclopentanol	86	>99(S)

cell (APG4), no reduction was observed, which would indicate the loss of the necessary coenzyme(s) and/or coenzyme regeneration system(s) during the treatment of the cells with acetone. Addition of coenzyme, NAD+, did not have a significant effect on the yield. Addition of 2-propanol resulted in only a small increase in the yield, but a significant improvement in the enantioselectivity was observed. Surprisingly, addition of both NAD+ and 2-propanol profoundly enhanced both chemical yield and enantiomeric excess. Addition of NADH, NADP+ or NADPH instead of NAD+ and addition of cyclopentanol instead of 2-propanol also gave enantiomerically pure alcohol in high yield.

The improvement in the enantioselectivity from 28% (R) to > 99% (S) was due to the suppression of every enzyme which reduces the substrate, followed by the stimulation of an S-directing enzyme by the addition of the coenzyme and an excess amount of 2-propanol, agents which push the equilibrium towards the reduction of the substrate.

It was confirmed, by separating the enzymes in the powder, that many S- and R-directing enzymes exist in the biocatalyst. The addition of coenzyme and cyclopentanol stimulates only one particula S-enzyme but not other S-enzymes and R-enzymes because the specific S-enzyme can oxidize cyclopentanol [concomitantly reducing NAD(P)+], while other S- or R-enzymes cannot use cyclopentanol as effectively[81]. This is a very interesting case where the reduction with a cell initially having both S- and R-directing enzymes was modified and resulted in excellent S enantioselectivity.

15.1.2.8
Cultivation Conditions of the Cell

The dependence of enantioselectivity in microbial transformations on the cultivation conditions of the microorganisms has also been investigated[82–86]. The enzymes induced during the growth phase and during starvation are certainly different, therefore the enantioselectivity of the product may be different when two competing enzymes with different enantioselectivities catalyze the reduction. Since the enzyme reducing the non-natural substrate is not usually known, cultivation conditions which induce the desired enzyme have to be found by trial and error.

For example, the effect of cultivation time and different carbon sources on the enantioselectivity of the reduction of sulcatone by some anaerobic bacteria has been investigated [83]. Another example is the investigation on the effect of the medium concentrations for cultivation of *Geotrichum candidum* IFO 4597 on the enantiose-lectivity of the reduction of acetophenone derivatives. The yield of *R*-alcohol (the minor enantiomer) increased with the medium concentration; therefore, the medium concentration was kept low, optimally to produce the *S*-enantiomer [82]. The effect of the aeration during cultivation on the enantioselectivity of bakers' yeast production of 3-hydroxyesters has also been reported [86]. Inducers such as a substrate analog may also induce the desired enzyme to improve the enantiose-lectivity.

15.1.2.9
Modification of Reaction Conditions: Incorporation of an Inhibitor

In the case of the observation of poor overall enantioselectivity due to the presence of two competing enzymes with different enantioselectivities, one of the most straight-forward methods to improve the enantioselectivity is the use of the inhibitor of the unnecessary enzyme(s). Ethyl chloroacetate, methyl vinyl ketone, allyl alcohol, allyl bromide, sulfur compounds, Mg^{2+}, Ca^{2+}, *etc.* have been reported as inhibitors of enzymes in yeast [87–97].

For example, the low enantioselectivity in the yeast reduction of β-keto ester was improved by addition of ethyl chloroacetate or methyl vinyl ketone as described in Fig. 15-13. The enzymes inhibited and those not inhibited were identified by enzymatic studies using purified enzymes [97]. The mechanism of the inhibition is reported to be non-competitive.

These inhibitors were also used to improve the enenatioselective reduction of

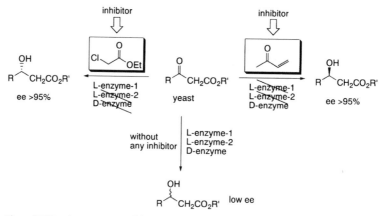

Figure 15-13. Improvement of the enantioselectivity by using an inhibitor of undesired enzymes [87, 88, 97].

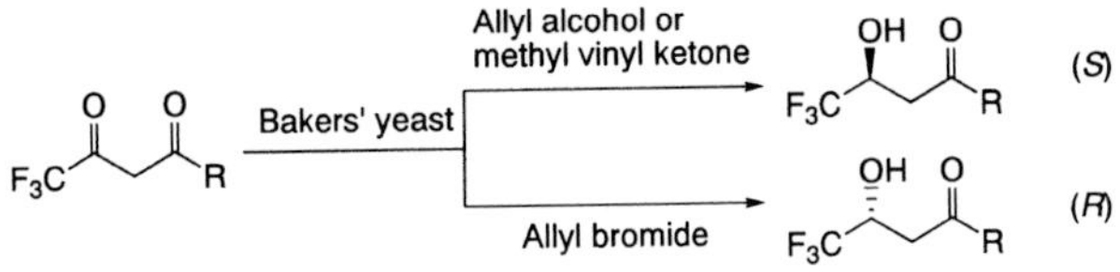

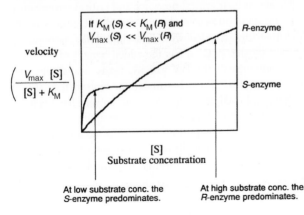

Figure 15-14. Stereochemical control using an inhibitor[89].

fluorinated diketones (Fig. 15-14). By applying a suitable inhibitor, both enantiomers of the alcohol can be obtained using only one kind of microorganism, namely bakers' yeast[89].

15.1.2.10
Modification of Reaction Conditions: Organic Solvent

Organic solvents have been used widely for esterifications and transesterifications using hydrolytic enzymes to shift the equilibrium towards esterification by avoiding hydrolysis. Organic solvents can also be used for reductions using dehydrogenases[98–109]. They can be used to control the overall enantioselectivity of the reduction, when there are more than two competing enzymes with different enantioselectivities, K_M and V_{max}.

Enzymatic reactions follow the Michaelis–Menten equation, therefore, the rate of the enzyme catalyzed reaction depends on the substrate concentration. When an organic solvent is introduced, most organic substrates usually dissolve in the organic phase, and the effective substrate concentration in the aqueous phase around the enzyme decreases. The change in substrate concentration by the addition of the organic phase causes the change in the enzyme species catalyzing the reduction. For example, as shown in Fig. 15-15, if the K_M for an S-directing enzyme is much smaller than that for an R-directing enzyme, and V_{max} for the S-directing enzyme is much smaller than that for the R-directing enzyme, then when the substrate concentration is low, the S-enzyme will dominate, whereas at high substrate concentration, the R-enzyme will dominate the biotransformation.

In fact, when the yeast reduction of ethyl 2-oxohexanoate was conducted in water,

Figure 15-15. Effect of the substrate concentration on enantioselectivity of the reduction with the system having both an S-enzyme with small K_M and small V_{max} and an R-enzyme with large K_M and large V_{max}.

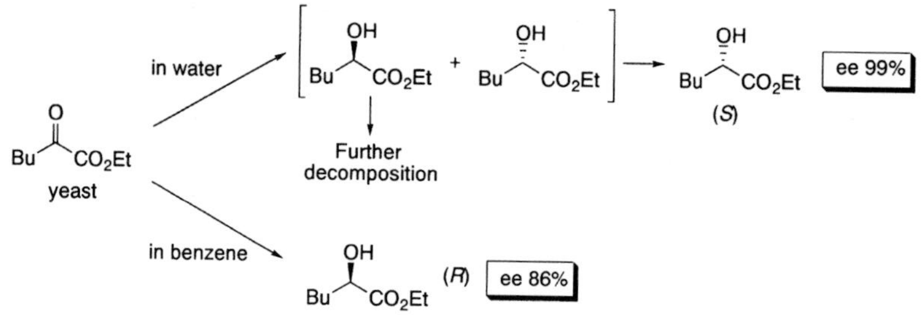

Figure 15-16. Stereochemical control by using an organic solvent.

Table 15-5. Mechanism of stereochemical control using benzene: kinetic parameters of yeast α-keto ester reductases (YKERs)[100, 110].

Enzyme	Enantioselectivity	K_M (mM)	k_{cat} (s^{-1})	V_{max} (U kg^{-1} yeast)
YKER-I	R	8.40	1.53	37.7
YKER-IV	R	0.142	4.59	41
YKER-V	S	5.72	27.8	649
YKER-VI	R	1.03	2.10	1774
YKER-VII	S	27.3	127	501

both *(R)*- and *(S)*-alcohols were produced and the *(S)*-alcohol was obtained as the major product as a result of the further enantioselective decomposition of the *(R)*-enantiomer (Fig. 15-16)[100, 109]. However, when the biotransformation was conducted in benzene, then the *(R)*-alcohol was formed selectively in high yield.

K_M and V_{max} for all enzymes existing in yeast and catalyzing the reduction were determined and it was found that an *R*-enzyme, YKER-IV, has a K_M which is smaller than other enzymes by an order of magnitude (Table 15-5), and, therefore, predominantly catalyzes the reduction in benzene[100, 110, 111].

15.1.2.11
Modification of Reaction Conditions: Use of a Supercritical Solvent

Supercritical fluids, materials above their critical pressure and critical temperature (Fig. 15-17), have been attracting attention as solvents with the advantages of gas-like low viscosities and high diffusivities coupled with their liquid-like solubilizing power[112]. Supercritical carbon dioxide (scCO$_2$) has the added benefit of an environmentally benign nature, nonflammability, low toxicity, ready availability, and ambient critical temperature (T_c = 31.0 °C) that is suitable for biotransformations. The attraction of combining natural catalysts with a "natural" solvent has been the driving force behind a growing body of literature on the stability, activity and specificity of enzymes in scCO$_2$. The first report on biotransformations in supercritical fluids was in 1985[113–115], and the benefit of using supercritical fluids for biotransformations has been demonstrated, e. g. through improved reaction rates, etc.[116, 117].

Recently the alcohol dehydrogenase from *Geotrichum candidum* was found to

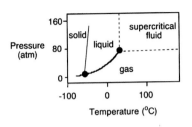

Figure 15-17.   Phase diagram of carbon dioxide.

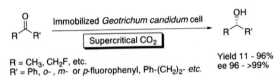

R = CH₃, CH₂F, etc.
R' = Ph, o-, m- or p-fluorophenyl, Ph-(CH₂)₂- etc.

Yield 11 - 96%
ee 96 - >99%

Figure 15-18. Reduction of fluoroketones by *Geotrichum candidum* IFO 5767 in supercritical CO_2[118].

catalyze the reduction of fluoroacetophenones *etc.* in scCO₂ at around 100 atm and 35 °C (Fig. 15-18)[118]. The enantioselectivity obtained was equivalent to the system using an organic solvent.

15.1.2.12
Modification of Reaction Conditions: Cyclodextrin

Cyclodextrin has also been used to control the enantioselectivity of bioreductions[119–121]. When added to a reaction mixture, the substrate can reside in the cyclodextrin, which decreases the effective substrate concentration around the enzyme and results in the domination of reactions involving enzymes with low K_M. The effect can be demonstrated by the reduction of ketopantoyl lactone by yeast. The enantioselectivity was improved from 73% to 93% by adding β-cyclodextrin to the reaction mixture. The improvement in enantioselectivity of the reduction in the presence of enzymes with different enantioselectivities and K_M values by decreasing the substrate concentration was confirmed by the ineffectiveness of α-cyclodextrin which is too small to include the substrate. It was also confirmed by dilution of the reaction mixture, which improved the enantioselectivity in the absence of cyclodextrin.

15.1.2.13
Modification of Reaction Conditions: Hydrophobic Polymer XAD

A decrease in the effective substrate concentration around the enzyme but not in the bulk can also be achieved using hydrophobic polymer XAD instead of using cyclodextrin or an organic solvent[122–126]. For example, the technique was used in the reduction of methyl benzyl ketone by *Zygosaccharomyes rouxii* for the synthesis of LY300164, a noncompetitive antagonist of the AMPA subtype of excitatory amino acid receptor[122]. The adsorption properties of the resin on both substrate and

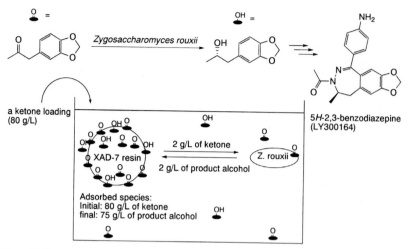

Figure 15-19. Decrease in the effective substrate concentration around the enzyme by using hydrophobic polymer XAD [122].

product allowed a ketone loading of 80 g L^{-1}, while limiting the effective solution concentration of both substrate and product to sublethal concentrations of 2 g L^{-1} (Fig. 15-19).

The hydrophobic resin has also been used for the purpose of controlling selectivity[123, 124]. Enantioselectivity, chemoselectivity and space-time yields of the yeast reduction of α,β-unsaturated carbonyl compounds were impressively enhanced. The distribution of substrates and products between the resin and the water phase showed that the improved selectivity could be attributed to the control of substrate concentration.

The powerful influence of the hydrophobic resin was also demonstrates in the *Geotrichum candidum* catalyzed reduction of simple aliphatic and aromatic ketones[126]. For example, the enantioselectivity of the reduction of 6-methylhept-5-en-2-one was improved from 27% ee *(R)* to 98% ee *(S)*.

15.1.2.14
Modification of Reaction Conditions: Reaction Temperature

Reaction temperature is one of the parameters that affects the enantioselectivity of a reaction[43–46]. For the oxidation of an alcohol, the values of k_{cat}/K_M were determined for the *(R)*- and *(S)*-stereodefining enantiomers; E is the ratio between them. From the transition state theory, the free energy difference at the transition state between *(R)*- and *(S)*-enantiomers can be calculated from E [Eq. (2)], and $\Delta\Delta G$ is in turn the temperature function [Eq. (3)]. The racemic temperature (T_r) can be calculated as shown in Eq. (4). With these equations, T_r for 2-butanol and 2-pentanol of the *Thermoanaerobacter ethanolicus* alcohol dehydrogenase was determined to be 26 °C and 77 °C, respectively.

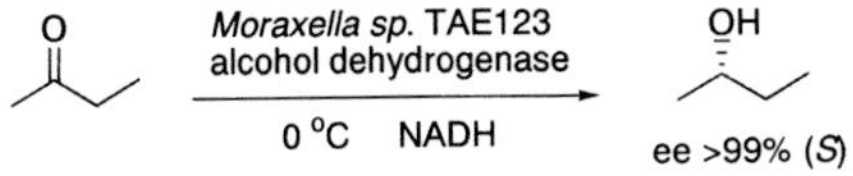

Figure 15-20. Reduction of 2-butanone by the alcohol dehydrogenase from *Moraxella sp.* TAE123 at 0 °C [17].

$$E = (k_{cat} / K_M)_R / (k_{cat} / K_M)_S \tag{1}$$

from transition state theory

$$- RT\ln(E) = \Delta\Delta G\ddagger \tag{2}$$
$$\Delta\Delta G\ddagger = \Delta\Delta H\ddagger - T\Delta\Delta S\ddagger \tag{3}$$

When

$$\Delta\Delta G\ddagger = 0, \; T_r = \Delta\Delta H\ddagger / \Delta\Delta S\ddagger \tag{4}$$

Since the transition state for alcohol oxidation and ketone reduction must be identical, the product distribution (under kinetic control) for reduction of 2-butanone and 2-pentanone is also predictable. Thus, one would expect to isolate *(R)*-2-butanol if the temperature of the reaction was above 26 °C. On the contrary, if the temperature is less than 26 °C, *(S)*-2-butanol should result. In fact, the reduction of 2-butanone and 2-pentanone at 37 °C resulted in 28 % ee *(R)*- and 44 % *ee* (S)-alcohol, respectively, as expected [43].

The temperature range that can be used for a biocatalytic reduction is very wide because alcohol dehydrogenases from various types of microorganisms (thermophilic and psychrophilic) are available. The extremely high stability of enzymes from thermophilic microorganisms are discussed in Sect. 15.1.1.4. On the other hand, conducting reactions at temperatures as low as 0 °C is also possible using an Antarctic psychrophile [17]. For example, the reduction of 2-butanone, which is an extremely challenging substrate for enantioselective reduction, with alcohol dehydrogenase from *Moraxella sp.* TAE123, at 0 °C afforded *(S)*-2-butanol in > 99 % ee (Fig. 15-20).

15.1.2.15
Modification of Reaction Conditions: Reaction Pressure

The effect of high hydrostatic pressure (400 bar) on microbial reductions of the ketones such as acetophenone, *etc.* has been examined using various strains of *Saccharomyces cerevisiae* and *Yarrowia lipolytica*. Higher enantioselectivities are generally achieved together with lower yields compared with the results obtained at atmospheric pressure as in the case of treatment of cells with high pressure homogenation [79]. Although the enantioselectivity obtained here is not as high as > 99 % ee, this finding added pressure as an adjustable parameter to control the enantioselectivity of the bioreduction.

15.1.3
Improvement of Dehydrogenases for use in Reduction Reactions by Genetic Methods

15.1.3.1
Overexpression of the Alcohol Dehydrogenase

Recent developments in molecular biology have contributed to the development of useful biocatalysts. Overexpression as well as rational and random mutations of many alcohol dehydrogenases have improved the function of enzymes so that they can be useful in organic synthesis [22, 127–134]. Examples of overexpressed enzymes are introduced here, and Sect. 15.1.3.2–15.1.3.8 will describe the improvement of catalytic functions achieved by using genetic methods. Although the non-genetic chemical modifications of enzymes can also be important in order to improve a biocatalyst [135], they are not mentioned here.

Example 1: The *Thermoanaerobacter ethanolicus* 39E adhB gene encoding the secondary alcohol dehydrogenase was overexpressed in *Escherichia coli* to form more than 10 % to total protein [136]. The recombinant enzyme was purified by heat treatment and precipitation with aqueous $(NH_4)_2SO_4$ and isolated in 67 % yield. Enzymes with mutation(s) around the active site residues were also created to examine the catalytically important zinc binding motif in the proteins.

Example 2: The gene encoding a phenylacetaldehyde reductase with a unique and wide substrate range was cloned from the genomic DNA of the styrene-assimilating *Corynebacterium* strain ST10 [137–139]. The enzyme was expressed in recombinant *E. coli* cells in sufficient quantity for practical use and purified to homogeneity by three column chromatography steps [140]. The amino acid residues assumed to be three catalytic and four structural zinc-binding ligands were characterized by site-directed mutagenesis of two zinc-binding centers within the enzyme [141].

Besides these examples, many other important enzymes for biocatalytic reductions, such as the NADPH-dependent carbonyl reductase from *Candida magnoliae* [142], the ketoreductase from *Zygosaccharomyces rouxii* [143], and the aldehyde reductase from *Sporobolomyces salmonicolor* AKU4429 [144], *etc.* have also been expressed in *E. coli etc.* and shown to be active.

The availability of sufficient quantities of enzymes for crystallization studies has led to the crystal structures been obtained for several dehydrogenases. For example, two tetrameric NADP⁺-dependent bacterial secondary alcohol dehydrogenases from the mesophilic bacterium *Clostridium beijerinckii* and the thermophilic bacterium *Thermoanaerobium brockii* have been crystallized in the apo- and the holo-enzyme forms, and their structures are available in the Protein Data Bank [145]. The crystal structure of the alcohol dehydrogenase from horse liver is also available [40–42].

15.1.3.2
Access to a Single Enzyme Within a Whole Cell: Use of Recombinant Cells

The advantages and disadvantages of using whole cell and isolated enzymes are described in Sect. 15.1.1.3. Here, genetic methods are used to build the systems with the advantages of both whole cells and isolated enzymes; the technology enables one to access essentially a single enzyme within a whole cell [127].

For example, to improve a low enantioselectivity due to the presence of plural enzymes in a cell with overlapping substrate specificities but different enantiose-lectivities, a recombinant cell with only the enzyme possessing the desired enantio-selectivity was used (Fig. 15-21). Isolation of the enzyme, of course, improves the enantioselectivity. However, the requirement of a laborious enzyme isolation process and expensive cofactor with its associated regeneration enzyme (if necessary) have limited the practical utility of isolated enzyme processes. However, once the gene encoding the enzyme with high enantioselectivity has been overexpressed in *E. coli*, then the essentially single enzyme system can be accessed within the whole cell. Since it is a whole cell system, it can be cultivated to supply an appropriate amount without involving a laborious process for the isolation of an enzyme. The fact that there is no coenzyme requirement is also a merit for the system. Because it has only one enzyme which transforms the substrate, the problems of overmetabolism or low selectivity are also resolved. Using *E. coli* expressing Gcy1p and *E. coli* expressing Gre3p, various β-keto esters and α-alkyl-β-keto esters were reduced with excellent enantio- (up to > 98 % ee) and diastereo-selectivities (> 98 % de) [128].

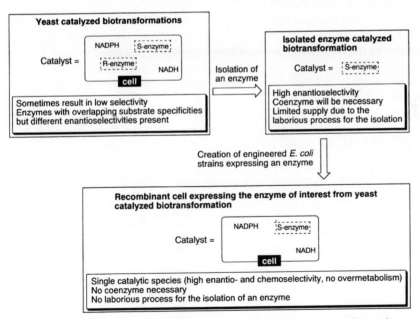

Figure 15-21. Advantages and disadvantages of whole cell, isolated enzymes and recombinant cell as biocatalysts.

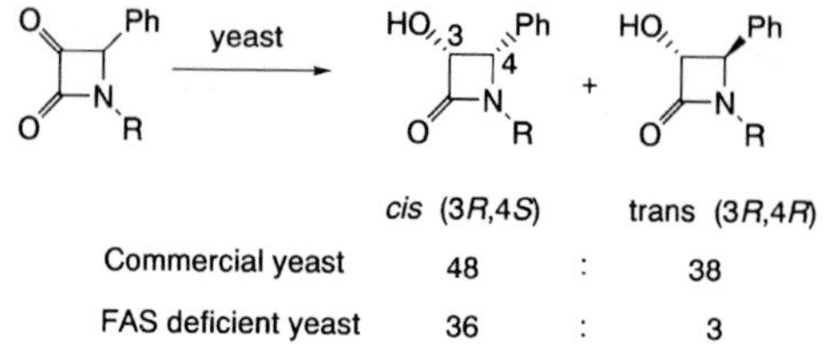

Figure 15-22. Use of FAS deficient yeast to improve the diastereoselectivity of a reduction [129].

	cis (3R,4S)		trans (3R,4R)
Commercial yeast	48	:	38
FAS deficient yeast	36	:	3

15.1.3.3.
Use of a Cell Deficient in an Undesired Enzyme

This is a similar approach to that described above. Use of a yeast strain deficient in fatty acid synthase (FAS) suppressed formation of the undesired *trans*-diastereomer of a β-lactam as shown in Fig. 15-22 [129].

15.1.3.4
Point Mutation for the Improvement of Enantioselectivity

Point mutation of enzymes has played an important role in determining those amino acid residues involved in catalytic activities. It has also been used to improve the enantioselectivity of dehydrogenases. For example, even a single point mutation of a secondary alcohol dehydrogenase from *Thermoanaerobacter ethanolicus* can change substantially the enantioselectivity for the reduction of 2-butanone and 2-pentanone as shown in Table 15-6 [45].

15.1.3.5
Broadening the Substrate Specificity of Dehydrogenase by Mutations

Developments in molecular biology enable us to change the substrate specificity of enzymes; the enzymes can be engineered to be more suitable for the requisite substrate. For example, variations have been made to the structure of the NAD⁺ dependent L-lactate dehydrogenase from *Bacillus stearothermophilus* (LDH) [130]. Two regions of LDH that border the active site (but are not involved in the catalytic

Table 15-6. Control of enantioselectivity by a single mutation (serine-39 to threonine) of the secondary alcohol dehydrogenase from *Thermoanaerobacter ethanolicus*[45].

Parameter	Wild type	Mutant (S39T)
k_{cat}/K_M (M^{-1} s^{-1}) for oxidation at 55 °C of:		
(R)-2-butanol	3.1×10^5	2.8×10^5
(S)-2-butanol	1.1×10^5	0.29×10^5
(R)-2-pentanol	0.87×10^5	3.5×10^5
(S)-2-pentanol	1.3×10^5	2.1×10^5
Ee of the reduction at 55 °C of:		
2-butanone	47(R)	81(R)
2-pentanone	20(S)	25(R)

Table 15-7. Broadening the substrate specifity of L-lactate dehydrogenase from *Bacillus stearothermophilus* by rational protein engineering[130].

Enzyme	R	k_{cat} (s^{-1})	K_M (mM)	k_{cat}/K_M $(M^{-1} s^{-1})$
Wild Type	CH_3	250	0.06	4 200 000
	$CH_2CH(CH_3)_2$	0.33	6.7	50
[102–105]GlnLysPro → MetValSer	CH_3	66	0.16	410 000
	$CH_2CH(CH_3)_2$	0.67	1.9	353
[236–237]AlaAla → GlyGly	CH_3	167	4	42 000
	$CH_2CH(CH_3)_2$	1.74	15.4	110
[102–105]GlnLysPro → MetValSer /[236–237]AlaAla → GlyGly	CH_3	32	4	8 000
	$CH_2CH(CH_3)_2$	18.5	14.3	1 300

reaction) were altered in order to accommodate substrates with hydrophobic side chains larger than that of the naturally preferred substrate, pyruvate. The mutations [102–105]GlnLysPro → MetValSer and [236–237]AlaAla → GlyGly were made to increase to tolerance for large hydrophobic substrate side chains as shown in Table 15-7. The five changes together produced a broader substrate specificity LDH, with a 55 fold improved k_{cat} for α-keto isocaproate [R = $CH_2CH(CH_3)_2$].

The substrate specificity of isocitrate dehydrogenase (IDH) has also been redesigned by genetic methods[131]. Despite the structural similarities between isocitrate (ISO) and isopropylmalate (IPM), wild type isocitrate dehydrogenase (IDH) exhibits a strong preference for its natural substrate (ISO). The substrate specificity of IDH was changed to that of isopropylmalate dehydrogenase (IPMDH) using a combination of rational and random mutagenesis. Three amino acids of IDH (S113, N115, V116) were changed and the chimeric enzyme ETV (S113E, N114T, V116V) showed

Table 15-8. Redesigning the substrate specificity of isocitrate dehydrogenase[131].

isocitrate (ISO)   isopropylmalate (IPM)

Enzyme	IDH position			k_{cat}/K_M IPM $(M^{-1} s^{-1})$	k_{cat}/K_M ISO $(M^{-1} s^{-1})$	k_{cat}/K_M IPM / k_{cat}/K_M ISO
	113	115	116			
Wild Type IPMDH	E	L	L	1.4×10	0	–
Wild Type IDH	S	N	V	1.7×10^{-6}	1.6×10	1.0×10^{-7}
EVG	E	V	G	1.1×10^{-5}	1.1×10^{-5}	1.0
ENA	E	N	A	1.5×10^{-5}	5.9×10^{-6}	2.5
ETV	E	T	V	1.8×10^{-4}	3.9×10^{-5}	4.6

Table 15-9. Elimination of the cofactor requirement by "blind" directed evolution[132].

Bacillus stearothermophillus lactate dehydrogenase	Cofactor (Fructose 1,6-bisphosphate)	$K_M$pyruvate (mM)
Wild	+	0.05
Wild	−	5
Mutated (R118C, 203L, N307S)	+	0.05
Mutated (R118C, 203L, N307S)	−	0.07

a preferred substrate specificity for IPM over ISO; [k_{cat}/K_MIPM] / [k_{cat}/K_MISO] of ETV was 4.6 while that of wild type IDH was 1.0×10^{-7}.

15.1.3.6
Production of an Activated Form of an Enzyme by Directed Evolution

One of the drawbacks of using alcohol dehydrogenases as catalysts for organic synthesis (comparing them with hydrolytic enzymes) is the cofactor requirement[132]. For example, *Bacillus stearothermophillus* lactate dehydrogenase is activated in the presence of fructose 1,6-bisphosphate[132]. The activator is expensive and representative of the sort of cofactor complications that are undesirable in industrial processes. Three rounds of random mutagenesis and screening produced a mutant which is almost fully activated in the absence of fructose 1,6-bisphosphate as shown in Table 15-9.

15.1.3.7
Change in the Coenzyme Specificity by Genetic Methods: NADP(H) Specific Formate Dehydrogenase

Formate/formate dehydrogenase is one of the most useful coenzyme regeneration systems as has been described in the Sect. 15.1.1.2. However, the known wild type formate dehydrogenases only accept NAD$^+$; NADP$^+$ is not the substrate. Multipoint site-directed mutagenesis was used to create a formate dehydrogenase which was able to accept NADP$^+$. This mutant enzyme was then coupled to the reduction using the alcohol dehydrogenase from *Lactobacillus sp* as shown in Fig. 15-23[22]. The activity of the NADP(H)-specific mutant (with NADP$^+$ as substrate) is about 60% of the activity of wild type formate dehydrogenase (with NAD$^+$ as substrate).

15.1.3.8
Use of a Mutant Dehydrogenase for the Synthesis of 4-Amino-2-Hydroxy Acids

The usefulness of a mutant dehydrogenase was demonstrated in a practical synthesis of 4-amino-2-hydroxy acids, which themselves are valuable as γ-turn mimics for investigations into the secondary structure of peptides[146]. Chemoenzymatic synthesis of these compounds were achieved by lipase catalyzed hydrolysis of a α-keto esters to the corresponding α-keto acids followed by reduction employing a lactate dehydrogenase in one pot. Wild type lactate dehydrogenase from either *Bacillus*

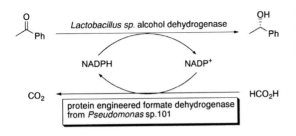

Figure 15-23. Recycling of NADPH with protein engineered formate dehydrogenase[22].

Table 15-10. The use of a mutant dehydrogenase for the synthesis of 4-amino-2-hydroxy acids[146].

Dehydrogenase	R	Reaction Time	Yield (%)
Wild type	a: CH$_3$	4 days	67
Staphylococcus epidermidis	b: CH(CH$_3$)$_2$	no reaction	–
lactate dehydrogenase	c: CH$_2$CH(CH$_3$)$_2$	no reaction	–
	d: CH$_2$Ph	no reaction	–
H205Q mutant of	a: CH$_3$	4 h	85
Lactobacillus delbrueckii bulgaricus	b: CH(CH$_3$)$_2$	5 h	90
D-hydroxyisocaproate dehydrogenase	c: CH$_2$CH(CH$_3$)$_2$	4 h	78
	d: CH$_2$Ph	5 h	85

stearothermophilus (BS-LDH) or *Staphylococcus epidermidis* (SE-LDH) could be used specifically to reduce the ketone of the alanine derived α-keto acid, **2a**, giving the *(S)*- and *(R)*-2-hydroxy acids, respectively, in good yields.

However, more bulky α-keto acids **2b–2d** were not substrates for these enzymes. In contrast, the genetically engineered H205Q mutant of *Lactobacillus delbrueckii bulgaricus* D-hydroxyisocaproate dehydrogenase proved to be an ideal catalyst for the reduction of all the α-keto acids **2a–2d**, giving excellent yields of the CBZ-protected (2R, 4S)-4-amino-2-hydroxy acid as a single diastereomer (Table 15-10). This genetically engineered oxidoreductase has great potential value in synthesis, not only due to its broad substrate specificity but also due to the high catalytic activity. For example, reduction of 1 mmol of **2a** took just 4 h with the H205Q mutant, whereas with SE-LDH the reaction required 4 days.

15.1.3.9

Catalytic Antibody

Nakayama and Schultz have developed antibodies to carry out the catalytic enantioselective reduction of an α-keto amide using NaBH$_3$CN as the reductant[147]. Monoclonal antibodies raised to phosphonate **3** were prepared (Fig. 15-24), and one antibody showed activity for the enantioselective reduction of a chiral keto amide **4**.

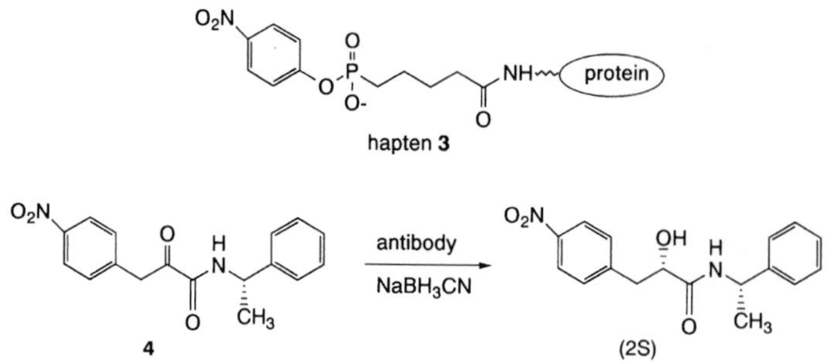

hapten 3

Figure 15-24. Reduction of a ketone by a catalytic antibody[147].

Reduction with the antibody gave the 2S product with a diastereomeric excess greater than 99% (opposite to the stereoselectivity of the uncatalyzed reaction which afforded the 2R product).

15.1.4
Reduction Systems with Wide Substrate Specificity

15.1.4.1
Bakers' Yeast

Many methods for asymmetric reduction have been developed and some of these are used for the synthesis of optically active alcohols on a preparative scale. Bakers' yeast is one of the most widely used microorganisms due to its commercial availability and its wide substrate specificity, which enables the non-expert in biochemistry to use the biocatalyst as a reagent for organic synthesis. Detailed reactions will not be described in this text since there are many reviews and original reports on this subject[1, 37–39, 148–162]. However, one of the most important and useful reactions using yeast, the reduction of a hydroxymethyl ketone, is featured here due to the excellent enantioselectivity obtained even on a large scale (Fig. 15-25).[163–166]. For example, 1-hydroxy-2-heptanone (50 g) was reduced to the corresponding (R)-diol in an optically pure form in 56% yield [Fig. 15-25 (b)][164]. Another example [Fig. 15-25 (c)] is the reduction of a sulphenyl hydroxyketone with yeast in the synthesis of a natural product[166]. Products isolated from the mandibular glands of the oriental hornet were synthesized using yeast reduction of an S-substituted hydroxyketone.

15.1.4.2
Rodococcus erythropolis

A carbonyl reductase isolated form *Rhodococcus erythropolis* accepts a broad range of substrates, including a variety of compounds useful for synthetic chemistry, as shown in Table 15–11[25]. Reduction of all the carbonyl compounds tested yielded (S)-configured hydroxyl compounds with high enantioselectivities.

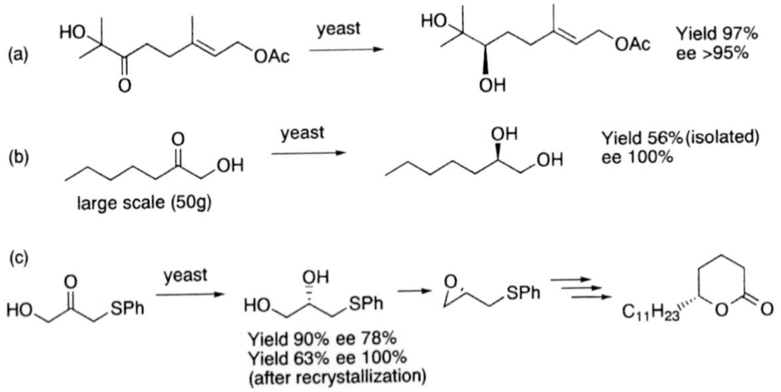
(a)
HO
OAc
O
yeast
HO
OH
OAc
Yield 97%
ee >95%
(b)
O
OH
large scale (50g)
yeast
OH
OH
Yield 56% (isolated)
ee 100%
(c)
HO
O
SPh
yeast
OH
HO
SPh
O
SPh
$C_{11}H_{23}$
O
O
Yield 90% ee 78%
Yield 63% ee 100%
(after recrystallization)

Figure 15-25. Reduction of hydroxyketones by bakers' yeast[163, 164, 166].

Table 15-11. Kinetic constants of the *R. erythropolis* carbonyl reductase[25].

Substrate	V_{max} (U mg⁻)	K_M (m*M*)	Substrate	V_{max} (U mg^{-1})	K_M (m*M*)
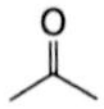	3.5	330	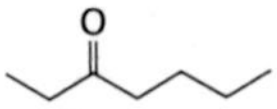	0.46	18
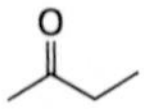	3.5	260	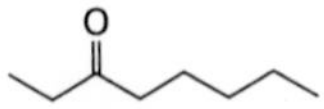	1.4	7.3
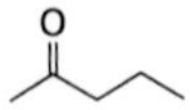	4.8	59	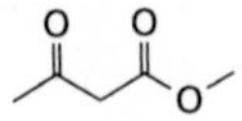	2.6	16
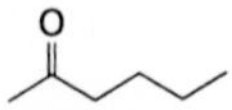	7.7	3.8	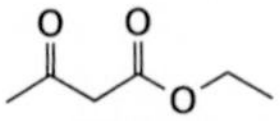	5.5	3.1
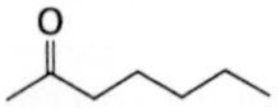	10.4	0.59	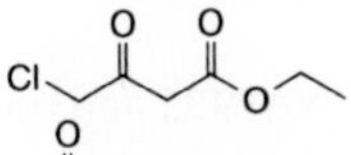	4.2	9.9
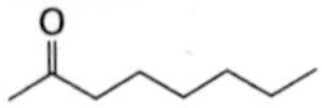	10.3	0.42	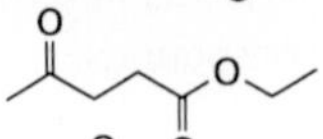	7.6	8.3
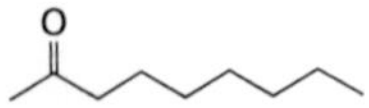	10.8	0.34	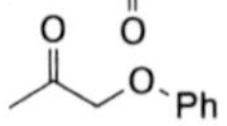	10.6	0.039
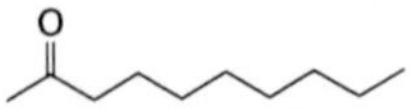	11.1	0.54	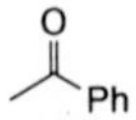	1.7	3.8

15.1.4.3
Pseudomonas sp. Strain PED and *Lactobacillus kefir*

The substrate specificities of the alcohol dehydrogenases from *Pseudomonas sp.* strain PED and *Lactobacillus kefir* have been investigated. It was reported that they reduce wide varieties of ketones[6, 7]. Both reactions use 2-propanol for the regeneration of coenzyme and produce (*R*)-alcohols as depicted in Table 15-12. However, they require different coenzymes. The alcohol dehydrogenase from the *Pseudomonas sp.* uses NADH and transfers to pro-*R* hydride of NADH to the *si*-face of carbonyl compounds as shown in Sect. 15.1.1.1. The mechanism is ordered bi–bi with the coenzyme binding first and released last. On the other hand, the enzyme from

Table 15-12. Enantioselectivities of the alcohol dehydrogenases from *Pseudomonas sp.* strain PED and *Lactobacillus kefir*[6, 7].

Product	ee (%)		Product	ee (%)	
	Pseudomonas sp. strain PED	*Lactobacillus kefir*		*Pseudomonas* sp. strain PED	*Lactobacillus kefir*
Ph⌒CF₃, OH	92	> 99	(structure: ester, O–CH₂C(O)–CH₂–CHCl with OH)	98	–
Ph⌒, OH	94	–	(branched chain with OH)	97	> 99
Ph, OH, O (keto)	86	–	Cl⌒⌒⌒ OH	93	> 97
Ph, OH, O–O (ester)	98	–	(furan) O⌒ OH	45	–
Ph (S,S) cyclopropyl OH	65	–	⌒⌒⌒⌒ OH	27	–
Ph⌒ cyclopropyl OH	92	–	cyclopropyl OH	–	> 97
pyridine OH	–	> 97	norbornyl OH	–	> 97
furan OH	–	95	Si≡ alkyne OH	–	94
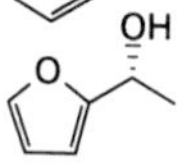			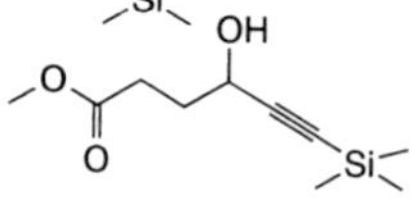 O= Si alkyne OH	–	97

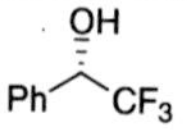

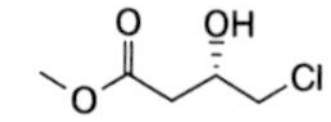

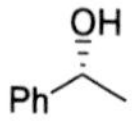

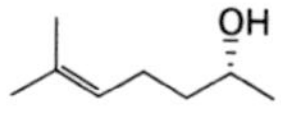

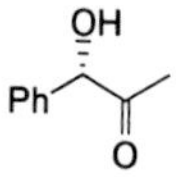

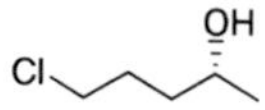

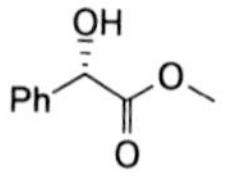

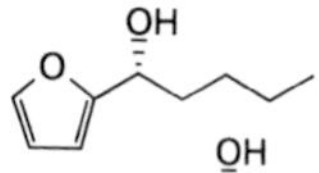

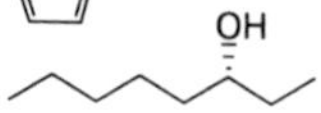

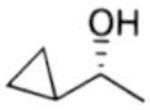

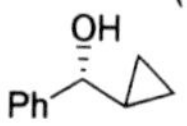

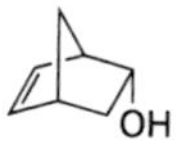

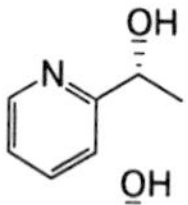

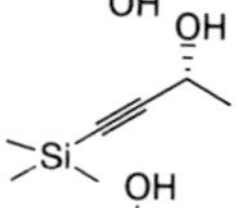

Lactobacillus kefir uses NADPH and transfers the *pro-R* hydride from the cofactor to the *si*-face of carbonyl compounds.

15.1.4.4

Thermoanaerobium brockii

The alcohol dehydrogenase from *Thermoanaerobium brockii* is very suitable for the reduction of aliphatic ketones[18, 19]. Even very simple aliphatic ketones can be reduced enantioselectively. An interesting substrate size-induced reversal of enantio-selectivity was observed. The smaller substrates (methyl ethyl, methyl isopropyl or methyl cyclopropyl ketones) were reduced to the *(R)*-alcohols, whereas higher ketones produced the *(S)*-enantiomers.

This example and the next one (Sect. 15.1.4.5) using *G. candidum* show that the biocatalytic reduction system is very beneficial for the reduction of aliphatic ketones over a non-enzymatic system where no report on highly enantioselective (> 99 % ee) reduction of unfunctionalized dialkyl ketones can be found, to the best of our knowledge.

Table 15-13. Asymmetric reduction of aliphatic ketones with the alcohol dehydrogenase from *Thermoanaerobium brockii*[18].

Product	Relative rate	ee (%)	Config.	Product	Relative rate	ee (%)	Config.
(structure, OH)	12.0	48	R	(structure, OH)	0.9	99	S
(structure, OH)	3.0	86	R	(structure, OH)	0.2	95	S
(structure, OH)	0.8	44	R	(structure, OH)	0.6	97	S
(structure, OH)	3.3	79	S	(structure, OH)	0.3	99	S
(structure, OH)	1.0	96	S	(structure, OH)	0.3	98	S
(structure, OH)	0.3	95	S	(structure, OH)	0.1	99	S
(structure, OH)	0.1	81	2 S, 3 R	(structure, OH, Cl)	1.5	98	S
(structure, OH)	0.9	97	S				

15.1.4.5
Geotrichum candidum

Reductions using an acetone powder of *G. candidum* (APG4), NAD⁺ and 2-propanol exhibit one of the widest substrate specificities together with very high enantiose-lectivities (Table 15-14) [20, 21]. Various ketones such as acetophenone derivatives can be reduced with APG4 with excellent enantioselectivities (> 99 % ee). The nature and electronegativity of substituents on the phenyl ring did not affect the enantioselectivity although the yield was slightly lower for *para* derivatives than for the corresponding *ortho* and *meta* derivatives.

Reduction by APG4 of several aromatic ketones having different length alkyl chains demonstrated the scope and limitations of the substrate specificity. The phenyl moiety of acetophenone can be replaced by a benzyl or even by a 2-phenyl-ethyl group with slightly better results in terms of chemical yield without any decrease in enantioselectivity. However, when the methyl moiety of actophenone was replaced by an ethyl, isopropyl or methoxymethyl group, the yield decreased dramatically, although the enantioselectivity remained high (> 99% ee). When the alkyl chain was elongated to a propyl or enlarged to a *t*-butyl group, the reaction was observed scarcely to proceed.

The versatility of the APG4 reduction system is further exemplified by the use of β-keto esters as substrates. 3-Oxobutyrates involving methyl, ethyl, *t*-butyl, or neo-pentyl esters are reduced to the *(S)*-hydroxyesters with > 99% ee and in quantitative yield. Moreover, simple aliphatic ketones from 2-octanone to 2-undecanone, as well

Table 15-14. Reduction of various ketones by the acetone powder of *G. candidum*, NAD⁺ and 2-propanol[20, 21].

Product	Yield (%)	ee (%)	Product	Yield (%)	ee (%)
OH (X)	X = H 89	> 99 (S)	OH (R, Ph)	R = Et 41	> 99 (S)
	o-F > 99	> 99 (S)		Pr 0	–
	m-F 95	> 99 (S)		i-Pr 12	99 (S)
	p-F 74	> 99 (S)		t-Bu 1	–
	o-Cl > 99	> 99 (S)		CH₂OMe 8	> 99 (R)
	m-Cl 95	99 (S)		CH₂Cl 80	98 (R)
	p-Cl 62	> 99 (S)			
	o-Br 97	> 99 (S)	OH O (O-R)	R = Me > 99	> 99 (S)
	m-Br 92	> 99 (S)		Et > 99	> 99 (S)
	p-Br 95	> 99 (S)		t-Bu > 99	> 99 (S)
	o-Me 96	> 99 (S)		neo-Pentyl > 99	> 99 (S)
	m-Me 86	> 99 (S)			
	p-Me 78	> 99 (S)	OH O	72	> 99 (S)
	o-MeO 84	> 99 (S)			
	m-MeO 90	> 99 (S)	OH (R)	R = me 87	> 99 (S)
	p-MeO 29	> 99 (S)		Et 87	> 99 (S)
	o-CF₃ 6	97 (S)		Pr 85	> 99 (S)
	m-CF₃ 96	> 99 (S)		Bu 60	> 99 (S)
	p-CF₃ 73	> 99 (S)			
	1',2',3',4',5'-F₅ 62	> 99 (S)	OH	90	99 (S)
OH (Ph)	96	> 99 (S)			
OH (Ph)	93	> 99 (S)	OH (Cl)	92	99 (S)

as 6-methyl-5-heptene-2-one and 5-chloro-2-pentanone are also reduced by the APG4 system to the corresponding (S)-2 alkanols giving high yields with 99 % ee.

In summary, a detailed investigation of substrate specificity for the acetone powder of a *G. candidum* system reveals that as long as there is a methyl group at the α-position of the carbonyl group, high yield and enantioselectivity can be obtained regardless of the substituent on the other side of the ketone moiety.

Apart from acetone-dried *G. candidum* IFO 4597, intact whole cells of various strains of *G. candidum* have been found to be useful for asymmetric reductions[75–78, 101, 126, 167–171]. For example, methyl 2-acetylbenzoate was reduced by *G. candidum* ATCC 34614, IFO 5767 or IFO 4597 as well as by other microorganisms such as *Mucor javanicus*, *Mucor heimalis*, *Endomyces magnusii*, *Endomyces reessii* and bakers' yeast to afford phthalide derivatives (Fig. 15-26) which have various pharmacological profiles such as relaxant, antiproliferative or antiplatelet effects, *etc.*[171].

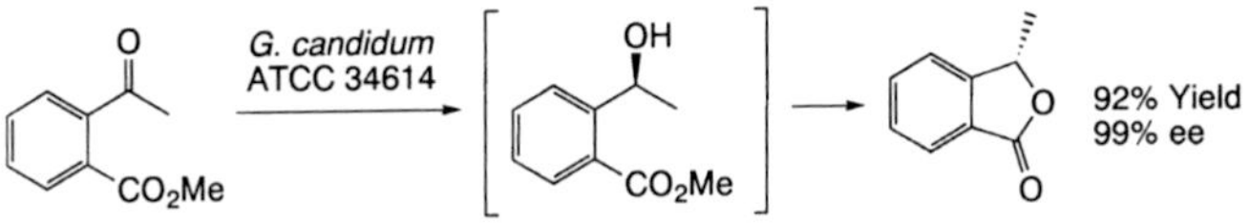

Figure 15-26. Asymmetric reduction by *G. candidum* ATCC 34614 for the synthesis of a bioactive phthalide derivative[171].

15.1.5
Reduction of Various Ketones

15.1.5.1
Reduction of Fluoroketones

The biocatalytic reduction of fluoroketones is useful in order to gain an insight into the enzyme recognition of fluorinated groups, and is also very important due to the high synthetic values of the products, optically active fluorinated alcohols[160, 172–185]. Sometimes the monofluorinated substrate can be a straightforward mimic of the unsubstituted counterpart, but with difluorinated and trifluorinated substrates, different recognition patterns compared with unfluorinated or monofluorinated substrates and with each other are often observed. For example, the enantioselectivity of yeast reduction is definitely affected by the fluorination pattern on the substrate[172]. One of the most prominent effects of the fluorination of a substrate is seen in the reduction of acetophenone derivatives by the acetone powder of *Geotrichum candidum* (APG4) as shown in Fig. 15-27[173, 174]. Reduction of methyl ketones afforded (*S*)-alcohols in excellent ee, whereas the reduction of trifluoromethyl ketones gave the corresponding alcohols of the opposite configuration, also in excellent ee. Monofluoroacetophenone and difluoroacetophenone were also reduced under the same conditions. The reduction proceeded quantitatively for both substrates. As expected, the stereoselectivity shifted from the acetophenone type to the trifluoroacetophenone type according to the number of fluorine substituents at the α-position as shown in Fig. 15-28.

The replacement of the methyl moiety with a trifluoromethyl group alters the bulkiness and electronic properties: the effect on the enantioselectivity has been examined. No inversion in stereochemistry was observed for the reduction of hindered ketones such as isopropyl ketone, while the stereoselectivity was inverted for the reduction of ketones with electron-withdrawing atoms such as chlorine. The mechanism for the inversion in stereochemistry was investigated in further studies. Several enzymes with different enantioselectivities were isolated; one of them

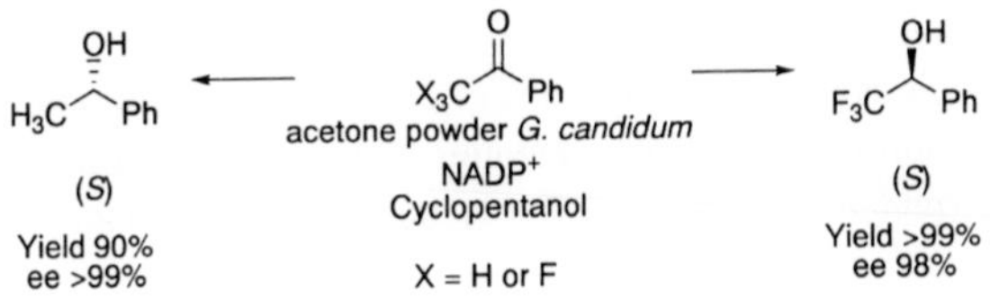

Figure 15-27. Reduction of acetophenone and trifluoroacetophenone by an acetone powder of *Geotrichum candidum*, NADP+ and cyclopentanol[173, 174].

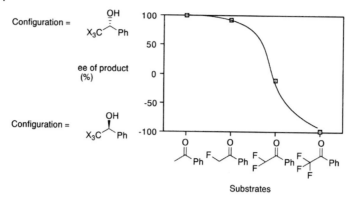

Figure 15-28. Effect of introducing a fluorine atom or atoms at the α-position of acetophenone on the stereoselectivity in the reduction by *G. candidum* acetone powder[174].

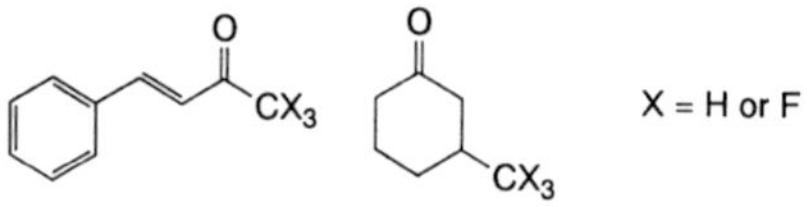

Figure 15-29. Substrates used for the examination of the stereodirecting effects of trifluoromethyl and methyl groups[175].

catalyzed the reduction of methyl ketones, and another, with the opposite enantiose-lectivity, catalyzed the reduction of trifluoromethyl ketones.

The differing abilities of trifluoromethyl and methyl groups to direct enantioselection in the reduction of carbonyl substrates has also been analyzed using various other microorganisms including different strains of *G. candidum, Hansenula anomala, Saccharomyces cervisiae, Streptomyces, etc.*[175]. The reduction of the cyclic ketone and enones shown in Fig. 15-29 was investigated. The differences in the electronic and steric properties of the trifluoromethyl and methyl residues resulted in different chemo- and enantioselectivities in the reduction of the phenylbutenones, while the cyclohexanones showed similar enantioselectivities.

Many synthetically valuable reactions involving reductions of fluoroketones have been reported as shown in Fig. 15-30[176–178]. Various monofluoroketones are reduced with yeast; some of them proceeded with high diastereoselectivity.

Chiral trifluoromethyl benzyl alcohols are useful synthons for ferroelectric liquid crystals. Therefore, Fujisawa *et al.* investigated the asymmetric reduction of the corresponding ketones using bakers' yeast[179, 180]. The enantioselectivity of the bakers' yeast reduction of trifluoroacetylbenzene derivatives was improved by the introduction of some functional groups at the *para*-position to give the corresponding *(R)*-trifluoromethyl substituted benzylic alcohols in high chemical and optical yields as shown in Fig. 15-31. The "enantio-controlling" functional group at the *para*-position was then used in further transformations.

Yeast and *G. candidum* acetone powder (APG4) are complementary to each other in the reduction of various trifluoromethyl biphenyl ketones. Yeast reduction affords the *(R)*-alcohol, whereas *G. candidum* reduction affords the *(S)*-alcohol (Fig. 15-32)[181].

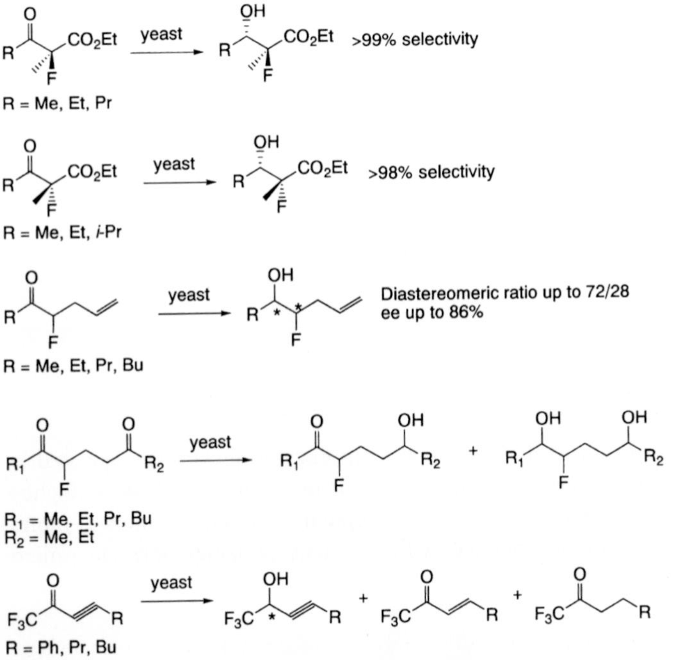

Figure 15-30.   Reduction of fluorinated ketones by yeast [176–178].

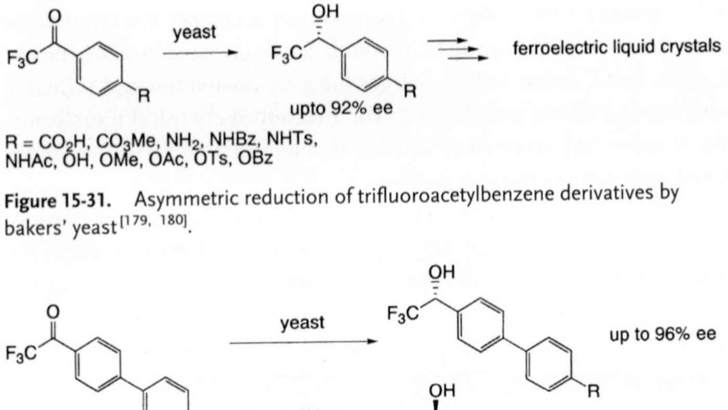

Figure 15-31.   Asymmetric reduction of trifluoroacetylbenzene derivatives by bakers' yeast [179, 180].

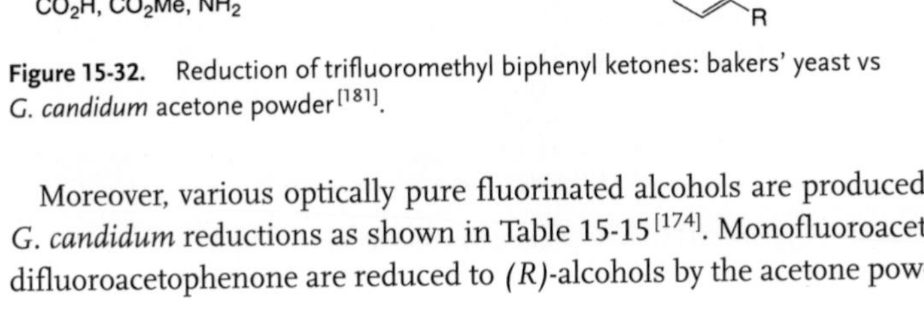

Figure 15-32.   Reduction of trifluoromethyl biphenyl ketones: bakers' yeast vs G. candidum acetone powder [181].

Moreover, various optically pure fluorinated alcohols are produced by employing G. candidum reductions as shown in Table 15-15 [174]. Monofluoroacetophenone and difluoroacetophenone are reduced to (R)-alcohols by the acetone powder, NAD+ and

Table 15-15. Synthesis of chiral fluorinated alcohols by the reduction with acetone powder and isolated enzymes of *Geotrichum candidum* IFO 4597[174].

Product		Yield (%)	ee (%)	Product	Yield (%)	ee (%)
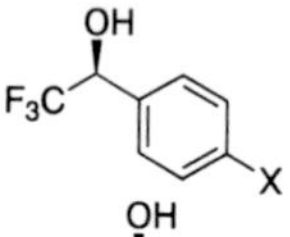 [structure: F_3C-C(OH)-C6H4-X]	X = H	84	98 (S)	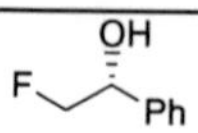 [structure: F...CH(OH)-Ph]	93	> 99 (R)
	X = Cl	81	> 99 (S)			
	X = Br	80	> 99 (S)	[structure: F-CH2...CH(OH)-Ph]	91	> 99 (S)[a]
[structure: F_3C-CH(OH)-CH2-Ph]		74	98 (S)	[structure: F,F...CH(OH)-Ph]	99	63 (R)
[structure: Cl,F,F...CH(OH)-Ph]		82	94 (S)	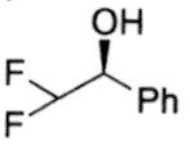 [structure: F,F,F...CH(OH)-Ph]	95	> 99 (S)[a]

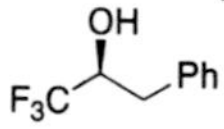

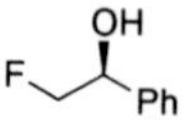

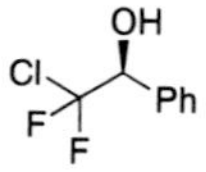

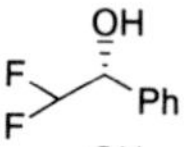

a The isolated enzyme was used for the reduction.

2-propanol, and to *(S)*-alcohols by a constituent enzyme previously separated by anion-exchange chromatography and using glucose-6-phosphate/glucose-6-phosphate dehydrogenase as the cofactor recycling system. Both enantiomers of monofluorophenylethanol can be obtained with excellent ee using only one microorganism.

15.1.5.2
Reduction of Fluoroketones Containing Sulfur Functionalities

As the demand for optically active fluorinated compounds increases, the importance of the development of asymmetric synthetic methods for fluorinated building blocks grows. On the other hand, sulfur functionalities such as phenylthio and dithianyl groups have been used as useful reactive units for a variety of chemical transformations. Therefore, various trifluoromethyl ketones containing a sulfur functionality have been reduced with various microorganisms [182–185].

For example, several microorganisms have been employed for the reduction of α,α,α,-trifluoromethyl α'-sulphenyl ketones (Fig. 15-33). Some of them produce the corresponding alcohols in high diastereo- and enantioselectivities; the high conver-

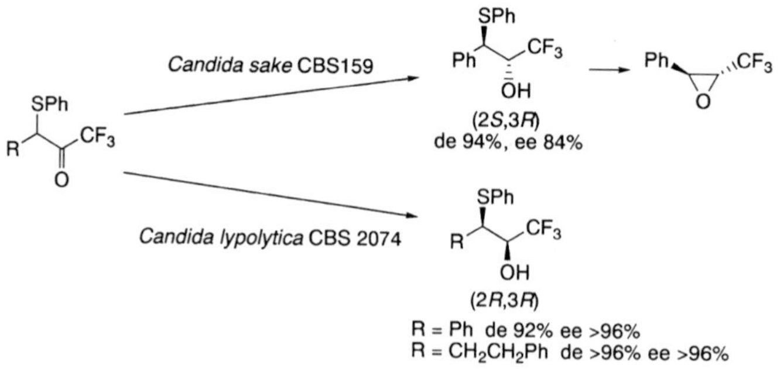

Figure 15-33. Reduction of sulphenyl ketones followed by epoxide formation [182].

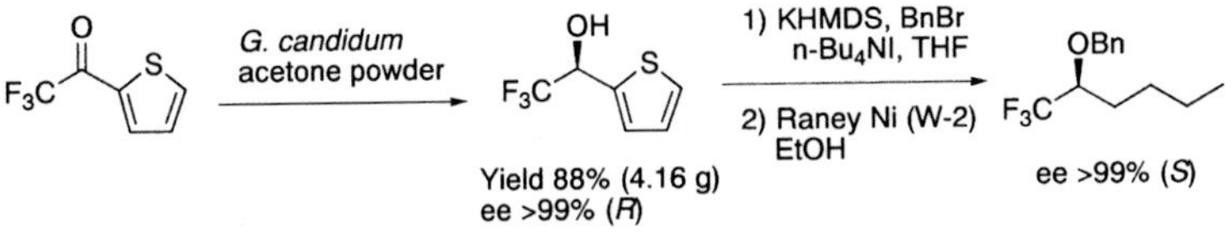

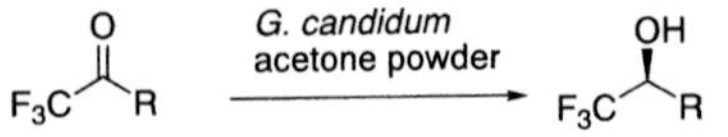

R	Yield (%)	ee (%)
SPh	90	98 (R)
CH$_2$CH$_2$CH$_2$SPh	78	>99 (S)
CH$_2$SC$_9$H$_{17}$	35	96 (R)
3-Thienyl	87	>99
1,3-Dithian-2-yl	42	>99

Figure 15-34. Asymmetric reduction of trifluoromethyl ketones containing a sulfur functionality by the acetone powder of *G. candidum*[183].

sion into a single enantiomer is secured by the racemization of starting ketones under the biotransformation conditions. Transformation of the resulting sulphenyl trifluoromethyl alcohols into trifluoromethyl epoxides was also achieved[182].

The acetone powder of *G. candidum* (APG4) has also been used for the reduction of sulfur containing trifluoromethyl ketones (Fig. 15-34)[183]. This reaction can be scaled up easily without the loss of enantioselectivity. For example, the reduction of trifluoro(2-thienyl)ethanone on the gram scale proceeded quantitatively and yielded the optically pure *(R)*-alcohol in 88% yield after purification (4.16 g, ee > 99%). The thienyl alcohol can be further transformed into a fluorinated aliphatic alcohol without racemization.

15.1.5.3
Reduction of Chloroketones

The reduction of chloroketones has been widely investigated since it can produce versatile chiral intermediates. For example, reduction of an α-chloroketone results in the formation of a chlorohydrin, which can easily be transformed into an epoxide on treatment with a base. On recently published example involves the reduction of 3,4-dichlorophenacylchloride by *Rhodotorula mucillaginosa* CBS 2378 or *Geotrichum candidum* CBS233.76 to give the *(R)*- or *(S)*-chlorohydrin with > 99% ee and > 98% ee, respectively, as shown in Fig. 15-35[186]. The *(S)*-enantiomer was transformed into the corresponding epoxide and then into a dichlorophenylbutanolide, an intermediate in the synthesis of (+)-*cis*-1S,4S-sertraline, which is an antidepressant drug of the selective serotonin reuptake inhibitor (SSRI) type.

There are also many other examples of the reduction of α-halomethyl ketones as shown in Table 15-16[187–189]. Various microorganisms are able to reduce fluoro-, chloro- and bromoketones[161, 190–192]. However, reduction of iodoacetophenone usually results in a poor yield, producing, mainly, acetophenone or phenylethanol.

Another example of the reduction of α-chloroketone involves dynamic kinetic resolution. The reduction of an α-chloroketo ester by *M. racemosus* and *R. glutinis* resulted in optically active *syn-* and *anti-*chlorohydrin, respectively, as shown in

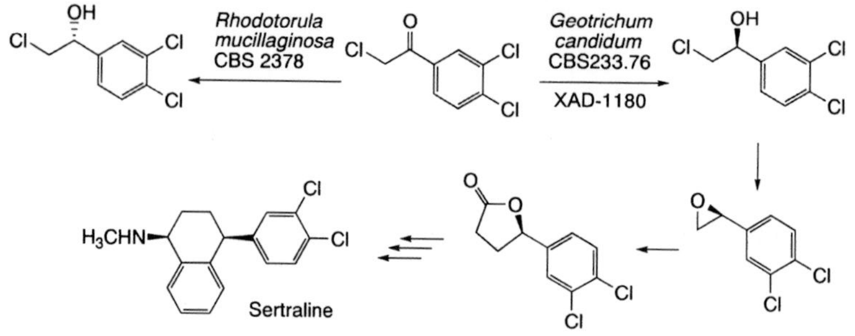

Figure 15-35. Reduction of a chloroketone followed by epoxidation for the synthesis of sertraline [186].

Table 15-16. Reduction of α-halogenated acetophenones.

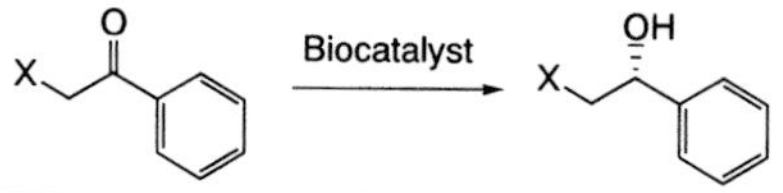

Catalyst	X	Yield[a] (%)	ee (%)	Reference
Cryptococcus macerans	Cl	80	100	187
	Br	95	93	187
Bakers' yeast	F	67	97	188
	Cl	37	90	188
	Br	9	97	188
	F	55	35	189
	Cl	6 (40)	68	189
	Br	0 (15)	–	189
Geotrichum candidum sp. 38	F	65	75	189
	Cl	86	87.4	189
	Br	15 (25)	94	189

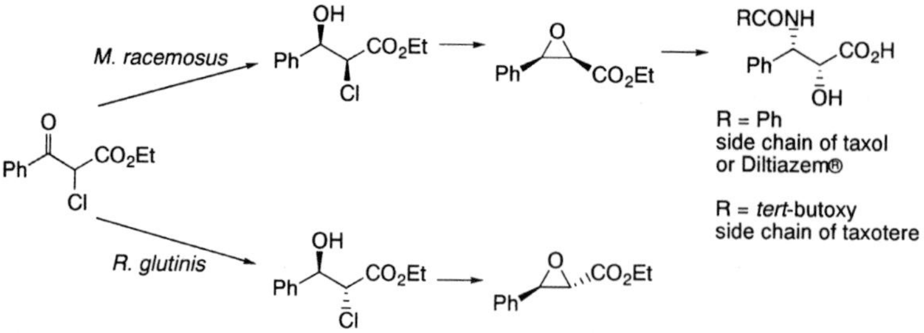

Figure 15-36. Enantio- and diastereo-selective reduction of a chloroketone [193, 194].

Fig. 15-36 [193]. The *syn*-isomer was transformed into the corresponding epoxide, followed by conversion into the side chain of taxol and taxotere [194].

One of the most studies α-chloroketones is ethyl 4-chloro-3-oxobutanoate. (*R*)- and (*S*)-enantiomers of the corresponding alcohol were produced by various micro-

Table 15-17. Comparison of various microorganisms for the reduction of ethyl 4-chloro-3-oxobutanoate.

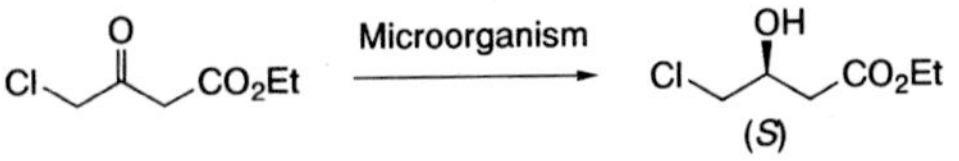

Microorganism	Yield (%)	ee (%)	Reference
Geotrichum candidum	98	96	170
Bakers' Yeast	100	90	90
Bakers' Yeast		55	61
Lactobacillus kefir	100	100	195
Candida magnoliae (recombinant and overexpressed in *Escherichia coli*)	88	100	142

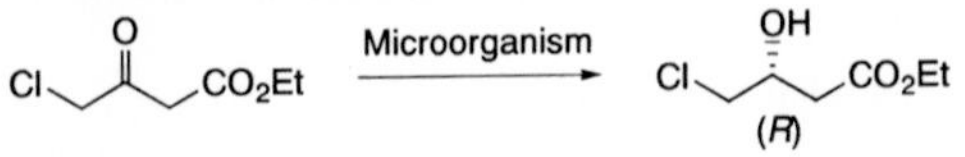

Microorganism	Yield (%)	ee (%)	Reference
Dancus carota	42	52	196
Sporobolomyces salmonicolor	95	86	197
Lactobacillus fermentum	70	98	195
Saccharomyces cerevisiae (FAS (β-keto reductase) negative)	55	16	63

organisms as shown in Table 15-17. The *(R)*-enantiomer is a promising chiral building block for the synthesis of L-carnitine, an essential factor for the β-oxidation of fatty acids in mitochondria.

As shown in Fig. 15-37, a chiral intermediate for a human immunodeficiency virus protease inhibitor (HIVPI) was also synthesized by the reduction of an α-chloroketone with a *Streptomyces* strain [198].

Another example of the reduction of chloroketone is the reduction of 5-chloro-2-pentanone by TBADH as shown in Fig. 15-38 [19]. Using this biotransformation in the synthetic pathway, a naturally occurring heterocycle isolated from the glandular secretion of the civet cat (*Viverra civetta*), was prepared.

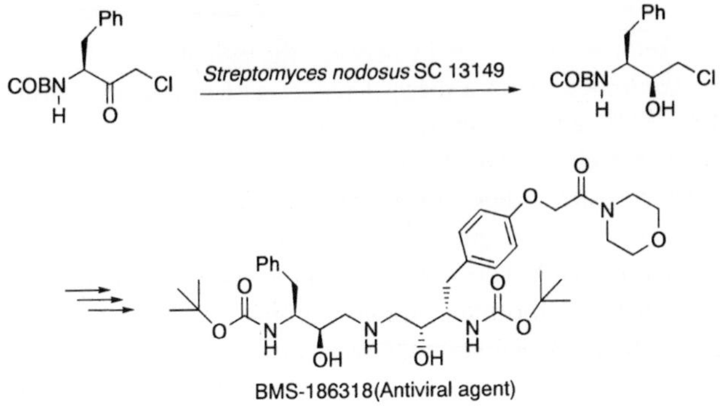

BMS-186318(Antiviral agent)

Figure 15-37. Synthesis of a chiral intermediate for an HIV-PI [198].

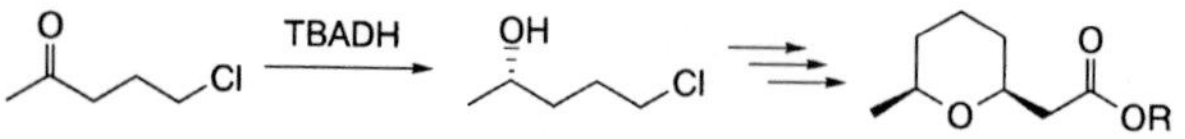

Figure 15-38. Reduction of 5-chloro-2-pentanone by TBADH for natural product synthesis[19].

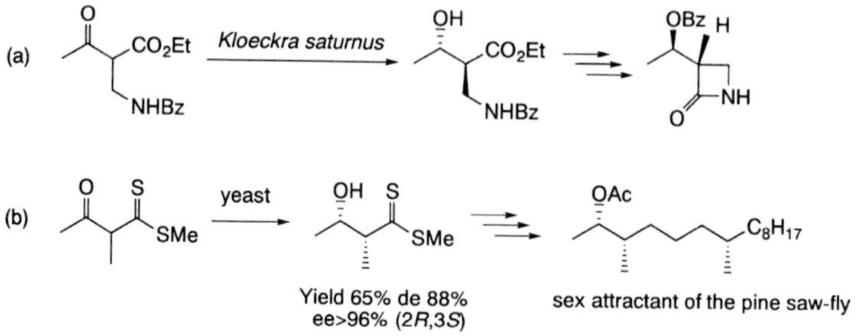

Yield 65% de 88%
ee>96% (2R,3S)

sex attractant of the pine saw-fly

Figure 15-39. Reduction of ketones containing sulfur or nitrogen functionality[199, 219].

15.1.5.4
Reduction of Ketones Containing Nitrogen, Oxygen, Phosphorus and Sulfur Functionalities

Ketones with useful heteroatomic functional groups containing nitrogen[199–212], oxygen[163, 213–217], phosphorus[218] and sulfur[154, 184, 219–227] have been reduced by biocatalysts. For example, an intermediate in the synthesis of β-lactam antibiotics was obtained by microbial reduction of a β-keto ester as shown in Fig. 5-39(a)[199], while yeast reduction of a β-keto dithioester afforded an easily separable mixture of β-hydroxy-dithioesters, the major component of which was converted enantiose-lectively into a sex attractant of the pine saw-fly as shown in Fig. 15-39(b)[219].

15.1.5.5
Reduction of Diketones

Regio- and enantioselective reduction of diketones can be achieved readily by using a biocatalyst[228–242]. As a result, optically active hydroxyketones and diols have been synthesized successfully.

For the reduction of α-diketones, the selectivity between the reduction to diol and to hydroxyketone can be controlled using a diacetyl reductase from *Bacillus stear-othermophilus* (Fig. 15-40)[233]. When a one-enzyme system was used for the coen-zyme recycling using endo-bicyclo[3.2.0]hept-2-en-6-ol (5), both carbonyl groups were reduced selectively to produce a diol. On the other hand, α-hydroxyketones were obtained using a two-enzyme system glucose 6-phosphate/glucose 6-phosphate dehydrogenase for coenzyme recycling. The synthetic potential of both systems has been illustrated by the synthesis of the male sex pheromone of the grape borer *Xylotrechus pyrrhoderus*, identified as a two-component mixture of the reduction products, **6** and **7**.

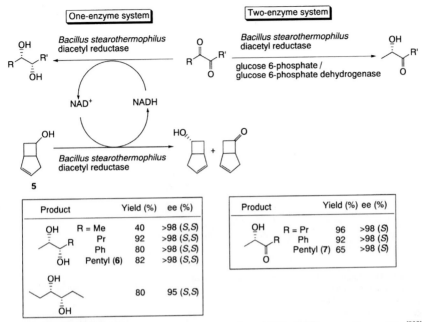

Figure 15-40. Reduction of α-diketones by diacetyl reductase from *Bacillus stearothermophilus*[233].

Regio- and enantioselective reduction of β-diketones may be carried out using biocatalysts. For example, a diketo ester **8** was reduced by the alcohol dehydrogenase from *Lactobacillus brevis*, to provide the corresponding hydroxyketo ester with 99.4 % ee in 78 % yield; this was used as an intermediate for the synthesis of dimeric metabolite vioxanthin of *Penicillium citreo-viride* in order to develop an assay system to monitor phenol oxidative coupling in lignan formation [Fig. 15-41(a)] [228]. Yeast reduction also proceeds regio- and enantioselectively with aliphatic diketones producing hydroxyketones with perfect selectivities as shown in Fig. 15-41(b) [232]. The yeast reduction also proceeds satisfactorily with 2,2-disubstituted cycloalkanediones, producing hydroxyketones with excellent enantio- and diastereoselectivities as shown in Fig. 15-41(c) [231].

15.1.5.6
Reduction of Diaryl Ketones

Bulky ketones such as diaryl ketones can be also reduced by biocatalysts. For example, a rice plant growth regulator, *(S)*-N-isonicotinoyl-2-amino-5-chlorobenzhydrol, was prepared by microbial reduction of 2-amino-5-chlorobenzophenone with *Rhodosporidium toruloides* followed by isonicotinoylation as shown in Fig. 15-42(a) [243]. A phosphodiesterase 4 inhibitor was also prepared by microbial reduction of a diaryl ketone **9** with *Rhodotorula pilimanae*, which was found by the screening of 310 microbial strains [Fig. 15-42(b)] [244].

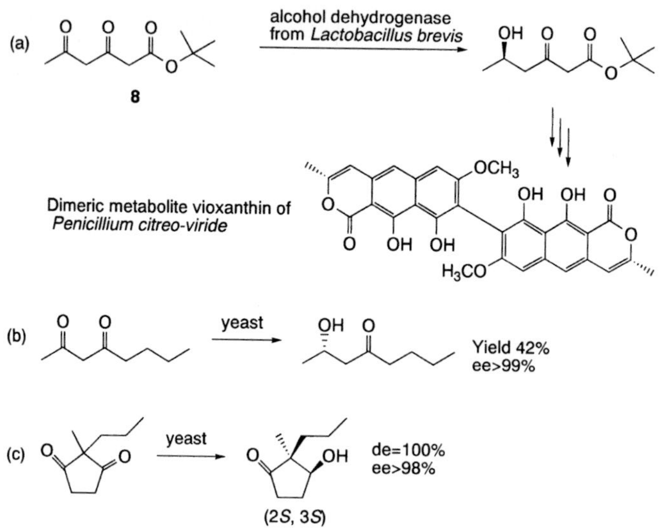

Figure 15-41. Regio- and enantioselective reduction of diketones [228, 231, 232].

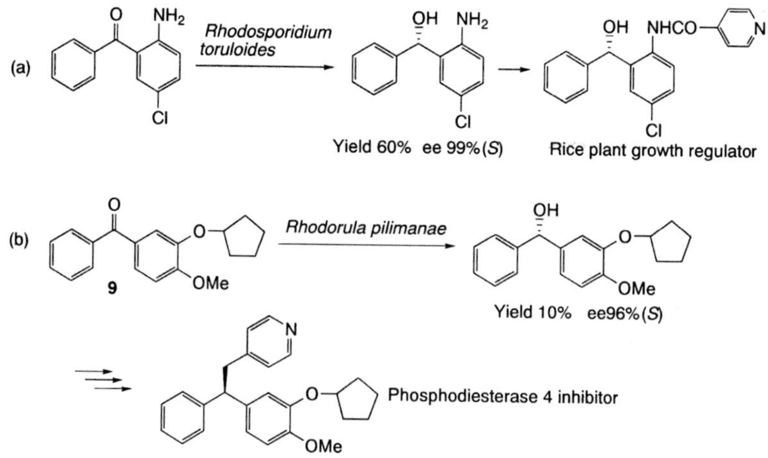

Figure 15-42. Reduction of diaryl ketones for the synthesis of bioactive compounds [243, 244].

15.1.5.7

Diastereoslective Reductions (Dynamic Resolution)

Enantio and diastereoselective reduction (dynamic resolution) of keto esters and ketones can be achieved using yeast and other microorganisms [55, 70, 74, 245–253]. As shown in Fig. 15-43, when the racemization rate of the keto ester is faster than that for the yeast reduction, and the product hydroxyester is not racemized under the reaction conditions, then the yeast reduction may proceed enantioselectively and

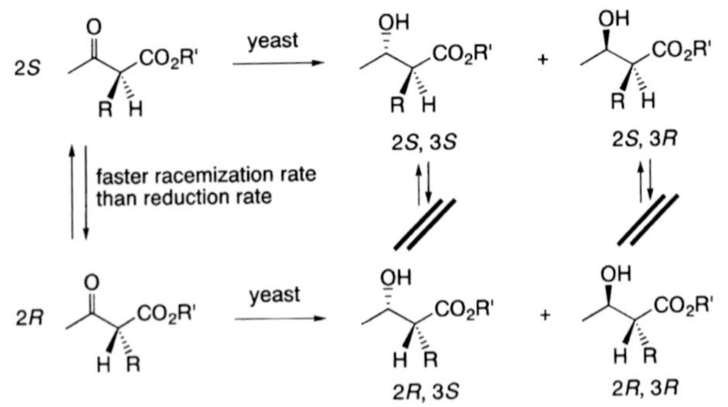

Figure 15-43. Diastereoselective reduction.

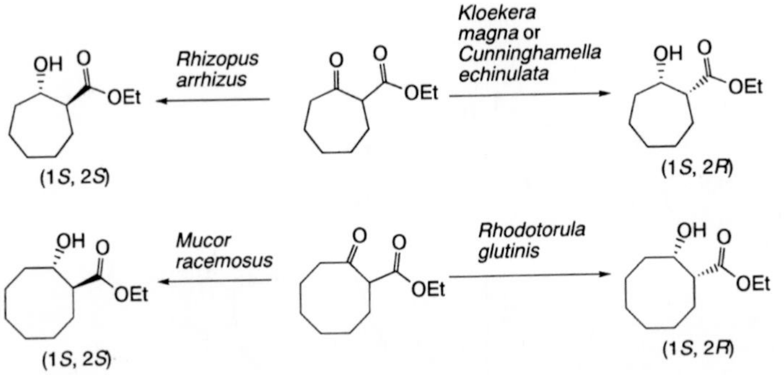

Figure 15-44. Diastereoselective reduction of cyclic keto esters [245].

diastereoselectively; thus only one stereoisomer out of the four possible ones can be obtained in one step. Actually, when bakers' yeast was used for the reduction of neopentyl 2-methyl-3-oxobutanoate (R = Me, R' = neopentyl), then the ratio of (2R, 3S) : (2S, 3R) : (2S, 3S) : (2R, 3R) products was found to be 96 : < 1 : 4 : < 1 [247]. When an enzyme was isolated from the yeast, then the diastereoselectivity was improved to > 99 : 1, and only a single isomer was obtained [248]. Another example is the large scale reduction of ethyl 2-methyl-3-oxobutanoate by *Klebsiella pneumoniae* IFO 3319 [70]. On a 200 L scale, 2 Kg of the substrate were converted into the (2R, 3S)-hydroxyester with 99% de, > 99% ee, and 99% chemical yield as shown in Table 15-2.

Enantio- and diastereoselective reduction of cyclic keto esters are also achieved using various microorganisms (Fig. 15-44) [245]. By selecting a suitable organism, *syn-* and *anti-*hydroxyesters may be synthesized enantio- and diastereoselectively.

15.1.5.8
Chemo-enzymatic Synthesis of Bioactive Compounds

Ketones with various functionalitis, containing F, Cl, N, S, O, etc., have been shown to be reduced by a biocatalyst, and by using the biocatalytic reduction as a key step, the chemoenzymatic synthesis of many bioactive compounds have been re-

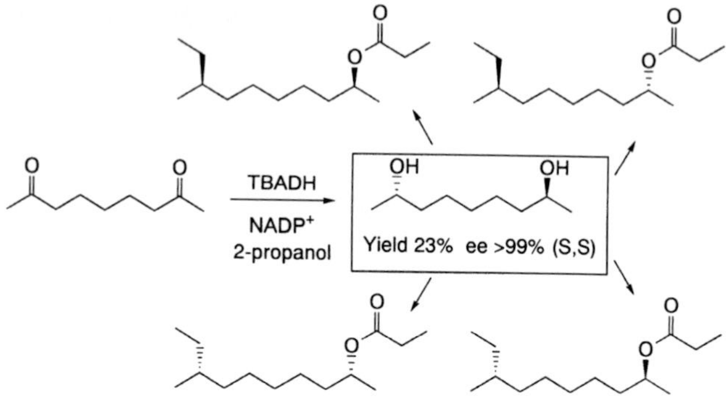

Figure 15-45. Synthesis of all four isomers of the western corn rootworm sex pheromone[234].

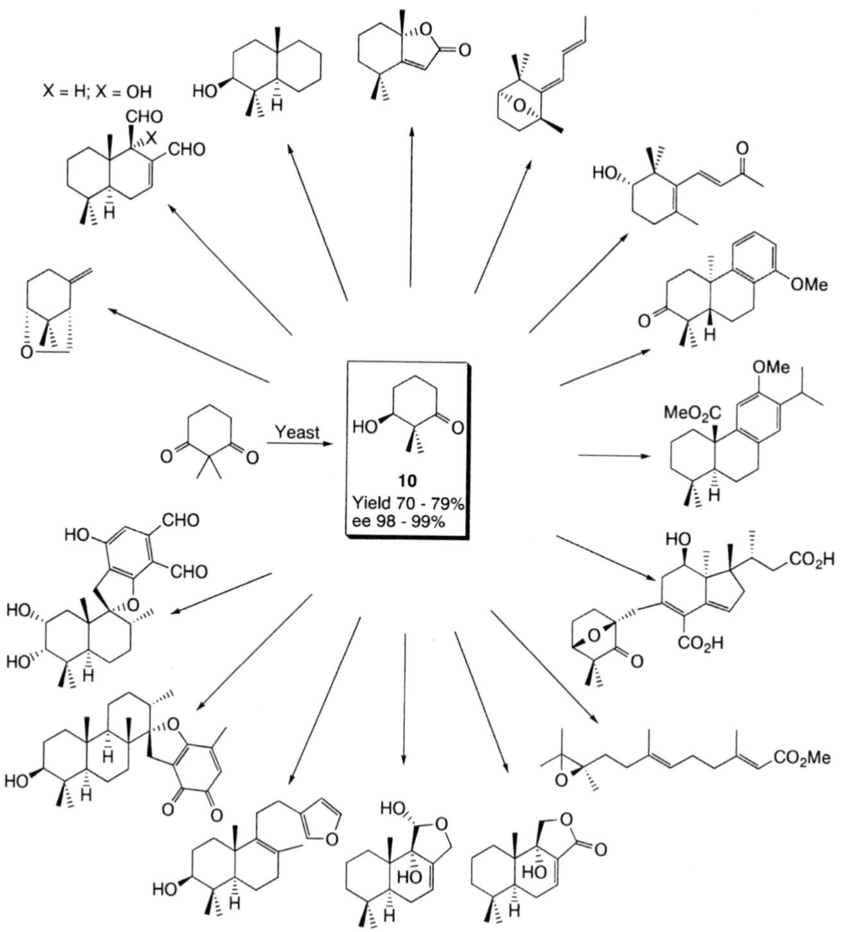

Figure 15-46. Synthesis of natural products from a key intermediate obtained by yeast reduction.

ported [122, 129, 199, 228–230, 234, 235, 243, 254–274]. For example, 2,8-nonandione can be reduced enantioselectively by TBADH to furnish the corresponding diol, from which all four isomers of 8-methyldec-2-yl propanoate, the western corn rootworm sex pheromone, were prepared (Fig. 15-45) [234].

One of the most versatile key intermediates discovered to date is the hydroxy-ketone **10** which is synthesized by the yeast reduction of the corresponding diketone [229, 230]. Starting with **10**, many terpenes have been enantioselectively synthesized by Mori *et al.*, as shown in Fig. 15-46.

15.2
Reduction of Various Functionalities

Kaoru Nakamura and Tomoko Matsuda

15.2.1
Reduction of Aldehydes

Many aldehyde reductases transform both aldehydes and ketones [138, 144, 275, 276]. For example, phenylacetaldehyde reductase from a styrene-assimilating *Corynebacterium* strain, ST-10, reduces aldehydes and ketones as shown in Table 15-18 [138]. Other aldehyde reductases such as one from *Sporobolomyces salmonicolor* also reduce aldehydes as well as ketones [144, 275].

Organometallic aldehydes can be reduced enantioselectively with dehydrogenases. For example, optically active organometallic compounds having planar chiralities were obtained by biocatalytic reduction of racemic aldehydes with yeast [277, 278] or HLADH [279] as shown in Fig. 15-47.

The dynamic resolution of an aldehyde is also possible as shown in Fig. 15-48 [280]. The racemization of the starting aldehyde and enantioselective reduction of a carbonyl group by bakers' yeast resulted in the formation of tertiary chiral carbon centers. The ee of the product was improved from 19% to 90% by changing the ester moiety from the isopropyl group to the neopentyl group.

Table 15-18. Examples of substrates of phenylacetaldehyde reductase from *Corynebacterium* strain, ST-10 [138].

Substrate (mM) (aldehyde)	Relative activity (%)	Substrate (mM) (ketone)	Relative activity (%)
Acetaldehyde (3)	0	Acetone (3)	0
n-Valeraldehyde (3)	181	2-Hexanone (3)	207
n-Hexyl aldehyde (3)	1220	2-Heptanone (3)	760
Phenylacetaldehyde (3)	100	Acetophenone (3)	35
3-Phenylpropionaldehyde (1)	364	4-Phenyl-2-butanone (3)	29

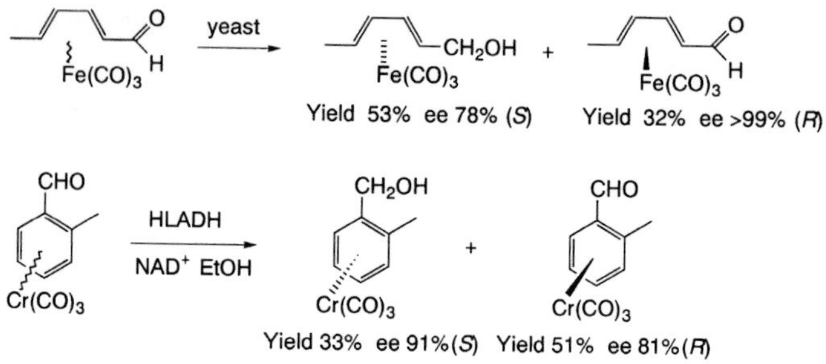
yeast
$Fe(CO)_3$
CH_2OH
Yield 53% ee 78% (S)
Yield 32% ee >99% (R)
CHO
HLADH
NAD+ EtOH
$Cr(CO)_3$
Yield 33% ee 91%(S)
Yield 51% ee 81%(R)

Figure 15-47. Reduction of organometallic aldehydes to produce alcohols with planar chiralities [277–279].

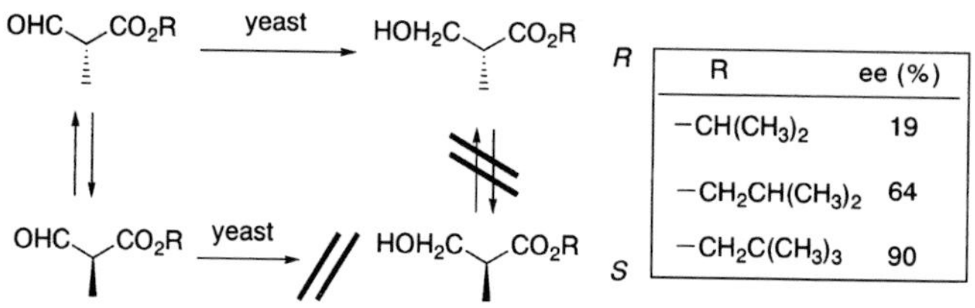
OHC CO2R
yeast
HOH2C CO2R
R
S
R ee (%)
−CH(CH3)2 19
−CH2CH(CH3)2 64
−CH2C(CH3)3 90

Figure 15-48. Reduction of aldehyde with dynamic resolution [280].

15.2.2
Reduction of Peroxides to Alcohols

Horseradish peroxidase has been used for the reduction of peroxide to alcohol [281–284]. The enzyme selectively recognizes sterically uncumbered *(R)*-alkyl aryl hydrogenperoxides, which allows kinetic resolution to provide *(R)*-alcohol and *(S)*-peroxide. However, poor enzyme recognition is observed with hydroperoxides possessing larger R_2 groups such as a propyl or an isopropyl moiety as shown in Fig. 15-49. This reaction can be performed on a preparative scale conveniently to provide optically pure hydroperoxides.

15.2.3
Reduction of Sulfoxides to Sulfides

Asymmetric synthesis of sulfoxides can also be achieved by biocatalytic reduction. One example is the reduction of alkyl aryl sulfoxides by intact cells of *Rhodobacter sphaeroides f. sp. denitrificans* [285]. In the reduction of methyl *p*-substituted phenyl sulfoxides, *(S)*-enantiomers were exclusively deoxygenated while enantiomerically pure *(R)*-isomers were recovered in good yield. For poor substrates such as ethyl phenyl sulfoxide, the repetition of the incubation after removing the toxic product was effective in enhancing the ee of recovered *(R)*-enantiomers to 100% as shown in Table 15-19.

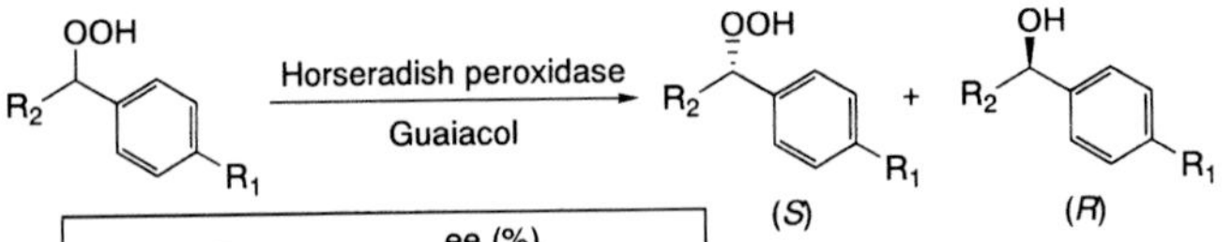

R$_1$	R$_2$	ee (%) (-)-(S)-ROOH	(+)-(R)-ROH
H	Me	>99	>99
Cl	Me	>95	>95
H	Et	93	95
H	Pr	<5	<5
H	i-Pr	15	14
etc.			

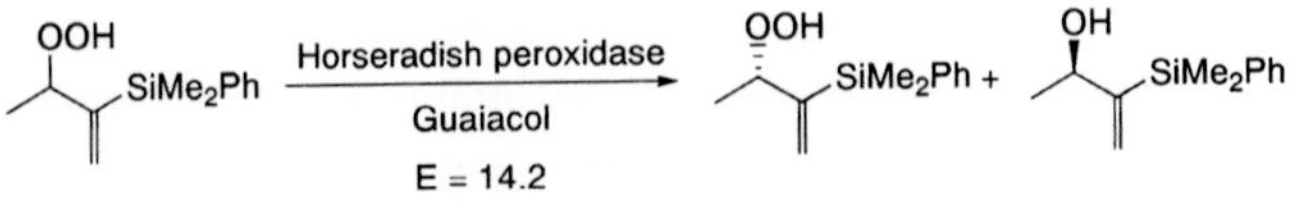

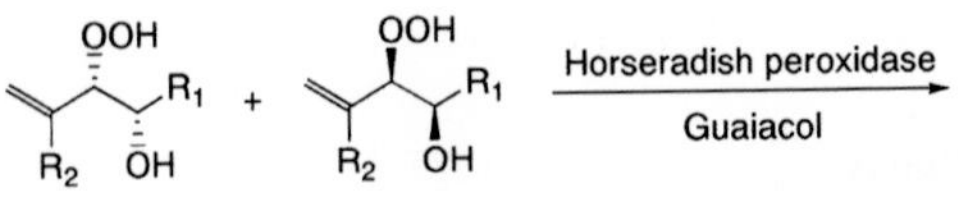

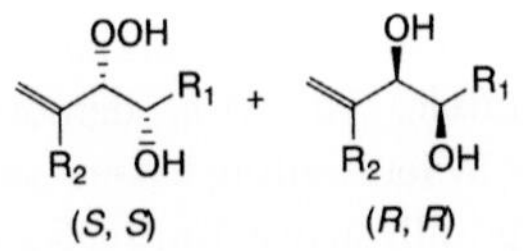

R$_1$	R$_2$	E
Me	H	10
Et	H	>200
i-Pr	H	>200
t-Bu	H	30
Me	Me	2

Figure 15-49. Reduction of peroxides to alcohols [281–284].

Table 15-19. Reduction of sulfoxide to obtain optically pure (R)-sulfoxide[285].

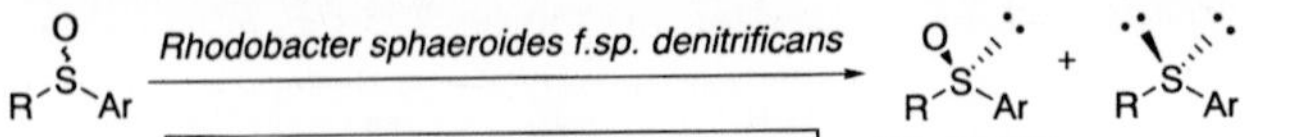

R	Ar	Yield (%)	ee (%)
Me	Ph	46	100
Me	p-Me-C$_6$H$_4$	40	100
Me	p-Br-C$_6$H$_4$	43	100
Me	p-MeO-C$_6$H$_4$	47	>99
Me	PhCH$_2$	41	90
Et	Ph	41	100
n-Pr	Ph	54	21

15.2.4
Reduction of Azide and Nitro Compounds to Amines

Bakers' yeast catalyzes the reduction of azides and nitro compounds to amines [286–291]. For example, it catalyzes chemoselective reduction of azidoarenes to arenamines as shown in Fig. 15-50 [286, 287]. Excellent yields are obtained for various aromatic compounds on reaction at room temperature. Aromatic nitro compounds

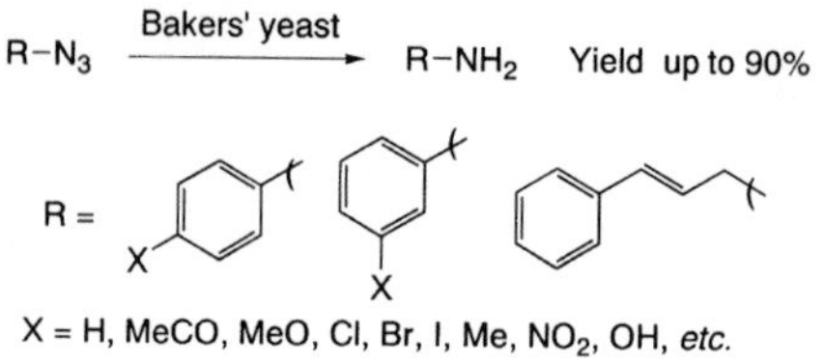

Bakers' yeast

R−N₃ $\xrightarrow{\text{Bakers' yeast}}$ R−NH₂   Yield up to 90%

Figure 15-50. Reduction of azide to amine by bakers' yeast [286, 287].

R =

X = H, MeCO, MeO, Cl, Br, I, Me, NO₂, OH, *etc.*

containing *o-*, *m-*, or *p*-electron withdrawing groups, such as carbonyl, halogen and nitro, were also selectively and rapidly reduced to their corresponding amino derivatives in good yields using bakers' yeast in basic solution [288].

  N-oxides can also be reduced. For example, the microbial deoxygenation of a series of aromatic and heteroaromatic *N*-oxide compounds, including quinoline *N*-oxides, isoquinoline *N*-oxides, 2-aryl-*2H*-benzotriazole 1-oxides, benzo[c]cinnoline *N*-oxide and azoxybenzene, has performed with bakers' yeast–NaOH to afford quinolines and pyridines [291].

15.2.5
Reduction of Carbon-Carbon Double Bonds

The reduction of carbon–carbon double bonds to single bonds has been studied with various substrates [51, 124, 292–306]. For example, Ohta *et al.* demonstrated that the reduction of a number of 1-nitro-1-alkenes by fermenting bakers' yeast was enantio-selective, resulting in the formation of optically active 1-nitroalkanes as shown in Fig. 15-51(a) [294, 295]. On the other hand, Fuganti *et al.* reduced α,β-unsaturated-δ-lactones to produce enantiomerically pure (+)-*(R)*-goniothalamins [Fig. 15-51(b)], which show CNS activity. They also performed the kinetic resolution of the corresponding embryotoxic epoxide with yeast [296].

  One of the most studied enzymes for the reduction of carbon–carbon double

(a)

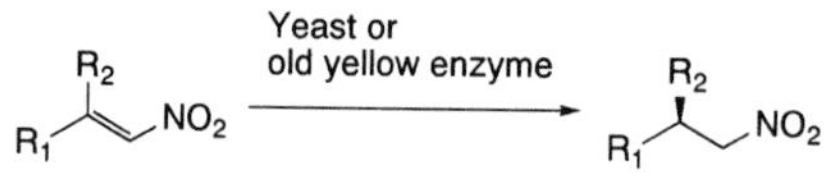

Yeast or old yellow enzyme

R_1	R_2	ee
Ph	H	-
Ph	Me	98
p-Cl-Ph	Me	89
p-Br-Ph	Me	94
Ph	Et	97
Ph	n-Pr	89
Ph	Hexyl	no reaction
Hexyl	Me (E)	83
Me	Hexyl (Z)	66
2-Thienyl	H	-
etc.		

(b)

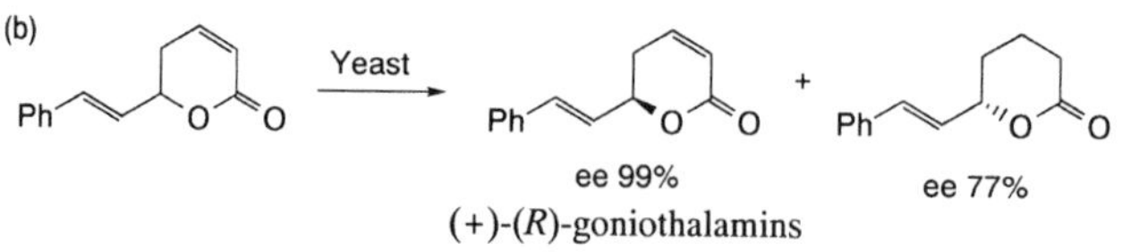

Ph $\xrightarrow{\text{Yeast}}$ Ph + Ph

ee 99%   ee 77%
(+)-*(R)*-goniothalamins

Figure 15-51. Examples of substrate specificity of yeast reduction of olefins [294–296].

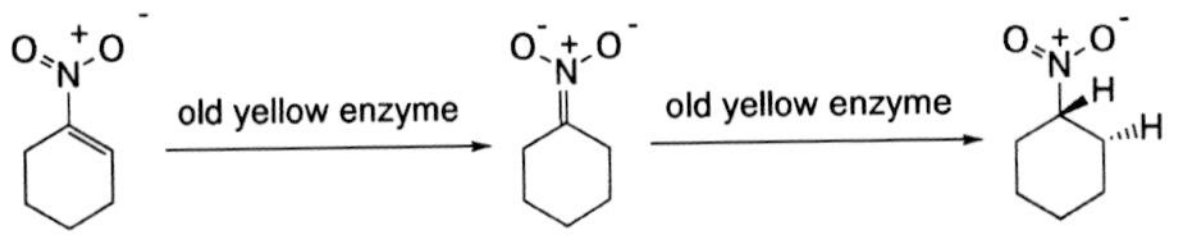

Figure 15-52. Mechanism of the reduction of nitro olefin by "old yellow enzyme" from yeast[292–294].

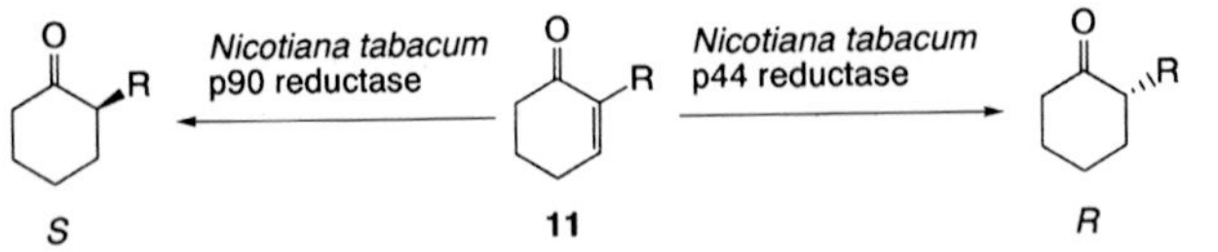

R	ee
Me	99
Et	98
n-Pr	95

R	ee
Me	>99
Et	>99
n-Pr	>99

Figure 15-53. Reduction of carbon–carbon double bonds by reductases from plant cell culture[298].

bonds is the "old yellow enzyme" from yeast[292–294] which has been shown efficiently to catalyze the NADPH-linked reduction of nitro olefins. The reduction of the nitro-olefin proceeds in a stepwise fashion (Fig. 15-52). The first step involves hydride transfer from the enzyme-reduced flavin to the β-carbon of the nitro-olefin which forms a nitronate intermediate that is freely dissociable from the enzyme. The second step, protonation of the nitronate at the α-carbon to form the final nitroalkane product, is also catalyzed by the enzyme.

Photosynthetic microorganisms and plant cell cultures are very important sources of enzymes for the reduction of olefins[51, 298]. For example, Hirata et al. found that reduction of enone **11** with *Nicotiana tabacum* p90 reductase and *Nicotiana tabacum* p44 reductase affords *(S)-* and *(R)*-alkylcyclohexanones, respectively, with excellent enantioselectivities as shown in Fig. 15-53. They also found two enone reductases from *Astasia longa*, a nonchlorophyllous cell line classified in *Euglenales*, and studied the mechanism. Both catalyzed enantiospecific *trans*-addition of hydrogen atoms to carvone from the *si*-face at the α-position and from the *re*-face at the β-position.

15.2.6
Transformation of α-Keto Acid to Amine

A dehydrogenase can also be used for the transformation of an α-keto acid to an amine (Fig. 15-54). The chiral intermediate for an antihypertensive drug was prepared by reduction of an α-keto acid with glutamate dehydrogenase from beef liver. The cofactor NADH was regenerated using glucose dehydrogenase from *Bacillus* sp. [307].

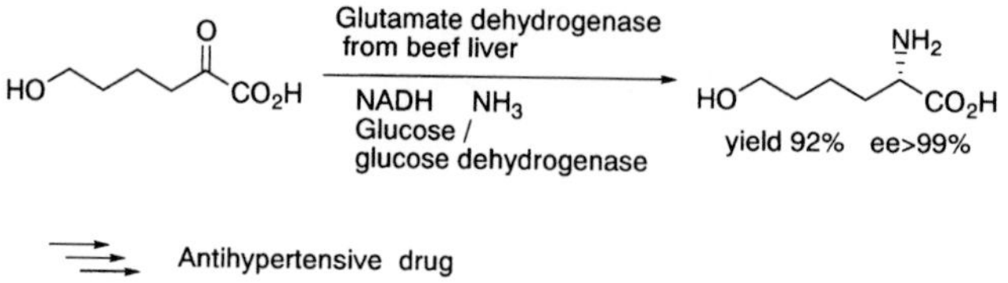

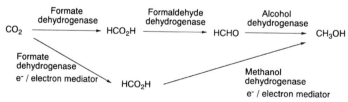

Glutamate dehydrogenase from beef liver

NADH NH₃
Glucose /
glucose dehydrogenase

yield 92% ee>99%

Antihypertensive drug

Figure 15-54. Reduction of α-keto acid to amine [307].

15.2.7
Reduction of Carbon Dioxide

15.2.7.1
Reduction of CO₂ to Methanol

Syntheses using CO_2 as a carbon source are attracting growing interest. The development of environmentally benign methods to utilize CO_2 is very important due to the abundance of CO_2. For this purpose, dehydrogenases have been successfully utilized. Formate, formaldehyde and alcohol dehydrogenases are used for the reduction of CO_2 to methanol as shown in Fig. 15-55 [33, 308–311].

For the efficient production of methanol, electrochemical methods have been used (Fig. 15-55) [33, 308, 310, 311]. Electrochemically, CO_2 was converted into formate by formate dehydrogenase with the aid of methyl viologen or pyrroloquinolinequinone as a mediator. Methanol dehydrogenase was used to reduce formate to formaldehyde and methanol with the same system [308, 310, 311].

An approach for the conversion of CO_2 into formic acid which combines a semiconductor photoelectrode with formate dehydrogenase is very interesting [33]. Electrons in the semiconductor can be produced with light of wavelengths shorter than 900 nm. Then, the photogenerated electrons were transferred to CO_2 through methyl viologen to produce formic acid as shown in Fig. 15-56 [33].

Another highly efficient process involves the immobilization of three enzymes in a silica sol-gel [309]. Since the process consists of a sequential reaction of *in situ* generated substrates with three different enzymes, the confinement of the system in a porous matrix resulted in an enhanced probability of the reactions as shown in Fig. 15-57 due to an overall increase in local concentration of reactants within the nanopores of the sol-gel processed glasses [309].

Formate
dehydrogenase

Formaldehyde
dehydrogenase

Alcohol
dehydrogenase

CO_2 ────────→ HCO_2H ────────→ HCHO ────────→ CH_3OH

Formate
dehydrogenase
e⁻ / electron mediator

HCO_2H

Methanol
dehydrogenase
e⁻ / electron mediator

Figure 15-55. Reduction of CO_2 to methanol with dehydrogenases.

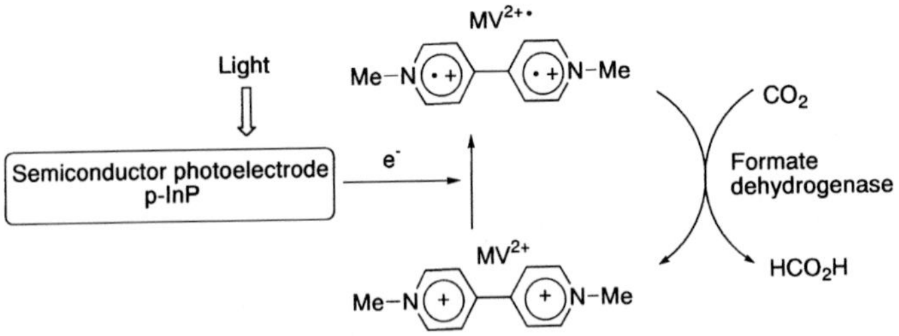

MV²⁺·

Light

Me—N⁺ ⊕ ⊕ N—Me

CO₂

Semiconductor photoelectrode
p-InP

e⁻

Formate
dehydrogenase

MV²⁺

Me—N⁺ ⊕ ⊕ N—Me

HCO₂H

Figure 15-56. Photoelectrochemical pumping of enzymatic CO_2 reduction [33].

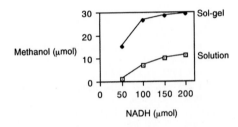

Figure 15-57. Effect of the confinement of the three enzymes in a porous matrix on methanol production [309].

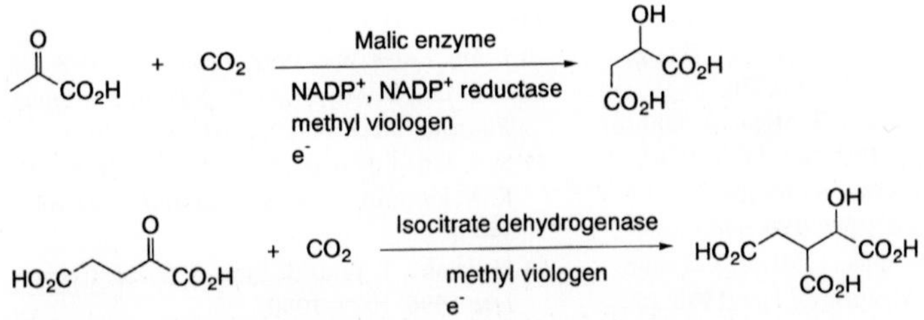

$$\underset{CO_2H}{\overset{O}{\parallel}} + CO_2 \xrightarrow[\substack{NADP^+, NADP^+ \text{ reductase} \\ \text{methyl viologen} \\ e^-}]{\text{Malic enzyme}} \underset{CO_2H}{\overset{OH}{|}} CO_2H$$

$$HO_2C \overset{O}{\underset{CO_2H}{\parallel}} + CO_2 \xrightarrow[\substack{\text{methyl viologen} \\ e^-}]{\text{Isocitrate dehydrogenase}} HO_2C \underset{CO_2H}{\overset{OH}{|}} CO_2H$$

Figure 15-58. Electrochemical reductive fixation of CO_2 [312, 313].

15.2.7.2
Reductive fixation of CO₂

Reductive fixation is an another important process. Malic enzyme and isocitrate dehydrogenase catalyze both the reduction of the carbonyl group in an α-keto acid and fixation of CO_2 at the α-position with the aid of an electric power source and an electron mediator (Fig. 15-58) [312, 313]. Uniquely, the reaction using isocitrate dehydrogenase does not require the use of NADP⁺. When CO_2 is reductively fixed in an organic molecule, the enzyme is oxidized; the oxidized enzyme is ultimately reduced back to its original form by methyl viologen cation radicals [312].

References

1 G. M. Whitesides, H. K. Chenault, H. Bertschy, H. Simon, A. S. Bommarius in: *Enzyme Catalysis in Organic Synthesis A Comprehensive Handbook* (Ed.: K. Drauz, H. Waldmann), Wiley-VCH, Weinheim, **1995**, p. 595–665.

2 E. Santaniello, P. Ferraboschi, A. Manzocchi in: *Enzymes in: Action* (Ed.: B. Zwanenburg, M. Mikolajczyk, P. Kielbasinski), Kluwer Academic Publishers, Dordrecht, **2000**, p. 95–115.

3 K. Faber, *Biotransformations in Organic Chemistry*, 2 d ed., Springer-Verlag, Berlin, **1995**, p. 145–180.

4 W. Hummel, M.-R. Kula, *Eur. J. Biochem.* **1989**, *184*, 1–13.

5 D. Schomburg, M. Salzmann, (GFB-Gesellschaft für Biotechnologische Forschung), *Enzyme Handbook*, Springer-Verlag, Berlin, D, **1990**.

6 C. W. Bradshaw, H. Fu, G.-J. Shen, C.-H. Wong, *J. Org. Chem.* **1992**, *57*, 1526–1532.

7 C. W. Bradshaw, W. Hummel, C.-H. Wong, *J. Org. Chem.* **1992**, *57*, 1532–1536.

8 K. Nakamura, T. Shiraga, T. Miyai, A. Ohno, *Bull. Chem. Soc. Jpn.* **1990**, *63*, 1735–1737.

9 K. Nakamura, S. Takano, K. Terada, A. Ohno, *Chem. Lett.* **1992**, 951–954.

10 K. Nakamura, T. Yoneda, T. Miyai, K. Ushio, S. Oka, A. Ohno, *Tetrahedron Lett.* **1988**, *29*, 2453–2454.

11 H. Dutler, J. L. Van Der Baan, E. Hochuli, Z. Kis, K. E. Taylor, V. Prelog, *Eur. J. Biochem.* **1977**, *75*, 423–432.

12 V. Prelog, *Pure Appl. Chem.* **1964**, *9*, 119–130.

13 J. B. Jones, *Tetrahedron* **1986**, *42*, 3351–3403.

14 J. B. Jones, J. F. Beck, in *Applications of Biochemical Systems in Organic Chemistry* (Ed.: J. B. Jones, C. J. Sih, D. Perslman), John Wiley and Sons, New York, **1976**, p. 107–401.

15 J. Davies, J. B. Jones, *J. Am. Chem. Soc.* **1979**, *101*, 5405–5410.

16 L. K. P. Lam, I. A. Gair, J. B. Jones, *J. Org. Chem.* **1988**, *53*, 1611–1615.

17 K. Velonia, I. Tsigos, V. Bouriotis, I. Smonou, *Bioorg. Med. Chem. Lett.* **1999**, *9*, 65–68.

18 E. Keinan, E. K. Hafeli, K. K. Seth, R. Lamed, *J. Am. Chem. Soc.* **1986**, *108*, 162–169.

19 E. Keinan, K. K. Seth, R. Lamed, *J. Am. Chem. Soc.* **1986**, *108*, 3474–3480.

20 K. Nakamura, T. Matsuda, *J. Org. Chem.* **1998**, *63*, 8957–8964.

21 K. Nakamura, K. Kitano, T. Matsuda, A. Ohno, *Tetrahedron Lett.* **1996**, *37*, 1629–1632.

22 K. Seelbach, B. Riebel, W. Hummel, M.-R. Kula, V. I. Tishkov, A. M. Egorov, C. Wandrey, U. Kragl, *Tetrahedron Lett.* **1996**, *37*, 1377–1380.

23 A. Liese, T. Zelinski, M.-R. Kula, H. Kierkels, M. Karutz, U. Kragl, C. Wandrey, *J. Mol. Catal. B: Enzymatic*, **1998**, *4*, 91–99.

24 G. Casy, T. V. Lee, H. Lovell, *Tetrahedron Lett.* **1992**, *33*, 817–820.

25 T. Zelinski, M.-R. Kula, *Bioorg. Med. Chem.* **1994**, *2*, 421–428.

26 C. H. Wong, D. G. Druckhammer, H. M. Sweers, *J. Am. Chem. Soc.* **1985**, *107*, 4028–4031.

27 D. C. Crans, C. M. Simone, J. S. Blanchard, *J. Am. Chem. Soc.* **1992**, *114*, 4926–4928.

28 S.-S. Lin, O. Miyawaki, K. Nakamura, *Biosci. Biotechn. Biochem.* **1997**, *61*, 2029–2033.

29 S.-S. Lin, T. Harada, C. Hata, O. Miyawaki, K. Nakamura, *J. Ferment. Bioeng.* **1997**, *83*, 54–58.

30 M. Utaka, T. Yano, T. Ema, T. Sakai, *Chem. Lett.* **1996**, 1079–1080.

31 S.-S. Lin, O. Miyawaki, K. Nakamura, *J. Biosci. Bioeng.* **1999**, *87*, 361–364.

32 K. Nakamura, R. Yamanaka, K. Tohi, H. Hamada, *Tetrahedron Lett.* **2000**, *41*, 6799–6802.

33 B. A. Parkinson, P. F. Weaver, *Nature*, **1984**, *309*, 148–149.

34 R. Yuan, S. Watanabe, S. Kuwabata, H. Yoneyama, *J. Org. Chem.* **1997**, *62*, 2494–2499.

35 A. E. Biade, C. Bourdillon, J. M. Laval, G. Mairesse, J. Moiroux, *J. Am. Chem. Soc.* **1992**, *114*, 893–897.

36 M. D. Leonida, S. B. Sobolow, A. J. Frey, *Bioorg. Med. Chem. Lett.* **1998**, *8*, 2819–2824.

37 S. Servi, *Synthesis*, **1990**, 1–25.

38 P. A. Levene, A. Walti, *Org. Synth.* **1943**, *Coll. Vol. II*, 545–547.

39 D. Seebach, M. A. Sutter, R. H. Weber, M. F. Züger, *Org. Synth.* **1990**, *coll. Vol. 7*, 215–220.

40 S. Ramaswamy, H. Eklund, B. V. Plapp, *Biochemistry* **1994**, *33*, 5230–5237.

41 E. Cedergren-Zeppezauer, *Biochemistry* **1983**, *22*, 5761–5772.

42 H. Eklund, *Pharmacol. Biochem. Behav.* **1983**, *18 Suppl. 1*, 73–81.

43 V. T. Pham, R. S. Phillips, L. G. Ljungdahl, *J. Am. Chem. Soc.* **1989**, *111*, 1935–1936.

44 C. Zheng, V. T. Pham, R. S. Phillips, *Catalysis Today* **1994**, *22*, 607–620.

45 A. E. Tripp, D. S. Brudette, J. G. Zeikus, R. S. Phillips, *J. Am. Chem. Soc.* **1998**, *120*, 5137–5141.

46 C. Heiss, M. Laivenieks, J. G. Zeikus, R. S. Phillips, *J. Am. Chem. Soc.* **2001**, *123*, 345–346.

47 D. Burdette, J. G. Zeikus, *Biochem. J.* **1994**, *302*, 163–170.

48 C. Zheng, R. S. Phillips, *J. Chem. Soc. Perkin Trans. 1*, **1992**, 1083–1084.

49 C. Zheng, V. T. Pham, R. S. Phillips, *Bioorg. Med. Chem. Lett.* **1992**, *2*, 619–622.

50 R. J. Lamed, J. G. Zeikus, *Biochem. J.* **1981**, *195*, 183–190.

51 K. Shimoda, T. Hirata, *J. Mol. Catal. B: Enzymatic*, **2000**, *8*, 255–264.

52 T. Hirata, S. Izumi, *Plant Tissue Culture Lett.* **1993**, *10*, 215–222.

53 T. Suga, H. Hamada, T. Hirata, *Plant Cell Reports* **1983**, *2*, 66–68.

54 A. Chadha, M. Manohar, T. Soundararajan, T. S. Lokeswari, *Tetrahedron: Asymm.* **1996**, *7*, 1571–1572.

55 Nakamura, K. Miyoshi, T. Sugiyama, H. Hamada, *Phytochemistry*, **1995**, *40*, 1419–1420.

56 H. Hamada, N. Nakajima, Y. Shisa, M. Funahashi, K. Nakamura, *Bioorg. Med. Chem. Lett.* **1994**, *4*, 907–910.

57 F. Baldassarre, G. Bertoni, C. Chiappe, F. Marioni, *J. Mol. Catal. B: Enzymatic* **2000**, *11*, 55–58.

58 D. Seebach, R. Imwinkelried, T. Weber in: *Modern Synthetic Methods 1986* (Ed.: R. Scheffold), Springer-Verlag, Berlin, **1986**, p. 125–259.

59 S. C. Stinson, *Chem. Eng. News*, **2000**, October 23, 55–80.

60 D. L. Lewis, A. W. Garrison, K. E. Wommack, A. Whittemore, P. Steudler, J. Melillo, *Nature*, **1999**, *401*, 898–901.

61 B.-n Zhou, A. S. Gopalan, F. VanMiddlesworth, W.-R. Shieh, C. J. Sih, *J. Am. Chem. Soc.* **1983**, *105*, 5925–5926.

62 C.-S. Chem, B.-n. Zhou, G. Girdaukas, W.-R. Shieh, F. VanMiddlesworth, A. S.

Gopalan, C. J. Sih, *Bioorg. Chem.* **1984**, *12*, 98–117.

63 W.-R. Shieh, A. S. Gopalan, C. J. Sih, *J. Am. Chem. Soc.* **1985**, *107*, 2993–2994.

64 T. Fujisawa, T. Itoh, T. Sato, *Tetrahedron Lett.* **1984**, *25*, 5083–5086.

65 K. Nakamura, K. Ushio, S. Oka, A. Ohno, S. Yasui, *Tetrahedron Lett.* **1984**, *25*, 3979–3982.

66 D. C. Demirjian, P. C. Shah, F. Morís-Varas, in *Biocatalysis From Discovery to Application* (Ed.: W.-D. Fessner), Springer-Verlag, Berlin, **2000**, p. 1–29.

67 Y. Yasohara, N. Kizaki, J. Hasegawa, S. Takahashi, M. Wada, M. Kataoka, S. Shimizu, *Appl. Microbiol. Biotechnol.* **1999**, *51*, 847–851.

68 S. Shimizu, H. Hata, H. Yamada, *Agric. Biol. Chem.* **1984**, *48*, 2285–2291.

69 M. Kataoka, S. Shimizu, Y. Doi, K. Sakamoto, H. Yamada, *Biotechnol. Lett.* **1990**, *12*, 357–360.

70 H. Miya, M. Kawada, Y. Sugiyama, *Biosci. Biotechn. Biochem.* **1996**, *60*, 95–98.

71 A. Shafiee, H. Motamedi, A. King, *Appl. Microbiol. Biotechnol.* **1998**, *49*, 709–717.

72 M. Wada, M. Kataoka, H. Kawabata, Y. Yasohara, N. Kizaki, J. Hasegawa, S. Shimizu, *Biosci. Biotech. Biochem.* **1998**, *62*, 280–285.

73 M. Wada, H. Kawabata, A. Yoshizumi, M. Kataoka, S. Nakamori, Y. Yasohara, N. Kizaki, J. Hasegawa, S. Shimizu, *J. Biosci. Bioeng.* **1999**, *87*, 144–148.

74 K. Nakamura, Y. Kawai, A. Ohno, *Tetrahedron Lett.* **1991**, *32*, 2927–2928.

75 R. Azerad, D. Buisson in: *Microbial Reagents in Organic Synthesis* (Ed.: S. Servi), Kluwer Academic Publishers, Netherlands, **1992**, p. 421–440.

76 D. Buisson, R. Azerad, C. Sanner, M. Larchevêque, *Biocatalysis* **1992**, *5*, 249–265.

77 D. Buisson, R. Azerad, C. Sanner, M. Larchevêque, *Tetrahedron: Asymm.* **1991**, *2*, 987–988.

78 D. Buisson, R. Azerad, C. Sanner, M. Larchevêque, *Biocatalysis* **1990**, *3*, 85–93.

79 G. Fantin, M. Fogagnolo, M. E. Guerzoni, R. Lanciotti, A. Medici, P. Pedrini, D. Rossi, *Tetrahedron: Asymm.* **1996**, *7*, 2879–2887.

80 C. T. Goodhue, J. P. Rosazza, G. P. Peruzzotti in: *Manual of Industrial Microbiology and Biotechnology* (Ed.: A. L. Demain, N. A. Solomon), American Society for Microbiology, Washington C. C., **1986**, p. 97–121.

81 M. Matsuda, T. Harada, N. Nakajima, K. Nakamura, *Tetrahedron Lett.* **2000**, *41*, 4135–4138.

82 K. Nakamura, T. Matsuda, A. Ohno, *Tetrahedron: Asymm.* **1996**, *7*, 3021–3024.

83 E. C. Tidswell, G. J. Salter, D. B. Kell, J. G. Morris, *Enzyme Microb. Technol.* **1997**, *21*, 143–147.

84 K. Nakamura, T. Miyai, K. Fukushima, Y. Kawai, B. R. Babu, A. Ohno, *Bull. Chem. Soc. Jpn.* **1990**, *63*, 1713–1715.

85 K. Ushio, K. Inoue, K. Nakamura, S. Oka, A. Ohno, *Tetrahedron Lett.* **1986**, *27*, 2657–2660.

86 A. C. Dahl, J. Ø. Madsen, *Tetrahedron: Asymm.* **1998**, *9*, 4395–4417.

87 K. Nakamura, Y. Kawai, A. Ohno, *Tetrahedron Lett.* **1990**, *31*, 267–270.

88 K. Nakamura, K. Inoue, K. Ushio, S. Oka, A. Ohno, *Chemistry Lett.* **1987**, 679–682.

89 A. Forni, I. Moretti, F. Prati, G. Torre, *Tetrahedron* **1994**, *50*, 11 995–12 000.

90 A. C. Dahl, M. Fjeldberg, J. Ø. Madsen, *Tetrahedron: Asymm.* **1999**, *10*, 551–559.

91 K. Nakamura, Y. Kawai, S. Oka, A. Ohno, *Bull. Chem. Soc. Jpn.* **1989**, *62*, 875–879.

92 K. Nakamura, Y. Kawai, S. Oka, A. Ohno, *Tetrahedron Lett.* **1989**, *30*, 2245–2246.

93 J.-H. Kim, W.-T. Oh, *Bull. Korean Chem. Soc.* **1992**, *13*, 2–3.

94 R. Hayakawa, K. Nozawa, M. Shimizu, T. Fujisawa, *Tetrahedron Lett.* **1998**, *39*, 67–70.

95 K. Ushio, J. Hada, Y. Tanaka, K. Ebara, *Enzyme Microb. Technol.* **1993**, *15*, 222–228.

96 R. Hayakawa, M. Shimizu, T. Fujisawa, *Tetrahedron: Asymm.* **1997**, *8*, 3201–3204.

97 K. Nakamura, Y. Kawai, N. Nakajima, A. Ohno, *J. Org. Chem.* **1991**, *56*, 4778–4783.

98 K. Nakamura, K. Inoue, K. Ushio, S. Oka, A. Ohno, *J. Org. Chem.* **1988**, *53*, 2589–2593.

99 K. Nakamura, S. Kondo, Y. Kawai, A. Ohno, *Tetrahedron Lett.* **1991**, *32*, 7075–7078.

100 K. Nakamura, S. Kondo, N. Nakajima, A. Ohno, *Tetrahedron*, **1995**, *51*, 687–694.

101 K. Nakamura, Y. Inoue, T. Matsuda, I. Misawa, *J. Chem. Soc. Perkin Trans. 1*, **1999**, 2397–2402.

102 O. Rotthaus, D. Krüger, M. Demuth, K. Schaffner, *Tetrahedron* **1997**, *53*, 935–938.

103 L. Y. Jayasinghe, D. Kodituwakku, A. J. Smallridge, M. A. Trewhella, *Bull. Chem. Soc. Jpn.* **1994**, *67*, 2528–2531.

104 M. North, *Tetrahedron Lett.* **1996**, *37*, 1699–1702.

105 C. Medson, A. J. Smallridge, M. A. Trewhella, *Tetrahedron: Asymm.* **1997**, *8*, 1049–1054.

106 A. M. Snijder-Lambers, E. N. Vulfson, H. J. Doddema, *Recl. Trav. Chim. Pays-Bas*, **1991**, *110*, 226–230.

107 K. Nakamura, T. Miyai, K. Inoue, S. Kawasaki, S. Oka, A. Ohno, *Biocatalysis* **1990**, *3*, 17–24.

108 K. Nakamura, Y. Inoue, A. Ohno, *Tetrahedron Lett.* **1995**, *36*, 265–266.

109 K. Nakamura, S. Kondo, Y. Kawai, A. Ohno, *Bull. Chem. Soc. Jpn.* **1993**, *66*, 2738–2743.

110 K. Nakamura, S. Kondo, Y. Kawai, N. Nakajima, A. Ohno, *Biosci. Biotechn. Biochem.* **1994**, *58*, 2236–2240.

111 K. Nakamura, S. Kondo, Y. Kawai, N. Nakajima, A. Ohno, *Biosci. Biotech. Biochem.* **1997**, *61*, 375–377.

112 A. J. Mesiano, E. J. Beckman, A. J. Russell, *Chem. Rev.* **1999**, *99*, 623–633.

113 T. W. Randolph, H. W. Blanch, J. M. Prausnitz, C. R. Wilke, *Biotechnol. Lett.* **1985**, *7*, 325–328.

114 D. A. Hammond, M. Karel, A. M. Klibanov, V. J. Krukonis, *Appl. Biochem. Biotechnol.* **1985**, *11*, 393–400.

115 K. Nakamura, Y. M. Chi, Y. Yamada, T. Yano, *Chem. Eng. Commun.* **1986**, *45*, 207–212.

116 T. Mori, Y. Okahata, *Chem. Soc., Chem. Commun.* **1998**, 2215–2216.

117 T. Mori, A. Kobayashi, Y. Okahata, *Chem. Lett.* **1998**, 921–922.

118 T. Matsuda, T. Harada, K. Nakamura, *Chem. Soc., Chem. Commun.* **2000**, 1367–1368.

119 K. Nakamura, S. Kondo, Y. Kawai, A. Ohno, *Tetrahedron: Asymm.* **1993**, *4*, 1253–1254.

120 K. Nakamura, S. Kondo, A. Ohno, *Bioorg. Med. Chem.* **1994**, *2*, 433–437.

121 T. Zelinski, M.-R. Kula, *Biocatal. Biotransform.* **1997**, *15*, 57–74.

122 B. A. Anderson, M. M. Hansen, A. R. Harkness, C. L. Henry, J. T. Vicenzi, M. J. Zmijewski, *J. Am. Chem. Soc.* **1995**, *117*, 12358–12359.

123 P. D'Arrigo, C. Fuganti, G. P. Fantoni, S. Servi, *Tetrahedron* **1998**, *54*, 15 017–15 026.

124 P. D'Arrigo, M. Lattanzio, G. P. Fantoni, S. Servi, *Tetrahedron: Asymm.* **1998**, *9*, 4021–4026.

125 P. D'Arrigo, G. P. Fantoni, S. Servi,

A. Strini, *Tetrahedron: Asymm.* **1997**, *8*, 2375–2379.

126 K. Nakamura, M. Fujii, Y. Ida, *J. Chem. Soc., Perkin Trans 1* **2000**, 3205–3211.

127 J. D. Stewart, K. W. Reed, C. A. Martinez, J. Zhu, G. Chen, M. M. Kayser, *J. Am. Chem. Soc.* **1998**, *120*, 3541–3548.

128 S. Rodríguez, K. T. Schoeder, M. M. Kayser, J. D. Stewart, *J. Org. Chem.* **2000**, *65*, 2586–2587.

129 M. M. Kayser, M. D. Mihovilovic, J. Kearns, A. Feicht, J. D. Stewart, *J. Org. Chem.* **1999**, *64*, 6603–6608.

130 H. M. Wilks, D. J. Halsall, T. Atkinson, W. N. Chia, A. R. Clarke, J. J. Holbrook, *Biochemistry* **1990**, *29*, 8587–8591.

131 S. A. Doyle, S.-Y. F. Fung, D. E. Koshland, Jr., *Biochemistry* **2000**, *39*, 14348–14355.

132 S. J. Allen, J. J. Holbrook, *Protein Eng.* **2000**, *13*, 5–7.

133 N. Bernard, K. Johnsen, J. L. Gelpi, J. A. Alvarez, T. Ferain, D. Garmyn, P. Hols, A. Cortes, A. R. Clarke, J. J. Holbrook, J. Delcour, *Eur. J. Biochem.* **1997**, *244*, 213–219.

134 G. Casy, T. V. Lee, H. Lovell, B. J. Nichols, R. B. Sessions, J. J. Holbrook, *Chem. Soc., Chem. Commun.* **1992**, 924–926.

135 T. Higuchi, Y. Imamura, M. Otagiri, *Biochim. Biophys. Acta* **1994**, *1199*, 81–86.

136 D. S. Burdette, F. Secundo, R. S. Phillips, J. Dong, R. A. Scott, J. G. Zeikus, *Biochem. J.* **1997**, *326*, 717–724.

137 N. Itoh, K. Yoshida, K. Okada, *Biosci. Biotech. Biochem.* **1996**, *60*, 1826–1830.

138 N. Itoh, R. Morihama, J. Wang, K. Okada, N. Mizuguchi, *Appl. Environ. Microbiol.* **1997**, *63*, 3783–3788.

139 N. Itoh, N. Mizucuhi, M. Mabuchi, *J. Mol. Catal. B: Enzymatic* **1999**, *6*, 41–50.

140 J.-C. Wang, M. Sakakibara, J.-Q. Liu, T. Dairi, N. Itoh, *App. Microbiol. Biotechnol.* **1999**, *52*, 386–392.

141 J. C. Wang, M. Sakakibara, M. Matsuda, N. Itoh, *Biosci. Biotech. Biochem.* **1999**, *63*, 2216–2218.

142 Y. Yasohara, N. Kizaki, J. Hasegawa, M. Wada, M. Kataoka, S. Shimizu, *Biosci. Biotech. Biochem.* **2000**, *64*, 1430–1436.

143 C. A. Costello, R. A. Payson, M. A. Menke, J. L. Larson, K. A. Brown, J. E. Tanner, R. E. Kaiser, C. L. Hershberger, M. J. Zmijewski, *Eur. J. Biochem.* **2000**, *267*, 5493–5501.

144 K. Kita, T. Fukura, K. Nakase, K. Okamoto, H. Yanase, M. Kataoka, S. Shimizu, *Appl. Environ. Microbiol.* **1999**, *65*, 5207–5211.

145 Y. Korkhin, F. Frolow, O. Bogin, M. Peretz, A. J. Kalb, Y. Burstein, *Acta Crystallogr.* **1996**, *D52*, 882–886.

146 A. Sutherland, C. L. Willis, *J. Org. Chem.* **1998**, *63*, 7764–7769.

147 G. R. Nakayama, P. G. Schultz, *J. Am. Chem. Soc.* **1992**, *114*, 780–781.

148 G. Fontana, P. Manitto, G. Speranza, S. Zanzola, *Tetrahedron: Asymm.* **1998**, *9*, 1381–1387.

149 S. Geresh, T. J. Valiyaveettil, Y. Lavie, A. Shani, *Tetrahedron: Asymm.* **1998**, *9*, 89–96.

150 G. Fantin, M. Fogagnolo, M. E. Guerzoni, A. Medici, P. Pedrini, S. Poli, *J. Org. Chem.* **1994**, *59*, 924–925.

151 T. Kometani, E. Kitatsuji, R. Matsuno, *J. Ferment. Bioeng.* **1991**, *71*, 197–199.

152 T. Kometani, Y. Morita, H. Furui, H. Yoshii, R. Matsuno, *J. Ferment. Bioeng.* **1994**, *77*, 13–16.

153 B. Adger, U. Berens, M. J. Griffiths, M. J. Kelly, R. McCague, J. A. Miller, C. F. Palmer, S. M. Roberts, R. Selke, U. Vitinius, G. Ward, *Chem. Soc., Chem. Commun.* **1997**, 1713–1714.

154 T. Fujisawa, B. I. Mobele, M. Shimizu, *Tetrahedron Lett.* **1992**, *33*, 5567–5570.

155 R. Chênevert, G. Fortier, R. B. Rhlid, *Tetrahedron* **1992**, *48*, 6769–6776.

156 K. Takabe, H. Hiyoshi, H. Sawada, M. Tanaka, A. Miyazaki, T. Yamada, T. Katagiri, H. Yoda, *Tetrahedron: Asymm.* **1992**, *3*, 1399–1400.

157 G. Fantin. M. Fogagnolo, A. Medici, P. Pedrini, S. Poli, F. Gardini, M. E. Guerzoni, *Tetrahedron: Asymm.* **1992**, *3*, 107–114.

158 M. Hamdani, B. D. Jeso, H. Deleuze, B. Maillard, *Tetrahedron: Asymm.* **1991**, *2*, 867–870.

159 R. Chenevert, G. Fortier, *Chem. Lett.* **1991**, 1603–1606.

160 A. Guerrero, F. Raja, *Bioorg. Med. Chem. Lett.* **1991**, *1*, 675–678.

161 S. Tsuboi, J. Sakamoto, T. Kawano, M. Utaka, A. Takeda, *J. Org. Chem.* **1991**, *56*, 7177–7179.

162 T. Kometani, H. Yoshii, E. Kitatsuji, H. Nishimura, R. Matsuno, *J. Ferment. Bioengl* **1993**, *76*, 33–37.

163 M. Kodama, H. Minami, Y. Mima, Y. Fukuyama, *Tetrahedron Lett.* **1990**, *31*, 4025–4026.

164 J. Barry, H. B. Kagan, *Synthesis* **1981**, 453–455.

165 T. Kometani, H. Yoshii, Y. Takeuchi, R. Matsuno, *J. Ferment. Bioeng.* **1993**, *76*, 414–415.

166 T. Fujisawa, T. Itoh, M. Nakai, T. Sato, *Tetrahedron Lett.* **1985**, *26*, 771–774.

167 K. Nakamura, S. Takano, A. Ohno, *Tetrahedron Lett.* **1993**, *34*, 6087–6090.

168 K. Nakamura, *J. Mol. Catal, B: Enzymatic,* **1998**, *5*, 129–132.

169 Z.-L. Wie, G.-Q. Lin, Z.-Y. Li, *Bioorg. Med. Chem.* **2000**, *8*, 1129–1137.

170 R. N. Patel, C. G. McNamee, A. Banerjee, J. M. Howell, R. S. Robison, L. J. Szarka, *Enzyme Microb. Technol.* **1992**, *14*, 731–738.

171 T. Kitayama, *Tetrahedron: Asymm.* **1997**, *8*, 3765–3774.

172 T. Kitazume, T. Yamazaki in: *Selective Fluorination in Organic and Bioorganic Chemistry, ACS Symposium Series No. 456* (Ed.: J. T. Welch), American Chemical Society, **1991**, p. 175–185.

173 K. Nakamura, T. Matsuda, T. Itoh, A. Ohno, *Tetrahedron Lett.* **1996**, *37*, 5727–5730.

174 T. Matsuda, T. Harada, N. Nakajima, T. Itoh, K. Nakamura, *J. Org. Chem.* **2000**, *65*, 157–163.

175 A. Arnone, R. Bernardi, F. Blasco, R. Cardillo, R. Resnati, I. I. Gerus, V. P. Kukhar, *Tetrahedron,* **1998**, *54*, 2809–2818.

176 T. Kitazume, T. Kobayashi, *Synthesis,* **1987**, 187–188.

177 T. Kitazume, Y. Nakayama, *J. Org. Chem.* **1986**, *51*, 2795–2799.

178 T. Kitazume, T. Sato, *J. Fluorine Chem.* **1985**, *30*, 189–202.

179 T. Fujisawa, K. Ichikawa, M. Shimizu, *Tetrahedron: Asymm.* **1993**, *4*, 1237–1240.

180 T. Fujisawa, T. Sugimoto, M. Shimizu, *Tetrahedron: Asymm.* **1994**, *5*, 1095–1098.

181 T. Fujisawa, Y. Onogawa, A. Sato, T. Mitsuya, M. Shimizu, *Tetrahedron,* **1998**, *54*, 4267–4276.

182 A. Arnone, G. Biagini, R. Cardillo, G. Resnati, J.-P. Bégué, D. Bonnet-Delpon, A. Kornilov, *Tetrahedron Lett.* **1996**, *37*, 3903–3906.

183 K. Nakamura, T. Matsuda, M. Shimizu, T. Fujisawa, *Tetrahedron* **1998**, *54*, 8393–8402.

184 Y. Ohtsuka, O. Katoh, T. Sugai, H. Ohta, *Bull. Chem. Soc. Jpn.* **1997**, *70*, 483–491.

185 A. Sakai, M. Bakke, H. Ohta, H. Kosugi, T. Sugai, *Chem. Lett.* **1999**, 1255–1256.

186 C. Barbieri, E. Caruso, P. D'Arrigo, G. P. Fantoni, S. Servi, *Tetrahedron: Asymm.* **1999**, *10*, 3931–3937.

187 M. Imuta, K. Kawai, H. Ziffer, *J. Org. Chem.* **1980**, *45*, 3352–3355.

188 M. d. Carvalho, M. T. Okamoto, P.J.S. Moran, J.A.R. Rodrigues, *Tetrahedron* **1991**, *47*, 2073–2080.

189 Z.-L. Wei, Z.,-Y. Li, G.-Q. Lin, *Tetrahedron* **1998**, *54*, 13059–13072.

190 P. Besse, T. Sokoltchik, H. Veschambre, *Tetrahedron: Asymm.* **1998**, *9*, 4441–4457.

191 S. Tsuboi, H. Furutani, M. H. Ansari, T. Sakai, M. Utaka, A. Takeda, *J. Org. Chem.* **1993**, *58*, 486–492.

192 J. Aleu, G. Fronza, C. Fuganti, V. Perozzo, S. Serra, *Tetrahedron: Asymm.* **1998**, *9*, 1589–1596.

193 O. Cabon, D. Buisson, M. Larchevêque, R. Azerad, *Tetrahedron: Asymm.* **1995**, *6*, 2199–2210.

194 O. Cabon, D. Buisson, M. Larchevêque, R. Azerad, *Tetrahedron: Asymm.* **1995**, *6*, 2211–2218.

195 F. Aragozzini, M. Valenti, E. Santaniello, P. Ferraboschi, P. Grisenti, *Biocatalysis* **1992**, *5*, 325–332.

196 Y. Akakabe, M. Takahashi, M. Kamezawa, K. Kikuchi, H. Tachibana, T. Ohtani, Y. Naoshima, *J. Chem. Soc., Perkin I Trans 1* **1995**, 1295–1298.

197 S. Shimizu, M. Kataoka, M. Katoh, T. Morikawa, T. Miyoshi, H. Yamada, *Appl. Environ. Microbiol.* **1990**, *56*, 2374–2377.

198 R. N. Patel, A. Banerjee, C. G. McNamee, D. B. Brzozowski, L. J. Szarka, *Tetrahedron: Asymm.* **1997**, *8*, 2547–2522.

199 C. Fuganti, S. Lanati, S. Servi, A. Tagliani, A. Bedeschi, G. Franceschi, *J. Chem. Soc., Perkin Trans 1.* **1993**, 2247–2249.

200 K. Nakamura, T. Kitayama, Y. Inoue, A. Ohno, *Tetrahedron* **1990**, *46*, 7471–7481.

201 M. Mehmandoust, D. Buisson, R. Azerad, *Tetrahedron Lett.* **1995**, *36*, 6461–6462.

202 R. Tanikaga, Y. Obata, K. Kawamoto, *Tetrahedron: Asymm.* **1997**, *8*, 3101–3106.

203 F. Molinari, E. G. Occhiato, F. Aragozzini, A. Guarna, *Tetrahedron: Asymm.* **1998**, *9*, 1389–1394.

204 C. Forzato, P. Nitti, G. Pitacco, E. Valentin, *Tetrahedron: Asymm.* **1997**, *8*, 1811–1820.

205 T. Izumi, K. Fukaya, *Bull. Chem. Soc. Jpn.* **1993**, *66*, 1216–1221.

206 D. Bailey, D. O'Hagan, U. Dyer, R. B. Lamont, *Tetrahedron: Asymm.* **1993**, *4*, 1255–1258.

207 D. W. Knight, N. Lewis, A. C. Share, D. Haigh, *J. Chem. Soc., Perkin Trans 1* **1998**, 3673–3683.

208 T. Hudlicky, G. Gillman, C. Andersen, *Tetrahedron: Asymm.* **1992**, *3*, 281–286.

209 T. Hudlicky, T. Tsunoda, K. G. Gadamasetti, J. A. Murry, G. E. Keck, *J. Org. Chem.* **1991**, *56*, 3619–3623.

210 G. Fantin, M. Fogagnolo, M. E. Guerzoni, E. Marotta, A. Medici, P. Pedrini, *Tetrahedron: Asymm.* **1992**, *3*, 947–952.

211 K. Nakamura, T. Kitayama, Y. Inoue, A. Ohno, *Bull. Chem. Soc. Jpn.* **1990**, *63*, 91–96.

212 K. Nakamura, Y. Inoue, J. Shibahara, S. Oka, A. Ohno, *Tetrahedron Lett.* **1988**, *29*, 4769–4770.

213 J. A. Macritchie, A. Silcock, C. L. Willis, *Tetrahedron: Asymm.* **1997**, *8*, 3895–3902.

214 G. Egri, A. Kolbert, J. Bálint, E. Fogassy, L. Novák, L. Poppe, *Tetrahedron: Asymm.* **1998**, *9*, 271–283.

215 N. W. Fadnavis, S. K. Vadivel, U. T. Bhalerao, *Tetrahedron: Asymm.* **1997**, *8*, 2355–2359.

216 K. Ishihara, N. Nakajima, S. Tsuboi, M. Utaka, *Bull. Chem. Soc. Jpn.* **1994**, *67*, 3314–3319.

217 V. Waagen, V. Partali, I. Hollingsæter, M. S. S. Huang, T. Anthonsen, *Acta Chem. Scand.* **1994**, *48*, 506–510.

218 E. Zymanczyk-Duda, B. Lejczak, P. Kafarski, J. Grimaud, P. Fischer, *Tetrahedron* **1995**, *51*, 11809–11814.

219 T. Itoh, Y. Yonekawa, T. Sato, T. Fujisawa, *Tetrahedron Lett.* **1986**, *27*, 5405–5408.

220 T. Cohen, S. Tong, *Tetrahedron*, **1997**, *53*, 9487–9496.

221 T. Sugai, Y. Ohtsuka, H. Ohta, *Chem. Lett.* **1996**, 233–234.

222 R. Tanikaga, N. Shibata, T. Yoneda, *J. Chem. Soc., Perkin Trans 1* **1997**, 2253–2258.

223 A. R. Maguire, L. L. Kelleher, *Tetrahedron Lett.* **1997**, *38*, 7459–7462.

224 A. Svatos, Z. Hunková, V. Kren, M. Hoskovec, D. Saman, I. Valterová, J. Vrkoc, B. Koutek, *Tetrahedron: Asymm.* **1996**, *7*, 1285–1294.

225 H. L. Holland, T. S. Manoharan, F. Schweizer, *Tetrahedron: Asymm.* **1991**, *2*, 335–338.

226 G. Fantin, M. Fogagnolo, A. Medici, P. Pedrini, S. Poli, F. Gardini, M. E. Guerzoni, *Tetrehedron: Asymm.* **1991**, *2*, 243–246.

227 R. Hayakawa, M. Shimizu, T. Fujisawa, *Tetrahedron Lett.* **1996**, *37*, 7533–7536.

228 D. Drochner, M. Müller, *Eur. J. Org. Chem.* **2001**, 211–215.

229 K. Mori, H. Mori, *Org. Synth.* **1993**, *Coll. Vol. 8*, 312–315.

230 K. Mori, S. Takayama, S. Yoshimura, *Liebigs Ann. Chem.* **1993**, 91–95.

231 D. W. Brooks, H. Mazdiyasni, P. G. Grothaus, *J. Org. Chem.* **1987**, *52*, 3223–3232.

232 H. Ohta, K. Ozaki, G. Tuchihashi, *Agric. Biol. Chem.* **1986**, *50*, 2499–2502.

233 O. Bortolini, G. Fantin, M. Fogagnolo, P. P. Giovannini, A. Guerrini, A. Midici, *J. Org. Chem.* **1997**, *62*, 1854–1856.

234 E. Keinan, S. C. Sinha, A. Sinha-Bagchi, *J. Org. Chem.* **1992**, *57*, 3631–3636.

235 K. Fuhshuku, N. Funa, T. Akeboshi, H. Ohta, H. Hosomi, S. Ohba, T. Sugai, *J. Org. Chem.* **2000**, *65*, 129–135.

236 R. Bel-Rhlid, A. Fauve, M. F. Renard, H. Veschambre, *Biocatalysis* **1992**, *6*, 319–337.

237 H. Ikeda, E. Sato, T. Sugai. H. Ohta, *Tetrahedron* **1996**, *52*, 8113–8122.

238 Y.-Y. Zhu, D. J. Burnell, *Tetrahedron: Asymm.* **1996**, *7*, 3295–3304.

239 S.-S. Lee, J.-L. Yan, K. C. Wang, *Tetrahedron: Asymm.* **1997**, *8*, 3051–3058.

240 P. Besse, J. Bolte, A. Fauve, H. Veschambre, *Bioorg. Chem.* **1993**, *21*, 342–353.

241 K. Nakamura, S. Kondo, Y. Kawai, K. Hida, K. Kitano, A. Ohno, *Tetrahedron: Asymm.* **1996**, *7*, 409–412.

242 M. Sakakibara, A. Ogawa-Uchida, *Biosci. Biotech. Biochem.* **1995**, *59*, 1300–1303.

243 M. Kato, K. Sasahara, K. Ochi, H. Akita, T. Oishi, *Chem. Pharm. Bull.* **1991**, *39*, 2498–2501.

244 M. Chartrain, J. Lynch, W.-B. Choi, H. Churchill, S. Patel, S. Yamazaki, R. Volante, R. Greasham, *J. Mol. Catal, B: Enzymatic* **2000**, *8*, 285–288.

245 S. Danchet, C. Bigot, D. Buisson, R. Azerad, *Tetrahedron: Asymm.* **1997**, *8*, 1735–1739.

246 K. Nakamura, T. Miyai, K. Nozaki, K. Ushio, S. Oka, A. Ohno, *Tetrahedron Lett.* **1986**, *27*, 3155–3156.

247 K. Nakamura, T. Miyai, A. Nagar, S. Oka, A. Ohno, *Bull. Chem. Soc. Jpn.* **1989**, *62*, 1179–1187.

248 K. Nakamura, Y. Kawai, N. Nakajima,

T. Miyai, S. Honda, A. Ohno, *Bull. Chem. Soc. Jpn.* **1991**, *64*, 1467–1470.

249 C. Abalain, D. Buisson, R. Azerad, *Tetrahedron: Asymm.* **1996**, *7*, 2983–2996.

250 G. Fantin, M. Fogagnolo, P. Giovannini, A. Medici, E. Pagnotta, P. Pedrini, A. Trincone, *Tetrahedron: Asymm.* **1994**, *5*, 1631–1634.

251 T. Kuramoto, K. Iwamoto, M. Izumi, M. Kirihata, F. Yoshizako, *Biosci. Biotech. Biochem.* **1999**, *63*, 598–601.

252 K. Nakamura, T. Miyai, Y. Kawai, N. Nakajima, A. Ohno, *Tetrahedron Lett.* **1990**, *31*, 1159–1160.

253 K. Nakamura, Y. Kawai, T. Miyai, A. Ohno, *Tetrahedron Lett.* **1990**, *31*, 3631–3632.

254 G. Gibbs, M. J. Hateley, L. McLaren, M. Welham, C. L. Willis, *Tetrahedron Lett.* **1999**, *40*, 1069–1072.

255 B. Das, P. Madhusudhan, A. Kashinatham, *Bioorg. Med. Chem. Lett.* **1998**, *8*, 1403–1406.

256 C. Gonzáles-Bello, M. K. Manthey, J. H. Harris, A. R. Hawkins, J. R. Coggins, C. Abell, *J. Org. Chem.* **1998**, *63*, 1591–1597.

257 R. N. Patel, R. L. Hanson, A. Banerjee, L. J. Szarka, *J. Am. Oil Chem. Soc.* **1997**, *74*, 1345–1360.

258 N. W. Fadnavis, S. K. Vadivel, M. Sharfuddin, U. T. Bhalerao, *Tetrahedron: Asymm.* **1997**, *8*, 4003–4006.

259 M. Amat, M.-D. Coll, J. Bosch, E. Espinosa, E. Molins, *Tetrahedron: Asymm.* **1997**, *8*, 935–948.

260 A. Sutherland, C. L. Willis, *Tetrahedron Lett.* **1997**, *38*, 1837–1840.

261 C. Fuganti, P. Grasselli, M. Mendozza, S. Servi, G. Zucchi, *Tetrahedron* **1997**, *53*, 2617–2624.

262 N. M. Kelly, R. G. Reid, C. L. Willis, P. L. Winton, *Tetrahedron Lett.* **1996**, *37*, 1517–1520.

263 H. Watanabe, T. Watanabe, K. Mori, *Tetrahedron* **1996**, *52*, 13939–13950.

264 M. Miyazawa, K. Tsuruno, H. Kameoka, *Tetrahedron: Asymm.* **1995**, *6*, 2121–2122.

265 T. Sugai, O. Katoh, H. Ohta, *Tetrahedron* **1995**, *51*, 11987–11998.

266 M. Zarevúcka, M. Rejzek, Z. Wimmer, D. Saman, L. Strcinz, *Tetrahedron* **1993**, *49*, 5305–5314.

267 S. Robin, F. Huet, A. Fauve, H. Veschambre, *Tetrahedron: Asymm.* **1993**, *4*, 239–246.

268 Y. J. Surh, S. S. Lee, *Biochem. Int.* **1992**, *27*, 179–187.

269 J.-X. Gu, Z.-Y. Li, G.-Q. Lin, *Tetrahedron: Asymm.* **1992**, *3*, 1523–1524.

270 G. Fronza, C. Fuganti, P. Grasselli, G. Pedrocchi-Fantoni, S. Servi, *Tetrahedron Lett.* **1992**, *33*, 5625–5628.

271 E. Santaniello, P. Ferraboschi, P. Grisenti, F. Aragozzini, E. Maconi, *J. Chem. Soc., Perkin Trans 1*, **1991**, 601–605.

272 E. Keinan, S. C. Sinha, A. Sinha-Bagchi, *J. Chem. Soc., Perkin Trans. 1*, **1991**, 3333–3339.

273 K. Pabsch, M. Petersen, N. N. Rao, A. W. Alfermann, C. Wandrey, *Recl. Trav. Chim. Pays-Bas*, **1991**, *110*, 199–205.

274 A. Kumar, D. H. Ner, S. Y. Dike, *Tetrahedron Lett.* **1991**, *32*, 1901–1904.

275 S. Shimizu, M. Kataoka in: *Biotransformations* (Ed.: K. Faber), Springer-Verlag, Berlin, **2000**, p. 109–123.

276 A. Muheim, R. Waldner, D. Sanglard, J. Reiser, H. E. Schoemaker, M.S.A. Leisola, *Eur. J. Biochem.* **1991**, *195*, 369–375.

277 J.A.S. Howell, M. G. Palin, G. Jaouen, S. Top, H. E. Hafa, J. M. Cense, *Tetrahedron: Asymm.* **1993**, *4*, 1241–1252.

278 J.A.S. Howell, M. G. Palin, H. E. Hafa, S. Top, G. Jaouen, *Tetrahedron: Asymm.* **1992**, *3*, 1355–1356.

279 C. Baldoli, P. D. Buttero, S. Maiorana, G. Ottolina, S. Riva, *Tetrahedron: Asymm.* **1998**, *9*, 1497–1504.

280 K. Nakamura, T. Miyai, K. Ushio, S. Oka, A. Ohno, *Bull. Chem. Soc. Jpn.* **1988**, *61*, 2089–2093.

281 W. Adam, U. Hoch, M. Lazarus, C. R. Saha-Möller, P. Schreier, *J. Am. Chem. Soc.* **1995**, *117*, 11898–11901.

282 W. Adam, C. Mock-Knoblauch, C. R. Saha-Möller, *Tetrahedron: Asymm.* **1997**, *8*, 1947–1950.

283 W. Adam, M. Lazarus, U. Hoch, M. N. Korb, C. R. Saha-Möller, P. Schreier, *J. Org. Chem.* **1998**, *63*, 6123–6127.

284 W. Adam, B. Boss, D. Harmsen, Z. Lukacs, C. R. Saha-Möller, P. Schreier, *J. Org. Chem.* **1998**, *63*, 7598–7599.

285 M. Abo, A. Okubo, S. Yamazaki, *Tetrahedron: Asymm.* **1997**, *8*, 345–348.

286 M. Baruah, A. Boruah, D. Prajapati, J. S. Sandhu, *Synlett.* **1996**, 1193–1194.

287 A. Kamal, B. Laxminarayana, N. L. Gayatri, *Tetrahedron Lett.* **1997**, *38*, 6871–6874.

288 W. Baik, J. L. Han, K. C. Lee, B. H. Kim, J.-T.

Hahn, *Tetrahedron Lett.* **1994**, *35*, 3965–3966.

289 W. Baik, D. I. Kim, H. J. Lee, W.-J. Chung, B. H. Kim, S. W. Lee, *Tetrahedron Lett.* **1997**, *38*, 4579–4580.

290 J. A. Blackie, N. J. Turner, A. S. Wells, *Tetrahedron Lett.* **1997**, *38*, 3043–3046.

291 W. Baik, D. I. Kim, S. Koo, J. U. Rhee, S. H. Shin, B. H. Kim, *Tetrahedron Lett.* **1997**, *38*, 845–848.

292 K. Stott, K. Saito, D. J. Thiele, V. Massey, *J. Biol. Chem.* **1993**, *268*, 6097–6106.

293 A. D. N. Vaz, S. Chakraborty, V. Massey, *Biochemistry* **1995**, *34*, 4246–4256.

294 Y. Meah, V. Massey, *Proc. Natl. Acad. Sci. USA* **2000**, *97*, 10733–10738.

295 H. Ohta, N. Kobayashi, K. Ozaki, *J. Org. Chem.* **1989**, *54*, 1802–1804.

296 C. Fuganti, G. Pedrocchi-Fantoni, A. Sarra, S. Servi, *Tetrahedron: Asymm.* **1994**, *5*, 1135–1138.

297 M. S. v. Dyk, E. v. Rendsburg, I. P. B. Rensburg, N. Moleleki, *J. Mol. Catal, B: Enzymatic* **1998**, *5*, 149–154.

298 T. Hirata, K. Shimoda, T. Gondai, *Chem. Lett.* **2000**, 850–851.

299 G. Fronza, C. Fuganti, M. Mendozza, R. S. Rallo, G. Ottolina, D. Joulain, *Tetrahedron* **1996**, *52*, 4041–4052.

300 K. Takabe, M. Tanaka, M. Sugimoto, T. Yamada, H. Yoda, *Tetrahedron: Asymm.* **1992**, *3*, 1385–1386.

301 G. Fronza, G. Fogliato, C. Fuganti, S. Lanati, R. Rallo, S. Servi, *Tetrahedron Lett.* **1995**, *36*, 123–124.

302 H. E. Högberg, E. Hedenström, J. Fägerhag, S. Servi, *J. Org. Chem.* **1992**, *57*, 2052–2059.

303 T. Sakai, S. Matsumoto, S. Hidaka, N. Imajo, S. Tsuboi, M. Utaka, *Bull. Chem. Soc. Jpn.* **1991**, *64*, 3473–3475.

304 D. L. Varie, J. Brennan, B. Briggs, J. S. Cronin, D. A. Hay, J. A. Rieck III, M. J. Zmijewski, *Tetrahedron Lett.* **1998**, *39*, 8405–8408.

305 B. Das, A. Kashinatham, P. Madhusudhan, *Tetrahedron Lett.* **1997**, *38*, 7457–7458.

306 K. Matsumoto, Y. Kawabata, J. Takahashi, Y. Fujita, M. Hatanaka, *Chem. Lett.* **1998**, 283–284.

307 R. N. Patel. *J. Am. Oil Chem. Soc.* **1999**, *76*, 1275-1281.

308 M. Kodaka, Y. Kubota, *J. Chem. Soc., Perkin Trans 2*, **1999**, 891–894.

309 R. Obert, B. C. Dave, *J. Am. Chem. Soc.* **1999**, *121*, 12192–12193.

310 S, Kuwabata, R. Tsuda, K. Nishida, H. Yoneyama, *Chem. Lett.* **1993**, 1631–1634.

311 S. Kuwabata, R. Tsuda, H. Yoneyama, *J. Am. Chem. Soc.* **1994**, *116*, 5437–5443.

312 K. Sugimura, S. Kuwabata, H. Yoneyama, *J. Am. Chem. Soc.* **1989**, *111*, 2361–2362.

313 K. Sugimura, S. Kuwabata, H. Yoneyama, *J. Electroanal. Chem.* **1990**, *299*, 241–247.

15.3
Reduction of C=N bonds

Andreas S. Bommarius

15.3.1
Introduction

Enantiospecific reduction of C=N bonds is of interest for the synthesis of α-amino acids and derivatives such as amines. While nonenzymatic reductive amination has been known since 1927[1], only recently have enzymatic procedures to L-amino acids became established. The reduction can be achieved by different enzymes following different mechanisms, e. g. by pyridoxalphosphate (PLP)-dependent transaminases (E. C. 2.6.1, discussed in Chapter 12.7) or by amino acid dehydrogenases (E. C. 1.4.1) using NADH or NADPH as the cofactor. The synthetic usefulness of the transaminase reaction is diminished by the location of the equilibrium (K_{eq} often is close to one), so that complex mixtures result, which are often laborious to separate (for solutions to this problem, see Chapter 12.7). For this reason, this chapter focuses on the reduction of C=N bonds by reductive amination with amino acid dehydrogenases, AADHs.

Reductive amination of α-keto acids to α-amino acids is similar to the reduction of C = O bonds to the corresponding α-hydroxy acids. In an equilibrium reaction, α-keto acids can be reductively aminated to α-amino acids or, *vice versa*, α-amino acids can be oxidatively deaminated:

$$
\begin{array}{c}
\text{R} \diagdown \diagup \text{COOH} \\
\text{O}
\end{array}
+ \text{ NADH } + \text{ NH}_3 + \text{H}^+ \quad \rightleftharpoons \qquad (1)
$$

$$
\begin{array}{c}
\text{R} \diagdown \diagup \text{COOH} \\
\text{NH}_2
\end{array}
+ \text{ NAD}^+ + \text{ H}_2\text{O}
$$

A very promising process route is the reductive amination of prochiral α-keto acids to α-amino acids with AADHs and the cofactor NADH and its regeneration by co-oxidation of formate to CO_2 by formate dehydrogenase (Fig. 15.3-1).

This asymmetric synthesis route possesses a number of advantages rendering it attractive in today's context of seeking environmentally benign processes:

- compact synthesis of α-keto acid substrates,
- formation of harmless and easily separable CO_2 as the only co-product,
- extreme enantioselectivity of amino acid dehydrogenases, and
- yields of up to 100% with respect to α-keto acid, resulting in no undesirable enantiomers and other by-products.

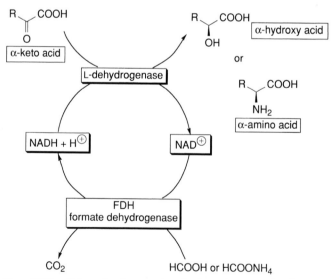

Figure 15.3-1. Schematic of enzymatic reductive amination with cofactor regeneration.

Table 15.3-1. List of NAD(P)$^+$-dependent amino acid dehydrogenases [5].

E.C. Number	Enzyme	Coenzyme	Sourcea
1.4.1.1	Alanine DH	NAD$^+$	B (*Bacillus, Streptomyces, Halobacterium*)
1.4.1.2	Glutamate DH	NAD$^+$	B, F, Y, P
1.4.1.3	Glutamate DH	NAD(P)$^+$	A, F, *Tetrahymena*
1.4.1.4	Glutamate DH	NAD$^+$	B, F, Y, *Chlorella*
1.4.1.7	Serine DH	NAD$^+$	P
1.4.1.8	Valine DH	NAD(P)$^+$	B (*Alcaligenes, Streptomyces*), P
1.4.1.9	ucine DH	NAD$^+$	B (*Bacillus, Clostridium*)
1.4.1.10	Glycine DH	NAD$^+$	B (*Mycobacterium*)
1.4.1.11	3,5-Diaminohexanoate DH	NAD$^+$	B (*Clostridium*)
1.4.1.12	2,4-Diaminopentanoate DH	NAD$^+$	B (*Clostridium*)
1.4.1.15	Lysine DH	NAD$^+$	Human, B (*Agrobacterium*)
1.4.1.16	Diaminopimelate DH	NADP$^+$	B (*Bacillus, Corynebacterium*)
1.4.1.20	Phenylalanine DH	NAD$^+$	B (*Brevibacterium, Bacillus, Rhodococcus*)
1.4.1.-	Tryptophan DH	NAD(P)$^+$	P

a Abbreviations: B: bacterium; F: fungi; Y: yeast; A: animal; P: plant; DH: dehydrogenase

With three exceptions (AlaDH from *Phormidium lapideum*, L-lysine-ε-dehydrogenase and meso-α,ε-diaminopimelate DH) all of the AADHs (Table 15.3-1) catalyze reduction of prochiral keto acids to the L-amino acids [(S)-configuration]. The natural function of L-AADHs is not known. The D-AADHs that have been found appear to be iron-sulfur membrane-associated flavoenzymes which seem to catalyze the oxidative reaction from keto acids to amino acids only; artificial dyes and the coenzyme Q analog serve as electron acceptors but not oxygen [2–4]. AADHs have been screened from a variety of organisms (Table 15.3-1), the most important enzymes for synthesis are alanine dehydrogenase (AlaDH, E.C. 1.4.1.1), phenylalanine dehydrogenase (PheDH, E.C. 1.4.1.20), and particularly leucine dehydrogenase (LeuDH, E.C. 1.4.1.9). The ubiquitous glutamate dehydrogenase (GluDH, E.C. 1.4.1.2.–4), however, is still the most studied member of the group.

Reviews on AADHs: Apart from early review articles on individual amino acid dehydrogenases by Schütte et al. (1985; LeuDH from *B. cereus*) [6], Ohshima et al. (1985a; LeuDH from *B. species*) [7] and Hummel et al. (1987; PheDH from *Rh. rhodocrous*) [8], comprehensive reviews have been published by Hummel and Kula (1989) [9], Ohshima and Soda (1989 and 1990) [5,10,11] and by Brunhuber and Blanchard (1994) [12].

15.3.2
Structural Features of Amino Acid Dehydrogenases (AADHs)

Most of the AADHs possess hexameric structure, although octamers, tetramers, dimers and even monomers have been found. The subunits are usually of similar size: for instance, most bacterial AADHs are hexamers with a molecular weight of around 49 000 per subunit.

Table 15.3-2. Identities of protein sequences of different amino acid dehydrogenases (in per cent) [22]. The data were calculated via BLAST search in the database 'Swissprot' [23].

Protein	LeuDH, B. cereus	LeuDH, B. sphaericus	PheDH, Rh. rhodocrous	PheDH, Th. intermedius	GluDH, C. symbiosum
LeuDH, *B. stearothermophilus*	82.5	79.9	32.0	45.6	12.6
LeuDH, *B. cereus*	–	76.9	31.5	44.5	13.4
LeuDH, *B. sphaericus*		–	31.7	41.8	14.0
PheDH, *Rh. rhodocrous*			–	26.4	12.4
PheDH, *T. intermedius*				–	14.2

15.3.2.1
Sequences and Structures

Several amino acid dehydrogenases have been screened from a variety of microorganisms, the preparatively most important are phenylalanine dehydrogenase (PheDH, from *Rhodococcus* sp. M4) and leucine dehydrogenase (LeuDH, from *Bacillus stearothermophilus* and *Bacillus cereus*). As of the end of February 2001, more than 20 gene and protein sequences for AADHs except GluDH (which more than triples the number) and 3D crystal structures from five different AADHs have been deposited (GluDH from *Clostridium symbiosum*[13], LeuDH from *B. sphaericus*[14], AlaDH from *Phormidium lapideum*[15], PheDH from *Nocardia* sp 239[16] and PheDH from *Rhodococcus* sp. M4[17,18]). Sequence homologies and similarities of 3D structures of the members of several organisms are so high that amino acid dehydrogenases can be termed a single superfamily, generated through divergent evolution[19–21] (Table 15.3-2).

Remarkable, on one hand, is the high degree of identity of the three leucine dehydrogenases, and on the other hand the sequence of glutamate dehydrogenase, which bears no homology to the other dehydrogenases. Although overall sequence homology varies from around 20% up to 80%, the residues essential for the three-dimensional structure of a subunit, for nicotinamide cofactor binding, and for catalysis have been conserved[20]. While a complex between NAD$^+$ and GluDH from *Clostridium symbiosum* left the overall conformation unaltered[24], a drastic conformational change (hinge movement) was observed on binding of the glutamate[13].

15.3.3
Thermodynamics and Mechanism of Enzymatic Reductive Amination

15.3.3.1
Thermodynamics

For reductive amination, basically no thermodynamic limitation exists: for the leucine/ketoleucine reaction at pH 11.0, K_{eq} equals 9×10^{12}[25], for phenylalanine/phenylpyruvate at pH 7.95 a K_{eq} of 2.5×10^7 has been reported[18], thus, the maximum degree of conversion is very close to 100%. Coupling of the reductive amination reaction with cofactor regeneration via the FDH/formate reaction, which is irreversible, further helps to pull the equilibrium towards the amino acid product.

15.3.3.2

Mechanism, Kinetics

As will be elucidated below, the mechanism of reductive amination and the geometry of the active center[13, 18, 19, 26, 27] cause the (S)-configured amino acid products of the reaction to be completely enantiomerically pure, an important criterion for a large-scale application.

The catalytic mechanism of AADHs has been studied most thoroughly with GluDH from *C. symbiosum*[13,24] and with PheDH from *Rhodococcus* M4 [18]. The mechanism was found to be remarkably similar in both cases so that the prediction by Stillman et al.[13] seems to have been borne out. In Fig. 15.3-2, the study on PheDH is illustrated [18]:

Following the scheme in Fig. 15.3-2, which depicts oxidative deamination, in a clockwise fashion starting from the top left, the α-N-protonated L-Phe molecule is stabilized by the ε-group of Lys66 at the carbonyl group as well as by the ε-group of Lys78 via a water molecule, the carbonyl group of Pro117 and the β-carboxyl group of Asp118 at the α-amino group. The first intermediate is the protonated imine after steps (2) and (3) in which Lys78 picks up the proton from the α-N-group of L-Phe and delivers a hydrogen to the *Si* face of the cofactor NAD+ with deprotonation of Lys78. Accompanied by another Lys78 protonation, the water molecule adds to the imine

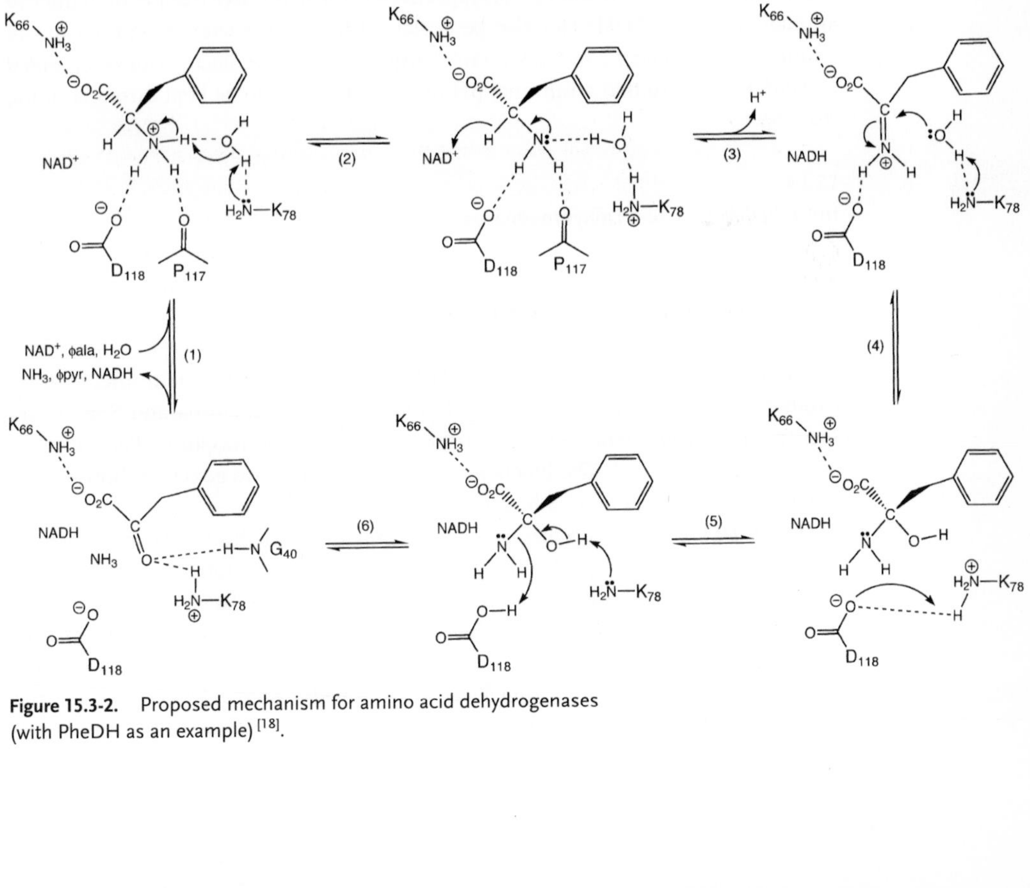

Figure 15.3-2. Proposed mechanism for amino acid dehydrogenases (with PheDH as an example) [18].

carbon to form the carbinolamine, the second intermediate [step (4)]. The Lys78 proton is picked up by Asp118 [step (5)] and in turn by the amino group [step (6)] of the substrate to liberate NH_3 and with the formation of phenylpyruvate. The keto group is stabilized by the protonated ε-sidechain of Lys78 as well as a by a proton from Gly40. The positioning of Lys78 and Gly40 also prevents the oxidation of phenyllactate, so that PheDH cannot act as a HicDH.

A similar mechanism had already been proposed for GluDH from *C. symbiosum*[13]; the only major difference seems to be the attribution of the initial deprotonation of the amino acid molecule to Asp165 (which corresponds to Asp117 on PheDH) instead of Lys125 (Lys78 in PheDH). The Lys125 in GluDH is known to have a low pK value[28], which causes this residue to act as a proton shuttle more easily.

The optimum degree of protonation and catalytically important amino acid residues can be determined from a log V_{max}-pH diagram[29]: on the acidic and alkaline side of the optimum pH, log V_{max} decreases nearly linearly with pH, the two slopes intersect at the optimum degree of protonation, which is also the optimum point of activity. The experimentally observed optimum pH value of 9.2–9.3 for LeuDH[30], corresponding to two pK values of around 8.7 and 10.0 for amino acid residues participating in the catalytic step, can be linked to lysine residues, corroborating the results of Rife and Cleland (1980)[26] and Sekimoto et al. (1993)[27] for the case of GluDH. Brunhuber et al. in their study of PheDH assigned their pK_as values of 8.1 and 9.4 to Asp118 and Lys78, respectively[18]. The influence of pH on reductive aminations with AADHs can also be explained by the dissociation equilibrium of ammonia (pK_a value 9.25). Only an uncharged ammonia molecule can be accepted by LeuDH[26, 30] so that a minimal pH of around 7.5 has to be kept throughout the reaction.

15.3.4
Individual Amino Acid Dehydrogenases

15.3.4.1
Leucine Dehydrogenase (LeuDH, E. C. 1.4.1.9)

Isolation and characterization of LeuDH has been pioneered by Hummel et al.[31] (from *B. sphaericus*), Schütte[6] (from *B. cereus*), and by Ohshima and Soda (from mesophilic *Bacillus sphaericus* and from moderately thermophilic *Bacillus stearothermophilus*[10, 20, 32]). The biochemical data for the last two enzymes, however, do not differ much, as Table 15.3-3 reveals.

The LeuDH from *B. stearothermophilus* as compared with the *B. sphaericus* enzyme has an extended pH range of activity (5.5–10 vs. 6.5–8.5), a higher heat stability (70 vs. 50 °C after a heat treatment of 5 min), a longer half-life (several months vs. six days at pH 7.2 and 6 °C), and much greater stability against organic solvents and denaturants[10].

LeuDH from *B. stearothermophilus* had already been cloned and overexpressed[20, 33] during early studies. Recently, the production of recombinant enzyme from *B. cereus* even on a large scale has been demonstrated[34, 35].

Table 15.3-3. Properties of LeuDH from *Bacillus sphaericus* and *Bacillus stearothermophilus* [10].

Source	B. sphaericus	B. stearothermophilus
M_r (kDa)	245 000	300 000
Subunit (M_r)	41 000	49 000
	hexamer	hexamer
Optimum pH: deamination	10.7	11.0
amination	9.0–9.5	9.0–9.5
Coenzyme	NAD (K_M 0.39 mM)	NAD (K_M 0.49 mM)
Substrate specificity (in % of L-leucine)		
Deamination: L-leucine	100 (K_M 1.0 mM)	100 (K_M 4.4 mM)
L-valine	74 (1.7)	98 (3.9)
L-isoleucine	58 (1.8)	73 (1.4)
L-norvaline	41 (3.5)	–
L-α-aminobutyrate	14 (10)	–
L-norvaline	10 (6.3)	–
D-leucine	0	0
Amination: α-ketoisocaproate	100 (0.31)	100
α-ketoisovalerate	126 (1.4)	167
α-ketovalerate	76 (1.7)	86
α-ketobutyrate	57 (1.7)	45
α-ketocaproate	46 (7.0)	–

The substrate specificity of LeuDHs, catalyzing mainly branched-chain α-keto acids to the α-amino acids, has been investigated by Zink and Sanwal (1962) [36] and subsequently by Schütte et al. (1985: *B. cereus*) [6], Ohshima and Soda (1989; *Bacillus stearothermophilus* and *Bacillus sphaericus*) [5], Nagata et al. (1990; *Bacillus DSM 7330*) [37], Misono et al. (1990; *Corynebacterium pseudodiphtheriticum*) [38] and by Bommarius et al. (1994; *Bacillus stearothermophilus*) [39]. In addition to the proteinogenic amino acids valine, leucine, and isoleucine, unnatural amino acids such as *tert*-leucine [40] or L-β-hydroxy-valine [41] can be synthesized.

The kinetic parameters of several leucine dehydrogenases show a similar pH-profile. The opposite tendency of V_{max} and K_M for all substrates is remarkable: the dimethyl-substituted substrates show K_M values above 10 mM (V_{max} values are between 0.2 and 30 % of the reactivity of 2-oxo-4-methyl-pentanoic acid, the base case), whereas K_M values below 1 mM are typical for good substrates ($V_{max} \approx 100\%$ compared with the base case, 2-oxo-4-methyl-pentanoic acid).

15.3.4.2
Alanine Dehydrogenase (AlaDH, E. C. 1.4.1.1)

AlaDH has been isolated and characterized from both mesophilic (*B. subtilis* and *B. sphericus*) [42] and thermophilic (*B. stearothermophilus*) [43] organisms. For cloning and purification of AlaDH, see ref. [44]. The narrow substrate specificity of AlaDH [42] renders the enzyme useful for synthesis of L-alanine and analogs only, such as [^{15}N]-L-alanine [45], 3-fluoro-L-alanine [46], and 3-chloro-L-alanine [47].

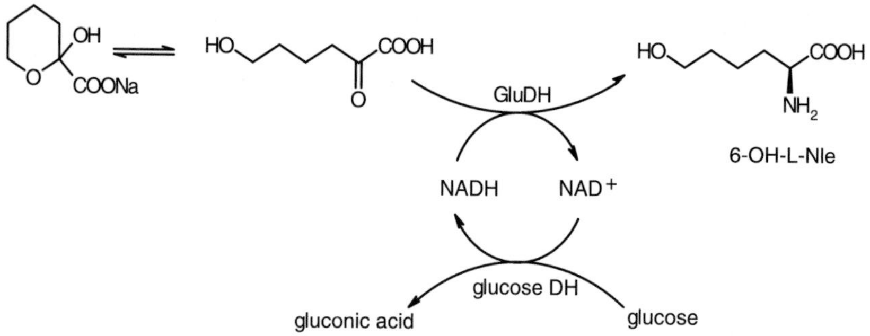

Figure 15.3-3. Synthesis of 6-hydroxy-L-norleucine with GluDH/glucose DH[51].

15.3.4.3
Glutamate Dehydrogenase (GluDH, E. C. 1.4.1.2–4)

GluDH has been investigated by the groups of Engel and Rice since the 1980s so that more is known about GluDH, especially from *C. symbiosum*, than about any other AADH. Although there is no sequence identity to other AADHs beyond random similarity (Table 15.3-2), site-directed mutagenesis of two amino acids residues, K89L and S380V, led to similar activity levels towards glutamate, norleucine and methionine and demonstrated the importance especially of the K89L mutation[48, 49]. Studies on GluDH from the same source define the knowledge base regarding conformational change of the enzyme upon binding of the substrate but not upon the preceding binding of the cofactor. These conformational changes also seem to be responsible in part for substrate specificity[50].

Just as with other AADHs, GluDH has potential as a catalyst in synthesis: beef liver GluDH was the best catalyst for the reductive amination of 2-keto-6-hydroxy-hexanoic acid Na salt to 6-hydroxy-L-norleucine, a potentially important building block for the vasopeptidase Vanlev (BMS) (Fig. 15.3-3)[51]. The reaction of 95 mM substrate (2 : 1 mixture of 2-keto-6-hydroxy-hexanoic acid Na salt in equilibrium with 2-hydroxy-tetrahydropyran-2-carboxylic acid) was complete in 3 h, resulting in an amino acid product of 89–92% chemical yield and >99% optical purity. As the keto acid substrate is very cumbersome to synthesize, an alternative way of providing the keto acid substrate was the separation of D,L-6-hydroxynorleucine, which can be prepared easily from 4-hydroxybutylhydantoin, by D-amino acid oxidase to L-amino acid and keto acid where the latter in turn was reduced by GluDH/NADH[51]. Both FDH/formate and glucose DH/glucose were employed for cofactor regeneration.

15.3.4.4
Phenylalanine Dehydrogenase (PheDH, E. C. 1.4.1.20)

An enzyme catalyzing the reductive amination of phenylpyruvate to the desired L-Phenylalanine was first found by Hummel et al.[52] in a strain of *Brevibacterium* and later in *Rhodococcus* sp.[8, 53]. Table 15.3-4 summarizes the microbiological and kinetic data[9].

Table 15.3-4. Comparison of PheDH from *Brevibacterium* and *Rhodococcus* species [9].

Parameter	*Brevibacterium*	*Rhodococcus*
Microbiological data:		
enzyme yield (U L^{-1}) after addition of 1% of		
L-phenylalanine	210	15 200
L-histidine	120	1800
L-phenylalaninamide	-	3500
L-isoleucine	0	0
D-phenylalanine	204	0
DL-phenylalanine	214	0
	9.0	9.25
Enzymological data:		
pH optimum		
reductive amination		
oxidative deamination	10	10
	0.11	0.16
K_M (mM)		
phenylpyruvate		
p-hydroxypyruvate	0.24	2.4
indolepyruvate	8.0	7.7
2-oxo-4-methylmercaptobutyrate	3.0	2.1
	100	100
V_{max} (relative to phenylpyruvate)		
phenylpyruvate		
p-hydroxypyruvate	96	5
indolepyruvate	24	3
2-oxo-4-methylmercaptobutyrate	59	33
	47	130
K_M (µM) NADH		
K_M (mM) NH$_4^+$	431	387
	4–8 h	10 d
Stability:		
stored at 4 °C ($t_{1/2}$)		
deactivation (% d^{-1}) under operation	26	5
Reference	8	53

Table 15.3-5. Substrate specifity of different PheDHs [39].

Substrate[a]	*Rhodococcus rhodocrous* V_{max} (U m L^{-1})	K_M (mM)	Rel. activity (%)	*B. sphaericus* Rel. activity (%)
Ketoisocaproate			4.2	
Keto-methionine[b]	50	2.1	33	6.0
Phenylpyruvate	150	0.16	= 100	= 100
p-OH-phenylpyruvate[b]	7.5	2.4	5	138
Indolepyruvate[b]	4.5	7.7	3	n. d.[c]
Keto-4-phenylbutyrate	96	0.01	64	1.9
Keto-5-phenylvalerate	46	0.65	30	1.5

a Conditions: pH 8.0, $T = 25$ °C, [S] = 0.1 M; comparison: LeuDH from *B. cereus*: 2-oxo-4-methyl-pentanoic acid = 100%, 2-oxo-4-phenylbutyrate = 0.2%; *B. sphaericus* data from [55]
b As in a except for a pH of 8.5 [8]
c Not determined

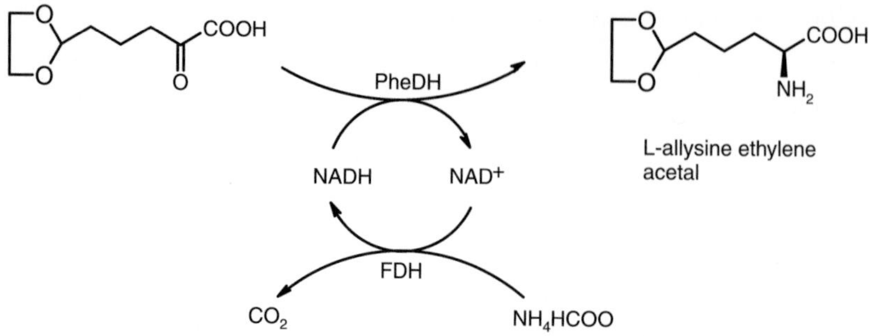

Figure 15.3-4. Synthesis of allysine ethylene acetal with PheDH/FDH [58].

Apart from L-phenylalanine, the homolog L-homophenylalanine (L-Hph), important as a component in ACE inhibitors, can be obtained from 2-keto-4-phenyl-butyrate with PheDH [54]. The substrate specificity of PheDH from *Bacillus sphaericus* has been investigated by Asano et al. [55]. Table 15.3-5 compares the activities of two PheDH from *Rhodococcus rhodocrous* [8] and *Bacillus sphaericus* [55] for the transformation of aromatic and aliphatic keto acids.

Sequencing, cloning, and heterologous expression of PheDH from *Rhodococcus* was first described by Brunhuber et al. [56]. A double mutation G124A/L307V was created by site-directed mutagenesis of PheDH from *Bacillus sphaericus* to change the substrate specificity from a PheDH closer to a LeuDH. This led to a mutant with decreased activity towards L-phenylalanine and enhanced activity towards almost all aliphatic amino acid substrates, thus confirming the predictions made from molecular modeling [57].

PheDH from *Thermoactinomyces intermedius* ATCC 33 205 was utilized recently to synthesize allysine ethylene acetal [(S)-2-amino-5-(1,3-dioxolan-2-yl)-pentanoic acid (2)] from the corresponding keto acid with regeneration of NAD^+ cofactor by FDH/formate [58] (Fig. 15.3-4); the specific activity towards the keto acid was 16% compared to the standard substrate phenylpyruvate.

The system was used in three different configurations: (i) the system with heat-dried cells from *Th. intermedius* (PheDH) and *C. boidinii* (FDH) yielded on average only 84 M% and could not be scaled up owing to lysis of the *Th. intermedius* cells; (ii) a similar system with recombinant PheDH from *E. coli* improved the yield to 91 M%; (iii) heat-dried *Pichia pastoris* containing endogeneous FDH and expressing recombinant PheDH from *Th. intermedius* yielded 98 M% with an optical purity of >98%.

Altogether, more than 200 kg of allysine ethylene acetal have been produced.

15.3.5
Summary of Substrate Specificities

The most comprehensive investigation of substrate specificity of LeuDH and PheDH has been conducted by Krix et al. (1997) [30]. Table 15.3-6 lists the relative rates of various substrates.

Table 15.3-6. Relative V_{max} values of keto acid substrates of various LeuDHs and PheDH [30].

Keto acid	B. stearo-thermophilus LeuDH	B. cereus LeuDH	B. sphaericus LeuDH	Rhodococcus Rhodocrous PheDH
Specific activity (U mg^{-1} of protein)	120	15.9	3.3	54.8
2-Oxobutyric acid	48	74	66	72
2-Oxo-3-methylbutyric acid	113	152	205	96
2-Oxo-3,3-dimethylbutyric acid	31	74	51	8
2-Oxopentanoic acid	63	81	102	157
2-Oxo-3-methylpentanoic acid	110	114	88	193
2-Oxo-4-methylpentanoic acid[a]	= 100	= 100	= 100	= 100
2-Oxo-3,3-dimethylpentanoic acid	2	11	5	4
2-Oxo-4,4-dimethylpentanoic acid	7	14	11	54
2-Oxohexanoic acid	15	63	75	250
2-Oxo-4-methylhexaoic acid	22	19	n.d.[b]	296
2-Oxo-4-ethylhexanoic acid	1	11	n.d.[b]	79
2-Oxo-4,4-dimethylhexanoic acid	0.5	1.2	0.2	146
2-Oxo-5,5-dimethylhexanoic acid	0.8	0.3	n.d.[b]	257
2-Oxo-3-cyclohexylpropanoic acid	0.8	0.1	0.3	140
2-Oxooctanoic acid	0.2	n.d.[b]	n.d.[b]	n.d.[b]
2-Oxo-3-(1-adamantyl)propanoic acid	0	n.d.[b]	n.d.[b]	16

a All V_{max} values refer to 2-oxo-4-methylpentanoic acid (= 100%), pH 8.5, T = 30 °C. Absolute activity of LeuDHs with 2-oxo-4-methyl-pentanoic acid (ketoisocaproic acid) were 120 U mg^{-1} (B. stearothermophilus), 15.9 U mg^{-1} (B. cereus) and 3.3 U mg^{-1} (B. sphaericus) as well as 54.8 U mg^{-1} with PheDH (Rh. rhodocrous).
b Not determined

LeuDHs from *B. cereus*, *B. sphaericus* and *B. stearothermophilus* display a remarkably similar substrate spectrum:

- LeuDHs accept 2-oxoacids with hydrophobic, aliphatic, branched and unbranched carbon side chains of up to six C atoms as well as some alicyclic keto acids as substrates, however, not the adamantyl group, where the geometric limit seems to be reached. 2-Oxo-3-methylpentanoic acid is the preferred substrate, the preferred chain length is C5.
- The keto acid substrate should have at least four C atoms; pyruvate is only converted at less than 3% of standard. Short-chain keto acids with branching at the C3 position are only preferred by the enzyme from *B. sphaericus*.
- The different amino acid dehydrogenases differentiate substrate side chains mainly based on steric parameters in the C3 and C4 position of branched ketoacids.
- Functionalized keto acids such as ketoglutarate are not accepted (activity <0.1% of the base case). Phenylpyruvate as a model compound of an aromatic substrate was inert [59].

A correlation of LeuDH activity with van-der-Waals volumes [60] or hydrophobicities [61] for different C atom configuration of side chains only yielded a moderate correlation [39, 61].

PheDH differs markedly from all LeuDHs, as it can convert not only aromatic substrates but also the aliphatic substrates typical for LeuDHs. Owing to the high intrinsic specific activity of PheDH from *Rhodococcus*, in many cases the enzyme actually registers higher specific activity with many sterically demanding α-keto acid substrates than LeuDH. The substrate specifity of PheDH from *Rhodococcus rhodocrous* and *Bacillus sphaericus* seems to vary more between the two PheDHs than the specificity between the different LeuDH species. PheDH from *B. sphaericus* mainly converts (substituted) phenylpyruvates whereas the enzyme from *Rhodococcus* sp. displays a fairly high degree of activity in the presence of a phenylalkyl group in the substrate.

15.3.6
Process Technology: Cofactor Regeneration and Enzyme Membrane Reactor (EMR)

15.3.6.1
Regeneration of NAD(P)(H) Cofactors

Enzymatic reductive amination with NADH as the cofactor can only be operated on a large scale if the cofactor is regenerated. Wandrey and Kula have developed a regeneration scheme using formate as the reductant of NAD^+ generated upon reductive amination (Fig. 15.3-1). The formate is oxidized irreversibly to CO_2 by formate dehydrogenase (FDH, E.C. 1.2.1.2) [62].

For soluble reactants and products, enzymes are preferentially immobilized in an enzyme-membrane reactor (EMR). To prevent the cofactor from penetrating through the membrane, it can be enlarged with polyethyleneglycol (PEG) [63].

L-leucine was produced in an EMR with LeuDH from both *B. sphaericus* [40] and *B. stearothermophilus* [64]. LeuDH has also been employed successfully for the synthesis of L-*tert*-leucine in batch processes [39] and in its continuous version [40b, 65]. L-*tert*-leucine is an important building block for several novel pharma developments [66, 67] as well as being on intermediate for templates for asymmetric synthesis [66]. L-Phe was produced in an EMR with PheDH starting from phenylpyruvate [Fig. 15.3-5, (i)] [68]. Owing to the instability and high cost of this compound, two additional processes were devised generating phenylpyruvate *in situ* (Fig. 15.3-5): (ii) intermittent oxidation of DL-phenyllactate with D- and L-hydroxyisocaproate DH (HicDH) [69], or (iii) hydrolysis of acetamidocinnamic acid (ACA) with ACA acylase [70]. For productivities of all processes, see Table 15.3-7.

Another regeneration scheme for NADH from NAD^+ utilizes glucose which is oxidized to gluconic acid with the help of glucose dehydrogenase (see Fig. 15.3-3 for an example) [51]. Regeneration to NADPH from $NADP^+$ can be afforded by glucose-6-phosphate dehydrogenase with glucose-6-phosphate as the substrate [71, 72]; the system, however, has not found widespread use yet, probably owing to the higher price of $NADP^+$ vs. NAD^+ and the cost associated with the generation of glucose-6-phosphate from glucose.

With the advantage of the potentially quantitative use of a keto acid substrate and with suitable processes of cofactor regeneration, reductive amination of keto acids is an interesting route to α-amino acids worthy of consideration in comparison with more established routes.

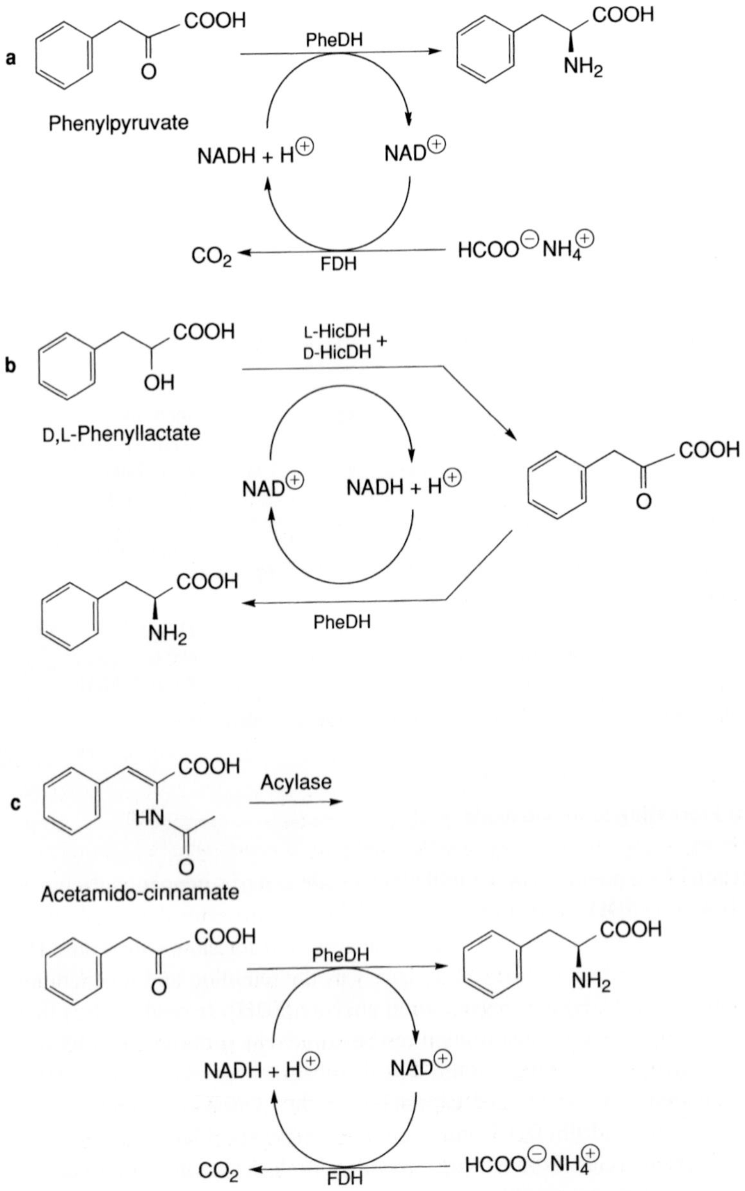

Figure 15.3-5. Enzymatic routes to L-phenylalanine via phenylpyruvate[9]. (i) Reductive amination of phenylpyruvate by PheDH with simultaneous NADH regeneration using FDH.
(ii) Oxidation of DL-phenyllactate with D- and L-2-hydroxy-4-methylpentanoate (HicDH) and simultaneous reductive amination of the phenylpyruvate formed in situ with PheDH. NADH is "substrate-coupled" regenerated from phenyllactate. (iii) In situ formation of phenylpyruvate by enzymatic deacetylation of N-acetamidoocinnamic acid by the respective acylase followed by simultaneous reductive amination with PheDH.

Table 15.3-7. Continuous production of L-amino acids with the aid of dehydrogenases in an enzyme membrane reactor[9].

AADH	Regeneration enzyme(s)	Precursor	Product	Product conc. (mM)	Degree of conversion	s.t.y. g/(Lxd)	Enzyme consumptions (U kg^{-1})	Ref.
LeuDH	FDH	oxomethyl-pentanoate	L-leu	80	80	250	300/300 (LeuDH,FDH)	40c
LeuDH	D-HmpDH L-HmpDH	DL-OH-methyl-pentanoate	L-leu	70	70	72	730/350/650 (LeuDH, D-HmpDH, L-HmpDH)	40a
LeuDH	D-HmpDH L-HmpDH	D,L-OH-methionine	L-met	240	60	143		40c
LeuDH	FDH	trimethyl-pyruvate	L-tle	425	85	640	1000/2000 (LeuDH, FDH)	40c
AlaDH	D-LDH L-LDH	D,L-lactate	L-ala	184	46	134	4700/2600 (LeuDH, FDH)	40a
PheDH	FDH	phenyl-pyruvate	L-phe	114	95	456	1500/150 (PheDH, FDH)	68
PheDH	D-HmpDH L-HmpDH	D,L-phenyl-lactate	L-phe	22	43	28		69
PheDH + ACA acylase	FDH	acetamid-ocinnamate	L-phe	70	88	277	1170/1770/400 (acylase, PheDH, FDH)	70

HmpDH: 2-hydroxy-4-methylpentanoate-DH; LDH: lactate dehydrogenase; L-tle: L-tert-leucine

15.3.6.2
Summary of Processing to Amino Acids

The production of L-*tert* leucine on a multi-100 kg scale and of L-neopentylglycine on a 30 kg scale with LeuDH from *B. stearothermophilus* demonstrates the suitability of enzymatic reductive amination on a large scale and even for slow substrates. The economics of the process is influenced decisively by the retention and regeneration of both production (AADH) and regeneration enzyme (FDH). If yields of less than 100% are acceptable enzyme consumption can be lowered by running the process in a continuous mode[40b]. Owing to the broad substrate specificity of AADHs, reductive amination can be utilized especially for the synthesis of hydrophobic amino acids. LeuDH and PheDH feature complementary specificities for aliphatic and aromatic L-amino acids. Both enzymes are enantioselective to the highest degree and stable in a coupled process with FDH. As non-enzymatic processes of reductive amination often lead to low yields and enantioselectivities[73–76], enzymatic schemes are superior to chemical ones. Additionally, enzymatic reductive aminations are conducted solely in water so that organic solvents can be avoided.

References

1 F. Knoop and H. Oesterlin, *Hoppe Seylers Z. Physiol. Chem.* **1927**, *170*, 186–211.

2 P. J. Olsiewski, G. J. Kaczorowski and C. Walsh, *J. Biol. Chem.* **1980**, *255*, 4487–94.

3 S. Nagata, N. Esaki, K. Tanizawa, H. Tanaka, K. Soda, *Agric. Biol. Chem.* **1985**, *49*, 1134–41.

4 M. Lobocka, J. Hennig, J. Wild, T. Klopotowski, *J. Bacteriol.* **1994**, *176*, 1500–10.

5 T. Ohshima, K. Soda, *Int. Ind. Biotech.* **1989**, *9*, 5–11.

6 H. Schütte, W. Hummel, H. Tsai, M.-R. Kula, *Appl. Microbiol. Biotechnol.* **1985**, *22*, 306–317.

7 T. Ohshima, S. Nagata, K. Soda, *Arch. Microbiol.* **1985**, *141*, 407–411.

8 W. Hummel, H. Schütte, E. Schmidt, C. Wandrey, M.-R. Kula, *Appl. Microbiol. Biotechnol.* **1987**, *26*, 409–416.

9 W. Hummel, M.-R. Kula, *Eur. J. Biochem.* **1989**, *184*, 1–13.

10 T. Ohshima, K. Soda, *TIBTECH* **1989**, *7*, 210–214.

11 T. Ohshima, K. Soda, *Adv. Biochem. Eng./Biotech.* **1990**, *42*, 187–209.

12 N. M. W. Brunhuber, J. S. Blanchard, *Crit. Rev. Biochem. Mol. Biol.* **1994**, *29(6)*, 415–467.

13 T. J. Stillman, P. J. Baker, K. L. Britton, D. W. Rice, *J. Mol. Biol.* **1993**, *234*, 1131–1139.

14 P. J. Baker, A. P. Turnbull, S. E. Sedelnikova, T. J. Stillman, D. W. Rice, *Structure* **1995**, *3*, 693–705.

15 P. J. Baker, Y. Sawa, H. Shibata, S. E. Sedelnikova, D. W. Rice, *Nature Struct. Biol.* **1998**, *5*, 561–567.

16 A. Pasquo, K. L. Britton, P. J. Baker, G. Brearley, R. J. Hinton, A. J. G. Moir, T. J. Stillman, D. W. Rice, *Acta Crystallogr.* **1998**, *D54*, 269–272.

17 J. L. Vanhooke, J. B. Thoden, N. M. W. Brunhuber, J. S. Blanchard, H. M. Holden, *Biochemistry* **1999**, *38*, 2326–2339.

18 N. M. W. Brunhuber, J. B. Thoden, J. S. Blanchard, J. L. Vanhooke, *Biochemistry* **2000**, *39(31)*, 9174–9187.

19 K. L. Britton, P. J. Baker, P. C. Engel, D. W. Rice, T. J. Stillman, *J. Mol. Biol.* **1993**, *234*, 938–45.

20 S. Nagata, K. Tanisawa, N. Esaki, Y. Sakamoto, T. Ohshima, H. Tanaka, K. Soda, *Biochemistry* **1988**, *27*, 9056–62.

21 H. Takada, T. Yoshimura, T. Ohshima, N. Esaki, K. Soda, *J. Biochem.* **1991**, *109*, 371–6.

22 A. S. Bommarius, Habilitation thesis, RWTH Aachen, **2000**.

23 a) M. C. Peitsch, *Bio/Technology* **1995**, *13*, 658–660; b) M. C. Peitsch, *Biochem. Soc. Trans.* **1996**, *24*, 274–279.

24 P. J. Baker, K. L. Britton, P. C. Engel, G. W. Farrants, K. S. Lilley, D. W. Rice, T. J. Stillman, *Proteins* **1992**, *12*, 75–86.

25 B. D. Sanwal, M. W. Zink, *Arch. Biochem. Biophys.* **1961**, *94*, 430–435.

26 J. E. Rife, W. W. Cleland, *Biochemistry* **1980**, *19*, 2328–33.

27 T. Sekimoto, T. Matsuyama, T. Fukui, K. Tanizawa, *J. Biol. Chem.* **1993**, *268*, 27039–27045.

28 D. Piszkiewicz, E. L. Smith, *Biochemistry* **1971**, *10*, 4538–44.

29 W. W. Cleland, *Enzyme Kinetics as a Tool for Determination of Enzyme Mechanisms. Investigation of Rates and Mechanisms of Reactions*, 4th ed., John Wiley & Sons, New York, **1986**, pp. 791–870.

30 G. Krix, A. S. Bommarius, K. Drauz, M. Kottenhahn, M. Schwarm, M.-R. Kula, *J. Biotechnol.* **1997**, *53*, 29–39.

31 W. Hummel, H. Schütte, M.-R. Kula, *Eur. J. Appl. Micobiol. Biotechnol.* **1981**, *12*, 22–27.

32 T. Ohshima, S. Nagata, K. Soda, *Arch. Microbiol.* **1985**, *141*, 407–411.

33 M. Oka, Y.-S. Yang, S. Nagata, N. Esaki, H. Tanaka, K. Soda, *Biotechology and Applied Biochem.* **1989**, *11*, 307–311.

34 M. B. Ansorge, M.-R. Kula, *Biotechnol. Bioeng.* **2000**, *68*, 557–562.

35 M. B. Ansorge, M.-R. Kula, *Appl. Microbiol. Biotechnol.* **2000**, *53(6)*, 668–673.

36 M. W. Zink, B. D. Sanwal, *Arch. Biochem. Biophys.* **1962**, *99*, 72–7.

37 S. Nagata, H. Misono, S. Nagasaki, N. Esaki, H. Tanaka, K. Soda, *J. Ferment. Biotechnol.* **1990**, *69*, 199–203.

38 H. Misono, K. Sugihara, Y. Kuwamoto, S. Nagata, S. Nagasaki, *Agric. Biol. Chem.* **1990**, *54*, 1491–1498.

39 A. S. Bommarius, K. Drauz, W. Hummel,

M.-R. Kula, C. Wandrey, *Biocatalysis* **1994**, *10*, 37–47.

40 a) C. Wandrey, B. Bossow, *Biotechnol. Bio-ind.* **1986**, *3*, 8–13; b) U. Kragl, D. Vasic-Racki, C. Wandrey, *Chem. Ing. Tech.* **1992**, *64*, 499–509; c) C. Wandrey in: *Enzymes as Catalysts in Organic Synthesis"* (Ed.: M. Schneider), D. Reidel, Dordrecht, **1986**, pp. 263–284.

41 R. L. Hanson, J. Singh, T. P. Kissik, R. N. Patel, L. J. Szarka, R. H. Mueller, *Bioorg. Chem.* **1990**, *18*, 116–30.

42 a) T. Ohshima, K. Soda, *Eur. J. Biochem.* **1979**, *100*, 29–39; b) H. Porumb, D. Vancea, L. Muresan, *J. Biol. Chem.* **1987**, *262*, 4610–4615; c) I. Vancurova, A. Vancura, J. Vole, *Arch. Microbiol.* **1988**, *150*, 438–440.

43 T. Ohshima, C. Wandrey, M. Sugiura, K. Soda, *Biotechnol. Lett.* **1985**, *7*, 871–876.

44 a) S. Kuroda, K. Tanizawa, H. Tanaka, K. Soda, *Biochemistry*, **1990**, *29*, 1009–1015; b) K. Soda, S. Nagata, H. Tanaka, JP 60 180 580, **1985**; c) K. Soda, S. Nagata, H. Tanaka, JP 60 180 590, **1985**.

45 A. Mocanu, G. Niac, A. Ivanof, V. Gorun, N. Palibroda, E. Vargha, M. Bologa, O. Barzu, *FEBS Lett.* **1982**, *143*, 153–156.

46 T. Ohshima, C. Wandrey, D. Conrad, *Bio-technol. Bioeng.* **1989**, *34*, 394–397.

47 Y. Kato, K. Fukumoto, Y. Asano, *Appl. Micro-biol. Biotechnol.* **1993**, *39*, 301–4.

48 X.-G. Wang, K. L. Britton, P. J. Baker, S. Martin, D. W. Rice, P. C. Engel, *Protein Eng.* **1995**, *8(2)*, 147–152.

49 X.-G. Wang, L. K. Britton, P. J. Baker, D. W. Rice, P. C. Engel, *Biochem. Soc. Trans.* **1996**, *24(1)*, 126S.

50 T. J. Stillman, A. M. B. Migueis, X.-G Wang, P. J. Baker, L. Britton, P. C. Engel, D. W. Rice, *J. Mol. Biol.* **1999**, *285*, 875–885.

51 R. L. Hanson, M. D. S., A. Banerjee, D. B. Brzozowski, B.-C. Chen, B. P. Patel, C. G. McNamee, G. A. Kodersha, D. R. Kro-nenthal, R. N. Patel, L. J. Szarka, *Bioorg Med. Chem.* **1999**, *7*, 2247–2252.

52 W. Hummel, N. Weiß, M.-R. Kula, *Arch. Microbiol.* **1984**, *137*, 47–52.

53 W. Hummel, E. Schmidt, C. Wandrey, M.-R. Kula, *Appl. Microbiol. Biotechnol.* **1986**, *25*, 175–185.

54 C. W. Bradshaw, C.-H. Wong, W. Hummel, M.-R. Kula, *Bioorg. Chem.* **1991**, *19*, 29–39.

55 Y. Asano, A. Yamada, Y. Kato, K. Yama-guchi, Y. Hibino, K. Hirai, K. Kondo, *J. Org. Chem.* **1990**, *55*, 5567–71.

56 N. M. W. Brunhuber, A. Banerjee, W. R. Ja-cobs Jr., J. S. Blanchard, *J. Biol. Chem.* **1994**, *269(23)*, 16203–16211.

57 S. Y. Seah, K. L. Britton, P. J. Baker, D. W. Rice, Y. Asano, P. C. Engel, *FEBS Lett.* **1995**, *370(1–2)*, 93–6.

58 R. L. Hanson, J. M. Howell, T. L. LaPorte, M. J. Donovan, D. L. Cazzulino, V. Zannella, M. A. Montana, V. B. Nanduri, S. R. Schwarz, R. F. Eiring, S. C. Durand, J. M. Waslyk, W. L. Parker, M. S. Liu, F. J. Oku-niewicz, Bang-Chi Chen, J. C. Harris, K. J. Natalie Jr., K. Ramig, S. Swaminathan, V. W. Rosso, S. K. Pack, B. T. Lotz, P. J. Bernot, A. Rusowicz, D. A. Lust, K. S. Tse, J. J. Venit, L. J. Szarka, R. N. Patel, *Enzyme Mi-crob. Technol.* **2000**, *26(5–6)*, 348–358.

59 T. Ohshima, C. Wandrey, M. Sugiura, K. Soda, *Biotechnol. Lett.* **1985**, *7 (12)*, 871–876.

60 A. Bondi, *J. Phys. Chem.* **1964**, *68 (3)*, 441–451.

61 J.-L. Fauchère, *QSAR of Oligopeptides (Amino-acid Side Chain Parameters and Some Specific QSAR Studies)*, QSAR Des. Bioact. Compd., Barcelona, Spanien **1984**.

62 H. Schütte, J. Flossdorf, H. Sahm, M.-R. Kula, *Europ. J. Biochem.* **1976**, *62*, 151–60.

63 M.-R. Kula, C. Wandrey, *Methods Enzymol.* **1988**, *136*, 34–45.

64 T. Ohshima, C. Wandrey, M.-R. Kula, K. Soda, *Biotechnol. Bioeng.* **1985**, *27*, 1616–1618.

65 R. Wichmann, C. Wandrey, A. F. Bück-mann, M.-R. Kula, *Biotechnol. Bioeng.* **1981**, *23*, 2789–2802.

66 A. S. Bommarius, M. Schwarm, K. Stingl, M. Kottenhahn, K. Huthmacher, K. Drauz, **1995**, *Tetrahedron: Asymm. 6*, 2851–2888.

67 M. Whittaker, C. D. Floyd, P. Brown, A.J.H. Gearing, *Chem. Rev.* **1999**, *99*, 2735–2776.

68 W. Hummel, H. Schütte, E. Schmidt, M.-R. Kula, *Appl. Microbiol. Biotechnol.* **1987**, *27*, 283–291.

69 E. Schmidt, D. Vasic-Racki, C. Wandrey, *Appl. Microbiol. Biotechnol.* **1987**, *26*, 42–48.

70 E. Schmidt, W. Hummel, C. Wandrey, *Proc. 4th Eur. Congr. Biotechnol.* **1987**, 189–191.

71 C.-H. Wong, G. M. Whitesides, *J. Am. Chem. Soc.* **1981**, *103*, 4890–99.

72 B. L. Hirschbein, G. M. Whitesides, *J. Am. Chem. Soc.* **1982**, *104*, 4458–60.

73 U. Groth, T. Huhn, B. Porsch, C. Schmeck,

U. Schöllkopf, *Liebigs Ann. Chem.* **1993**, 715–719.

74 U. Groth, C. Schmeck, U. Schöllkopf, *Liebigs Ann. Chem.* **1993**, 321–323.

75 Z. X. Shen, J. F. Qian, W. J. Qiang, Y. W. Zhang, *Chin. Chem. Lett.* **1992**, *3 (4)*, 237–238.

76 G. C. Cox and L. M. Harwood, *Tetrahedron: Asymm.* **1994**, *5*, 1669–1672.

16
Oxidation Reactions

16.1
Oxygenation of C-H and C=C Bonds

Sabine Flitsch, Gideon Grogan and D. Ashcroft

16.1.1
Introduction

Reactions catalyzed by oxygenase enzymes (mono or dioxygenases) are interesting for applications in organic synthesis. There are numerous examples of such reactions in biological systems, yet there are few chemical reagents or catalysts that can compete with biocatalysts. Examples of monoxygenases catalyzed biotransformations are shown in Figure 16.1-1. These include heteroatom oxygenation, aromatic hydroxylation, Bayer-Villiger oxidation, double-bond epoxidation and hydroxylation of non-activated hydrocarbon atoms. The latter can occur with regio-, stereo-,

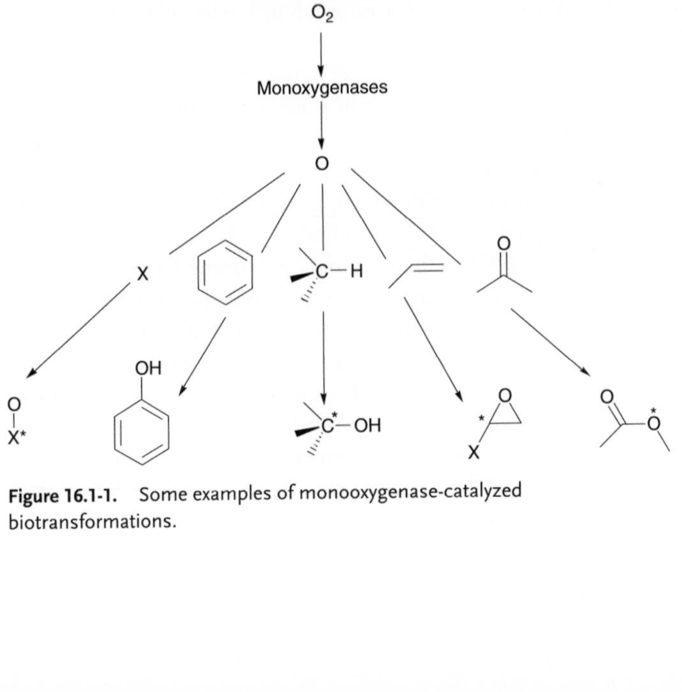

Figure 16.1-1. Some examples of monooxygenase-catalyzed biotransformations.

and in some cases enantioselectivity that is difficult to achieve using conventional chemistry. There is a large body of literature describing the exploitation of these oxygenase enzymes for synthetic applications, and the current chapter will only give a few representative examples of what has been done.

There are few reports on biotransformations using isolated oxygenase enzymes because of several problems with cell free enzymes. Many of the oxygenases are membrane-bound, and require a complex set of co-factors and co-proteins. Despite the fact that a large number of their genes have been identified, the enzymes themselves are difficult to isolate and quite unstable. Thus, oxidative bioconversions, especially on an industrial scale, generally use whole-cell bioconversion techniques, which makes them less accessible for use by organic chemists in organic synthesis laboratories. However, very recently new techniques have been developed for the isolation, cloning and over-expression of oxygenases in heterologous expression systems, and the number of reports using isolated systems is increasing. These new developments will be discussed at the end of the chapter, and point to the possibility of overcoming technical difficulties that have hampered the application of oxygenase systems in biocatalysis in the past.

A number of excellent reviews with comprehensive coverage on the literature of biooxidations have appeared in journals and books [1–8]. In this chapter we will only try to highlight some of these biotransformation reactions, in particular hydroxylation of non-activated carbon atoms and double-bond epoxidation reactions.

16.1.2
Hydroxylating Enzymes

Hydroxylation reactions in nature are generally catalyzed by monooxygenase, a subclass of the oxidoreductase enzyme group. These enzymes are very important and ubiquitous proteins found in almost all living cells, ranging from bacterial to mammalian. One of the most important groups of this type of enzyme is the cytochrome P450 family. These are heme-dependent monooxygenases whose essential role, among other functions, is to ensure detoxification of exogenous compounds by rendering these very often lipophilic molecules water soluble, thus facilitating their excretion. Because of this essential function, mammalian monooxygenases have been thoroughly studied in the context of drug metabolism [1, 9–14].

The majority of the cytochrome P450 systems reported to date are multicomponent, requiring the involvement of additional proteins for transport of reducing equivalents from NAD(P)H to the terminal cytochrome P450 component. Increasing attention is given to microbial monooxygenases, in particular in their application for biotransformations. One of the earliest and most important industrial applications of microbiologically mediated bioconversions is the 11-α-hydroxylation of progesterone using *Rhizopus arrhizus* cells (Figure 16.1-2) [15]. Microbial monooxygenases tend to be soluble enzymes that can be purified fom cell free extracts, and a number of crystal structures of microbial monooxygenases are now available [16, 17]. The first X-ray structure for P450 was that of cytochrome P450$_{cam}$, which was isolated from *Pseudomonas putida*, and catalyzes the 5-*exo* hydroxylation of its natural substrate D-camphor to 5-*exo* hydroxycamphor as shown in Figure 16.1-2.

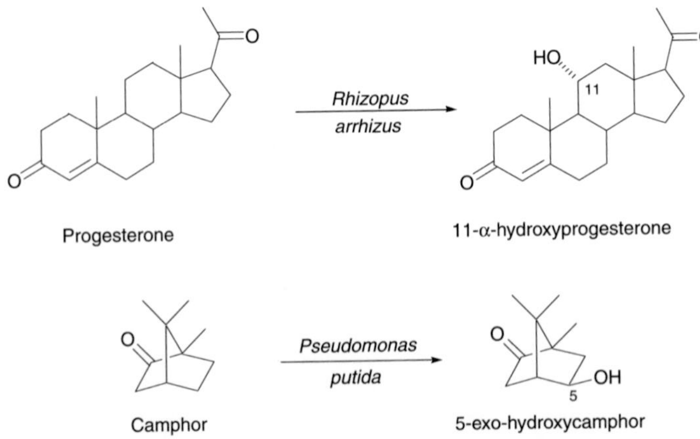

Figure 16.1-2. Regio- and stereoselective microbiological hydroxylation of progesterone and D-camphor.

The P450$_{cam}$ enzyme has served as a model system for general studies of cytochrome P450 enzymes in terms of structure, function and mechanism[16, 18]. The exquisite regio- and stereoselectivity can be explained by the active site geometry of the P450$_{cam}$, which shows several van der Waals interactions with hydrophobic side chains and a key hydrogen bond between tyrosine 96 and the carbonyl oxygen of the substrate. Removal of the tyrosine-96 hydroxyl group by site-directed mutagenesis or removal of the carbonyl oxygen by using camphane results in loss of selectivity[19].

There are also an increasing number of non-P450 type biohydroxylases. Examples are the n-octane ω-hydroxylase of *Pseudomonas oleovorans* and the n-decane hydroxylase of *Pseudomonas denitrificans*, which have been shown to be also responsible for epoxidation of 1-octene and for O-demethylation of heptyl methyl ether[20–24].

Similarly, the progesterone 9-α-hydroxlase from *Norcardia* sp., one of the first microbial hydroxlases obtained in crude form, was shown not to be a cytochrome P450 protein[25]. Interestingly, this enzyme allows for the functionalization of the steroid skeleton, thus opening the way to production of the C-11-oxygenated corticosteroids[26] as shown in Figure 16.1-3.

Other very important examples of non-cytochrome P450 enzymes are methane monooxygenases. These appear to be more reactive than P450 oxygenases and are able to catalyze the conversion of methane to methanol, chemically one of the most difficult steps[27]. These enzymes have been shown to reductively activate dioxygen for incorporation into a wide variety of hydrocarbon substrates, including alkanes, alkenes and alicyclic or aromatic hydrocarbons. The enzyme harbors a hydroxy-bridged dinuclear iron cluster in its active site, and its structure has been determined by X-ray crystallography[28].

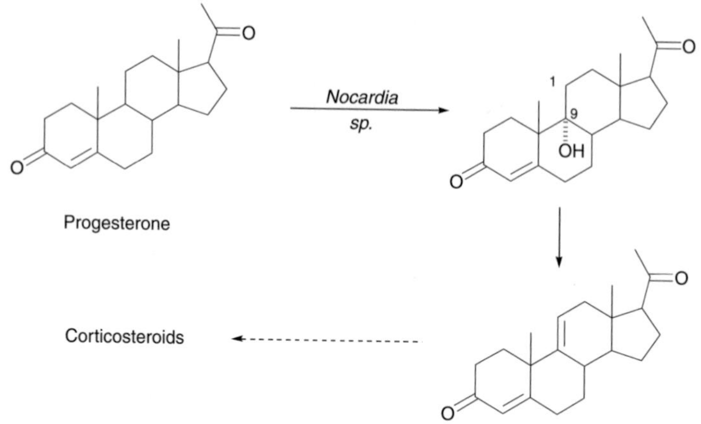

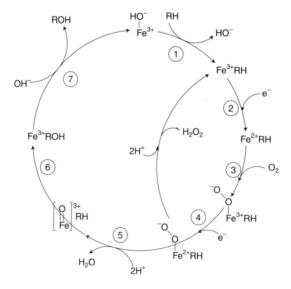

Figure 16.1-3. Use of 9-α-hydroxylation of progesterone as a way to corticosteroids.

16.1.2
Hydroxylating Enzymes

The active site of P450 monooxygenases contains an iron-heme center that is directly involved in the oxidation process by activating molecular oxygen. The catalytic cycle by which cytochrome P450-mediated alkane hydroxylation occurs is by now well studied[1, 18]. The reaction cycle of cytochrome $P450_{cam}$ is outlined in Figure 16.1-4. This mechanism involves (i) reversible substrate binding which converts the six-coordinate, low spin form of the protein to the penta-coordinate high-spin form, (ii) electron reduction of the ferric substrate-enzyme complex by flavoprotein NADPH-

Figure 16.1-4. The catalytic cycle of cytochrome P450 enzymes.

cytochrome P450 reductase leading to the ferrous enzyme, (iii) binding of molecular oxygen to give the six-coordinate iron-dioxygen intermediate, (iv) + (v) reduction of this species with a second electron and addition of two protons, thus leading to an activated oxygen intermediate, (vi) insertion of an oxygen atom into the substrate, and (vii) release of the iron atom in its original ferrous state. Apart from the products of steps (iv) and (v), all these intermediates have now been investigated by crystallography using trapping and cryocrystallography methods [18].

The exact mechanistic details of oxygen insertion into the C-H bond are still the subject of intense discussion. One of the most popular proposals appears to be that of the so-called "rebound mechanism", which proceeds by an initial hydrogen abstraction from the alkane (RH) by the active oxygen intermediate to form a radical R and a hydroxo-iron species as intermediates. The radical then rebounds on the hydroxy group and generates the enzyme-product species [29]. Alternative proposals involve cationic intermediates [30] or two-state reactivity with multiple electromer species for epoxidations [31].

It should be noted that the same cytochrome P450 enzyme is able to achieve reactions as different as double-bond epoxidation or heteroatom demethylation. Thus it appears clear that the chemo-, regio- and stereoselectivity of the reaction is a function of the nature and the fit of the substrate, or, more properly, of its transition state with the protein, rather than being governed by enzyme reaction specificity.

It is obvious that all these very powerful enzymes are of high synthetic value for the organic chemist since they prove to be able to achieve, at normal temperature and in aqueous media, reactions which are very difficult, if not impossible, to perform using conventional chemistry. One additional bonus offered by these biological tools is their generally high selectivity. It is thus understandable that a variety of oxygenative biotransformations have been explored using numerous substrates. We will focus in the following pages on two such particularly interesting reaction types, namely the hydroxylation of non-activated carbon atoms and the stereospecific epoxidation of "*isolated*" double bonds.

16.1.4
Hydroxylation of Non-Activated Carbon Atoms

16.1.4.1
Hydroxylation of Monoterpenes

Because of their involvement in the flavor and fragrance industry, monoterpenes are one type of natural compounds which have been considered as interesting substrates for biohydroxylation studies. For instance, geraniol, nerol and linalool were studied by different groups and were shown to lead to the 8-hydroxylated products with the fungus *Aspergillus niger* [32] as well as with four strains of *Botrytis cinerea* [33]. Interestingly, the same 8-hydroxylated products were found starting from the corresponding acetates, which were shown to be hydrolyzed to the starting alcohol by the fungus *Aspergillus niger* prior to hydroxylation [34]. This C-8 regioselectivity has also been observed in hydroxylations of geraniol and nerol with reconstituted

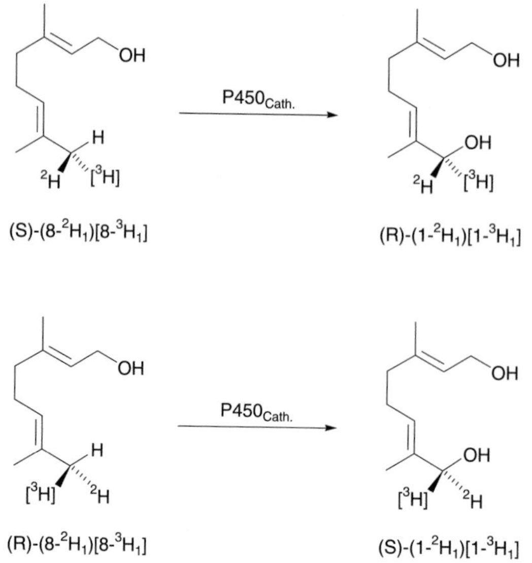

Figure 16.1-5. Retention of configuration during the hydroxylation at C-8 of isotopically labeled geraniol.

hydroxylating enzyme systems from rabbit liver[35–39] and from plant cells like *Vinca rosea* as well as *Catharanthus roseus* (L.) G. DON [40].

In this last case, incubation of different ^{13}C- and ^{2}H-labeled geraniols revealed that hydrogen abstraction is completely regioselective in favor of the CH$_3$ group *trans* to the chain at C-6, i.e. at position C-8. An intramolecular isotope effect of K_H/K_D 8.0 was determined, suggesting that the hydrogen abstraction is one of the major rate determining steps. Furthermore, Fretz and Woggon [41] have studied incubation of the (R)(8-^{2}H$_1$) (8-^{3}H$_1$) and (S)(8-^{2}H$_1$) (8-^{3}H$_1$) geraniols (Fig. 16.1-5). This resulted in the formation of the chiral 8-hydroxy products, thus indicating clearly retention of configuration during the allylic hydroxylation process. Interestingly, in neither of these cases has an allylic radical rearrangement (migration of the double bond) been observed.

Bioconversions of (+)-limonene, the major constituent of citrus essential oils, have also been studied in recent years in order to afford biotechnological routes to interesting products of natural source valuable for the perfume and/or flavor industries. For instance, (+)-limonene was shown to be transformed by *Pseudomonas gladioli* to (+)-α-terpineol (one of the most commonly used products in fragrances and flavors i.e. lemon, nutmeg, orange, ginger, peach and spices), which is resistant to further degradation by the bacterium, and to (+)-perillic acid, which is further metabolized [42]. The corresponding levorotatory limonene antipode is known as being the primary olefinic constituent of the volatile oils of immature *Mentha piperita* (peppermint), *Mentha spicata* (spearmint) and *Perrilla frutescens* leaves, whereas (-)-menthone, (-)-carvone and (-)-perillyl aldehyde, respectively, are the major oxygenated compounds. The enzymatic hydroxylation of (-)-limonene at C-3, C-6 and C-7 to the corresponding derivatives has been studied using light membrane preparations from leaves of each of these plants. It has thus been shown that they lead to the

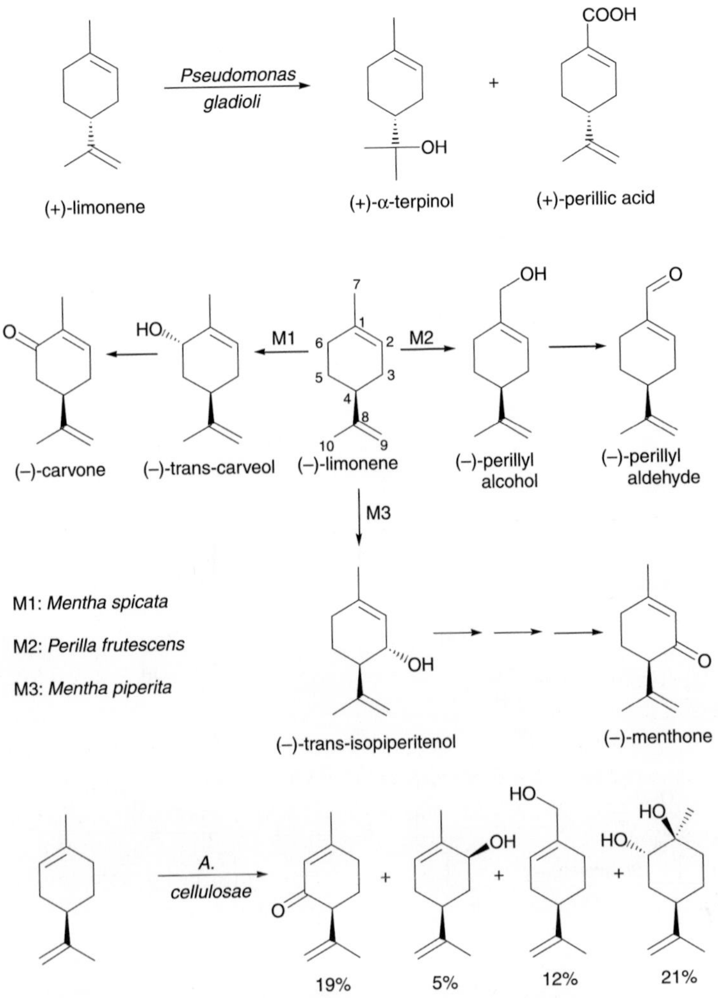

Figure 16.1-6. Bioconversion of limonene using various biocatalysts.

corresponding oxygenated compounds in a mutually exclusive manner in each species. This suggests very strongly that different forms of cytochrome P450 are present in each type of plant, each one showing an exclusive regiochemistry of oxygen insertion[43] as shown in Fig. 16.1-6.

Another terpene interesting for flavor chemistry is β-ionone (Fig. 16.1-7). Its microbiologically mediated transformation has been explored in order to afford a mixture of derivatives that is utilized as an essential oil of tobacco, used for tobacco flavoring at the ppm level[44, 45]. One of the best microorganisms capable of converting β-ionone to the desired mixture of its useful derivatives was an *Aspergillus niger* strain. This process has been recently improved using bioconversion in the presence of organic solvents and immobilization techniques. Thus, this fungus

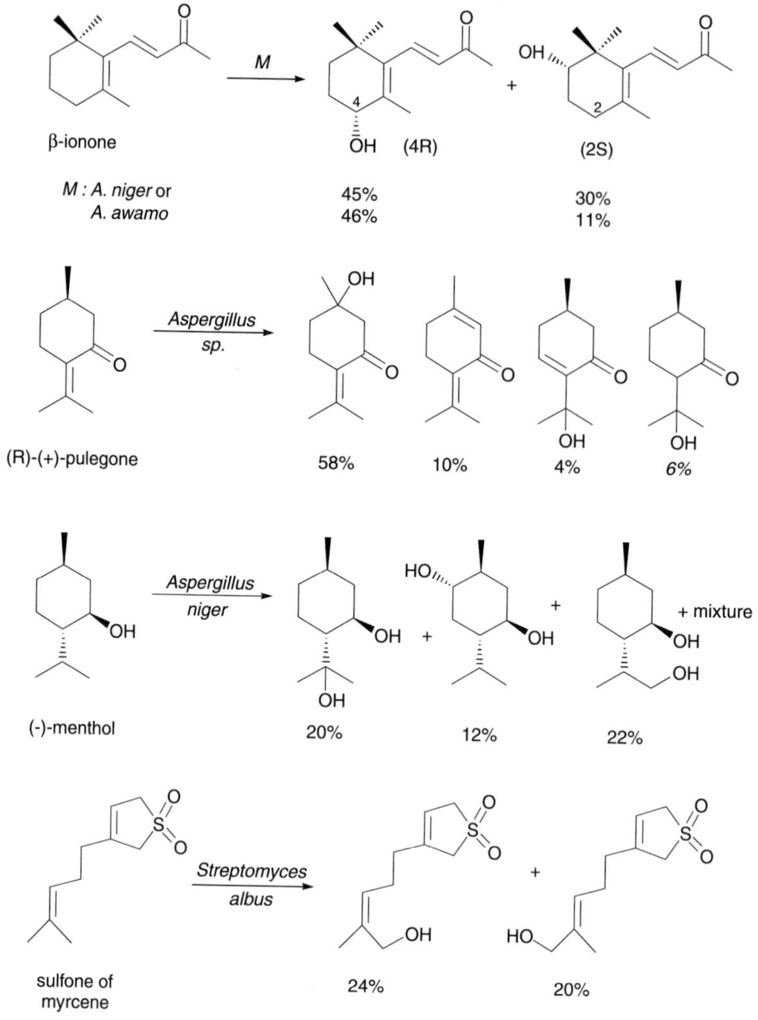

β-ionone

M : A. niger or
A. awamo

(4R)
45%
46%

(2S)
30%
11%

(R)-(+)-pulegone

58%   10%   4%   6%

(-)-menthol

20%   12%   22%   + mixture

sulfone of
myrcene

24%   20%

Figure 16.1-7. Biohydroxylations of various monocyclic terpenes.

could be repeatedly used for microbial conversion of β-ionone in the presence of isooctane for more than 480 h[46]. Another *Aspergillus* strain, *A. awamori*, has also been shown recently to achieve hydroxylation of β-ionone. This led to a mixture of two alcohols, the major product being a building block usable for further synthesis of abscisic acid (an important phytohormone) analogs[47].

Other monocyclic terpenes like for instance R-(+)-pulegone (a mint-like odor monoterpene ketone which constitutes the main component of *Mentha pulegium* essential oil)[48] menthols, terpinolene and carvotanacetone[49] have been investigated. All these substrates proved to be transformed by various *Aspergillus* strains, including *A. niger*, leading essentially to monohydroxylation. Also, monocyclic

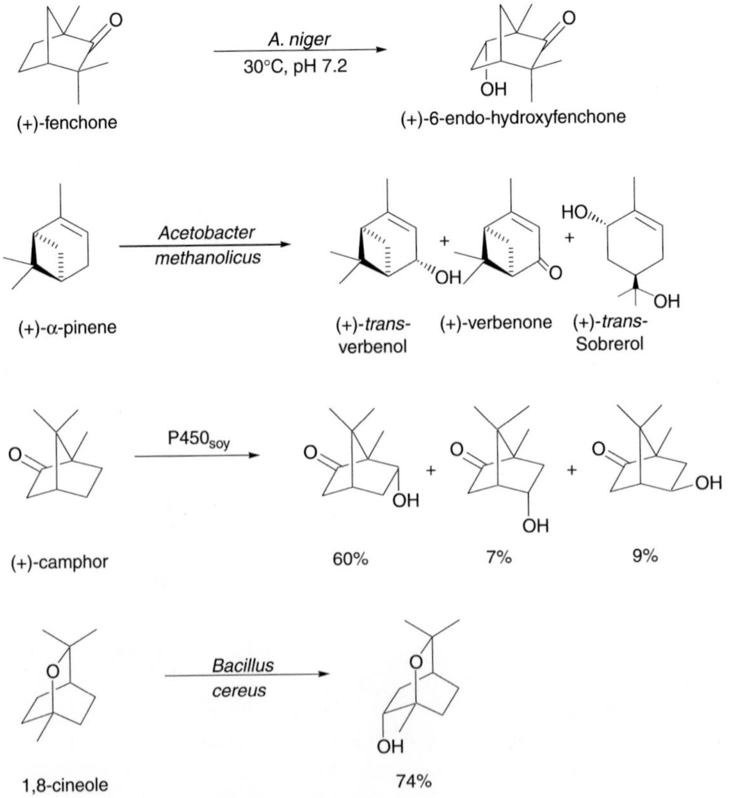

Figure 16.1-8. Biohydroxylation of various bridged bicyclic terpenes.

sulfoxide derivatives of the linear terpenes myrcene and ocimene were shown to be good substrates for several bacteria and fungi, whereas they were themselves only very poorly transformed[50].

Finally, some bicyclic monoterpenes have also been recently described to be subject to microbiological hydroxylation (Fig. 16.1-8). For instance (+)-fenchone was transformed to (+)-6-endo-hydroxyfenchone by A. niger[51], and α-pinene led to the predominant metabolites (+)-trans-verbenol, (+)-verbenone and (+)-trans-sobrerol by the action of several strains of the methylotrophic species Acetobacter methanolicus[52]. (+)-Camphor is transformed to the major metabolite 6-endo-hydroxy-camphor by cytochrome P450soy enriched intact cells of Streptomyces griseus[53], and 1,8-cineole (the major component of the oil from leaves of Eucalyptus radiata var.) is hydroxylated to 6-(R)-exo-hydroxy-1,8-cineole by the bacterium Bacillus cereus following a high yielding (74%) and stereospecific route[54]. Interestingly, this same bacterium had been previously described to be able to achieve hydroxylation of 1,4-cineole yielding good yields of essentially pure 2-(R)-endo and 2-(R)-exo-hydroxy-1,4-cineole[55].

16.1.4.1

Hydroxylation of Monoterpenes

Similarly to monoterpenes, several studies aimed at the microbiological hydroxylation of sesquiterpenes have been described, and a review has summarized the most
interesting results obtained in this area[56]. Since then, additional examples have
been published. Thus, three germacrone-type sesquiterpenoids, (+)-germacrone-
4,5-epoxide, germacrone and (+)-curdione, were described as being transformed by
Aspergillus niger[57]. The interesting feature of these results is the fact that they
essentially led to hydroxylated guaiane-type sesquiterpenoids (together with allylic
alcohols and spirolactone) which arise from transannular cyclization of the carbon

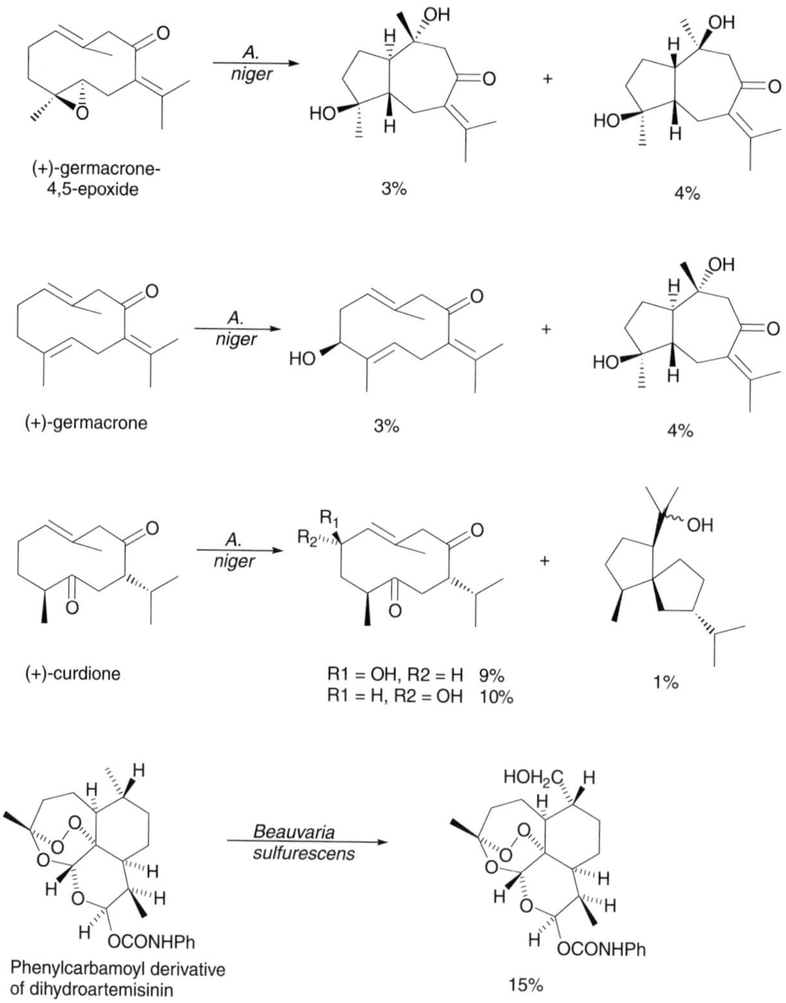

Figure 16.1-9. Biohydroxylation of various higher terpenes.

skeleton as shown in Fig. 16.1-9. The biohydroxylation of an *N*-phenylcarbamoyl derivative of dihydroartemisinine by the fungus *Beauveria sulfurescens* has been described [58]. This allowed the preparation of new and novel derivatives of artemisinine, a drug known to be active against *Plasmodium falciparum*, the strain responsible for malaria, which claims more than one million lives a year. The C-10-*N*-phenylcarbamoyl derivative of dihydroartemisinine, a highly oxygenated sesquiterpene, is thus converted into its 14-hydroxymethyl derivative in 15% yield. Although this yield can be considered as rather modest, this biohydroxylation is interesting since it allows us to prepare derivatives which retain the peroxide group required for biological activity of these drugs. Also, it emphasizes the possibility to favor biohydroxylation by introducing an amide or urethane group into a substrate, as already observed on other model substrates (Fig. 16.1-9) [59, 60].

A series of biotransformations of 6β-santonin and of some of its derivatives, achieved by *Curvularia lunata* and *Rhizopus nigricans* cultures, have also been described (Fig. 16.1-10) [61]. Depending on the strain used and on the starting substrate, several metabolites were obtained, including products resulting from hydroxylation as well as double bond and/or carbonyl reduction. The same group also described biotransformations of several 1,6-difunctionalized eudesmanes leading to 12-hydroxy derivatives which are interesting intermediates for the synthesis of 6,12-eudesmanolides [62]. Similarly, starting from 6β-acetoxyeudesmanone, biohy-

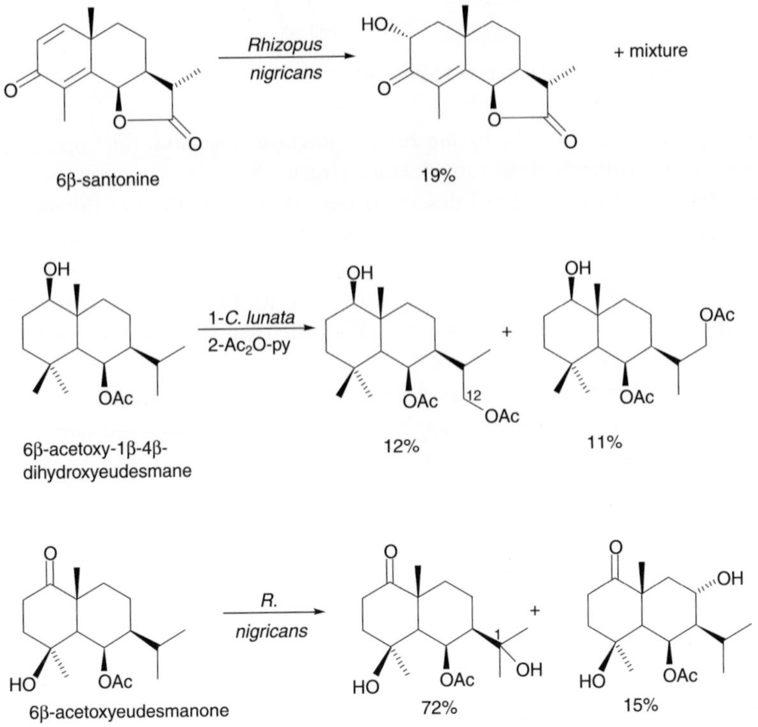

Figure 16.1-10. Further examples of biohydroxylations of bicyclic natural compounds.

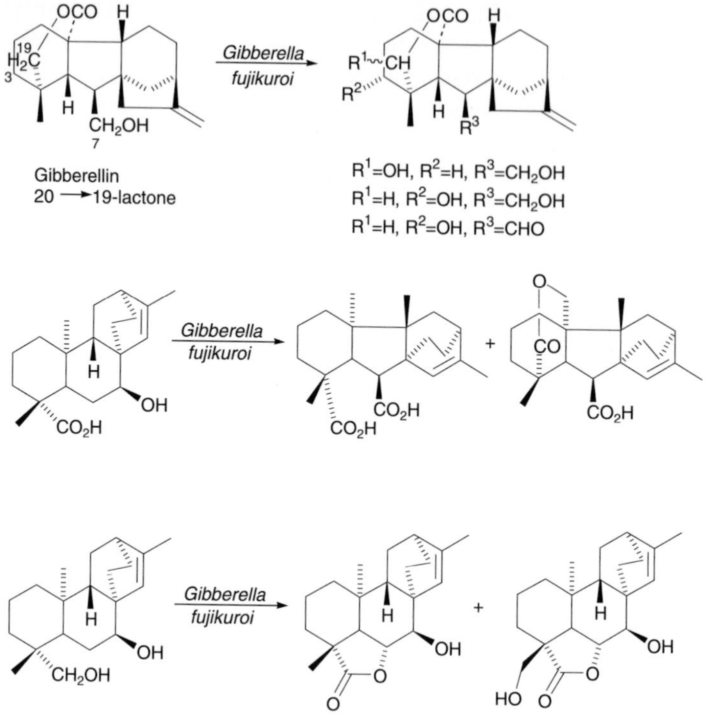

Figure 16.1-11. Some examples of diterpenes hydroxylation by *Gibberella fujikuroi*.

droxylation was achieved at C-11 by the fungus *Rhizopus nigricans*, thus opening another way to the synthesis of the same lactone targets [63].

Several diterpenes have also been described recently to be subject to microbiological hydroxylations. Thus, as shown in Fig. 16.1-11, a compound prepared from gibberellin A_{13} was transformed by the fungus *Gibberella fujikuroi*, affording three metabolites, two of these arising from hydroxylation at its 3α- and 19-positions [64]. Similarly, the same fungus was shown to transform isoatisene derivatives into rearranged isoatisagibberellin derivatives [65].

Sclareol, a natural product first isolated from the essential oil of *Salvia sclarea* L. (Labiatae) in 1931, is used for diverse applications in the perfumery and flavoring industries and in folk medicine. This diterpene has been described recently to be hydroxylated by three strains, i.e. *Cunninghamella* sp., *Septomyxa affinis* [66] and *Mucor plumbeus* [67, 68], leading essentially to hydroxylation reactions on the A ring of this compound (Fig. 16.1-12). Some of these metabolites could be used for further synthesis of some biologically active targets or as mammalian metabolism models.

Amazingly, as in the case of sciareol, the presence of an oxygenated function on the C ring position of grindelic acid again orients the hydroxylation process towards the A ring, since mainly 3β-hydroxylation is observed [69]. This is to be compared with the results obtained in studying the behavior of *Rhizopus* and *Aspergillus* strains as potential hydroxylating species for kaurene sesquiterpenes [70] (Fig. 16.1-13). Inter-

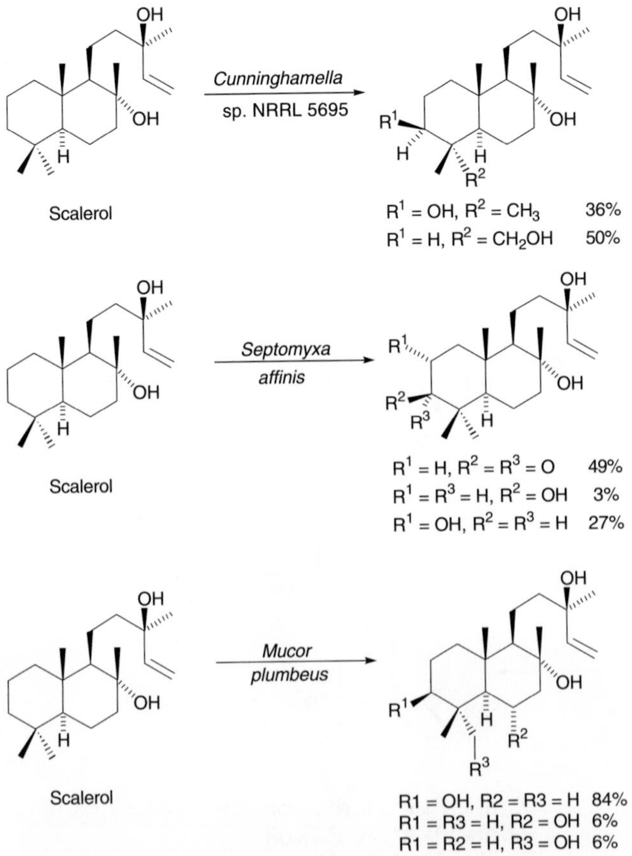

Figure 16.1-12. Hydroxylation of natural sclareol using different microorganisms.

estingly, the fact that the starting substrate bears an oxygenated function on the A ring now orients the oxidations catalysed by *R. nigricans* toward the C ring (and in particular to the C-13 position). This nicely resembles the results previously observed in the steroid family by Jones and coworkers [71], and once more emphasizes the role of a preexisting oxygenated function which operates as an anchoring and thus a site-directing entity inside the hydroxylating active site.

By modifying the location of this function on the starting material, it should therefore be possible to orient the hydroxylation locus differently. This is nicely exemplified by hydroxylation of stemodine, a diterpene bearing a preexistent OH group at position 2 of the A ring. Hydroxylation by *C. elegans* and by *Polyangium cellulosum* then orients the hydroxylation process toward positions 17 and 19 [72].

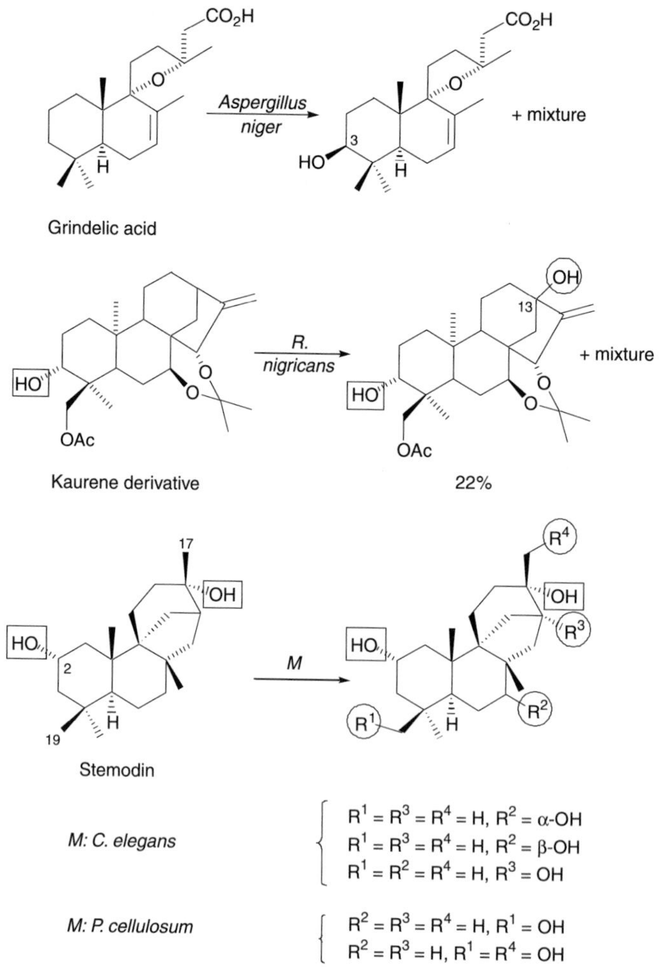

Grindelic acid → *Aspergillus niger* → + mixture

Kaurene derivative → *R. nigricans* → 22% + mixture

Stemodin → M →

M: *C. elegans*

$R^1 = R^3 = R^4 = H, R^2 = \alpha\text{-OH}$
$R^1 = R^3 = R^4 = H, R^2 = \beta\text{-OH}$
$R^1 = R^2 = R^4 = H, R^3 = OH$

M: *P. cellulosum*

$R^2 = R^3 = R^4 = H, R^1 = OH$
$R^2 = R^3 = H, R^1 = R^4 = OH$

Figure 16.1-13. Effect of a preexisting oxygenated function on the orientation of a biohydroxylation process.

16.1.4.3
Hydroxylation of Steroids

Because of their utmost importance as bioactive molecules, steroids have been the most thoroughly studied family as far as microbiological hydroxylations are concerned.

The most important features and references have been put together by Holland in his important monograph [73]. At the present time, one could presumably almost consider that one or even several strains are known which are able to introduce a hydroxyl group at every carbon atom of the steroidal framework. Obviously, however, further work will have to be achieved in order to improve the selectivities and yields

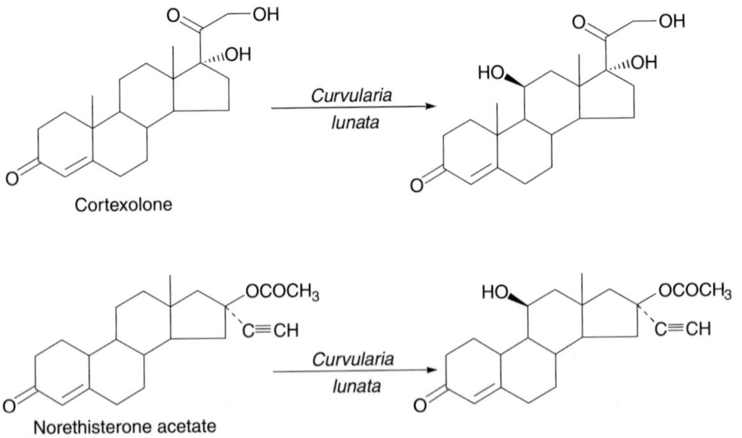

Figure 16.1-14. Examples of steroid hydroxylations by *Curvularia lunata*.

of these bioconversions. Thus, the course of the 11β-hydroxylation of cortexolone by *Curvularia lunata*, a hydroxylation of considerable commercial importance, has been more recently reexamined. The work described by Chen and Wey[74] focused on the improvement of this process by studying the characteristics of mycelial growth as well as the role of substrate addition time and dissolved oxygen tension. These studies have provided some more insight into the fundamental aspects of this biotransformation. The 11β-hydroxylation of norethisterone acetate by the same microbial strain has also been described[75] (Fig. 16.1-14).

16.1.4.4
Miscellaneous Compounds

Although being initially the major part of literature concerning microbiological hydroxylations, natural compounds of the terpene or steroid families have not been the only ones to be studied in this context. Indeed, more recently studies have focused on using biohydroxylations to provide interesting synthetic intermediates, in particular chiral intermediates.

Linear or branched-chain alkanes have been shown previously to undergo hydroxylation by various microorganisms, and can lead for instance to fatty acids, hydroxyacids or α-dicarboxylic acids of commercial importance. *Pseudomonas oleovorans* is one of the strains capable of achieving such transformations, but this process suffers from the fact that the monocarboxylic acids formed in the first step are submitted to β-oxidation and thus used as a source of energy and carbon. Such problems can now be overcome using heterologous expression of P450 genes in microorganism, which will be discussed later on.

An interesting aspect of biohydroxylations is that the biocatalyst can in principle generate one single enantiomer starting from a prochiral substrate – a reaction which could be defined as "*enantiogenic*"[76]. This enantiogenic hydroxylation can lead to enantiopure compounds by stereospecific attack of one single enantiotopic

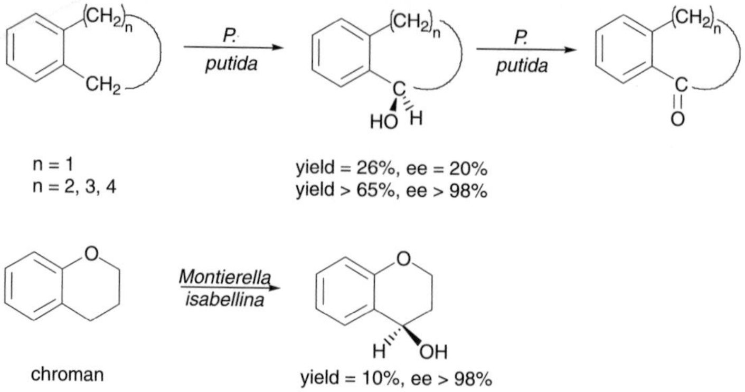

n = 1   yield = 26%, ee = 20%
n = 2, 3, 4   yield > 65%, ee > 98%

chroman   yield = 10%, ee > 98%

Figure 16.1-15.   Examples of regio- and stereoselective benzylic hydroxylation.

face of the starting substrate. Biohydroxylation reaction can therefore be used to prepare high value chiral synthetic intermediates from low value prochiral starting materials, and there are now a number of reports of such reactions.

Thus, benzocycloalkenes have been described to undergo bacterial hydroxylation by the *Pseudomonas putida* strain UV4. As shown in Fig. 16.1-15, this yielded exclusively hydroxylation at the benzylic position, and also one single enantiomer, i.e. the (R)-alcohol. The biotransformation of benzocyclobutene proved, however, to be different from that observed for higher benzocycloalkenes, presumably because of its particular chemical reactivity [77]. A similar result has been observed by Holland and coworkers in the course of chroman biotransformation by the fungus *Mortierella isabellina* [78], which leads, although in low yield, to the benzylic (R)-alcohol.

Another interesting example of asymmetric synthesis from a prochiral substrate is the preparation of (S)-naproxen, a non-steroidal anti-inflammatory drug. It has been shown that several strains are able to regioselectively oxidize one of the enantiotopic methyl groups of the isopropyl moiety. This allows the preparation of the corresponding acid, which is obtained with high enantiomeric purity (Fig. 16.1-16).

Next to *Aspergillus niger* the fungus *Beauveria bassiana* (previously classified as *Sporotrichum sulfurescens* and *B. sulfurescens*) is one of the most frequently used fungal biocatalyst [2]. In particular, hydroxylations of piperidine and pyrrolidine derivatives have been studied by several groups, and interesting regio- and ster-

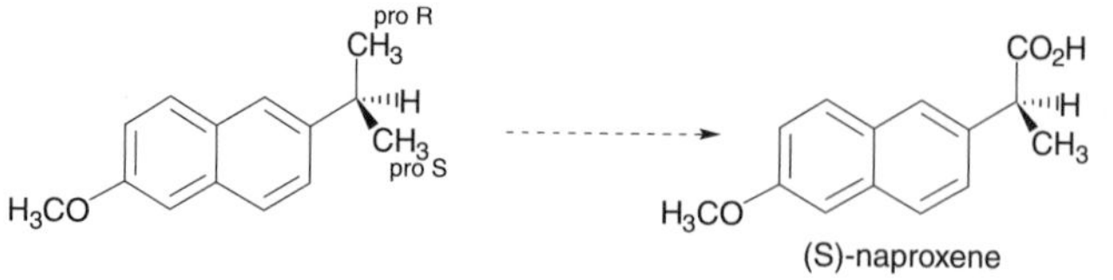

Organism	ee (%)
Cordyceps militaris	99
Graphinium fructicola	96
Exophiala mansonni	68
Exophiala jeanselmei	66

Figure 16.1-16.
Selective hydroxylation of one enantiotopic methyl group as an approach to optically pure Naproxene.

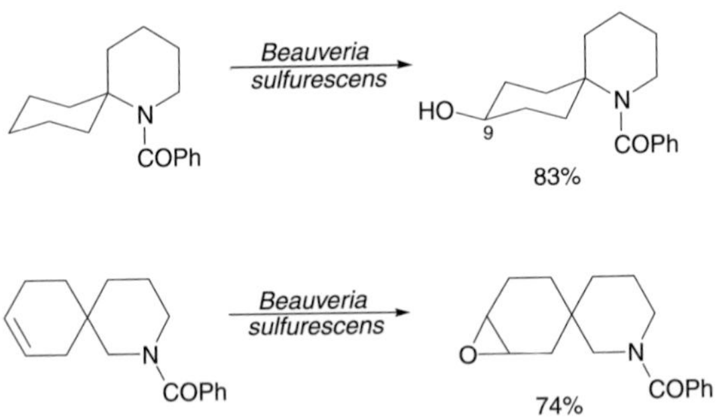

Figure 16.1-17. Hydroxylation versus epoxidation of two spiro-bicyclic amides.

eoselectivities have been reported [79–83]. It should be noted that the ring nitrogen generally needs to be protected for a successful biohydroxylation. This can be used to advantage since the choice of protecting group can influence the regio- and stereochemistry of the hydroxylation [84]. Some examples of hydroxylations of spirobicyclic amides are shown in Fig. 16.1-17. Similarly to previous results described by Fonken and coworkers [85, 86] and by Furstoss and coworkers [87, 88], these led to good yields of hydroxylated products. In all these cases, the regioselectivity of the reaction is partly or even exclusively oriented toward the C-9 carbon atom, a result which could have been predicted on the basis of the previously described results. Interestingly, a similar substrate bearing a double bond at carbon C-9 led to the corresponding epoxide.

Because they constitute partial structures of various higher terpenes and/or steroids, enantiomers of different substituted hexahydronaphthalenones are pivotal intermediates in the total synthesis of these target compounds. Therefore, several differently substituted octalone derivatives have been studied for microbiological hydroxylations. These substrates were prepared in optically active form by chemical synthesis from (S)-(+)-Wieland-Miescher's ketone and were submitted for screening with nine strains known to hydroxylate polyterpenic or steroidal substrates. Thus, submitted to a culture of *Rhizopus arrhizus*, these substrates led to allylic hydroxylation at the B ring, as shown in Fig. 16.1-18 [89, 90]. Similar results were obtained by Azerad and coworkers [91] in the same series, starting from differently substituted octalones. These authors have investigated the biotransformation of their substrates with a variety of fungal strains. For most of these strains, the (R)-enantiomer of hydronaphthalenone led to the 8-hydroxyenone as the main product, i.e. again a product of allylic hydroxylation, which is quite disappointing since this product is easily accessible by (electro)chemical oxidation. However, the fungus *Mucor plumbeus* produced another hydroxylated metabolite, the 6α-hydroxyl derivative. Interestingly, the S-enantiomer of the starting substrate only led to the 8-hydroxyenone in this last case. Introduction of an additional methyl group on the carbon framework of the starting octaenone also led to different regioselectivities of the hydroxylation.

Hydrindane derivatives, which bear a five-membered B ring (instead of a six-membered ring in the decalones derivatives) have also been examined for biohydroxylation by the fungus *Rhizpous arrhizus* [92]. All the hydroxylations observed now occur at position 3, to the α,β-unsaturated ketone. This can be considered as

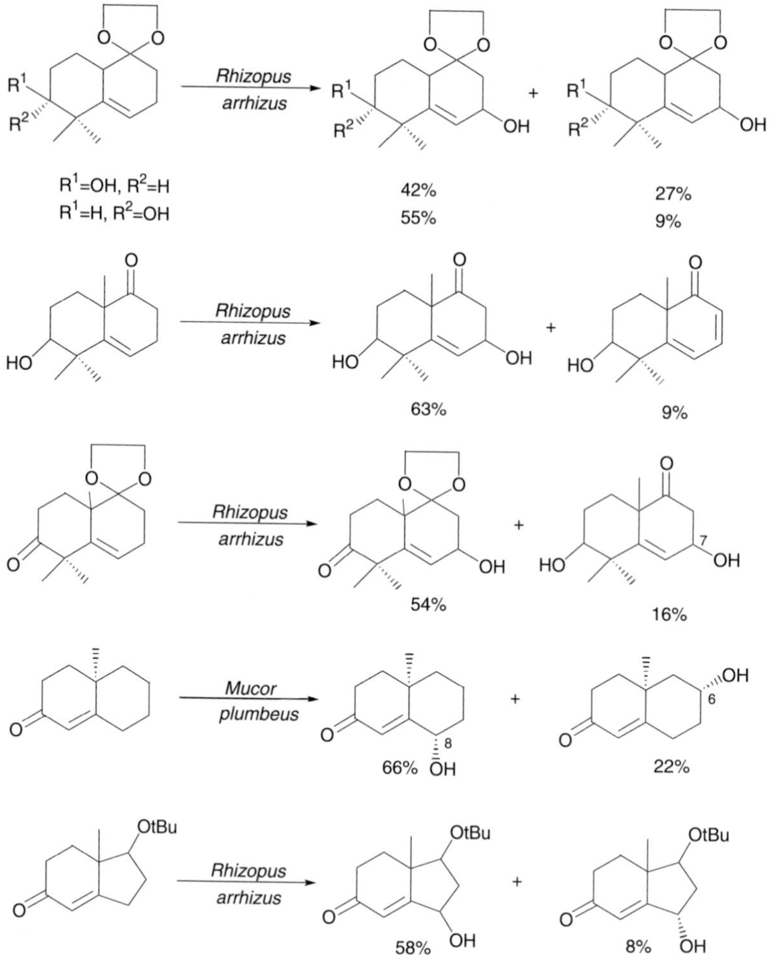

Figure 16.1-18. Microbiological hydroxylation of differently substituted octalones.

being formally analogous regioselectivity as compared to the results obtained on decalones. In this case, however, reactivity is identical in both antipodal series, and led almost quantitatively but with moderate or low stereoselectivity to the formation of the epimeric alcohols. Interestingly, these biohydroxylations prove to be complementary to lead tetraacetate oxidation of these substrates, which affords the 6-acetyl substituted products.

Structurally much more complex molecules have also been submitted to regioselective enzymatic hydroxylation. Two such examples have been described involving milbemycin, a sixteen-membered macrolide which exhibits broad-spectrum insecticidal and acaricidal activity, and monensin, a carboxylic polyether antibiotic[93, 94]. Milbemycin (Fig. 16.1-19) was thus regioselectively hydroxylated at the 13β position (followed eventually by a C-29 hydroxylation) to afford the 13β,29-

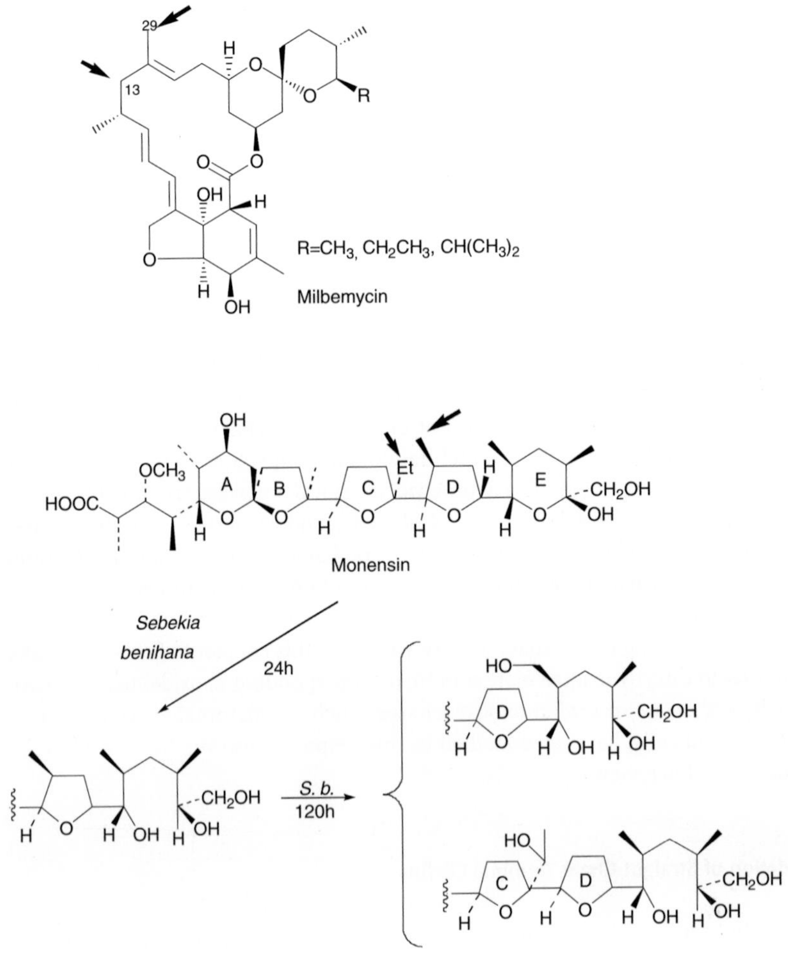

Figure 16.1-19. Regioselective hydroxylation of structurally complex substrates.

hydroxylated product by a strain isolated from solonized brown Mallee oil (collected in Adelaide in Australia) and identified as *Streptomyces cavourensis*. It is interesting to emphasize here the high regioselectivity observed for this hydroxylation of a rather complex and multifunctional compound.

Even more complex is the structure of monensin, a compound which has been extensively used as an anticoccidial agent for poultry and shown to improve the efficiency of feed utilization in ruminant animals. When submitted to a culture of *Sebekia bevihana*, monensin was first quantitatively converted by enzymatic reduction of the δ-hydroxy-ketone (which is in equilibrium with its hemiketal tautomeric form) and was regioselectively further hydroxylated at the C-29 methyl group as well as at the nearby ethyl group substituent[94].

All the previously described examples exemplify the ability of various monooxygenase enzymes to achieve, often with good to reasonable yields and interesting regioselectivities, the hydroxylation of non-activated carbon atoms which are inaccessible using conventional chemistry. This thus allows one-step syntheses of these metabolites, which can in certain cases be of high enantiomeric purity. Another type of oxygenation reaction which is of interest is the stereoselective epoxidation of double bonds, the essential aim being in this matter the access to epoxides of high enantiomeric purity. This will be the subject of the following part of the discussion.

16.1.5
Epoxidation of Olefins

As discussed previously, monooxygenases provide highly activated oxygen intermediates that can oxidize a wide range of functional groups. One of the most studied among these has been the epoxidation of olefins [95–97]. This epoxidation is particularly interesting when applied to prochiral double bonds. Spectacular success has been obtained in the field of asymmetric chemical epoxidation, notably using Sharpless epoxidation catalysts for allyl alcohols and Jacobsen catalysts for aryl olefins, which has made epoxides key intermediates in the synthesis of chiral compounds. However, these chemical catalysts often have a limited "substrate" range, and biocatalysts can provide access to complementary structural motifs.

Without attempting to be exhaustive, we will try in this chapter to focus on results allowing us to directly oxidize olefins to their corresponding epoxide, using microbial cells. Other sources of monooxygenases, such as mammalian cells (microsomes) or plant cells, have been studied in this respect. However, these will not be considered in this review.

16.1.5.1
Epoxidation of Straight-Chain Terminal Olefins

One of the earliest observations implicating the formation of epoxides during microbial olefin metabolism was the report by Bruyn in 1954 that *Candida lipolytica* grown on 1-hexadecene produced 1-hexadecanediol (about 5% of the hydrocarbon consumed was accounted for as the diol) [98]. Molecular ^{18}O was shown to be incorporated into this diol, and the 1,2-epoxide was identified as one of the by-products of this metabolism [99, 100]. Several further reports confirmed that enzymatic systems are able to achieve epoxidations. For instance, Van der Linden showed in 1963 that *Pseudomonas aeruginosa* grown on *n*-heptane and resuspended in a buffer solution produced the epoxide from 1-octene (Fig. 16.1-20) [101]. This led the authors to conclude that this epoxide was formed by enzymes already present in the alkane-grown cells and that epoxidation might be catalyzed by the same hydroxylases that would normally oxidize alkanes. A similar conclusion was reached by Maynert and coworkers [102], who demonstrated that epoxides are obligatory intermediates in the metabolism of simple olefins in rat liver microsomes.

However, the real breakthrough in the study of enzymatic epoxidations is due to Abbot and coworkers [103] and to May and coworkers [104], who established unequivo-

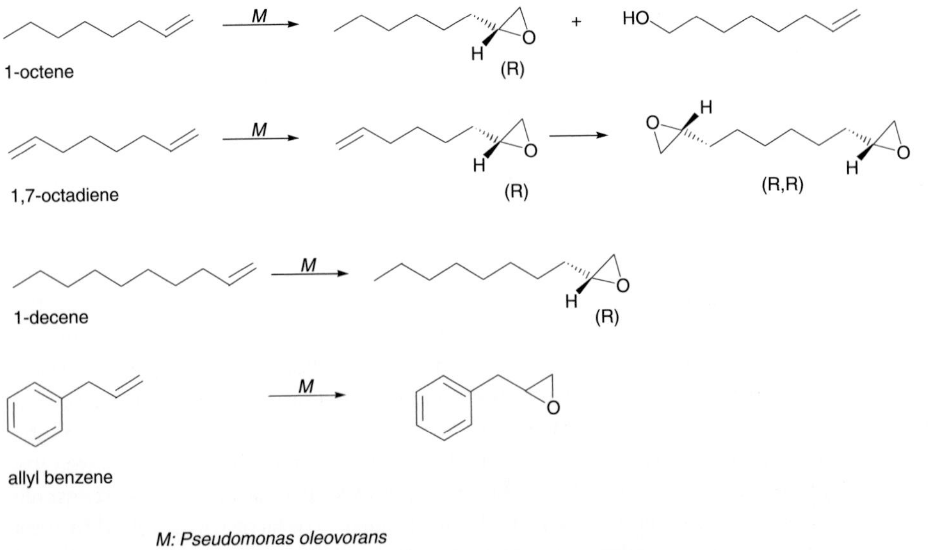

M: Pseudomonas oleovorans

Figure 16.1-20. Stereospecific epoxidation of straight-chain terminal olefins.

cally that epoxides are formed from terminal olefins by the bacterial strain *Pseudomonas oleovorans* (Fig. 16.1-20). They showed that 1-octene is epoxidized to 1,2-epoxyoctane of (*R*)-configuration (*ee* 70 %) or hydroxylated to 7-octen-1-ol. The 1,7-diene is exclusively epoxidized, affording 7,8-epoxy-1-octene, which can be further processed to the corresponding diepoxide[105]. It was shown later that this monoepoxidation was stereospecific, leading to the *R*(+)-7-epoxide showing an *ee* of about 80%. Furthermore, the diepoxide was shown to be essentially of (*R,R*) configuration. This interestingly indicates that the configuration of the monoepoxide formed at one end of the molecule profoundly affects the stereochemical course of the reaction. Indeed, the authors showed that when starting from racemic monoepoxide, the diepoxide was essentially formed from the (*R*)-monoepoxide. Interestingly it was observed that olefins bearing an allylic (or homoallylic) hydroxyl were not epoxidized, but were converted instead to the corresponding saturated ketones.

One of the most useful characteristics of this work is the fact that these epoxides could be routinely produced at yields approaching (at best) 1 g L^{-1} after simple overnight shaking using whole-cell or even crude cell-free systems. Thus, these results clearly opened the way to a new type of biotransformation which should be very useful for organic synthesis.

The enzymatic system involved in hydroxylation reactions of long-chain alkanes had been previously studied by Coon and coworkers, who isolated an enzyme system from *P. oleovorans* that catalyzes co-hydroxylation of alkenes and fatty acids[20, 106–115]. This was resolved into three protein components: rubredoxin (an iron-sulfur protein of molecular weight 19 000), an NADH-rubredoxin reductase (a flavoprotein of molecular weight 55 000) and an "ω-hydroxylase" (characterized as being a non-heme iron protein, with one iron atom and one cysteine per polypeptide chain). Interestingly, it was shown that this same enzyme system is responsible for the conversion of terminal olefins to their corresponding 1,2-epoxides [104]. This leads to a competition between the two types of biotransformations, which results in a specific pattern for each type of substrate. Thus, further investigation demonstrated that this monooxygenase can produce epoxyalkanes with from six to twelve carbon atoms containing terminal alkenes. As a result of the influence of carbon chain length on epoxidation versus hydroxylation it was shown that hydroxylation predominates for the "*short*" substrates propylene and 1-butene, but that epoxidation activity falls off much less readily than hydroxylation for "*long*" substrates. For the "*medium*" length substrates, like for instance 1-octene, both reactions do occur. Thus, this substrate is epoxidized to 1,2-epoxyoctane or hydroxylated to 7-octen-1-ol, while for 1-decene epoxidation largely predominates. Interestingly, the epoxidation reaction exhibits a specificity far different from that expected for chemical reactivity. Indeed, terminal olefins are epoxidized exclusively even in the presence of more highly substituted (electron-rich) double bonds. Thus, cyclic and internal olefins were not epoxidized. This indicated that the substrate specificity pattern observed severely moderates the inherent reactivity of the activated oxygen species involved in these transformations. Methyl imidoesters as well as sodium cyanide were found to be inhibitors of enzymatic epoxidation, and the potency of a homologous series of imidoester inhibitors was examined.

In the reaction with dienes, 1,5-hexadiene to 1,11-dodecadiene were epoxidized while dienes with a smaller number of carbon atoms were hydroxylated to the corresponding unsaturated alcohols [116]. The reactivity was shown to be maximal for octadiene (leading to 0.3 to 0.4 g of diepoxyoctane per liter) and falls off rapidly as the carbon chain is shortened, but decreases only slightly as the chain is lengthened. In a further study, it was shown that a very efficient conversion of 1,7-octadiene to 7,8-epoxy-1-octene and 1,2–7,8-diepoxyoctane could be obtained by incorporating a high concentration of cyclohexane into the conventional fermentation medium. Thus, a 90% yield of product was achieved within 72 h, instead of a 18.5% yield in the absence of cyclohexane, when a 20% (v/v) amount of cyclohexane was used. Clearly, this is an early example of the use of organic solvents applied to microbial transformations [117].

A similar result was obtained later on using the 1-octene substrate itself as the organic phase (20% v/v), leading to comparable results (70% *ee*) [118]. Interestingly it also has been shown in the course of this work that, when *n*-hexadecane (which is not metabolized by the cells) is used as a solvent, racemic epoxide is enantioselectively degraded by the "ω-hydroxylation" enzymatic system of P. *oleovorans*, leading to an enrichment in (*S*)-1,2-epoxyoctane.

Further work by Wynberg and coworkers was aimed at even increasing the yield of 1,2-epoxyoctane using an optimized two-phase system and a cell renewal procedure [119]. Thus, yields up to 150 mg 1,2-epoxyoctane per mL 1-octene and up to 20–25 mg 1,2-epoxyoctane per mL culture was obtained. Some other substrates were tested in this optimized system. Of these, 1-decene was converted into (*R*)-1,2-epoxydecane (60% o. p.), while allylbenzene was converted to the corresponding epoxide. However, no effort was made to determine the absolute configuration and the optical purity of this product.

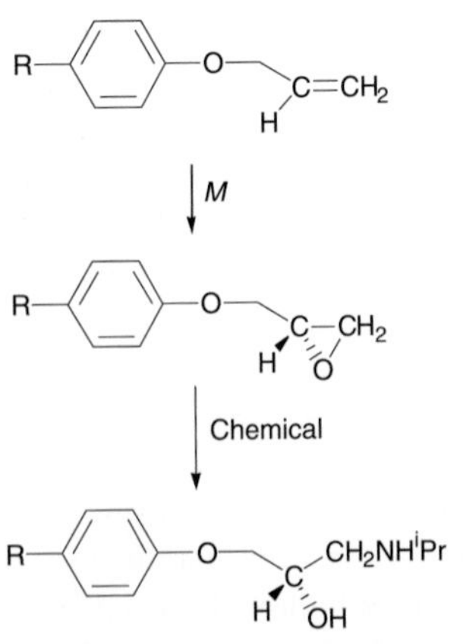

Organism M	Metropolol o.p. (%)
R. equi NCIB 12035	95.4
P. putida NCIB 9571	98
P. oleovorans AT CC 29347	98.4
P. aeruginosa NCIB 8704	98.8

R = CH₂CH₂OCH₃ : Metropolol
R = CH₂CONH₂ : Atenolol (*M: P. oleovorans* o.p. = 97%)

Figure 16.1-21. Microbiological epoxidation as a way to optically pure β-blocker drugs.

All these results led to an interesting application for asymmetric organic synthesis. Thus, *P. oleovorans* has been used, among some other microorganisms, for stereospecific epoxidation of some arylallylethers into (+)-arylglycidyl ethers (Fig. 16.1-21). These intermediates were chemically converted into (S)-(–)-3-substituted-1-alkylamino-2-propanols, which are the physiologically active components of the β-adrenergic receptor blocking drugs. This method has been used to synthesize (S)-(–)-Metoprolol and (S)-(–)-Atenolol with enantiomeric purities of 95.4–98.8% and 97% respectively[120]. These applications are of great industrial interest, since it has been shown that (S)-(–)-Metoprolol is 270–380 times more active than its antipode[121].

Microorganisms screened for epoxidation activity were selected from bacteria belonging to the genera *Rhodococcus*, *Mycobacterium*, *Nocardia* and *Pseudomonas*. Species of *Pseudomonas* gave the best activities, but there were variations between the individual members, and *P. oleovorans* was the most active organism. The activity was further enhanced by carrying out the transformation in the presence of a cosubstrate such as glucose.

This pioneering work on microbial epoxidation of straight-chain terminal olefins has triggered several further studies aimed at preparing enantiopure epoxides via biotransformations. Thus, a number of alkene-utilizing microorganisms have been described in the literature. In the context of aliphatic substrates these efforts have been developed essentially along two lines: epoxidation of long-chain olefins and epoxidation of short (C₁-C₄) chain compounds. Thus, for instance, it was shown that *Corynebacterium equi* (IFO 3730) grown on *n*-octane is able to oxidize 1-hexadecene to give the corresponding optically pure (R)-(+)-epoxide (41% yield based on consumed substrate)[122, 123]. This strain also assimilated other terminal olefins and produced the corresponding epoxides from substrates which have a carbon chain longer than fourteen, although in very low yields (less than 1%). Production of 7,8-epoxy-1-octene from 1,7-octadiene by non-growing *Pseudomonas putida* species using two-phase transformation has also been achieved[124]. Similarly, a gaseous hydrocarbon-assimilating microorganism *Nocardia corallina* B-276 grown on 1-alkenes (C_3, C_4 and C_{13}-C_{18}) was described as being able to produce the corresponding 1,2-epoxy-alkanes. One of the products, 1,2-epoxytetradecane, was shown to be optically active. Glucose-grown cells could also transform styrene and C_2-C_{18} 1-alkenes to their epoxyalkanes[125]. Similarly, production of epoxides from C_6-C_{10} 1-alkenes and styrene was shown to be enhanced by using *n*-hexadecane as an additional solvent, while this led to a decreased rate for epoxidation of longer chain 1-alkenes[126]. Epoxidation of unsaturated fatty acids such as palmitoleic acid by *Bacillus megaterium* has also been reported[127]. Here again, experiments indicated that epoxidation and hydroxylation were catalyzed by the same soluble cytochrome P450-dependent enzymatic system.

16.1.5.2
Short-Chain Alkenes

Short-chain alkenes are another type of substrates which have been studied for microbiological epoxidation during the last thirty years. In this context, an extensive study has been conducted by De Bont and coworkers in order to prepare epoxides from gaseous olefins. Thus, a *Mycobacterium* sp. (E 20) was isolated from soil and shown to excrete ethylene oxide when grown on ethylene[128, 129]. Studies carried out using $^{18}O_2$ showed that a monooxygenase was involved in these epoxidations, as proved by incorporation of only one ^{18}O into the product. Another *Mycobacterium* (Py 1) was also shown to achieve this reaction. Experiments were performed in a gas-solid reactor to prevent accumulation of the toxic ethylene oxide in the immediate vicinity of the biocatalyst[130]. An experimental set-up, allowing for automatic gas chromatography analysis of circulation gas in a batch-reactor system, was also described allowing on-line monitoring of the microbial oxidation of the gaseous alkenes propene and 1-butene (Fig. 16.1-22)[131]. Optimization was achieved by studying the influence of various organic solvents on the retention of immobilized cell activity[132]. High activity retention was favored by a low polarity in combination with a high molecular weight. Using chiral gas chromatography (at that time recently described by Schurig and Bürkle[133]), eleven strains of alkene-utilizing bacteria were screened with respect to the stereospecific epoxidation of propene, 1-butene and 3-chloro-1-propene. The results obtained showed that seven of these bacteria strongly resembled each other, in that they all produced 1,2-epoxypropane and 1,2-epoxybutane mainly in the (*R*)-form (93 and 85% *ee* respectively). Several of these strains were also able to epoxidize stereoselectively 1-chloro-2,3-epoxypropane, thus leading to the synthetically very useful (*S*)-epichlorohydrin (*ee* > 95%). Stereoselective epoxidation of 4-bromo-1-butene and of 3-buten-1-ol was similarly studied using three strains. The results showed that the epoxides were again obtained predominantly in the (*R*)-form but that their enantiomeric purity depended on both

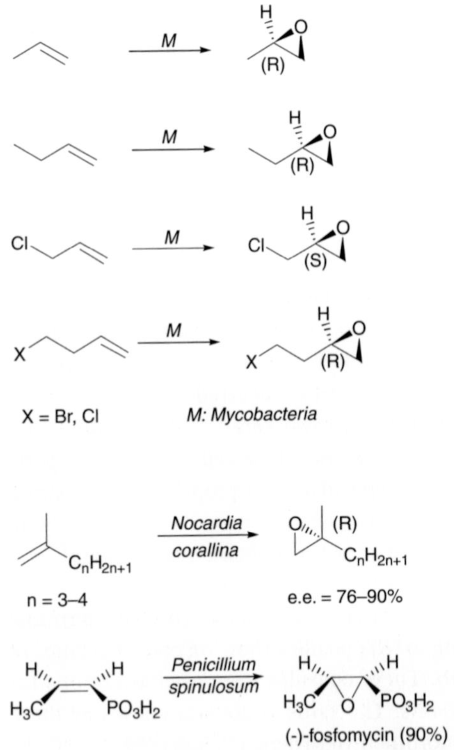

Figure 16.1-22. Short-chain alkene epoxidation.

the strain used and on the substrate studied[87]. Inactivation of the alkene oxidation enzymatic system by the produced epoxide was also investigated in view of setting up a biotechnological procedure for producing these epoxides[134]. Modeling the effects of mass transfer on the kinetics of propene epoxidation was also achieved by the same authors[135, 136], and they showed that product inhibition can be reduced by absorbing the epoxide in the gas phase in cold di-n-octyl phthalate[137].

In addition to the Mycobacterium species, several other strains have been reported to achieve epoxidation of olefins. Thus, three distinct types of methane-grown methylotrophic bacteria (*Methylosinus trichosporium*, *Methylobacterium capsulatus* and *Methylobacterium organophilum*) were shown by Hou and coworkers[138] to be able to oxidize terminal C_2 to C_4 n-alkenes to their corresponding 1,2-epoxides, which accumulated extracellularly. Results from inhibition studies indicated, as in the case of the previously discussed ω-hydroxylation system of *P. oleovorans*, that the same monooxygenase enzyme was responsible for the hydroxylation of methane and the epoxidation of alkenes. Further work achieved by the same group showed that whole cells of *Methylosinus* sp. CRL 31, immobilized by adsorption on glass beads, were able to convert propylene to propylene oxide for several hours until the reduced NAD cofactor was depleted. This could be regenerated by periodic addition of methanol. These authors also observed that attempts to immobilize the cells by covalent binding or entrapment in polyacrylamide gel led to complete loss of propylene epoxidation activity[139]. However, no mention is made in this work of the enantiomeric purities of the obtained epoxides. Further studies carried out by Subramanian[140] revealed that these were nearly racemic compounds, and also that the major problem of these biotransformations was again product (epoxide) inhibition. More information about the reaction mechanism of the epoxidation achieved with whole-cell *M. trichosporium* was gained by Okura and coworkers[141], who showed that the configuration of the double bond was retained during the epoxidation of *cis*-2-butene. This result was further confirmed by studies of the epoxidation of 1,2-deuterated-*cis*-propene. A concerted insertion of oxygen was postulated to account for this result[142]. Oxidation of propylene to propylene oxide by *Methylococcus capsulatus* (Bath) was studied in order to optimize the biotransformation for a possible industrial production. However, the high rates obtained could only be sustained for 3–4 min before loss of biocatalytic activity occurred[143].

Similar results were obtained by Wyngard and coworkers[144] in the course of a study aimed at exploring how immobilization of the whole cells on solid supports would influence the rate and duration of the epoxidation of propylene by the strain *Nocardia corallina* B-276 initially isolated by Furuhashi and coworkers[125, 126]. Here again the results suggested that entrapment in a hydrophobic matrix might be a favorable system, but that loss of activity was quite rapid with time. The same *Nocardia* strain has been shown to be able to epoxidize branched chain terminal olefins in an asymmetric manner leading to (*R*)-epoxides showing optical purities of 76–90% depending on the chain length. These epoxides were used as chirons for further synthesis of prostaglandin ω-chains. The same strain was shown by these authors to be also able to epoxidize trifluoromethylethylene (75% *ee*)[145].

Some newly isolated *Xanthobacter* sp. were recently shown to be able to accumulate 1,2-epoxyethane from ethene or, when grown on propene, to accumulate 2,3-epoxybutane from *cis*- or *trans*-2-butene but with apparently low yields[146]. Similarly, *Rhodococcus rhodochrous*, a propane-oxidizing strain, was shown to produce 1,2-epoxyalkanes from short-chain terminal alkenes. Interestingly, its oxygenase enzyme appeared to be capable of tolerating high levels of product without inhibition[147].

Finally, a very useful and industrially interesting epoxidation which deserves special attention is the stereospecific epoxidation of *cis*-propenylphosphonate. Eighteen species of *Penicillium*, one of *Oidium* and one of *Paecilomyces* were found to effect this reaction, which affords directly (-)-fosfomycin, a broad-spectrum antibiotic. Using the strain *Penicillium spinulosum* MB 2843 at optimum culture conditions, a 90% efficiency (based on olefin charged, 0.5 g L^{-1}) was obtained after 6 days, leading to a product claimed to be optically pure[148].

16.1.5.3

Terpenes

Besides the extensive studies aimed at preparing optically active epoxides starting from short or long straight-chain alkenes, another area of investigation has been the microbiological epoxidation of various natural substrates, essentially in the terpene and steroid area. Interestingly enough, it appears that terminal olefins (and only

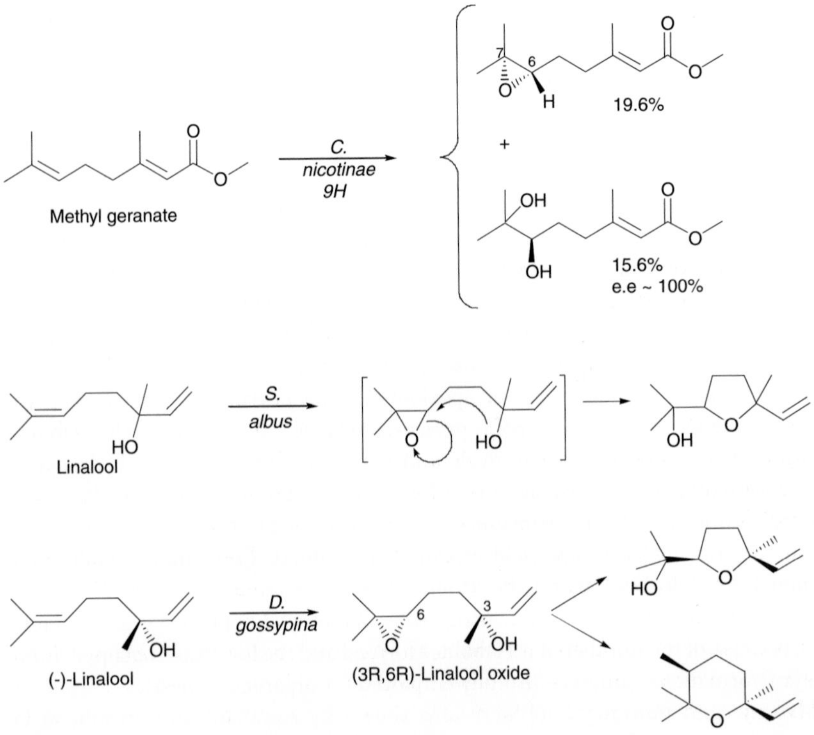

Figure 16.1-23. Some examples of olefinic terpene epoxidation.

these) are epoxidized almost exclusively by bacteria, and lead to accumulation of the corresponding epoxide in the culture. On the other hand, more substituted double bonds are often preferentially oxidized by higher organisms like fungi. The product is generally the corresponding vicinal diol arising from further metabolism (hydrolysis) of the primarily formed epoxide. Numerous publications describe microbial transformations of various terpenes [149, 150]. However, there are few cases of an accumulation of intermediates in sufficient amounts for further use in synthesis [151].

One of the first examples of such a transformation has been described by Marumo and coworkers (Fig. 16.1-23) [152]. Their investigations, aimed at preparing optically active insect juvenile hormone, showed that methylgeranate was metabolized by the fungus *Colletotrichum nicotianae*, leading to 19.6% of S(−)-methyl-6,7-epoxygeranate and to 15.6% of R(+)methyl-6,7-dihydroxygeranate after 9 h incubation. Longer incubation times (24 h) produced only the optically pure glycol with an isolated yield as high as 85%, showing that the first epoxidation step had to be stereospecific. Unfortunately, this analytical study was not pursued on a preparative scale, and no accurate results concerning the stereochemical and kinetic aspects of these interesting biotransformations have been described.

A similar microbial oxidation of the isoprene double bond has been studied by

Veschambre and coworkers starting from linalool[153]. Thus *Streptomyces albus*, a strain which synthesized nigericine, transforms each enantiomer of linalool, as well as the racemic compound, into a mixture (10–20% yield) of two diastereoisomeric linalool oxides. In this case, the epoxide formed primarily is trapped by an intramolecular cyclization. Based on the reported proportions of these products, one can deduce that the *ee* of the formed epoxide was about 35%. Further work achieved using several other microorganisms showed that *Beauveria sulfurescens* gave similar yields (15–20% analytical) of an equimolar mixture of linalool oxides[154]. *Botyris cinerea*, a fungus which participates in the formation of flavors in sweet wines, was also checked for linalool biotransformation. This led to several metabolites including linalool oxides, presumably arising from prior epoxidation of the olefinic bond[155]. Interestingly, these products were also detected in the *Carica papaya* fruit flavor, together with the diastereoisomeric epoxides[156]. It was also observed by Abraham and coworkers[157] that (–)-linalool is processed by *Diplodia gossypina* exclusively to a mixture of *trans*-(3R,6R)-linalool oxide and to the corresponding tetrahydropyran. These were proposed to arise by intramolecular cyclization of the intermediate 6(S)-epoxide. Some other similar substrates have been studied in the course of this study, but they generally led to low yield mixtures of products. Comparable results were obtained from linalool using the strain *Streptomyces cinnamonensis*[158]. Similar transformations were observed starting from 2-methyl-2-heptene-6-one[159]. However, because of the number of metabolites formed and the low yields obtained, these biotransformations cannot be usefully employed for organic synthesis.

Myrcene and *trans*-nerolidol were also shown by Abraham and Stumpf to be transformed by two fungi (*Diplodia gossypina* and *Corynespora cassiicola* respectively) into a mixture of several products including vicinal diols arising from oxidation of the isoprenyl double bond. These were shown to be further degraded, presumably via an acyloin-splitting mechanism[160]. During the course of the fermentation, the diol occurred at first in the culture medium followed by the nordiols and the trialcohols. So, the formation of these compounds from diols seemed to be very likely. Some other related substrates were also studied in the same context, and it was shown that both strains revealed a pronounced and almost opposite substrate selectivity. Much more impressive is the result obtained by the same group[161], who conducted a broad screen of 800 various microorganisms using both the (S)(–)- and the (R)(+)-limonene enantiomers as a starting substrate, as well as some other terpenes which were tested with the best suited strains (Fig. 16.1-24). The most interesting results were observed with *Diplodia gossypina* (ATCC 10936), which afforded 380 mg of a diol which was found to be the (1R, 2R, 4S)-8-p-menthen-1,2-diol from 1 g of (S)(–)-limonene. Similarly, *Corynespora cassiicola* (DSM 62474) was described to yield 1.1g of (1R, 2R)-3-p-menthen-1,2-diol from 1.8g of α-terpene. (R)(+)-limonene was shown to afford (1S, 2S, 4R)-p-8-menthene-1,2-diol.

Because of the interest of these products in flavor chemistry, the preparative-scale transformation of this enantiomer by the fungus *Diplodia gossypina* has been undertaken: thus 1300 g were transformed, yielding 900 g of the (1S, 2S) diol showing high optical purity[162]. Interestingly, these strains convert the substrates fast with only negligible amounts of side products. Also, it is noteworthy that the

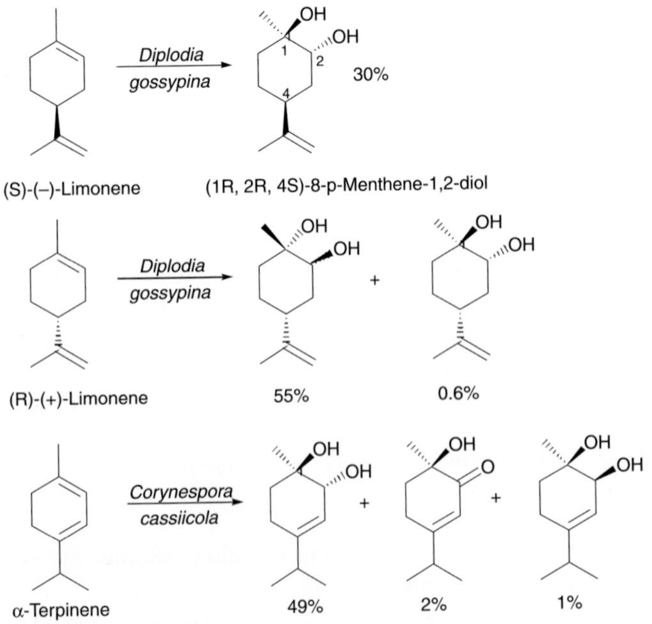

(S)-(–)-Limonene   (1R, 2R, 4S)-8-p-Menthene-1,2-diol

(R)-(+)-Limonene   55%   0.6%

α-Terpinene   49%   2%   1%

Figure 16.1-24. Stereoselective oxidation of monocyclic terpenes.

obtained diols are almost exclusively of *trans* configuration. No indication is provided concerning the determination and the values of the obtained products' optical purities. It was suggested that these *trans*-diols were formed via an intermediate epoxide, which could be further cleaved enzymatically to the obtained diols. Surprisingly, both these microorganisms were shown not to attack 3,3,5,5-tetra-methyllimonene[163]. However, geranylacetone, nerylacetone, *trans*-nerolidol, *cis*-nerolidol, farnesol and 2,5-dimethyl-1,3-hexadiene were transformed by these strains to the corresponding glycols in yields of up to 70%[164, 165] and interesting optical purities of up to 98%. Using (+)-*trans*-nerolidol as a substrate, the strain *Nocardia alba* DSM 43 130 was shown to be lacking an epoxide hydrolase, thus leading to a 27% yield of the corresponding (S)-epoxide which accumulates in the culture medium[157].

Also, the ability of the monensin-producing organism *Streptomyces cinnamonensis* to convert the *cis* and *trans* isomers of nerolidol has been investigated[158]. However, here again this led to a low-yield mixture of several products.

Much more useful in that sense are the results obtained by Furstoss and coworkers in the course of their study of biooxygenation of geraniol derivatives (Fig. 16.1-25). Indeed, it has been described in a first paper that, if the N-phenylcarbamate of geraniol is used instead of geraniol itself, its transformation by the fungus *Aspergillus niger* leads to a 49% isolated yield of the 6,7-dihydroxylated product. Moreover, this diol proved to be of (6S) absolute configuration and was shown to possess an enantiomeric excess of about 95%[76, 166]. This diol, which is a very versatile substrate for further organic synthesis, can thus be obtained without problem in

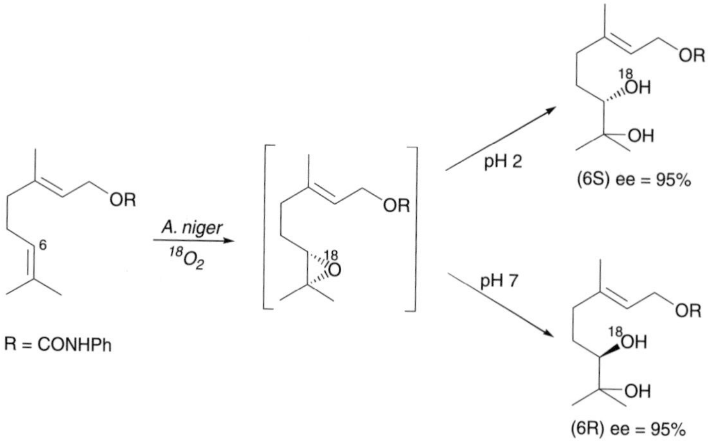

Figure 16.1-25. Stereoselective pH-dependent oxidation of geraniol N-phenyl carbamate.

gram-scale quantities (1 g substrate treated for 36 h in 1 L culture afforded 550 mg pure diol). Further work aimed at exploring the influence of the culture conditions showed that a unique stereochemical control could be achieved simply by modulating the pH of the medium. Thus, although when the culture was at pH 2 the diol of (S)-configuration was obtained, at pH 6–7 the diol of opposite (R) absolute configuration was isolated in similar yields and with an *ee* again as high as 95%. This interestingly showed that the fungus *A. niger* not only is able to convert the substrate across the pH 2–7 range, but that the (6S)-epoxide must be the primarily formed metabolite. This can then be further hydrolyzed in acidic medium (following the classical acid-catalysis mechanism) to afford the (6S)-diol or, at pH 6, be hydrolyzed enzymatically to the (6R)-diol by attack on the less substituted oxirane carbon atom [167]. Experiments conducted in the presence of ^{18}O confirmed this hypothesis. When the incubation was carried out at pH 2, the distribution of the ^{18}O label in the obtained diol was 95% on C-6 and 5% on C-7. This ratio was inverted at pH 7. These results show clearly that, whatever the pH, molecular oxygen is involved in these oxygenations but only one labeled oxygen atom is incorporated into the diol, leading to an epoxide which is differently hydrolyzed, depending on the pH of the medium.

Very interestingly as far as organic synthesis is concerned, these biooxygenations can be conveniently performed on a scale of several grams (5 g), thus allowing easy preparation of either enantiopure diol. These can be conveniently used as "*chirons*" for the synthesis of various natural or non-natural products. For instance they can be cyclized to the optically pure linalool oxides [168] or the corresponding tetrahydropyranols [169].

Biooxygenation of some other similar compounds, i.e. 7-geranyloxycoumarin, citronellyl N-phenylcarbamate and sulcatol N-phenylcarbamate were studied (Fig. 16.1-26) [170–172]. The reaction was shown to be operative in all these cases, leading, for instance to either enantiomer of marmin (a member of the umbelliferone family). Moreover, this result opens the way to an easy preparation of either

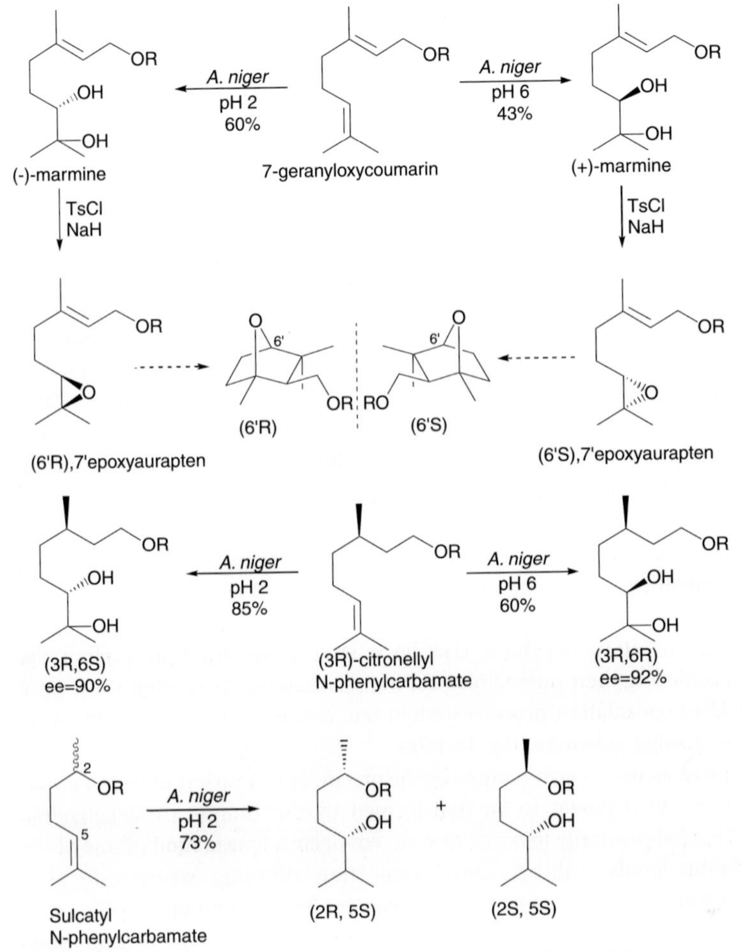

Figure 16.1-26. Application of the pH-dependent oxidation of geranyl derivatives to the synthesis of some natural products.

enantiomer of 6',7'-epoxyaurapten and of 3',6'-epoxyaurapten, both these compounds being natural products isolated from various sources. Similar results were obtained from both commercially available citronellol enantiomers, leading to the corresponding diols showing *ee*'s as high as 90 and 92%.

Bioconversions conducted at pH 2 on racemic sulcatol *N*-phenylcarbamate led to a 73% yield of a 1/1 mixture of the two expected diastereoisomeric diols, which can be readily separated by flash chromatography. They both show *ee*'s > 95%, indicating that the first (epoxidation) step again occurred in a highly stereospecific manner. Interestingly in this case, it was also possible to avoid hydrolysis of the intermediate epoxide by changing the preculture conditions and performing the reaction at neutral pH. This intermediate can thus be obtained directly with high enantiomeric purity. Using this chiron allows the four-step synthesis of optically pure pityol, a

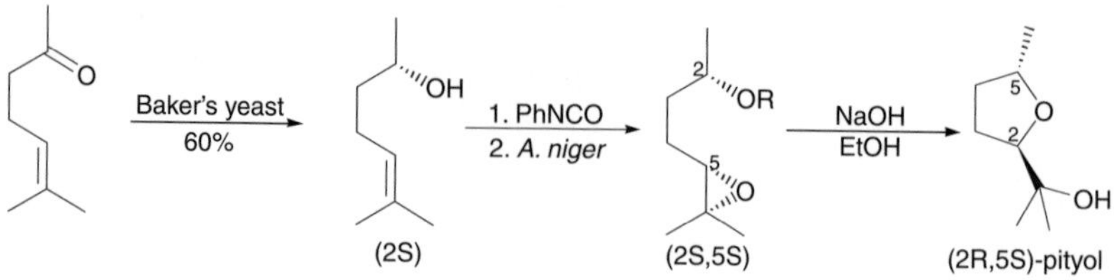

Figure 16.1-27. A four-step synthesis of (2*R*,5*S*)-pityol using microbiologically mediated steps.

male-specific attractant of the bark beetle *Pityophtorus pityographus*. Thus, prochiral 6-methyl-hept-5-en-2-one was reduced with baker's yeast to the corresponding alcohol (60% yield, 98.5% *ee*). This was converted to its *N*-phenylcarbamate, which was subsequently subjected to epoxidation using *A. niger*, thus affording a 50% preparative yield of the corresponding enantiopure epoxide. In a final step, treatment of the epoxycarbamate with an alcoholic NaOH solution led to the natural (2*R*,5*S*)-pityol (7.5% overall yield, 100% *ee*, 98% *de*) (Fig. 16.1-27).

16.1.5.4
Cyclic Sesquiterpenes

Various cyclic sesquiterpenes have also been studied in order to explore the possibility of achieving their microbiological transformations. Very often these were shown to lead to epoxidation processes when one (or several) double bonds were present in the starting substrate (Fig. 16.1-28).

Thus germacrone, which is thought to be the precursor of a variety of bicarbocyclic sesquiterpenoids, was shown to be transformed by the fungus *Cunninghamella blakesleena*. This led primarily to regio- and stereoselective epoxidation of one of the intracyclic double bonds of this prochiral triene, thus affording two epoxides. The third product isolated from this experiment was due to subsequent epoxidation of the remaining intracyclic double bond. Interestingly, the exocyclic olefinic bond conjugated to the carbonyl function appeared resistant to oxidation[173].

Valencene, another olefinic sesquiterpene, has been studied in the same context using microorganisms isolated from soil[174]. It was observed that these biotransformations led in reasonable yields to a mixture of three main metabolites, including an epoxide and nootkatone, an interesting flavoring compound.

The microbial transformation of humulene, a substrate showing a structure similar to that of germacrone, was studied by Abraham and Stumpf using a screen of about 300 strains[175]. This led the authors to select the fungi *Diplodia gossypina* and *Chaetonium cochlioides* for preparative scale experiments. It was thus observed that the main reaction path starts with the epoxidation of the 1,2-double bond, as shown by direct biotransformation of this monoepoxide obtained by chemical synthesis. This is then further oxidized to yield a multitude of products including diepoxides and hydroxy-epoxides (Fig. 16.1-28).

Comparable results were obtained from caryophyllene, a compound similar to humulene. Again, the biotransformation of this substrate with cultures of *Chaeton-*

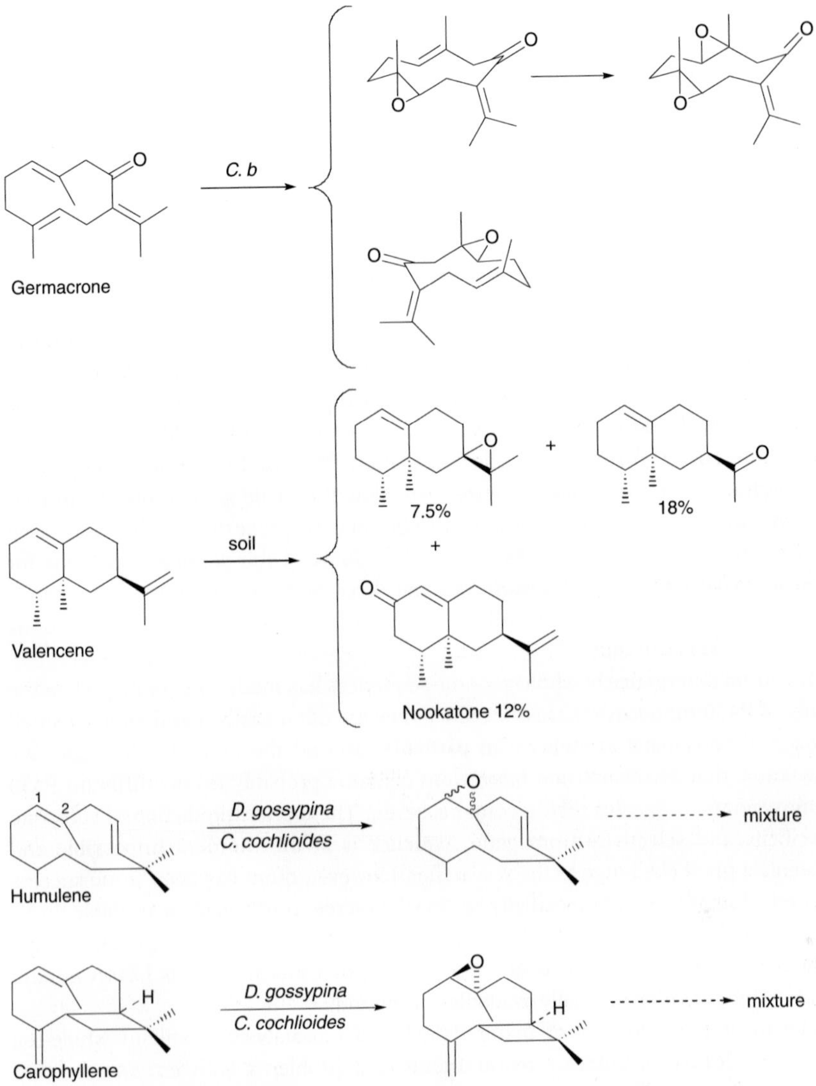

Figure 16.1-28. Epoxidation steps in the course of sesquiterpene biotransformations.

ium cochlioides as well as of *Diplodia gossypina* give a broad spectrum of products, resulting from an initial epoxidation of the 1–2 double bond followed by additional epoxidation or hydroxylation processes (Fig. 16.1-28) [176, 177].

16.1.6
Conclusions, Current and Future Trends

This review has illustrated the very broad range of biohydroxylations and epoxidations that can be achieved using monooxygenase enzymes. In fact, one can propose that almost all organic compounds are potential substrates for these enzymes. Since each substrate can lead to many different oxidized products, the range of compounds that can be generated is clearly enormous.

Finding new enzymes with novel substrate specificities and selectivities of reaction has in the past been achieved by screening organisms and substrates and has very much been down to good luck. Current and future work is focused on finding methods to make this process faster and more rational and predictable. This is now possible because of new technologies in genetics, molecular biology and structural biology, of which a few highlights are discussed below.

More and more P450 monoxygenases have been sequenced and cloned into heterologous expression systems. This can have the advantage of higher turnover yields because of higher expression of the enzyme in the host or because higher cell mass can be obtained when using easy growing organisms such as *E. coli* as hosts[178, 179]. Heterologous expression can also overcome problems of loss of product because of further metabolic degradation[180] as in the case of the *alk* gene of *P oleovorans*. Such expression systems also allow the facile generation of chimeric enzymes and mutants with more desirable biocatalytic properties, such as increased activity towards a particular substrate[181, 182]. Some of the popular organisms for biohydroxylations such as *Beauvaria bassiana* also might contain several endogenous P450 enzymes that can interfer with selectivity of one enzyme and make predictions of reactions very difficult[183].

The rapid emergence of whole genome sequences has made a major impact on the study of P450 monooxygenases[184], since they are often easily identifiable by small conserved consensus sequences, in particular around the heme binding site. We now know that *Mycobacterium tuberculosis* contains probably twenty different P450 monooxygenases; *Bacillus subtilis* contains seven. The *a priori* prediction of substrate specificity and selectivity from gene sequence is at the moment impossible and presents a great challenge to the researcher. However, there has been some success in prediction of substrate specificity by "in silico screening" based on available three-dimensional structures of P450-monooxygenases[185]. Thus, substrate docking algorithms were used to predict substrate suitability for P450cam and its L244A mutant from a library of commercially available compounds.

The most practical way of using P450-based biocatalysts is still in whole-cell systems, because of cofactor requirements and problems with enzyme stability. However, some P450 monooxygenases, such as the P450cam, can be isolated in sufficient quantities and reconstituted for cell-free preparative scale biotransformations[182]. This might be particularly useful for substrates that cannot penetrate cell walls, are toxic to the organism or are unstable in the organism. One solution for overcoming co-factor requirements might be the use of electrochemical methods, and is has indeed been shown that P450cam can be immobilized on an electrode and can take up electrons from the electrode[186].

Another novel area of intense research is the application of mutagenesis (random and directed) to obtain desired changes in substrate specificity. Thus P450cam, which is highly selective for camphor and closely related analogs, was subjected to site-specific mutagenesis, changing the tyrosine in position 96 to a phenylalanine[182], which resulted in about a 20-fold increase in the reactivity towards naphthalene. The P450 monooxygenase was independently subjected to random mutagenesis by Arnold and co-workers[187], and mutants were screened for increased activity towards naphthalene. Similar improvements to those observed by specific mutagensis were obtained. However, interestingly, the mutations that were found to be responsible for improved activity were not at position 96, but were distant from the active site of the enzyme. Such a "directed evolution" approach has great promise in quickly generating desired biohydroxylation catalysts, provided that a suitable screening system for the product can be found. The method has also been recently used by the same group on P450BM3[188].

In conclusion, the application of biocatalysts in biohydroxylations and epoxidations is rapidly expanding in terms of practicality, substrate range and selectivity. A vast diversity of P450 genes is generated by genomics programmes and mutagenesis. Methods for screening such oxidation catalysts are becoming more rapid, and one can forsee a future where designer biooxidation catalysts, tailored for a specific substrate and even for selectivity of reaction, can be generated within short time spans using a combination of rational and screening methods.

16.1.7
Cis Hydroxylation of Aromatic Double Bonds

16.1.7.1
Introduction

The microbial dioxygenation of aromatic compounds **1** has been known for over thirty years through the pioneering efforts of D. Gibson et al., who characterized the metabolic pathway of toluene degradation[189]. In lower organisms, the chiral *cis* glycol intermediates **2** are rapidly oxidized by dihydrodiol dehydrogenase, involving rearomatisation to the diol **3**, which is further oxidized by ring cleavage dioxygenase to give dicarboxylic acid **4**, which can be channeled into the organism's normal metabolic pathways (Scheme 16.1-1) [190–192].

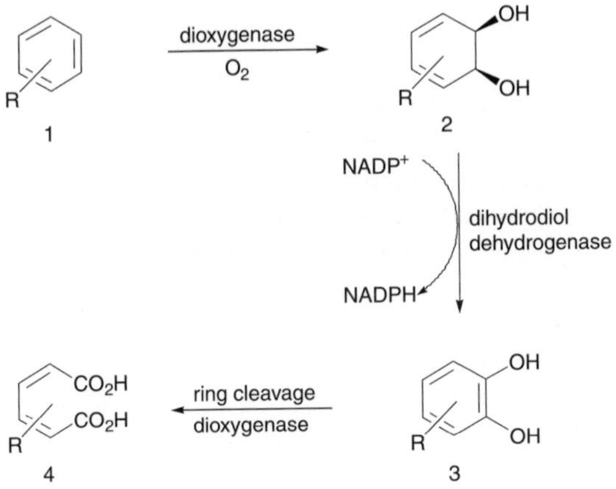

Scheme 16.1-1. Oxidative degradation of aromatic compounds by microorganisms.

The use of certain strains of *Pseudomonas putida,* most notably the mutant 39 D with blocked dehydrogenase activity [193], allows accumulation of the chiral glycols in the fermentation medium associated with high stereospecificity while the substrate tolerance remains high with respect to ring substituents. The enzymology of dioxygenases has been surveyed [190], and refinement of the mechanistic details of the dioxygenases continues [194], but only those enzymes and applications of relevance to the preparative biotransformations will be considered here.

16.1.7.2
Preparation of *cis* Dihydrodiols

An impressive number of substituted aromatic compounds **5** have been converted by mutant strains of *Pseudomonas putida* into the corresponding chiral *cis* glycols **6** with often excellent stereoselectivity [195, 196]. The remarkable substrate range and selectivity of this dioxygenase system for the aromatic ring have been demonstrated by the conversion of a series of substituted benzenes and of alkenyl benzenes with the side chain double bond being left intact (Scheme 16.1.2) [197, 198]. An analogous product was obtained from *para*-fluorotoluene, but the dihydrodiols from *para*-chloro- and *para*-bromotoluenes were found to be racemic [199]. Unlike the substrates shown in Scheme 16.1-2, benzoic acid, toluic acid, and their halogenated analogs, for example **7**, undergo enzymatic dioxygenation by *Alcaligenes eutrophus* B 9 and two strains of *Pseudomonas,* for example JT 103, mainly at the 1,2-position (Scheme 16.1-3) [200]. However, with other strains of *Pseudomonas putida,* for example JT 106, enantiospecific *cis* 2,3-dihydroxylation is possible, too [201].

The structure of the substrates is not necessarily restricted to monocyclic aromatic compounds such as those shown in Scheme 16.1-2. The dioxygenase activity of *Pseudomonas putida* and *Beijerinckia* species has been used exclusively for the synthesis of *cis* dihydrodiols from polycyclic [202] and heterocyclic [203] derivatives. Such products have been obtained from naphthalene, anthracene, phenanthrene, benz[a]pyrene, benz[a]anthracene, and methylsubstituted benz[a]anthracenes, and

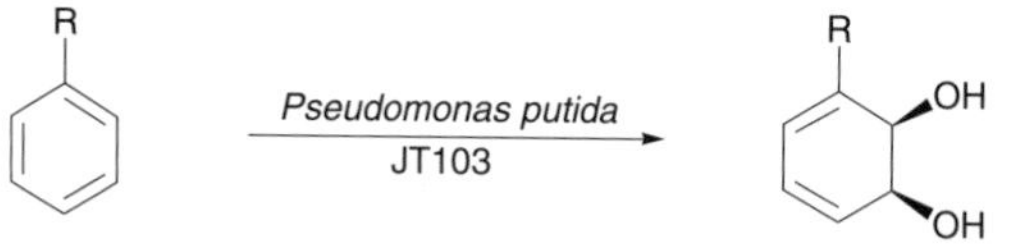

R = H, Me, Et, *n*Pr, *n*Bu, *t*Bu, EtO, *n*PrO, halogen, CF_3, Ph, $PhCH_2$
$PhCO$, $CH(OH)CH_3$, $COCH_3$, $CH_2=CH$, $CH_2=CHCH_2$, $CH_3CH=CH$
$CH3C=CHCH3$, $HC\equiv C$, CF_3, CN, CO_2R, $SiMe_3$

Scheme 16.1-2. Synthesis of *cis* diols by *Pseudomonas putida.*

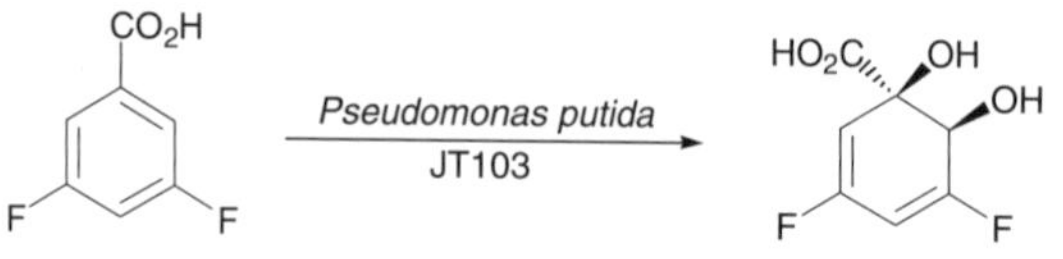

Scheme 16.1-3. *Cis*-hydroxylation of aromatic carboxylic acids by *Pseudomonas putida* JT 103.

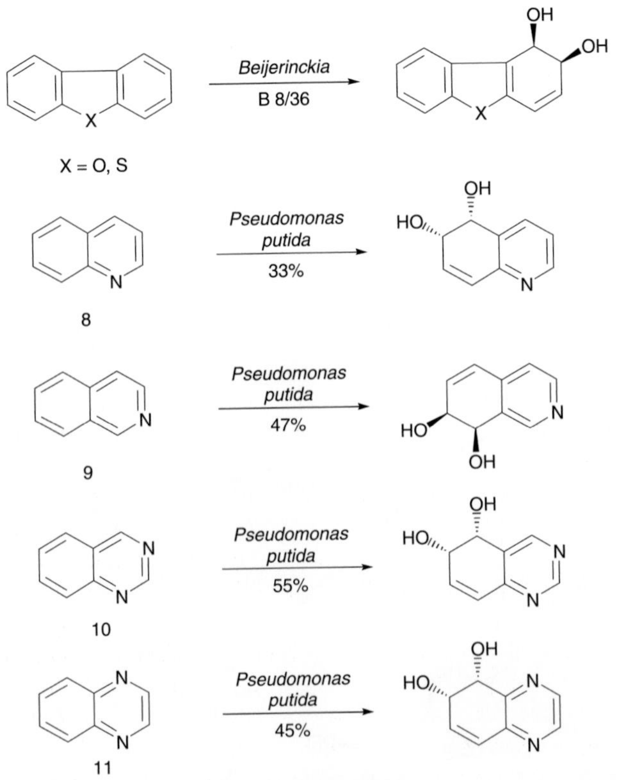

Scheme 16.1-4.
Dioxygenation of con-
densed aromatics by
Beijerinckia and *Pseudomonas putida*.

many of the enzymes responsible have been identified and characterized[191]. Benz[a]anthracene, for example, is converted to three *cis* dihydrodiol regioisomers by *Beijerinckia* B 8/36[204]. This organism has also been reported to produce dihydrodiols from dibenzofuran[205] and dibenzothiophene (Scheme 16.1-4)[206]. The ability of a *Pseudomonas putida* mutant to metabolize heteroaromatic compounds is demonstrated by the bioconversion of quinoline **8**, isoquinoline **9**, quinazoline **10**, and quinoxaline **11**[207]. Attack occurred exclusively in the carbocyclic ring (Scheme 16.1-4).

The impact of the genetic revolution has been greater in the area of dioxygenase-catalyzed reactions than in many other areas of bioconversion, largely because of the bacterial origin of the enzymes concerned. The bacterial oxidation of aromatic double bonds to *cis* diols in an enantiospecific manner leads to highly interesting synthons for organic chemistry. For example, the diene may be subjected to Diels-Alder reactions, as well as Michael-type addition reactions. Alternatively, oxidative cleavage of the cyclohexadiene ring leads to open chain products, which further react to yield cyclopentanoids. The large synthetic potential of chiral *cis* glycols is illustrated in Scheme 16.1-5.

T. Hudlicky et al. efficiently synthesized the prostaglandin PGE$_2$ **12** through an oxidative ring cleavage of the methyl-substituted diol **6** (R = CH$_3$)[208], the vinyl

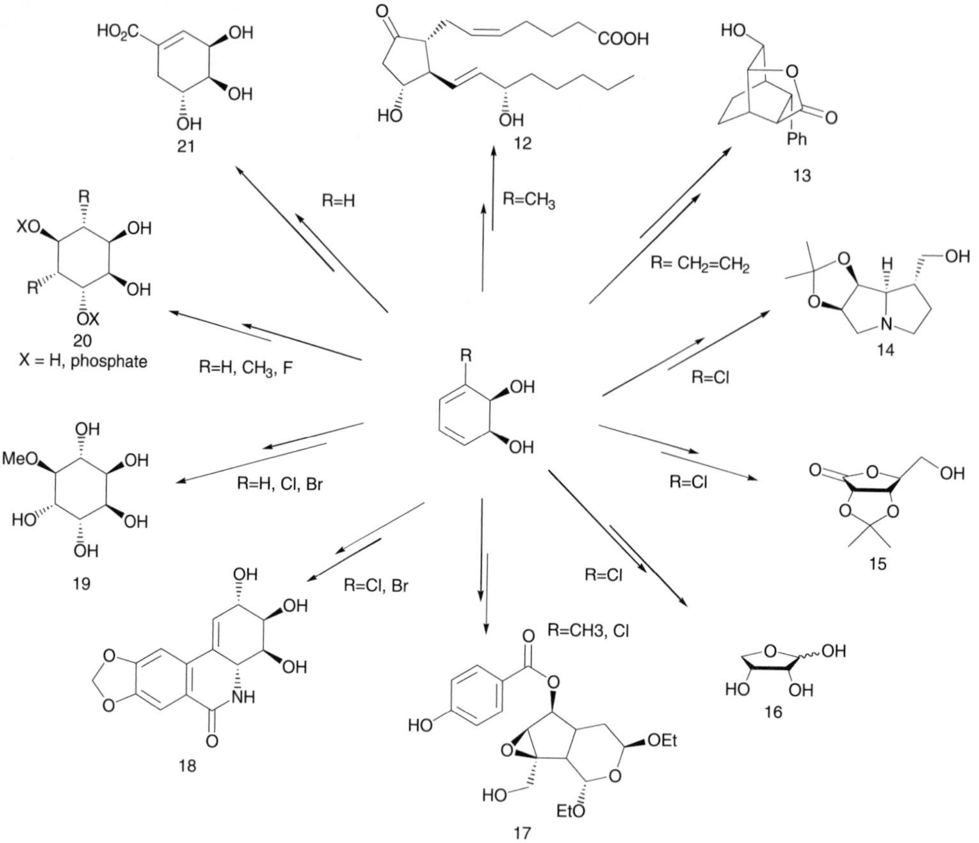

Scheme 16.1-5. Syntheses of natural products from substituted cyclohexadienediols.

derivative **6** (R = CH = CH$_2$) was used for the construction of the plant metabolite (-)-zeylena **13** [209], and the chloro-substituted diol for the synthesis of the alkaloid trihydroxyheliotridane **14** [210] and the carbohydrates L-ribonolactone **15** [211] and D-erythrose **16** [212]. Hudlicky et al. also prepared the sesquiterpene specionin **17** [213], an antifeedant to the spruce budworm, and the narcissus alkaloid lycoricidine **18** [214] in only nine steps.

Biologically active polyols like pinitol **19** [215, 216], D-myo-inositol **20** [217], conduritol C [218] and conduritol E [219] were obtained from diol **6** in both enantiomeric forms in only a few steps using this approach.

Futhermore C. R. Johnson et al. synthesized (–)-shikimic acid **21**, the biosynthetic precursor of the benzene moiety of aromatic amino acids [220].

In the case of the cyclohexadienediols **6**, the current development promises to complement the traditional and rather arduous use of carbohydrates as starting materials from the chiral pool. The popularity of diol-based methods will, therefore, be directly proportional to their ready commercial availability and to the operational ease of their transformations for the stereocontrolled introduction of further functionalities.

Several supply houses are now providing some simple chiral diols of type **6**, and further applications will assuredly follow.

References

1 P R Ortiz De Montellano (ed) *Cytochrome P450*. 2nd edn. 1995, Plenum Press: New York.

2 G J Grogan, H L Holland, *J. Mol. Catalysis B:Enzymatic*, **2000**, 9, 1–32.

3 H L Holland, *Curr. Opin. Chem. Biol.*, **1999**, 3, 22–27.

4 H L Holland, H K Weber, *Curr. Opin. Biotechnol.*, **2000**, 11, 547–553.

5 H L Holland, in *Biotechnology, Volume 8a:Biotransformations*, H.-J. Rehm and G. Reed (eds), 1998, Wiley-VCH, Weinheim.

6 H L Holland, *Stereoselective hydroxylation reactions*, in *Stereoselective Biocatalysis*, R N Patel (ed), 2000, Marcel Dekker: New York. p. 131–152.

7 R Azerad, *Regio- and stereoselective microbial hydroxylation of terpenoid compounds*, in *Stereoselective Biocatalysis*, R N Patel (ed), 2000, Marcel Dekker: New York. p. 152–180.

8 H L Holland, *Steroids*, **1999**, 64, 178–186.

9 M J Coon, A H Conney, R W Eastbrook, H V Gelboin, J R Gillette, P J O'Brien (eds) *Microsomes, Drug Oxidation and Chemical Carcinogenesis*. Vols. 1 and 2. 1980, Academic Press: New York.

10 F P Guengerich (ed) *Mammalian Cytochromes P450*. Vols. 1 and 2. 1987, CRC Press: Boca Raton, Fl.

11 J Seidegard, J W Depierre, *Biochim. Biophys. Acta*, **1983**, 695, 251–271.

12 J Meijer, J W Depierre, *Chemico-biological Interactions*, **1988**, 64, 207.

13 P F Hollenberg, *Faseb J.*, **1992**, 6, 686–694.

14 C S Yang, J F Brady, J-Y Hong, *Faseb J.*, **1992**, 6, 737–744.

15 J Peterson, H C Murray, S H Eppstein, L M Reineke, A Weintraub, P D Meister, H M Leigh, *J. Am. Chem. Soc.*, **1952**, 74, 5933–5936.

16 T L Poulos, R Raag, *Faseb J.*, **1992**, 6, 674–679.

17 J K Yano, L S Koo, D J Schuller, H Li, P R Ortiz de Montellano, T L Poulos, *J. Biol. Chem.*, **2000**, 275, 31086–31092.

18 I Schlichting, J Berendzen, K Chu, A M Stock, S A Maves, D E Benson, R M Sweet, D Ringe, G A Petzko, S G Sligar, *Science*, **2000**, 287, 1615–1622.

19 R Raag, T L Poulos, *Biochemistry*, **1991**, 30, 2674–2684.

20 J Peterson, M J Coon, *J. Biol. Chem.*, **1968**, 243, 329–334.

21 M Kusunose, J Matsumoto, K Ichihara, E Kusunose, E Nozaki, *J. Biochem.*, **1967**, 61, 665–667.

22 M Kusunose, K Ichihara, E Kusunose, J Nozaki, J Matsumoto, *Agric. Biol. Chem.*, **1967**, 21, 990.

23 A G Katopodis, K Wimalasena, J Lee, S W May, *J. Am. Chem. Soc.*, **1984**, 106, 7928–7935.

24 H Fu, M Newcomb, C H Wong, *J. Am. Chem. Soc.*, **1991**, 113, 5878–5880.

25 A Strijewski, *Eur. J. Biochem.*, **1982**, 128, 125–135.

26 K Kieslich, *Bull. Soc. Chem. Fr.*, **1980**, 119–1123.

27 J D Lipscomb, L Que, *JBIC*, **1998**, 3, 331–336.

28 A C Rosenzweig, C A Frederick, S J Lippard, P Nordlund, *Nature (London)*, **1993**, 366, 537–543.

29 M Filatov, N Harris, S Shaik, *Angew. Chem., Int. Ed. Engl.*, **1999**, 38, 3510–3512.

30 M Newcomb, R Shen, S-Y Choi, P H Toy, P F Hollenberg, A D Vaz, M J Coon, *J. Am. Chem. Soc.*, **2000**, 122, 2677–2686.

31 S P de Visser, F Ogliaro, N Harris, S Shaik, *J. Am. Chem. Soc.*, **2001**, 123, 3037–3047.

32 K M Madyastha, N S R K Murthy, *Tetrahedron Lett.*, **1988**, 29, 579–580.

33 H L Holland, E Riemland, *Can. J. Chem.*, **1985**(63), 1121–1126.

34 K M Madyastha, N S R K Murthy, *Appl. Microbiol. Biotechnol.*, **1988**, 28, 324–329.

35 T D Meehan, C J Coscia, *Biochem. Biophys. Res. Commun.*, **1973**, 53, 1043–1048.

36 K M Madyastha, T D Meehan, C J Coscia, *Biochemistry*, **1976**, 15, 1097–1102.

37 H J Licht, C J Coscia, *Biochemistry,* **1978**, *17,* 5638–5646.

38 H J Licht, C J Coscia, in *Microsomes, Drug Oxidation and Chemical Carcinogenesis,* M. J. Coon, et al. (eds), 1980, Academic Press, New York. p. 24.

39 K M Madyastha, C J Coscia, *J. Biol. Chem.,* **1979**, *254,* 2419–2427.

40 H Fretz, W D Woggon, *Helv. Chim. Acta,* **1986**(69), 1959–1970.

41 H Fretz, W D Woggon, R Voges, *Helv. Chim. Acta,* **1989**, *72,* 391–400.

42 K R Cadwallader, R J Braddock, M E Parish, D P Higgins, *J. Food Sci.,* **1989**(54), 1241–1245.

43 F Karp, C A Mihaliak, J L Harris, R Croteau, *Arch. Biochem. Biophys.,* **1990**, *276,* 219–226.

44 K Sode, K Kajiwara, E Tamiya, I Karube, Y Mikami, N Hori, T Yanagimoto, *Biocatalysis,* **1987**, *1,* 77–86.

45 V Krasnobajew, D Helmlinger, *Helv. Chim. Acta,* **1982**, *65,* 1590–1601.

46 K Sode, I Karube, R Araki, Y Mikami, *Biotechnol. Bioeng.,* **1989**, *33,* 1191–1195.

47 H Kakeya, T Sugai, H Ohta, *Agric. Biol. Chem.,* **1991**, *55*(1873–1876).

48 M Ismailialaoui, B Benjilali, D Buisson, R Azerad, *Tetrahedron Lett.,* **1992**, *33,* 2349–2352.

49 Y Asakawa, H Takahashi, M Toyota, Y Noma, *Phytochemistry,* **1991**, *30,* 3981–3987.

50 W-R Abraham, H-A Arfmann, *Tetrahedron,* 48, 6681–6688.

51 M Miyazawa, K Yamamoto, Y Noma, H Kameoka, *Chem. Express.,* **1990**, *4,* 237–240.

52 H D Repp, U Stottmeister, M Dorre, L Weber, G Haufe, *Biocatalysis,* **1990**, *4,* 75.

53 F S Sariaslani, L R McGee, M K Trower, F G Kitson, *Biochem. Biophys. Res. Commun.,* **1990**, *170,* 456–461.

54 W G Liu, J P N Rosazza, *Tetrahedron Lett.,* **1990**, *31,* 2833–2836.

55 W G Liu, A Goswami, R P Steffek, R L Chapman, F S Sariaslani, J J Steffens, J P N Rosazza, *J. Org. Chem.,* **1988**, *53,* 5700–5704.

56 V Lamare, R Furstoss, *Tetrahedron,* **1990**, *46,* 4109–4132.

57 Y Asakawa, H Takahashi, M Toyota, *Phytochemistry,* **1991**, *30,* 3993–3997.

58 Y Hu, R J Highet, D Marion, H Ziffer, *J. Chem. Soc., Chem. Commun* **1991**, 1176–1177.

59 R Furstoss, A Archelas, J D Fourneron, B Vigne, in *Organic Synthesis. An interdisciplinary Challenge 5th IUPAC Symposium,* H. P. J Streith, G Schill, (eds) 1985, Blackwell Scientific Publications. p. 215–226.

60 B Vigne, A Archelas, J D Fourneron, R Furstoss, *Nouv. J. Chim.,* **1987**, 297–298.

61 Y Amate, A Garcia-Granados, A Martinez, A Saenz de Buruaga, J L Breton, M E Onorato, J M Arias, *Tetrahedron,* **1991**, *47,* 5811–5818.

62 A Garcia-Granados, A Martinez, M E Onorato, F Rivas, J M Arias, *Tetrahedron,* **1991**, *47,* 91–102.

63 A Garcia-Granados, A Martinez, M E Onorato, F Rivas, J M Arias, *Tetrahedron Lett.,* **1991**, *32,* 5383–5384.

64 M S Ali, C A Davis, J R Hanson, *Phytochemistry,* **1991**(30), 3967–3969.

65 B M Fraga, R Guillermo, *Phytochemistry,* **1987**, *36,* 2521–2524.

66 S A Kouzi, J D McChesney, *Helv. Chim. Acta,* **1990**, *73,* 2157–2164.

67 G Aranda, J Y Lallemand, A Hammoumi, R Azerad, *Tetrahedron Lett.,* **1991**, *32,* 1783–1786.

68 G Aranda, M S Elkortbi, J Y Lallemand, A Neuman, A Hammouni, I Facon, R Azerad, *Tetrahedron,* **1991**, *47,* 8339–8350.

69 A J Aladesanmi, J J Hoffmann, *Phytochemistry,* **1991**, *30,* 1847–1848.

70 A Garcia-Granados, A Martinez, M E Onorato, M L Ruiz, J M Sanchez, J M Arias, *Phytochemistry,* **1990**, *29,* 121–126.

71 J C Sih, J P Rosazza, in *Applications of Biochemical Systems in Organic Chemistry,* C.J.S. J B Jones, D Perlman (eds) 1976, John Wiley Interscience, New York, Chapter 3.

72 F A Badria, C D Hufford, *Phytochemistry,* **1991**, *30,* 2265–2268.

73 H L Holland, *Organic Synthesis with Oxidative Enzymes.* 1992, New York: VCH. 1–304.

74 K C Chen, H C Wey, *Enzyme Microb. Technol.,* **1990**, *12,* 305–308.

75 P Chosson, H Vidal, A Aumelas, F Couderc, *FEMS Microbiol. Lett.,* **1991**, *83,* 17–22.

76 J D Fourneron, A Archelas, R Furstoss, *J. Org. Chem.,* **1989**, *54,* 4686–4689.

77 D R Boyd, N D Sharma, P J Stevenson, J Chima, D J Gray, H Dalton, *Tetrahedron Lett.,* **1991**, *32,* 3887–3890.

78 H L Holland, T S Manoharan, F Schweizer, *Tetrahedron: Asymmetry,* **1991**, *2,* 335–338.

79 R A Johnson, H C Murray, L M Reineke, G S Fonken, *J. Org. Chem.,* **1969**, *34,* 2279.

80 W Carruthers, J D Prail, S M Robert, A J Willetts, *J. Chem. Soc. Perkin Trans. 1*, 1990, 2854.

81 S J Aitken, G Grogan, C S-Y Chow, N J Turner, S L Flitsch, *J. Chem. Soc., Perkin Trans. 1*, 1998, 3365–3370.

82 S L Flitsch, S J Aitken, C S-Y Chow, G Grogan, A Staines, *Bioorganic Chemistry*, 1999, 27, 81–90.

83 Z Li, H-J Feiten, D Chang, W A Duetz, J B van Beilen, B Witholt, *J. Org. Chem.*, *pre-published on web at time of writing*, 2001.

84 A de Raadt, H Griengl, H-J Weber, *Chem-Eur. J.*, 2001, 7, 27–31.

85 R A Johnson, M E Herr, H C Murray, G S Fonken, *J. Org. Chem.*, 1970, 35, 622–626.

86 G S Fonken, R A Johnson, *Chemical Oxidations with Microorganisms*. 1972, New York: Marcel Dekker Inc. 1–27.

87 A Archelas, S Hartmans, J Tramper, *Biocatalysis*, 1988, 1, 283–292.

88 R Furstoss, A Archelas, J D Fourneron, B Vigne (eds) *Enzymes as Catalysts in Organic Synthesis*. NATO ASI, M P Schneider (ed), Vol. 178. 1986. 361–370.

89 J Ouazzani, S Arseniyadis, R Alvarez-Manzaneda, E Cabrera, G Ourisson, *Tetrahedron Lett.*, 1991, 32, 647–650.

90 J Ouazzani, S Arseniyadis, R Alvarez-Manzaneda, E Cabrera, G Ourisson, *Tetrahedron Lett.*, 1991, 32, 1983–1986.

91 A Hammamouni, G Revial, J D'Angelo, J P Girault, R Azerad, *Tetrahedron Lett.*, 1991, 32, 651–654.

92 S Arseniyadis, J Ouazzani, R Rodriguez, A Rumbero, G Ourisson, *Tetrahedron Lett.*, 1991, 32, 3573–3576.

93 K Nakagawa, K Sato, T Okazaki, A Torikata, *J. Antibiot.*, 1991, 44, 803–805.

94 F Vaufrey, A M Delort, G Jeminet, G Dauphin, *J. Antibiot.*, 1990, 43, 1189–1191.

95 S W May, *Enzyme Microb. Technol.*, 1979, 1, 15–22.

96 C A G M Weijers, A De Haan, J A M D Bont, *Microbiol. Sci.*, 1988, 5, 156–159.

97 K Furuhashi, *Chemical Economy and Engineering Review*, 1989, 18, 21–27.

98 J K Bruyn, *Ned. Akad. Wet. (Amsterdam)*, 1954, 54, 41.

99 T Ishikura, J W Foster, *Nature (London)*, 1961, 192, 892.

100 M J Klug, A J Markovetz, *J. Bacteriol.*, 1968, 1115.

101 A C Van Der Linden, *Biochim. Biophys. Acta*, 1963, 77, 157–159.

102 E W Maynert, R L Foreman, T Watabe, *J. Biol. Chem.*, 1966, 241, 5162.

103 B J Abbott, C T Hou, *Appl. Microbiol.*, 1973, 26, 86–91.

104 S W May, B J Abbott, *J. Biol. Chem.*, 1973, 248, 1725–1730.

105 S W May, M S Steltenkamp, R D Swartz, C J McCoy, *J. Am. Chem. Soc.*, 1976, 98, 7856–7858.

106 J A Peterson, D Basu, M J Coon, *J. Biol. Chem.*, 1966, 241, 5162–5164.

107 J A Peterson, M Kusnose, E Kusnose, M J Coon, *J. Biol. Chem.*, 1967, 242, 4334–4340.

108 E J McKenna, M J Coon, *J. Biol. Chem.*, 1970, 245, 3882–3889.

109 E T Lode, M J Coon, *J. Biol. Chem.*, 1971, 246, 791–802.

110 T Ueda, E T Lode, M J Coon, *J. Biol. Chem.*, 1972, 247, 2109–2116.

111 A Benson, K Tomoda, J Chang, G Matsueda, E T Lode, M J Coon, K T Yasunobu, *Biochem. Biophys. Res. Commun.*, 1971, 42, 640.

112 R R Boyer, E T Lode, M J Coon, *J. Biochem. Biophys. Res. Commun.*, 1971, 44, 925–930.

113 T Ueda, M J Coon, *J. Biol. Chem.*, 1972, 247, 5010–5016.

114 R T Ruettinger, S T Olson, R F Boyer, M J Coon, *Biochem. Biophys. Res. Commun.*, 1974, 57, 1011–1017.

115 R T Ruettinger, G R Griffith, M J Coon, *Arch. Biochem. Biophys.*, 1977, 183, 528–537.

116 S M May, R D Schwartz, B J Abbott, O R Zaborsky, *Biochim. Biophys. Acta*, 1975, 403, 524–255.

117 R D Schwartz, C J McCoy, *Appl. Environ. Microbiol.*, 1977, 34, 47–49.

118 M J De Smet, B Witholt, H Wynberg, *J. Org. Chem.*, 1981, 46, 3128–3131.

119 M J De Smet, J Kingma, H Wynberg, B Witholt, *Enzyme Microb. Technol.*, 1983, 5, 352–360.

120 S L Johnstone, G T Phillips, B W Robertson, P D Watts, M A Bertola, H S Koger, A F Marx, in *Biocatalysis in Organic Media*, J T C Laane, M D Lilley (eds) 1987, Elsevier, Amsterdam, p. 387–392.

121 N Toda, S Hayashi, Y Hatano, H Olunishi, M Miyazaki, *J. Pharmacol. Exp. Ther.*, 1978, 207, 311.

122 H Ohta, H Tetsukawa, *J. Chem. Soc., Chem. Commun*, 1978, 849–850.

123 H Ohta, H Tetsukawa, *Agric. Biol. Chem.*, **1979**, *43*, 2099–2104.

124 S Harbron, B W Smith, M D Lilley, *Enzyme Microb. Technol.*, **1986**, *8*, 85–88.

125 K Furuhashi, A Taoka, S Uchida, I Karube, S Suzuki, *Eur. J. Appl. Microbiol. Biotechnol.*, **1981**, *12*, 39–45.

126 K Furuhashi, M Shintani, M Takagi, *Appl. Microbiol. Biotechnol.*, **1986**, *23*, 218–223.

127 R T Ruettinger, A J Fulco, *J. Biol. Chem.*, **1981**, *256*, 5728–5734.

128 J A M De Bont, W Harder, *FEMS Microbiol. Lett.*, **1978**, *3*, 89–93.

129 J A M De Bont, M M Attwood, S B Primrose, W Harder, *FEMS Microbiol. Lett.*, **1979**, 183–188.

130 J A M De Bont, C G Van Ginkel, J Tramper, K C A M Luyben, *Enzyme Microb. Technol.*, **1983**, *5*, 55–59.

131 L E S Brink, J Tramper, K Van't Riet, K H A M Luyben, *Anal. Chim. Acta*, **1984**, *163*, 207–217.

132 L E S Brink, J Tramper, *Biotechnol. Bioeng.*, **1985**, *27*, 1258–1269.

133 V Schurig, W Burkle, *J. Am. Chem. Soc.*, **1982**, *104*, 7573–7580.

134 A Q H Habets-Crutzen, J A M D Bont, *Appl. Microbiol. Biotechnol.*, **1985**, *22*, 428–433.

135 L E S Brink, J Tramper, *Enzyme Microb. Technol.*, **1986**, *8*, 281–288.

136 L E S Brink, J Tramper, *Enzyme Microb. Technol.*, **1986**, *8*, 334–340.

137 L E S Brink, J Tramper, *Enzyme Microb. Technol.*, **1987**, *9*, 612–618.

138 C T Hou, R Patel, A I Laskin, N Barnabe, *Appl. Environ. Microbiol.*, **1979**, *38*, 127–134.

139 C T Hou, *Appl. Microbiol. Biotechnol.*, **1984**, *19*, 1–4.

140 V Subramanian, *J. Ind. Microbiol.*, **1986**, *1*, 119–127.

141 S Aono, M Ono, I Okura, *J. Mol. Catal.*, **1989**, *49*, L65-L67.

142 M Ono, I Okura, *J. Mol. Catal.*, **1990**, *61*, 113–122.

143 S H Stanley, H Dalton, *Biocatalysis*, **1992**, *6*, 163–175.

144 L B Wyngard, R P Roach, O Miyawaki, K A Egler, G E Klinzing, *Enzyme Microb. Technol.*, **1985**, *7*, 503–509.

145 O Takahashi, K Furuhashi, M Fukumasa, T Hirai, *Tetrahedron Lett.*, **1990**, *31*, 7031–7034.

146 C G Van-Ginkel, H G J Welten, J A M D Bont, *Appl. Microbiol. Biotechnol.*, **1986**, *24*, 334–337.

147 N R Woods, J C Murrell, *Biotechnol. Lett.*, **1990**, *12*, 409–414.

148 R F White, J Birnbaum, R T Meyer, J T Broeke, J M Chemerda, A L Demain, *Appl. Microbiol.*, **1971**, *22*, 55–60.

149 V Krasnobajew, in *Biotechnology: Biotransformations, Vol 6A, Chapter 4.* 1984, Verlag Chemie, Weinheim.

150 B B Mukherjee, G Kraidman, I D Mill, *Appl. Microbiol.*, **1973**, *25*, 447–453.

151 J R Devi, P K Bhattacharyya, *Ind. J. Biochem. Biophys.*, **1977**, *14*, 288–291.

152 K Imai, S Marumo, *Tetrahedron Lett.*, **1976**, *15*, 1211–1214.

153 L David, H Veschambre, *Tetrahedron Lett.*, **1984**, *25*, 543–546.

154 L David, H Veschambre, *Agric. Biol. Chem.*, **1985**, *49*, 1487–1489.

155 G Bock, I Benda, P Schreier, *J. Food Sci.*, **1986**, *51*, 659–662.

156 P Winterhalter, D Katzenberger, P Schreier, *Phytochemistry*, **1986**, *25*, 1347–1350.

157 W R Abraham, B Stumpf, H A Afrmann, *J. Essent. Oil Res.*, **1990**, *2*, 251–257.

158 D S Holmes, D M Ashworth, J A Robinson, *Helv. Chim. Acta*, **1990**, *73*, 260–271.

159 E Schwab, A Bernreuther, P Puapoomcharoen, K Mori, P Schreier, *Tetrahedron: Asymmetry*, **1991**, *2*, 471–479.

160 W R Abraham, B Stumpf, *Z. Naturforsch*, **1986**, *42C*, 559–566.

161 W R Abraham, B Stumpf, K Kieslich, *Appl. Microbiol. Biotechnol.*, **1986**, *24*, 24–30.

162 W R Abraham, J M R Hoffmann, K Kieslich, G Reng, B Stumpf, in *Enzymes in Organic Synthesis*, S. C. R Porter, Editor. 1985, Pitman Press: London. p. 146–157.

163 W R Abraham, B Stumpf, K Kieslich, S Reif, M R Hoffmann, *Appl. Microbiol. Biotechnol.*, **1986**, *24*, 31–34.

164 W R Abraham, H A Arfmann, B Stumpf, P Washausen, K Kieslich, in *Bioflavor*, P. Schreier (ed) 1988, Walter de Gruyter, New York, p. 399–414.

165 H A Arfmann, W R Abraham, K Kieslich, *Biocatalysis*, **1988**, *2*, 59–67.

166 R Furstoss, in *Microbial Reagents in Organic Synthesis*, S. Servi (ed) 1992, p. 333–346.

167 X M Zhang, A Archelas, R Furstoss, *J. Org. Chem.*, **1991**, *56*, 3814–3817.

168 A Meou, N Bouanah, A Archelas, X M

Zhang, R Guglielmetti, R Furstoss, *Synthesis*, **1980**, 752–753.

169 A Meou, X M Zhang, A Archelas, R Gugleilmetti, R Furstoss, *Synthesis*, **1991**, 681–682.

170 X M Zhang, A Archelas, A Meou, R Furstoss, *Tetrahedron: Asymmetry*, **1991**, *2*, 247–250.

171 X M Zhang, A Archelas, R Furstoss, *Tetrahedron: Asymmetry*, **1992**, *3*, 1373–1376.

172 A Archelas, R Furstoss, *Tetrahedron Lett.*, **1992**, *33*, 5241–5242.

173 H Hikino, C Konno, T Nagashima, T Kohama, T Takemoto, *Tetrahedron Lett.*, **1971**, *4*, 337–340.

174 R S Dhavlikar, G Albroscheit, *Dragoco Report*, **1973**, *12*, 251–258.

175 W R Abraham, L Ernst, B Stumpf, H A Arfman, *J. Ess. Oil Res.*, **1989**, *1*, 19–27.

176 W R Abraham, L Ernst, H A Arfmann, *Phytochemistry*, **1990**, *29*, 757–763.

177 W-R Abraham, L Ernst, B Stumpf, *Phytochemistry*, **1990**, *29*, 115–120.

178 M S Shet, C W Fisher, R W Estabrook, *Arch. Biochem. Biophys.*, **1997**, *339*, 218–225.

179 A Parikh, E M J Gillam, F P Guengerich, *Nature Biotechnology*, **1997**, *15*, 784–788.

180 O Favrebulle, T Schouten, J Kingma, B Witholt, *Biotechnology*, **1991**, *9*, 367–371.

181 D J Fraser, Y Q He, G Harlow, R., R Halpert J, *Mol. Pharmacol*, **1999**, *55*, 241–247.

182 S M Fowler, P A England, A C G Westlake, D R Rouch, D P Nickerson, C Blunt, D Braybrook, S West, L-L Wong, S L Flitsch, *J. Chem. Soc. Chem. Commun.*, **1994**, 2761–2762.

183 H L Holland, T A Morris, P J Nava, M Zabic, *Tetrahedron*, **1999**, *55*, 7441–7460.

184 D R Nelson, *Archives of Biochemistry and Biophysics*, **1999**, *369(1)*, 1–10.

185 J J De Voss, O Sibbensen, Z Zhang, P R Ortiz De Montellano, *J. Am. Chem. Soc.*, **1997**, *119*, 5489–5498.

186 J Kazlauskaite, A C G E Westlake, L-L Wong, H A O Hill, *Chem.Commun.*, **1996**, 2189–2190.

187 J Hyun, Z Lin, F H Arnold, *Nature (London)*, **1999**, *399*, 670–673.

188 E T Farinas, U Schwaneberg, A Glieber, F H Arnold, *Adv. Synth. Catal.*, **2001**, *343*, 601–606.

189 D T Gibson, M Hensley, H Yoshioka, T S Mabry, *Biochemistry*, **1970**, *9*, 1626–1630.

190 M Nozaki, *Top. Curr. Chem.*, **1979**, *78*, 145–186.

191 C E Cerniglia, *Adv. Appl. Microbiol.*, **1984**, *30*, 31–71.

192 P R Wallnofer, G Engelhardt, *Biotechnology*, **1984**, *6a*, 277–327.

193 D T Gibson, J R Koch, R E Kallio, *Biochemistry*, **1968**, *7*, 3795–3802.

194 C J Batie, E LaHaie, D P Ballou, *J. Biol. Chem.*, **1987**, *262*, 1510–1518.

195 D W Ribbons, C T Evans, J T Rossiter, S C J Taylor, S D Thomas, D A Widdowson, D J Williams, *Adv. Appl. Biotechnol. Ser.*, **1990**, *4*, 213–245.

196 D T Gibson, B Gschwendt, W K Yeh, V M Kobal, *Biochemistry*, **1973**, *12*, 1520–1528.

197 G Bestetti, E Galli, C Benigni, F Orsini, F Pelizzoni, *Appl. Microbiol. Biotechnol.*, **1989**, *30*, 252–256.

198 J A Schofield. 1989: U. S. Patent 4,8889,804, 26 December.

199 H L Holland, in *Organic Synthesis with Oxidative Enzymes.* 1992, VCH Publishers Inc., New York. p. 199.

200 W Reineke, H J Knackmuss, *Biochim. Biophys. Acta*, **1978**, *542*, 412–423.

201 S J C Taylor, D W Ribbons, A M Z Slawin, D A Widdowson, D J Williams, *Tetrahedron Lett.*, **1987**, *28*, 6391–6392.

202 T Hudlicky, M E Deluca, *Tetrahedron Lett.*, **1990**, *31*, 13–16.

203 L P Wackett, L D Kwart, D T Gibson, *Biochemistry*, **1988**, *27*, 1360–1367.

204 D M Jerina, P J van Bladeren, H Yagi, D T Gibson, V Mahadevan, A S Neese, M Koreeda, N D Sharma, D R Boyd, *J. Org. Chem.*, **1984**, *49*, 3621–3628.

205 C E Cerniglia, J C Morgan, D T Gibson, *Biochem. J.*, **1979**, *180*, 175–185.

206 A L Laborde, D T Gibson, *Appl. Environ. Microbiol.*, **1977**, *34*, 783–790.

207 D R Boyd, R A S McMordie, H P Porter, H Dalton, R O Jenkins, O W Howarth, *J. Chem. Soc. Chem. Commun.*, **1987**, 1722–1724.

208 T Hudlicky, H Luna, G Barbieri, L D Kwart, *J. Am. Chem. Soc.*, **1988**, *110*, 4735–4741.

209 T Hudlicky, G Seoane, T Pettus, *J. Org. Chem.*, **1989**, *54*, 4239–4243.

210 T Hudlicky, H Luna, J D Price, F Rulin, *J. Org. Chem.*, **1989**, *55*, 4683–4687.

211 T Hudlicky, J D Price, *Synlett.*, **1990**, 159–160.

212 T Hudlicky, H Luna, J D Price, F Rulin, *Tetrahedron Lett.*, **1989**, *30*, 4053–4054.

213 T Hudlicky, M Natchus, *J. Org. Chem.*, **1992**, *57*, 4740–4744.

214 T Hudlicky, H F Olivo, *J. Am. Chem. Soc.*, **1992**, *114*, 9694–9696.

215 T Hudlicky, J D Price, F Rulin, T Tsunoda, *J. Am. Chem. Soc.*, **1990**, *112*, 9439–9440.

216 S Ley, F Sternfeld, *Tetrahedron*, **1989**, *45*, 3463–3476.

217 S Ley, M Parra, A J Redgrave, F Sternfeld, *Tetrahedron*, **1990**, *46*, 4995–5026.

218 T Hudlicky, J D Price, H Luna, C M Andersen, *Synlett.*, **1990**, 309–310.

219 T Hudlicky, H Luna, H F Olivo, C Andersen, T Nugent, J D Price, *J. Chem. Soc. Perkin Trans. 1*, **1991**, 2907–2917.

220 C R Johnson, J P Adams, M A Collins, *J. Chem. Soc. Perkin Trans. 1*, **1993**, 1–2.

16.2
Oxidation of Alcohols

Andreas Schmid, Frank Hollmann, and Bruno Bühler

16.2.1
Introduction

The enzymatic oxidation of alcohols is catalyzed by different oxidoreductases. Here, examples of dehydrogenases, oxidases, and peroxidases are discussed. Single enzymes were selected based on representative or demanding reactions that are catalyzed, or because of interesting reaction engineering solutions applied. Reactions catalyzed by whole microbial cells are described in a separate chapter. A focus is put on presenting or introducing enzyme catalysts and their substrate spectra in order to give the reader a basis for designing his or her own, new reactions with sterically or electronically similar compounds or with such compounds which are compatible with a certain reaction mechanism.

Biocatalysis usually exploits advantageous features of enzymes such as chemoselectivity, regioselectivity, enantioselecivity and substrate spectrum of a certain broadness as depicted in Fig. 16.2-1. These points are addressed in examples in the following chapters.

16.2.2
Dehydrogenases as Catalysts

16.2.2.1
Regeneration of Oxidized Nicotinamide Coenzymes

Regeneration of $NAD(P)^+$ from $NAD(P)H$ is a redox reaction involving the transfer of two electrons and a proton (successively or at once as hydride ion H⁻) to a suitable acceptor. Most commonly these acceptors are carbonyl functions, molecular oxygen or the anode. Apart from a few exceptions the direct hydride transfer is slow or disadvantageous so that catalytic procedures have to be applied. Here we selected representative examples to give an overview. Excellent review articles are available, too [1–3, 10].

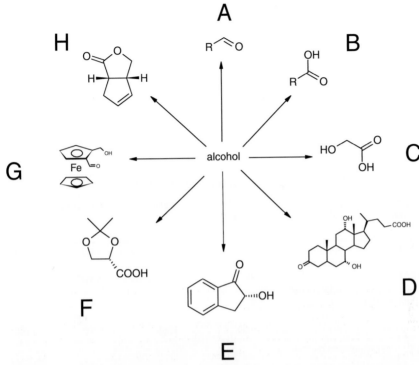

Figure 16.2-1. Enzyme-catalyzed oxidations of alcohols. Reactions are grouped according to the feature mainly exploited in the preparative application. A-C: Chemoselectivity (e.g. Sects. 16.2.2.3, 16.2.2.6, and 16.2.2.11); C, D: Regioselectivity (e.g. Sects. 16.2.2.9, 16.2.2.10, and 16.2.3.4); E, F: Enantioselectivity (e.g. Sects. 16.2.5.2 and 16.2.6.4); G: Non-natural substrates (e.g. Sect. 16.2.2.3); H: Complex structures from simple starting materials (e.g. Sect. 16.2.2.3).

16.2.2.2
Dehydrogenases as Regeneration Enzymes

Today, the utilization of a dehydrogenase-catalyzed reduction reaction is still the most widespread approach for the regeneration of oxidized NAD(P)$^+$. Its principle is displayed in Fig. 16.2-2.

Most commonly, alcohol dehydrogenase (E. C. 1.1.1.1) from yeast (YADH), horse liver (HLADH), or *Thermoanaerobium brockii* (TBADH) as well as glutamate dehydrogenase (E. C. 1.4.1.2.) or lactate dehydrogenase (E. C. 1.1.1.27) are used for NAD(P)$^+$ regeneration (Table 16.2-1). Thus, the reduction equivalents are transferred to an aldehyde or ketone as terminal electron acceptor yielding the corresponding alcohols.

The drawbacks of this approach result from the necessity to use a second enzyme, whose optimal reaction conditions may differ significantly from those of the actual production enzyme, and the presence of cosubstrates and coproducts. Furthermore,

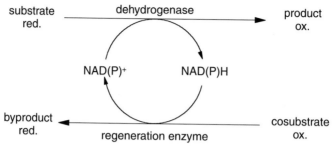

Figure 16.2-2.   Enzymatic regeneration of oxidized NAD(P)$^+$.

Table 16.2-1.   Comparison of commonly used dehydrogenases for NAD(P)$^+$ regeneration.

Regeneration enzyme	Cosubstrate/ coproduct	Specific Activity [U mg^{-1}]	Stability	Coenzyme	E'$_0$ [V] vs. NHE^a [1]
YADH	Acetaldehyde/ ethanol	300	Low, sensitive to O$_2$	NAD$^+$	– 0.199
TBADH	Acetone/ isopropanol	30–90	Thermostable	NADP$^+$	– 0.286
Glutamate DH	α-Ketoglutarate/ glutamate	40	High	NAD$^+$ and NADP$^+$	– 0.121
Lactate DH	Pyruvate/lactate	1000	High	NAD$^+$	– 0.185

a NHE: normal hydrogen electrode.

1 H. K. Chenault, G. M. Whitesides, *Appl. Biochem. Biotech.* **1987**, *14*, 147–197.

the thermodynamical driving force is low because the formal redox potential of the cosubstrate/coproduct couple is often close to that of the NADH/NAD$^+$ couple. Some of these problems can be addressed using the following regeneration concepts:

16.2.2.2.1 Enzyme-Coupled Regeneration

Since dehydrogenase catalysis is reversible, the production enzyme can be used to perform the regeneration reaction of NAD(P)$^+$ using a suitable cosubstrate as electron acceptor. In this case, the regeneration enzyme in Fig. 16.2-2 is identical with the production dehydrogenase. However, conversion rates in this set-up tend to be low because of a given reaction equilibrium, which requires an efficient method to withdraw the products and coproducts.

16.2.2.2.2 Intrasequential Regeneration

One elegant way of *in situ* product removal is to use the product of a first dehydrogenase reaction as substrate for a subsequent enzymatic reaction, thus recycling the oxidized nicotinamide coenzyme (Fig. 16.2-3). Various NAD(P)-dependent enzymes can be applied as regeneration enzymes in this cascade reaction.

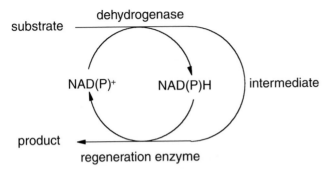

Figure 16.2-3. Intrasequential regeneration of NAD(P)$^+$. The strategy applied is the synthetic coupling of a dehydrogenase-catalyzed oxidation and a regeneration reaction yielding the final product and NAD(P) regeneration.

If the regeneration enzyme is a second dehydrogenase, an overall redoxisomerization takes place. But also monooxygenases are reported as regeneration enzymes thus yielding an overall double oxidation of the substrate (see Sect. 16.2.2.6.1).

16.2.2.3
Molecular Oxygen as Terminal Acceptor

The application of molecular oxygen as oxidant is favorable for several reasons. It is cheap and easily applicable. Furthermore, the high redox potentials of the O_2/H_2O or O_2/H_2O_2 couples (in acidic solution + 1.23 V and + 0.682 V, respectively) result in a strong thermodynamic driving force for the regeneration reaction. Since direct oxidation of NAD(P)H by molecular oxygen is very slow[4], the electron transfer has to be accelerated via enzymatic or chemical techniques.

NADH oxidases (NADH dehydrogenases, E.C. 1.6.99.x) from several organisms have been characterized in recent years[5]. Two types of NADH oxidases can be distinguished, namely those reducing molecular oxygen to water and those performing the reduction to hydrogen peroxide. Interestingly, few examples are found in literature employing NADH oxidases for the regeneration of NAD$^+$, probably because of stability reasons. However, an NADH oxidase from *Thermus aquaticus* was reported to be stable at 80 °C for at least 1 h[6], which might allow small scale applications.

FMN reductase (NAD(P)H dehydrogenase (FMN), E.C. 1.6.8.1) catalyzes the transhydrogenation from NAD(P)H to FMN[7], yielding the oxidized nicotinamide coenzyme and FMNH$_2$, which reacts spontaneously with molecular oxygen (Fig. 16.2-4). The reaction might be coupled to the catalase reaction in order to decrease the degree of enzyme inactivation over longer reaction times. Compared to the non-catalyzed hydride transfer from NAD(P)H to FMN[8], up to 1000-fold increases in the transhydrogenation rate are reported, which is not very high when applied synthetically. In this respect also the operational stability of FMN reductase has to be optimized. Besides the native substrate, cheaper alloxazine-based analogs are also accepted[9].

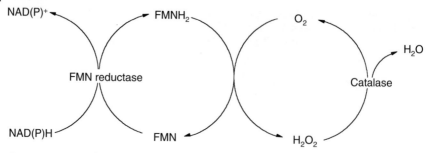

Figure 16.2-4. Transhydrogenation catalyzed by FMN reductase.

Among the chemical mediator systems especially *o-quinones* are capable of accepting the hydride equivalent from reduced nicotinamides. The oxidized mediators are regenerated by molecular oxygen. Since these mediators can also be recycled electrochemically, they are discussed in the following chapter.

16.2.2.4
Electrochemical Regeneration

A very elegant method to regenerate NAD(P)$^+$ from NAD(P)H is to use the anode as terminal electron acceptor. The most common approaches are summarized in Figure 16.2-5.

Direct electrochemical NAD(P)H oxidation (Fig. 16.2-5 A)
The easiest way to oxidize NAD(P)H is to withdraw the excess electrons anodically. Although the formal potential of the NADH/NAD$^+$ couple is $-$ 320 mV [- 324 mV for NADP] vs NHE[10], overpotentials as large as 1 V are required to achieve significant oxidation rates at bare electrodes[11, 12]. The number of enzymes, substrates, and

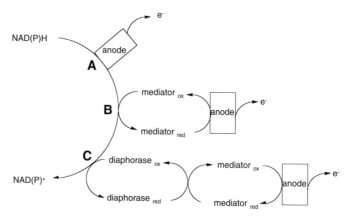

Figure 16.2-5. Electrochemical regeneration of NAD(P)$^+$. A: direct anodic oxidation; B: indirect electrochemical oxidation; C: diaphorase-accelerated indirect electrochemical oxidation.

products that can withstand this oxidizing power is limited. In addition, direct oxidation is often accompanied by electrode fouling, which is attributed to the formation of NAD dimers or stable adducts [12, 13].

Indirect electrochemical NAD(P)H oxidation (Fig. 16.2-5 B,C)

The high overpotentials needed for NAD(P)H oxidation can be considerably lowered by the use of redox mediators. Organic compound (such as *ortho-* and *para*-substituted quinones [14–18], diimines [20], and organic dyes [21–22]) undergoing two-electron transfer processes were found to be ideal fpr NAD(P)⁺ regeneration. Amongst these, 1,10-phenanthroline-5,6-diones [23, 24] are probably the most potent mediators. Furthermore, quinoid mediators can be generated in the surface of carbon electrodes by oxidative pretreatment [19].

Besides these hydride acceptors, single-electron-transfer mediators (e. g. transition metal complexes [25, 22], viologene derivatives [26, 27], ferrocenes [28], heteropolyanions [29], conducting polymers [30] or ABTS [31] are also capable of oxidizing NAD(P)H. Examples for one- and two-electron acceptors are listed in Table 16.2-2.

These mediators have been applied mostly freely diffusing but also immobilized at the electrode surface. A great variety of immobilization techniques have been used for the preparation of these modified electrodes – the mediator molecules have, for example, been directly adsorbed onto electrode surfaces, incorporated into conducting polymers or covalently linked to functional groups on electrode surfaces.

Often the electron transfer between the reduced nicotinamide coenzyme and the mediator is rather slow because of kinetic limitations. In many of these cases electron transfer catalyzed by *diaphorase* (E. C. 1.6.99.x) results in a drastic enhancement of the reaction rate (Fig. 16.2-5 C). Diaphorase-catalyzed NAD(P)⁺ regeneration was reported for example with methylene blue [32], PQQ [33] (under aerobic conditions), ferrocene [28], *N*-methyl-*p*-aminophenol [34], *N*,*N*-dimethylindoaniline, 2,6-dichlorophenol indophenol (DCIP), $[Fe(CN)_6]^{2-}$ [35], viologenes or several quinoid structures [36].

Many of the quinone-based mediators react in their reduced states with molecular oxygen. This aerobic regeneration has the advantage that no additional electrochemical equipment is necessary to perform NAD(P)⁺ regeneration. On the other hand, reactive oxygen species are generated, which might inactivate enzymes and which therefore need to be removed from the reaction mixture.

It should be mentioned at this point that most of the mediators described here were developed for analytical purposes. Only a few systems were applied to electrochemically driven dehydrogenase-catalyzed oxidations. This is partially because some systems exhibit moderate half-life times.

In conclusion it can be said that, for each individual case, a mediator with a good performance and stability under the given production conditions has to be found.

Table 16.2-2. Selection of frequently used mediators for indirect electrochemical regeneration of NAD(P)$^+$.

One-electron acceptors

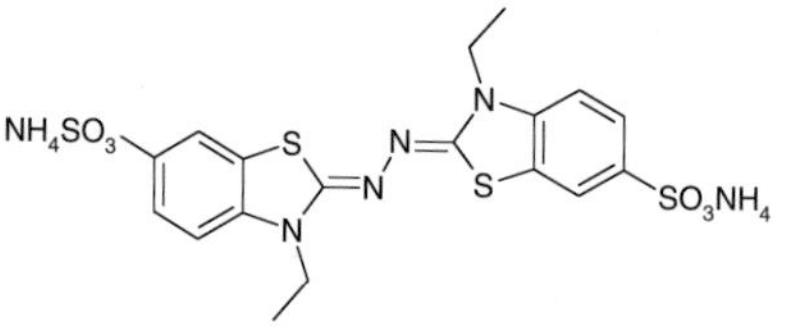

2,2'-azino-bis-(3-ethylbenzothiazoline-6-sulfonic acid)-diammonium salt (ABTS)

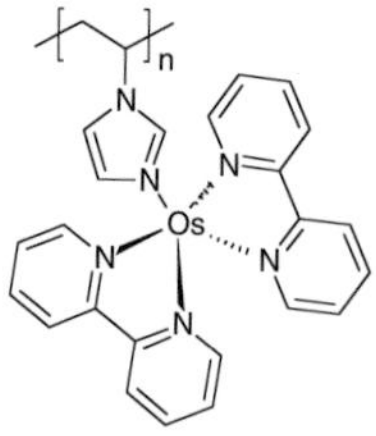

[Os(bpy)$_2$(PVI)$_{10}$Cl]$^+$

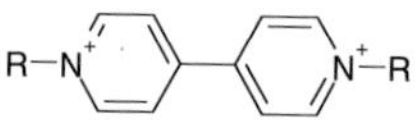

R = Me: methyl viologene
R = Bz: benzyl viologene

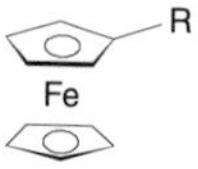

Ferrocenes

Two-electron acceptors

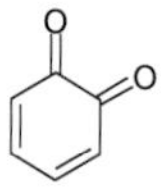

ortho-quinone (and various derivatives)

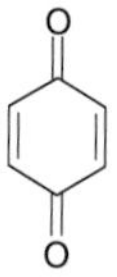

para-quinone (and various derivatives)

(BF$_4^-$)
N-methyl-1,10-phenanthroline-5,6-dione

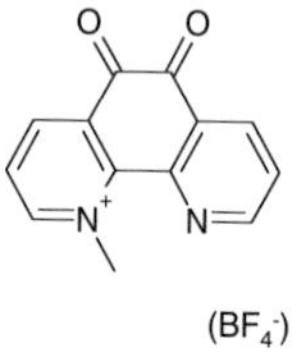

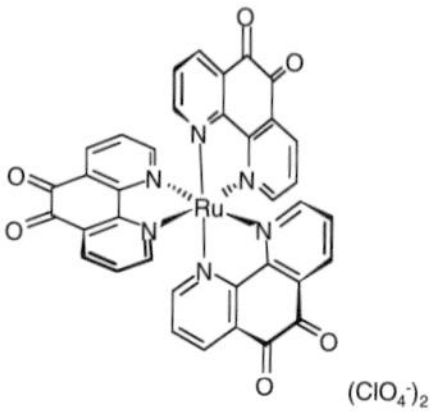

[Ru(PDOn)$_3$]$^{2+}$

(ClO$_4^-$)$_2$

16.2.2.5
Photochemical Regeneration

Various methods for photosensitized oxidation of NAD(P)H have been developed[37]. Photochemical methods are based either on the light-induced excitation of a mediator enabling it to oxidize NAD(P)H (reductive quenching mechanism) or on the light-induced excitation of the already reduced mediator, thus facilitating its re-oxidation (oxidative quenching mechanism) (Fig. 16.2-6).

For reductive quenching, photosensitizers such as tin porphyrins[38], methylene blue[39], and other dyes[40] are reported (Fig. 16.2-7). Ruthenium(II) *tris* bipyridine complexes in combination with viologenes are used for oxidative quenching. After

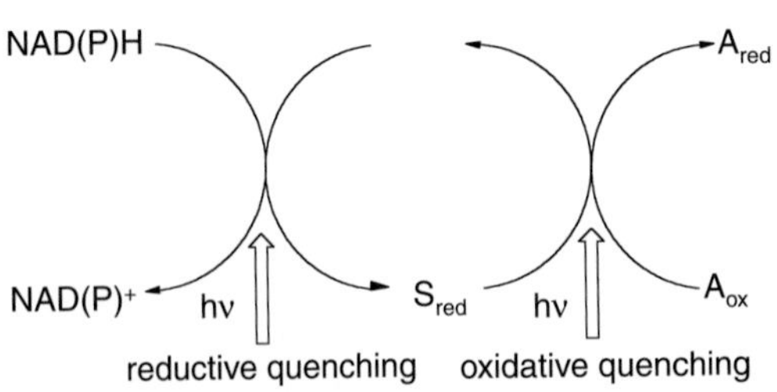

Figure 16.2-6. Electron transfer from NAD(P)H to acceptors (A) via photosensitizers (S) facilitated by photochemical activation.

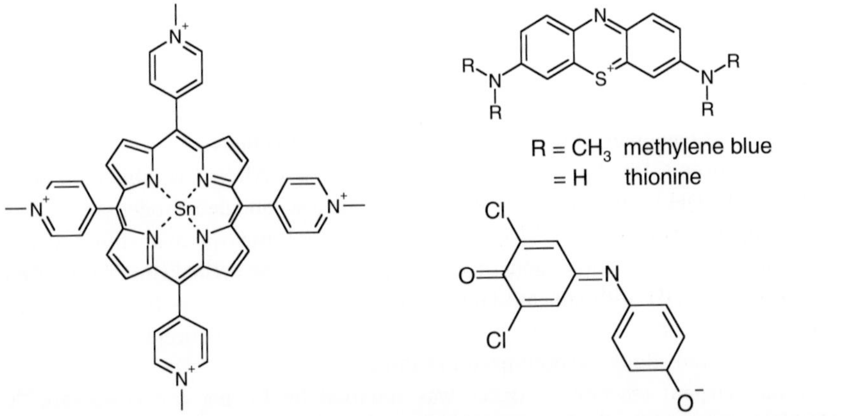

NAD(P)H

NAD(P)$^+$ hv   S$_{red}$ hv   A$_{ox}$

A$_{red}$

reductive quenching   oxidative quenching

R = CH$_3$ methylene blue
= H   thionine

Sn(II)-meso-tetramethylpyridinium porphyrin   2,6-dichlorophenol indophenol (DCP/P)

Figure 16.2-7. Photosensitizers used for photochemical regeneration of NAD(P)$^+$ from NAD(P)H.

the oxidation of NAD(P)H, the reduced Ru complex is excited by light. The resulting powerful reduction agent transforms methyl viologene into the radical cation. The electrons from NAD(P)H are usually transferred to molecular oxygen, protons or the anode [38, 40, 41].

Next to soluble photosensitizers, semi-conductors were reported for NAD$^+$ regeneration [42]. The advantage of these photochemical systems is that some of them utilize visible light, pointing towards the possibility of using sunlight for driving organic reactions. Disadvantageous, however, are the still low performances (TTN and TF of the photosensitizers and coenzymes) and the fact that photoexcitation results in the formation of strong oxidizing agents and the formation of free reactive radicals. Therefore, photochemical regeneration has not become one of the standard procedures, yet [37].

16.2.2.6
Oxidations Catalyzed by Alcohol Dehydrogenase from Horse Liver (HLADH)

HLADH is certainly one of the most prominent and widely used oxidoreductases. The NAD-dependent enzyme is a dimer consisting of two almost identical subunits, which both contain two zinc atoms [43, 44]. The 3-dimensional structure was eluci-dated via X-ray analysis [45, 46].

HLADH exhibits a unique combination of a very broad tolerance for primary and secondary alcohols (or aldehydes and ketones in the reductive direction) with an almost invariable and predictable stereospecificity [47, 48]. HLADH exhibits tolerance to many organic solvents [49] and is active even in water-saturated organic sol-vents [50, 42]. Even though HLADH exhibits a rather poor specific activity in the range of 1–2 U mg^{-1}, it is commercially available at reasonable prices ($ 570/1000 U, Sigma 2001) and, more importantly, is fairly stable even in oxygen-containing media [48]. Also because of that, HLADH has been studied extensively during the last few decades.

16.2.2.6.1 Regeneration of NAD⁺ in HLADH-catalyzed Reactions

Various concepts for the enzymatic regeneration of NAD⁺ in combination with isolated HLADH have been reported, ranging from a second dehydrogenase such as glutamate dehydrogenase [51, 52] to enzyme-coupled or intrasequential approaches.

A Baeyer-Villiger monooxygenase was applied to oxidize cyclic ketones produced *in situ* by HLADH with concomitant regeneration of NAD⁺ (Fig. 16.2-8) [53]. Even though yields and enantiomeric excesses are moderate, this concept has synthetic significance and should be optimized in future.

A very elegant reaction sequence was reported by Tanaka and coworkers [54]. HLADH was used for the kinetic resolution of a series of racemic β-hydroxysilanes yielding one enantiomer in *ee* values ranging from 20 to 97 % in reasonable yields and the corresponding β-ketosilane. This β-ketosilane hydrolyzes spontaneously and drives the regeneration of NAD⁺ catalyzed by HLADH (Fig. 16.2-9).

Other NAD⁺ regeneration approaches are based on the transfer of hydride either to PQQ (catalyzed by diaphorase) [33], directly to flavins [55–57], or to flavins via FMN reductase catalysis [58]. Direct hydride transfer to flavins has the advantage that the alloxazine acceptor can be chosen freely, e. g. cheap riboflavin instead of FAD. On the

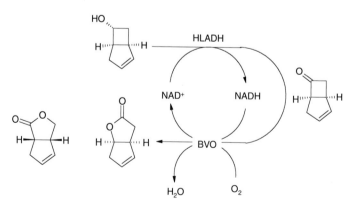

Figure 16.2-8.   Intrasequential regeneration of NAD with HLADH and a Baeyer-Villiger monooxygenase (BVO) from *Acinetobacter calcoaceticus*.

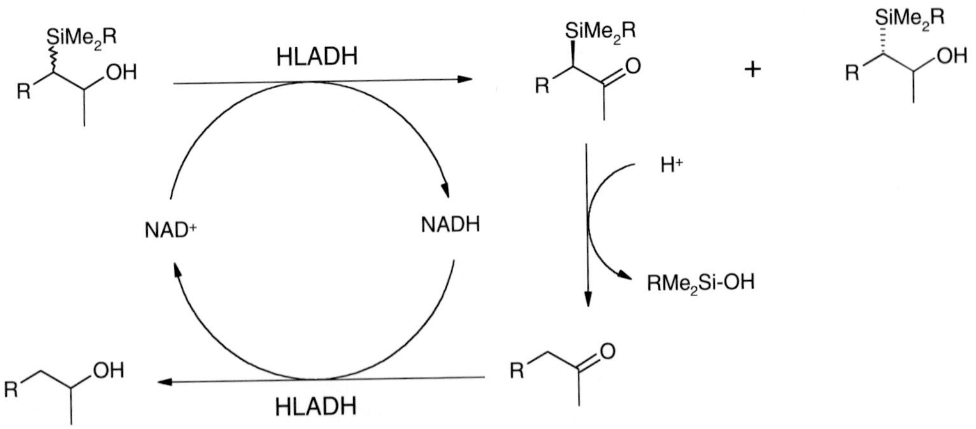

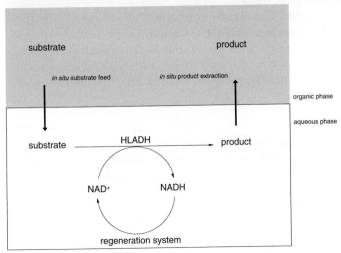

Figure 16.2-9. Intrasequential NAD⁺ regeneration for HLADH-driven kinetic racemate resolution of β-hydroxysilanes.

other hand, the spontaneous hydride transfer suffers from sluggish kinetics ($k = 0.2$ M^{-1} s^{-1}; turnover rates ranging between 0.06 and 1.8 h^{-1}) [59], which is app. 1000-fold slower than the values reported for enzymatic regeneration. For this reason, high excesses of the acceptor have to be applied in order to achieve acceptable regeneration rates. Introduction of FMN reductase accelerates this reaction remarkably.

Electrochemical methods utilizing quinoid mediators [23, 24] or ferrocenes [28] as well as photochemical [42] methods have also been applied to regenerate NAD⁺ in combination with HLADH. Especially the electrochemical variants utilizing quinoid shuttle systems proved to be very efficient, with mediator performances as high as 130 catalytic cycles per hour and quantitative yields.

Figure 16.2-10. HLADH-catalyzed oxidations in two-liquid phase systems (in the case of buffer-saturated organic solvents, the aqueous phase is limited to a layer around HLADH).

Table 16.2-3. Synthetic application of HLADH in organic solvents.

Substrate(s)	Product(s)	Solvent	Remarks/Ref.
Geraniol	Geranial	Hexane	Plugged-flow reactor for continuous production [2]
Cinnamylalcohol	Cinnamylaldehyde	Isopropyl ether	[3]
		Hexane	[4]
Racemic		Ethyl acetate, chloroform, Isopropyl ether, butyl acetate	HLADH immobilized on glass beads [5]
		Hexane	HLADH in polyacrylamide particles [6]

2 R. Lortie, I. Villaume, M. D. Legoy, D. Thomas, *Biotech. Bioeng.* **1989**, *33*, 229–232.
3 T. Kawamoto, A. Aoki, K. Sonomoto, A. Tanaka, *J. Ferm. Bioeng.* **1989**, *67*, 361–362.
4 J. R. Matos, C.-H. Wong, *J. Org. Chem.* **1986**, *51*, 2388–2389.

5 J. Grundwald, B. Wirz, M. P. Scollar, A. M. Klibanov, *J. Am. Chem. Soc.* **1986**, *108*, 6732–6734.
6 C. Gorrebeck, M. Spanghoe, G. Lanens, G. L. Lemiere, R. A. Dommisse, J. A. Lepoivret, F. C. Adlerweireldt, *Rec. Trav. Chim. Pays-Bas* **1991**, *110*, 231–235.

16.2.2.6.2 HLADH in Organic Media

Several applications of HLADH in organic/aqueous media have been reported (Table 16.2-3). The concept of these two liquid-phase reaction systems is shown schematically in Fig. 16.2-10. This approach is especially suitable for substrates and products with low solubility in aqueous media. Furthermore, the organic phase serves as a sink for products, thus decreasing problems resulting from product inhibition or back reactions.

16.2.2.6.3 Kinetic Resolution of Alcohols using HLADH

Because of its high enantioselectivity, HLADH has found widespread applications in the kinetic resolution of racemic alcohols and α-amino alcohols. Total turnovers of up to 10^8 for HLADH and 800 for NAD were reported with 90% residual activity, yielding the corresponding aldehydes in enantiomeric excesses up to 96%. The α-hydroxy aldehydes were metabolized *in situ* by an aldehyde dehydrogenase to the corresponding α-hydroxy acids (Fig. 16.2-11) [51].

Examples of further kinetic resolutions of racemates via regioselective oxidation using HLADH are given in Fig. 16.2-12.

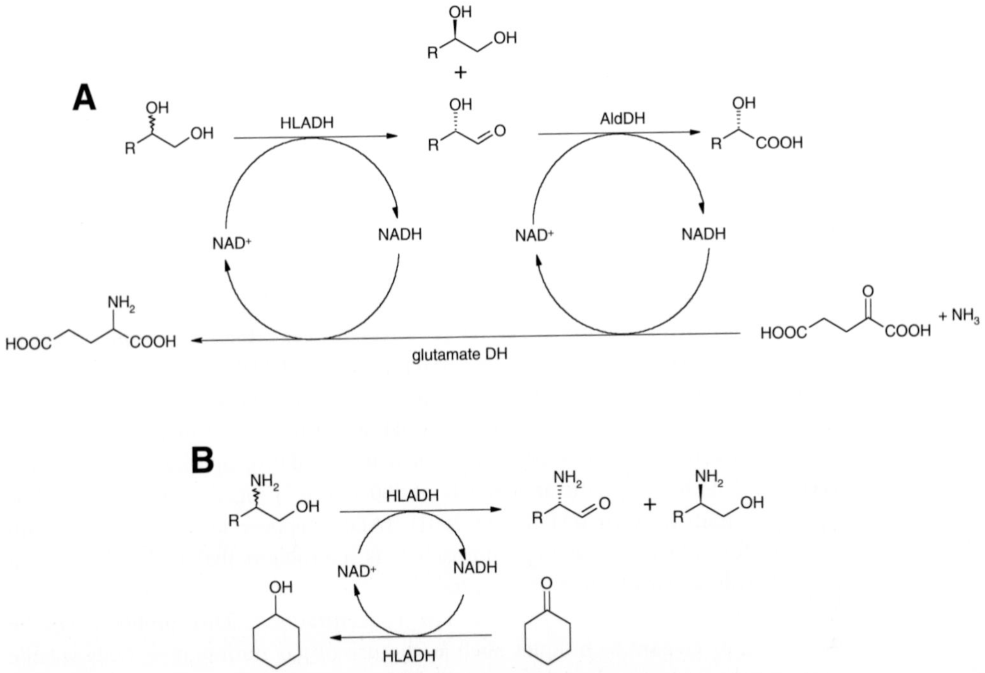

Figure 16.2-11. HLADH as enantioselective catalyst in the kinetic resolution of *vic*-diols (A) and α-amino alcohols (B). R = CH₂OH, CH₂F, CH₂Cl, CH₂Br, CH₃, CH=CH₂, C₂H₅, CH₂NH₂, (CH₃)₂.

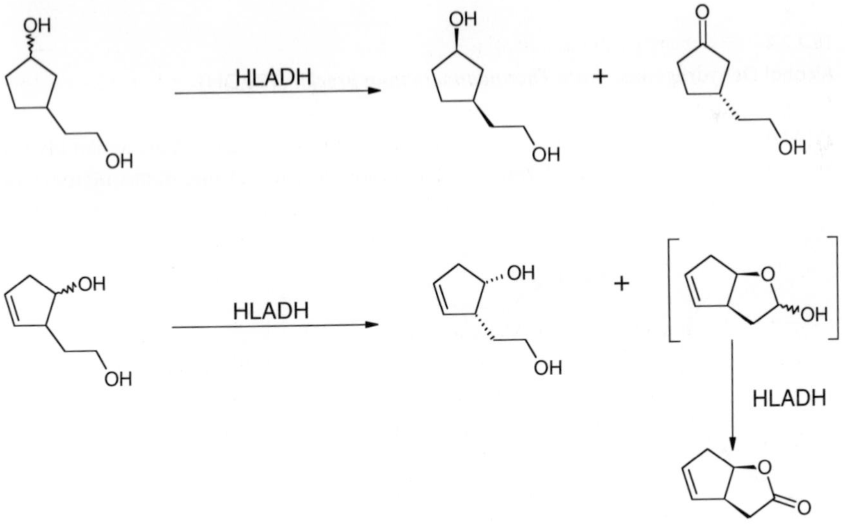

Figure 16.2-12. Chemo- and stereoselective oxidations of *sec*-alcohols.

16.2.2.6.4 HLADH for the Oxidation of *meso*-Compounds

Probably the most prominent application of HLADH is the oxidation of *meso*-diols to homochiral lactones. Both 1,4- and 1,5-diols are accepted as substrates (Table 16.2-4). The overall 4-electron oxidations proceed via two successive steps (tandem oxidation). The enantiomeric excesses often exceed 97%.

16.2.2.7
Alcohol Dehydrogenase from Yeast (YADH)

Even though the primary sequences differ significantly, YADH exhibits almost the same quaternary structure as HLADH[60]. Nevertheless, far fewer applications in biocatalytic processes are known for YADH than for HLADH. In part this is due to its low overall stability and its low resistance towards organic solvents[61]. Furthermore the substrate spectrum of YADH is limited to primary alcohols and 2-hydroxyalkanes[62]. It has been used in a few oxidative applications[63, 64]. On account of its high specific activity (about 300 U mg^{-1}) together with its very low price (less than 1.2 $/1000 U, Sigma, 2001), YADH has been used as a regeneration enzyme for NADH[65]. In this approach it is a problem that both ethanol and acetaldehyde as cosubstrate and coproduct of the regeneration reaction inactivate YADH and also other enzymes at low concentrations. This problem can be addressed by elegant techniques such as the use of gas membranes. Only volatile compounds such as ethanol or acetaldehyde can pass into the gas phase. This concept has been applied for lactate dehydrogenase (Fig. 16.2-13)[66, 67]. Hazardous acetaldehyde is removed and even recycled to form ethanol by treatment with sodium borohydride in the gas phase. Cycle numbers of over 10 000 are reported.

16.2.2.8
Alcohol Dehydrogenase from *Thermoanaerobium brockii* (TBADH)

TBADH is a NADP-dependent dehydrogenase with remarkable thermostability up to 65 °C[68]. Neither HLADH nor YADH are able to convert linear secondary alco-

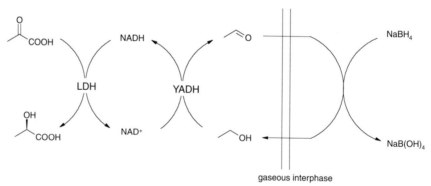

Figure 16.2-13. Regeneration of NADH with YADH. Acetaldehyde diffuses through the gaseous interphase into the second liquid phase where it is regenerated chemically to ethanol.

Table 16.2-4. Examples of HLADH-catalyzed enantioselective oxidations of *meso*-diols.

Meso-diol	Lactone	Yield [%]	ee [%]	References
		99	> 97	[7]
		90	95	[8]
		ND[a]	> 97	[9]
		68	> 97	[10]
		90	> 97	[10]
		55	99	[11]
		95	"100"	[8]
		> 99	> 99	[7]
X = CH₂, O		70	99	[12]
Fe	Fe	81	86	[13]

a ND: not determined.

7 G. Hilt, B. Lewall, G. Montero, J. H. P. Utley,
E. Steckhan, *Liebigs Ann./Recueil* **1997**, 2289–2296.

8 T. Osa, Y. Kashiwagi, Y. Yanagisawa, *Chem. Lett.*
1994, 367–370.

9 K. Mori, M. Amaike, J. E. Oliver, *Liebigs Ann.
Chem.* **1992**, 1179.

10 Y. Yamazaki, K. Hosono, *Tetrahedron Lett.* **1989**, 30,
5313–5314.

11 M.-E. Gourdel-Martin, C. Comoy, F. Huet, *Tetrahedron: Asym.* **1999**, 10, 403–404.

12 R. N. Patel, M. Liu, A. Banerjee, S. L., *Ind. J.
Chem.* **1992**, 31B, 832–836.

13 Y. Yamazaki, K. Hosono, *Tetrahedron Lett.* **1988**, 29,
5769–5770.

hols. TBADH fills this gap: its activity is highest for secondary alcohols, being low for
primary alcohols [48]. Because of this rather narrow substrate spectrum, TBADH is
mostly used for the regeneration of NADPH. Only a few synthetic applications are
reported [24, 69]. Figure 16.2-14 gives one example where YADH was used simultaneously as an oxidizing enzyme and a NADPH regeneration enzyme (intrasequential
cofactor regeneration).

16.2.2.9
Glycerol Dehydrogenase (GDH, E. C. 1.1.1.6)

GDH was isolated from various bacterial strains, especially from *Schizosaccharomyces pombe* [70, 71] and *Cellulomonas* sp. [72, 73].

It displays a somewhat complementary substrate specificity to HLADH. While
HLADH oxidizes *meso*-diols with secondary hydroxyl groups rather badly, they are
readily oxidized by GDH to the corresponding (S)-α-hydroxyketones [1]. Furthermore,
the natural substrate glycerol is transformed to achiral dihydroxy acetone by GDH
while HLADH produces optically active (S)-glyceraldehyde. In many cases GDH
seems to prefer secondary hydroxyl groups (Table 16.2-5), although this rule of
thumb has some exceptions.

In aqueous buffers GDH exhibits only low enantioselectivity, e. g. for the kinetic
resolution of 1-phenyl-1,2-ethanediol (which is most probably due to spontaneous
racemization via enolization) [74]; furthermore, it suffers from pronounced product
inhibition, accounting for low yields. Both problems (product inhibition and

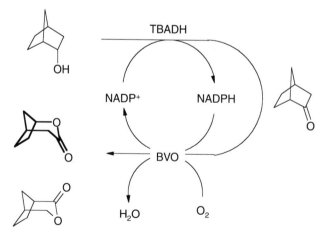

Figure 16.2-14.
Intrasequential
regeneration of NADP
with TBADH and a
Baeyer-Villiger mono-
oxygenase (BVO) from
*Acinetobacter calcoaceti-
cus.*

Table 16.2-5. Alcohol oxidations catalyzed by glycerol dehydrogenase[14].

Substrate	Product

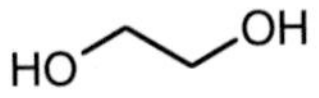

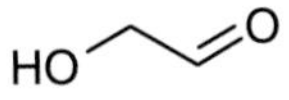

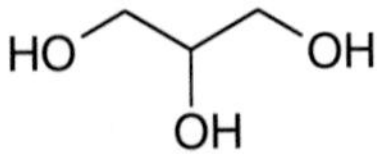

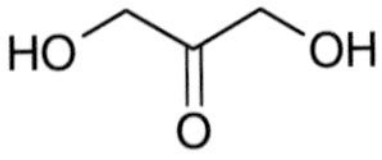

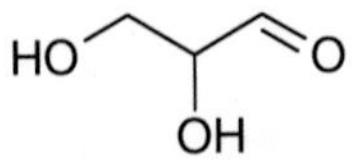

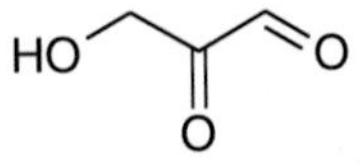

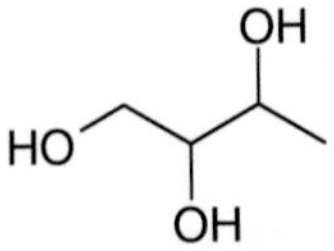

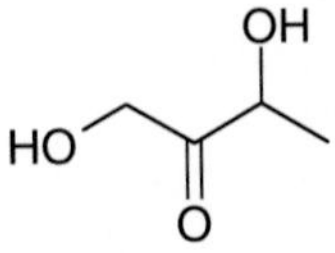

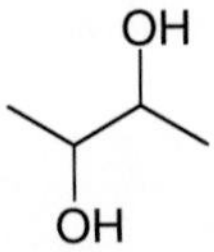

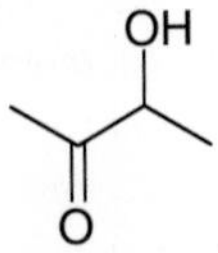

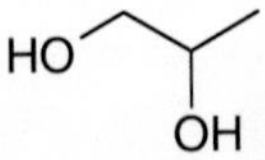

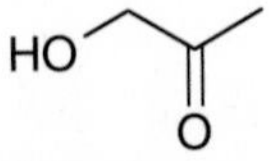

14 J. H. Marshall, J. W. May, J. Sloan, *J. Gen. Microbiol.* **1985**, *131*, 1581–1588.

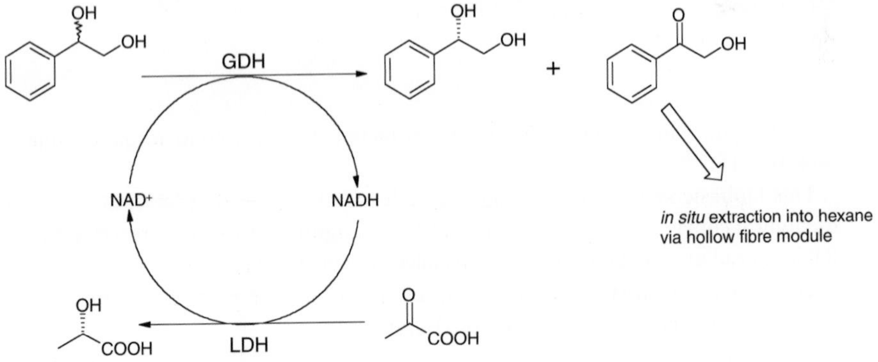

Figure 16.2-15. Deracemization of *rac* 1-phenyl-1,2-ethandiol coupled to *in situ* product extraction via a hollow fiber module.

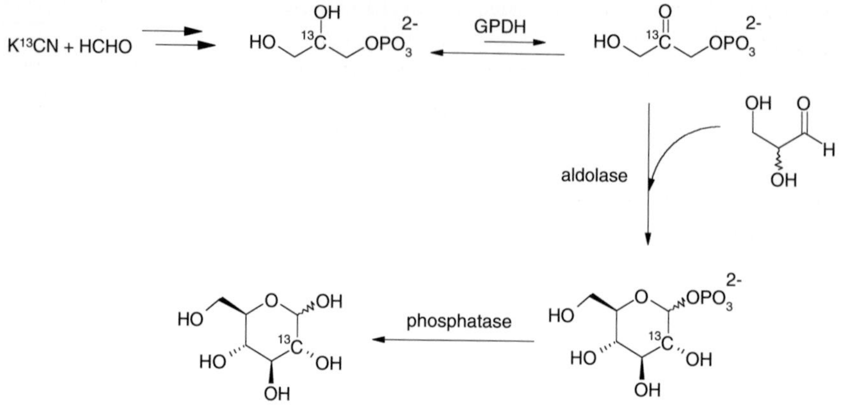

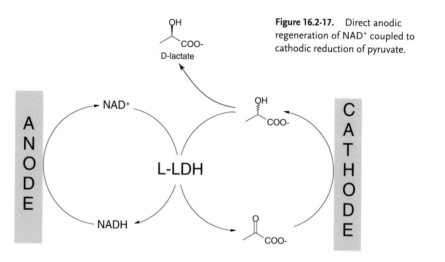

Figure 16.2-16. Synthesis of ^{13}C-labeled sugars in a tandem reaction of GPDH, aldolase, and phosphatase.

Figure 16.2-17. Direct anodic regeneration of NAD$^+$ coupled to cathodic reduction of pyruvate.

racemization) can be solved by *in situ* extraction into a second (organic) phase (Fig. 16.2-15) [74].

This biphasic system yielded higher *ee* values (99% instead of 58%) at maximal theoretical conversions (50% instead of 38%) in significantly shorter reaction times (60 h instead of 170 h) compared to the solely aqueous system.

Since GDH contains autooxidizable thiol groups, it is necessary to perform such reactions in media essentially free from oxygen.

16.2.2.10
Glycerol-3-phosphate Dehydrogenase (GPDH, E. C. 1.1.1.8)

GPDH has been isolated from various organisms. The enzyme from rabbit muscle is commercially available. Its synthetic applications are limited because of its very narrow substrate spectrum it almost exclusively accepts L-glycerol-3-phosphate[75, 76]. The product 3-hydroxyacetone phosphate, however, is an essential substrate of aldolases and therefore can serve as a building block in the enzymatic synthesis of non-native sugars and polyols. Although the redox equilibrium of GPDH favors the reduced substrates even more than in the case of GDH, it has been employed in the synthesis of radioactively labeled carbohydrates starting from K^{13}CN and formaldehyde (Fig. 16.2-16)[77, 78]. Depending on the substrates, single- or double-labeled glucose, fructose or sorbose are available by the sequence outlined in Fig. 16.2-16.

16.2.2.11
Lactate Dehydrogenase (LDH, E.C. 1.1.1.27)

LDH was used to catalyze the deracemization of lactate in a very elegant electrochemical approach. The driving force of the endergonic reaction was supplied by anodic regeneration of NAD$^+$ and cathodic reduction of pyruvate (Fig. 16.2-17)[79, 80]. Thus, both LDH products were removed efficiently, avoiding product inhibition. The electrochemical reduction of pyruvate leads to racemic lactate, producing 50% of the desired product and 50% of "new" substrate for LDH. An

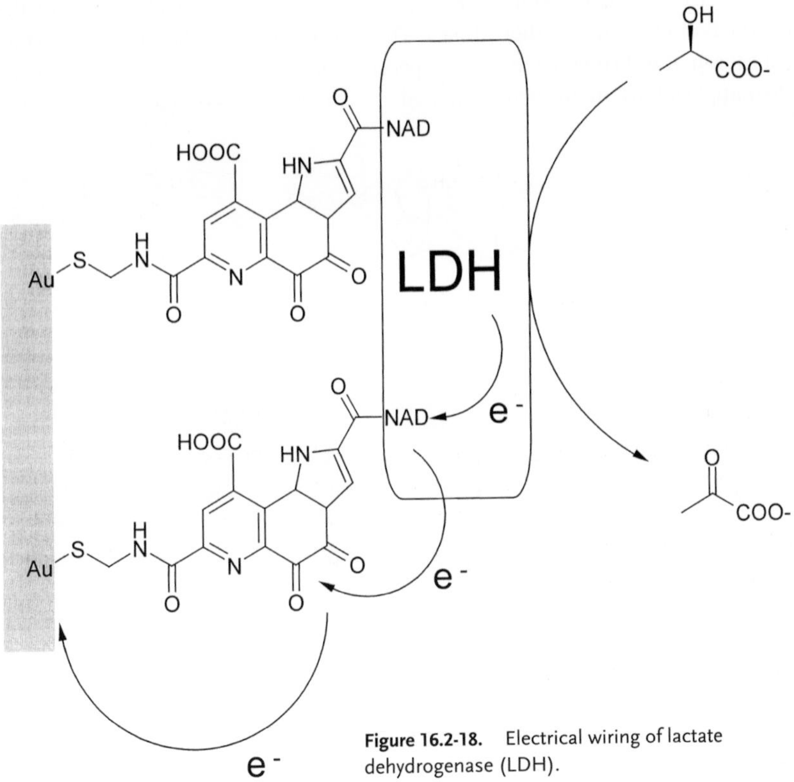

Figure 16.2-18. Electrical wiring of lactate dehydrogenase (LDH).

interesting approach to direct "electrical wiring" of LDH to an electrode was reported recently (Fig. 16.2-18) [81].

NAD was covalently linked via a PQQ spacer to a gold electrode. This modified electrode is capable of binding LDH over the exposed nicotinamide groups. Upon oxidation of lactate to pyruvate the excess electrons tunnel from NADH in the active site to PQQ and eventually to the anode. Thus, a kind of electrical linkage between the enzyme and the electrode is established. The enzymes were crosslinked, as LDH is a homotetramer and might dissociate during the reaction. This approach is not only useful for electrochemical biosensors but might be transferred to other oxidoreductase reactions.

16.2.2.12
Carbohydrate Dehydrogenases

Many so-called polyol dehydrogenases have been reported in literature, for example various glucose dehydrogenases, mannitol dehydrogenase, fructose dehydrogenase, and uridine-5'-diphosphoglucose dehydrogenase. Glucose dehydrogenase (E.C. 1.1.1.47) was applied for the production of D-gluconic acid in a plug-flow reactor with direct electrochemical regeneration of NAD^+ [82]. Glucose-6-phosphate dehydrogenase (E.C. 1.1.1.49) is a common regeneration enzyme for NADPH [69]. Most polyol dehydrogenases are not specific for their native substrate, but also catalyze the oxidoreduction of various carbohydrates. Thus, they can be applied for the production of (non-)natural sugars which are especially valuable in the sweetener industry. Yet their applications are limited compared to the polyol oxidases (see Sect. 16.2.3)

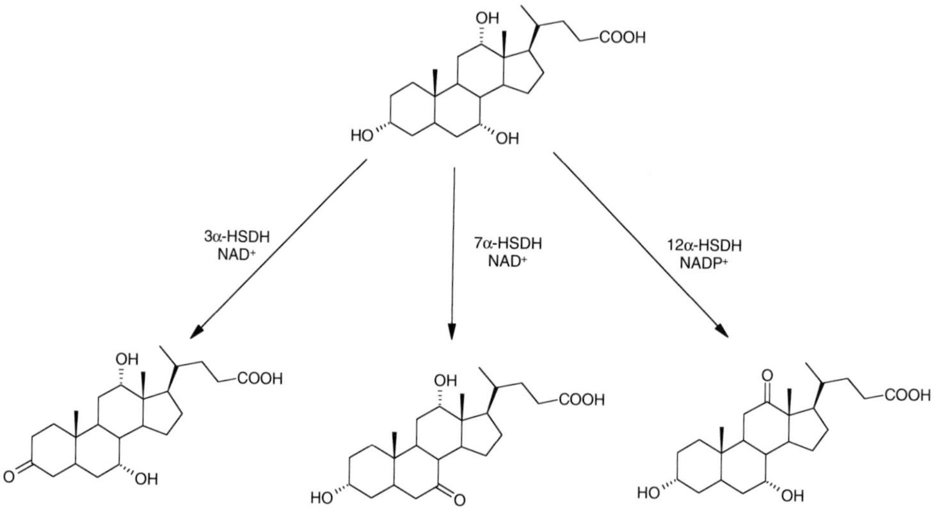

Figure 16.2-19. Regioselective oxidation of cholic acid by hydroxysteroid dehydrogenases.

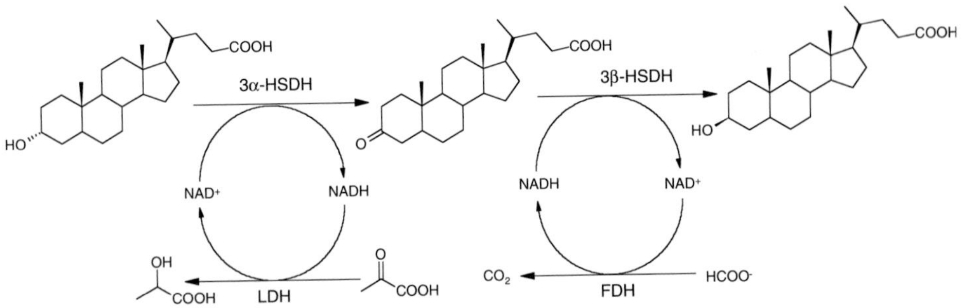

Figure 16.2-20. 3α- and 3β-hydroxysteroid dehydrogenase (HSDH) catalyzed stereoinversion in steroids.

16.2.2.13
Hydroxysteroid Dehydrogenases (HSDH)

The hydroxysteroid dehydrogenases comprise another group of synthetically interesting dehydrogenases. For many hydroxylated positions of the steroid backbone, individual NAD(P)$^+$ dependent dehydrogenases exist, which selectively oxidize the respective residue.

For example, the three hydroxy groups of cholic acid in the 3-,7-, and 12-positions can all be oxidized regioselectively (Fig. 16.2-19) [83–85].

In addition to the regioselective oxidation of the hydroxy groups in virtually every position, a discrimination of the absolute stereochemistry can be achieved by various α- or β-selective HSDHs. Thus, the stereoinversion of various steroids was achieved by successive oxidation at position 3 with 3α-HSDH and subsequent reduction with 3β-HSDH (Fig. 16.2-20) [84]. Hydroxy functions in other positions were not modified, and the products at the end of the sequence were essentially pure. Because of the low solubility of the reactants, biphasic systems with ethyl (butyl) acetate as organic solvents were used as reaction media.

16.2.2.14
Other Dehydrogenases

In addition to the alcohol dehydrogenases mentioned above, ADHs from various other sources were examined, especially with respect to increased stability, resistance to organic solvents, and catalytic properties.

A NAD$^+$ dependent ADH isolated from *Sulfolobus solfataricus* was found to exhibit better thermostability than HLADH [$t^{1}/_{2}$ (60 °C) = 20 h] together with a distinctive preference for (S)-alcohols (complementary to HLADH) [86]. The enzyme has a broad substrate specificity that includes linear and branched primary alcohols and linear and cyclic secondary alcohols [48]. The highly purified enzyme exhibits a specific activity of 4 U mg^{-1} (for benzyl alcohol at 65 °C) [87, 88]. To date, this enzyme is not commercially available.

Hummel *et al.* established a new route to enantiomerically pure alcohols by the

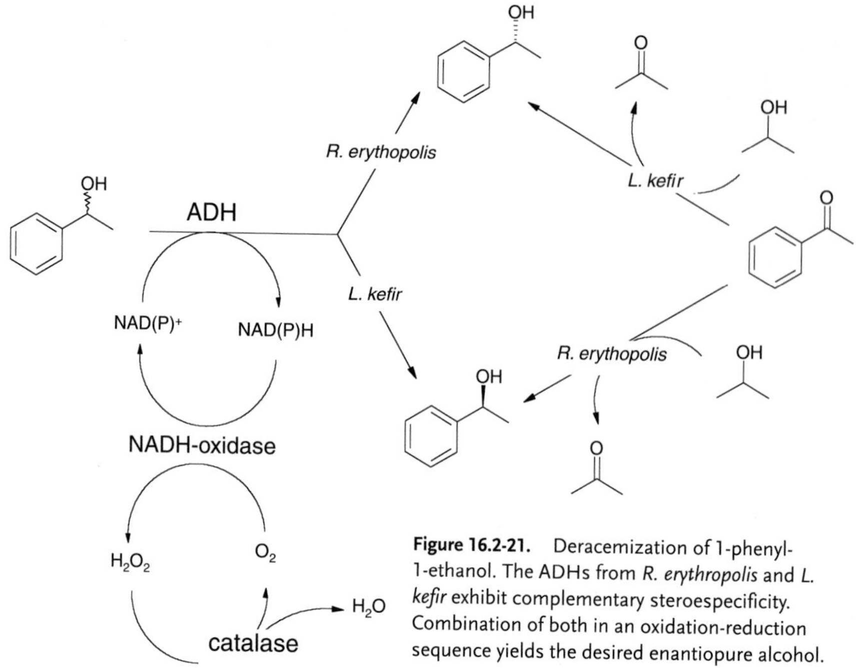

Figure 16.2-21. Deracemization of 1-phenyl-1-ethanol. The ADHs from *R. erythropolis* and *L. kefir* exhibit complementary stereospecificity. Combination of both in an oxidation-reduction sequence yields the desired enantiopure alcohol.

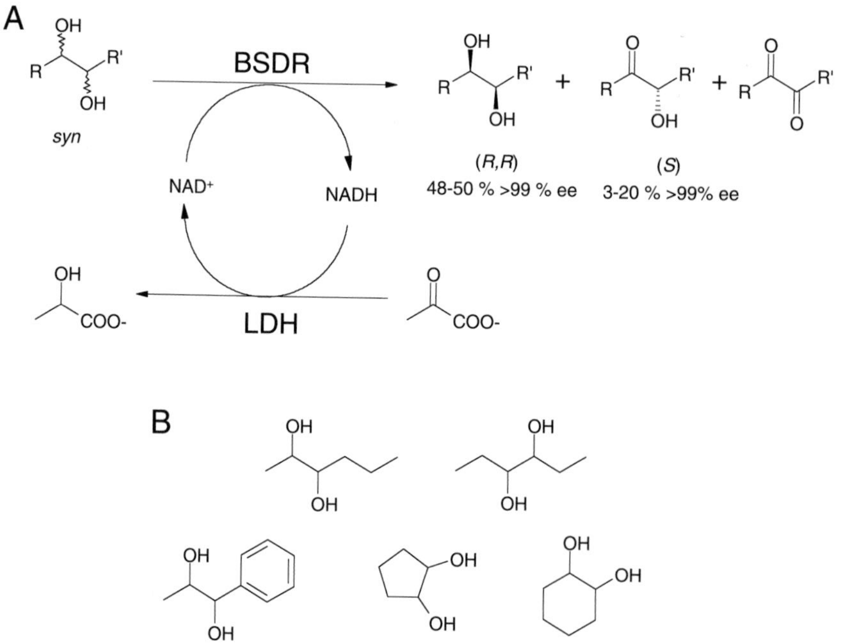

Figure 16.2-22. Kinetic resolution of racemic *syn*-diols by *Bacillus stearothermophilus* diacetyl reductase (BSDR). A: reaction with LDH-catalyzed regeneration of NAD⁺; B: selection of *syn*-diols applied.

combination of a (R)-specific, NADP-dependent ADH from *Lactobacillus kefir* and a (S)-specific, NAD-dependent ADH from *Rhodococcus erythropolis*[89]. In a first step, a kinetic resolution yielded 50% of the desired alcohol. Subsequently the ketone was reduced with the suitable ADH, finally yielding the desired optically pure enantiomer in 100% yield (Fig. 16.2-21).

Recently, diacetyl reductase (Acetoin reductase, E.C. 1.1.1.5) from *Bacillus stearothermophilus* (BSDR) was reported to be a powerful catalyst in the oxidative kinetic resolution of *vic*-diols (Fig. 16.2-22)[90]. All *syn*-diols tested yielded the enantiopure (R,R) diols in almost maximum theoretical yields, α-hydroxy ketones were largely further oxidized to the corresponding diketones. Oxidation of *vic-anti* diols only gave *ee* values in the range of 62–76%.

16.2.3
Oxidases as Catalysts

16.2.3.1
General Remarks

Oxidases utilize molecular oxygen as terminal electron acceptor. This can be considered as aerobic regeneration of the prosthetic group of the oxidase. At first glance, this seems to offer a simpler enzymatic oxidation procedure compared to the coenzyme-dependent dehydrogenases or monooxygenases. However, with few exceptions such as cytochrome c oxidase[91], some NADH oxidases[92] or laccases[93], which reduce molecular oxygen directly to water in an overall four-electron transfer step, O_2 reduction generally leads to hydrogen peroxide (transfer of two electrons) or to the superoxide radical anion (transfer of one electron) as primary reduction products.

16.2.3.2
Methods to Diminish/Avoid H_2O_2 formation

Autoregeneration of oxidases with concomitant catalase-catalyzed disproportionation of hydrogen peroxide is a simple and effective regeneration method (Fig. 16.2-23); it is quite commonly used with oxidase reactions.

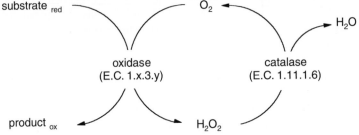

Figure 16.2-23. Coupling of oxidase autoregeneration and catalase for dismutation of hydrogen peroxide.

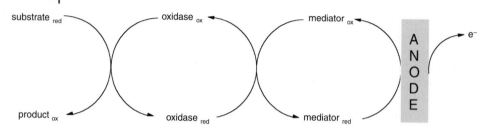

Figure 16.2-24. Indirect electrochemical regeneration of an oxidase.

Hydrogen peroxide, however, is highly reactive and irreversibly inhibits enzyme activity (also catalase) even in low concentrations.

Hydrogen peroxide can be avoided if excess electrons are transferred to the anode. However, direct electron transfer between enzymes and solid electrodes is usually very slow because the enzymatic active sites are often deeply buried within the protein shell and therefore inaccessible for the electrode (the tunneling probability of electrons is a function of distance). In order to accelerate the electron transfer, low molecular weight redox active substances can be used to shuttle the electrons between the enzyme and the electrode. This indirect electrochemical enzyme regeneration is represented schematically in Figure 16.2-24.

For the anaerobic electrochemical regeneration of a given oxidase, a suitable mediator can be chosen from various organometallic complexes, especially ferrocenes [94–102], but also bipyridine/phenanthroline, terpyridine, or hexacyano complexes [103, 104]. Also, quinoid salts such as TTF/TCNQ (tetrathiofulvalene/tetracyanoquinodimethane) [105, 106] as well as benzoquinones [107] and redox dyes such as phenazine and phenothiazine derivatives (MPMS, thionin, azure A, and azure C) [108] proved to be useful redox agents for indirect electron transfer. Even incorporation of oxidases into conducting polymers made of polypyrrole or polythiophene derivatives proved to function for electrochemical regeneration [109].

It should be mentioned at this point that most of the research in the field of electrochemical oxidase regeneration concentrates on analytical applications, inspired by the search for electrochemical biosensors [110].

However, it was demonstrated that indirect electrochemical methods are suitable for prolonging oxidase operational stability. In a particular example, glucose oxidase (E. C. 1.1.3.4) was immobilized on a carbon felt anode and regenerated with the benzoquinone/hydroquinone redox couple (Fig. 16.2-25) [107]. Thus, the operational stability of glucose oxidase could be increased at least 50 times compared to the use of molecular oxygen as oxidant. Productivities as high as $100 \text{ g h}^{-1} \text{ L}^{-1}$ were reached.

One disadvantage of the electrochemical methods is the need for rather elaborate equipment. Recently, Baminger *et al.* proposed a novel concept of enzymatic regeneration of a range of redox mediators including quinones and various redox dyes [93]. Instead of reoxidizing these mediators via the anode, laccases are employed. Laccases (E. C. 1.10.3.2) are multi-copper oxidases [111] that are found in various trees and fungi [112, 113]. Laccases catalyze the oxidation of various structurally diverse

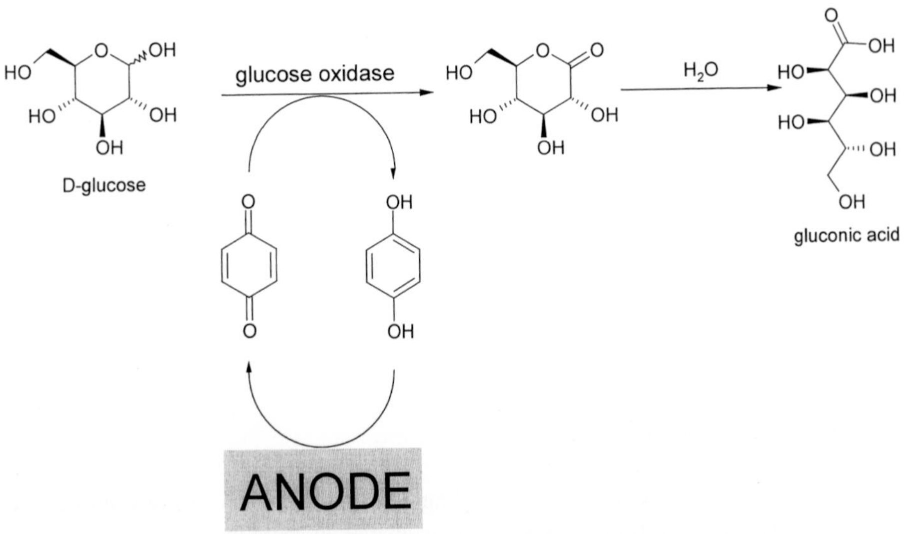

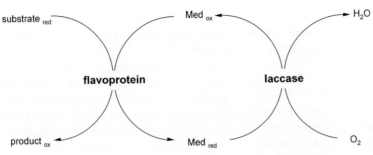

Figure 16.2-25. Indirect electro-enzymatic oxidation of glucose using glucose oxidase.

Figure 16.2-26. Laccase-based regeneration concept for oxidized flavoproteins (oxidases).

substances with concomitant reduction of molecular oxygen to water[114], thus avoiding the generation of hazardous hydrogen peroxide (Fig. 16.2-26).

This regeneration concept was tested with pyranose oxidase (P2O, E.C. 1.1.3.10)[93]. Interestingly, it was found that P2O shows higher affinity for some mediators than for O_2 (K_M value for 1,4-benzoquinone is 120 mM compared to 650 mM for O_2) with otherwise comparable activities yielding a 6 times higher k_{cat}/K_M value. Preparative scale biotransformations could be performed with two-fold volumetric productivities. The TTNs were 1.1×10^6 for P2O, with a residual activity of 85%, and 800 for 1,4-benzoquinone. Similar results were obtained with the enzyme cellobiose-dehydrogenase (E.C. 1.1.99.18), which is incapable of autoregeneration, in combination with ABTS or DCIP and laccase.

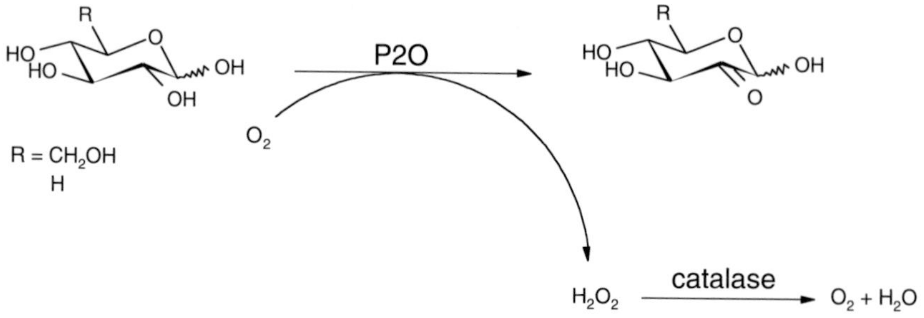

Figure 16.2-27. Oxidation of carbohydrates specifically at C-2 by pyranose oxidase (P2O).

16.2.3.3
Pyranose Oxidase (P2O, E. C. 1.1.3.10)

P2O is common among wood-degrading basidiomycetes[115]. It has been isolated and characterized from various microorganisms[116]. Although the substrate specificity varies to some extent among the P2Os isolated from different fungi, P2Os have some properties in common, such as the homotetrameric structure with covalently bound FAD. The main metabolic role of P2O appears to be as a constituent of the fungal ligninolytic system that provides the lignin-degrading lignin peroxidase and manganese peroxidase with hydrogen peroxide[117].

Natural substrates of P2O are probably D-glucose, D-galactose, and D-xylose, which are abundant in lignocellulose and which are oxidized to the corresponding 2-keto sugars. In addition, P2O exhibits significant activity with a number of other carbohydrates[118]. During such oxidations, electrons are transferred to molecular oxygen, yielding hydrogen peroxide. In addition, benzoquinones, 2,6-dichloroindophenol, as well as ABTS were reported to function as electron acceptors[93, 119]. Interestingly, up to 11-fold increased reactivity (compared to molecular oxygen as electron acceptor) was found.

P2O is currently used in various analytical applications, e. g., in clinical chemistry for the determination of 1,5-anhydro-D-glucitol, an important marker for glycemic control in diabetes patients[116], or in amperometric biosensors for the detection of monosaccharides[120, 121]. For the last two decades, P2O has received increased attention as the key catalyst in several biotechnological applications. Only a few can be mentioned here.

The essential structural requirements of substrates for P2O are the six-membered ring of pyranoid saccharides and an equatorially orientated 2-OH group[122]. In some cases regioselective oxidation at C-3 was observed[118]. The general reaction scheme is given in Fig. 16.2-27. Table 16.2-6 gives a selection of preparative oxidations reported with P2O.

The "Cetus process"
P2O is involved in the so-called "Cetus process", in which D-fructose is produced from cheap D-glucose (Fig. 16.2-28).

Table 16.2-6. Substrates and oxidation products of pyranose oxidase.

Substrate	Product	Yield [%][a]	Activity [%][a]	References
	Oxidation of monosaccharides			
D-Glucose	D-Glucosone	100	100	[15]
D-Allose	D-Ribo-hexos-2-ulose	94	40	[15]
D-Galactose	D-Lyxo-hexos-2-ulose	70	8	[15]
D-Ribose	D-Erythro-pentos-2-ulose	5	very low	[15]
3d-D-Glucose	3d-D-Erythro-hexos-2-ulose	100	96	[15]
6d-D-Glucose	6d-D-Ribo-hexos-2-ulose	100	92	[15]
1,5-Anhydro-D-glucitol	1,5-Anhydro-D-fructose	100	75	[15]

Table 16.2-6. (cont.).

Substrate	Product	Yield [%][a]	Activity [%][a]	References
	Oxidation of Disaccharides			
 Allolactose	 Allolactulose	100	ND[b]	[16]
 Meliobiose	 Meliobiulose	100	ND[b]	[16]
 Gentiobiose	 Gentiobiulose	100	ND[b]	[16]
 Isomaltose	 Palatinose	100	ND[b]	[16]

a expressed as percentage of yield and activity of D-glucose oxidation; **b** ND: not determined.

15 S. Freimund, A. Huwig, F. Giffhorn, S. Köpper, *Chem. Eur. J.* **1998**, *4*, 2442–2455.

16 C. Leitner, P. Mayr, S. Riva, J. Volc, K. D. Kulbe, B. Nidetzky, D. Haltrich, *J. Mol. Cat. B: Enzymatic* **2001**, *11*, 407–414.

Hydrogen peroxide is not merely dismutated by catalase, but used as substrate in a second enzyme cascade reaction producing propylene oxide [123–125]. In an alternative process [126] the reduction step was performed enzymatically using aldose reductase and formate dehydrogenase for NADH regeneration. Thus, essentially glucose free D-fructose was obtained.

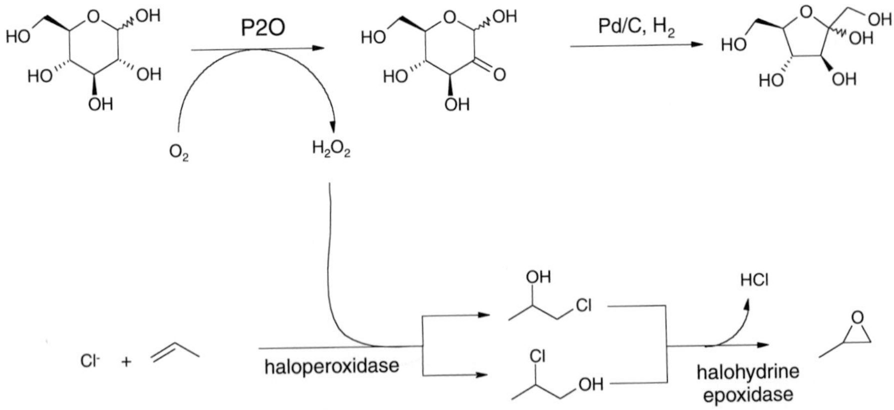

Figure 16.2-28. Isomerization of D-glucose to D-fructose with pyranose oxidase (P2O) and coupling of hydrogen peroxide to a synthetic reaction (Cetus process).

Table 16.2-7. Kinetic resolution of some racemic 2-hydroxy acids to the (R)-2-hydroxy acids and the corresponding 2-keto acids [17].

Substrate	Yield [%]	ee [%]
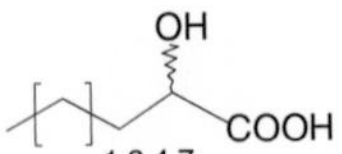 OH, COOH, 1,2,4,7	49–50	> 98
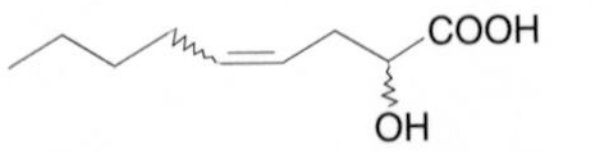 COOH, OH, cis/trans	50	> 99
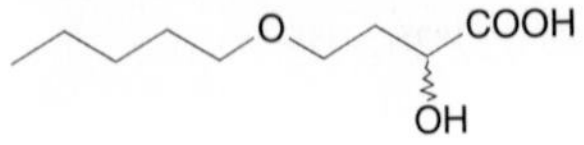 O, COOH, OH	47	86

17 W. Adam, M. Lazarus, B. Boss, C. R. Saha-Möller, H.-U. Humpf, P. Schreier, *J. Org. Chem.* **1997**, *62*, 7841–7843.

16.2.3.4
Glycolate Oxidase (E. C. 1.1.3.15)

Glycolate oxidase is a peroxisomal enzyme that is found in the leaves of many green plants and in the liver of mammalians. The enzyme isolated and for economic reasons only partially purified from spinach (*Spinacia oleracea*) was applied to the enantioselective oxidation of various 2-hydroxy acids yielding the corresponding 2-keto acid and the remaining (R) alcohol [127]. Enantiopure 2-hydroxy acids are valuable building blocks in the synthesis of glycols [128], haloesters [129] or epoxides [130]. Unless the steric demand of the substituents close to the alcohol function is too big, the oxidation proceeds smoothly to the full theoretical conversion with enantiomeric excesses of the alcohols usually in the range of 98–99% (Table 16.2-7).

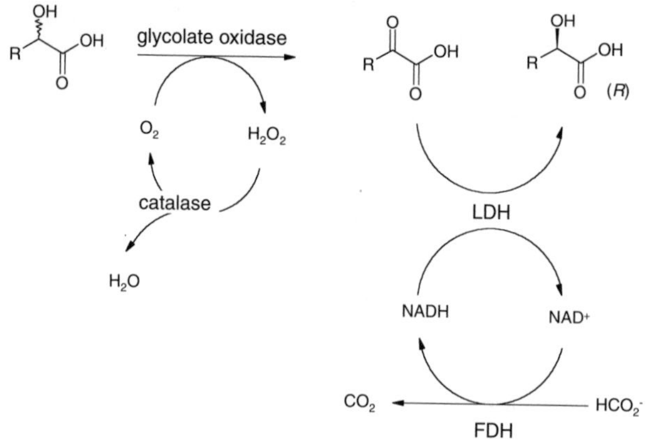

Figure 16.2-29. Deracemization of racemic 2-hydroxy acids in a combination of glycolate oxidase and lactate dehydrogenase (LDH).

Table 16.2-8. Conversion of racemic 2-hydroxy acids into (R)-2-hydroxy acids by the combined action of glycolate oxidase and D-lactate dehydrogenase[18].

Substrate	Oxidase [U]	Dehydrogenase [U]	Reaction time [h]	Yield [%]	ee [%]
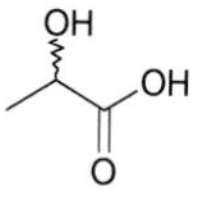	2	450	66	100	> 99
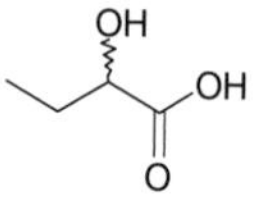	2	900	210	100	94

One unit (U) is defined as the amount of enzyme which converts 1 μmol of substrate per minute.

18 W. Adam, M. Lazarus, C. R. Saha-Möller, P. Schreier, *Tetrahedron Asymmetry* **1998**, 9, 351–355.

Kinetic resolutions have a maximum yield of only 50%. Therefore, a second enzymatic process was added after completion of the glycolate oxidase-catalyzed kinetic resolution[131]. By addition of D-lactate dehydrogenase (E.C. 1.1.1.28) together with formate dehydrogenase for NADH regeneration, enantiospecific reduction of the 2-keto acid was achieved. Overall, a quantitative transformation (deracemization) of the racemic 2-hydroxy acid into the corresponding (R)-2-hydroxy acid was achieved (Fig. 16.2-29).

Unfortunately, this process cannot be performed in a more elegant and more efficient one-pot synthesis. On the one hand, the pH optima for the three enzymes are not compatible with each other, and on the other, lactate dehydrogenase is air sensitive. In addition to this, glycolate oxidase also catalyzes the reverse reaction under aerobic conditions, thus lowering the ee-value. Therefore, the reaction mixture is filtered (glycolate oxidase can be reused) and, after pH adjustment, the second enzymatic transformation is performed. Table 16.2-8 shows some results of this procedure.

Glycolate oxidase has been studied thoroughly not only for specific oxidation of

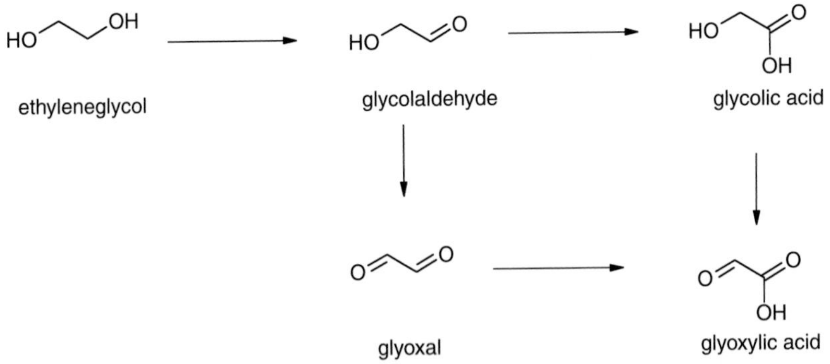

Figure 16.2-30. Sequential oxidation of ethylene glycol to glycolic acid.

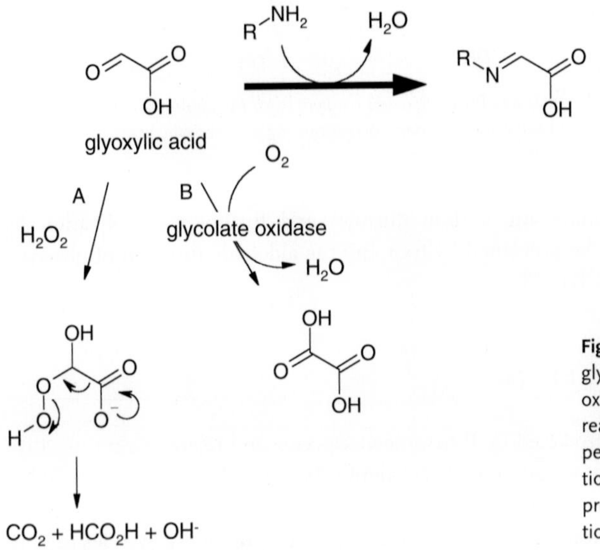

Figure 16.2-31. Synthesis of glyoxylic acid by glycolate oxidase. The undesired side-reactions (A) with hydrogen peroxide and (B) overoxidation by glycolate oxidase are prevented by *in situ* formation of an imine.

(S)-2-hydroxy propionic acid (lactate) [132] and for the kinetic resolution of racemic 2-hydroxy acids [127, 131], but also for selective oxidations of 1,2-diols such as ethylene glycol (Fig. 16.2-30).

Reports on the specific conversion of glycolic acid into glyoxylic acid are numerous. Isobe *et al.* Introduced an *in vivo* system utilizing *Alcaligenes* sp. isolated from media containing 1,2-propanediol. By carefully adjusting the pH, a yield of 95 % was obtained [133].

DiCosimo and coworkers optimized the *in vitro* production of glyoxylic acid from glycolic acid with glycolate oxidase from spinach [134]. Improvements in operational stability as well as in productivity were achieved by enzyme immobilization either onto a solid matrix [135] or in permeabilized, metabolically inactive cells of *Pichia pastoris* or *Hansenula polymorpha*, containing overexpressed glycolate oxidase from spinach together with catalase. The undesired oxidation of glyoxylic acid by hydrogen

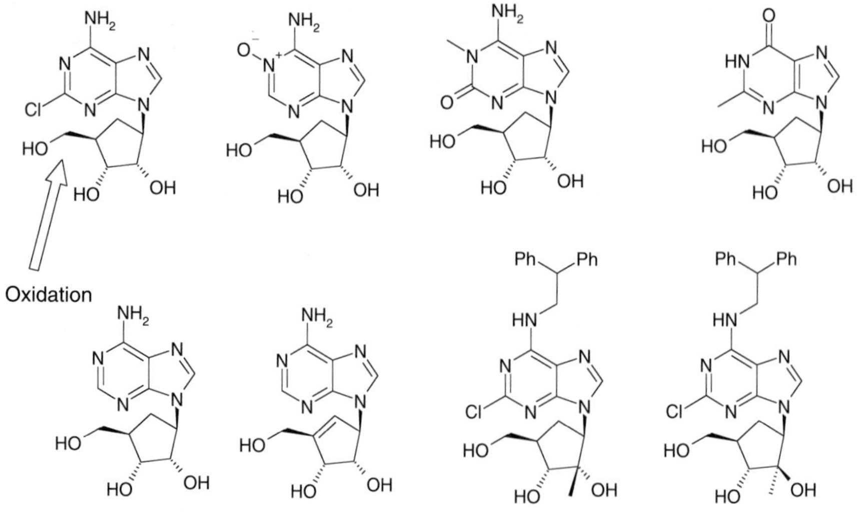

Figure 16.2-32. Non-natural substrates for nucleoside oxidase from *Pseudomonas* sp. These compounds are converted selectively to their corresponding 5'-carboxylic acids.

peroxide (yielding formate and carbon dioxide) and further metabolization by glycolate oxidase could be prevented by trapping the aldehyde function of glyoxylic acid as imine (Fig. 16.2-31) [136].

16.2.3.5
Nucleoside Oxidase (E. C. 1.1.3.28)

Nucleoside oxidase is produced by *Pseudomonas* species and related Gram negative bacteria [137]. The hetero-tetramer with covalently bound FAD oxidizes the 5'-hydroxyl group of purine and pyrimidine nucleosides to the corresponding carboxylic acids. It has found application in the analytical determination of nucleosides (e. g. in assessing food freshness) [138]. At Glaxo Wellcome R&D it found attention as key step in the production of anti-inflammatory compounds [139–141]. Several non-natural substrates were selectively converted on multi-gram scale into their 5'-carboxylic acids (Fig. 16.2-32).

The operational stability of the enzyme was improved by immobilization onto a solid matrix and especially by substitution of molecular oxygen as the primary electron acceptor by stoichiometric amounts of hydroquinone.

16.2.3.6
Glucose Oxidase (E. C. 1.1.3.4)

The most prominent of the alcohol oxidases is glucose oxidase. The dimeric flavoenzyme catalyzes the oxidation of β-D-glucose to D-glucono-δ-lactone, a reaction that has attracted the attention of generations of analytical chemists because of its possible applicability in glucose sensors for diabetes control[142]. The reaction of the stoichiometrically formed hydrogen peroxide with various dyes can be used as the analytical signal[143]. More elegant variants (that at the same time avoid the formation of hazardous hydrogen peroxide) utilize anaerobic, electrochemical regeneration with a suitable mediator. Thus, the catalytic current becomes the analytical signal. Several approaches have been reported, e.g. the utilization of freely diffusible quinones[107], the incorporation of glucose oxidase in a conducting polymer (produced from 1,4-hydroquinones and soybean peroxidase), or the immobilization of several mediators in the vicinity of the prosthetic redox center[98, 99].

Because of the high substrate specificity of glucose oxidase, which almost exclusively accepts glucose (other substrates such as D-maltose, D-xylose, or L-sorbose are converted with less than 6% of the activity on glucose[144, 145]), this oxidase has not found any synthetic application, but it is frequently used in the food industry to remove traces of molecular oxygen from vacuum sealed products. Immobilized glucose oxidase is also used for the deoxygenation of juices and beer[146].

16.2.3.7
Alcohol Oxidase (E.C. 1.1.3.13)

The aliphatic alcohol oxidase, a FAD-dependent enzyme, catalyzes the oxidation of primary short-chain alcohols to the corresponding aldehydes. Dioxygen can be replaced by synthetic acceptors such as dichlorophenolindophenol or phenazine methosulfate[147].

By utilizing an alcohol oxidase from *Pichia pastoris* or *Candida* sp.[148], almost complete conversion of ethylene glycol into glyoxal (Fig. 16.2-30) was observed. These enzymatic routes were shown to be superior in terms of reaction conditions and yields compared to the chemical variants that make use of metal catalysts or even nitric acid for the oxidation of ethylene glycol.

Recently, aliphatic alcohol oxidase was applied as dehydrated enzyme in a gas-solid bioreactor[149]; an excess amount of catalase was added to prevent oxidase inactivation.

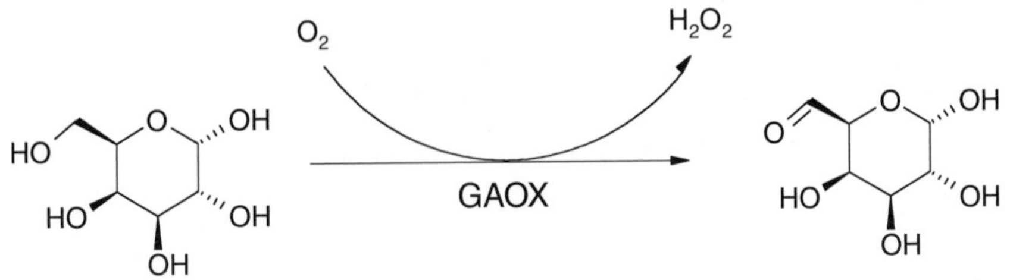

Figure 16.2-33. Galactose oxidase (GAOX) catalyzed oxidation of α-D-galactose to *meso*-galactohexodialose.

Table 16.2-9. Substrates and products of galactose oxidase.

Substrate	Product	References
D-Galactose	*meso*-Galactohexodialdose	
UDP-[14C]-Galactose	UDP-[14C]-Galacturonic acid	[19]
D,L-Threitol	D-Threose + L-Threitol	[20]
Xylitol	L-Xylose	[20]
D,L-Glucitol	L-Glucose + D-Glucitol	[20]
D,L-Galactitol	L-Galactose + D-Galactitol	[20]

Table 16.2-9. (cont.).

Substrate	Product	References
HO—CH(OH)—CH(OH)—OH	HO—CH(OH)—CH=O L(-)Glyceraldehyde	[21]
HO—CH(OH)—CH(OH)—CH₂Cl	HO—CH(OH)—CH(OH)—CH₂Cl + O=CH—CH(OH)—CH₂Cl (S)-Halodiol + (R)-Aldehyde	[21]

19 S. S. Basu, G. D. Dotson, C. R. H. Raetz, *Anal. Biochem.* **2000**, *280*, 173–177.
20 D. G. Drueckhammer, W. J. Hennen, R. L. Pederson, D. F. Barbas, C. M. Gautheron, T. Krach, C. H. Wong, *Synthesis* **1991**, *7*, 499–525.
21 A. M. Klibanov, B. N. Alberti, M. A. Marletta, *Biochem. Biophys. Res. Commun.* **1982**, *1982*, 108.

Table 16.2-10. Substrates and products in the kinetic resolution of allylic alcohols with cholesterol oxidase [22].

Substrate (R = H, OH)	Product
(cyclohexenyl methyl allylic alcohol)	(enone) + (allylic alcohol)
(decalin system, R)	(enone, R)
(decalin system, R)	(enone, R)
(decalin system, R)	No product detected

22 S. Dieth, D. Tritsch, J.-F. Biellmann, *Tetrahedron Lett.* **1995**, *36*, 2243–2246.

16.2.3.8
Galactose Oxidase (GAOX, E. C. 1.1.3.9)

Galactose oxidases belong to the group of copper-dependent oxidases. For the GAOX from *Dactylium dendroides* the existence of covalently bound pyrroloquinoline quinone (PQQ) could be shown [145]. It catalyzes the specific oxidation of the hydroxyl group in position 6 of galactose (Fig. 16.2-33) [150].

The enzyme regeneration can be performed aerobically or utilizing mediators

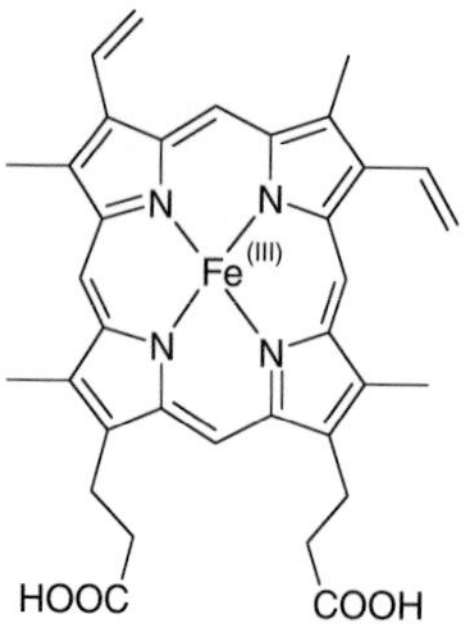

Figure 16.2-34. Ferric protoporphyrin IX as prosthetic group in most peroxidases.

such as ferrocene[102], tetracyano-iron-1,10-phenanthroline, or cobalt *tert*-pyridine complexes[103].

GAOX stereospecifically oxidizes a broad range of substrates (Table 16.2-9). In synthetic applications, the oxidation of racemic or *meso*-polyols such as D,L-threitol or xylitol to the non-native sugars are of special interest[151, 152]. In addition to the monosaccharides represented in Table 16.2-9, GAOX also converts di- or oligo-saccharides[153].

16.2.3.9
Cholesterol Oxidase (ChOX, E. C. 1.1.3.6)

ChOX from *Rhodococcus erythropolis* was applied for the kinetic resolution of racemic mono- and bicyclic allyl alcohols (Table 16.2-10)[154]. Although the substrates tested were much smaller than the native substrate cholest-4-en-3β-ol, reasonable enantio-selectivities (*E*) in the range of 7–20 were found for the (*S*) alcohols.

Both enantiomers of the alcohol (entry 1) were oxidized with moderate enantiose-lectivities (*E* = 7) for the (*S*) enantiomer. For bicyclic alcohols, the position of the hydroxyl group with respect to the methyl group is essential. Only at a relative *trans* configuration of both substituents significant oxidation occurred.

By utilizing organic redox dyes as primary electron acceptors and concomitant reoxidation at a glassy carbon electrode, amperometric biosensors for cholesterol based on cholesterol oxidase were developed[108].

16.2.4
Peroxidases as Catalysts

16.2.4.1
Introduction

Peroxidases (E. C. 1.11.1.7) are ubiquitously found in plants, microorganisms, and animals. Most peroxidases studied so far contain ferric protoporphyrin IX (proto-heme, Fig. 16.2-34) as the prosthetic group[155]. However, some peroxidases also contain selenium (glutathione peroxidase)[156], vanadium (bromoperoxidase)[157],

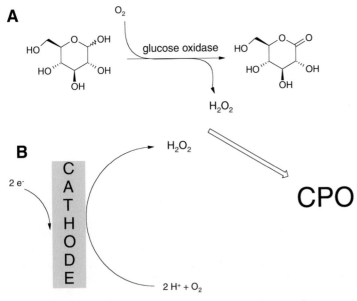

Figure 16.2-35. Methods of generating appropriate hydrogen peroxide concentrations for chloroperoxidase reactions, (A) enzymatically with glucose oxidase and (B) electrochemically by cathodic reduction of molecular oxygen.

manganese (manganese peroxidase) [158], and flavin (flavoperoxidase) [159] as prosthetic groups.

Most peroxidases accept a variety of peroxides, such as hydrogen peroxide or alkyl hydroperoxides, as oxidizing agents. The mechanism includes the activation of oxygen in a high valence iron-oxo species [155, 160].

16.2.4.2
Methods to Generate H_2O_2

At a first glance, utilization of cheap hydrogen peroxide as electron acceptor seems appealing. The major drawback, however, is the sometimes rapid inactivation of peroxidases by their substrate. For example, chloroperoxidase (CPO, E.C. 1.11.1.10) exhibits a half-life time of 38 min even at an H_2O_2 concentration of 50 µM [161].

Several approaches to controlling hydrogen peroxide at a constant low concentration have been reported. In aqueous/organic emulsions, the use of *tert*-butyl hydroperoxide is beneficial. On the one hand, the peroxide concentration is limited according to the partition coefficient, and on the other hand, *tert*-butanol was shown to exert a stabilizing effect on CPO [162].

The slow continuous addition of hydrogen peroxide results in better CPO performance [163], which can be even further improved by sensor-controlled addition of H_2O_2 [162], increasing the CPO total turnover number for indole oxidation more than 20-fold to ca. 860 000.

Table 16.2-11. Chloroperoxidase-catalyzed oxidation of some alcohols to the corresponding aldehydes.

Substrate	Yield [%]	Remarks and reference
	94	H_2O_2 or *tert*-butyl hydroperoxide as oxidants [23]
	95	H_2O_2 or *tert*-butyl hydroperoxide as oxidants [23]
	92	H_2O_2 or *tert*-butyl hydroperoxide as oxidants [23]
	quantitative	3 times higher activity with *tert*-butyl hydroperoxide in biphasic systems compared to H_2O_2 in buffer [24]
	81	[25]
	95	Production in gram-scale; low, non-enzymatic *cis/trans* isomerization observed [25]
	99	[25]
	97	[25]
	50 (40% *ee*)	Production in gram-scale, low yield with *cis*-isomer [25]
	46 (45% *ee*)	[25]
	92	[26]
	74	Quantitative conversion; significant amounts of acid as the product of overoxidation were found [26]

23 S. Hu, L. P. Hager, *Biochem. Biophys. Res. Commun.* **1998**, *253*, 544–546.

24 B. K. Samra, M. Andersson, P. Adlercreutz, *Biocat. Biotransf.* **1999**, *17*, 381–391.

25 E. Kiljunen, L. T. Kanerva, *J. Mol. Cat. B: Enzymatic* **2000**, *9*, 163–172.

26 M. P. J. van Deurzen, F. van Rantwijk, R. A. Sheldon, *J. Carbohydr. Chem.* **1997**, *16*, 299–309.

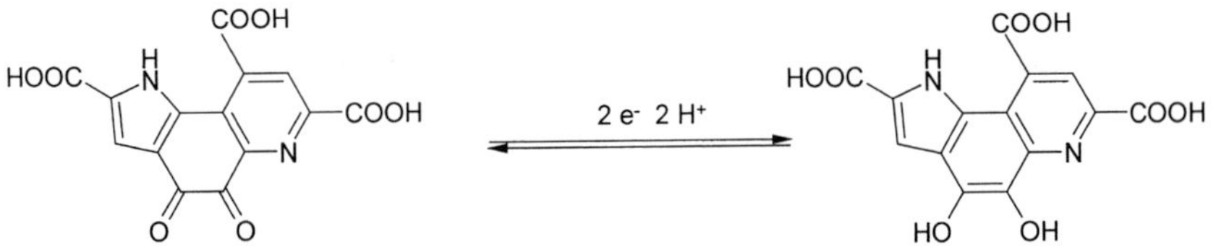

Figure 16.2-36. Pyrroloquinoline quinone (PQQ) in its oxidized and reduced form as prosthetic group for most quinoprotein dehydrogenases.

However, external H_2O_2 addition still has the disadvantage that locally high concentrations occur at the entry points, resulting in CPO inactivation at these hot spots. This can be circumvented via *in situ* generation of hydrogen peroxide. Two promising approaches have been reported so far: (i) another enzymatic reaction producing H_2O_2 e.g. with glucose oxidase[164], and (ii) electrochemical reduction of molecular oxygen (Fig. 16.2-35)[161, 165]. In both approaches, drastic increases of the number of CPO catalytic cycles up to 1.1×10^6 were achieved.

16.2.4.3
Chloroperoxidase (CPO, E. C. 1.11.1.10)

Publications on CPO-catalyzed oxidations of alcohols are rare. However, some selective oxidations of aliphatic, allylic, propagylic and benzylic alcohols to the aldehyde stage have been reported (Table 16.2-11).

16.2.4.4
Catalase (E. C. 1.11.1.6)

Most commonly, catalase is applied for the dismutation of hydrogen peroxide[166]. On reaction of catalase with one molecule of hydrogen peroxide, the intermediate high valence iron-oxo species is generated. This species, however, is a potent oxidant and readily reacts not only with a second molecule of hydrogen peroxide (yielding water and molecular oxygen) but has been reported to oxidize various other compounds such as methanol or nitrite[166].

Klibanov and coworkers enlarged the substrate spectrum by including a variety of alcohols that were oxidized to the corresponding aldehydes. Depending on the substrate and the reaction medium, high enantioselectivities are reported[167].

The generation of reactive catalase in its oxidized stage can also be achieved by direct electrochemical oxidation (transfer of electrons from ferric protoporphyrin IX to the electrode). Thus, catalase immobilized on graphite electrodes has been used for the hydrogen peroxide-free oxidation of phenol[168].

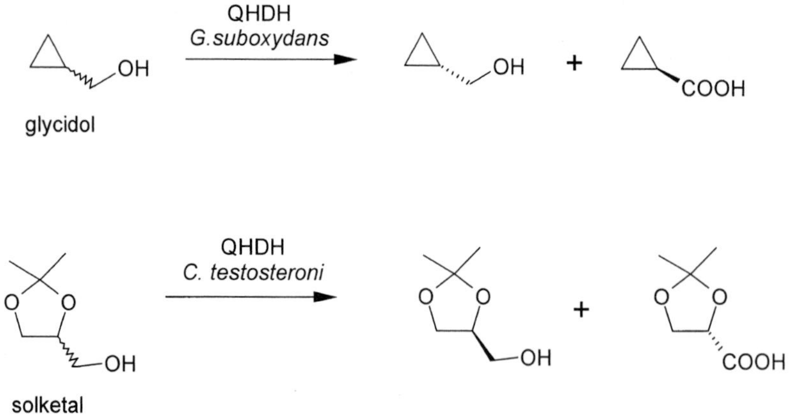

glycidol

solketal

Figure 16.2-37. Resolution of alcohols by enantioselective oxidation using quinohemoprotein dehydrogenases (QHDH) from different microorganisms.

16.2.5
Quinoprotein Dehydrogenases (QDH)

16.2.5.1
General Remarks

Quinoproteins constitute a class of dehydrogenases distinct from the nicotinamide- and flavin-dependent oxidoreductases [169]. They use different quinone cofactors to convert a vast variety of alcohols and amines into their corresponding carbonyl products [170]. Proteins containing the cofactor pyrroloquinoline quinone (PQQ) (Fig. 16.2-36) form the largest and best-characterized sub-group.

QDHs are independent from classical coenzymes like NAD(P)$^+$. The substrate electrons are preferentially transferred to organic acceptors (quinones) and non-native redox mediators such as phenazine derivatives, DCPIP, Wursters blue [171], ferrocene [101], ferricyanide [172, 173], osmium complexes [174], or direct contact to an electrode [175].

One advantage of the PQQ-dependent dehydrogenases over the NAD(P)-dependent dehydrogenases is the more positive redox potential of the PQQ/PQQH$_2$ couple (+ 90 mV/pH 7 [176] compared to – 320 mV [177, 178]).

Similarly to the flavin-dependent reactions, several mechanisms have been discussed, including covalent substrate-PQQ intermediates or hydride transfer [179–181]. The most important QDHs are methanol (alcohol) dehydrogenase (E. C. 1.1.99.8) and glucose dehydrogenase (E. C. 1.1.99.17), which will be discussed briefly.

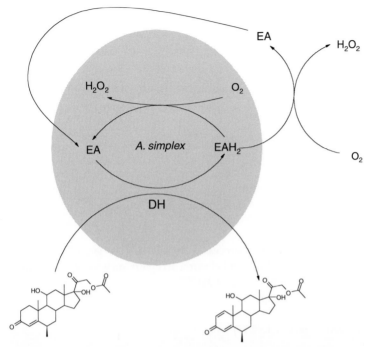

Figure 16.2-38. Δ¹-dehydrogenation of 6-α-methyl-hydrocortisone-21-acetate with polyurethane-entrapped *Arthrobacter simplex* cells in buffer-saturated 1-decanol. The dehydrogenase (DH) activity is largely increased on addition of quinoid electron acceptors (EA).

16.2.5.2
Methanol Dehydrogenase (E. C. 1.1.99.8)

In addition to PQQ, the methanol dehydrogenases from *Comamonas testosteroni* and *Gluconobacter suboxydans* contain a heme group, which is indicated in their synonym quinohemoprotein dehydrogenase.

The regeneration of these enzymes has been achieved by anodic reoxidation of ferricyanide[173], Os-modified anodes[174], or even direct contact to the anode[175].

Quinohemoprotein dehydrogenases (from *Comamonas testosteroni* and *Gluconobacter suboxydans*) have been reported to oxidize the alcohols solketal and glycidol (Fig. 16.2-37) enantioselectively[172].

Alcohol oxidases from various strains, and especially NAD(P) dependent dehydrogenases (except HLADH together with thio-NAD$^+$ [182]), were found to be extremely inefficient for the oxidations in Figure 16.2-37, a fact, which is attributed to the significantly lower redox potential of the NAD(P)$^+$/NAD(P)H redox system[172].

The QDH from *C. testosteroni* was further characterized[183]. It oxidizes stereospecifically the (R) enantiomer of secondary alcohols. Both, k_{cat}/K_M and E increased with the substrate chain length. *In vitro*, ferricyanide was used as sacrificial electron acceptor. *In vivo*, the excess electrons are most probably transferred to molecular

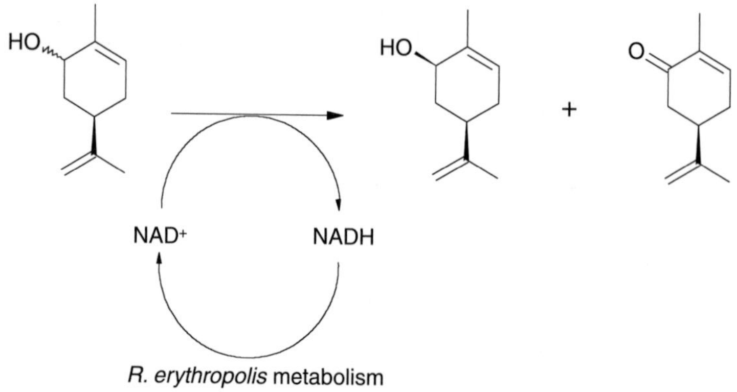

R. erythropolis metabolism

Figure 16.2-39. Enantiospecific oxidation of racemic carveol to (−)-carvone and (−)-cis carveol using whole cells of Rhodococcus erythropolis.

oxygen via the respirator chain. This process is considerably accelerated (by a factor of 12) upon addition of external quinoid electron acceptors such as vitamin K (that are capable of autoregeneration) (Fig. 16.2-38) [184].

16.2.5.3
Glucose Dehydrogenase (E. C. 1.1.99.17)

So far, a membrane-bound [185] and a soluble glucose dehydrogenase [186] have been identified. The latter oxidizes a wide range of mono- and disaccharides [186]. In addition to cytochrome b_{562}, regeneration with artificial acceptors such as DCPIP or ferrocene [187, 188] is effective and unproblematic, as no autoregeneration with molecular oxygen (producing reactive O-species) is possible. It has commercial interest as a component of glucose test strips for diabetes control [189].

16.2.6
Whole-Cell Oxidations

16.2.6.1
Stereoselective Oxidation of (−)-Carveol to (−)-Carvone [190]

By using whole cells of *Rhodococcus erythropolis* DCL14, a racemic mixture of (−)-carveol was converted to (−)-carvone and (−)-*cis*-carveol (Fig. 16.2-39). The system was optimized using the two-liquid concept, in which a second organic phase serves as substrate and product reservoir. (−)-Carvone is an important flavor compound.

The enzyme responsible for this bioconversion, catalyzed by wild-type cells of *Rhodococcus erythropolis* DCL14, is carveol dehydrogenase [191]. A high enantiose-lectivity and no further conversion of (−)-carvone was obtained. Carveol dehy-drogenase has a broad substrate specificity and prefers substituted cyclohexanols as substrates [191]. The regeneration of the cofactor NAD$^+$ was accomplished by the use of living cells.

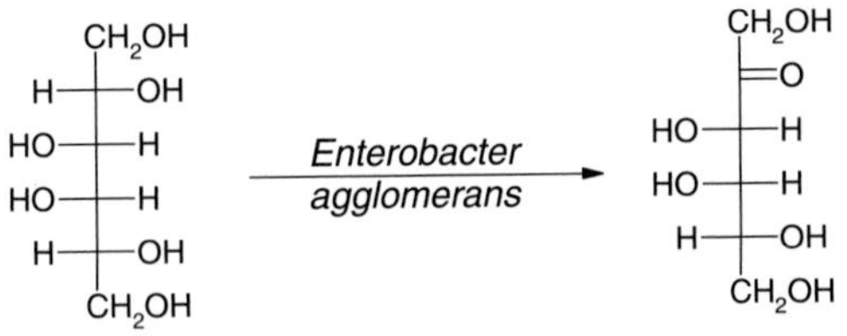

Figure 16.2-40. Production of the low calorie sweetener tagatose from D-galactitol by whole cells of *Enterobacter agglomerans*.

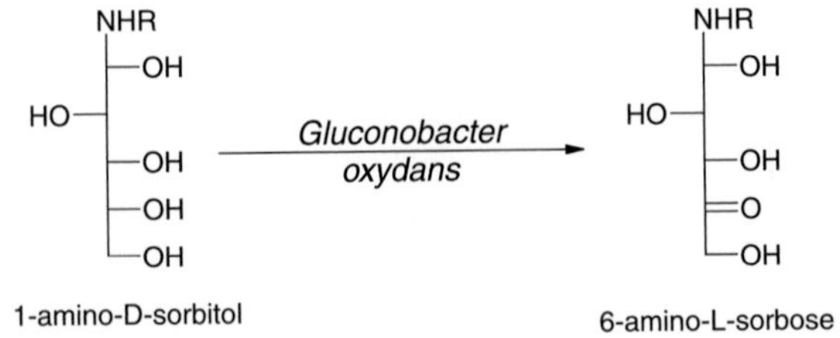

Figure 16.2-41. Oxidation of N-protected 1-amino-D-sorbitol to 6-amino-L-sorbose using *Gluconobacter oxydans*.

The use of a two-liquid phase system consisting of a 1:1 mixture of phosphate buffer and dodecane resulted in an increase of the initial (–)-*trans*-carveol conversion rate by 70% (to 26 nmol per minute and per mg protein). The production was increased from 4.3 to 208 µmol (–)-carvone formed per mg protein as compared to the aqueous system. A simple downstream process consisting of phase separation, methanol extraction, evaporation, and separation of (–)-*cis*-carveol and (–)-carvone over a silica gel column, was developed.

In another study, *Rhodococcus globerulus* PWD8 was found to oxidize D-limonene regio- and enantioselectively via (+)-*trans*-carveol to (+)-carvone[192].

16.2.6.2
Sugar Dehydrogenases Applied in Whole Cells

Cofactor regeneration by the cell metabolism is the main advantage of whole cells in polyalcohol oxidations. The induction of whole-cell biocatalyst activity is dependent on the nature of the growth substrate. An example is the production of the low calorie carbohydrate sweetener tagatose from D-galactitol (Fig. 16.2-40). As biocatalysts, wild-type strains of *Enterobacter agglomerans* and *Gluconobacter oxydans* DSM 2343, in which sugar dehydrogenases catalyze the reaction of interest, were described[193, 194].

In the case of *Enterobacter agglomerans*, cells growing on 1% glycerol plus 1% erythritol resulted in the best biocatalytic performance. In 30 h, galactitol (50 g/L) was converted with a tagatose yield of 86%. Immobilization and storage at – 20 °C are possible.

With *Gluconobacter oxydans*, growing cells were found to be more effective than resting cells. Furthermore, galactitol adaptation gave a notable increase in tagatose yield.

Another example is the oxidation of 1-amino-D-sorbitol (N-protected) to 6-amino-L-sorbose (Fig. 16.2-41)[195]. This reaction was published as a step in the synthesis of

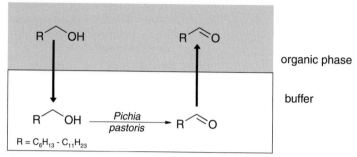

Figure 16.2-42. Selective oxidation of linear and branched aliphatic alcohols to the corresponding aldehydes using *P. pastoris* in aqueous/organic reaction mixtures.

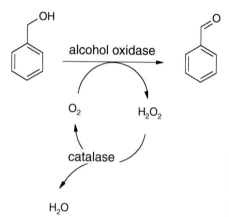

Figure 16.2-43. Oxidation of benzylic alcohol to benzaldehyde using whole cells of *P. pastoris* in organic/aqueous emulsions or with purified alcohol oxidase. *In vitro* hydrogen peroxide was removed by catalase.

1-desoxynojirimycin. Derivatives of 1-desoxynojirimycin are pharmaceuticals for the treatment of carbohydrate metabolism disorders (e. g. diabetes mellitus). Suspended whole cells of *Gluconobacter oxydans* were used as the biocatalyst, in which D-sorbitol dehydrogenase is responsible for this biotransformation.

To prevent undesired follow-up reactions of 6-amino-L-sorbose in water, the amino group has to be protected by, for example, a benzyloxycarbonyl group (R). Cells are produced by fermentation on sorbitol and used for the bioconversion step as resting cells in water without added nutrients. The biotransformation is carried out by Bayer in a 10 000 L reactor with 90 % yield.

16.2.6.3
Oxidation of Aromatic and Aliphatic Alcohols to Corresponding Aldehydes and Acids

Flavin-containing alcohol oxidase combined with catalase in peroxisomes of *Pichia pastoris* naturally catalyzes the oxidation of methanol to formaldehyde. *In vivo* and *in vitro* applications are possible. The alcohol oxidase has a broader substrate specificity than the subsequent enzymes of the methanol degradation pathway. Therefore,

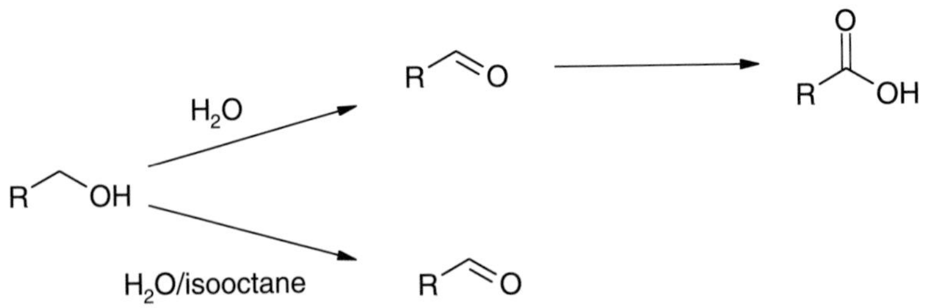

Figure 16.2-44. Oxidation of alcohols by whole cells of *Acinetobacter*. In aqueous media the oxidation proceeds until the acid stage, whereas the aldehyde is accumulated in the presence of organic solvents.

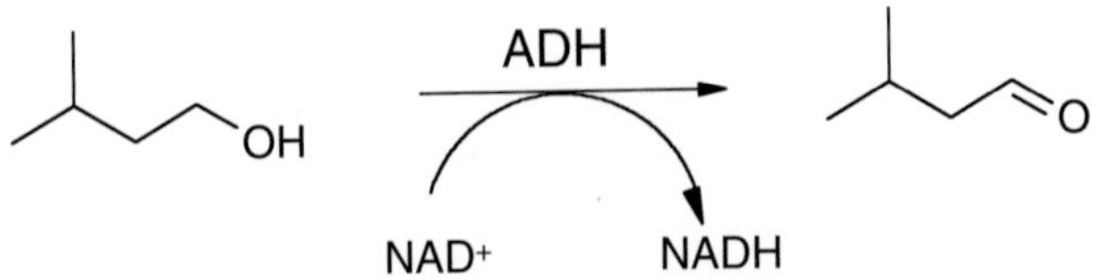

Figure 16.2-45. Preparation of isovaleraldehyde using an alcohol dehydrogenase (ADH) in whole cells of *Gluconobacter oxydans*.

products other than formaldehyde are not degraded further. The spectrum of alcohols oxidized by whole cells of *Pichia pastoris* includes aliphatic C_1-C_5 alcohols (saturated, unsaturated or branched). In biphasic media, *Pichia pastoris* also oxidizes C_6-C_{11} alcohols, phenylethyl alcohol and 3-phenyl-1-propanol [196].

Up to 70 g/L acetaldehyde was produced from ethanol [197–199]. Here, competitive product inhibition was partially overcome by high Tris buffer concentrations. Tris is able to bind acetaldehyde and markedly improve reaction yields. In a biphasic system consisting of 97 % hexane and 3 % aqueous phase, hexanol (11 g/L) was converted to hexanal (Fig. 16.2-42) within 24 h at a yield of 96 % [196]. In another example, benzyl alcohol was oxidized by whole cells of *Pichia pastoris* and purified alcohol oxidase (Fig. 16.2-43) [200]. For this reaction the importance of solute partitioning in the biphasic reaction system was studied [201]. With immobilized cells in organic (xylene)/aqueous media, benzaldehyde concentrations up to 30 g/L were reached in the organic phase [201]. With purified alcohol oxidase, up to 45 g/L benzaldehyde was produced within 8 h and with an enzyme concentration of 0.94 g/L.

Dehydrogenases of Acinetobacter and Gluconobacter strains catalyze the oxidation of various alcohols to corresponding aldehydes and acids *in vivo*. Substrates tested

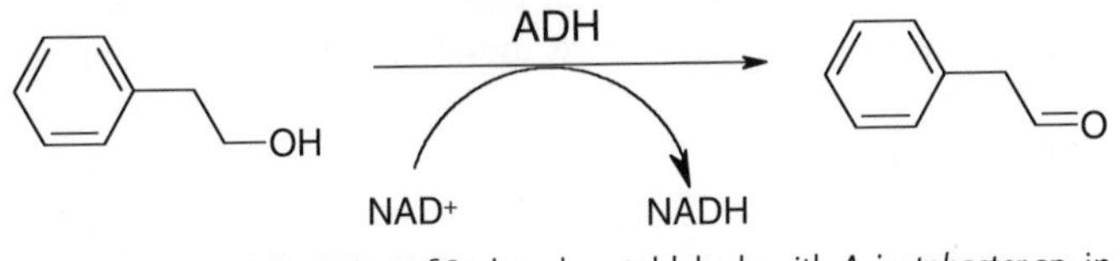

Figure 16.2-46. Preparation of 2-phenyl acetaldehyde with *Acinetobacter* sp. in organic/aqueous emulsions.

Table 16.2-12. Oxidations catalyzed by *Acinetobacter* sp. in aqueous and biphasic media [27].

Substrate	Water		Water/isooctane (vol/vol 1/1)	
	Acid yield [%]	Time [h]	Aldehyde yield [%]	Time [h]
	> 97	3	74	1
	> 97	3	90	1
	> 97	3	87	1
	> 97	24	72	4
	25	24	< 5	24
	> 97	3	90	45 min
	> 97	2	93	45 min
	> 97	8	77	45 min
racemic	40 ((S)-alcohol: 95 % ee)	24	< 5	24

27 R. Gandolfi, N. Ferrara, F. Molinari, *Tetrahedron Lett.* **2001**, *42*, 513–514.

include ethanol, propanol, butanol, 2-methyl-1-butanol, 3-methyl-1-butanol, 1-penta-nol, 1-hexanol, geraniol, 2-phenylethanol, 2-phenylthioethanol, cinnamyl alcohol, benzyl alcohol and (R,S)-2-phenyl-1-propanol (Tables 16.2-12 and 16.2-13) [202, 203]. The molecular structure of substrates and products as well as physicochemical conditions significantly influence bioconversions of short-chain aliphatic and aro-matic alcohols into acids [204, 205]. Yields depend on the toxicity of the alcohol (different inhibitory concentrations for different alcohols), since product inhibition is often the major limiting factor [204]. Furthermore, dissolved oxygen concentrations and pH conditions are important factors for improving such bioconversions. Depending on strain and substrate (specificity of dehydrogenases), the reaction is directed to aldehyde *or* acid accumulation (Fig. 16.2-44). In principle, acid accumula-tion is favored in aqueous media, whereas aldehydes preferentially accumulate in biphasic media [203].

Table 16.2-13. Oxidations catalyzed by *Gluconobacter asaii* in aqueous and biphasic systems [27].

Substrate	Water		Water/isooctane (vol/vol 1/1)	
	Acid yield [%]	Time [h]	Aldehyde yield [%]	Time [h]
 ~~~~OH	> 97	4	93	45 min
Y~~OH	> 97	4	90	1
~~OH	> 97	3	91	45 min
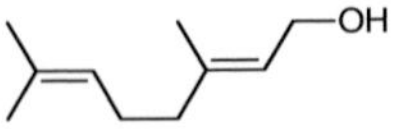 >=~~=~-OH	16	24	29	5
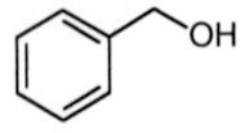 Ph-OH (benzyl)	< 5	24	< 5	24
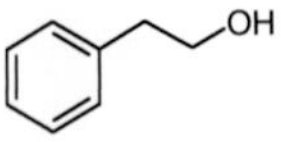 Ph-~OH	> 97	5	85	2
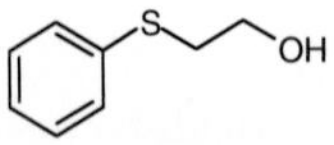 Ph-S-~OH	> 97	5	96	1
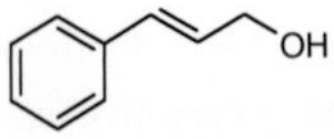 Ph-~=~OH	20	24	24	4
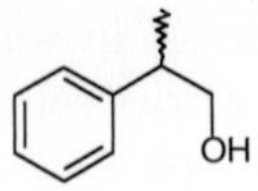 Ph-(CH)-CH2OH racemic	33	24	< 5	24

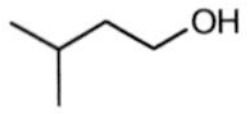

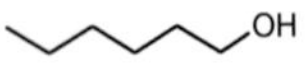

27 R. Gandolfi, N. Ferrara, F. Molinari, *Tetrahedron Lett.* **2001**, *42*, 513–514.

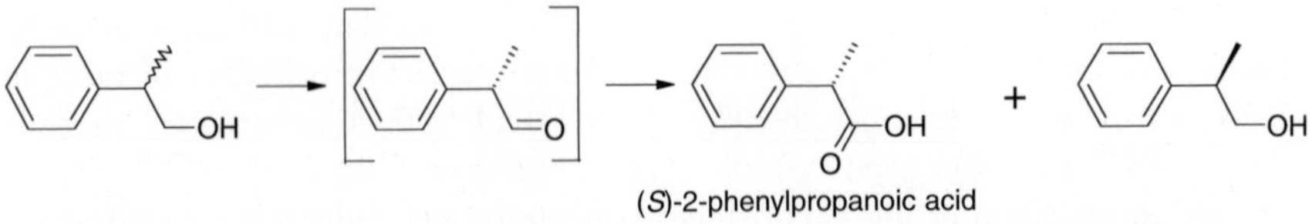

(S)-2-phenylpropanoic acid

Figure 16.2-47. Resolution of racemic (R,S)-2-phenylpropionic alcohol with whole cells of *Gluconobacter oxydans* yielding (S)-2-phenylpropanoic acid and (R)–2-phenylpropionic alcohol.

An example of aldehyde formation is the production of isovaleraldehyde by *Gluconobacter oxydans* R (Fig. 16.2-45) [202, 206]. Glycerol-grown *Gluconobacter oxydans* slowly oxidizes 3-methyl-1-butanol to isovaleraldehyde, with yields of over 90%. The product was recovered by bisulphite trapping or cold traps [202]. Extractive bioconversion in a hollow-fiber membrane bioreactor allowed continuous produc-

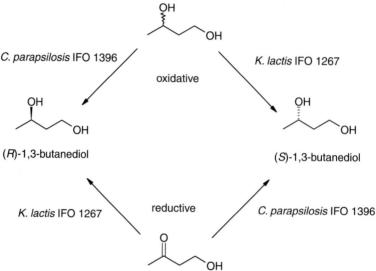

Figure 16.2-48. Preparation of both enantiomers of 1,3-butanediol with whole cells of
K. lactis and *C. parapsilosis* either by enantioselective oxidation of 1,3-butanediol
(oxidative) or enantioselective reduction of 4-hydroxybutanone (reductive).

tion of isovaleraldehyde at overall productivities of 2–3 g L^{-1} h^{-1}[206]. Yields between
72 and 90% were reached.

Another example of a synthesis is the production of phenylacetaldehyde using
Acinetobacter strains (Fig. 16.2-46) [207, 208]. Different two-liquid phase systems were
tested for their ability to remove the aldehyde into the organic phase before its
further conversion to acid. In an optimized two-liquid-phase process, in which
isooctane (at a volume fraction of 50%) was used as the organic carrier solvent,
product concentrations of 9 g/L were reached in 4 h of reaction, corresponding to a
yield of 90% [208]. The production strain *Acinetobacter* sp. ALEG showed satisfactory
long-term stability, being able to perform the transformation with 80% of the
original activity after 3 days of contact with the solvent.

Besides the multigram-scale production of different aliphatic carboxylic acids by
biocatalytic alcohol oxidation, especially the enantioselective oxidation of racemic
2-phenyl-1-propanol to (S)-2-phenylpropanoic acid with *Gluconobacter oxydans*
(Fig. 16.2-47) is another good example of acid production from alcohols [209].

After optimization of the parameters temperature, pH, substrate concentration,
and agitation speed using a simplex sequential method, the resolution involving two
oxidation steps yielded 45% product with an *ee* of 98%.

16.2.6.4
Enantiospecific Reactions

Two ways of producing *optically pure 1,3-butanediol via microbial resolution* have been
reported: the oxidation of a racemic mixture of 1,3-butanediol yielding one enantio-

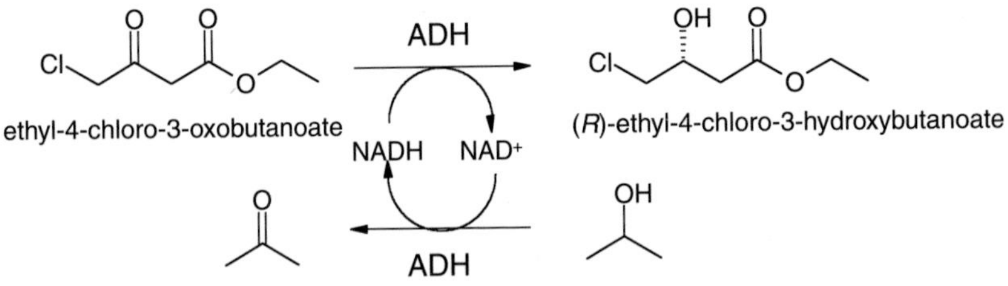

Figure 16.2-49. Asymmetric reduction of ethyl-4-chloro-3-oxobutanoate catalyzed by an alcohol dehydrogenase (ADH) in recombinant *E. coli*. The necessary reduction equivalents were derived from the oxidation of isopropanol with the same enzyme.

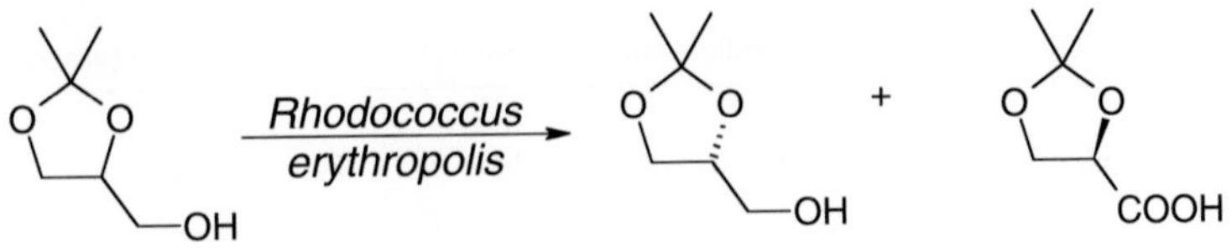

Figure 16.2-50. Enantioselective oxidation of isopropylideneglycerol utilizing *Rhodococcus erthyropolis*.

mer and 4-hydroxy-2-butanone, and the reduction of the 4-hydroxy-2-butanone yielding one enantiomer of 1,3-butanediol (Fig. 16.2-48). (R)-1,3-butanediol is an important chiral synthon for the synthesis of various optically active compounds such as azetidinone derivatives, which are intermediates in the production of antibiotics, pheromones, fragrances, and insecticides.

From a screening procedure, *Kluyveromyces lactis* IFO 1903 and *Candida parapsilosis* IFO 1396 were found to be effective in the enantioselective oxidation of (R)-1,3-butanediol and (S)-1,3-butanediol, respectively, and in the asymmetric reduction of 4-hydroxy-2-butanone to (R)-1,3-butanediol and (S)-1,3-butanediol, respectively [210].

The equilibria between ketones and alcohols are catalyzed by secondary alcohol dehydrogenases. The secondary alcohol dehydrogenase of *C. parapsilosis* IFO 1396

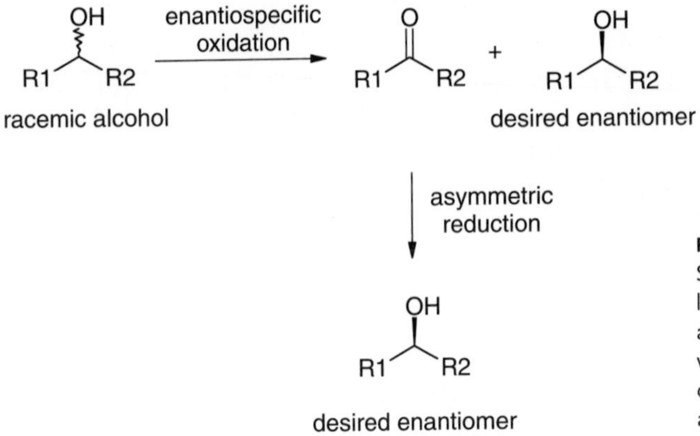

Figure 16.2-51. Stereoinversion catalyzed by two different alcohol dehydrogenases via enantiospecific oxidation followed by an asymmetric reduction.

was purified and characterized as an NAD⁺-dependent dehydrogenase with a broad substrate specificity (secondary alcohols > primary alcohols) [211]. The alcohol dehydrogenase gene of *C. parapsilosis* was cloned and expressed in recombinant *E. coli* JM109, which showed more than twofold higher specific alcohol dehydrogenase activity than *C. parapsilosis* [212].

Resting cells of *C. parapsilosis* were used for the large-scale (2000 L) production of (R)-1,3-butanediol (94% *ee*) from racemic 1,3-butanediol. After down-stream processing 3.1 kg product was isolated (overall yield: 15.5%), and a chemical purity of

Table 16.2-14. Biocatalytic stereoinversions with *Geotrichum candidum* [28].

Substrate	Without allyl alcohol			With allyl alcohol (33 mM)		
	Yield [%]	*ee* [%]	Configuration	Yield [%]	*ee* [%]	Configuration
	96	99	(R)	94	98	(R)
	65	92	(R)	57	86	(R)
	99	3	(S)	100	0	–
	90	16	(R)	85	23	(R)
	97	89	(R)	97	96	(R)
	99	2	(R)	89	21	(R)
	89	21	(R)	55	94	(R)
	95	79	(R)	74	96	(R)
	77	97	(R)	54	99	(R)

28 K. Nakamura, Y. Inoue, T. Matsuda, A. Ohno, *Tetrahedron Lett.* **1995**, *36*.

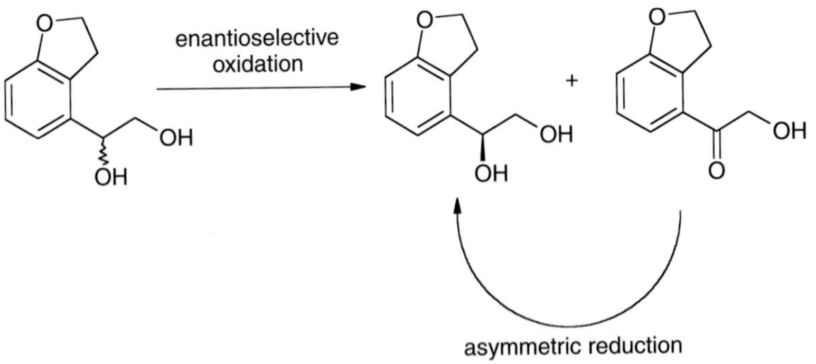

Figure 16.2-52. Synthetic application of the stereoinversion concept using *Candida* sp. and *Pichia* sp.

98.8 % was reached [213]. Resting cells of recombinant *E. coli* were reported to produce (*R*)-1,3-butanediol (93.5 % *ee*, 94.7 % yield) from the racemate without any additive to regenerate NAD+ from NADH [212].

In another application, recombinant *E. coli* produced 36.6 g/L ethyl-(*R*)-4-chloro-3-hydroxybutanoate (99 % *ee*) from 40 g/L ethyl-4-chloro-3-oxo-butanoate [210]. Here, the secondary alcohol dehydrogenase served as both synthetic (asymmetric reduction) and regenerating (NADH-regeneration via isopropanol oxidation) enzyme (Fig. 16.2-49).

Enzymatic resolution of (R/S) isopropylideneglycerol [214, 215]
Whole cells of *Rhodococcus erythropolis* were used for the selective oxidation of the (*S*)-enantiomer of isopropylideneglycerol (Fig. 16.2-50). With a 50 % conversion of the racemate, an *ee* value of over 98 % was reached for (*R*)-isopropylideneglycerol and of over 90 % for (*R*)-isopropylideneglyceric acid.

(*R*)-Isopropylideneglycerol is a useful C_3-synthon in the synthesis of (*S*)-β-blockers; e.g. (*S*)-metoprolol. (*R*)-Isopropylideneglyceric acid can also be used as starting material for the synthesis of biologically active compounds.

16.2.6.5
Stereoinversions using Microbial Redox Reactions [216]

Racemic mixtures of secondary alcohols can be resolved completely by enantiospecific enzyme-catalyzed oxidation resulting in one enantiomer of the alcohol and the ketone followed by asymmetric enzyme-catalyzed reduction of the ketone (Fig. 16.2-51). For oxidation and reduction, two separate microorganisms [217–219] or two different enzymes in a single microorganism [220–222] may be used.

An example of a suitable biocatalyst is *Geotrichum candidum*, harboring both an oxidizing and a reducing enzyme activity. Table 16.2-14 shows the catalytic performance of *Geotrichum candidum* towards different substrates [220] when the biocatalyst is incubated for 24 h with 27 mM substrate. Allyl alcohol effectively shifts the stereoselectivity of the reduction. It is presumed to inhibit enzyme(s) that reduce aryl

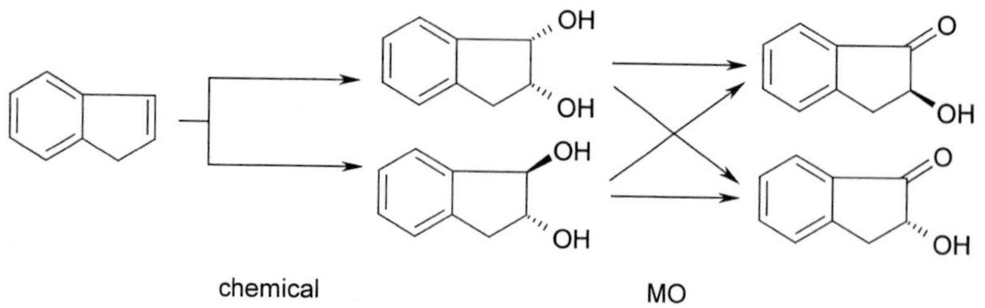

Figure 16.2-53. Chemoenzymatic synthesis of 2-hydroxy-1-indanone. The racemic *syn* and *anti* diols were prepared by chemical dihydroxylation of indane. Asymmetric induction was achieved by microbial oxidation (MO) of these diols.

Table 16.2-15. Substrate spedificity of *Arthrobacter* and *Pseudomonas* strains [29].

Taxonomy	Substrate specificity (no substrates)
Arthrobacter sp. strain 1HB	*cis*-(1*S*, 2*R*)-diol > *trans*-(1*S*, 2*S*)-diol >> (*cis*-(1*R*, 2*S*)-diol, *trans*-(1*R*, 2*R*)-diol)
Arthrobacter sp. strain 1HE	*cis*-(1*S*, 2*R*)-diol >> *trans*-(1*S*, 2*S*)-diol >> (*cis*-(1*R*, 2*S*)-diol, *trans*-(1*R*, 2*R*)-diol)
Pseudomonas aeruginosa strain IN	*trans*-(1*R*, 2*R*)-diol > *cis*-(1*S*, 2*R*)-diol > *cis*-(1*R*, 2*S*)-diol >> (*trans*-(1*S*, 2*S*)-diol)

29 Y. Kato, Y. Asano, *J. Mol. Cat. B: Enzymatic* **2001**, *13*, 27–36.

Table 16.2-16. Microbial stereoselective oxidation of *cis*- and *trans*-1,2-indandiols [29].

Strain	Substrate	Product	Reaction time [h]	Yield [%]	*ee* [%]
Arthrobacter sp. 1HB	*Cis*	R	4	46	> 99.9
	Trans	S	12	35	> 99.9
Arthrobacter sp. 1HB	*Cis*	R	4	47	> 99.9
	Trans	S	24	8	> 99.9
P. aeruginosa IN	*Cis*	R	5	7	82.5
	Trans	R	24	40	> 99.9

29 Y. Kato, Y. Asano, *J. Mol. Cat. B: Enzymatic* **2001**, *13*, 27–36.

methyl ketone to (*S*)-1-arylethanol. The inhibition of yeast reductases by allyl alcohols has been reported [223].

Another example is the deracemization of (*RS*)-1-{2',3'-dihydrobenzo[*b*]furan-4'-yl}-ethane-1,2-diol by biocatalytic stereoinversion (Fig. 16.2-52) [224]. In order to find an appropriate biocatalyst to accomplish such a deracemization, different microorganisms were screened. Several microorganisms belonging to the genera *Candida* and *Pichia* allowed yields of 60–70 % with 90–100 % enantiomeric excess. Substrate dissolved in DMF was added to the biotransformation mixture consisting of resting cells suspended in phosphate buffer (pH 7). The presence of glucose generally increased the yield but lowered the enantiomeric excess. Different microorganisms can be suitable for a given stereoinversion and the optimal biocatalyst should be chosen by screening.

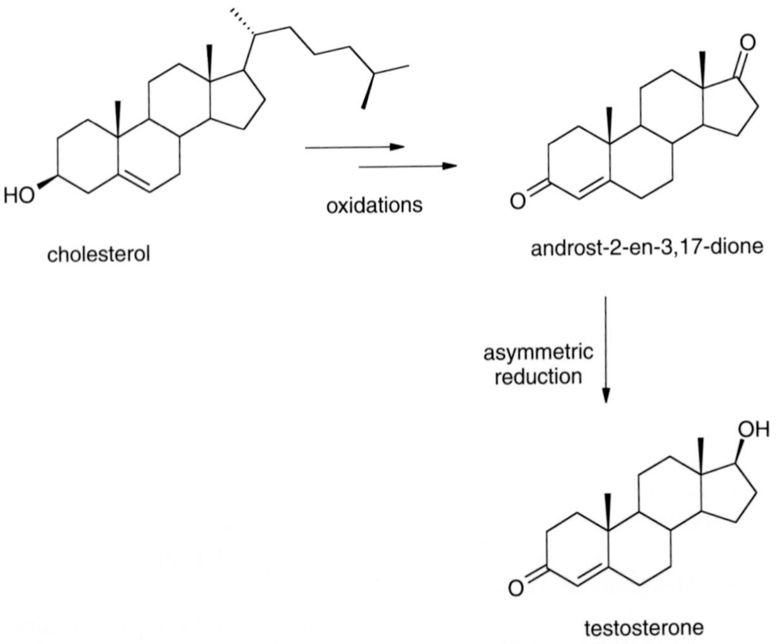

Figure 16.2-54. Selective oxidation of cholesterol to testosterone by whole cells of *Mycobacterium* sp NRRL B-3805.

Stereoselective oxidation of racemic 1,2-indandiols[225]

Kato *et al.* described the stereoselective microbial synthesis of both enantiomers of 2-hydroxy-1-indanone, selecting *cis-* or *trans*-diol as the substrate (Fig. 16.2-53). *Cis*-1-amino-2-indanol is an important synthon in organic chemistry (for example in the synthesis of the leading HIV protease inhibitor Crixivan) and can easily be synthesized from optically active 2-hydroxy-1-indanone[226].

Microorganisms degrading indane derivatives were screened for stereoselective oxidation of racemic *cis-* or *trans*-1,2-indandiol. Three promising strains specifically oxidizing the benzylic hydroxyl group were found (see Table 16.2-15).

All strains produced inducible enzymes responsible for the oxidation reaction, recognizing the stereochemistry of the 1- or 2-positions of the diol regardless of their *cis* and *trans* geometry. By using the resting cells of the strains, both enantiomers of 2-hydroxy-1-indanone were synthesized in enantiomerically pure form simply by selecting *cis-* or *trans*-1,2-indandiol as the substrate. Growth conditions were optimized to promote cell growth and the formation of 1,2-indanediol-oxidizing activity. The biocatalyst activity was optimally induced with 0.05% indanol. Carefully choosing appropriate carbon and nitrogen sources is crucial for optimal biocatalyst activity and cell growth.

Table 16.2-16 shows the stereoselective oxidation of racemic *cis*-diol or *trans*-diol into optically active 2-hydroxy-1-indanone at a 2 mL scale with 50 mg dry cells per ml.

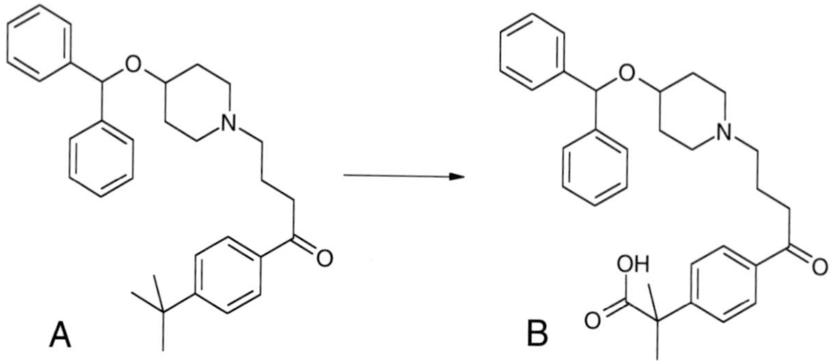

Figure 16.2-55. Regioselective three-step oxidation of ebastine (A) to carebastine (B) using *Cunninghamella blakesleeana*.

Production of testosterone from cholesterol using Mycobacterium sp. [227]
In this multistep reaction the microbial degradation of sterol side chains combined with the reduction of an intermediate thereof is used to accumulate testosterone from cholesterol. A cholesterol-assimilating and androst-2-en-3,17-dione-accumulating mutant of *Mycobacterium* sp. NRRL B-3805 oxidizes cholesterol through multiple steps of the sterol side chain degradation pathway, also involving alcohol oxidations, to androst-2-en-3,17-dione (Fig. 16.2-54). This multistep oxidation is followed by the reduction of androst-2-en-3,17-dione to testosterone by the NADH requiring activity of 17β-hydroxysteroid dehydrogenase. This activity is dependent on the presence of glucose as the carbon source. After the glucose in the fermentation culture is completely consumed, most testosterone is oxidized to androst-2-en-3,17-dione. Adding a larger amount of glucose prevents this oxidation.

On a 2.5 L scale a yield of 51% was reached in 120 h of cultivation. Here, the initial substrate concentration amounted to 0.1% (w/v).

Microbial oxidation of ebastine [228]
Ebastine is a new generation antihistaminic drug with fewer side-effects. The microbial three-step oxidation of ebastine, using whole cells of the mold *Cunninghamella blakesleeana* as biocatalysts, involves an alcohol and an aldehyde oxidation step and results in the formation of carebastine, which is the pharmacologically active compound [229]. The initial step in the oxidation of ebastine is hydroxylation by a cytochrome P-450-dependent monooxygenase to the corresponding alcohol. The two consecutive oxidations are catalyzed by oxidoreductases, which are not further characterized, and lead via the aldehyde to the corresponding carboxylic acid carebastine (Figure 16.2-55).

Growth in a complex medium containing soybean-peptone and yeast extract is necessary for biocatalyst activity. A component of soybean-peptone, genistein, is thought to act as an inducer of cytochrome P-450 enzymes. Growing cells provide a higher yield than resting cells. Addition of 1% poly(vinyl alcohol) was found to prevent pellet formation and thereby to guarantee constant mass transfer rates.

From a 3 L batch fermentation, 270 mg carebastine was isolated (yield: 45%).

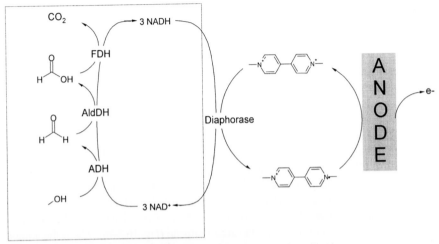

Figure 16.2-56. Enzymatic three-step oxidation of methanol to carbon dioxide in the anodic compartment of a biofuel cell.

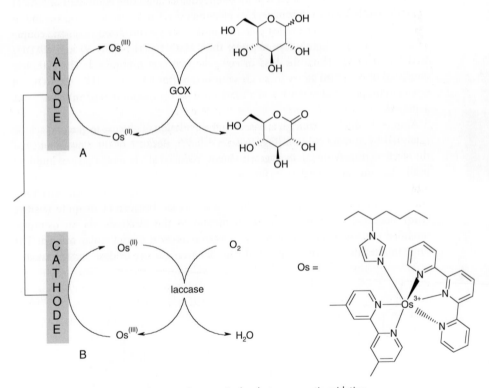

Figure 16.2-57. Mediated electron transfer steps in the electroenzymatic oxidation of glucose (A) and reduction of O_2.

Therefore, after 24 h of cultivation, 600 mg ebastine was added and the incubation was continued for 68 h.

16.2.7
Miscellaneous

16.2.7.1
Biofuel Cells

In recent years, biofuel cells have gained tremendous attention. The use of methanol instead of dihydrogen as the oxidizable substance offers special advantages as it is readily available and easy to store and handle. At the same time, the theoretical cell voltage of an MeOH/O_2 cell (1.19 V) is near that of H_2/O_2 (1.23 V).

Whitesides and coworkers recently developed a biofuel cell based on the step-wise enzymatic oxidation of methanol to carbon dioxide (Fig. 16.2-56)[230]. In the anodic compartment of the biofuel cell, methanol is oxidized to carbon dioxide in three steps: by an alcohol dehydrogenase, an aldehyde dehydrogenase, and ultimately formate dehydrogenase. In each of these enzymatic steps, one equivalent of NADH is produced. NADH itself transfers its electrons via diaphorase to viologene and in the end to the anode. The redox potential of the reduced/oxidized viologene couple (– 0.55 V) is only slightly less negative than MeOH/CO_2 (– 0.64 V) and NADH/NAD$^+$ (– 0.59 V). Thus, the loss in cell potential was minimized. The catholyte consisted of platinum gauze in an O_2-saturated buffer (O_2 + 4e$^-$ + 4H$^+$ → 2H_2O). An open-circuit potential of 0.8 V and a maximum power output of 0.67 mW cm^{-2} was achieved.

Another biofuel cell concept is based on the oxidation of glucose to gluconolactone catalyzed by glucose oxidase (Fig. 16.2-57)[231, 232]. Because of the slow kinetics of the electron transfer to O_2, dioxygen is usually reduced at a potential several hundred millivolts more negative than its formal potential, thus lowering the power density of a fuel cell. Utilizing laccase to catalyze this reaction can circumvent that. ABTS is a suitable mediator between the electrode and laccase because of its quite positive redox potential[233]. Wiring laccase reduction to the electrode via an osmium-modified electrode also facilitates the electroreduction of molecular oxygen. The same modification serves as the conductor between glucose oxidase and the anode.

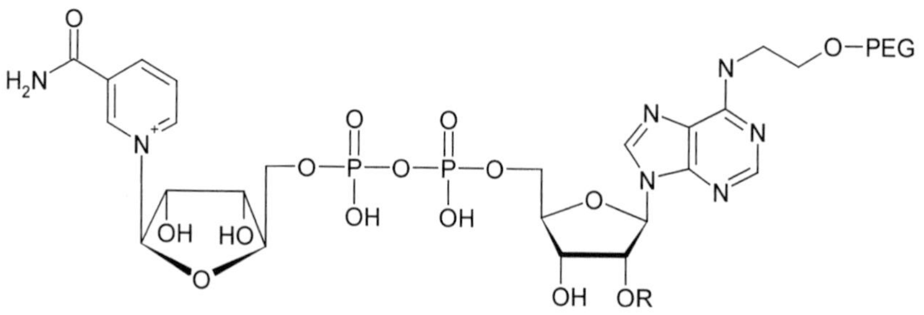

Figure 16.2-58. NAD modified with polyethylene glycol (PEG).

Table 16.2-17. Kinetic constants of different dehydrogenases for NAD(P)$^+$ and PEG-NAD(P)$^+$.

NAD$^+$-dependent enzymes	Native cofactor	PEG-bound cofactor	
	K_M [μM]	K_M [μM]	V_{max} [as % of NAD$^+$]^a
FDH	15	82	57
Glutamate DH	175	444	53
YADH	154	1310	64
HLADH	62	1150	72
LDH	182	142	21
3α-HSDH	29	647	66
Glucose DH	96	2030	3
NADP$^+$-dependent enzymes	**K_M [μM]**	**K_M [μM]**	**V_{max} [as % of NADP$^+$]**
Glutamate DH	160	425	96
Malic enzyme	5	12	86
TBADH	13	28	84

a 100% correspond to V_{max} values of the dehydrogenases determined with native coenzymes.

A miniaturized cell was constructed which exhibited a power output of 0.137 mW cm^{-2}. After 72 h of operation, 75% of the initial power output was still present.

Even though biofuel cells are generally considered to be in their infancy[234], their potential, which is based on non-hazardous, easy-to-handle substrates and electrolytes (especially the moderate temperatures compared to those of conventional fuel cells: 80–1000 °C) cannot be neglected. Even photosynthetic biofuel cells (converting light energy into electrical energy) have been shown to work in principle[235].

16.2.7.2
Biomimetic Analogs to Nicotinamide Coenzymes

For large-scale applications of NAD(P)-dependent enzymes, continuous-flow reactors with ultrafiltration membranes have been proposed[236]. In order to retain low molecular weight nicotinamide cofactors in the reactor, charged membranes have been used, retarding the overall negatively charged nicotinamide coenzymes by electrostatic repulsion[237, 238]. Retention rates of approx. 99% and TTNs (NAD) of up to 10 000 were reported.

Another approach makes use of polymer-modified NAD [modification with polyethylene glycol (PEG; MW = 20 000)], thus retaining it on account of its drastically increased size (Fig. 16.2-58)[239–241]. The polymer modification usually leads to a drastically increased K_M value, whereas the V_{max} value is generally over 50% of that of low molecular weight NAD(P) (Table 16.2-17).

Another area of research deals with synthetic analogs of NAD(P) coenzymes. Besides the lower costs, these analogs may offer better stability or easier regeneration and may add new functionalities to known enzyme systems (e.g. thio-NAD together with HLADH[182]). Some artificial redox coenzymes were developed mimicking the "shape" of native nicotinamide coenzymes (Fig. 16.2-59)[242, 244]. Activity with various NAD-dependent enzymes was found, even though the activity was only

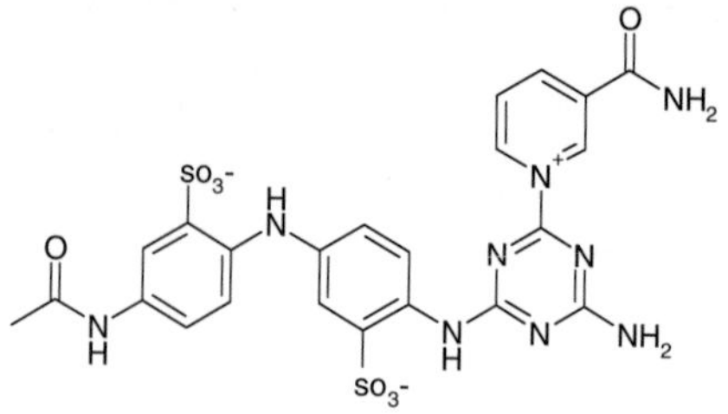

Figure 16.2-59. Synthetic analogs of NAD.

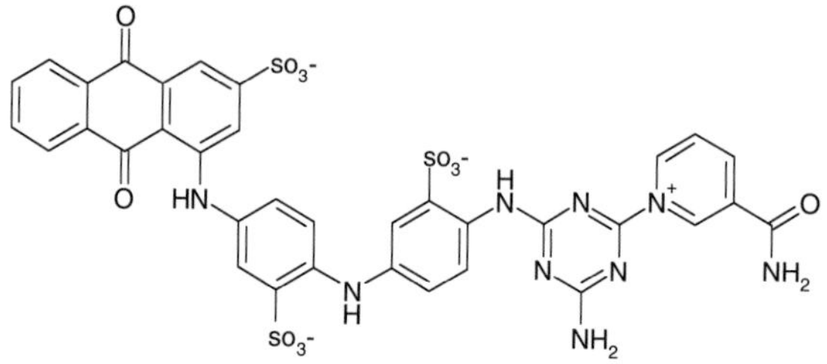

CL4

blue N-3

in the region of less than 10% of that with the native cofactor. However, it was shown that these analogs could have at least some potential.

References

1 L. G. Lee, G. M. Whitesides, *J. Org. Chem.* **1986**, *51*, 25–36.

2 P. Adlercreutz, *Biocat. Biotransf.* **1996**, *14*, 1–30.

3 M. D. Leonida, *Curr. Med. Chem.* **2001**, *8*, 345–369.

4 C. Ricci, *Acta Vitaminol. Enzymol.* **1971**, *25*, 65–69.

5 Y. Nishiyama, V. Massey, K. Takeda, S. Kawasaki, J. Sato, T. Watanabe, Y. Niimura, *J. Bacteriol.* **2001**, *183*, 2431–2438.

6 E. Sanjust, N. Curreli, A. Rescigno, J. V. Bannister, C. D., *Biochem. Mol. Biol. Int.* **1997**, *41*, 555–562.

7 G. A. Michaliszyn, S. S. Wing, E. A. Meighen, *J. Biol. Chem.* **1977**, *252*, 7495–7499.

8 J. B. Jones, K. E. Taylor, *Can. J. Chem.* **1976**, *54*, 2969 and 2974.

9 B. Nefsky, M. DeLuca, *Arch. Biochem. Biophys.* **1982**, *216*, 10–16.

10 H. K. Chenault, G. M. Whitesides, *Appl. Biochem. Biotech.* **1987**, *14*, 147–197.

11 P. N. Bartlett, P. Tebbutt, M. J. Whitaker, *Progr. React. Kinet.* **1991**, *16*, 55.

12 I. Katakis, E. Dominguez, *Mikrochim. Acta* **1997**, *126*, 11–32.

13 H. Jaegfeldt, T. Kuwana, G. Johansson, *J. Am. Chem. Soc.* **1983**, *105*, 1805–1814.

14 C. Degrand, L. Miller, *J. Am. Chem. Soc.* **1980**, *102*, 5728–5732.

15 A. S. N. Murthy, J. Sharma, *Talenta* **1998**, *45*, 951–956.

16 S. I. Bailey, I. M. Ritchie, *Electrochim. Acta* **1985**, *30*, 3–12.

17 H. R. Zare, S. M. Golabi, *J. Electroanal. Chem.* **1999**, *464*, 14–23.

18 H. R. Zare, S. M. Golabi, *J. Solid State Electrochem.* **2000**, *4*, 87–94.

19 W. B. Nowall, W. G. Kuhr, *Anal. Chem.* **1995**, *67*, 3583–3588.

20 C. R. Raj, T. Ohsaka, *Electrochem. Commun.* **2001**, *3*, 633–638.

21 B. Gründig, G. Wittstock, U. Rüdel, B. Strehlitz, *J. Electroanal. Chem.* **1995**, *395*, 143–157.

22 J. R. Komoschinski, E. Steckhan, *Tetrahedr. Lett.* **1988**, *29*.

23 G. Hilt, T. Jarbawi, W. R. Heineman, E. Steckhan, *Chem. Eur. J.* **1997**, *3*, 79–88.

24 G. Hilt, B. Lewall, G. Montero, J. H. P. Utley, E. Steckhan, *Liebigs Ann./Recueil* **1997**, 2289–2296.

25 H. Ju, D. Leech, *Anal. Chim. Acta* **1997**, *345*, 51–58.

26 B. W. Carlson, L. L. Miller, P. Neta, J. Grodkowski, *J. Am. Chem. Soc.* **1984**, *106*, 7233–72339.

27 A. Malinauskas, T. Ruzgas, L. Gorton, *J. Coll. Interf. Sci.* **2000**, *224*, 325–332.

28 T. Osa, Y. Kashiwagi, Y. Yanagisawa, *Chem. Lett.* **1994**, 367–370.

29 M. Sadakane, E. Steckhan, *Chem. Rev.* **1998**, *98*, 219–237.

30 N. F. Atta, A. Galal, E. Karagözler, H. Zimmer, J. Rubinson, H. B. Mark, *J. Chem. Soc., Chem. Commun.* **1990**, 1347–1349.

31 J. Botzem, *Dissertation, Univerity Bonn, Germany* **2000**.

32 D. Schwartz, M. Stein, K.-H. Schneider, F. Giffhorn, *J. Biotech.* **1994**, *33*, 95–101.

33 S. Itoh, T. Terasaka, M. Matsumiya, M. Komatsu, Y. Ohshiro, *J. Chem. Soc., Perkin Trans. I* **1992**, 3253–3254.

34 Y. Ogino, K. Takagi, K. Kano, T. Ikeda, *J. Electroanal. Chem.* **1995**, *396*, 517–524.

35 G. F. Hall, A. P. F. Turner, *Electroanal.* **1994**, *6*, 217–220.

36 K. Takagi, K. Kano, T. Ikeda, *J. Electroanal. Chem.* **1998**, *445*, 211–219.

37 I. Willner, D. Mandler, *Enz. Microb. Tech.* **1989**, *11*, 467–483.

38 J. Handman, A. Harriman, G. Porter, *Nature* **1984**, *307*, 534–535.

39 M. Julliard, J. Le Petit, *J. Photochem. Photobiol.* **1982**, *36*, 283.

40 R. P. Chambers, J. R. Ford, J. H. Allender, W. H. Baricos, W. Cohen, *Enz. Eng.* **1974**, *2*, 195.

41 R. Ruppert, E. Steckhan, *J. Chem. Soc. Perkin Trans. II* **1989**, *112*, 13–23.

42 T. Kawamoto, A. Aoki, K. Sonomoto, A. Tanaka, *J. Ferm. Bioeng.* **1989**, *67*, 361–362.

43 J. A. Lepoivre, *Janssen Chim. Acta.* **1984**, *2*, 20.

44 A. Plant, *Pharm. Manufac. Rev.* **1991**, *March*, 5.

45 E. S. Cedergen-Zeppezauer, I. Andersson, S. Ottonello, *Biochem.* **1985**, *24*, 4000–4010.

46 S. Ramaswamy, H. Eklund, B. V. Plapp, *Biochem.* **1994**, *33*, 5230–5237.

47 J. B. Jones, I. J. Jackovac, *Can. J. Chem.* **1982**, *60*, 19.

48 W. Hummel, New Alcohol Dehydrogenases for the Synthesis of Chiral Compounds. In *Adv. Biochem. Eng. Biotech.*, T. Scheper (ed), Springer, Berlin, 1997, Vol. 58, pp. 147–179.

49 M. Andersson, H. Holmberg, P. Adlercreutz, *Biocat. Biotransf.* **1998**, *16*, 259–273.

50 R. Lortie, I. Villaume, M. D. Legoy, D. Thomas, *Biotech. Bioeng.* **1989**, *33*, 229–232.

51 C.-H. Wong, J. R. Matos, *J. Org. Chem.* **1985**, *50*, 1992–1994.

52 J. R. Matos, C.-H. Wong, *J. Org. Chem.* **1986**, *51*, 2388–2389.

53 A. J. Willetts, C. J. Knowles, M. S. Levitt, S. M. Roberts, H. Sandey, N. F. Shipston, *J. Chem. Soc. Perkin Trans. I* **1991**, 1608–1610.

54 Y. Tsuji, T. Fukui, T. Kawamoto, A. Tanaka, *Appl. Microbiol. Biotech.* **1994**, *41*, 219–224.

55 R. N. Patel, M. Liu, A. Banerjee, L. Szarka, *Ind. J. Chem.* **1992**, *31B*, 832–836.

56 C. Hertweck, W. Boland, *J. prakt. Chem.* **1997**, *339*, 754–757.

57 M.-E. Gourdel-Martin, C. Comoy, F. Huet, *Tetrahedron: Asym.* **1999**, *10*, 403–404.

58 D. G. Drueckhammer, R. V. W., C.-H. Wong, *J. Org. Chem.* **1985**, *50*, 5387–5389.

59 L. G. Lee, G. M. Whitesides, *J. Am. Chem. Soc.* **1985**, *107*, 6999–7008.

60 A. J. Ganzhorn, D. W. Green, A. D. Hershey, R. M. Gould, B. V. Plapp, *J. Biol. Chem.* **1987**, *262*.

61 A. M. Snijder-Lamberts, E. N. Vulfson, H. J. Doddema, *Rec. Trav. Chim. Pays-Bas* **1991**, *110*, 226–230.

62 K. Nakamura, T. Miyai, K. J., N. Nakajima, A. Ohno, *Tetrahedron Lett.* **1990**, *31*, 1159–1160.

63 F. Yang, A. J. Russel, *Biotech. Bioeng.* **1993**, *43*, 232–241.

64 J. M. Laval, J. Moiroux, C. Bourdillon, *Biotech. Bioeng.* **1991**, *38*, 788–796.

65 H. G. Davis, *Best Synthetic Methods/Biotransformations in Preparative Organic Chemistry*, Academic Press, London, 1989, p. 165.

66 P. van Eikeren, D. J. Brose, D. C. Much-

more, J. B. West, *Ann. N. Y. Acad. Sci.* **1990**, *613*, 796–801.

67 P. van Eikeren, D. J. Brose, D. C. Muchmore, R. H. Colton, *Ann. N. Y. Acad. Sci.* **1992**, *672*, 539–551.

68 E. Keinan, S. C. Sinha, A. Sinha-Bagchi, *J. Chem. Soc. Perkin Trans. I* **1991**, *12*, 3333–3339.

69 F. Bastos, A. G. dos Santos, J. Jones, E. G. Oestreicher, G. F. Pinto, L. M. C. Paiva, *Biotech. Techniques* **1999**, *13*, 661–664.

70 J. H. Marshall, J. W. May, J. Sloan, *J. Gen. Microbiol.* **1985**, *131*, 1581–1588.

71 Y.-C. Kong, J. W. May, J. H. Marshall, *J. Gen. Microbiol.* **1985**, *131*, 1571–1579.

72 H. Nishise, A. Nagao, Y. Tani, H. Yamada, *Agric. Biol. Chem.* **1984**, *48*, 1603–1609.

73 H. Nishise, S. Maehashi, H. Yamada, Y. Tani, *Agric. Biol. Chem.* **1987**, *51*, 3347–3353.

74 A. Liese, M. Karutz, J. Kamphuis, C. Wandrey, *Biotech. Bioeng.* **1996**, *51*, 544–550.

75 G. Klöck, K. Kreuzberg, *Biochim. Biophys. Acta* **1989**, *991*, 347–352.

76 A. Nilsson, L. Adler, *Biochim. Biophys. Acta* **1990**, *1034*, 180–185.

77 C.-H. Wong, G. M. Whitesides, *J. Org. Chem.* **1983**, 3199–3205.

78 J. R. Dürrwachter, D. G. Drückhammer, K. Nozaki, H. M. Sweeres, C.-H. Wong, *J. Am. Chem. Soc.* **1986**, *108*, 7812–7818.

79 A.-E. Biade, C. Bourdillon, J.-M. Laval, G. Mairesse, J. Moiroux, *J. Am. Chem. Soc.* **1992**, *114*, 893–897.

80 A. Anne, C. Bourdillon, S. Daninos, J. Moiroux, *Biotech. Bioeng.* **1999**, *64*, 101–107.

81 A. Bardea, E. Katz, A. F. Bückmann, I. Willner, *J. Am. Chem. Soc.* **1997**, *199*, 9114–9119.

82 A. Fassouane, J.-M. Laval, J. Moiroux, C. Bourdillon, *Biotech. Bioeng.* **1990**, *35*, 935–939.

83 S. Riva, R. Bovara, P. Pasta, G. Carrea, *J. Org. Chem.* **1986**, *51*, 2902–2906.

84 S. Riva, R. Bovara, L. Zetta, P. Pasta, G. Ottolina, G. Carrea, *J. Org. Chem.* **1988**, *53*, 88–92.

85 G. Carrea, R. Bovara, R. Longhi, S. Riva, *Enz. Microb. Tech.* **1985**, *7*, 597–600.

86 C. A. Raia, S. D'Auria, M. Rossi, *Biocat.* **1994**, *11*, 143–150.

87 S. Ammendola, C. A. Raia, C. Caruso, L. Camardella, S. Dauria, M. Derosa, M. Rossi, *Biochem.* **1992**, *31*, 12514–

12523.

88 C. A. Raia, A. Giordano, M. Rossi, Alcohol Dehydrogenase from Sulfolobus solfataricus. In *Methods in Enzymology*. M. W. W. Adams (ed), Academic Press, San Diego, 2001, Vol. 331, pp. 176–195.

89 W. Hummel, B. Riebel, *Ann. N. Y. Acad. Sci.* **1996**, *799*, 713–716.

90 O. Bortolini, E. Casanova, G. Fantin, A. Medici, S. Poli, S. Hanau, *Tetrahedron: Asymmetry* **1998**, *9*, 647–651.

91 D. D. Hoskin, H. R. Whiteley, B. Mackler, *J. Biol. Chem.* **1962**, *237*, 2647–2651.

92 H. L. Schmidt, W. Stoecklein, J. Danzer, P. Kirch, B. Limbach, *Eur. J. Biochem.* **1986**, *156*, 149–155.

93 U. L. R. Baminger, C. Galhaup, C. Leitner, K. D. Kulbe, D. Haltrich, *J. Mol. Cat. B: Enzymatic* **2001**, *11*, 541–550.

94 P. N. Bartlett, M. J. Whitaker, J. Green, J. Frew, *J. Chem. Soc., Chem. Commun.* **1987**, 1603–1604.

95 R. M. Baum, *Chem. Eng. News* **1987**, *3*, 24–26.

96 Y. Degani, A. Heller, *J. Phys. Chem.* **1987**, *91*, 1285–1289.

97 Y. Degani, A. Heller, *J. Am. Chem. Soc.* **1988**, *110*, 2615–2620.

98 A. Heller, *Acc. Chem. Res.* **1990**, *23*, 128–134.

99 W. Schuhmann, T. J. Ohara, H. L. Schmidt, A. Heller, *J. Am. Chem. Soc.* **1991**, *113*, 1394–1397.

100 A. E. G. Cass, G. Davis, M. J. Green, H. A. O. Hill, *J. Electroanal. Chem.* **1985**, *190*, 1237–1243.

101 G. Davis. In *Biosensors*. A. P. F. Turner, J. Karube, G. S. Wilson (eds), Oxford University Press, Oxford, 1987, pp. 247–256.

102 A. Petersen, E. Steckhan, *Bioinorg. Med. Chem.* **1999**, *7*, 2203–2208.

103 J. M. Hohnson, H. B. Halsall, W. R. Heineman, *Biochem.* **1985**, *27*, 1579–1585.

104 A. L. Crumbliss, H. A. O. Hill, D. J. Page, *J. Electroanal. Chem.* **1986**, *206*, 327–331.

105 P. D. Hale, T. A. Skotheim, *Synth. Meth.* **1989**, *28*, C853-C858.

106 S. Zhao, R. B. Lennox, *Anal. Chem.* **1991**, *63*, 1174–1178.

107 C. Bourdillon, L. R., M. Laval, *Biotech. Bioeng.* **1988**, *31*, 553–558.

108 T. Nakaminami, S. Kuwabata, H. Yoneyama, *Anal. Chem.* **1997**, *69*, 2367–2372.

109 C. G. J. Koopal, B. de Ruiter, R. J. M. Nolte, *J. Chem. Soc., Chem. Commun.* **1991**, 1691–1692.

110 R. D. Scheller, F. Schubert, D. Pfeiffer, *Spektr. Wiss.* **1992**, 9, 99–103.

111 E. I. Solomon, M. J. Baldwin, M. D. Lowery, *Chem. Rev.* **1992**, 92.

112 U. A. Germann, G. Mueller, P. E. Hunziger, K. Lerch, *J. Biol. Chem.* **1988**, 263, 885.

113 B. Reinhammar. In *Copper Proteins and Copper Enzymes.* R. Lonntie (ed), CRC Press, Boca Raton, 1984, Vol. III, pp. 1–35.

114 C. F. Thurston, *Microbiology* **1994**, 140, 19.

115 H.-J. Danneel, M. Ullrich, F. Giffhorn, *Enzyme Microb. Technol.* **1992**, 14, 898–903.

116 F. Griffhorn, *Appl. Microbiol. Biotechnol.* **2000**, 54, 727–740.

117 G. Daniel, J. Volc, E. Kubatova, *Appl. Environ. Micobiol.* **1994**, 60, 2524–2532.

118 S. Freimund, A. Huwig, F. Giffhorn, S. Köpper, *Chem. Eur. J.* **1998**, 4, 2442–2455.

119 C. Leitner, J. Volc, D. Haltrich, *Appl. Environ. Micobiol.* **2001**, 67, 3636–3644.

120 M. Petrivalsky, P. Skladal, L. Macholan, J. Volc, *Coll. Czech. Chem. Commun.* **1994**, 59, 1226.

121 H. Liden, J. Volc, G. Marko-Varga, L. Gorton, *Electroanalysis* **1998**, 10, 223–230.

122 H. W. Ruelius, R. M. Kerwin, F. W. Janssen, *Biochim. Biophys. Acta* **1968**, 167, 493–500 and 501–510.

123 K. Soda, K. Yonaha. In *Biotechnology.* J. F. Kennedy (ed), VCH, Weinheim, 1987, Vol. 7a, pp. 606.

124 T. E. Liu, B. Wolf, J. Geigert, S. L. Neidleman, D. J. Chin, D. S. Hirnao, *Carbohydr. Res.* **1983**, 113, 151–157.

125 J. Geigert, *Carbohydr. Res.* **1983**, 113, 159–162.

126 C. Leitner, W. Neuhauser, J. Volc, K. D. Kulbe, B. Nidetzky, D. Haltrich, *Biocat. Biotransf.* **1998**, 16, 365–382.

127 W. Adam, M. Lazarus, B. Boss, C. R. Saha-Möller, H.-U. Humpf, P. Schreier, *J. Org. Chem.* **1997**, 62, 7841–7843.

128 V. Prelog, M. Wilhelm, D. B. Bright, *Helv. Chim. Acta* **1954**, 37, 221–224.

129 J. B. Lee, I. M. Downie, *Tetrahedron* **1967**, 23, 359–363.

130 K. Mori, T. Takigawa, T. Matsuo, *Tetrahedron* **1979**, 35, 933–940.

131 W. Adam, M. Lazarus, C. R. Saha-Möller, P. Schreier, *Tetrahedron Asymmetry* **1998**, 9, 351–355.

132 A. Eisenberg, J. E. Seip, J. E. Gavagan, M. S. Payne, D. L. Aton, R. DiCosimo, *J. Mol. Cat. B: Enzymatic* **1997**, 2, 223–232.

133 K. Isobe, H. Nishise, *J. Mol. Cat. B: Enzymatic* **1999**, 75, 265–271.

134 J. E. Seip, S. K. Fager, J. E. Gavagan, L. W. Gosser, D. L. Anton, R. DiCosimo, *J. Org. Chem.* **1993**, 58, 2253–2259.

135 J. E. Seip, S. K. Fager, J. E. Gavagan, D. L. Anton, R. DiCosimo, *Bioinorg. Med. Chem.* **1994**, 2, 371–378.

136 J. E. Gavagan, S. K. Fager, J. E. Seip, D. S. Clark, M. S. Payne, D. L. Anton, R. DiCosimo, *J. Org. Chem.* **1997**, 62, 5419–5427.

137 Y. Isono, T. Sudo, M. Hoshino, *Agric. Biol. Chem.* **1989**, 53, 1663–1669 and 1671–1677.

138 Y. Isono, M. Hoshino. In *US Patent Appl.*; 5,156,955, 1992.

139 M. Mahmoudian, B. A. M. Rudd, B. Cox, C. S. Drake, R. M. Hall, P. Stead, M. J. Dawson, M. Chandler, D. G. Livermore, N. J. Turner, G. Jenkins, *Tetrahedron* **1998**, 54, 8171–8182.

140 M. Mahmoudian, *Biochem. Biotransf.* **2000**, 18, 105–118.

141 M. Gregson, B. E. Ayres, G. B. Ewan, F. Ellis, J. Knight. In *PCT Int. Appl.*, WO 9417090, 1994.

142 P. Abel, *Spektr. Wiss.* **1992**, 9, 115.

143 Y. Saito, *Talanta* **1987**, 34, 667–669.

144 R. L. Kelley, C. A. Reddy, *Methods Enzymol.* **1988**, 161, 307–316.

145 R. L. Kelley, C. A. Reddy, *J. Bacteriol.* **1986**, 166.

146 M. R. Kula, *ChiuZ* **1980**, 14, 61–70.

147 G. A. Hamilton in *Techniques of Chemistry.* J. B. Jones, C. J. Sih, D. Perlman (eds), Wiley, New York, 1976, Vol. 10, pp. 875–972.

148 K. Isobe, H. Nishise, *Biosci. Biotech. Biochem.* **1994**, 58, 170–173.

149 N. Hidaka, T. Matsumoto, *Ind. Eng. Chem. Res.* **2000**, 39, 909–915.

150 G. A. Hamilton, P. K. Adolf, J. De Jersey, G. C. DuBois, G. R. Dyrkacz, R. D. Libby, *J. Am. Chem. Soc.* **1978**, 100, 1899–1912.

151 R. L. Root, J. R. Durrwachter, C. H. Wong, *J. Am. Chem. Soc.* **1985**, 107, 2997–2999.

152 D. G. Drueckhammer, W. J. Hennen, R. L. Pederson, D. F. Barbas, C. M. Gautheron, T. Krach, C. H. Wong, *Synthesis* **1991**, 7, 499–525.

153 R. A. Schlegel, C. M. Gerbeck, R. Montgomery, *Carbohydr. Res.* **1968**, 7, 193–199.

154 S. Dieth, D. Tritsch, J.-F. Biellmann, *Tetrahedron Lett.* **1995**, *36*, 2243–2246.

155 W. Adam, M. Lazarus, C. R. Saha-Möller, O. Weichold, U. Hoch, D. Häring, P. Schreier, Biotransformations with Peroxidases. In *Adv. Biochem. Eng. Biotech.* K. Faber (ed), Springer, Berlin, Heidelberg, 1999, Vol. 63, pp. 74–104.

156 L. Flohé.; CIBA Foundation Symposium, 1979.

157 E. de Boer, Y. van Kooyk, M. G. M. Tromp, H. Plat, R. Wever, *Biochim. Biophys. Acta* **1986**, *869*, 48.

158 M. Kuwahara, J. K. Glenn, M. A. Morgan, M. H. Gold, *FEBS LETT.* **1984**, *169*, 247.

159 M. I. Dolin, *J. Biol. Chem.* **1957**, *225*, 557.

160 M. P. J. van Deurzen, F. van Rantwijk, R. A. Sheldon, *Tetrahedron* **1997**, *53*, 13 183–13 220.

161 A. Liese, S. Lütz, I. Schröder. "The Relevance of Reaction Engineering in Electroenzymatic Oxidations"; BioTrans 2001, 2001, Darmstadt, Germany.

162 K. Seelbach, M. P. J. van Deurzen, F. van Rantwijk, R. A. Sheldon, U. Kragl, *Biotech. Bioeng.* **1997**, *55*, 283–288.

163 M. P. J. van Deurzen, B. W. van Groen, F. van Rantwijk, R. A. Sheldon, *Biocatalysis* **1994**, *10*, 247–255.

164 F. van de Velde, N. D. Lourenço, M. Bakker, R. A. Sheldon, *Biotech. Bioeng.* **2000**, *69*, 286–291.

165 S. Lütz, E. Steckhan, C. Wandrey, A. Liese.; DE 100 54082.1: Germany, 2000.

166 G. R. Schonbaum, B. R. Chance, Catalase. In *The Enzymes.* P. D. Boyer (ed), Academic Press, New York, 1976, pp. 363–408.

167 E. Magner, A. M. Klibanov, *Biotech. Bioeng.* **1994**, *46*, 175–179.

168 E. Horozova, N. Dimcheva, Z. Jordanova, *Bioelectrochem.* **2000**, *53*, 11–16.

169 J. A. Duine, J. J. Frank, *Trends Biochem. Sci.* **1981**, *2*, 278–280.

170 J. A. Duine, *Eur. J. Biochem.* **1991**, *200*.

171 J. A. Duine, J. Frank, *Biochem. J.* **1980**, *187*, 213–219.

172 A. Geerlof, J. B. A. van Tol, J. A. Jongejan, J. A. Duine, *Biosci. Biotech. Biochem.* **1994**, *58*, 1028–1036.

173 W. Somers, R. T. M. van den Dool, G. A. H. de Jong, J. A. Jongejan, J. A. Duine, J. P. van der Lugt, *Biotech. Techn.* **1994**, *8*, 407–412.

174 E. C. A. Stigter, G. A. H. de Jong, J. A. Jongejan, J. A. Duine, J. P. van der Lugt, W. A. C. Somers, *Enz. Microb. Tech.* **1996**, *18*, 489–494.

175 T. Ikeda, D. Kobayashi, F. Matsushita, *J. Electroanal. Chem.* **1993**, *361*, 221–228.

176 J. A. Duine, J. J. Frank, P. E. G. Verwiel, *Eur. J. Biochem.* **1981**, *118*, 395–399.

177 K. Burton, T. H. Wilson, *Biochem. J.* **1953**, *54*, 86.

178 J. A. Olsen, C. B. Anfinsen, *J. Biol. Chem.* **1953**, *202*, 841.

179 C. Anthony, *Biochem. J.* **1996**, *320*, 697–711.

180 A. J. J. Olsthoorn, J. A. Duine, *Biochem.* **1998**, *37*, 13 854–13 861.

181 A. Oubrie, H. J. Rozeboom, K. H. Kalk, J. J. Olsthoorn, J. A. Duine, B. W. Dijkstra, *EMBO J.* **1999**, *18*, 5187–5194.

182 R. J. Kazalauskas, *J. Org. Chem.* **1988**, *53*, 4633–4635.

183 E. C. A. Stigter, J. P. van der Lugt, W. A. C. Somers, *J. Mol. Cat. B: Enzymatic* **1997**, *2*, 291–297.

184 H. M. Pinheiro, J. M. S. Cabral, P. Adlercreutz, *Biocat.* **1993**, *7*, 83–96.

185 J. A. Duine, J. Frank, J. K. van Zeeland, *FEBS Lett.* **1979**, *108*, 443–446.

186 K. Matsushita, E. Shinagawa, O. Adachi, M. Ameyama, *Biochem.* **1989**, *28*, 6276–6280.

187 P. Dokter, J. J. Frank, J. A. Duine, *Biochem. J.* **1986**, *239*, 163–167.

188 P. Dokter, J. E. Wielink, M. A. van Kleef, J. A. Duine, *Biochem. J.* **1988**, *254*, 131–138.

189 J. Hoenes, V. Unkrig. Method for the calorimetric determination of an analyte with a PQQ-dependent dehydrogenase; US Patent 5.484.708, 1996.

190 C. S. R. Tecelao, F. v. Keulen, M. M. R. d. Fonseca, *Journal of Molecular Catalysis B: Enzymatic* **2001**, *11*, 719–724.

191 M. J. van der Werf, C. v. d. Ven, F. Barbirato, M. H. Eppink, J. A. M. d. Bont, W. J. v. Berkel, *J. Biol. Chem.* **1999**, *274*, 26 296–26 304.

192 W. A. Duetz, A. H. M. Fjaellman, S. Ren, C. Jourdat, B. Witholt, *Appl. Environ. Microbiol.* **2001**, *67*, 2829–2832.

193 S. Muniruzzaman, H. Tokunaga, K. Izumori, *J. Ferm. Bioeng.* **1994**, *78*, 145–148.

194 M. Manzoni, M. Rollini, S. Bergomi, *Process Biochem.* **2001**, *36*, 971–977.

195 G. Kinast, M. Schedel, *Angew. Chem.* **1981**, *93*, 799–800.

196 W. D. Murray, S. J. B. Duff, *Appl. Microbiol. Biotech.* **1990**, *33*, 202–205.

197 S. J. B. Duff, W. D. Murray, *Biotech. Bioeng.* **1988**, *31*, 44–49.

198 S. J. B. Duff, W. D. Murray, *Biotech. Bioeng.* **1988**, *31*, 790–795.

199 S. J. B. Duff, W. D. Murray, R. P. Overend, *Enz. Microb. Tech.* **1989**, *11*, 770–775.

200 S. J. B. Duff, W. D. Murray, *Biotech. Bioeng.* **1989**, *34*, 153–159.

201 K. Kawakami, T. Nakahara, *Biotech. Bioeng.* **1994**, *43*, 918–924.

202 F. Molinari, R. Villa, M. Manzoni, F. Aragozzini, *Appl. Microbiol. Biotech.* **1995**, *43*, 989–994.

203 R. Gandolfi, N. Ferrara, F. Molinari, *Tetrahedron Lett.* **2001**, *42*, 513–514.

204 J. Svitel, P. Kutnik, *Lett. Appl. Microbiol.* **1995**, *20*, 365–368.

205 J. Svitel, E. Sturdik, *Enz. Microb. Tech.* **1995**, *17*, 546–550.

206 F. Molinari, F. Aragozzini, J. M. S. Cabral, D. M. F. Prazeres, *Enz. Microb. Tech.* **1997**, *20*, 604–611.

207 M. Manzoni, F. Molinari, A. Tirelli, F. Aragozzini, *Biotech. Lett.* **1993**, *15*, 341–346.

208 F. Molinari, R. Villa, F. Aragozzini, R. Leon, D. M. F. Prazeres, *Tetrahedron Asym.* **1999**, *10*, 3003–3009.

209 F. Molinari, R. Gandolfi, F. Aragozzini, R. Leon, D. M. F. Prazeres, *Enz. Microb. Tech.* **1999**, *25*, 729–735.

210 A. Matsuyama, H. Yamamoto, N. Kawada, Y. Kobayashi, *J. Mol. Cat. B: Enzymatic* **2001**, *11*, 513–521.

211 H. Yamamoto, N. Kawada, A. Matsuyama, Y. Kobayashi, *Biosci. Biotech. Biochem.* **1995**, *59*, 1769–1770.

212 H. Yamamoto, N. Kawada, A. Matsuyama, Y. Kobayashi, *Biosci. Biotech. Biochem.* **1999**, *63*, 1051–1055.

213 A. Matsuyama, Y. Kobayashi, *Biosci. Biotech. Biochem.* **1994**, *58*, 1148–1149.

214 M. A. Bertola, H. S. Koger, G. T. Phillips, A. F. Marx, V. P. Claassen. A process for the preparation of (R)- and (S)-2,2-R1,R2–1,3-dioxolane-4-methanol: EP, 1987.

215 A. Liese, K. Seelbach, C. Wandrey, Oxidase of *Rhodococcus erythropolis*. In *Industrial Biotransformations*. A. Liese, K. Seelbach, C. Wandrey (eds), Wiley-VCH, Weinheim, 2000, pp. 163–164.

216 A. J. Carnell, *Adv. Biochem. Eng. Biotech.* **1999**, *63*, 57–72.

217 G. Fantin, M. Fogagnolo, P. P. Giovannini, A. Medici, P. Pedrini, *Tetrahedron Asym.* **1995**, *6*, 3047–3053.

218 E. Takahashi, K. Nakamichi, M. Furui, *J. Ferm. Bioeng.* **1995**, *80*, 247–250.

219 S. Shimizu, S. Hatori, H. Hata, H. Yamada, *Enz. Microb. Tech.* **1987**, *9*, 411–416.

220 K. Nakamura, Y. Inoue, T. Matsuda, A. Ohno, *Tetrahedron Lett.* **1995**, *36*.

221 M. Takemoto, K. Achiwa, *Tetrahedron Asym.* **1995**, *6*, 2925–2958.

222 J. Hasegawa, M. Ogura, S. Tsuda, S. I. Maemoto, H. Kutsuki, T. Ohashi, *Agric. Biol. Chem.* **1990**, *54*, 1819–1828.

223 K. Nakamura, K. Inoue, K. Ushio, S. Oka, A. Ohno, *Chem. Lett.* **1987**, *4*, 679–682.

224 A. Goswami, K. D. Mirfakhrae, R. N. Patel, *Tetrahedron Asym.* **1999**, *10*, 4239–4244.

225 Y. Kato, Y. Asano, *J. Mol. Cat. B: Enzymatic* **2001**, *13*, 27–36.

226 H. Kajiro, S. Mitamura, A. Mori, T. Hiyama, *Bull. Chem. Soc. Jpn.* **1999**, *72*, 1093–1100.

227 W.-H. Liu, C.-K. Lo, *J. Ind. Microbiol. Biotechnol.* **1997**, *19*, 269–272.

228 H. Schwartz, A. Liebig-Weber, H. Hochstätter, H. Böttcher, *Appl. Microbiol. Biotech.* **1996**, *44*, 731–735.

229 M. Matsuda, M. Sakashita, Y. Mitsuki, T. Yamaguchi, T. Fujii, Y. Sekine, *Arzneimittel-Forschung* **1994**, *44*, 55–59.

230 G. T. R. Palmore, H. Bertschy, S. H. Bergens, G. M. Whitesides, *J. Electroanal. Chem.* **1998**, *443*, 155–161.

231 S. C. Barton, H.-H. Kim, G. Binyamin, Y. Zhang, A. Heller, *J. Am. Chem. Soc.* **2001**, *123*, 5802–5803.

232 T. Chen, S. C. Barton, G. Binyamin, Z. Gao, Y. Zhang, H.-H. Kim, A. Heller, *J. Am. Chem. Soc.* **2001**, *123*, 8630–8631.

233 G. Tayhas, R. Palmore, H.-H. Kim, *J. Electroanal. Chem.* **1999**, *464*, 110–117.

234 J. St-Pierre, D. P. Wilkinson, *AIChE J.* **2001**, *47*, 1482–1486.

235 S. Tsujimura, A. Wadano, K. Kano, T. Ikeda, *Enz. Microb. Tech.* **2001**, *29*, 225–231.

236 A. F. Bückmann, G. Carrea, *Adv. Biochem. Eng. Biotechnol.* **1989**, *39*, 97–152.

237 B. Nidetzky, W. Neuhauser, D. Haltrich, K. Kulbe, *Biotech. Bioeng.* **1996**, *52*, 387–396.

238 J. M. Obon, M. J. Almagro, A. Manjon, J. L. Iborra, *J. Biotechnol.* **1996**, *50*, 27–36.

239 E. Steckhan, S. Herrmann, R. Ruppert, J. Thoemmes, C. Wandrey, *Angew. Chem. Int. Ed. Engl.* **1990**, *29*, 388–390.

240 P. Pasta, G. Carrea, N. Gaggero, G. Grogan, A. Willetts, *Biotech. Lett.* **1996**, *18*, 1123–1128.

241 G. Ottolina, G. Carrea, S. Riva, *Enz. Microb. Tech.* **1990**, *12*, 596–602.

242 R. J. Ansell, C. R. Lowe, *Appl. Microbiol. Biotech.* **1999**, *51*, 703–710.

243 S. Dilmaghanian, C. V. Stead, R. J. Ansell, C. R. Lowe, *Enz. Microb. Tech.* **1997**, *20*, 165–173.

244 R. J. Ansell, D. A. P. Small, C. R. Lowe, *J. Mol. Cat. B: Enzymatic* **1999**, *6*, 111–123.

16.3
Oxidation of Phenols

Andreas Schmid, Frank Hollmann, and Bruno Bühler

16.3.1
Introduction

Several classes of oxidoreductases accept phenols and their derivatives as substrates for oxidation reactions. A broad range of products can be obtained depending on the substrates and enzymes applied (Fig. 16.3-1). Several monooxygenases catalyze the hydroxylation of the aromatic ring specifically *ortho* or *para* to the existing phenolic alcohol function (Fig. 16.3-1 A). Oxidases can be used to catalyze the stereospecific benzylic hydroxylation of aliphatic side chains to (*R*) or (*S*) alcohols and the further oxidation of benzylic alcohols to corresponding ketones or aldehydes; furthermore, elimination to (*Z*) or (*E*) alkenes can be obtained if desired (Fig. 16.3-1 B). Laccases and peroxidases generate phenoxy radicals which – depending on the reaction conditions – can react further with phenols to structurally complex dimers or conducting polymers (Fig. 16.3-1 C). Even nitration reactions are reported (Fig. 16.3-1 D). Thus, enzymatic modification opens up new possibilities for synthetic chemistry with aromatic compounds under mild and non-toxic conditions.

16.3.2
Oxidases

16.3.2.1
Vanillyl-alcohol oxidase (E. C. 1.1.3.38)

The enzyme vanillyl-alcohol oxidase (VAO, E.C. 1.1.3.38) was examined in detail with respect to mechanism, structural properties, and biotechnological applications by van Berkel and coworkers, giving an excellent example of how detailed biochemical studies provide a basis for preparative biocatalytic applications (for recent reviews see [1, 2]). The homooctamer with a monomer mass of 65 kDa was isolated and purified from *Penicillium simplicissimum*. The catalytic mechanism of VAO-catalyzed oxidation of *para*-alkyl phenols was studied in detail [3–5]. After initial hydride abstraction from the Cα atom, a binary complex of the intermediate *para*-quinone methide and reduced FAD reacts with molecular oxygen, regenerating the

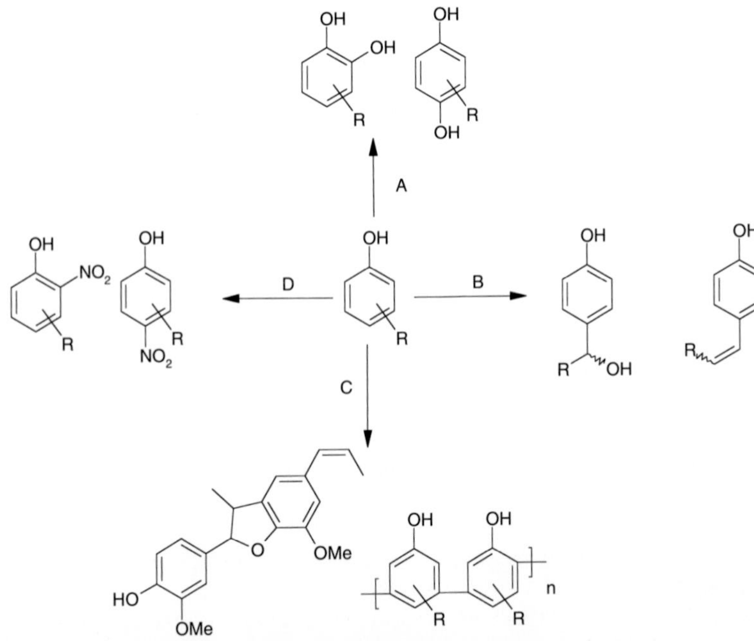

Figure 16.3-1. Enzyme-catalyzed oxidations of phenols. A: *ortho-* and *para-*hydroxylations catalyzed by monooxygenases (Sects. 16.3.3.2 and 16.3.6.2); B: oxidation at the benzylic position catalyzed by oxidases (Sects. 16.3.2.1 and 16.3.5); C: coupling reactions catalyzed by peroxidases and laccases (Sects. 16.3.4.1 and 16.3.2.2); D: nitration reactions catalyzed by peroxidases (Sect. 16.3.4.3).

oxidized prosthetic group. Depending on the nature of the aliphatic side chain, the *para*-quinone methide is hydroxylated to (chiral) benzylic alcohols (short aliphatic side chains) or rearranges yielding benzylic alkenes (long aliphatic side chains) (Fig. 16.3-2). Table 16.3-1 shows a selection of reactions catalyzed by VAO as well as the kinetic constants thereof[3, 6].

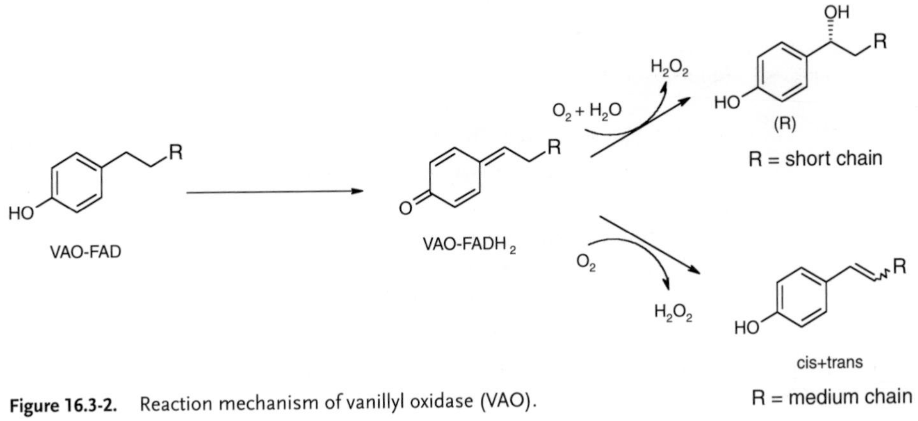

Figure 16.3-2. Reaction mechanism of vanillyl oxidase (VAO).

Table 16.3-1. Substrate spectrum and kinetic constants of vanillyl oxidase.

Substrate	Product(s)[a]	K_M [µM]	k_{cat} [s^{-1}]	k_{cat}/K_M [10^{-3}] [s^{-1} M^{-1}]
	76% alcohol 24% alkene	9	2.5	280
	68% alcohol 32% alkene	4	4.2	1050
	90% alcohol 10% alkene	6	4.9	820
	20% alcohol 80% alkene	16	1.3	81
	26% alcohol 74% alkene	72	0.5	7
	1% alcohol 99% alkene	2	1.2	600
	100% alkene	8	0.3	38
	100% alkene	42	< 0.001	< 0.02
	40% 60%	65	1.4	21
	16% alcohol 60% ketone 24% alkene	77	0.5	7
	4% alcohol 2% ketone 94% alkene	94	0.7	7

Table 16.3-1.   (cont.).

Substrate	Product(s)[a]	K_M [μM]	k_{cat} [s^{-1}]	k_{cat}/K_M [10^{-3}] [s^{-1} M^{-1}]
(structure)	100% ketone	222	0.7	3
(structure)	100% ketone	4.9	13.0	2700
(structure)	(structure)	4.8	6.5	1400
(structure)	(structure)	290	5.4	19
(structure)	(structure)	240	1.3	5.4
(structure)	(structure)	65	5.3	82

a Beside s the structure shown the products formed include benzylic alcohols, benzylic alkenes and benzylic ketones.

VAO exhibits a remarkable activity towards 4-alkylphenols, bearing aliphatic side chains of up to seven carbon atoms. The maximum chain-length of 7 is in accordance with structural data obtained from X-ray crystallography[7]. Short-chain 4-alkylphenols are mainly hydroxylated at the Cα position, whereas medium-chain 4-alkylphenols are dehydrogenated to 1-(4'-hydroxyphenyl)alkenes (Fig. 16.3-2)[6]. The hydroxylation reaction is highly stereospecific, producing the (R)-enantiomer with ee values of up to 94%[8]. Furthermore, VAO also catalyzes the further oxidation of the alcohols to the corresponding ketones. Here, the VAO-catalyzed oxidation of (S)-alcohols is far more efficient than the oxidation of (R)-alcohols, promoting a possible application in kinetic resolution reactions. Substrates with more space-consuming alkyl side chains are dehydrogenated by the action of VAO. With para-methyl phenols (e.g. cresol), a very low conversion rate is found which is due to the formation of a stable intermediate formed through a nucleophilic attack of the reduced FAD on the para-quinone methide, yielding a covalent bond[2]. Since the rate-limiting hydrolysis of this intermediate is acid-catalyzed, the pH optimum of the reaction shifts from alkaline to acidic values. The formation of such a covalent

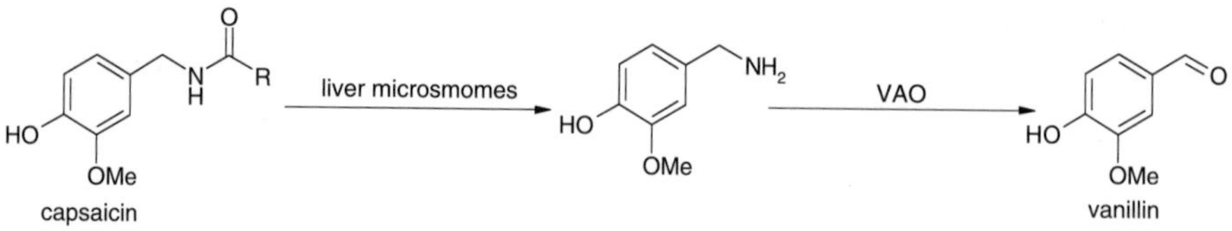

Figure 16.3-3. Potential biotechnological production route to vanillin from natural components with vanillyl oxidase.

intermediate is supposed to be more unlikely with increasing length of the aliphatic side chain, because of increasing steric hindrance.

Much attention has been paid to the shift from hydroxylation to dehydrogenation with increasing length of the side chain. The product ratio between alcohols and alkenes is strongly influenced by the extent of hydratation of the intermediate, *para*-quinone methide. Thus, by using organic media with a low water content the overall alkene yield could be significantly increased. The same is true for monovalent anions such as Cl⁻, Br⁻, or SCN⁻, which bind to the active site, thereby decreasing the water concentration at the active site [9]. By enzyme engineering based on the three-dimensional structure [7], the ratio between hydroxylation products and dehydrogenation products could be shifted either in favor of the alcohols, when Asp170 was exchanged with Glu, or in favor of the alkenes, when Asp170 was exchanged with Ser [10]. Double mutants of VAO (D170S/T457E and D170A/T457E) were produced based on the same rational approach, thus inverting the stereospecificity of the VAO-catalyzed hydroxylation of 4-ethyl phenol from (R) to (S) (ee = 80%) [11].

The VAO-catalyzed production of vanillin is of special synthetic interest. In particular, a route starting from capsaicin that is readily available from red hot pepper has some biotechnological potential. Here, vanillylamine is obtained by hydrolysis of capsaicin using rat liver microsomes and further oxidized by VAO (Fig. 16.3-3). Furthermore, a one-pot synthesis using carboxylesterase for capsaicin hydrolysis is proposed [12].

16.3.2.2
Laccase (E.C. 1.10.3.2)

Recently, laccases found some interest for synthetic application. Laccases are widely distributed in plants and fungi [13]. The copper-containing enzymes are some of the few oxidases so far reported to reduce molecular oxygen to water (aside from cytochrome c oxidase and others). This ability was recently exploited in a novel regeneration concept for flavin-dependent enzymes (see Chapter 16.2) [14].

Purified laccase oxidizes various phenolic compounds via hydrogen abstraction. The resulting phenoxy radical undergoes various dimerization and oligomerization reactions. Even though the synthetic potential of such reactions has to be considered as moderate, in some cases interesting products (such as complex coumaran type compounds) can be obtained in reasonable yields from simple phenols [15].

Laccases alone are not able to oxidize benzyl alcohols. Bourbonnais and Paice [16]

Table 16.3-2. Laccase/ABTS-catalyzed oxidations to corresponding aldehydes.

Catalyzed reaction	Yield [%]	Literature
	94	[1]
	92	[1]
	92	[1]
	98	[1]
	90	[1]
	92	[2]
	98	[2]
	89	[2]

1 A. Potthast, T. Rosenau, C. L. Chen, J. S. Gratzl, J. Mol. Cat. A.: Chemical 1996, 108, 5–9.

2 A. Potthast, T. Rosenau, C. L. Chen, J. S. Gratzl, J. Org. Chem. 1995, 60, 4320–4321.

were the first to report that laccase in the presence of a specific compound, usually called a "mediator", is able to catalyze the oxidation of benzyl alcohols. Mostly ABTS (2,2'-azino-bis(3-ethylbenzothiazoline-6-sulfonic acid), HOBT (1-hydroxybenzotria-zole) [17], and NHAA (N-hydroxyacetanilide) [18] have been used as mediators so far.

The actual role of the mediator is not yet fully understood, although Potthast et al. recently found evidence that laccase produces reactive radical species of ABTS and

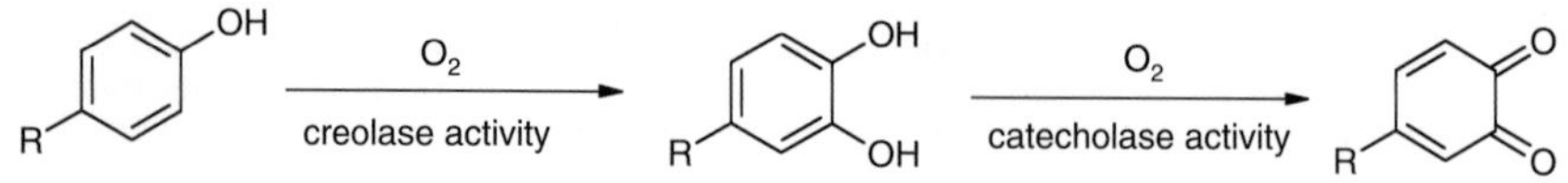

Figure 16.3-4. Oxidation of phenols catalyzed by tyrosinase displaying so-called creolase and catecholase activities.

HOBT, which perform the actual oxidations[17]. Nevertheless, some preparative oxidations of various benzylic alcohols are reported (Table 16.3-2).

It should be pointed out here that the laccase-mediator system still is far from being economically feasible.

16.3.3
Monooxygenases

16.3.3.1
Tyrosinase (E.C. 1.10.3.1)

Tyrosinases (synonyms: phenol oxidases, poly-phenolases or polyphenol oxidases) are copper-containing monooxygenases, which catalyze two consecutive reactions with molecular oxygen as cosubstrate, namely the *ortho*-hydroxylation of phenols and the oxidation of the resulting catechols to *ortho*-quinones (Fig. 16.3-4).

The initial (phenol-hydroxylating) activity is usually referred to as creolase activity, whereas the second (catechol-oxidizing) activity is most commonly called catecholase activity[19]. The classification of tyrosinases (polyphenol oxidases) is somewhat ambiguous; enzymes exhibiting monophenol oxidase activity are classified as E.C. 1.14.18.1., but those with catechol oxidase activity as E.C. 1.10.3.2. However, many enzymes exhibit both activities, and a more appropriate classification of all two-electron-accepting copper monooxygenases as E.C. 1.14.18.1 was proposed[20].

In animals, tyrosinase is involved in the formation of melamines, and in plants, tyrosinase leads to the well-known browning of open surfaces of fruits[21].

Much attention has been paid to the mechanism[20, 22]. In the active site, two copper(I) ions bind molecular oxygen. Upon binding of the phenolic substrate, the *ortho*-position is attacked electrophilically by one of the activated oxygen atoms. The resulting copper-bound catechol serves as an internal electron donor and leaves the active site as *ortho*-quinone. Figure 16.3-5 illustrates this mechanism.

In order to prevent rapid quinone polymerization in aqueous media, the quinones are usually reduced to the catechols (most commonly by ascorbic acid) (Fig. 16.3-6).

Several tyrosinase-catalyzed oxidations of phenols have been reported; some of these are presented in Table 16.3-3.

Tyrosinase was reported to hydroxylate and oxidize tyrosine residues in proteins[23], which is important in the production of moisture-resistant adhesives. In fact, tyrosinase has been used for the production of synthetic glues with similar compositions to those of naturally occurring adhesives such as mussel glue[24].

An interesting cascade reaction was reported by Waldmann *et al.*[25, 26]. Tyr-

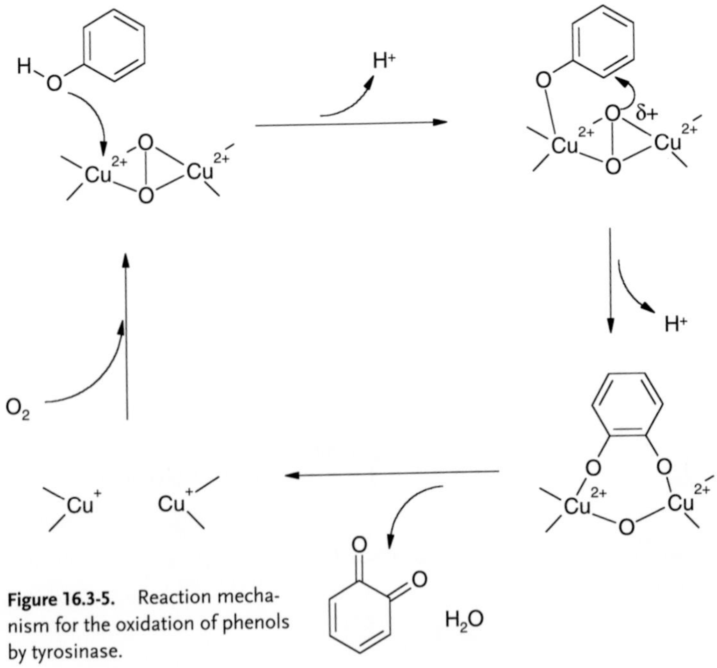

Figure 16.3-5. Reaction mechanism for the oxidation of phenols by tyrosinase.

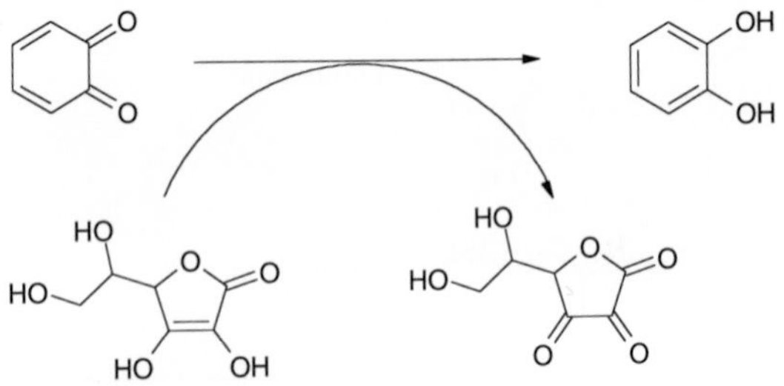

Figure 16.3-6. Ascorbic acid-driven reduction of quinones.

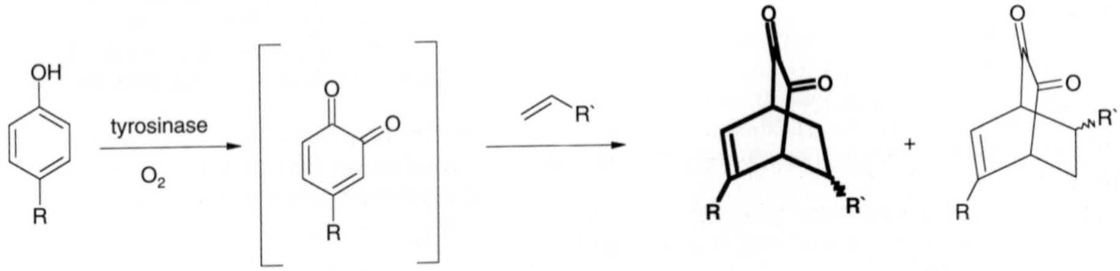

Figure 16.3-7. Chemoenzymatic Diels-Alder reactions. *Ortho*-quinones (dienes), derived from phenols by oxidation with tyrosinase, spontaneously react with dienophils.

osinase, immobilized on glass beads, was used to oxidize several phenols in chloroform as the organic medium. The products of the enzymatic oxidation step, the *ortho*-quinones, served *in situ* as dienes in a Diels-Alder reaction (Fig. 16.3-7). Table 16.3-4 summarizes some phenols (dienes after enzymatic oxidation) and dienophiles with which such a reaction cascade was observed.

Table 16.3-3. Oxidations of phenols catalyzed by tyrosinase.

Substrate	Product	References and remarks
R = OCH₃, OC₂H₅, CH₃, C(CH₃)₃, Halogen, etc.		Electron-rich phenols are preferred [3]
R = H, CH₃		[4, 5]
		L-DOPA production [6]
		Possible agent in melanoma treatment [7, 8]
		Coumestans [9]
		Phenoxazones [10]

3 S. Passi, M. Nazzaro-Porro, *Brit. J. Dermatol.* **1981**, *104*, 659.

4 M. Jimenez, F. Garcia-Carmona, F. Garcia-Canovas, J. L. Iborra, J. A. Lozano, F. Martinez, *Arch. Biochem. Biophys.* **1984**, *235*, 438.

5 M. Jimenez, F. Garcia-Carmona, F. Garcia-Canovas, J. L. Iborra, J. A. Lozano, *Int. J. Biochem.* **1985**, *17*, 891.

6 G. M. Carvalho, T. L. M. Alves, D. M. G. Freire, *Appl. Biochem. Biotech.* **2000**, *84–86*, 791–800.

7 M. E. Rice, B. Moghaddam, C. R. Creveling, K. R. Kirk, *Anal. Chem.* **1987**, *59*, 1534.

8 R. S. Phillips, J. G. Fletscher, R. L. Von Tersch, K. L. Kirk, *Arch. Biochem. Biophys.* **1990**, *276*, 65.

9 U. T. Bhalearo, C. Muralikrishna, G. Pandey, *Synth. Commun.* **1989**, *19*, 1303.

10 O. Toussaint, K. Lerch, *Biochem.* **1987**, *26*, 8567.

By this reaction sequence, highly functionalized bicyclo-[2.2.2]-octenes can be obtained from simple phenols and alkenes as starting materials. The overall yields reported are usually satisfactory (> 70%). The Diels-Alder products are racemic, probably because the Diels-Alder reaction proceeds in the bulk organic phase without involvement of tyrosinase.

Table 16.3-4. Substrates for the reaction cascade including tyrosinase catalyzed oxidation of phenols and a Diels-Alder-reaction [11, 12].

Phenols	Dienophiles

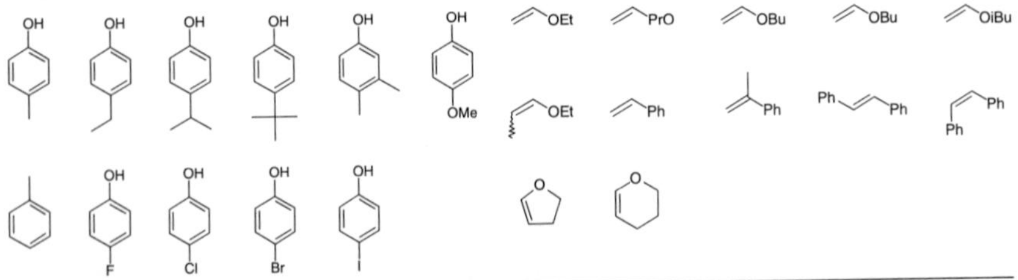

Via a tyrosinase catalyzed reaction the phenols are transformed to dienes, which subsequently react with the dienophiles in a Diels-Alder-reaction as shown in Figure 16.3-7.

11 G. H. Müller, H. Waldmann, *Tetrahedron Lett.* **1996**, *37*, 3833–3836.

12 G. H. Müller, A. Lang, D. R. Seithel, H. Waldmann, *Chem. Eur. J.* **1998**, *4*, 2513–2522.

16.3.3.2
2-Hydroxybiphenyl-3-monooxygenase (HbpA, E. C. 1.14.13.44)

The flavin-dependent, homotetrameric HbpA is the first enzyme in the biodegradation pathway of 2-hydroxybiphenyl in *Pseudomonas azelaica* HBP1 [27]. HbpA catalyzes the selective *ortho*-hydroxylation of a broad range of phenols to the corresponding catechols, utilizing NADH as cofactor (Fig. 16.3-8 and Table 16.3-5).

Compared to the chemical synthesis of *ortho*-substituted catechols (*ortho*-hydroxylation and aromatization procedures) [28–31], such an enzymatic approach is superior with respect to the number of steps involved as well as simplicity, selectivity, and yield. The resulting *ortho*-substituted catechols are valuable building blocks [32].

HbpA is an excellent example of *in vivo* as well as *in vitro* biocatalysis. Since the desired catechols are rapidly degraded via the *P. azelaica meta*-cleavage pathway by two catechol-2,3-dioxygenases, the gene coding for HbpA was expressed in *E. coli* JM109, which served as a biocatalyst accumulating the desired products [33]. Drawbacks such as inhibition by substrate and product can be overcome by continuous substrate feeding and *in situ* recovery of the catechol products with solid adsorbents

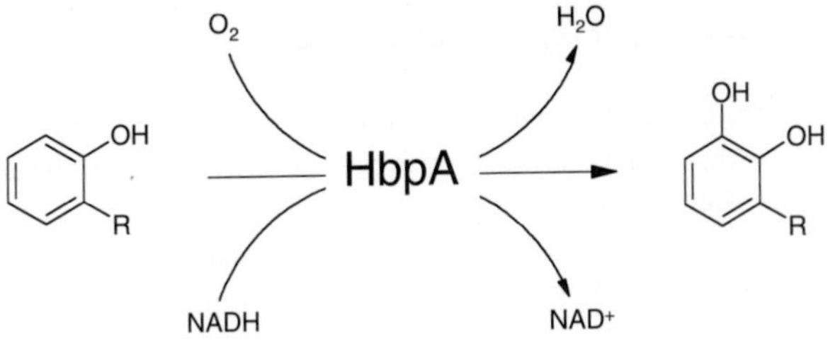

R = Ph, 2'-OH-Ph, 2,3-(OH)₂Ph, F, Cl, Br, Me, Et, Pr, *i*-Pr, But

Figure 16.3-8. Reaction scheme for the *ortho*-hydroxylation of phenol derivatives catalyzed by 2-hydroxybiphenyl-3-monooxygenase (HbpA).

Table 16.3-5. Substrates and relative activities of 2-hydroxybiphenyl-3-monooxygenase (HbpA) [13].

Substrate	Product	Relative activity [%][a]
		100
		34
		49 Native substrate
		24
		36
		10
		20
		33

a Relative activities were determined polarographically with whole cells of recombinant *E. coli* containing HbpA. 100 % corresponds to the HbpA-dependent specific oxygen uptake rate of whole cells incubated with 2,2'-dihydroxybiphenyl.

13 A. Schmid, H.-P. E. Kohler, K.-H. Engesser, *J. Mol. Cat. B: Enzymatic* **1998**, 5, 311–316.

in such a way that substrate and product concentrations can be kept below toxic levels [32]. Thus, several 3-substituted catechols were produced in gram amounts with satisfactory to high yields (Table 16.3-6).

The *in vivo* processes are based on a recombinant *E. coli* as catalyst [33]. Optimized space-time yields of up to 0.39 g L^{-1} h^{-1} for the formation of 3-phenyl catechol from 2-phenyl phenol can be reached [34].

The enzyme itself was purified and characterized in detail [27, 35]. Based on this knowledge and via directed evolution, HbpA characteristics were modified (Meyer, Schmid and Witholt, unpublished results) yielding HbpA variants with improved

Table 16.3-6. Preparative-scale production of 3-substituted catechols using *E. coli* JM101 containing 2-hydroxybiphenyl-3-monooxygenase [14].

Product	Product recovered [g]	Molar yield [%]
HO, OH — 3-phenylcatechol (biphenyl)	8.1	94
HO, OH — 3-ethylcatechol	2.1	95
HO, OH — 3-propylcatechol	0.6	71
HO, OH — 3-isopropylcatechol	2.2	77
HO, OH — 3-isobutylcatechol	1.7	85
HO, OH — Cl (3-chlorocatechol)	0.9	71
HO, OH — Br (3-bromocatechol)	2.1	71

14 M. Held, W. Suske, A. Schmid, K. Engesser, H. Kohler, B. Witholt, M. Wubbolts, *J. Mol. Cat. B: Enzymatic* 1998, 5, 87–93.

catalytic properties and changed substrate spectrum. For example, a new mutant with drastically decreased unproductive NADH oxidation and concomitant formation of hydrogen peroxide was developed. This so-called uncoupling reaction is quite common amongst flavin-dependent monooxygenases, and represents the major mechanism of autoregeneration amongst oxidases. Furthermore, the activity toward several substrates that are poorly converted by native HbpA, such as 2-*sec*-butylphenol (30 % activity increase), 2-*tert*-butylphenol (fivefold activity increase) or guaiacol (more than eightfold increase in K_M/k_{cat}), could be improved [36]. The HbpA substrate spectrum could be enlarged even more via directed evolution. Recently, an HbpA mutant was found that initiated the production of indigo starting from indole. It is assumed that HbpA converts indole into the 2,3-epoxide, which spontaneously dimerizes to indigo (Fig. 16.3-9) [37].

In vitro application of HbpA (and monooxygenases in general) offers some advantages over whole-cell biotransformations. For example, toxic effects on cell metabolism and further metabolization of the desired product can be avoided, and experimentally demanding *in vivo* set-ups are not necessary (beneficial for organic chemists). The major challenge in *in vitro* biotransformations is the efficient

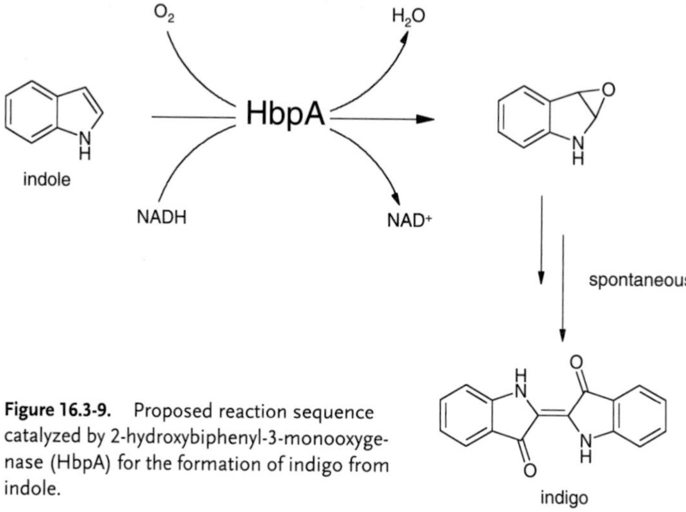

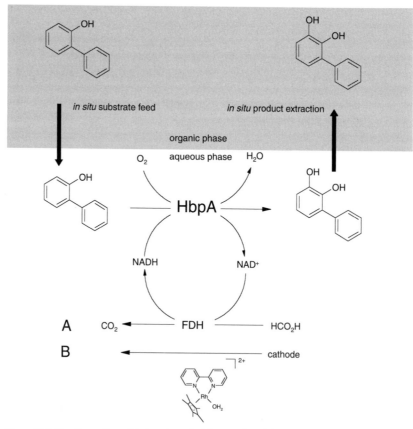

Figure 16.3-9. Proposed reaction sequence catalyzed by 2-hydroxybiphenyl-3-monooxygenase (HbpA) for the formation of indigo from indole.

Figure 16.3-10. Formation of 3-phenylcatechol from 2-phenylphenol catalyzed by partially purified 2-hydroxybiphenyl-3-monooxygenase (HbpA) in organic aqueous emulsions. Regeneration of NADH was achieved *in situ* with formate dehydrogenase (FDH) (A) or indirectly electrochemically with [Cp*Rh(bpy)(H₂O)]²⁺ (B).

Table 16.3-7. Substrates and products of peroxidase – catalyzed oxidative di- and oligomerizations of phenols.

Substrate	Products	References and applications
		Alkaloid synthesis [15]
		Alkaloid synthesis [15]
		Antimicrobial compounds [16]
		Phytoalexin activity [17]
		Cancer theraphy [18]

regeneration of reduced nicotinamide coenzymes. The general strategies are described in Chapter 7. Furthermore, the production enzyme must be easily available in large amounts. HbpA was obtained in gram amounts from recombinant *E. coli* in a one-step operation via expanded bed adsorption chromatography [38]. Limitations

Table 16.3-7.   (cont.).

Substrate	Products	References and applications
		Melanin synthesis[19]
		Racemic[20]
		Racemic[20]
		Quest Int. Naarden, The Netherlands, R = arrabinoxylan, carbohydrate gel which retains water

15 A. R. Krawczyk, E. Lipkowska, J. T. Wrobel, *Coll. Czech. Chem. Commun.* **1991**, *56*, 1147.
16 A. Kobayashi, Y. Koguchi, H. Kanzaki, S. I. Kajiyama, K. Kawazu, *Biosci. Biotech. Biochem.* **1994**, *58*, 133.
17 D. M. X. Donelly, F. G. Murphy, J. Polonski, T. Prangé, *J. Chem. Soc. Perkin Trans. I* **1987**, 2719.

18 A. E. Goodbody, T. Endo, J. Vukovic, J. P. Kutney, L. S. L. Choi, M. Misawa, *Planta Med.* **1988**, 136.
19 M. d'Ischia, A. Napolitano, K. Tsiakas, G. Prota, *Tetrahedron* **1990**, *46*, 5789.
20 M. M. Schmitt, E. Schüler, M. Braun, D. Häring, P. Schreier, *Tetrahedron Lett.* **1998**, *39*, 2945–2946.

due to low solubility of substrates and products can be overcome in biphasic reaction systems (Fig. 16.3-10). HbpA exhibits significant activity in the presence of various organic solvents such as 1-decanol, hexadecane or heptane[39].

Thus, the synthetic *in vitro* application of HbpA was done via an emulsion process. Several regeneration strategies for NADH were reported (Fig. 16.3-10).

In the emulsion process, a high 3-phenylcatechol concentration in the organic phase and the same or higher productivities (up to 0.45 g L^{-1} h^{-1}) as in the *in vivo* process were achieved[40]. Here, formate dehydrogenase and formate served as the coenzyme regeneration system (Fig. 16.3-10 A). The benefits of this regeneration

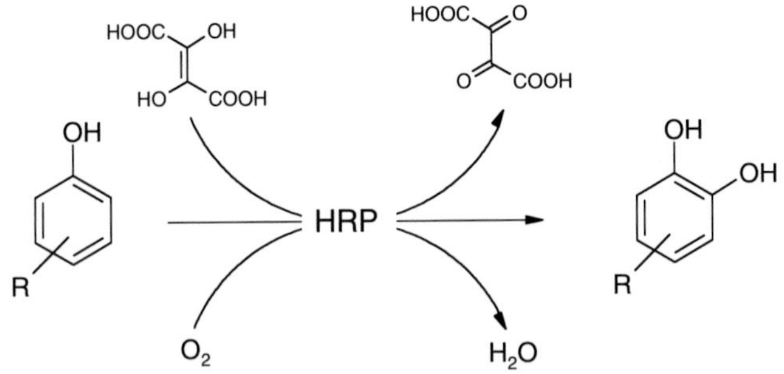

Figure 16.3-11. Hydroxylation of phenols to catechols catalyzed by horseradish peroxidase (HRP).

system are described in Chapter 16.6. Even electrical power could be used as a source of reduction equivalents (Fig. 16.3-10 B) [41].

16.3.4
Peroxidases

16.3.4.1
Oxidative Coupling Reactions

Phenols are typical substrates for peroxidases. Quite similarly to the laccase-mechanism (described earlier in this chapter), peroxidases catalyze phenol oxidations via hydrogen abstraction. The radicals thus generated leave the active site and

Table 16.3-8. Selected hydroxylation reactions of phenols catalyzed by horseradish peroxidase.

Substrate	Product	Literature
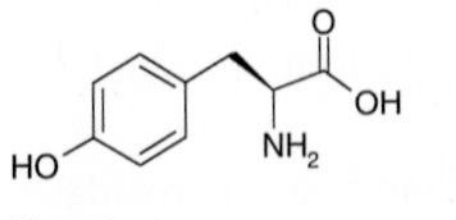 Tyrosine	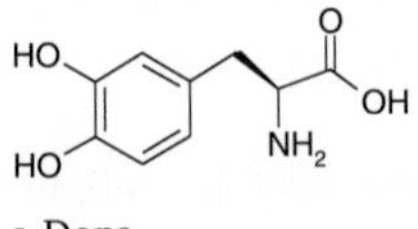 L-Dopa	[21]
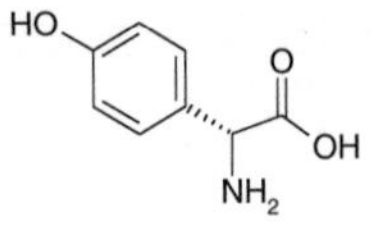	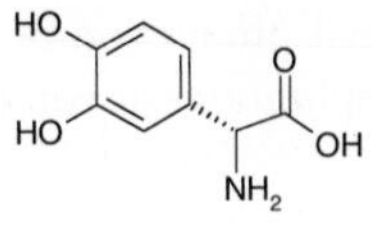	[21]
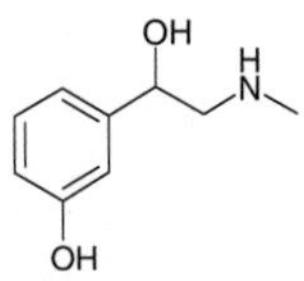	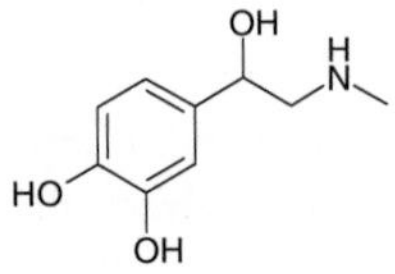 Adrenaline	[21]

21 A. M. Klibanov, Z. Berman, B. N. Alberti, *J. Am. Chem. Soc.* **1981**, *103*, 6263–6364.

Table 16.3-9. Selected nitration reactions of phenols catalyzed by soybean peroxidase.

Substrate	Product(s), Yield [%]	
	ortho	*para*

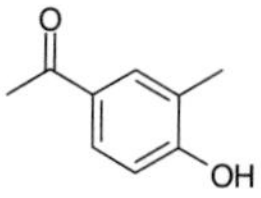

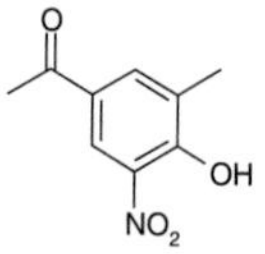

58   27

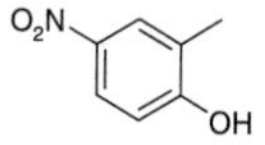

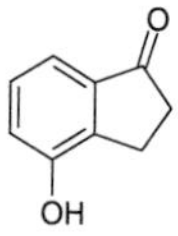

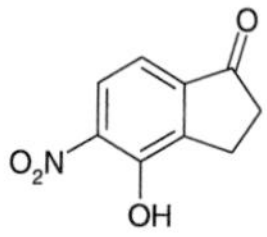

22   25

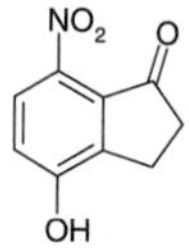

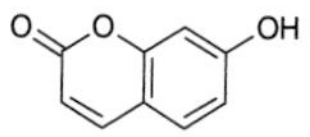

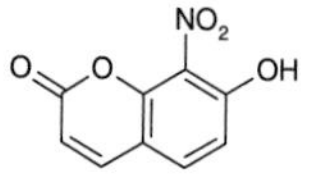

41   20

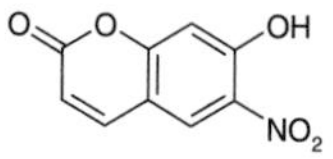

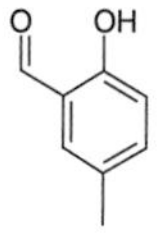

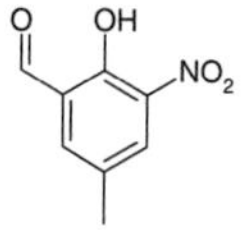

25   –

react with other aromatic compounds (depending on the reaction conditions) to form dimeric and polymeric products[42]. A selection of dimeric products is presented in Table 16.3-7.

Recently, peroxidases, especially horseradish (HRP) and soybean peroxidase, found increasing interest in resin manufacturing. The peroxidase-catalyzed coupling of phenols[43], catechols[44], hydroquinones[45], or anilines[46, 47] is a potential substitute for the conventional production of phenolic resins using toxic formaldehyde[48]. The resins find applications as conductive polymers[45, 49].

16.3.4.2
Hydroxylation of Phenols

As early as 1961, Mason and coworkers reported that HRP, in the presence of dihydrofumaric acid as cofactor, catalyzes the hydroxylation of arenes (Fig. 16.3-11)[50].

Also lignin peroxidase was found to catalyze the oxidation of phenol, cresol, and tyrosine[51].

Table 16.3-10. Oxidation reactions of arylamines catalyzed by peroxidases.

Substrate	Product	References and remarks
NH_2 (aniline)	NO_2 (nitrobenzene)	Bromoperoxidase [22]
Aminopyrrolonitrin	Pyrrolonitrin	Chloroperoxidase [23]
NH_2, R (aminoarene)	NO_2, R (nitroarene)	Chloroperoxidase [24] R = o-, m-, p-Cl; p-CH$_3$; p-COOH

22 N. Itoh, N. Morinaga, T. Kouzai, *Biochem. Mol. Biol.* **1993**, *29*, 785–791.

23 S. Kirner, K.-H. van Pee, *Angew. Chem. Int. Ed.* **1994**, *33*, 352.

24 V. N. Burd, K.-H. van Pee, *Bioorg. Khim.* **1998**, *24*, 462–464.

16.3.4.3
Nitration of Phenols

Khmelnitsky and coworkers recently reported a rather unusual application of soybean peroxidase. In the presence of nitrite and hydrogen peroxide, phenols are nitrated. The nitration of tyrosine has been reported earlier [52, 53]. The substrate spectrum was enlarged by various phenolic compounds (Table 16.3-9). Thus, such an enzymatic nitration represents an alternative to chemical nitration (especially for acid-labile phenols, which cannot by nitrated chemically).

Other peroxidases such as HRP or CPO were also able to perform such reactions.

Another approach to the production of nitroarenes with peroxidases is based on the CPO (or bromoperoxidase)-catalyzed oxidation of arylamines. Table 16.3-10 gives a selection of peroxidase-catalyzed conversions of aniline derivatives to corresponding nitroarenes.

For example, aniline was converted into nitrobenzene by a bromoperoxidase from *Pseudomonas putida* [54], and aminopyrrolonitrin was converted into the antibiotic pyrrolonitrin by a CPO from *P. pyrrocinia* [55].

Table 16.3-11. Substrates and products of 4-cresol-oxidoreductase [25, 26].

Substrate	Product	Substrate	Product

25 W. McIntire, D. J. Hopper, T. P. Singer, *Biochem. J.* **1985**, *228*, 325–335.

26 W. McIntire, D. J. Hopper, J. C. Craig, E. T. Everhart, E. V. Webster, M. J. Causer, T. P. Singer, *Biochem. J.* **1984**, *224*, 617–621.

16.3.5
Other Oxidoreductases

16.3.5.1
4-Cresol-oxidoreductase (PCMH, E.C. 1.17.99.1)

This enzyme shares structural and mechanistic properties with VAO [11]. In contrast to VAO it is not an oxidase as regeneration of the covalently bound FAD with molecular oxygen is not possible. It is a flavocytochrome enzyme. The reduction equivalents from the substrate are transferred to a type c cytochrome [56, 57]. In

Table 16.3-12. Oxidations of 4-alkylphenols catalyzed by 4-ethylphenol oxidoreductase[27].

> 98 % e. e.

Substrate	Relative conversion rate [%][a]
p-Cresol	44
4-Ethylphenol	100
4-Propylphenol	112
4-Butylphenol	114
4-Pentylphenol	116
4-Heptylphenol	52
4-Nonylphenol	14

a 100 % corresponds to the 4-ethylphenol conversion rate.

27 C. D. Reeve, M. A. Carver, D. J. Hopper, *Biochem. J.* **1990**, *269*, 815–819.

addition to a cytochrome c / cytochrome c oxidase regeneration system[58], chemical reoxidation agents such as phenazine methosulfate, dichlorophenol indophenol[59], and ferrocenes[60–62] have been used.

The reaction mechanism is quite similar to the one of VAO and also includes an intermediate, the *para*-quinone methide. Like VAO, 4-cresol-oxidoreductase also exhibits a high enantioselectivity for (S)-1-(4'-hydroxyphenyl)alkylalcohols[59].

This enzyme accepts a broad range of substrates; *para*-methylphenols are preferably oxidized to the corresponding aldehydes, whereas the oxidation of *para*-alkylphenols results in the formation of significant amounts of (S)-alcohols (Table 16.3-11)[59, 63].

16.3.5.2
4-Ethylphenol Oxidoreductase

4-Ethylphenol oxidoreductase from *Pseudomonas putida* JD1 is structurally almost identical to 4-cresol oxidoreductase, but catalyzes the hydroxylation of *para*-alkylphenols with longer aliphatic chains (Table 16.3-12). The hydroxylation reactions enantioselectively produce (R)-alcohols[64, 65]. The regeneration properties of this enzyme are quite similar to 4-cresol oxidoreductase[61].

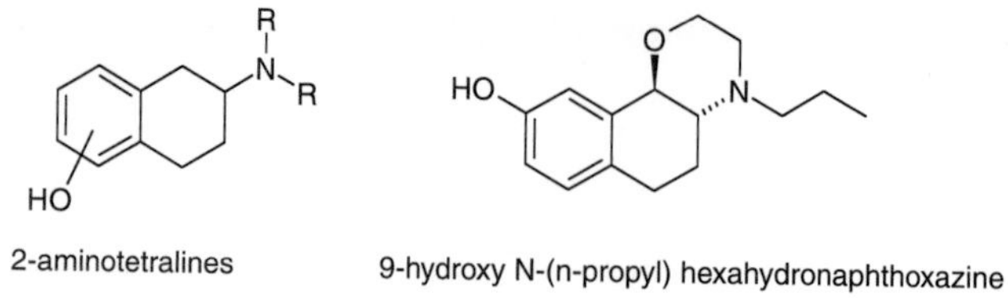

2-aminotetralines   9-hydroxy N-(n-propyl) hexahydronaphthoxazine

Figure 16.3-12. Substrates for phenol oxidase from *Mucuna pruriens*.
5-, 6-, or 7-Hydroxylated 2-aminotetralins with R = H or C_3H_7 and
9-hydroxy-*N*-(*n*-propyl)-hexahydronaphthoxazine are substrates for the
phenol oxidase.

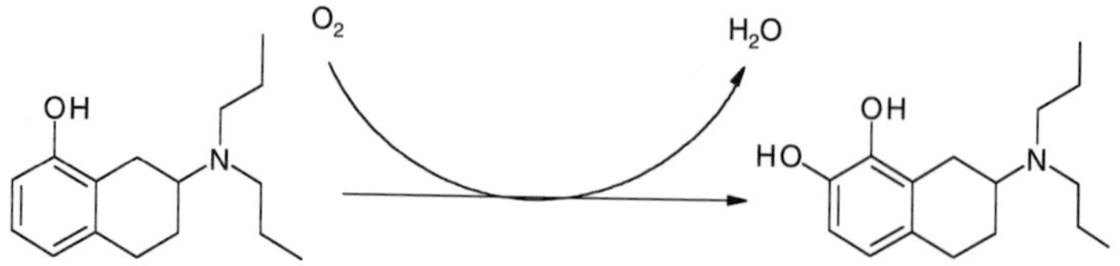

Figure 16.3-13. Formation of 7,8-dihydroxy *N*-(di-*n*-propyl)-2-aminotetralin
with *Mucuna*-phenoloxidase. Quinone formation is prevented *in situ* with
ascorbate as reductant.

16.3.6
In vivo Oxidations

16.3.6.1
Phenoloxidase of *Mucuna pruriens*

Like other phenoloxidases, this enzyme has a low substrate specificity and is able to
ortho-hydroxylate a whole range of *para*-substituted monocyclic phenols. The catechols produced belong to groups of fine chemicals and pharmaceuticals[66]. Furthermore, also bi- and tri-cyclic phenols were converted into catechols (Figure16.3-
12)[67]. 2-Aminotetralines, on the basis of their dopaminergic properties, are compounds of pharmaceutical interest.

Phenoloxidase (monophenol monooxygenase, E.C. 1.14.18.1) introduces one
atom of molecular oxygen into the substrate and was used in alginate-entrapped cells
or in partially purified form. The pharmaceutical 7,8-dihydroxy-*N*-(di-*n*-propyl)-
2-aminotetralin was produced continuously using a phenol oxidase suspension in
dialysis tubing in an airlift fermenter coupled to an aluminium oxide column for
selective product isolation (Figure 16.3-13)[68]. A product concentration of 130 mg/L
and a yield of 25 % were reached.

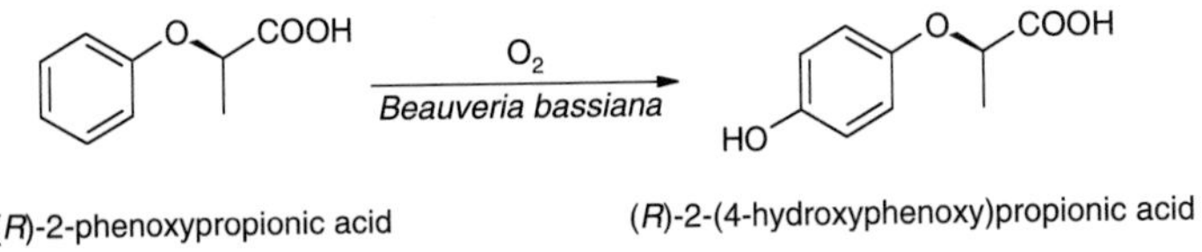

Figure 16.3-14. Regioselective *para*-hydroxylation of (R)-2-phenoxypropionic acid catalyzed by *Beauveria bassiana* (HPOPS process).

16.3.6.2
Monohydroxylation of (R)-2-Phenoxypropionic Acid and Similar Substrates [69, 70]

The product is a frequently used intermediate for the synthesis of enantiomerically pure aryloxyphenoxypropionic acid type herbicides. The enzyme catalyzing the hydroxylation of the phenolether is an oxidase, which is not further characterized. The biocatalyst *Beauveria bassiana* was found by an extensive screening of microorganisms for regioselective hydroxylation of (R)-2-phenoxypropionic acid and for substrate tolerance. This fungal strain was improved by random mutagenesis and screening, which resulted in strain LU 700. The hydroxylation is not growth-associated and the *ee* is increased during oxidation from 96% for the substrate to 98% for the product. After process optimization, a productivity of 7 g L^{-1} d^{-1} was reached. The biotransformation is carried out in a 120 000 L reactor at BASF in Germany.

The biocatalyst has a broad substrate spectrum. A compound needs the structural elements of a carboxylic acid and an aromatic ring system to be a substrate for the oxidase. Hydroxylation primarily takes place at the *para* position if it is free. If an alkyl group is in the *para* position, only the side chain is oxidized. In systems with more than one ring, the most electron-rich ring is hydroxylated.

16.3.6.3
Biotransformation of Eugenol to Vanillin [71]

The biotechnological production of vanillin is of interest because there is a large demand for vanillin originating from so called "natural" sources. Possible strategies for the biotechnological production of vanillin are reviewed by Priefert *et al.* [72].

One synthetically interesting strategy is the production of vanillin from eugenol. Here, a part of a catabolic pathway is used to accumulate an intermediate of this pathway. This was achieved by the knock-out of the enzyme catalyzing the further conversion of the putative product.

For the accumulation of vanillin from eugenol, the catabolism of eugenol in *Pseudomonas* sp. Strain HR199 (DSM7063) was used. In order to prevent further degradation of vanillin, the gene enconding vanillin dehydrogenase, responsible for the oxidation of vanillin to vanillic acid, was inactivated by insertion mutagenesis.

In a non-optimized biotransformation using growing cells in an aqueous mineral salts medium containing gluconate as a source of carbon and energy and 6.5 mM eugenol, vanillin accumulated up to a concentration of 2.9 mM, corresponding to a

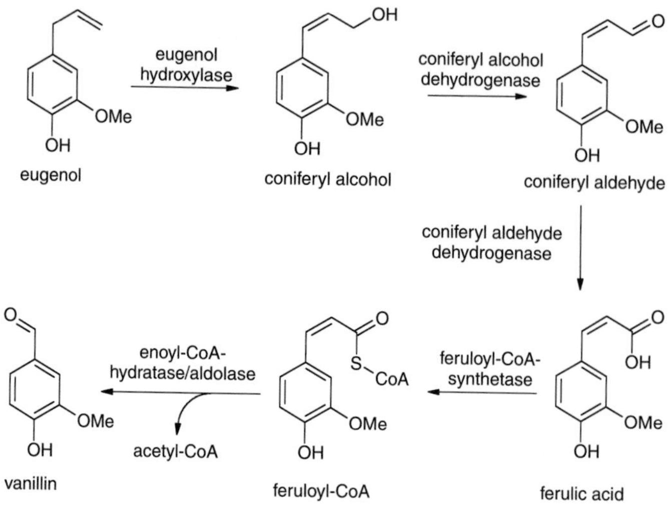

Figure 16.3-15. Multistep biotransformation of eugenol to vanillin catalyzed by whole cells of *Pseudomonas* sp. HR 199.

molar yield of 44.6%. The major drawback of the process is the degradation of vanillin by the action of coniferyl aldehyde dehydrogenase when coniferyl aldehyde is depleted from the medium.

References

1 R. H. H. van den Heuvel, M. W. Fraaije, A. Mattevi, C. Laane, W. J. H. van Berkel, *J. Mol. Cat. B: Enzymatic* **2001**, *11*, 185–188

2 R. H. H. van den Heuvel, C. Laane, W. J. H. van Berkel, *Adv. Synth. Cat.* **2001**, *343*, 515–520.

3 M. W. Fraaije, C. Veeger, W. J. H. van Berkel, *Eur. J. Biochem.* **1995**, *234*, 271–277.

4 M. W. Fraaije, W. J. H. van Berkel, *J. Biol. Chem.* **1997**, *272*, 18 111–18 116.

5 M. W. Fraaije, R. H. H. van den Heuvel, J. C. Roelofs, W. J. H. van Berkel, *Eur. J. Biochem.* **1998**, *253*, 712–719.

6 R. H. H. van den Heuvel, M. W. Fraaije, C. Laane, W. J. H. van Berkel, *J. Bacteriol.* **1998**, *180*, 5646–5651.

7 A. Mattevi, F. M. W., A. Mozzarelli, A. Olivi, W. J. H. van Berkel, *Structure* **1997**, *5*, 907–920.

8 F. P. Drijfhout, M. W. Fraaije, H. Jongejan, W. J. H. van Berkel, M. C. R. Franssen, *Biotech. Bioeng.* **1998**, *59*, 171–177.

9 R. H. H. van den Heuvel, J. Partridge, C. Laane, P. J. Halling, W. J. H. van Berkel, *FEBS Lett.* **2001**, *503*, 213–216.

10 R. H. H. van den Heuvel, F. M. W., W. J. H. van Berkel, *FEBS LETT.* **2000**, *481*, 109–112.

11 R. H. H. van den Heuvel, F. M. W., M. Ferrer, A. Mattevi, W. J. H. van Berkel, *PNAS* **2001**, *97*, 9455–9460.

12 R. H. H. van den Heuvel, M. W. Fraaije, C. Laane, W. J. H. van Berkel, *J. Argi. Food Chem.* **2001**, *49*, 2954–2958.

13 H. P. Call, I. Mücke, *J. Biotech.* **1997**, *53*, 163–202.

14 U. L. R. Baminger, C. Galhaup, C. Leitner, K. D. Kulbe, D. Haltrich, *J. Mol. Cat. B: Enzymatic* **2001**, *11*, 541–550.

15 T. Shiba, X. Ling, T. M., C.-L. Chen, *J. Mol. Cat. B: Enzymatic* **2000**, *10*, 605–615.

16 R. Bourbonnais, M. G. Paice, *Febs Lett.* **1990**, *267*, 99–102.

17 A. Potthast, T. Rosenau, K. Fischer, *Holzforschung* **2001**, *55*, 47–56.

18 M. Amann.; Proceedings of the International Symposium an Wood and Pulping Chemistry, 1997, Montreal, Canada.

19 D. Kertesz, D. Zito, *Biochim. Biophys. Acta* **1965**, *96*, 447.

20 S. G. Burton, *Catalysis Today* **1994**, *22*, 459–487.

21 D. Strack, W. Schliemann, *Angew. Chem.* **2001**, *113*, 3907–3911.

22 P. Capdeville, M. Maumy, *Tetrahedron Lett.* **1982**, *23*, 1573–1576.

23 K. Marumo, J. H. Waite, *Biochim. Biophys. Acta* **1986**, *872*, 98.

24 H. Yamamoto, H. Tanisho, S. Ohara, A. Nishida, *Int. J. Biol. Macromol.* **1992**, *14*, 66.

25 G. H. Müller, H. Waldmann, *Tetrahedron Lett.* **1996**, *37*, 3833–3836.

26 G. H. Müller, A. Lang, D. R. Seithel, H. Waldmann, *Chem. Eur. J.* **1998**, *4*, 2513–2522.

27 W. A. Suske, M. Held, A. Schmid, T. Fleischmann, M. G. Wubbolts, H.-P. E. Kohler, *J. Biol. Chem.* **1997**, *272*, 24 257–24265.

28 F. Chioccara, P. Gennaro, G. la Monica, R. Sebastino, B. Rindone, *Tetrahedron* **1991**, *47*, 4429–4434.

29 D. H. R. Barton, D. M. X. Donnelly, P. J. Guiry, J.-P. Finet, *J. Chem. Soc. Perkin Trans. I* **1994**, 2921ff.

30 A. Feigenbaum, J.-P. Pete, A. Poquet-Dhimane, *Tetrahedron Lett.* **1988**, *29*, 73–74.

31 K. A. Parker, K. K. A., *Journal of Organic Chemistry* **1987**, *52*, 674–676.

32 M. Held, W. Suske, A. Schmid, K. Engesser, H. Kohler, B. Witholt, M. Wubbolts, *J. Mol. Cat. B: Enzymatic* **1998**, *5*, 87–93.

33 A. Schmid, H.-P. E. Kohler, K.-H. Engesser, *J. Mol. Cat. B: Enzymatic* **1998**, *5*, 311–316.

34 M. Held, A. Schmid, H.-P. E. Kohler, W. A. Suske, B. Witholt, M. G. Wubbolts, *Biotech. Bioeng.* **1999**, *62*, 641–648.

35 W. A. Suske, W. J. H. van Berkel, H.-P. E. Kohler, *J. Biol. Chem.* **1999**, *274*, 33 355–33365.

36 A. Meyer, A. Schmid, M. Held, A. H. Westphal, M. Röthlisberger, H.-P. E. Kohler, W. J. H. van Berkel, B. Witholt, **2001**, submitted.

37 A. Schmid, **2001**.

38 J. Lutz, B. Krummenacher, B. Witholt, A. Schmid. "2-Hydroxybiphenyl 3-Monooxygenase: Large Scale Preparation and Cell Free Application in Emulsions"; BioTrans 2001, 2001, Darmstadt, Germany.

39 A. Schmid, J. Lutz, V. V. Mozhaev, L. Khmelnitsky, B. Witholt, *J. Mol. Cat. B: Enzymatic*, submitted.

40 A. Schmid, I. Vereyken, M. Held, B. Witholt, *J. Mol. Catal. B: Enzymatic* **2001**, *11*, 455–462.

41 F. Hollmann, A. Schmid, E. Steckhan, *Angew. Chem.* **2001**, *113*, 190–193.

42 W. Adam, M. Lazarus, C. R. Saha-Möller, O. Weichold, U. Hoch, D. Häring, P. Schreier, Biotransformations with Peroxidases. In K. Faber (ed), *Adv. Biochem. Eng. Biotech.*, Springer, Berlin, Heidelberg, 1999, Vol. 63, pp. 74–104.

43 H. Kurioka, H. Uyama, S. Kobayashi, *Polymer J.* **1998**, *30*, 526–529.

44 S. Dubey, D. Singh, R. A. Misra, *Enz. Microb. Tech.* **1998**, *23*.

45 P. Wang, S. Amarasinghe, J. Leddy, M. Arnold, J. S. Dordick, *Polymer* **1998**, *39*, 123–127.

46 J. A. Akkara, P. Salapu, D. L. Kaplan, *Ind. J. Chem.* **1992**, *31B*, 855–858.

47 J. Y. Shan, S. K. Cao, *Polym. Adv. Technol.* **2000**, *11*, 288–293.

48 P. W. Kopf, Encyclopedia of Polymer Science and Engineering; Wiley, New York, 1986, Vol. 11; pp. 45–95.

49 S. Kobayashi, I. Kaneko, H. Uyama, *Chem. Lett.* **1992**, 393.

50 D. R. Buhler, H. S. Mason, *Arch. Biochem. Biophys.* **1961**, *2*, 224.

51 M. W. Schmall, L. S. Gorman, J. S. Dordick, *Biochim. Biophys. Acta* **1989**, *999*, 267.

52 H. Shibata, Y. Kono, S. Yamashita, Y. Sawa, H. Ochiai, K. Tanaka, *Biochim. Biophys. Acta* **1995**, *1230*, 45–50.

53 A. van der Vliet, J. P. Eiserich, B. Halliwell, C. E. Cross, *J. Biol. Chem.* **1997**, *272*, 7617–7625.

54 N. Itoh, N. Morinaga, T. Kouzai, *Biochem. Mol. Biol.* **1993**, *29*, 785–791.

55 S. Kirner, K.-H. van Pee, *Angew. Chem. Int. Ed. Engl.* **1994**, *33*, 352.

56 W. McIntire, D. E. Edmondson, T. P. Singer, D. J. Hopper, *J. Biol. Chem.* **1980**, *255*, 6553–6555.

57 A. L. Bhattacharyya, G. Tollin, W. McIntire, T. P. Singer, *Biochem. J.* **1985**, *228*, 337–345.

58 W. McIntire, C. Bohmont. In de Gruyter, *Flavins and Flavoproteins*, Berlin, 1987, pp. 677–686.

59 W. McIntire, D. J. Hopper, J. C. Craig, E. T. Everhart, E. V. Webster, M. J. Causer, T. P. Singer, *Biochem. J.* **1984**, *224*, 617–621.

60 H. A. O. Hill, B. N. Oliver, D. J. Page, D. J. Hopper, *J. Chem. Soc., Chem. Commun.* **1985**, 1469–1471.

61 B. Brielbeck, M. Frede, E. Steckhan, *Biocatalysis* **1994**, *10*, 49–64.

62 E. Steckhan, Electroenzymatic Synthesis. In *Top. Curr. Chem.*; Springer-Verlag: Berlin; Heidelberg, 1994; Vol. 170; pp. 84–111.

63 W. McIntire, D. J. Hopper, T. P. Singer, *Biochem. J.* **1985**, *228*, 325–335.

64 C. D. Reeve, M. A. Carver, D. J. Hopper, *Biochem. J.* **1989**, *263*, 431–437.

65 C. D. Reeve, M. A. Carver, D. J. Hopper, *Biochem. J.* **1990**, *269*, 815–819.

66 N. Pras, H. J. Wichers, A. P. Bruins, T. M. Malingre, *Plant Cell, Tissue and Organ Culture* **1988**, *13*, 15–26.

67 N. Pras, G. E. Booi, D. Dijkstra, A. S. Horn, T. M. Malingre, *Plant Cell, Tissue and Organ Culture* **1990**, *21*, 9–15.

68 N. Pras, S. Batterman, D. Dijkstra, A. S. Horn, T. M. Malingre, *Plant Cell, Tissue and Organ Culture* **1990**, *23*, 209–215.

69 C. Dingler, W. Ladner, G. A. Krei, B. Cooper, B. Hauer, *Pesticide Science* **1996**, *46*, 33–35.

70 B. Cooper, W. Ladner, B. Hauer, H. Siegel. Verfahren zur fermentativen Herstellung von 2-(4-hydroxyphenoxy-)propionsäure, 1992, EP0465494B1.

71 J. Overhage, H. Priefert, J. Rabenhorst, A. Steinbüchel, *Appl. Microbiol. Biotech.* **1996**, *52*, 820–828

72 H. Priefert, J. Rabenhorst, A. Steinbüchel, *Appl. Microbiol. Biotech.* **2001**, *56*, 296–314.

16.4
Oxidation of Aldehydes

Andreas Schmid, Frank Hollmann, and Bruno Bühler

16.4.1
Introduction

To date, few reports on synthetic enzymatic oxidations of aldehydes have been published. Preparative applications reported include bioconversions of natural products such as retinal (Fig. 16.4-1 A) and various aliphatic and unsaturated aldehydes (Fig. 16.4-1 B). A broad range of aromatic acids can be obtained from their corresponding aldehydes (Fig. 16.4-1 C). Another reported reaction type is the production of olefins from aldehydes by oxidative removal of formic acid from the substrate (Fig. 16.4-1 D).

16.4.2
Alcohol Dehydrogenases

Alcohol dehydrogenases are generally applied for the interconversion of alcohols and aldehydes. Yet, these enzymes have also attracted interest due to their ability to oxidize aldehydes[1]. HLADH was shown to oxidize butanal[2]. This reaction, however, shows no potential for synthetic application unless a very efficient NAD^+ regeneration system is applied (Fig. 16.4-2). The catalytic activity of HLADH for the reduction of the aldehyde is more than 100 times higher than that for aldehyde oxidation (examined for benzaldehyde)[3]. As a result, the initially formed NADH is

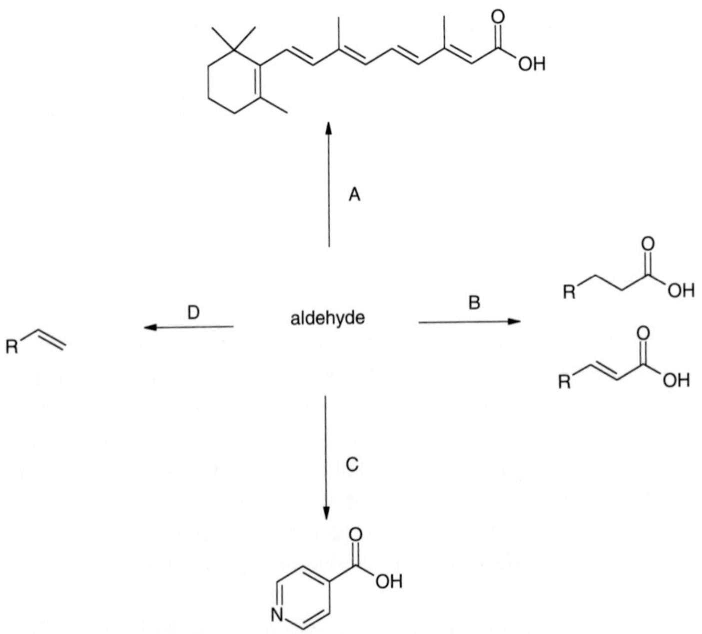

Figure 16.4-1. Selected enzymatic oxidations of aldehydes. A: oxidation of complex natural products such as retinal; B: oxidation of aliphatic and α,β-unsaturated aldehydes; C: oxidation of (hetero)arylic aldehydes; D: oxidative cleavage of the aldehyde-carbon atom yielding terminal alkenes.

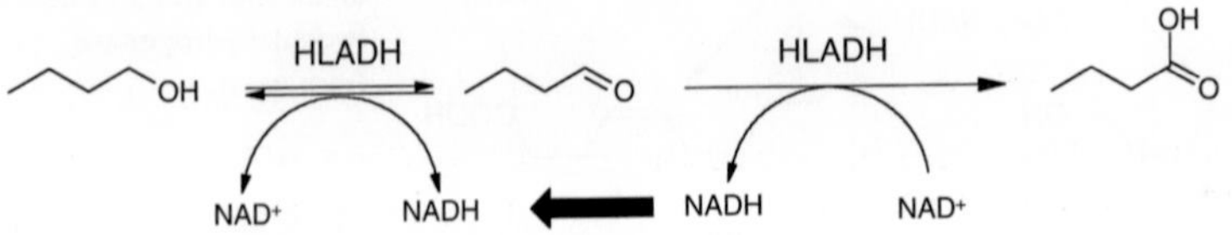

Figure 16.4-2. Oxidation activity for aldehydes exhibited by horse liver alcohol dehydrogenase (HLADH). Only minor amounts of acid are produced because of the higher HLADH activity for aldehyde reduction.

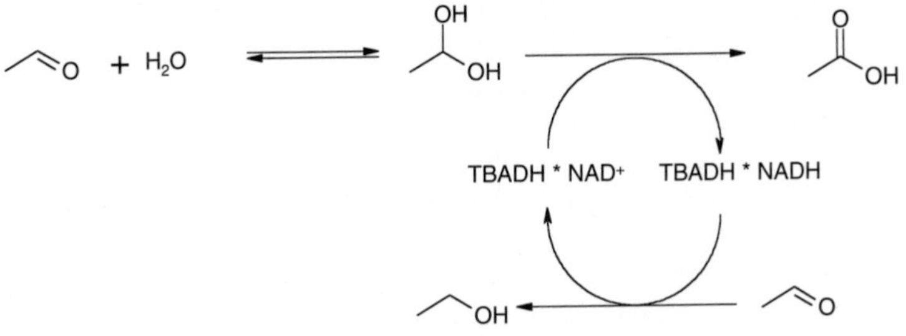

Figure 16.4-3. Aldehyde dismutase acitivity of *Thermoanaerobium brockii* alcohol dehydrogenase (TBADH). A high affinity of the TBADH-NAD⁺ complex for hydrated acetaldehyde is proposed, explaining the stochiometric acetaldehyde dismutation.

used for aldehyde reduction, yielding a dynamic equilibrium between alcohol and aldehyde.

TBADH also exhibits the so-called aldehyde dismutase activity[4]. In contrast to HLADH, stochiometric dismutation of acetaldehyde into one equivalent of ethanol and acetic acid has been reported. A *gem*-diol mechanism was proposed for this reaction (Fig. 16.4-3).

16.4.3
Aldehyde Dehydrogenases

Several aldehyde dehydrogenases have been reported for biocatalytic applications.

Recently, aldehyde dehydrogenase (E. C. 1.2.1.5) from yeast was applied to oxidize (Z,Z)-nona-2,4-dienal[5]. Recycling of NAD$^+$ was achieved *in situ* by addition of an alcohol dehydrogenase, reducing (Z,Z)-nona-2,4-dienal to the corresponding alcohol. Since both reactions are stochiometrically linked via NAD, this corresponds to an overall dismutation of the aldehyde (Fig. 16.4-4). This concept was extended to industrially relevant metabolites of linoleic acid (detergents and polymer building-blocks) (Fig. 16.4-5). No isomerization of the double bonds and yields up to 90% were reported[5].

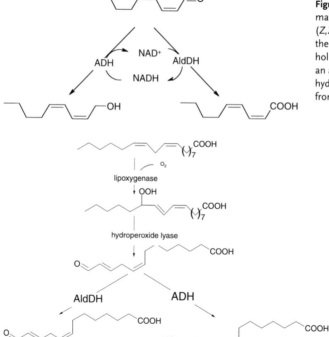

Figure 16.4-4. Enzymatic transformation of (Z,Z)-nona-2,4-dienal to the corresponding alcohol and acid catalyzed by an alcohol and an aldehyde dehydrogenase from yeast.

Figure 16.4-5. Enzymatic cleavage of linoleic acid to ω-hydroxy and dicarboxylic acids.

Table 16.4-1. Kinetic constants of bovine kidney aldehyde dehydrogenase for different substrates [1].

Substrate	V_{max} [%][a]	K_M [μM]
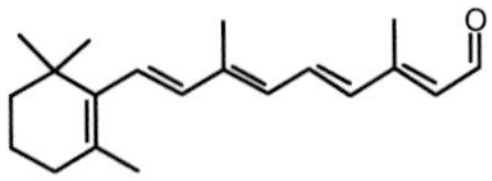	100	9.1
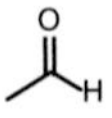	758	1
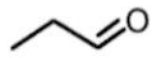	855	1.5
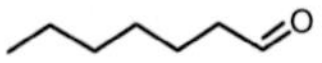	1960	30
	1683	33.9
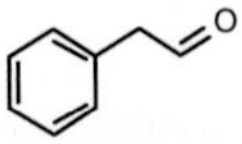	3026	8.2

a The V_{max} values are relative to retinal as substrate.

1 P. V. P. Bhat, L., Wang, X. L., *Biochem. Cell Biol.* **1996**, *74*, 695–700.

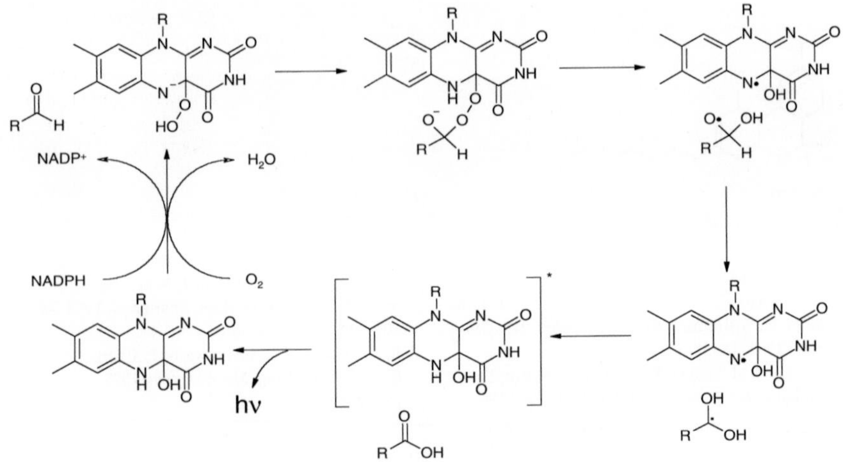

Figure 16.4-6. Mechanism proposed for light emission in the course of the luciferase reaction.

Another NAD⁺-dependent aldehyde dehydrogenase (from bovine kidney) was characterized with respect to its activity toward retinal and other aldehydes (Table 16.4-1) [6].

Table 16.4-2. Oxidation of aldehydes to corresponding carboxylic acids catalyzed by P450 monooxygenases.

O_2, NAD(P)H H_2O, NAD(P)$^+$

Substrate	Reference
Aliphatic aldehydes	[2, 3]
	[3]
	[4]
	[5]
	[6]
	[6]
Losartan	[7]

2 Y. Terelius, C. Norsten-Höög, T. Cronholm, M. Ingelman-Sundberg, *Biochem. Biophys. Res. Commun.* **1991**, *179*, 689–694.

3 K. Watanabe, T. Matsunaga, S. Narimatsu, I. Yamamoto, H. Yoshimura, *Biochem. Biophys. Res. Commun.* **1992**, *188*, 114–119.

4 S. Tomita, M. Tsujita, Y. Matsuo, T. Yubisui, Y. Chikawa, *Int. J. Biochem.* **1993**, *25*, 1775–1754.

5 K. Watanabe, T. Matsunaga, I. Yamamoto, H. Yashimura, *Drug. Metab. Dispos.* **1995**, *23*, 261–265.

6 K. Watanabe, S. Narimatsu, T. Matsunaga, I. Yamamoto, H. Yoshura, *Biochem. Pharmacol.* **1993**, *46*, 405–411.

7 R. A. Stearns, P. K. Chakravarty, R. Chen, S.-H. L. Chiu, *Drug. Metab. Dispos.* **1995**, *23*, 207–215.

16.4.4

Monooxygenases

16.4.4.1

Luciferase (E.C. 1.14.14.3)

Probably the most prominent oxidation reaction of aldehydes is the well-known luciferase reaction. The flavin-dependent luciferase is present in a number of marine and terrestrial species [7, 9]. Light of about 490 nm (blue-green) is emitted as a by-

Table 16.4-3. Oxidations and subsequent decarboxylations of aldehydes catalyzed by P450 monooxygenases.

Substrate	Reference
	[8]
	[8]
	[9]
	[10]

8 E. S. Roberts, A. D. N. Vaz, M. J. Coon, *Proc. Natl. Acad. Sci USA* **1991**, *88*, 8963–8966.
9 A. D. N. Vaz, E. S. Roberts, M. J. Coon, *J. Am. Chem. Soc.* **1991**, *113*, 5886–5887.
10 A. D. N. Vaz, K. J. Kessel, M. J. Coon, *Biochem.* **1994**, *33*, 13651–13661.

product of the oxidation of aliphatic aldehydes. Excited flavin species are discussed as emitters (Fig. 16.4-6) [9, 10].

16.4.4.2
Cytochrome P450BM-3

The oxidation of an aldehyde to the corresponding carboxylic acid with P450 systems is reported for various substrates (Table 16.4-2). In some cases oxidative decarboxylation is observed yielding formic acid and an olefin, one carbon atom shorter than the substrate (Table 16.4-3).

Several ω-oxo fatty acids are transformed to the corresponding α,ω -dicarboxylic acids, whereas ω-formylesters of fatty acids are decarboxylated to the ω-hydroxy fatty acids and carbon dioxide [11]. For several ω-oxo fatty acids turnover frequencies (measured as O_2 consumption) between 1.8 to 25 s^{-1} were found. Many P$_{450}$ systems are multi-component enzymes with small protein cofactors such as putidaredoxin performing the electron mediation between NAD(P)H and the active site of the enzyme. Vilker and coworkers recently were able to show that NADPH can be omitted from the catalytic cycle by direct electrochemical reduction of putidar-

Table 16.4-4. Kinetic constants of xanthine oxidase [11].

Substrate	K_M [mM]	V_{max} [s^{-1}]
(formaldehyde structure)	161.5	22.2
(acetaldehyde structure)	130	100
(propionaldehyde structure)	430	23.3
(butyraldehyde structure)	142	2.4
(pyridine-2-carbaldehyde structure)	0.36	3.4
(pyridine-3-carbaldehyde structure)	0.046	2.7
(pyridine-4-carbaldehyde structure)	1.7	4.2
(2-hydroxybenzaldehyde structure, OH)	1.03	7.7
(dihydroxybenzaldehyde structure, HO, OH)	0.068	15.7
(indole-3-carbaldehyde structure, CHO)	0.085	1.8
(O= ... COOH structure)	1	1
(O= ... OPO$_3^{2-}$, OH structure)	2	0.1

11 F. F. Morpeth, *Biochim. Biophys. Acta* **1983**, 744, 328–334.

edoxin [12–14], thus oxidizing styrene or camphor. Other approaches utilize Co sepulchrate as reducing agent, which can be regenerated either chemically (via Zn) [15] or electrochemically [16, 17].

16.4.5
Oxidases

16.4.5.1
Xanthine Oxidase (E.C. 1.1.3.22)

Xanthine oxidase was examined for its catalytic applicability for the oxidation of aldehydes as early as 1967[18]. In addition to O_2, xanthine oxidase was reported to accept e.g. methylene blue, PMS or ferricyanide[19] as electron acceptors. Table 16.4-4 gives kinetic data for some substrates[20].

16.4.6
Oxidations with Intact Microbial Cells[21]

Burkholderia cepacia was reported to transform aromatic aldehydes into the corresponding acids. Vanillin, *para*-hydroxybenzaldehyde, and syringaldehyde were converted to corresponding acids with high yields of 94%, 92%, and 72%, respectively (Fig. 16.4-7)[22].

The acid produced is not further metabolized as long as the aldehyde still is accessible to the cells. The enzyme responsible for aldehyde oxidation in *Burkholderia cepacia* was not further characterized. However, the gene of an NAD-dependent vanillin dehydrogenase of *Pseudomonas* sp. strain HR199 was cloned and characterized[23]. Recombinant *E. coli* containing this vanillin dehydrogenase transformed vanillin to vanillate at a clearly higher rate than *Burkholderia cepacia*.

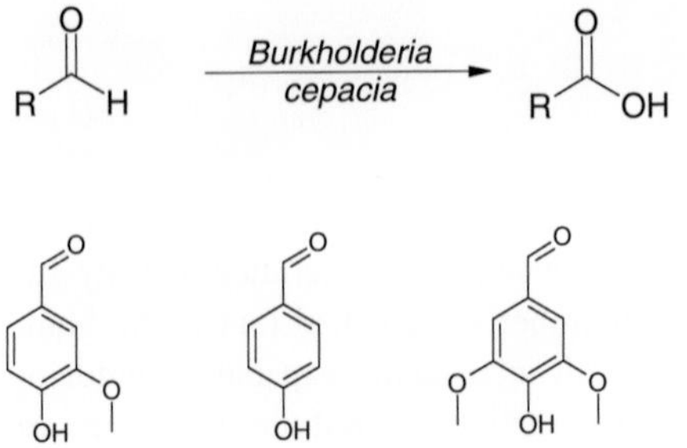

Figure 16.4-7. Oxidation of aromatic aldehydes by *Barkholderia cepacia* TM1.

References

1 L. P. Olson, J. Luo, Ö. Almarsson, T. C. Bruice, *Biochemistry* **1996**, *35*, 9782–9791.

2 G. T. M. Henehan, N. J. Oppenheimer, *Biochemistry* **1993**, *32*, 735–738.

3 G. L. Shearer, K. Kim, K. M. Lee, C. K. Wang, B. V. Plapp, *Biochemistry* **1993**, *32*, 11186–11194.

4 S. Trivic, V. Leskova, G. W. Winston, *Biotech. Lett.* **1999**, *21*, 231–234.

5 A. Nunez, T. A. Foglia, G. J. Piazza, *Biotechnol. Appl. Biochem.* **1999**, *29*, 207–212.

6 P. V. Bhat, L. Poissant, X. L. Wang, *Biochem. Cell Biol.* **1996**, *74*, 695–700.

7 T. O. Baldwin, M. M. Ziegler. In *Chemistry and Biochemistry of Flavoenzymes*, CRC

Press, Boca Raton, 1992, Vol. III, pp. 467–530.

8 A. Palfey, V. Massey, Flavin-Dependent Enzymes. In *Comprehensive Biological Catalysis.* M. Sinnott (ed), Academic Press, San Diego, London, 1998, Vol. III, pp. 83–154.

9 C. T. Walsh, Y.-C. J. Chen, *Angew. Chem.* **1988**, *100*, 342–352.

10 P. Macheroux, S. Gishla, *Nachr. Chem. Tech. Lab.* **1985**, *33*, 785.

11 S. C. Davis, Z. Sui, J. A. Peterson, P. R. Ortiz de Montellano, *Arch. Biochem. Biophys.* **1996**, *328*, 35–42.

12 M. P. Mayhew, V. Reipa, M. J. Holden, V. L. Vilker, *Biotechnol. Prog.* **2000**, *16*, 610–616.

13 V. Reipa, M. Mayhew, V. L. Vilker, *PNAS* **1997**, *94*, 13 554–13 558.

14 V. L. R. Vilker, Vytas; Mayhew, Martin; Holden, Marcia J., *J. Am. Oil Chem. Soc.* **1999**, *76*, 1283–1289.

15 U. Schwaneberg, D. Appel, J. Schmitt, R. D. Schmid, *J. Biotech.* **2000**, *84*, 249–257.

16 R. W. Estabrook, K. M. Faulkner, M. Shet, C. W. Fisher, Application of Electrochemistry for P450-Catalyzed Reactions. In *Methods in Enzymology*, Academic Press. San Diego, London, Boston, New York, Sydney, Tokyo, Toronto, 1996, Vol. 272, pp. 44–51.

17 K. M. Faulkner, M. S. Shet, C. W. Fisher, R. W. Estabrook, *PNAS* **1995**, *92*, 7705–7709.

18 F. Dastoli, S. Price, *Arch. Biochem. Biophys.* **1967**, *118*, 163–165.

19 G. Pelsey, A. M. Klibanov, *Biochim. Biophys. Acta* **1983**, *742*, 352–357.

20 F. F. Morpeth, *Biochim. Biophys. Acta* **1983**, *744*, 328–334.

21 M. Tanaka, Y. Hirokane, *J. Biosci. Bioeng.* **2000**, *90*, 341–343.

22 S. Adachi, M. Tanimoto, M. Tanaka, R. Matsuno, *Chem. Eng. J.* **1992**, *49*, B17-B21.

23 H. Priefert, J. Rabenhorst, A. Steinbüchel, *J. Bacteriol.* **1997**, *179*, 2595–2607.

16.5
Baeyer-Villiger Oxidations

Sabine Flitsch and Gideon Grogan

16.5.1
Introduction

The enzymatic Baeyer-Villiger oxidation continues to receive attention from synthetic organic chemists as it offers advantages of regio- and enantioselectivity still rarely exhibited by reagents such as *meta*-chloroperbenzoic acid (*m*-CPBA). Some recent advances have resulted in abiotic catalytic reagents capable of inducing modest enantioselectivity in the Baeyer-Villiger reaction [1–3], but these reactions are outside the scope of this section. The most encouraging examples of enantioselective Baeyer-Villiger reactions are still those catalyzed by microorganisms and enzymes and the extensive research in this area over the last decade has been covered in a number of recent reviews [4–7].

16.5.1.1
Steroidal Substrates

It had been known for many years that Baeyer-Villiger-type processes occur during the catabolic transformations of natural compounds. In 1953, it was described that the C17 side chain of steroids can be cleaved by several microorganisms including

Fusarium, Penicillium, Cylindrocarpon, Aspergillus and *Gliocladium* species [8–10]. One example reported was the conversion of progesterone into $\Delta^{1,4}$-androstadien-3,17-di-one in 84% yield as illustrated in Fig. 16.5-1 [8].

Since these reports, many others describing the microbiological Baeyer-Villiger oxidation of various steroids have been published [11–14]. Interestingly, it has been shown that depending on the microbial strain used, further oxidation may occur leading to incorporation of an oxygen atom into the D-ring, thus affording the corresponding lactone. In general, these oxidations are restricted to this ring. This selectivity may be due to the fact that the A-ring bears an α, β-unsaturated ketone moiety, which appears to display a different reactivity compared with the other carbonyl functions [15]. Introduction of a Δ^1 double bond also often occurs during these processes. Other examples involving oxidation of the A ring have been described with a *Glomerella fusaroides* strain [16] and with *Gymnoascus reesii* [17]. Thus, eburicoic acid affords a 30% yield of A-secoacid whereas the steroidal alkaloid tomatidine leads to the corresponding ketone as the major product, but a smaller amount of A-seco acid is also obtained. This could well be due to hydrolysis of the lactone which would be formed from Baeyer-Villiger oxidation of the parent ketone Fig. 16.5-2.

The mechanism of these reactions has been studied by several groups. Fonken and coworkers [18] first showed using 21-^{14}C labelled progesterone, that the testosterone acetate formed during degradation of progesterone by *Cladosporium resinae* is not an artefact but is indeed an intermediate in the degradation pathway. Further work by Prairie and Talalay [19] using the strain *Penicillium liliacinum* established the involve-ment of two enzymes, a Δ^1-dehydrogenase and an NADPH-dependent oxygenase. They also showed that $^{18}O_2$ molecular oxygen is incorporated as the ring oxygen atom of testololactone. Rahim and Sih [20] succeeded in showing that an oxygenase (requiring the presence of oxygen) as well as an esterase were involved in the degradation of the progesterone side-chain. In other studies using the 17α-labelled substrate, Singh and Rahkit [21] showed that retention of the deuterium label at the C17 position occurs and that the molecular oxygen is incorporated into the product (Fig. 16.5-3). More recently, a gene from *Rhodococcus rhodochrous* has been cloned and expressed [22], which encodes for a steroid monooxygenase that inserts an atom of oxygen between the C17 and C20 carbons of progesterone, forming testosterone acetate.

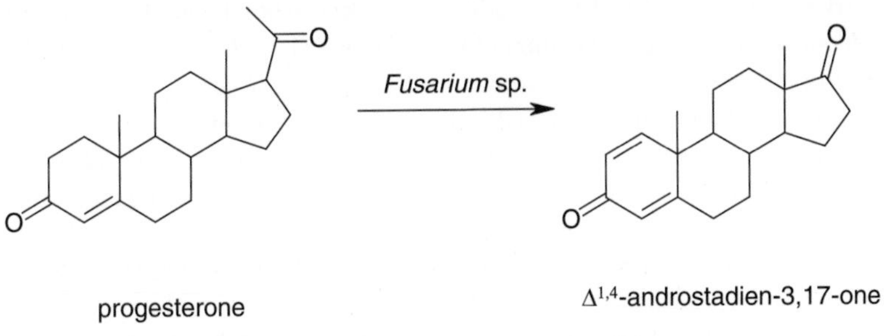

Figure 16.5-1. Biotransformation of progesterone using *Fusarium* spp.

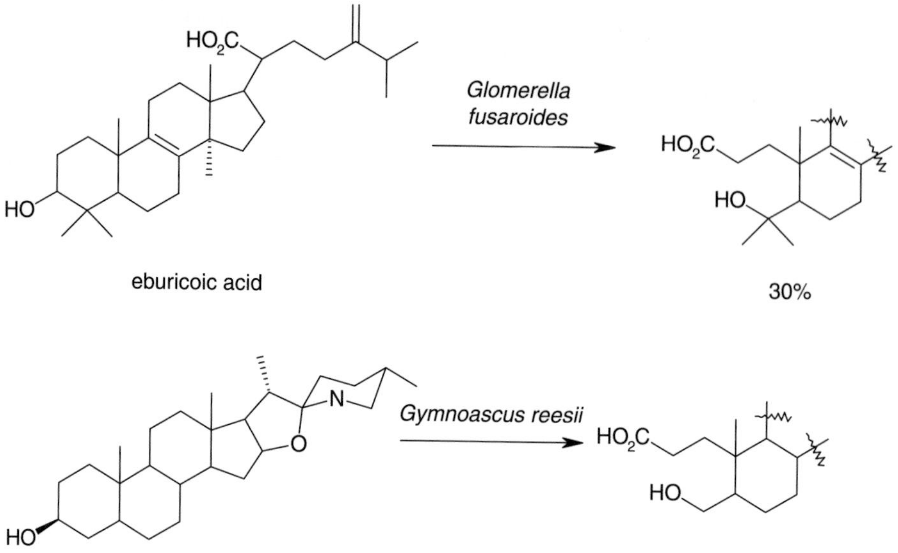

eburicoic acid

Glomerella fusaroides

30%

Gymnoascus reesii

Figure 16.5-2. A-ring cleavage by *Glomerella fusaroides* and *Gymnoascus reesii*.

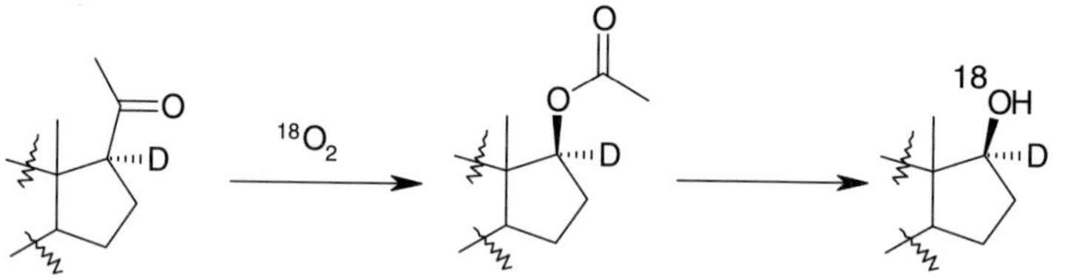

Figure 16.5-3. Retention of the deuterium label and oxygen incorporation during the side-chain degradation of progesterone.

All these results led to the conclusion that a process similar to the Baeyer-Villiger oxidation must occur during these degradations. The general scheme for the formation of testololactone from progesterone can thus be described, as shown in Fig. 16.5-4. It involves four successive steps; first a Baeyer-Villiger oxidation of the steroid sidechain leading to a testosterone acetate, secondly an esterase hydrolysis, thirdly oxidation of the C17 hydroxyl leading to the corresponding 3,17-dione and finally a second Baeyer-Villiger oxidation of this diketone at the D-ring leading to the corresponding δ-lactone. It has been shown in the fungus *Cylindrocarpon radicicola* that one bifunctional enzyme is involved in these transformations, which is able to catalyze oxygenative esterification of 20-ketosteroids as well as oxygenative lactonisation of 17-ketosteroids [23, 24]. It is noteworthy that all the above investigations into steroid substrates for lactonization were conducted on single enantiomers and thus, no reference to the enantioselectivity of the processes had been recorded.

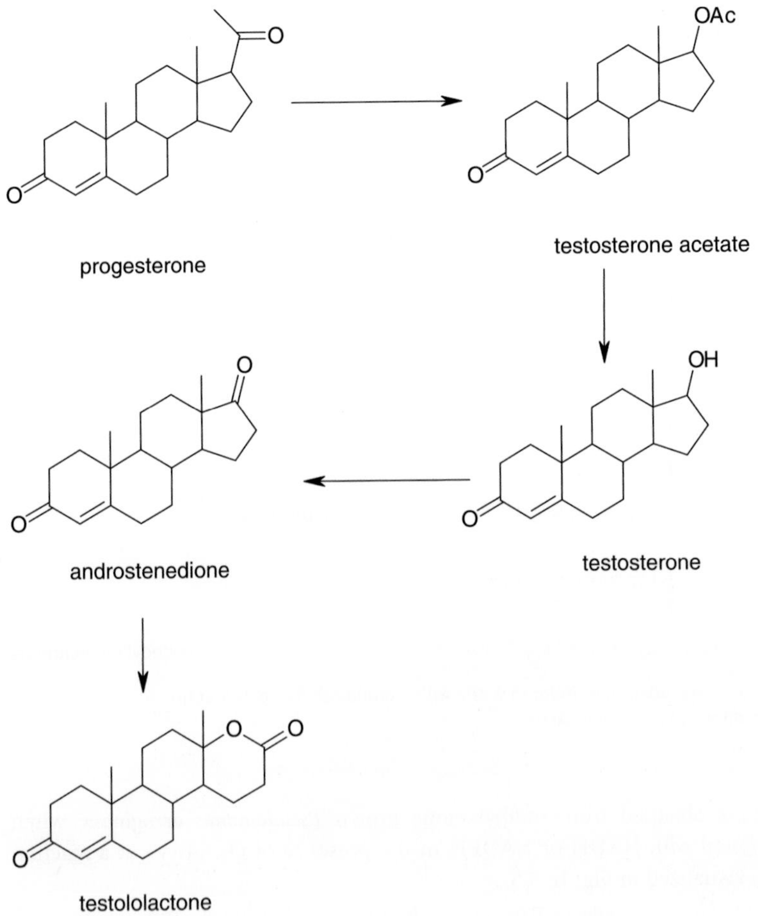

progesterone

testosterone acetate

testosterone

androstenedione

testololactone

Figure 16.5-4. Mechanism of the biotransformation of progesterone into testololactone.

16.5.1.2
Aliphatic Substrates

Baeyer-Villiger oxidation has also been reported for aliphatic ketones. Several strains able to grow on various aliphatic or alicyclic substrates have been isolated, and it has been shown that their degradation often involves a Baeyer-Villiger oxidation. For example, it has beeen observed that *Pseudomonas multivorans, Pseudomonas aeruginosa, Pseudomonas cepacia* and *Nocardia* sp. are able to grow on tridecan-2-one[25–28].

Forney and Markovetz isolated undecyl acetate directly from growing cultures of *Pseudomonas aeruginosa*. They showed that all early intermediates in the pathway arise biologically and sequentially from their precursors, indicating involvement of a Baeyer-Villiger type oxidation. In a further study they also showed that cell-free

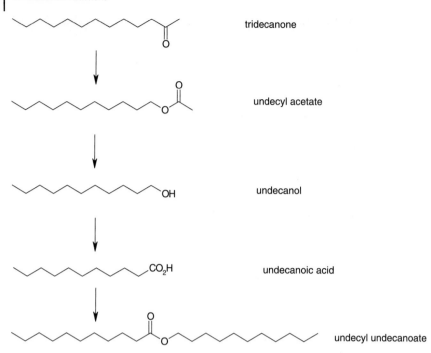

Figure 16.5-5. Degradation of tridecan-2-one with a crude cell-free preparation from a *Pseudomonas aeruginosa* strain.

preparations obtained from methylketone grown *Pseudomonas aeruginosa*, when supplemented with NADH or NADPH in the presence of O_2, carry out a reaction sequence visualized in Fig. 16.5-5.

Using *Pseudomonas cepacia* grown on tridecan-2-one, Markovetz and coworkers[28] later showed that experiments conducted with $^{18}O_2$ led to 84% incorporation of ^{18}O into the C – O – C linkage, rather than into the carbonyl function, indicating the occurrence of a Baeyer-Villiger type process. They also observed that the undecyl esterase involved in the degradation process is able to hydrolyze both aliphatic and aromatic acetate esters. They also reported that this enzyme is strongly inhibited by organophosphates such as tetraethylpyrophosphate (TEPP), as well as by other esterase inhibitors like *p*-chloromercuribenzoate[27].

A similar degradation pathway was described for oxidation of tetradecane and 1-tetradecene with *Penicillium* sp.[29]. Similar mechanisms were proposed for the degradation of other aliphatic substrates such as butan-2-one[28], acetol[30], acetophenone[31] and 1-phenylethanol[32]. Interestingly, cell extracts of *Nocardia* sp. LSU 169 grown on butan-2-one were also shown to be capable of oxidizing tridecan-2-one. Generally, the Baeyer-Villiger reaction was followed by an esterase catalyzed hydrolysis[33].

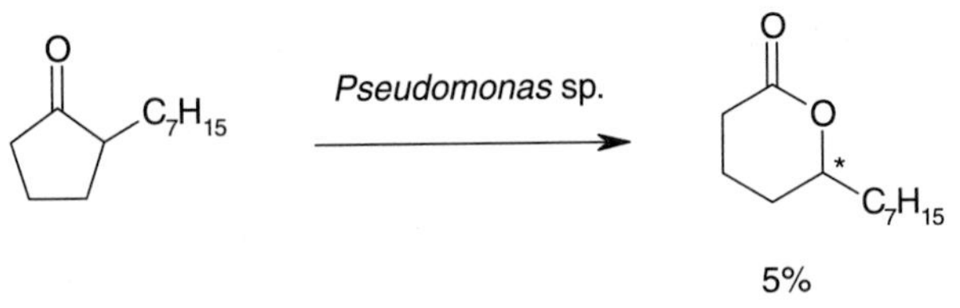

Figure 16.5-6.
Degradation of 2-heptyl-cyclopentanone by a *Pseudomonas* sp.

5%

16.5.1.3
Alicyclic Substrates

Baeyer-Villiger oxidation is also a common feature during the catabolic degradation of a variety of other compounds, including monocyclic, bicyclic or polycyclic molecules. For monocyclic compounds, one of the first reports describing formation of a lactone from racemic α-substituted cyclopentanone by various *Pseudomonas* sp. was by Shaw[34]. This could be regarded as the first indication that these reactions were to prove of interest for asymmetric synthesis since the lactone product displayed some optical activity (Fig. 16.5-6).

Further studies showed that other substrates such as cyclopentanol[35], cyclohexane[36–39], cyclohexanol[40–42], cyclohexan-1,2-diol[43–45], cycloheptanone[46] and, more recently, cyclododecane[47] were degraded via analogous pathways. These were studied using bacterial strains including *Pseudomonas* sp. NCIMB 9872[35, 48], *Nocardia globerula* CL1[40], *Acinetobacter* TD 63[43], *Acinetobacter calcoaceticus* NCIMB 9871[39], *Xanthobacter* sp.[38] and *Rhodococcus ruber*[47].

All these degradation pathways were shown to involve a Baeyer-Villiger oxidation of a cycloalkanone that led to formation of the corresponding lactone. Further degradation then occured via hydrolysis of this lactone by a lactone hydrolase which has, in some cases, been isolated. As an example, the reaction sequence for the degradation of cyclopentanol by *Pseudomonas* sp. NCIMB 9872[35] is shown in Fig. 16.5-7.

A pathway for the degradation of (–)-menthol and menthane-3,4-diol by a bacterium classified as a *Rhodococcus* sp. was proposed by Shukla and coworkers. Again, the proposed scheme involves formation of the corresponding lactone by a Baeyer-Villiger process[49]. Interestingly, an identical process has been shown to occur in the degradative pathway of menthol and menthone in peppermint (*Mentha piperita*) rhizomes[50]. *Rhodococcus erythropolis* DCL 14[51] has also been reported to degrade menthone in addition to 1-hydroxy-2-oxo-limonene and dihydrocarvone via an enzymatic Baeyer-Villiger reaction.

Some other monocyclic compounds bearing ketonic side chains have also been shown to undergo degradation processes involving Baeyer-Villiger type oxidation. For example, oxidation of β-ionone by *Lasioplodia theobromae*[52] affords, among other products, the alcohols shown in Fig. 16.5-8. In this case, the loss of two carbons from the sidechain has been attributed to a contribution of Baeyer-Villiger oxidation followed by ester hydrolysis and reduction.

Similar results were described by Nespiak and coworkers[53] in the course of their study of cyclopentyl ketones by *Acremonium roseum* (Fig. 16.5-9). When R = CH$_3$ or

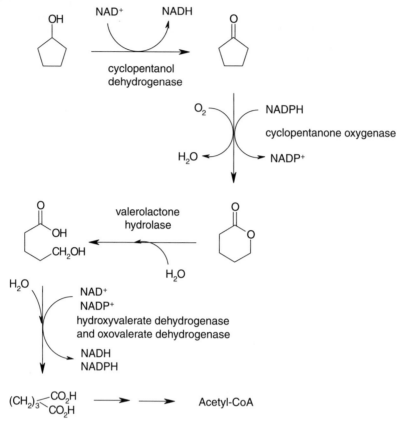

Figure 16.5-7. Reaction sequence for the oxidation of cyclopentanol by *Pseudomonas* sp. NCIMB 9872.

C_2H_5, the alcohol formed via Baeyer-Villiger oxidation and ester hydrolysis was the only product isolated after 2 days. However, higher esters (R = n-or i-C_3H_7, R = n-butyl) have also, if not predominantly, some amount of allylic oxidation product. In this study, it was shown that the (S)-enantiomer of the substrate methyl ester was oxidized more rapidly than the (R)-isomer and that the reaction proceeded with retention of configuration at the chiral center. Thus, by using short incubation times (2 days) the racemic substrate led to the (S)-alcohol, but the optical purity was low (around 20%). Although interesting, this apparent enantioselectivity could also be due eventually to an enantioselective hydrolysis of the intermediate ester or to some other catabolic pathway. However, the butyl ketone led to the (R)-alcohol showing 100% optical purity.

An extensive study by Fuganti and coworkers [54–57] showed that the metabolism of 4-(4-hydroxyphenyl)butan-2-one ("raspberry ketone") by the fungus *Beauveria bassiana* unexpectedly yielded tyrosol, through insertion of oxygen via a Baeyer-Villiger reaction and subsequent acetate hydrolysis [54] (Fig. 16.5-10). Only a narrow range of

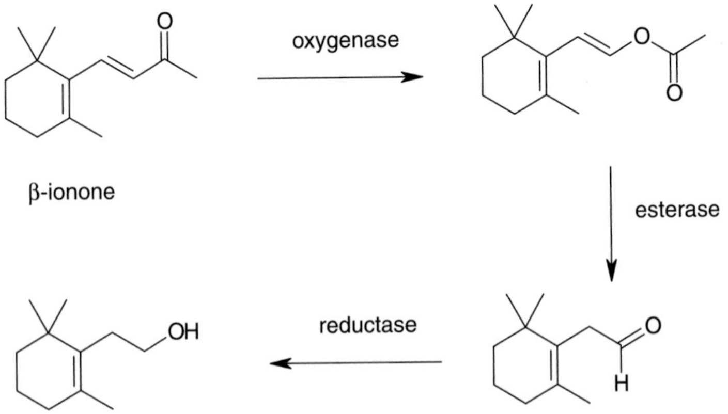

β-ionone

Figure 16.5-8. Biotransformation of β-ionone by *Lasiodiploida theobromae.*

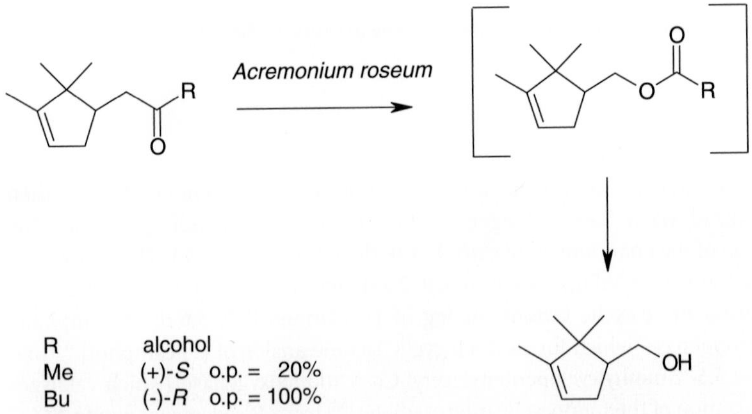

R	alcohol	
Me	(+)-*S*	o.p. = 20%
Bu	(-)-*R*	o.p. = 100%

Figure 16.5-9. Transformation of cyclopentyl ketones by *Acremonium roseum.*

substrates was converted in this manner, however[57]. The authors were able to show via deuterium incorporation experiments, that the configuration of the migrating carbon-carbon bond was retained[56], this being a defining characteristic of the peracid-catalyzed Baeyer-Villiger process.

Camphor and its analogs are the most studied bicyclic substrates for the biological Baeyer-Villiger reaction[58–67]. It has been shown by Gunsalus and coworkers that, in the early steps of D-(+)-camphor oxidation by *Pseudomonas putida* C1, both alicyclic rings are cleaved by lactonization reactions: Thus, the conversion of (+)-camphor to 5-keto-1,2-campholide involves three reactions; hydroxylation, oxidation and lactonization. Using a different *Pseudomonas* strain, the non-hydroxylated campholide has been isolated, suggesting that in this case lactonization occurs prior to hydroxylation[58]. It was also shown, by analysis of extracted metabolites, that an analogous, enantiocomplementary pathway existed for the metabolism of L-(−)-camphor. Several further studies have been devoted to clarifying these steps. Interestingly, it has been shown that *P. putida* does not express lactone hydrolases that are active towards

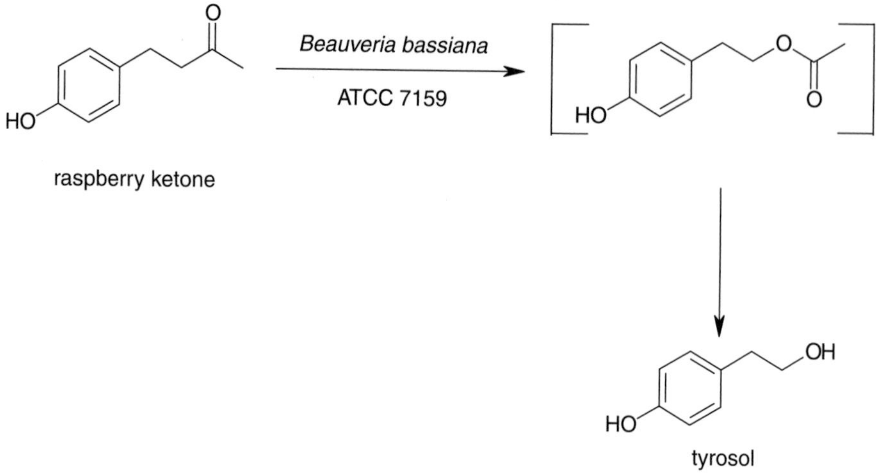

Figure 16.5-10. Biotransformation of raspberry ketone to tyrosol by *Beauveria bassiana* ATCC 7159.

the lactone intermediate. The intermediate bicyclic lactone is unstable under reaction conditions, and spontaneously opens to a cyclopentenone. This is then again oxidized via a Baeyer-Villiger reaction to the corresponding lactone. The degradation of the enantiomers of camphor is shown in Fig. 16.5-11. Three enzymes catalyzing the Baeyer-Villiger reaction, i.e. 2,5-diketocamphane 1,2-monooxygenase [which forms the bicyclic lactone analog of (+)-camphor][64], 3,6-diketocamphane 1,6-monooxygenase [which forms the bicyclic lactone analog of (–)-camphor][67] and 2-oxo-Δ^3–4,5,5-trimethylcyclopentenylacetyl-Co-A monooxygenase (which catalyzes the lactonization of the monocyclic intermediate)[63] from *Pseudomonas putida* ATCC 17453 have been purified to homogeneity and thoroughly characterized.

Enzymatic Baeyer-Villiger oxygenations are not restricted to microbial cells. It has been shown that (+)-camphor, a major constituent of the volatile oil of immature sage (*Salvia officinalis* L.) leaves, is converted into a water soluble metabolite via enzymatic lactonization to 1,2-campholide, followed by conversion into the β-D-glucoside-6-O-glucose ester of the corresponding hydroxy acid [68, 69].

The oxidation of racemic fenchone by a *Corynebacterium* sp. [70] (reclassified as *Mycobacterium rhodochrous*), an organism which grows at the expense of either (+)- or (-)- camphor, has also been reported. This was shown to lead, in a 45% yield, to a 90/10 mixture of 1,2 and 2,3-fencholides, as shown in Fig. 16.5-12. This result contrasts with the chemical oxidation of fenchone with peracetic acid, where 2,3-fencholide is the major product in a 40/60 mixture. Accumulation of these lactones is *a priori* surprising as compared with the total degradation of the structurally similar camphor substrate. However this may simply be due to the fact that this lactone, unlike that formed from camphor, is chemically stable in the medium. Of course, one has also to assume that, here again, the strain is devoid of any lactone hydrolase. This bioconversion was the first gram-scale preparative report

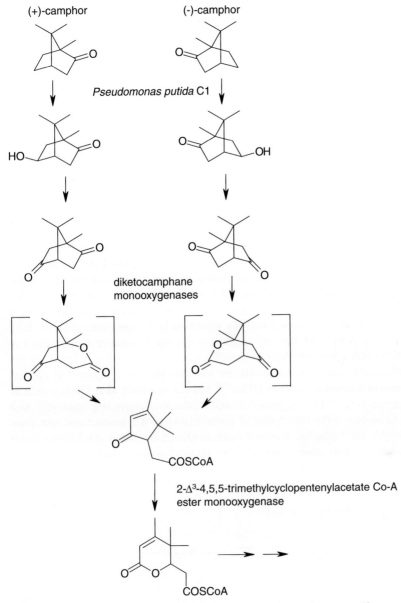

Figure 16.5-11. Metabolism of both enantiomers of camphor by *Pseudomonas putida* C1 (= NCIMB 10007 = ATCC 17453).

of a non-steroidal product, yet no indication of any enantioselectivity for this reaction was presented.

Similarly, it has been shown that 1,8-cineole and 6-oxo-cineole are degraded via the scheme shown in Fig. 16.5-13 [71]. As in the case of camphor, the first step involves a

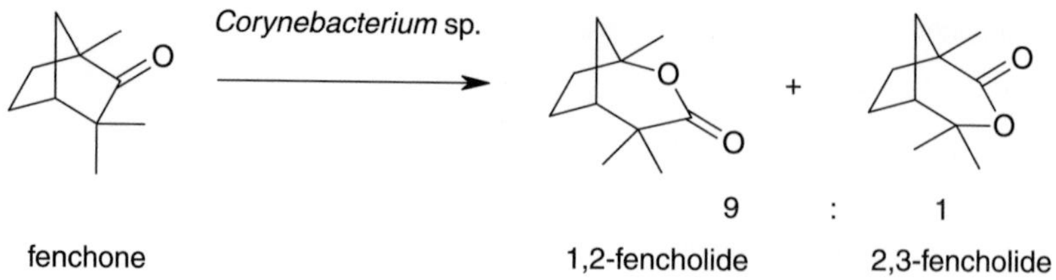

Figure 16.5-12. Oxidation of racemic fenchone to the corresponding fencholides.

hydroxylation, followed by oxidation of the alcohol to form 6-oxocineole. This is then processed via a Baeyer-Villiger reaction leading to a lactone which is spontaneously opened to the hydroxy acid.

16.5.1.4
Polycyclic Molecules

Enzymatic Baeyer-Villiger reactions have also been described in the metabolic processing of larger, polycyclic non-steroidal molecules (Fig. 16.5-14). This is the case for the biosynthesis of aflatoxin B1 where it has been demonstrated that formation of versiconal acetate intermediate from averufin occurs via such a process[72]. Similarly, aflatoxin G1 was shown to be formed from aflatoxin B[73] whereas degradation of the anthraquinone questin to desmethylsulochrin was shown to imply a Baeyer-Villiger process[74].

Furthermore, biological Baeyer-Villiger reactions have been reported in the biosynthesis of polyketides such as DTX-4[75] and the aureolic acid antibiotics such as mithramycin[76, 77]. The oxygenase MtmOIV from *Streptomyces argillaceus* responsible for cleavage of the fourth ring of premithramycin B is unique amongst those responsible for biological Baeyer-Villiger reactions, in that it displays sequence homology not with other "Baeyer-Villiger monooxygenases" (*vide infra*), but with flavin-type hydroxylases encoded in polyketide synthase gene clusters from other *Streptomyces* spp. [76].

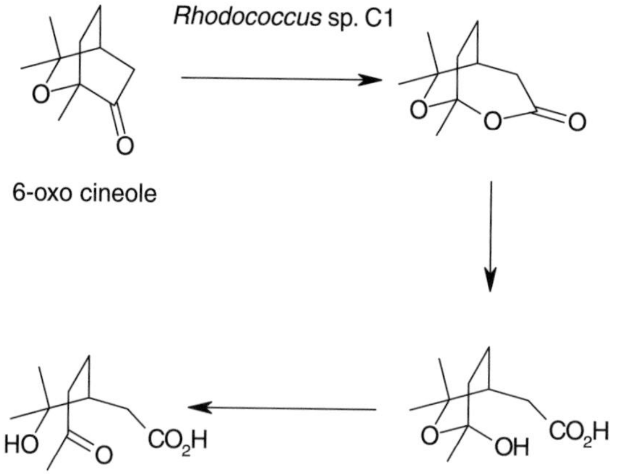

Figure 16.5-13. Degradation of 6-oxo-cineole by a *Rhodococcus* sp.

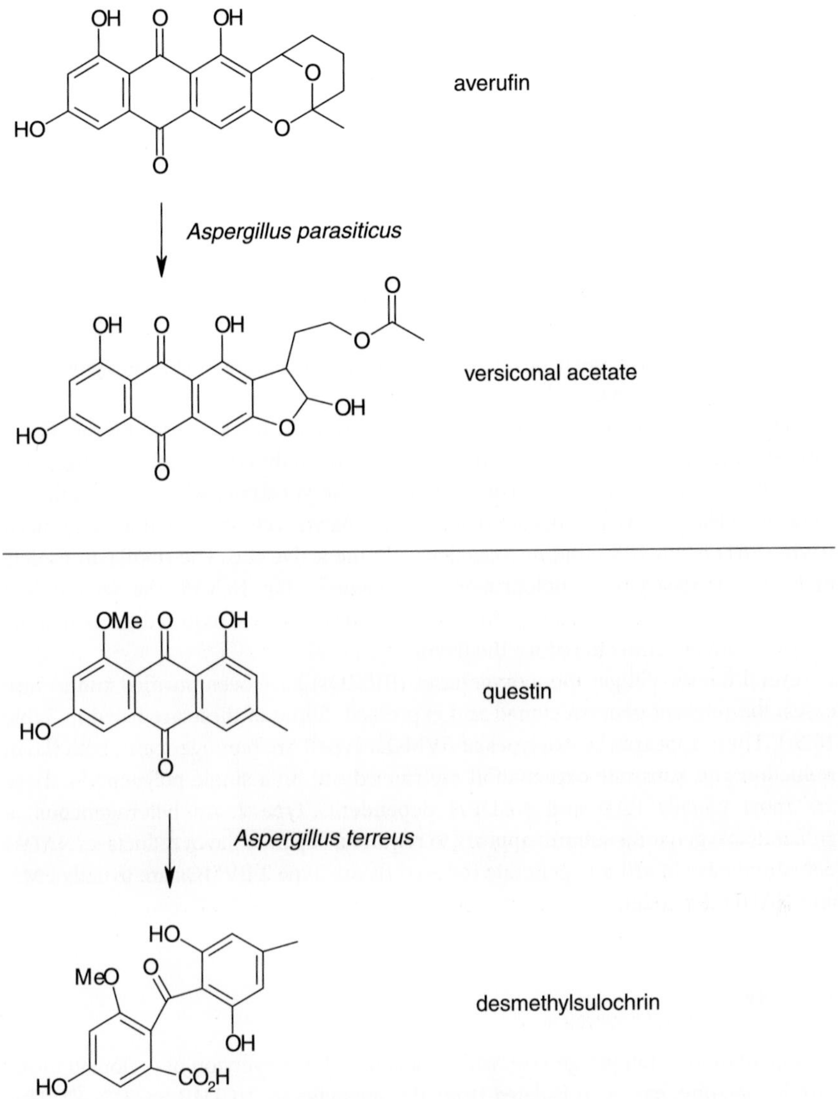

averufin

Aspergillus parasiticus

versiconal acetate

questin

Aspergillus terreus

desmethylsulochrin

Figure 16.5-14. Involvement of enzymatic Baeyer-Villiger processes in the degradation of non-steroidal polycyclic compounds.

16.5.2
Baeyer-Villiger Monooxygenases

The reactions described above illustrate that there are numerous metabolic routes wherein biological Baeyer-Villiger reactions have been implicated. The synthetic potential of the enzymatic Baeyer-Villiger reaction has dictated that intensive

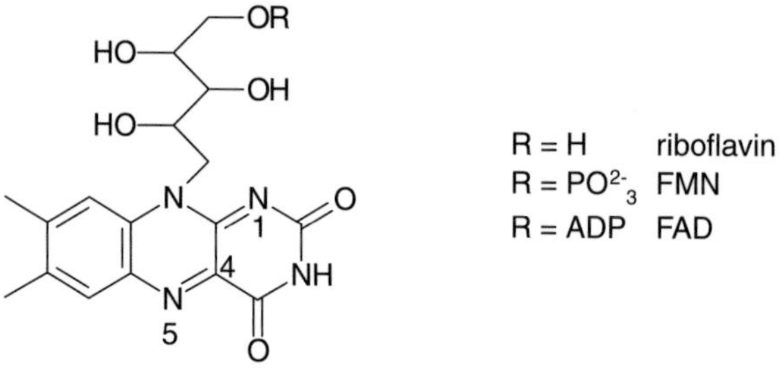

Figure 16.5-15. Riboflavin derivatives: the coenzymically active forms of flavoprotein.

R = H   riboflavin
R = PO$^{2-}_3$ FMN
R = ADP   FAD

research efforts have been devoted to studying the nature of the enzymes that catalyze these reactions.

Enzymes that catalyze the Baeyer-Villiger reaction are a subset of the flavin monooxygenases. In the mechanism of oxidation catalyzed by such enzymes one atom of molecular oxygen is incorporated into the substrate, whereas the other is reduced to H_2O. Two cofactors are required for catalytic activity. The first is a reduced flavin (FAD or FMN) bound non-covalently in the active site. The riboflavin moiety of flavin monooxygenase holoproteins is shown in Fig. 16.5-15; the second is a reduced nicotinamide cofactor (NADPH or NADH), which is required to furnish the enzyme with electrons to reduce the flavin.

Several Baeyer-Villiger monooxygenases (BVMOs) have been purified and in rare cases, the relevant genesm cloned and expressed. Some of these are listed in Table 16.5-1. There appear to be two types of BVMOs. Type 1 are homogeneous; both flavin reduction and substrate oxygenation are carried out on a single polypeptide, these are most usually FAD and NADPH dependent. Type 2 are heterogeneous, a substrate oxygenating subunit appears to require a separate flavin reductase/NADH dehydrogenase in order to generate reduced flavin. Type 2 BVMOs are usually FMN and NADH dependent.

16.5.2.1
Type 1 BVMOs

Cyclopentanone monooxygenase, which catalyzes the conversion of cyclopentanone to valerolactone, has been isolated from *Pseudomonas* sp. NCIMB 9872 [46, 78]. This has been shown to be made up of three identical subunits, each using one FAD equivalent, and to be NADPH dependent. Cyclohexanone monooxygenases have been purified from *Acinetobacter calcoaceticus* NCIMB 9871 and *Nocardia globerula* CL1 [40] and *Rhodococcus coprophilus* [79]. These enzymes were shown to be single polypeptides and to be FAD and NADPH dependent. Tridecanone monooxygenase from *Pseudomonas cepacia* is a dimer of two identical subunits, however, but is also FAD plus NADPH dependent [80]. A cyclohexanone monooxygenase from *Xantho-bacter* sp. is unusual in that it is dependent on FMN, but NADPH as a nicotinamide cofactor [82]. Steroid monooxygenase from *Rhodococcus rhodochrous* [83] and mono-cyclic monoterpene ketone monooxygenase from *Rhodococcus erythropolis* DCL 14 [51]

Table 16.5-1. Characteristics of various Baeyer-Villiger monooxygenases.

Enzyme and source	Number of proteins	Subunit structure	Cofactor specificity	Native molecular mass × 1000 Da	Mole of flavin/ mole of protein	Optimum pH	Reference
Cyclopentanone monooxygenase Pseudomonas NCIMB 9872	1	3–4 identical subunits	NADPH	200 (54–58 each)	1 FAD per subunit	7.7	[48]
Cyclohexanone monooxygenase Acinetobacter calcoaceticus NCIMB 9871	1	single polypeptide	NADPH	59	1 FAD	9.0	[81]
Cyclohexanone monooxygenase Nocardia globurela CL 1	1	single polypeptide	NADPH	53	1 FAD	8.4	[81]
2-Tridecanone monooxygenase Pseudomonas cepacia	1	2 identical subunits	NADPH	123 (55 each)	1 FAD	7.8–8.0	[80]
2-Oxo-Δ^3-4,5,5-trimethylcyclopentenyl acetyl Co-A monooxygenase Pseudomonas putida ATCC 17453	1	2 identical subunits	NADPH	106	1 FAD	9.0	[63]
2,5-Diketocamphane 1,2-monooxygenase Pseudomonas putida ATCC 17453	2	2 identical substrate oxygenating subunits + NADH dehydrogenase	NADH	78 (39 each)	1 FMN per subunit	7.2	[64]
3,6-Diketocamphane 1,6-monooxygenase Pseudomonas putida ATCC 17453	2	2 identical substrate oxygenating subunits + NADH dehydrogenase	NADH	72 (36 each)	1 FMN per subunit	–	[67]
Steroid monooxygenase Cylindrocarpon radicicola ATCC 11011	1	2 identical subunits	NADPH	115 (56 each)	1 FAD per subunit	7.8	[83]
Cyclohexanone monooxygenase Xanthobacter sp.	1	single polypeptide	NADPH	50	FMN	8.8	[82]
Monocyclic monoterpene ketone monooxygenase Rhodococcus erythropolis DCL 14	1	single polypeptide	NADPH	60	FAD	9.0	[51]
Cyclohexanone monooxygenase Rhodococcus coprophilus	1	single polypeptide	NADPH	58	FAD	–	[79]
Steroid monooxygenase Rhodococcus rhodochrous	1	single polypeptide	NADPH	60	FAD	–	[83]

are also Type 1 BVMOs as is the 2-oxo-Δ^3-4,5,5-trimethylcyclopentenyl acetyl Co-A monooxygenase from *Pseudomonas putida* ATCC 17453[63].

Cyclohexanone monooxygenase (CHMO, E.C. 1.14.13.X) is by far the most studied Type 1 BVMO and has been used extensively for as a model for mechanistic studies and as a catalyst in synthesis (*vide infra*). CHMO was purified from *Acinetobacter calcoaceticus* NCIMB 9871 grown on cyclohexanol as the sole carbon source, by Trudgill and coworkers[81]. It was found to be active as a monomer and to contain one non-covalently bound FAD molecule per monomer. The gene was cloned and the protein expressed in *Escherichia coli*[84] and more recently in *Saccharomyces cerivisiae*[85]. Each subunit is a polypeptide of 542 amino acids and, although no definitive structure of a BVMO has yet been published, a potential flavin binding site at the *N*-terminus was identified, in addition to a potential NADP binding site. Analysis of the sequence reveals that the N-terminus of the enzyme bears strong homology with the FAD binding domain of other flavoproteins such as glutathione reductase from *Escherichia coli*.

16.5.2.2
Type 2 BVMOs

The diketocamphane monooxygenases (DKCMOs) from *Pseudomonas putida* ATCC 17453 involved in camphor degradation, are FMN plus NADH dependent and are heterogeneous, consisting of two identical substrate oxidizing polypeptides and an NADH dehydrogenase. The enzymes have been purified, extensively characterized[64, 67] and their N-terminal amino acid sequences determined[86]. These data showed the oxygenating subunits of the DKCMOs to have homology with the NADH plus FMN dependent luciferase of *Vibrio harveyi*[87], an enzyme which catalyzes the Baeyer-Villiger oxidation of dodecanal to dodecanoic acid with the release of a photon of light. The application of the DKCMOs enzymes to synthesis has also been investigated (*vide infra*) and, whilst the genes encoding these proteins have not been identified, preliminary X-ray crystallographic data on 3,6-diketocamphane-1,6 monooxygenase has been reported[88].

16.5.2.3
Mechanism of the Enzymatic Baeyer-Villiger Reaction

The mechanism of the enzymatic Baeyer-Villiger oxidation, with reference to CHMO, has been studied by the group of Walsh[73, 89] who proposed the scheme shown in the top cycle in Fig. 16.5-16. The tricyclic isoalloxazine ring is the center of catalysis. Initially, the exogenous reductant NAD(P)H acts as the electron donor to afford the reduced flavin. This can be readily reoxidized by both one-electron or two electron processes in the presence of O_2 to yield a 4-*a*-hydroperoxyflavin. This intermediate undergoes an O-O bond fission upon nucleophilic attack on an electrophilic ketone substrate, a mechanism similar to the chemical Baeyer-Villiger oxidation of ketones by peracids. This initially affords the 4-*a*-hydroxyflavin which, by loss of H_2O regenerates the starting FAD for a subsequent catalytic cycle.

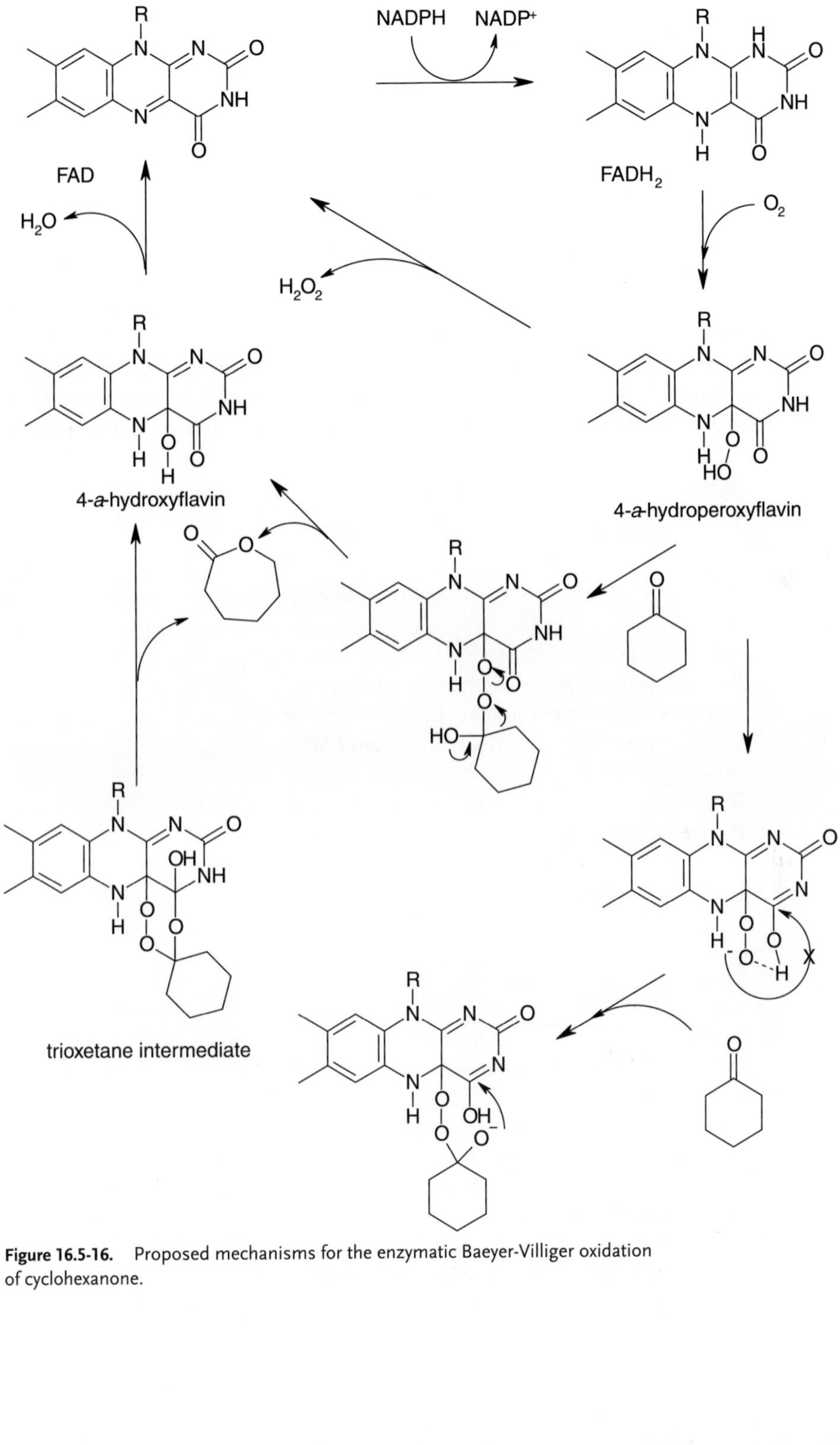

Figure 16.5-16. Proposed mechanisms for the enzymatic Baeyer-Villiger oxidation of cyclohexanone.

However, the FAD-4-a-OOH can also break down directly via liberation of H_2O_2.

A variation on this model has recently been proposed by Kelly et al. [7] who suggested that the hydroxy group of the Criegee intermediate could not be immobilized in such a mechanism, and that unreasonable steric constraints would be imposed for many of the substrates transformed reported for these enzymes. A new tautomer of the the flavin hydroperoxide was proposed as part of an alternative scheme (lower cycle, Fig. 16.5-16) in which an intermediate trioxane decomposes to yield the lactone and flavin hydrate.

In addition to ketone substrates, the 4-a-hydroperoxyflavin can also react by nucleophilic attack on other molecules. Thus, boronic acid substrates were transformed into the corresponding alcohols via the intermediate borate esters as hydrolytically labile initial enzyme products [90, 91].

However, the 4-hydroperoxyflavin, acting in these cases as an electrophile, is also able to oxidize other nucleophilic substrates and in particular heteroatoms such as sulfur [91], selenium [91, 92], nitrogen [73] and phosphorous [91]. Indeed, CHMO oxygen-

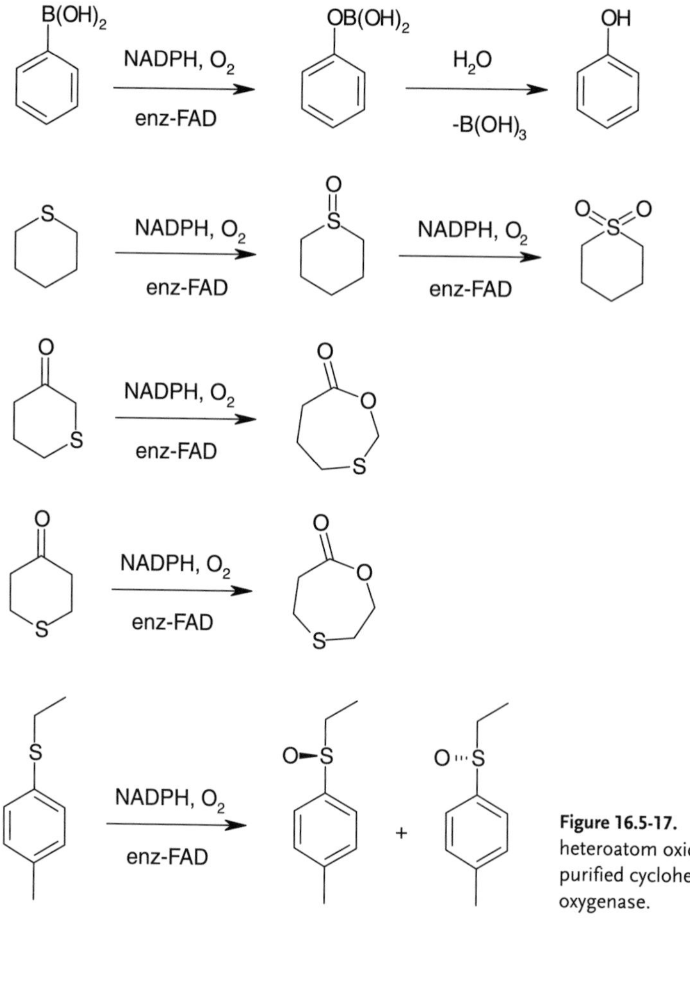

Figure 16.5-17. Studies on heteroatom oxidation using purified cyclohexanone mono-oxygenase.

ates trimethyl phosphite to trimethyl phosphate, sulfides to sulfoxides (one equivalent) or sulfones (two equivalents). If 3- or 4-thiocyclohexanones were used as substrates, these were converted exclusively into the lactone products, showing that Baeyer-Villiger oxidation is preferred in these cases (Fig. 16.5-17). The synthetic applications of heteroatom, notably sulfur, oxidation by BVMOs have been thoroughly explored and and reviewed [93].

These results illustrate that reactions performed by BVMOs are similar to those of peroxide containing reagents (hydrogen peroxide, alkyl hydroperoxides or peracids), which are able to deliver either a formally nucleophilic or a formally electrophilic oxygen atom to a substrate. Indeed, whereas Baeyer-Villiger oxidation or boronic acid oxygenation involve initial attack of a nucleophilic oxygen, the sulfide, selenide or phosphite ester oxygenations require the transfer of an electrophilic oxygen to a nucleophilic electron pair of the substrate. Interestingly, no epoxidation of olefinic double bonds by BVMOs have been reported however [73].

The substrate selectivity of CHMO was first explored by Trudgill and coworkers [70, 81], who demonstrated that the enzyme processes C4-C8 cyclic ketones. The migratory aptitude of the enzymatic oxygen insertion process was probed initially with two types of substrates. First, in an attempt to explore the stereochemical mode of these reactions, Schwab et al. studied the Baeyer-Villiger oxidation of (2R)-deuterated cyclohexanone. Detailed NMR multinuclear spectroscopic studies led to the conclusion that CHMO catalyzes the conversion of cyclohexanone to ε-caprolactone with complete retention of configuration at the migrating carbon center [94], a result identical to the chemical route (Fig. 16.5-18). To eliminate the possibility of an enolization and/or rearrangement route, 2,2,6,6-tetradeuterocyclohexanone was also incubated with the enzyme. The fact that no loss of deuterium was observed by GC again militates in favor of a mechanism similar to that proposed for chemical Baeyer-Villiger oxidation.

In an elegant further study [95], these authors confirmed their preliminary proposal of the (R)- absolute configuration of the starting 2-deuterocyclohexanone as well as the occurrence of a total retention of configuration of the CHMO catalyzed Baeyer-

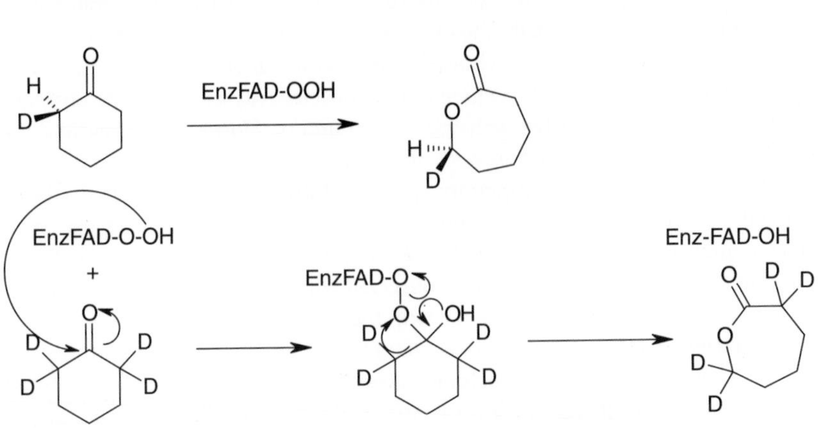

Figure 16.5-18. Stereochemical studies using deuterated cyclohexanone.

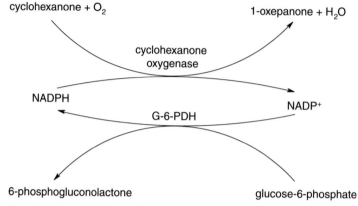

Figure 16.5-19. NADPH recycling in the course of CHMO catalyzed oxidation.

Villiger reaction. Interestingly, they described for the first time that the efficient conversion of ketone into lactone could be brought about in the presence of a catalytic amount of NADPH, with cofactor recycling accomplished by the glucose-6-phosphate dehydrogenase as shown in Fig. 16.5-19.

In order to test enzyme regio- and enantioselectivity rigorously, Schwab and co-workers also studied the asymmetric substrate 2-methylcyclohexanone. A "virtual racemate" made up of equivalent quantities of $(2R)$-2-[methyl-^{2}H$_3$]methylcyclohexanone and of $(2S)$-[methyl-^{13}C]methylcyclohexanone (each one prepared by different methods) was studied, using a multinuclear NMR technique. The conclusions from this experiment were two fold. First, they confirmed that 6-methyl-ε-caprolactone is the only reaction product, thus indicating a total regioselectivity of oxygen insertion into the "more substituted" carbon-carbon bond. Second, these results showed for the first time a two-fold rate difference between transformation of the two enantiomers of cyclohexanone. This was an interesting result that suggested that CHMO could show far greater discrimination toward enantiomers of a substrate that bore a far bulkier C2 substituent. Finally, the measurement of the reaction kinetics for each one of the substrate enantiomers showed that, after about 50 % reaction, there is a progressive decrease in both the degree of enantioselectivity as well as the absolute rate of lactonization of the two substrate enantiomers. However, the reasons for these diminishing rates of reaction were not clear.

Further work exploring the migratory aptitude of different substituents has been described[91]. Phenylacetone is converted into benzyl acetate, showing exclusive benzyl migration in accordance with the chemical reaction achieved with trifluoroacetic acid. Different results were observed with phenacetaldehyde, where an inverted preference is seen as compared with the peracidic reactions.

Using purified CHMO from *A. calcoaceticus* NCIMB 9871, Taschner and coworkers[96, 97] showed that several prochiral substrates including some 4-substituted cyclohexanones were efficiently converted into their corresponding lactones, each of them showing very high enantiomeric purities (Fig. 16.5-20). Thus, CHMO prove to

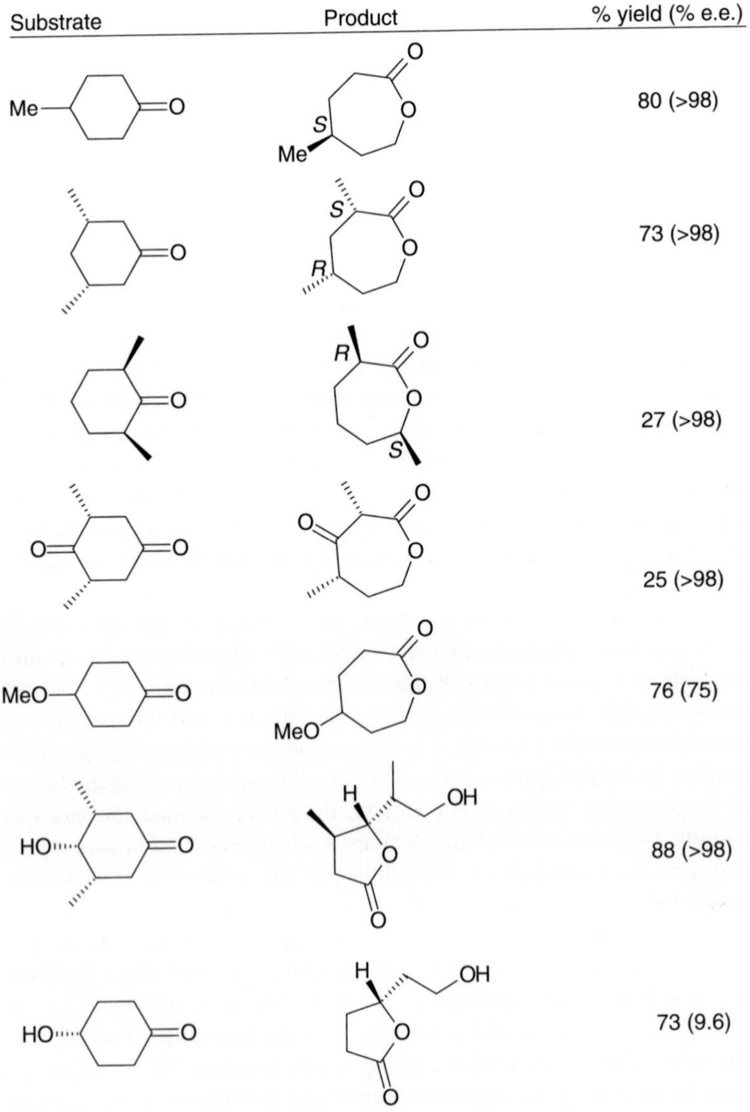

Substrate	Product	% yield (% e.e.)
		80 (>98)
		73 (>98)
		27 (>98)
		25 (>98)
		76 (75)
		88 (>98)
		73 (9.6)

Figure 16.5-20. CHMO catalyzed oxidation of various prochiral substrates.

be extremely effective at discriminating between the two sides of the carbonyl function of such prochiral substrates. However, the presence of an alcohol or methyl ether function at position 4 leads unexpectedly to products of lower ee values.

16.5.3

Synthetic Applications

With the exception of steroid type substrates, the results described up to now have dealt with small-scale analytical studies. However, in view of the potential of BVMOs for regio- and even enantioselective transformations of various substrates, studies into the scale-up of these transformations began in earnest soon after these earlier investigations, followed by considerations of their application in chiral organic synthesis.

In this context, Abril et al.[98] examined a variety of readily available ketones in order to determine the substrate selectivity, regioselectivity and enantioselectivity of CHMO immobilized in a polyacryalamide gel. They also used the NADPH recycling system previously desribed by Schwab for *in situ* regeneration of this cofactor. These experiments showed that 2-norbornanone, L- and D-fenchone, (+)-camphor and (+)-dihydrocarvone are processed by CHMO. In a typical experiment, 10.2 g of racemic 2-oxabicyclo[3.2.1]octan-3-one were obtained from 11.4 g of 2-norbornanone, using 1.7 g of NADP cofactor. The authors concluded that the enzyme did not display a useful degree of enantioselectivity, therefore offering no major advantages over chemical oxidation.

One major drawback of employing CHMO as a catalyst is the necessity to regenerate the expensive nicotinamide cofactor NADPH. One strategy for circumventing this problem is use of whole-cell preparations of microorganisms for Baeyer-Villiger oxidations. One early example of this technique involved the oxidation of 2,2,5,5-tetramethyl-1,4-cyclohexanedione to the optically pure (*S*)-ketol by *Curvularia lunata* described by Azerad and coworkers[99]. They showed that during the fungal reaction of the dione, as shown in Fig. 16.5-21, the already formed (*S*)-ketol was isomerized to its five-membered isomer. Moreover, when submitted to appropriate culture conditions, the racemic ketol afforded the (*S*)-lactone (81% ee) as well as the unchanged (*R*)-lactol of 97% ee The remaining substrate could then be further treated by *m*-chloroperbenzoic acid to afford the (*R*)-hydroxylactone enantiomer.

Extensive studies have been performed on the microbial Baeyer-Villiger oxidation of bicyclic [3.2.0] ketones and analogues. These studies were prompted by the important findings of Furstoss and coworkers[100], who determined that the oxygenation of bicyclo[3.2.0]hept-2-en-6-one using *Acinetobacter* sp. TD 63 led to *two* regioisomeric lactones in equal quantities and almost quantitative yield. The first arises from the "normal" oxygen insertion mode into the more substituted carbon-carbon bond, whereas the second is the result of an oxygen insertion into the less substitiuted bond leading to the so-called "abnormal" lactone. Moreover, both these lactones were of high optical purity i.e. showing a 98% ee for the (−)-(1*S*, 5*R*) isomer and a 95% enantiomeric excess for the (−)-(1*R*, 5*S*) enantiomer. These results appeared to suggest that biological Baeyer-Villiger oxidations could indeed be used for the large-scale preparation of optically active lactones.

In the case of bicyclo[3.2.0]hept-2-en-6-one, each one of the substrate enantiomers reacts with a different and divergent regioselectivity for the oxygen atom insertion. This result is noteworthy since it describes for the first time such an almost perfect

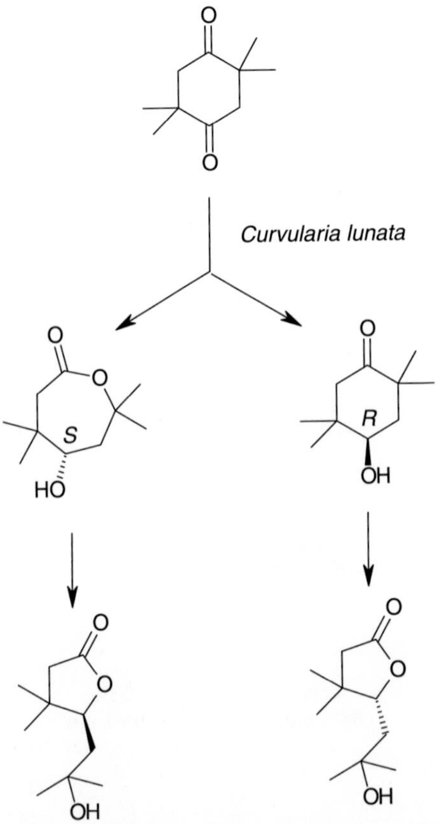

Figure 16.5-21. Baeyer-Villiger oxidation of 2,2,5,5-tetramethyl-1,4-cyclohexane-dione by *Curvularia lunata*.

regio- vs. enantioselectivity for the Baeyer-Villiger oxygenation. A more complete study[101], aimed at exploring the synthetic potential of these reactions, confirmed that this *enantiodivergent* selectivity is not restricted to one particular substrate but is a general phenomenon within a series of similar compounds. Two strains of bacteria, *Acinetobacter* sp. TD 63 and *A. calcoaceticus* NCIMB 9871 were used throughout this study and led to almost identical results. In most cases, both "normal" and "abnormal" lactones were obtained in approximately 1:1 ratios and with almost quantitative yields. Also, it was observed as shown in Fig. 16.5-22 that the "abnormal" lactone, which is not accessible using conventional Baeyer-Villiger oxidation, always shows very high ee values, whereas the enantiomeric purity of the "normal" lactone is somewhat lower for the substrate bearing a saturated six-membered ring. Both of these lactones are interesting chiral synthons; the "normal" one being an important chiron for prostaglandin synthesis. It is noteworthy that all lactones of a particular type are formed from the same enantiomer of the starting ketone: thus, the substrate enantiomer bearing an (S)-configuration at the bridge-head carbon atom α to the carbonyl group leads to the "normal" lactones, whereas the (R)- configuration affords the "abnormal" ones.

Similar results were obtained in the course of a study conducted on bicyclic

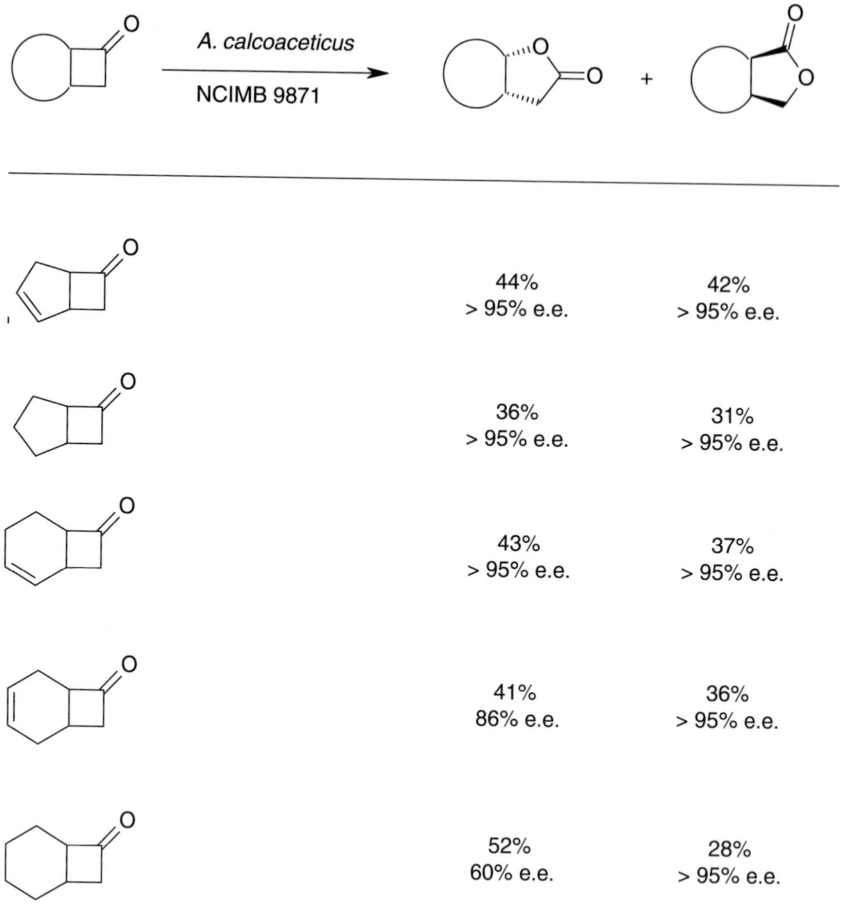

Figure 16.5-22. Oxidation of various [*n*.2.0] bicyclic ketones with *Acinetobacter calcoaceticus* NCIMB 9871.

substrates bearing an oxygen atom in the five or six-membered ring[102] (Fig. 16.5-23). Here again, equivalent ratios as well as high ee values were obtained for both the 'normal' and "abnormal" lactones. Since the lactones are unreported in the literature in their optically inactive form, detailed studies using circular dichroism were conducted in order to attribute the absolute configuration of the products.

Whilst the whole-cell approach has proved invaluable, the associated problems of overmetabolism and side reactions can be encountered. Another way to counter the problems of high cost in using isolated BVMOs is to use an NADH dependent enzyme, as NADH retails at approximately one tenth of the cost of NADPH. The Type 2 DKCMOs from *Pseudomonas putida* ATCC 17453 (≡ NCIMB 10007) are NADH dependent, and Grogan et al. were successful in applying a complement of these enzymes, termed MO1, to the transformation of bicyclo[3.2.0]hept-2-en-6-one, to yield another enantiodivergent mix of lactones enantiomeric to those obtained

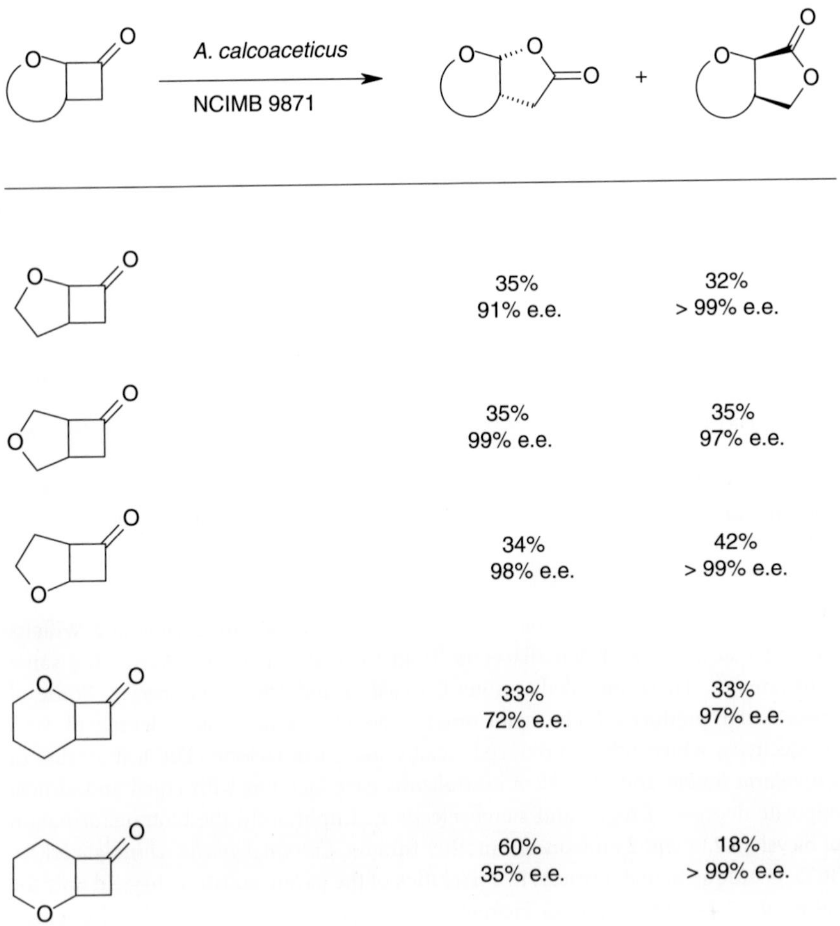

A. calcoaceticus

NCIMB 9871

35%	32%
91% e.e.	> 99% e.e.

35%	35%
99% e.e.	97% e.e.

34%	42%
98% e.e.	> 99% e.e.

33%	33%
72% e.e.	97% e.e.

60%	18%
35% e.e.	> 99% e.e.

Figure 16.5-23. Oxidation of various oxo-[n.2.0] bicyclic ketones with *Acinetobacter calcoaceticus* NCIMB 9871.

with *A. calcoaceticus* NCIMB 9871/TD 63. The use of NADH dependent enzymes is also important in this context, as it allows use of the NAD dependent formate dehydrogenase/sodium formate recycling strategy for cofactor regeneration[103], reducing costs still further. Interestingly, the separated isoenzymes, 2,5-diketocamphane 1,2-monooxygenase and 3,6-diketocamphane 1,6-monooxygenase were shown to have different selectivities for this transformation, compromising the result obtained with MO1[104] (Fig. 16.5-24). Further transformations of this ketone by luminescent bacteria containing NADH dependent luciferases (also Type 2 BVMOs) have also been reported[105], although characterization of cell-free systems employing these enzymes has not been investigated further.

The biotransformation of bicyclo[3.2.0]hept-2-en-6-one using whole cell suspensions of the fungus *Cylindrocarpon destructans* gave not only different ratios of both lactones depending on the degree of conversion, but also no enantioselectivity was

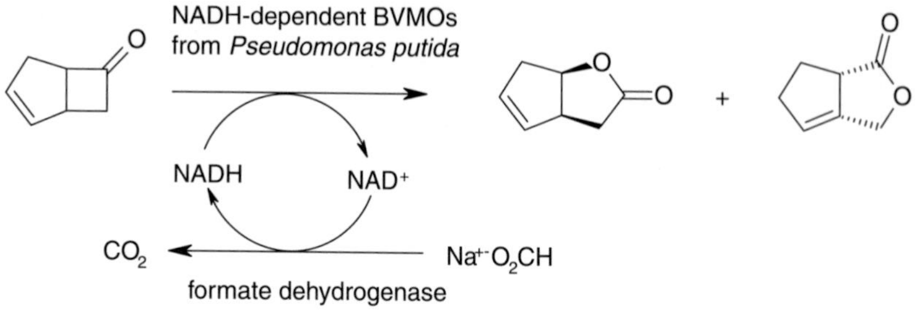

'MO1'	63%, 60% e.e.	37%, 95% e.e.
2,5-DKCMO	57%, 82% e.e.	43%, 100% e.e.
3,6-DKCMO	17%, 10% e.e.	13%, 72% e.e.

Figure 16.5-24. Biotransformation of bicyclo[3.2.0]hept-2-en-6-one by NADH dependent BVMOs from camphor grown *Pseudomonas putida* ATCC 17453.

observed[106]. Further fungal biotransformations described by Carnell and Willetts showed that a series of dematiaceous fungi were also able to lactonize the same substrate[107]. These included various *Curvularia* and *Dreschlera* species. Some of these fungi produced both regioisomeric lactones with a high degree of stereoselectivity, whilst others produced mostly the 3-oxa lactone. The test strains of *Curvularia lunata* and *Dreschlera australiensis* gave lactones with equal and almost opposite degrees of regio- and stereoselectivity. Importantly, the biotransformation of bicyclo[3.2.0]hept-2-en-6-one by another fungus, *Cunninghamella echinulata* NRRL 3655, is unique in that it results in a *resolution* of the parent substrate to yield only the "abnormal" (−)-(1*R*, 5*S*)-3-oxa lactone in 30% yield and 95% ee[108]. This chiral synthetic intermediate has been used to synthesize both single enantiomer cyclosarkomycin[108] and the marine brown algae pheremones (+)-multifidene and (+)-viridiene[109] (Fig. 16.5-25).

Further reports by Furstoss and coworkers concerned Baeyer-Villiger oxidation of α-substituted cyclopentanones[110]. Using the same two *Acinetobacter* strains used previously, this study aimed to explore the possibility of synthesising optically active δ-lactones bearing aliphatic chains, these compounds being of particular interest as chiral synthons. This study showed that various lactones of (*S*) configuration can be obtained in fair yields with moderate to excellent ee values depending on the chain length and on the conversion ratio. Using *Acinetobacter calcoaceticus* NCIMB 9871 it was, however, necessary to run these biotransformations in the presence of tetraethylpyrophosphate (TEPP), a well known inhibitor of hydrolases. This was necessary in order to avoid hydrolytic degradation of the δ-lactones formed. The use of this inhibitor was, however, unnecessary when using the *Acinetobacter* sp. TD 63 strain which is known to lack a lactone hydrolase. One interesting application of this study was the preparative two-step synthesis of both enantiomers of 5-hexadecanolide, a

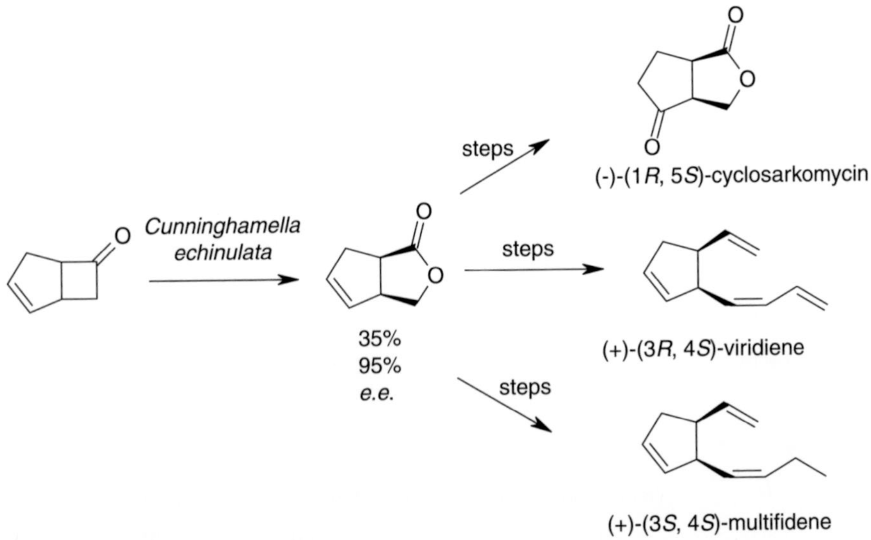

(-)-(1R, 5S)-cyclosarkomycin

(+)-(3R, 4S)-viridiene

(+)-(3S, 4S)-multifidene

35%
95%
e.e.

Figure 16.5-25. Biotransformation of bicyclo[3.2.0]hept-2-en-6-one by *Cunninghamella echinulata* NRRL 3655 and synthetic targets.

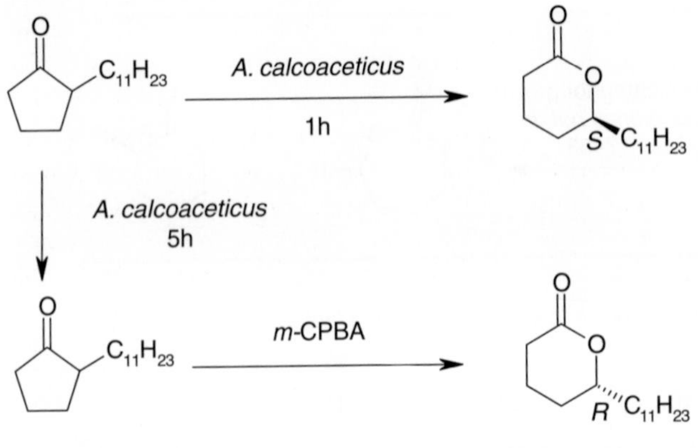

Figure 16.5-26. Baeyer-Villiger oxidation of α-undecylcyclopentanone: synthesis of either enantiomer of hexadecanolide.

pheromone isolated from the oriental hornet *Vespa orientalis*. As shown in Fig. 16.5-26, Baeyer-Villiger oxidation of racemic undecylcyclopentanone with *A. calcoaceticus* NCIMB 9871 led to a 25% isolated yield of (S)-5-hexadecanolide showing an ee of 74%. Interestingly, a 30% yield of remaining (R)-2-undecylcyclopentanone of 95% optical purity can also be isolated using a longer incubation time, thus allowing direct access, via chemical Baeyer-Villiger oxidation, to the (R)-(+)-5-hexadecanolide known to be the sole bioactive enantiomer.

The biotransformation of α-substituted cycloalkanones using the BVMOs from camphor grown *Pseudomonas putida* has also been investigated in depth. Whilst the NADPH dependent activity corresponding to 2-oxo-Δ^3-4,5,5-trimethylcyclopentenylacetyl-Co-A monooxygenase (and termed MO2) resolved a series of α-alkyl cyclopentanones with good selectivity, poorer resolution of these compounds was per-

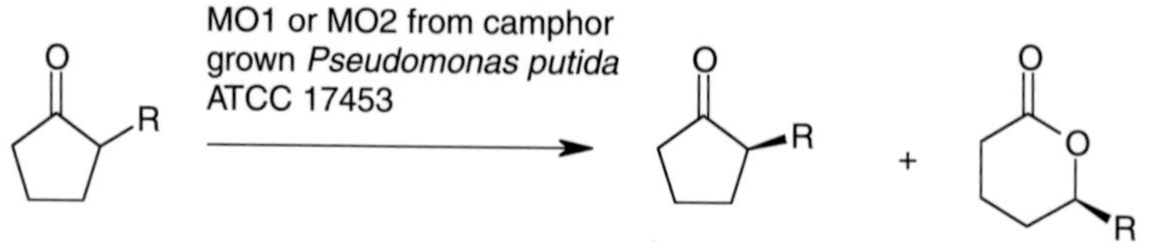

MO1

R	Yield ketone	e.e. ketone	Yield lactone	e.e. lactone
C_4H_9	14	9	16	58
C_6H_{13}	48	48	34	74
C_8H_{17}	35	22	11	90

MO2

R	Yield ketone	e.e. ketone	Yield lactone	e.e. lactone
C_4H_9	26	-	40	95
C_6H_{13}	51	75	35	92
C_8H_{17}	44	59	29	95

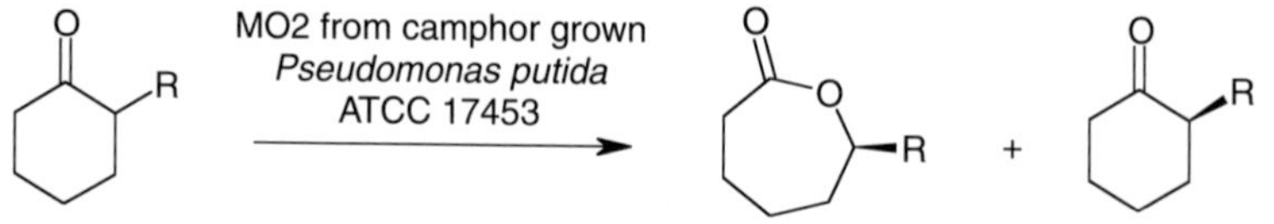

MO2

R	Yield ketone	e.e. ketone	Yield lactone	e.e. lactone
C_6H_{13}	30	65	36	72
C_8H_{17}	49	61	34	77
CH_2CO_2Et	43	89	30	93
CH_2CH_2OAc	13	75	34	83

Figure 16.5-27. Biotransformation of 2-substituted monocyclic ketones by BVMOs from camphor grown *Pseudomonas putida* ATCC 17 453.

formed by the NADH dependent MO1 complement[104] (Fig. 16.5-27). An extension to this study revealed that MO2 could be used to resolve a series of α-substituted cyclohexanones wherein the subsituents consisted of esters, acetates and common protecting groups[111]. This led to the development of a chemoenzymatic synthesis of (R)-(+)-lipoic acid incorporating a BVMO catalyzed resolution as the key step (Fig. 16.5-28). Interestingly, the preferred selectivity of cyclopentanone monooxygenase from *Pseudomonas* sp. NCIMB 9872, is opposite to that of MO2, and in a

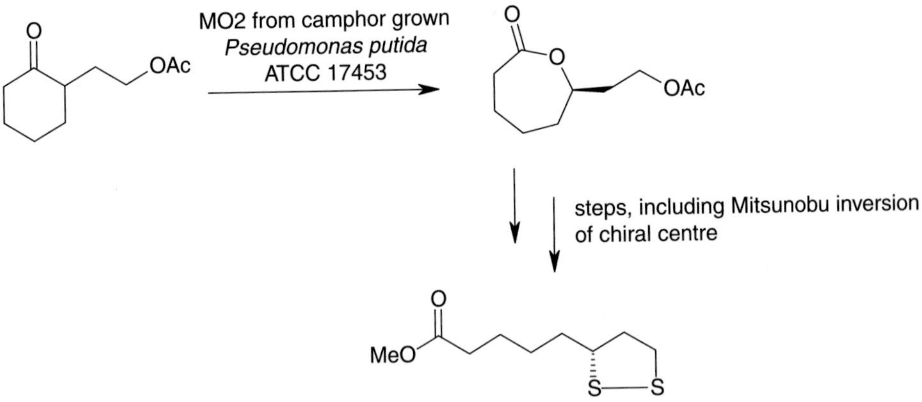

Figure 16.5-28. Chemoenzymatic synthesis of (+)-lipoic acid incorporating a BVMO catalysed resolution as the key step.

separate investigation, it was suggested that this enzyme be used in the place of MO2 to eliminate the need for the Mitsonobu inversion in the chemoenzymatic synthesis [112].

The biological Baeyer-Villiger oxidation has also been applied, in a variety of forms, to the production of optically active lactones from prochiral 3-substituted cyclobutanones. A series of cyclobutanones was subjected to oxidation by *Acinetobacter* sp. and to the MO1 and MO2 enzyme preparations derived from camphor-grown *Pseudomonas putida* ATCC 17453 [113]. The results are summarized in Fig. 16.5-29. In general, the reactions performed with *Acinetobacter* sp. displayed better enantioselectivities, but the value of a multi-biocatalyst approach was illustrated by the fact that certain BVMOs from *P. putida* displayed opposite enantioselectivity. A further series of cyclobutanone substrates was oxidized by *Acinetobacter* sp. and by the fungus *Cunninghamella echinulata* [114] (Fig. 16.5-30). The lactonization of 3-(4'-chlorobenzyl)-cyclobutanone was performed by this fungus to yield (R)-lactone of 99% ee in 30% yield, which was used in a chemoenzymatic synthesis of baclofen [115], a lipophilic derivative of γ-aminobutyric acid. The *Cunninghamella* strain was also used to oxidize 3-(benzyloxymethyl)-cyclobutanone to the optically pure (R)-(-)-γ-butyrolactone, which was used in enantiodivergent chemoenzymatic syntheses of (R)- and (S)-proline [116].

The oxidation of either enantiomer of menthone and dihydrocarvone by *Acinetobacter* sp. were also reported [117]. (–)-Menthone is not metabolized but (+)-menthone leads to the expected lactone, whereas both enantiomers of dihydrocarvone are oxidized. Thus (–)-dihydrocarvone leads to the expected lactone, whereas (+)-dihydrocarvone afforded the unexpected 'abnormal' lactone product (Fig. 16.5-31). Both enantiomers of dihydrocarvone are also transformed by MMKMO [51] from *Rhodococcus erythropolis* DCL 14, which in contrast to *Acinetobacter* sp., also transforms both enantiomers of menthone.

Taschner and coworkers described the oxidation of *cis*-3,5-dimethylcyclohexanone by whole-cell preparations of *A. calcoaceticus* NCIMB 9871 [118], which led directly to

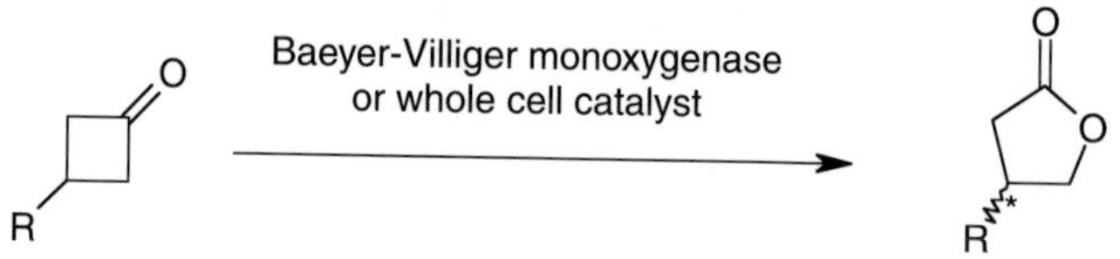

Baeyer-Villiger monoxygenase or whole cell catalyst

A. calcoaceticus NCIMB 9871

R	Conversion	Yield lactone	e.e. lactone
Bu	95	68	(S)-,17%
Bu^i	98	56	(R)-, 84%
CH$_2$Ph	100	57	(R)-, 82%
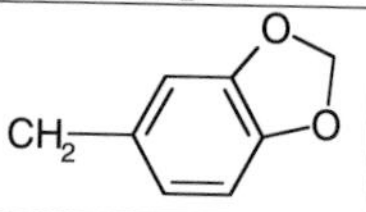 CH$_2$-[benzodioxole]	100	83	(R)-, 95%
CH$_2$OCH$_2$Ph	100	89	(S)-, 55

MO1

R	Conversion	Yield lactone	e.e. lactone
Bu	100	nd	(R)-, 69
Bu^i	78	nd	(R)-, 91
CH$_2$Ph	58	40	(S)-, 15
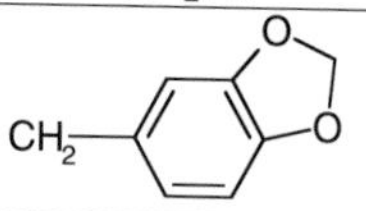 CH$_2$-[benzodioxole]	48	38	(R)-, 7
CH$_2$OCH$_2$Ph	98	74	(S)-, 74

MO2

R	Conversion	Yield lactone	e.e. lactone
Bu	93	nd	(R)-, 54
Bu^i	97	nd	(R)-, 85
CH$_2$Ph	37	26	(S)-, 20
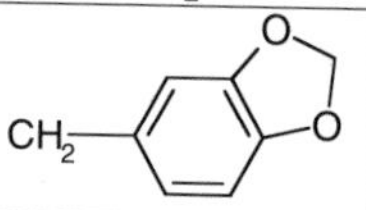 CH$_2$-[benzodioxole]	71	6	(S)-, 14
CH$_2$OCH$_2$Ph	95	nd	(R)-, 90

Figure 16.5-29. Biotransformation of prochiral 3-substituted cyclobutanones using BVMOs.

the corresponding optically active lactone and thence to the hydroxyacid, which was converted into the methylester by reaction with diazomethane. This methylester, which was shown to be optically active, is a key intermediate in the synthesis of the polyether antibiotic ionomycin.

In addition, several bridged bicyclic compounds have been examined as potential substrates (Fig. 16.5-32). In contrast to the regiodivergent behaviour of the [n.2.0] bicyclic compounds, in these cases, only one lactone product is usually obtained. This high selectivity compares favorably with the chemical Baeyer-Villiger oxidation of compounds of this type, which often afford regiomixtures[119]. In addition, the

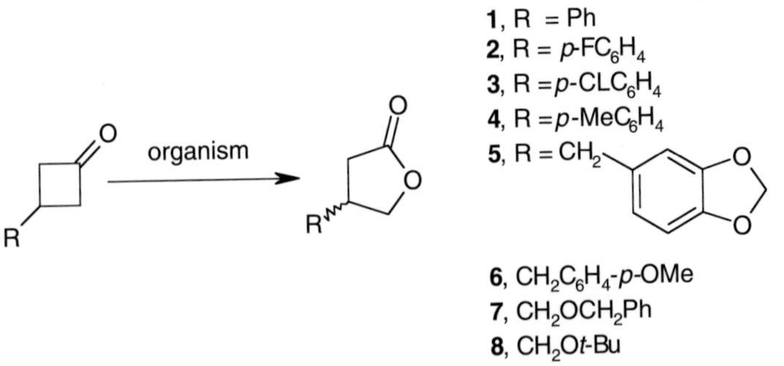

1, R = Ph
2, R = p-FC$_6$H$_4$
3, R = p-CLC$_6$H$_4$
4, R = p-MeC$_6$H$_4$
5, R = CH$_2$

6, CH$_2$C$_6$H$_4$-p-OMe
7, CH$_2$OCH$_2$Ph
8, CH$_2$Ot-Bu

Ketone	Microorganism	Yield	e.e.
1	C. echinulata	65	(R)-, 98
	A. calcoaceticus	70	(R)-, 43
	Acinetobacter TD63	84	(R)-, 47
2	C. echinulata	80	98
	A. calcoaceticus	89	19
	Acinetobacter TD63	92	5
3	C. echinulata	30	(R)-, 98
	A. calcoaceticus	88	(S)-, 85
	Acinetobacter TD63	15	(S)-, 89
4	C. echinulata	4	nd
	A. calcoaceticus	73	(S)-, 91
	Acinetobacter TD63	61	(S)-, 93
5	C. echinulata	68	(S)-, 91
	A. calcoaceticus	70	(S)-, 100
	Acinetobacter TD63	64	nd
6	C. echinulata	68	(S)-, 91
	A. calcoaceticus	83	(S)-, 96
	Acinetobacter TD63	94	(S)-, 94
7	C. echinulata	74	(R)-, 98
	A. calcoaceticus	89	(S)-, 55
	Acinetobacter TD63	90	(R)-, 25
8	C. echinulata	25	98
	A. calcoaceticus	43	89
	Acinetobacter TD63	15	88

Figure 16.5-30. Biotransformation of prochiral cyclobutanones by three whole-cell preparations.

obtained bridgehead lactones are often described to be of high optical purity. The benzyloxy derivative is known to be an important intermediate for prostaglandin synthesis. The residual fluorinated bicyclic ketone of high enantiomeric excess was used to synthesize an antiviral carbocyclic nucleoside[120]. In this last case, detailed studies showed that the first formed product is the corresponding alcohol (about 80% conversion) and that over the next 3 h period, the alcohol concentration decreased, the amount of ketone rose and the production of lactone started[121]. This observation led to an elegant closed-loop recycling procedure, as shown in Fig. 16.5-33, where the alcohol dehydrogenase from *Thermoanaerobium brockii* was used in conjunction with the purified monooxygenase from *A. calcoaceticus* NCIMB 9871. In

rac-menthone → A. calcoaceticus or Acinetobacter TD 63 → (-)-menthone + (+)

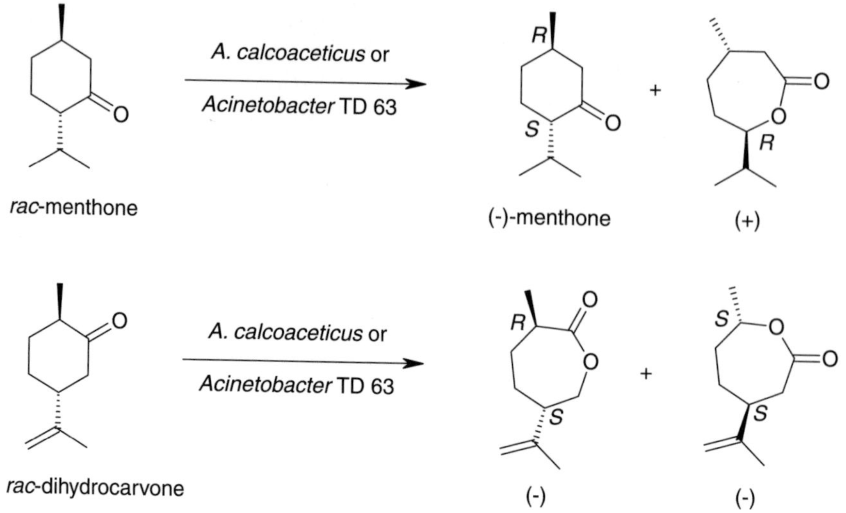

rac-dihydrocarvone → A. calcoaceticus or Acinetobacter TD 63 → (-) + (-)

Figure 16.5-31. Oxidation of dihydrocarvone enantiomers with *Acinetobacter calcoaceticus* NCIMB 9871 and *Acinetobacter* sp. TD63.

this case, the substrate alcohol also serves as a co-substrate for the NADPH recycling reaction. Thus, *endo*-bicyclo[2.2.1]heptan-2-ol was transformed using catalytic amounts of NADP. An analogous recycling loop was set up using the NAD dependent alcohol dehydrogenase from *Pseudomonas* sp. NCIMB 9872 and the NADH dependent MO1 isozyme complement from *Pseudomonas putida* ATCC 17 453, for the oxidation of 7- *endo*-methylbicyclo[3.2.0]hept-2-en-6-ol [122].

A further series of prochiral bicyclic [2.2.1] substrates have also been studied by Taschner and coworkers and lead generally to lactones of high enantiomeric purity. One of these is a valuable precursor for chorismic acid synthesis [97].

The transformation of a series of norbornanone derivatives (Fig. 16.5-34) was studied by Roberts and coworkers who determined that both the MO1 complement of NADH dependent BVMOs from *Pseudomonas putida* ATCC 17 453 and the NADPH dependent fraction MO2 were successful in the resolution of hydroxy, acetoxy and benzyloxy norbornanones [123]. Interestingly 25DKCMO and 36DKCMO when separate, displayed notably different reactivity toward the hydroxy and acetoxy derivative, again emphasizing their complementary nature as potential individual biocatalysts. The benzyloxy lactone is an intermediate in the synthesis of the insect antifeedant azadirachtin.

Further studies also been performed on the bicyclo[3.2.0]heptan-6-one series of compounds [124, 125]. These results are summarised in Fig. 16.5-35. Oxidation of this ketone with *Pseudomonas* NCIMB 9872 gave the (1S, 5R)-lactone of low optical purity (23 % ee) with only small amounts (5 %) of the isomeric lactone, whereas its oxidation with an *Acinetobacter* sp. gave these lactones in a 9:1 ratio and a modest yield, a result quite different from the one described previously. However, oxidation of 7-*endo*-methylbicyclo[3.2.0]hept-2-en-6-one using either *Pseudomonas* sp. or *Acinetobacter* sp. produced optically pure (ee > 96 %) of both lactones in equal quantities

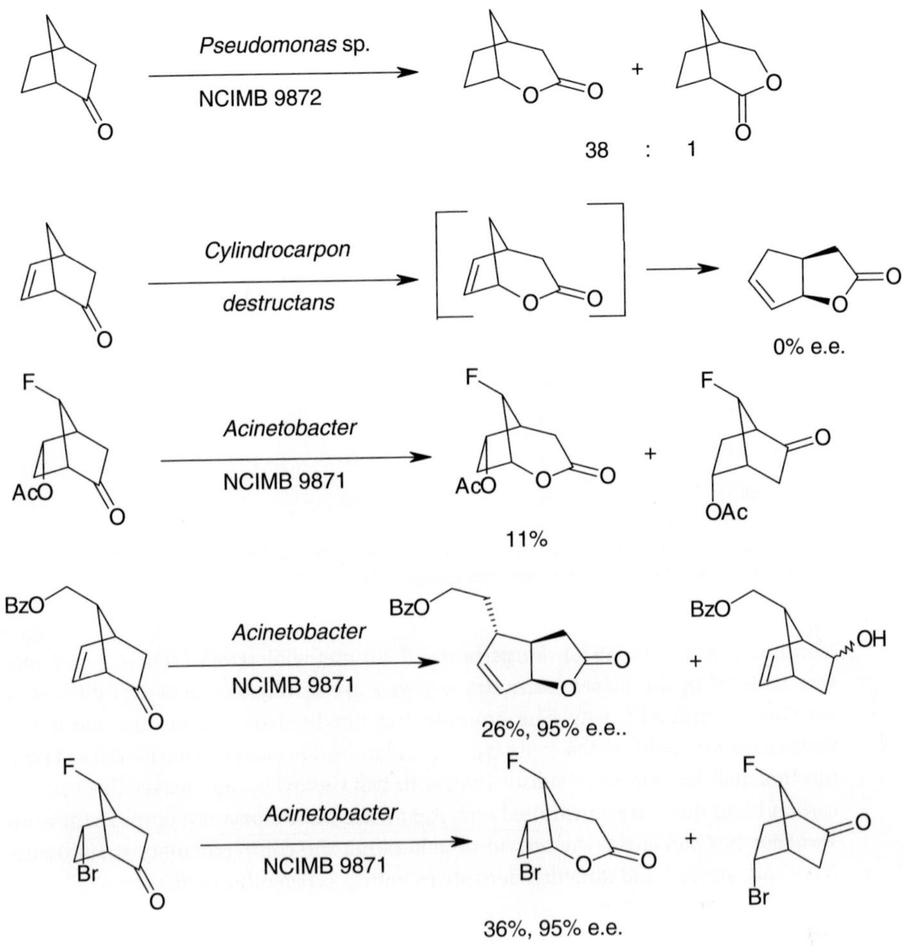

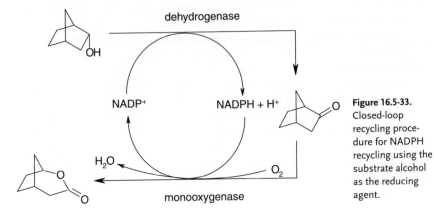

Figure 16.5-32. Baeyer-Villiger oxidation of various [2.2.1] bicyclic substrates.

Figure 16.5-33. Closed-loop recycling procedure for NADPH recycling using the substrate alcohol as the reducing agent.

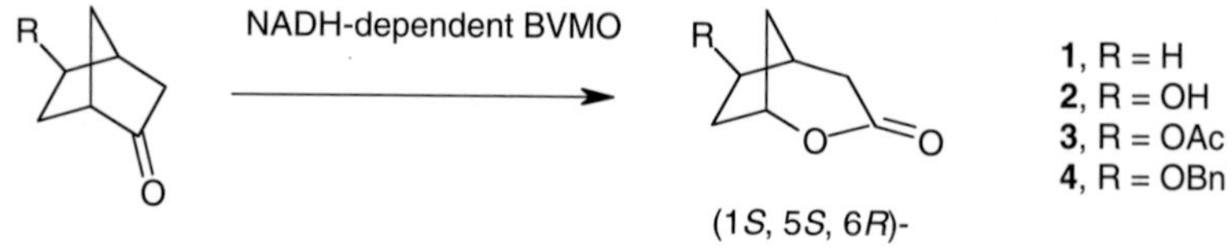

NADH-dependent BVMO

(1*S*, 5*S*, 6*R*)-

1, R = H
2, R = OH
3, R = OAc
4, R = OBn

Enzyme	Substrate	Conversion (%)	Lactone e.e. (%)
25DKCMO	1	20	60
36DKCMO	1	48	>90
25DKCMO	2	0	-
36DKCMO	2	33	>95
25DKCMO	3	35	>95
36DKCMO	3	0	-
'MO1'	4	39	>95

Figure 16.5-34. Biotransformation of norbornanone derivatives using NADH dependent BVMOs from camphor grown *Pseudomonas putida* ATCC 17 453.

(combined yields 50–55%). Surprisingly, 7,7-dimethylbicyclo[3.2.0]hept-2-en-6-one was oxidized by the *Acinetobacter* strain to give exclusively one lactone of 29% ee, a very low enantioselectivity. The bromohydrin obtained from this substrate led to similar results, yielding the same type of oxidation. This can be considered as being the "normal" lactone since substitution with two methyl groups makes this carbon-carbon bond the more substituted one. Again, the MO1 isozymic complement from *Pseudomonas putida* was successful in generating the complementary enantiomers from *endo*-methyl and dimethyl derivatives with good enantiomeric excess [103].

16.5.4
Models for the Action of Baeyer-Villiger Monooxygenases

The results of biological Baeyer-Villiger oxidations have been, in some cases unpredictable and surprising, and, in the continued absence of a structure of one of these enzymes, several groups have attempted to explain the various observations of selectivity with an increasingly complex series of models.

Initially, some workers proposed that enantiodivergent biotransformations of the type witnessed in the oxygenation of bicyclo[3.2.0]hept-2-en-6-one by, for instance CHMO and 25DKCMO could be due to the presence in either of these preparations of two separate enzymatic activities. Whilst this was once and indeed still is, a reasonable assumption in the light of results obtained with whole-cell preparations, the use of highly purified preparations of the two named enzymes to effect this biotransformation [104, 126] have eliminated this possibility in these cases. The phenomenon of enantiodivergence has therefore been addressed with respect to one enzyme active site.

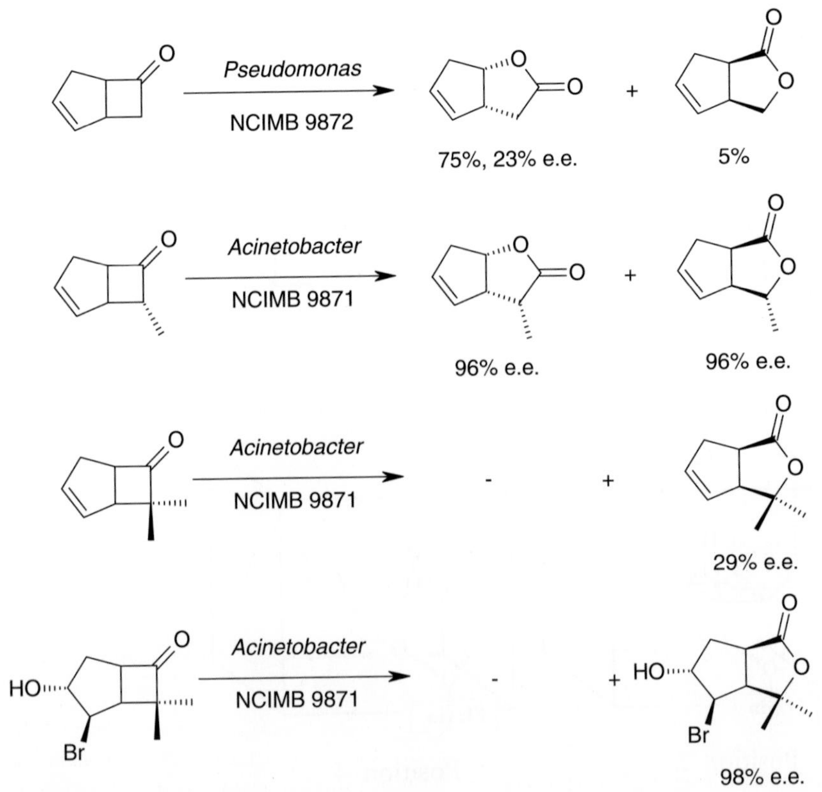

Figure 16.5-35. Baeyer-Villiger oxidation of various [n.2.0] bicyclic compounds.

The first model was proposed by Furstoss and coworkers, based on steric and stereoelectronic considerations. In this model, shown in Fig. 16.5-36, the 4-a-hydroxyperflavin is considered as being the oxygen transfer agent, according to the hypothesis of Walsh and coworkers[84]. The enantioselectivity of the reaction would be due to a different positioning of each intermediate in the active site. It is supposed, primarily, that the attack of the hydroperoxyflavin should take place on the least hindered face of the ketone. On the other hand, the migrating C-C bond of the peroxidic intermediate should be antiperiplanar to the peroxidic bond and to a non-bonded electron pair of the hydroxide group, as suggested for chemical Baeyer-Villiger oxidations. Thus, the cycloalkyl part of the (S,S)-enantiomer of the ketone (the one leading to the "normal" lactone) could be accommodated in only one region of the active site (position 1). Position 2 would never be adopted due to some steric hindrance with the active site (dotted cube). Similarly, in the case of the (R,R)-enantiomer, position 4 would be favored over position 3 leading to the "abnormal" lactone. This model was augmented by further work by the inclusion of results obtained with both monocyclic monterpene[117], 3-substituted cyclobutanone substrates[113] and α-substituted cyclohexanones[127].

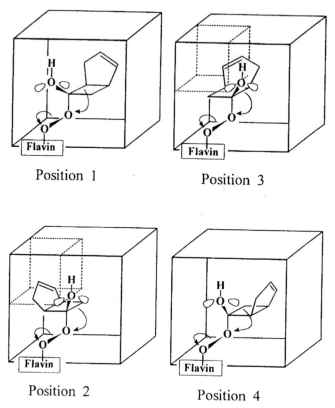

Position 1   Position 3

Position 2   Position 4

Figure 16.5-36. Furstoss model for the active site of cyclohexanone monooxygenase from *Acinetobacter calcoaceticus* NCIMB 9871.

Taschner and coworkers proposed a similar model based on two other fla-voenzymes; the human and *E. coli* glutathione reductase. The FAD binding domain of glutathione reductase and *p*-hydroxybenzoate hydroxylase have been shown to resemble each other closely via comparison of their respective X-ray crystal struc-tures. Extrapolating this information to CHMO leads to the proposal that the hydroperoxide is attached to the *re*-face of the isoalloxazine ring and that the ketone substrates approach the hydroperoxide from the direction of the dimethylbenzene moiety[97]. Further stereochemical and stereoelectronic considerations lead to a hypothesis explaining the observed stereoselectivities.

In the model of Furstoss and coworkers, stereoselectivity of CHMO is determined by the differentiation of groups of different sizes in the active site. A different model, proposed by Kelly and coworkers[128–130], extends Taschner's idea that the source of stereoselectivity might be the flavin cofactor itself. It was suggested that the stereoselectivity of oxygen insertion arises solely as a result of the flavin face, *re*- or *si*-, from which the hydroperoxide attacks. This would lead to two distinct Criegee intermediates of opposing absolute configuration (Fig. 16.5-37). Hence it was

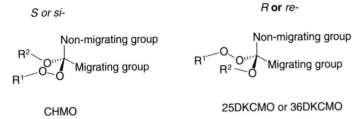

Figure 16.5-37. Schematic representation of enantiomeric Criegee intermediates for the enzymatic Baeyer-Villiger reaction.

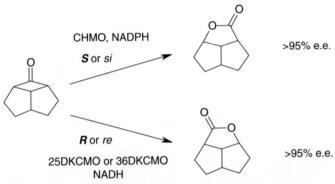

Figure 16.5-38. Enantioselective Baeyer-Villiger oxidation of a tricyclic ketone by Type 1 and Type 2 BVMOs.

demonstrated that for the tricyclic ketone shown in Fig. 16.5-38 for which attack from only the *exo*-face is possible, pure preparations of BVMOs always resulted in lactones of >95% ee Interestingly, all Type 1, FAD plus NADPH dependent BVMOs yield lactone from the (R)-configuration of the intermediate, and all Type 2, NADH plus FMN dependent BVMOs yield lactone from the (R)-intermediate. Substrate interaction with the topology of the active site must also be considered however, as the enantiocomplementary DKCMOs, both proposed to catalyze oxygen insertion via (R)-Criegee intermediates, catalyze complementary resolutions of racemic camphor[67].

This additional dependence on active site topology for selectivity in CHMO was carefully considered by Ottolina et al. [131], who developed a sophisticated cubic space model for the active site of CHMO (Fig. 16.5-39). This group was able to show that, for example, for the biotransformation of 7-*endo*-methylbicyclo[3.2.0]hept-2-en-6-one, of the eight possible intermediates in oxidation, the only two "allowed" by the model were the two which led to the lactones observed by experiment. The model was successfully applied to a series of other ketones and also predicts the stereoselectivity of sulfur oxidation by this enzyme[132]. The group of Colonna established in a series of reports that CHMO was able to catalyze the oxidation of a range of alkylaryl sulfides, benzyl alkyl sulfides, functionalized sulfides and 1,3-dithioacetals with absolute configuration and enantiomeric excesses being highly dependent

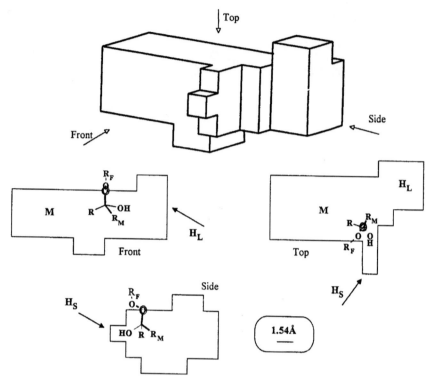

Figure 16.5-39. Cubic space filling model of the active site of cyclohexanone monooxygenase from *Acinetobacter calcoaceticus* NCIMB 9871, based on the results of the oxidations of a series of bicyclic ketones. The catalytic oxygen is circled. The main (M) hydrophobic large (H$_L$) and hydrophobic small (H$_S$) pockets are depicted. The correct arrangements of the Criegee intermediate are also shown.

on the structure of the substrate[93]. This group has also recently reported the first asymmetric oxidation of tertiary amines using CHMO[133].

The ability of BVMOs to oxidize sulfur was also exploited by Beecher and Willetts in order to construct space filling cubic models of the active site of the DKCMO enzymes from *Pseudomonas putida* ATCC 17453 (Fig. 16.5-40). They note that the more relaxed enantiospecificity of 36DKCMO, at least in terms of sulfoxidation, appears to be due to an overall larger 3D cubic space available in the active site[134]. 36DKCMO appears to be the best candidate for a first X-ray structure of a BVMO, as preliminary crystal data have been reported[88].

16.5.5
Conclusion and Outlook

It is apparent from the many application of BVMOs in synthesis, that these enzymes currently represent the most valuable method of effecting the enantioselective

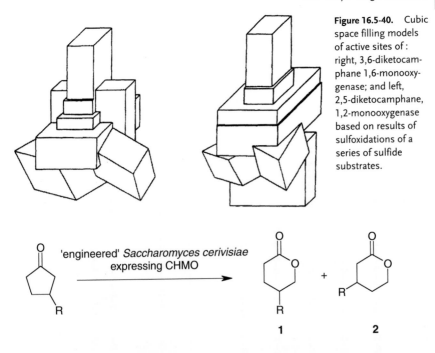

Figure 16.5-40. Cubic space filling models of active sites of: right, 3,6-diketocamphane 1,6-monooxygenase; and left, 2,5-diketocamphane, 1,2-monooxygenase based on results of sulfoxidations of a series of sulfide substrates.

R	Ratio 1:2 Combined yield lactones	e.e. lactone 1 (%)	e.e. lactone 2 (%)
Me	13:87 95%	9	36
Et	80:20 80%	33	19
n-Pr	83:17 44%	33	60
n-Bu	99:1 34%	38	-
n-Oct	99:1 19%	16	-

Figure 16.5-41. Biotransformation of 3-alkylcyclopentanones by "engineered" *Saccharomyces cerivisiae* expressing CHMO.

Baeyer-Villiger reaction. The primary sources of BVMO enzymes carry associated disadvantages that must now be addressed, although recent biotechnological advances suggest that BVMOs will be more accessible to the synthetic organic chemist in the future.

Type 1 BVMOs

1	Steroid monooxygenase *Cylindrocarpon radicicola*	2	**A**–E–W–**A**–E–E–**F**–**D**–**V**–**L**–V–V–**G**–**A**–**G**–**A**–**G**–
2	CHMO *Rhodococcus coprophilus*	2	**A**–Q–T–I–H–**G**–V–**D**–**A**–V–V–I–**G**–**A**–**G**–**F**–**G**–**G**–**I**–**Y**–**A**–**V**–**H**–**K**–
3	CHMO *Acinetobacter* NCIMB 9871	1	M–**S**–Q–L–M–**D**–**F**–**D**–A–I–V–I–**G**–**G**–**G**–**F**–**G**–**G**–**L**–**Y**–**A**–**V**–**K**–**K**–
4	CPMO *Pseudomonas* NCIMB 9872	14	–N–**S**–V–N–**D**–K–L–**D**–V–L–L–I–**G**–**A**–**G**–**F**–
5	Steroid monooxygenase *Rhodococcus rhodochrous*	1	M–N–G–Q–H–P–R–V–V–V–A–A–P–D–A

Type 2 BVMOs

6	2,5-DKCMO *Pseudomonas putida*	1	–M–Q–A–**G**–F–F–G–**T**–**P**–**Y**–D–L–**P**–**T**–R–**T**–**A**–R–Q–M–
7	3,6-DKCMO *Pseudomonas putida*	1	A–M–E–T–G–L–I–F–H–**P**–**Y**–M–Y–**P**–G–K–S–**A**–**A**–Q–

Figure 16.5-42. N-terminal amino acid sequence alignment of Type 1 BVMOs (**1–5**) and Type 2 BVMOs (**6** and **7**). Conserved residues are marked in bold.

The *Acinetobacter* strain from which CHMO is derived is a Class II pathogen as defined by the Advisory Committee on Dangerous Pathogens (ACDP), and hence, may only be handled in suitably equipped microbiological facilities. One solution to this problem has been the cloning and expression of the gene encoding CHMO in *Saccharomyces cerivisiae*[85]. In a series of reports by Stewart and coworkers[135–137], the "designer yeast" was shown to catalyze many of the reactions which had previously been shown to be catalyzed by either whole cells of *Acinetobacter* sp. or CHMO in addition to some new ones (Fig. 16.5-41). Recently, a similar strategy has seen whole-cell preparations of *Escherichia coli* expressing recombinant CHMO for the same purpose[138]. It remains to be seen whether constraints on the use of genetically engineered microorganisms of this type will render these strains as "difficult" to manipulate as the wild-type strains.

The use of purified enzyme would circumvent the need for whole-cell containment procedures, and indeed, amounts of CHMO are now available from Fluka[139]. However, the attendant costs associated with cofactor recycling must be addressed if this approach is to prove viable. The recent production of a formate dehydrogenase suitable for use in NADP/NADPH recycling systems[140] should prove attractive in this regard, as should the further investigation of NADH dependent enzymes. The practicalities associated with the industrial scale up of biological Baeyer-Villiger reactions are currently being investigated[141].

New sources of enzyme will also become important and with the advent of genomic science, paralogs of genes that encode CHMO-like proteins are being identified amongst whole bacterial genomes, most recently those of *Pseudomonas aeruginosa*[142] and *Mycobacterium tuberculosis*[143]. The availability of gene and amino acid sequence data for BVMOs will prove useful in identifying more new activities in this manner. BVMOs of the same Type (1 or 2) exhibit sequence homology within their *N*-terminal amino acid sequences although homology between types is not conserved[130] (Fig. 16.5-42). In the future, the "tailoring" of enzyme characteristics by either rational redesign or so-called "directed" evolution approaches could also doubtless be applied to BVMOs. Fundamental to these studies would be the development of an efficient, rapid screen for BVMO activity. Rational redesign would require more knowledge of the 3D structure of these enzymes. This is one reason why the acquisition of a complete X-ray crystal structure of a BVMO must be considered of fundamental importance to the ongoing development of this area.

References

1 A. Gusso, C. Baccin, F. Pinna, G. Strukul, *Organometallics* **1994**, *13*, 3442–3451.

2 C. Paneghetti, R. Gavagnin, F. Pinna, G. Strukul, *Organometallics* **1999**, *18*, 5057–5065.

3 C. Bolm, G. Schlingoff, F. Bienewald, *J. Mol. Catal. A* **1997**, *117*, 347–350.

4 A. J.Willetts, *Trends. Biotechnol.* **1997**, *15*, 55–62.

5 S. M. Roberts, P. H. W. Wan, *J. Mol. Catal. B. Enz.* **1998**, *4*, 111–136.

6 D. R. Kelly, P. Wan, J. Tang in: *Biotransformations*, Vol. 8a in the series *Biotechnology*. (volume Ed. D. R. Kelly; series Ed. H.-J. Rehm, G. Reed, A. Puhler, P. J. W. Stadler), Wiley VCH, Weinheim, 1998, pp. 535–588.

7 D. R. Kelly, *Chem. Oggi.* **2000**, *18*, 33–39 and 52–56.

8 E. Vischer, A. Wettstein, *Experientia* **1953**, *9*, 371–372.

9 J. Fried, R. W. Thoma, A. Klingsberg, *J. Am. Chem. Soc.* **1953**, *75*, 5764–5765.

10 D. J. Petersen, S. H. Eppstein, P. D. Meister, H. C. Murray, H. M. Leigh, A. Weintraub, L. M. Reineke, *J. Am. Chem. Soc.* **1953**, *75*, 5768.

11 J. Fried, R. W. Thoma, D. Perlman, J. R. Gerke, *Recent Prog. Hormone. Res.* **1955**, *11*, 149.

12 H. C. Murray, *Ind. Microbiol.* **1976**, 79.

13 C. K. A. Christoph, *Adv. Appl. Microbiol.* **1977**, *22*, 29.

14 H. L. Holland, *Synthesis with Oxidative Enzymes*, VCH Publishers Inc. NY, **1992**, chap. 5.

15 F. Viola, O. Caputo, G. Balliano, L. Delprino, L. Cattel, *J. Steroid Biochem.* **1983**, *19*, 1451–1458.

16 A. I. Laskin, P. Grabowich, C. de Lisle Meyers, J. Fried, *J. Med. Chem.* **1964**, *7*, 406–409.

17 V. Gaberc-Porekar, H. E. Gottlieb, M. Mervic, *J. Steroid Biochem.* **1983**, *19*, 1509–1511.

18 G. S. Fonken, H. C. Murray, L. M. Reineke, *J. Am. Chem. Soc.* **1960**, *82*, 5507–5508.

19 R. L. Prairie, P. Talalay, *Biochemistry* **1963**, *2*, 203–208.

20 M. A. Rahim, C. J. Sih, *J. Biol. Chem.* **1966**, *241*, 3615–3623.

21 K. Singh, S. Rahkit, *Biochim. Biophys. Acta* **1967**, *144*, 139–144.

22 S. Morii, S. Sawamoto, Y. Yamauchi, M. Miyamoto, M. Iwami, E. Itagaki, *J. Biochem.* **1999**, *126*, 624–631.

23 E. Itagaki, *J. Biochem.* **1986**, *99*, 825–832.

24 E. Itagaki, *J. Biochem.* **1986**, *99*, 815–824.

25 F. W. Forney, A. J. Markovetz, *J. Bacteriol.* **1968**, *96*, 1055–1964.

26 F. W. Forney, A. J. Markovetz, *Biochem. Biophys. Res. Commun.* **1969**, *37*, 31–38.

27 A. C. Shum, A. J. Markovetz, *J. Bacteriol.* **1974**, *118*, 890–897.

28 L. N. Britton, J. M. Brand, A. J. Markovetz, *Biochim. Biophys. Acta* **1974**, *369*, 45–49.

29 J. E. Allen, A. J. Markovetz, *J. Bacteriol.* **1970**, *103*, 426–434.

30 S. Hartmans, J. A. M. De Bont, *FEMS Microbiol. Lett.* **1996**, *36*, 155–158.

31 R. E. Cripps, *Biochem. J.* **1975**, *152*, 233–241.

32 R. E. Cripps, P. W. Trudgill, J. G. Whateley, *Eur. J. Biochem.* **1978**, *86*, 175–186.

33 E. F. Eubanks, F. W. Forney, D. Larson, *J. Bacteriol.* **1974**, *120*, 1133–1143.

34 R. Shaw, *Nature* **1966**, 1369.

35 M. Griffin, P. W. Trudgill, *Biochem. J.* **1972**, *129*, 595–603.

36 M. S. Anderson, R. A. Hall, M. Griffin, *J. Gen. Microbiol.* **1980**, *120*, 89–94.

37 L. A. Stirling, R. J. Watkinson, *J. Gen. Microbiol.* **1977**, *99*, 119–125.

38 M. K. Trower, R. M. Buckland, R. Higgins, M. Griffin, *Appl. Environ. Microbiol.* **1985**, *49*, 1282–1289.

39 A. M. Magor, J. Warburton, M. K. Trower, M. Griffin, *Appl. Environ. Microbiol.* **1986**, *52*, 665–671.

40 D. B. Norris, P. W. Trudgill, *Biochem. J.* **1971**, *121*, 363–370.

41 N. A. Donoghue, P. W. Trudgill, *Eur. J. Biochem.* **1975**, *60*, 1–7.

42 J. R. Murray, T. A. Scheikowski, I. C. Macrae, *Anyonie van Leeuwenhoek* **1974**, *40*, 17–24.

43 J. F. Davey, P. W. Trudgill, *Eur. J. Biochem.* **1977**, *74*, 115–127.

44 Y. Yugari, *Bikens's J.* **1961**, *4*, 197–207.

45 N. A. Donoghue, P. W. Trudgill, *Biochem. Soc. Trans.* **1973**, *1*, 1287–1290.

46 Y. Hasegawa, K. Hamano, H. Obata, T. Tokuyama, *Agric. Biol. Chem.* **1982**, *46*, 1139–1143.

47 J. D. Schumacher, R. M. Fakoussa, *Appl. Microbiol. Biotechnol.* **1999**, *52*, 85–90.

48 M. Griffin, P. W. Trudgill, *Eur. J. Biochem.* **1976**, *63*, 199–209.

49 O. P. Shukla, R. C. Bartholomus, I. C. Gunsalus, *Can. J. Microbiol.* **1987**, *33*, 489–497.

50 R. Croteau, V. K. Sood, B. Renstrom, R. Bhushan, *Plant Physiol.* **1984**, *76*, 489–497.

51 M. J. van der Werf, *Biochem. J.* **2000**, *347*, 693–791.

52 V. Krasnobajew, D. Helmlinger, *Helv. Chim. Acta* **1982**, *65*, 1590–1601.

53 A. Siewinski, J. Dmochowska-Gladysz, T. Kolek, A. Zabza, K. Derdzinski, A. Nespiak, *Tetrahedron* **1983**, *39*, 2265–2270.

54 C. Fuganti, M. Mendozza, D. Joulain, J. Minut, G. Pedrocchi-Fantoni, V. Pierigianni, S. Servi, G. Zucchi, *J. Agric Food Chem.* **1996**, *44*, 3616–3619.

55 F. Donzelli, C. Fuganti, M. Mendozza, G. Pedrocchi-Fantoni, S. Servi, G. Zucchi, *Tetrahedron: Asymmetry*, **1996**, *11*, 3129–3134.

56 G. Fronza, C. Fuganti, G. Pedrocchi-Fantoni, V. Perozzo, S. Servi, G. Zucchi, D. Joulain, *J. Org. Chem.* **1996**, *61*, 9362–9367.

57 C. Fuganti, J. Minut, G. Pedrocchi-Fantoni,

S. Servi, *J. Mol Catal. B. Enz.*, **1998**, *4*, 47–52.

58 H. E. Conrad, J. Hedegaard, I. C. Gunsalus, E. J. Corey, H. Uda, *Tetrahedron Lett.* **1965**, 561–565.

59 H. E. Conrad, K. Lieb, I. C. Gunsalus, *J. Biol. Chem.* **1965**, *240*, 4029–4037.

60 H. E. Conrad, R. Dubus, M. J. Namtvedt, I. C. Gunsalus, *J. Biol. Chem.* **1965**, *240*, 495–503.

61 P. W. Trudgill, R. Dubus, I. C. Gunsalus, *J. Biol. Chem.* **1966**, *241*, 4288–4297.

62 C. A. Yu, I. C. Gunsalus, *J. Biol. Chem.* **1969**, *244*, 6149–6152.

63 H. J. Ougham, D. G. Taylor, P. W. Trudgill, *J. Bacteriol.* **1983**, *153*, 140–152.

64 D. G. Taylor, P. W. Trudgill, *J. Bacteriol.* **1986**, *165*, 489–497.

65 P. J. Chapman, G. Meerman, I. C. Gunsalus, *Biochem. Biophys. Res. Commun.* **1965**, *20*, 104–108.

66 R. A. Chandler, R. F. Galligan, I. C. Macrae, *Aust. J. Biol. Sci.* **1973**, *26*, 999–1003.

67 K. H. Jones, R. T. Smith P. W. Trudgill, *J. Gen. Microbiol.* **1993**, *139*, 797–805.

68 R. Croteau, H. El-Bialy, S. El-Hindawi, *Arch. Biochem. Biophys.* **1984**, *228*, 667–680.

69 R. Croteau, H. El-Bialy, S. S. Dehal, *Plant Physiol.* **1987**, *84*, 643–648.

70 P. W. Trudgill in: *Microbial Degradation of Organic Compounds* (Ed.: D. T. Gibson), Marcel Dekker, NY, **1984**, Chapter 6, pp. 131–180.

71 D. R. Williams, P. W. Trudgill, D. G. Taylor, *J. Gen. Microbiol.* **1989**, *135*, 1957–1967.

72 C. A. Townsend, S. B. Christensen, S. G. Davis, *J. Am. Chem. Soc.* **1982**, *104*, 6154–6155.

73 C. T. Walsh, Y.-C. J. Chen, *Angew. Chem. Intl. Ed. Engl.* **1988**, *103*, 333–343.

74 I. Fujii, Y. Ebizuka, U. Sankawa, *J. Biochem.* **1988**, *103*, 878–883.

75 J. L. C. Wright, T. Hu, J. L. McLachlan, J. Needham, J. Walter, *J. Am. Chem. Soc.* **1996**, *118*, 8757–8758.

76 L. Prado, E. Fernandez, U. Weissbach, G. Bianco, L. M. Quiros, A. F. Brana, C. Mendez, J. Rohr, J. A. Salas, *Chem. Biol.* **1999**, *6*, 19–30.

77 J. Rohr, C. Mendez, J. A. Salas, *Bioorg. Chem.* **1999**, *27*, 41–54.

78 M. Griffin, P. W. Trudgill, *Biochem. Soc. Trans.* **1973**, 1255–1258.

79 M. A. Wright, I. N. Taylor, M. J. Lenn, D. R. Kelly, J. G. Mahdi, C. J. Knowles, *FEMS Microbiol. Lett.* **1994**, *116*, 67–72.

80 L. N. Britton, A. J. Markovetz, *J. Biol. Chem.* **1977**, *252*, 8561–8566.

81 N. A. Donoghue, D. B. Norris, P. W. Trudgill, *Eur. J. Biochem.* **1976**, *63*, 175–192.

82 M. K. Trower, R. M. Buckland, R. Griffin, *Eur. J. Biochem.* **1989**, *181*, 199–206.

83 M. Miyamoto, J. Matsumoto, T. Iwaya, E. Itagaki, *Biochim. Biophys. Lett.* **1995**, *1251*, 115–120.

84 Y.-C. J. Chen, O. P. Peoples, C. T. Walsh, *J. Bacteriol.* **1988**, *170*, 781–789.

85 J. D. Stewart, K. W. Reed, C. A. Martinez, J. Zhu, G. Chen, M. M. Kayser, *J. Am. Chem. Soc.* **1998**, *120*, 3541–3548.

86 J. Beecher, G. Grogan, S. M. Roberts, A. Willetts, *Biotechnol. Lett.* **1996**, *18*, 571–576.

87 D. H. Cohn, R. C. Ogden, J. N. Abelson, T. O. Baldwin, K. H. Nealson, M. I. Simon, A. J. Mileham, *Proc. Nat. Acad. Sci.* **1993**, *80*, 120–128.

88 E. J. McGhie, M. N. Isupov, E. Schroder, J. A. Littlechild *Acta Crystallogr. Sect. D.* **1998**, *54*, 1035–1038.

89 C. C. Ryerson, D. P. Ballou, C. Walsh, *Biochemistry* **1982**, *21*, 2644–2655.

90 J. A. Latham, C. Walsh, *J. Chem. Soc. Chem. Commun.* **1986**, 527–528.

91 B. P. Branchaud, C. T. Walsh, *J. Am. Chem. Soc.* **1985**, *107*, 2153–2161.

92 J. A. Latham, B. P. Branchaud, Y.-C. Chen, C. Walsh, *J. Chem. Soc. Chem. Commun.* 1986, 528–530.

93 S. Colonna, N. Gaggero, P. Pasta, G. Ottolina, *J. Chem. Soc. Chem. Commun.* **1996**, 2302–2307.

94 J. M. Schwab, *J. Am. Chem. Soc.* **1981**, *103*, 1876–1879.

95 J. M. Schwab, W.-B. Li, L. P. Thomas, *J. Am. Chem. Soc.* **1988**, *110*, 6892–6893.

96 M. J. Taschner, D. J. Black, *J. Am. Chem. Soc.* **1988**, *110*, 6892–6893.

97 M. J. Taschner, L. Peddada, P. Cyr, Q.-Z. Chen, D. J. Black, *Microbial Reagents in Organic Synthesis* (NATO Asi Series, Serie C) **1992**, *381*, 347–360.

98 O. Abril, C. C. Ryerson, C. Walsh, G. M. Whitesides, *Bioorg. Chem.* **1989**, *17*, 41–52.

99 J. Ouzzani-Chahdi, D. Buisson, R. Azerad, *Tetrahedron Lett.* **1987**, *28*, 1109–1112.

100 V. Alphand, A. Archelas, R. Furstoss, *Tetrahedron Lett.* **1989**, *30*, 3663–3664.

101 V. Alphand, R. Furstoss, *J. Org. Chem.* **1992**, *57*, 1306–1309.

102 F. Petit, R. Furstoss, *Tetrahedron: Asymmetry* **1993**, *4*, 1341–1352.

103 G. Grogan, S. M. Roberts, A. J. Willetts, *J. Chem. Soc. Chem. Commun.* **1993**, 699–701.

104 R. Gagnon, G. Grogan, M. S. Levitt, S. M. Roberts, P. W. H. Wan, A. J. Willetts, *J. Chem. Soc., Perkin Trans. 1* **1994**, 2537–2543.

105 R. Villa, A. Willetts, *J, Mol. Catal. B: Enz.* **1997**, *2*, 193–197.

106 K. Konigsberger, G. Braunegg, K. Faber, H. Griengl, *Biotechnol. Lett.* **1990**, *12*, 509–514.

107 A. Carnell, A. Willetts, *Biotechnol. Lett.* **1992**, *14*, 17–21.

108 L. Andrau, J. Lebreton, P. Viazzo, V. Alphand, R. Furstoss, *Tetrahedron Lett.* **1997**, *38*, 825–826.

109 J. Lebreton, V. Alphand, R. Furstoss, *Tetrahedron Lett.* **1996**, *37*, 1011–1014.

110 V. Alphand, A. Archelas, R. Furstoss, *Biocatalysis* **1990**, *3*, 73–83.

111 B. Adger, M. T. Bes, G. Grogan, R. McCague, S. Pedragosa-Moreau, S. M. Roberts, R. Villa, P. W. H. Wan, A. J. Willetts, *J. Chem. Soc. Chem. Commun.* **1995**, 1563–1564.

112 B. Adger, M. T. Bes, G. Grogan, R. McCague, S. Pedragosa-Moreau, S. M. Roberts, R. Villa, P. W. H. Wan, A. J. Willetts, *Bioorg. Med. Chem.* **1997**, *5*, 253–261.

113 R. Gagnon, G. Grogan, E. Groussain, S. Pedragosa-Moreau, P. F. Richardson, S. M. Roberts, A. J. Willetts, V. Alphand, J. Lebreton, R. Furstoss, *J. Chem. Soc., Perkin Trans. 1* **1995**, 2527–2528

114 V. Alphand, C. Mazzini, J. Lebreton, R. Furstoss, *J. Mol. Cat. B: Enz.* **1998**, *5*, 219–221.

115 C. Mazzini, J. Lebreton, V. Alphand, R. Furstoss, *Tetrahedron Lett.* **1997**, *38*, 1195–1196.

116 C. Mazzini, J. Lebreton, V. Alphand, R. Furstoss, *J. Org. Chem.* **1997**, *62*, 5215–5218.

117 V. Alphand, R. Furstoss, *Tetrahedron: Asymmetry*, **1992**, *3*, 379–382.

118 M. J. Taschner, Q.-Z. Chen, *Bioorg. Med. Chem. Lett.* **1991**, *1*, 535–538.

119 P. Hamley, A. B. Holmes, D. R. Marshall, J. W. M. MacKinnon, *J. Chem. Soc., Perkin Trans. 1* **1991**, 1793–1802.

120 M. S. Levitt, R. F. Newton, S. M. Roberts, A. J. Willetts, *J. Chem. Soc. Chem. Commun.* **1990**, 619–620.

121 A. J. Willetts, C. J. Knowles, M. S. Levitt, S. M. Roberts, H. Sandey, N. F. Shipston, *J. Chem. Soc., Perkin. Trans. 1* **1991**, 1608–1610.

122 G. Grogan. S. Roberts, A. Willetts, *Biotechnol. Lett.* **1992**, *14*, 1125–1130.

123 R. Gagnon, G. Grogan, S. M. Roberts, R. Villa, A. J. Willetts, *J. Chem. Soc., Perkin Trans. 1* **1995**, 1505–1511

124 A. J. Carnell, S. M. Roberts, V. Sik, A. J. Willetts, *J. Chem. Soc. Chem. Commun.* **1990**, 1438–1439.

125 A. J. Carnell, S. M. Roberts, V. Sik, A. J. Willetts, *J. Chem. Soc., Perkin Trans. 1* **1991**, 2385–2389.

126 N. F. Shipston, M. J. Lenn, C. J. Knowles, *J. Microbiol. Methods* **1992**, *15*, 41–52

127 V. Alphand, R. Furstoss, S. Pedragosa-Moreau, S. M. Roberts, A. J. Willetts, *J. Chem. Soc., Perkin Trans. 1* **1996**, 1867–1872.

128 D. R. Kelly, C. J. Knowles, J. G. Mahdi, M. A. Wright, I. N. Taylor, *J. Chem. Soc., Chem. Commun.* **1995**, 729–730.

129 D. R. Kelly, C. J. Knowles, J. G. Mahdi, M. A. Wright, I. N. Taylor, D. E. Hibbs, M. B. Hursthouse, A. K. Mish'al, S. M. Roberts, P. W. H. Wan, G. Grogan. A. J. Willetts, *J. Chem. Soc., Perkin Trans. 1* **1995**, 2057–2066.

130 D. R. Kelly, C. J. Knowles, J. G. Mahdi, M. A. Wright, I. N. Taylor, A. K. Mish'al, S. M. Roberts, P. W. H. Wan, G. Grogan, S. Pedragosa-Moreau, A. J. Willetts, *J. Chem. Soc., Chem. Commun.* **1996**, 2333–2334.

131 G. Ottolina, G. Carrea, S. Colonna, A. Ruckemann, *Tetrahedron: Asymmetry* **1996**, *7*, 1123–1136.

132 G. Ottolina, P. Pasta, G. Carrea, S. Colonna, S. Dallavalle, H. L. Holland, *Tetrahedron: Asymmetry* **1995**, *6*, 1375–1386.

133 G. Ottolina, S. Bianchi, B. Belloni, G. Carrea, B. Danieli, *Tetrahedron Lett.* **1999**, *40*, 8483–8486.

134 J. Beecher, A. Willetts, *Tetrahdron: Asymmetry* **1998**, *9*, 1899–1916.

135 M. M. Kayser, G. Chen, J. D. Stewart, *J. Org. Chem.* **1998**, *63*, 7103–7106.

136 M. M. Kayser, G. Chen, J. D. Stewart, *Synlett* **1999**, 153–158.

137 J. D. Stewart, *Curr. Opin. Biotechnol.* **2000**, *4*, 363–368.

138 G. Chen, M. M. Kayser, M. D. Mihovilovic, M. E. Mrstik, C. A. Martinez, J. D. Stewart, *New. J. Chem.* **1999**, *23*, 827–832.

139 Fluka catalogue **2000**.

140 K. Seelbach, B. Riebel, W. Hummel, M. R. Kula, V. I. Tishkov, A. M. Egorov, C. Wandrey, U. Kragl, *Tetrahedron. Lett.* **1996**, *9*, 1377–1380.

141 M. C. Hogan, J. M. Woodley, *Chem. Eng. Sci.* **2000**, *55*, 2001–2008.

142 PEDANT: http://pedant.mips.biochem. mpg.de/

143 C. A. Rivera-Marrero, M. A. Burroughs, R. A. Masse, F. O. Vannberg, D. L. Leimbach, J. Roman, J. J. Murtagh, *Microb. Pathogenesis* **1998**, *25*, 307–316.

16.6
Oxidation of Acids

Andreas Schmid, Frank Hollmann, Bruno Bühler

16.6.1
Introduction

At a first glance, synthetically relevant oxidations of carboxylic acids, except for oxidations at positions other than the carboxylate group, can hardly be found in literature. However, some preparative applications in whole cell catalysis were reported and will be discussed in the following (Fig. 16.6-1 A,B,C). *In vitro*, the high thermodynamic driving force for the oxidation of formate and pyruvate [$E°$ (formate/CO_2)= -0.42 V[1]; $E°$ (pyruvate/(acetate, CO_2)) = -0.70 V[2]] are used for the regeneration of coenzymes such as NAD(P)H or, indirectly, ATP (Fig. 16.6-1 D,E).

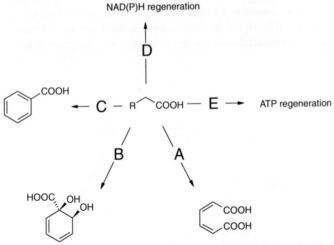

Figure 16.6-1. Synthetic and preparative applications of oxidations of acids. A, B: Oxidations of benzoic acid initiated by dihydroxylation (Sects. 16.6.4.2 and 16.6.4.3); C: oxidative decarboxylation (Sect. 16.6.4.1); D,E : energy coupling for the regeneration of coenzymes (Sects. 16.6.2, 16.6.3).

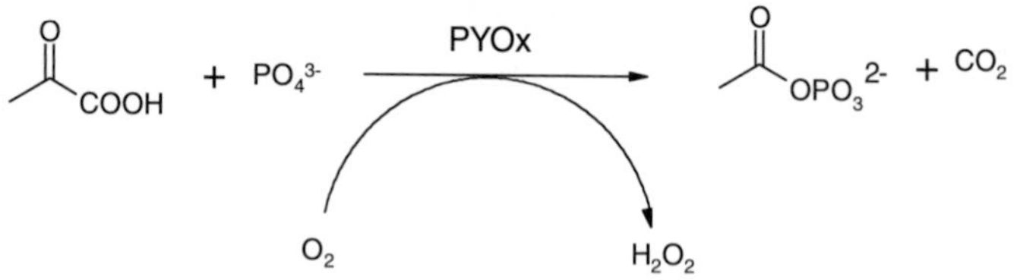

Figure 16.6-2. Oxidative phosphorylation of pyruvate by pyruvate oxidase (PYOx).

16.6.2
Pyruvate Oxidase (PYOx, E. C. 1.2.3.3)

PYOx from *Lactobacillus plantarum*[3, 4] or *Streptococcus sanguis*[5] catalyzes the decarboxylative phosphorylation of pyruvate to acetylphosphate, or the homologous arsenylation (Fig. 16.6-2).

Acetylphosphate is an important substrate for the enzyme acetate kinase (E. C. 2.7.2.1), which catalyzes the phosphorylation of various nucleotide diphosphates such as ADP, GDP, TDP, IDP, or UDP to the activated triphosphates[6–8]. This reaction can be applied to regenerate ATP in ATP-dependent enzymatic *in vitro* reactions (Fig. 16.6-3).

In a recent example, PYOx-catalyzed regeneration of ATP was coupled to *in vitro* protein biosynthesis (e. g. for human lymphotoxin)[9]. Under aerobic conditions, no external regeneration system for PYOx has to be applied; catalase however has to be added in order to destroy harmful hydrogen peroxide. An alternative to this autoregeneration approach (Fig. 16.6-3 A) was reported by Steckhan and coworkers for cases where hydrogen peroxide formation has to be prevented (Fig. 16.6-3 B)[10].

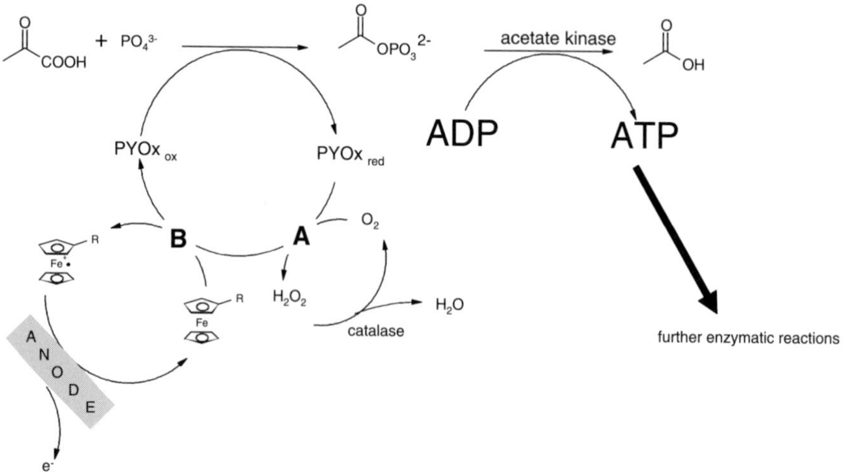

Figure 16.6-3. Decarboxylative phosphorylation of pyruvate by pyruvate oxidase as driving force for the regeneration of ATP; A: aerobic regeneration; B: indirect electrochemical regeneration.

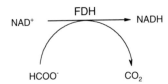

Figure 16.6-4. Regeneration of NADH using the formate dehydrogenase (FDH) reaction.

Here, the anode, together with the mediation by ferrocene, removes excess electrons from the PYOx active site.

Another possible application of the PYOx-catalyzed production of acetylphosphate lies within the *in vitro* regeneration of acetyl-CoA [11].

16.6.3
Formate Dehydrogenase (FDH, E. C. 1.2.1.2)

Probably the most prominent oxidation of a carboxylic acid is catalyzed by the enzyme formate dehydrogenase (FDH, E. C. 1.2.1.2). FDH was isolated from various bacteria, yeasts, and plants, where its physiological role is the regeneration of NADH [12].

FDH catalyzes the oxidation of formate to carbon dioxide, concomitant with the reduction of NAD^+ to NADH (Fig. 16.6-4). Because of the favorable thermodynamic equilibrium of the reaction and the volatility of the reaction product, the enzyme is commonly applied for *in situ* regeneration of NADH during asymmetric synthesis of chiral compounds [13].

FDH from *Candida boidinii* is mostly used as regeneration enzyme. It found industrial application at Degussa-Hüls AG in a leucine dehydrogenase-catalyzed reductive amination of 2-keto acids yielding various amino acids (e.g. *tert*-leucine) [14–16]. Native FDH is very selective for NAD^+. Recently a new FDH was developed by site-directed mutagenesis that shows all advantages of the NAD^+-dependent enzymes and additionally accepts $NADP^+$ as substrate [17]. The activity of the mutant with $NADP^+$ is about 60% of the wild-type FDH with NAD^+ [18].

16.6.4
Oxidations with Intact Microbial Cells

16.6.4.1
Production of Benzaldehyde from Benzoyl Formate or Mandelic Acid

Benzaldehyde can be produced from benzoyl formate with whole cells of *Pseudomonas putida* ATCC 12633 as biocatalyst [19, 20] (Fig. 16.6-5). Alternatively, but less effectively, mandelic acid can be used as starting material. A pH of 5.4 was found to be optimal for benzaldehyde accumulation. At this proton concentration, partial inactivation of the benzaldehyde dehydrogenase isoenzymes and activation of the benzoyl formate decarboxylase are reported. Fed-batch cultivation prevented substrate inhibition. *In situ* product removal is necessary to prevent product inhibition.

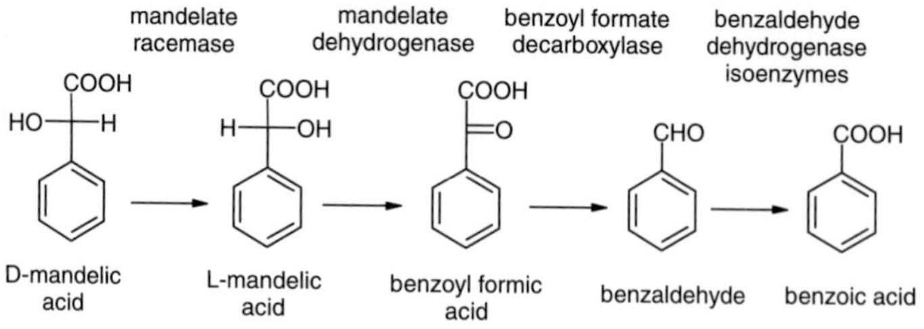

Figure 16.5-5. Degradation of tridecan-2-one with a crude cell-free preparation from a *Pseudomonas aeruginosa* strain.

Activated charcoal served as a solid-phase adsorption device[20]. Thus, benzaldehyde and thiophene-2-carboxaldehyde were obtained from benzoyl formic acid and thiophene-2-glyoxylic acid respectively, in final concentrations of up to 4.8 g L^{-1} and molar yields exceeding 85%.

16.6.4.2
Microbial Production of *cis,cis*-Muconic Acid from Benzoic Acid

Significant effort was put into the oxidation of benzoic acid to *cis,cis*-muconic acid via a multi-step reaction catalyzed by whole microbial cells[21–24]. *Cis,cis*-muconic acid is used as raw material for the synthesis of resins and polymers (precursor of adipic acid). Furthermore, it is widely used as building block in the synthesis of pharmaceuticals and agrochemicals.

As biocatalyst, growing cells of a mutant *Arthrobacter* strain (lacking *cis,cis*-muconate derivatization activity) was used. The reaction cascade (Fig. 16.6-6) is initiated by a dioxygenation of the benzylic ring followed by decarboxylation yielding catechol, which is transformed to the product via dioxygenase-catalyzed ring cleavage.

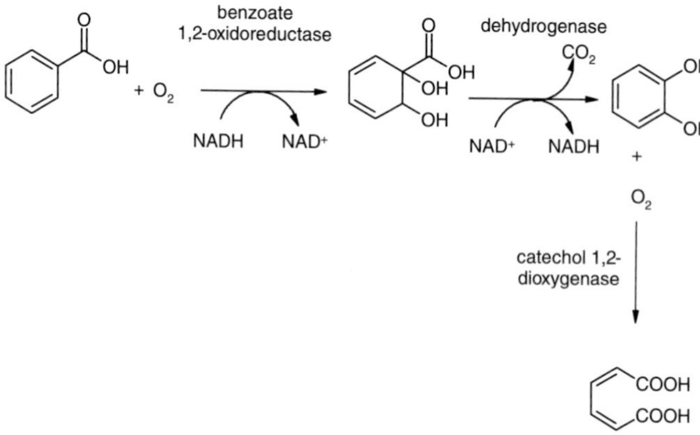

Figure 16.6-6. Sequential oxidation of benzoate to (*cis,cis*)-muconic acid catalyzed by *Arthrobacter* sp.

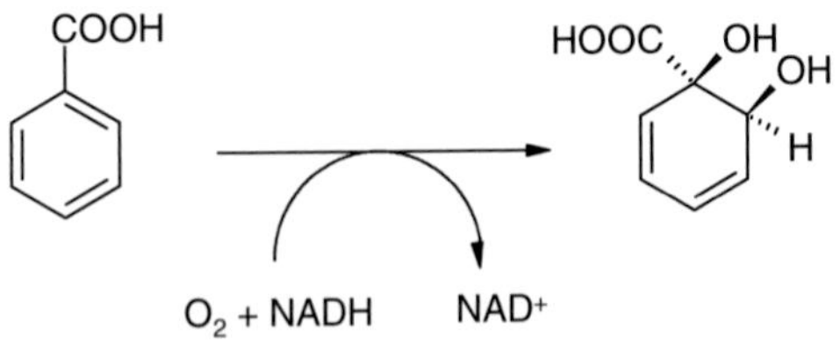

Figure 16.6-7. Dioxygenation of benzoate to corresponding *cis*-1,2-diols.

Benzoic acid was fed continuously to the fermentation medium. The space-time yield of the process including downstream processing amounts to 70 g L^{-1} d^{-1}.

16.6.4.3
Biotransformation of Substituted Benzoates to the Corresponding *cis*-Diols

Enantiopure 1,2-*cis*-dihydroxycyclohexa-3,5-diene carboxylic acids have considerable synthetic potential as building blocks in chiral synthesis. Such *cis*-diols can be produced from benzoic acid derivatives by the action of toluate-1,2-dioxygenase of *Pseudomonas putida* mt-2[25] or homologous enzymes of a different origin (Fig. 16.6-7).

Growing cells or recombinant *Pseudomonas oleovorans* GPo12 containing toluate-1,2-dioxygenase efficiently transform a whole range of *meta*- and *para*-substituted benzoates to the corresponding *cis*-diols, which are not further degraded by the *Pseudomonas* host. In the *ortho* position only hydrogen and fluorine were accepted as substituents. Toluate-1,2-dioxygenase activity is induced by *ortho*-toluate or the substrates themselves.

Similar reactions were reported for the broad-substrate-specific benzoate dioxygenase of *Rhodococcus* sp. strain 19070[26]. Recombinant *E. coli* containing this enzyme transform benzoate and anthranilate to catechol and 2-hydro-1,2-dihydroxybenzoate, respectively.

References

1 D. D. Woods, *Biochem. J.* **1936**, *30*, 515.

2 K. Burton, *Ergeb. Physiol.* **1957**, *49*, 275.

3 B. Sedewitz, K. H. Schleifer, F. Götz, *J. Bacteriol.* **1984**, *160*, 273–278.

4 B. Sedewitz, K. H. Schleifer, F. Götz, *J. Bacteriol.* **1984**, *160*, 462–465.

5 J. Carlsson, U. Kujala, *FEMS Microbiol. Lett.* **1985**, *25*, 53–56.

6 H. Vigenschow, H.-M. Schwarm, K. Knobloch, *Biol. Chem.* **1986**, *367*, 951–956.

7 K. Suzuki, H. Nakajima, K. Imahori, *Methods Enzmol.* **1982**, *90*, 179–185.

8 J. S. Nishimura, M. J. Griffith, *Methods Enzmol.* **1981**, *71*, 311–316.

9 D.-M. Kim, J. R. Swartz, *Biotech. Bioeng.* **1999**, *66*, 180–188.

10 E. Steckhan. Kontinuierliche enzymatische Synthesen enantiomerenreiner organischer Zwischenprodukte durch elektrochemische Aktivierung von Redoxenzymen in elektrochemischen Enzymmembranreaktoren – Final report for the period 01. 03. 1998 to 31. 08. 2000 on the Research Project 11556 N/1; AiF: Bonn, 2000.

11 U. M. Billhardt, P. Stein, G. M. Whitesides, *Bioorg. Chem.* **1989**, *17*, 1–12.

12 V. O. Popov, V. S. Lamzin, *Biochem. J.* **1994**, *301*, 625–643.

13 M.-R. Kula, U. Kragl, Dehydrogenases in synthesis of chiral compounds. In *Stereoselective Biocatalysis*. R. N. Patel, (ed) Marcel Dekker, New York, 1999, pp. 839–866.

14 A. Liese, K. Seelbach, C. Wandrey. In *Industrial Biotransformations*, Wiley-VCH, Weinheim, 2000, pp. 125–128.

15 A. S. Bommarius, M. Schwarm, K. Drauz, *J. Mol. Cat. B: Enzymatic* 1998, *5*, 1–11.

16 U. Kragl, D. Vasic-Racki, C. Wandrey, *Bioproc. Eng.* 1996, *14*, 291–297.

17 K. Seelbach, B. Riebel, W. Hummel, M.-R. Kula, V. I. Tishkov, A. M. Egorov, C. Wandrey, U. Kragl, *Tetrahedron Lett.* 1996, *37*, 1377–1380.

18 V. I. Tishkov, A. G. Galkin, G. N. Marchenko, Y. D. Tsyganov, H. M. Egorov, *Biotech. Appl. Biochem.* 1993, *18*, 201–207.

19 J. Simmonds, G. K. Robinson, *Enz. Microb. Tech.* 1997, *21*, 367–374.

20 J. Simmonds, G. K. Robinson, *Appl. Microbiol. Biotech.* 1998, *50*, 353–358.

21 S. Mizuno, N. Yoshikawa, M. Seki, T. Mikawa, Y. Iamada, *Appl. Microbiol. Biotech.* 1988, *28*, 20–25.

22 N. Yoshikawa, S. Mizuno, K. Ohta, M. Suzuki, *J. Biotech.* 1990, *14*, 203–210.

23 N. Yoshikawa, O. Ohta, S. Mizuno, H. Ohkishi, Production of *cis,cis*-muconic acid from benzoic acid. In *Industrial Application of Immobilized Biocatalysts*. A. Tanaka, T. Tosa, T. Kobayashi (eds), Marcel Dekker, New York, 1993, pp. 131–147.

24 A. Liese, K. Seelbach, C. Wandrey, Oxygenase of *Arthrobacter* sp. In *Industrial Biotransformations*. A. Liese, K. Seelbach, C. Wandrey (eds), Wiley-VCH, Weinheim, 2000, pp. 137–138.

25 M. G. Wubbolts, K. N. Timmis, *Appl. Environ. Microbiol.* 1990, *56*, 569–571.

26 S. Haddad, D. M. Eby, E. L. Neidle, *Appl. Environ. Microbiol.* 2001, *67*, 2507–2514.

16.7
Oxidation of C-N Bonds

Andreas Schmid, Frank Hollmann, and Bruno Bühler

16.7.1
Introduction

Enzymatic oxidations of carbon-nitrogen bonds are as diverse as the substances containing this structural element. Mainly amine and amino acid oxidases are reported for the oxidation of C-N bonds. The steroespecificity of amine-oxidizing enzymes can be exploited to perform resolutions and even deracemizations or stereoinversions (Fig. 16.7-1 A). Analogous to the oxidation of alcohols, primary amines are oxidized to the corresponding imines, which can hydrolyze and react with unreacted amines (Fig. 16.7-1 B). In contrast to ethers, internal C-N bonds are readily oxidized, yielding substituted imines. This can be exploited for the production of substituted pyridines (Fig. 16.7-1 C). Furthermore, pyridines can be oxidized not only to N-oxides but also to α-hydroxylated products (Fig. 16.7-1 D).

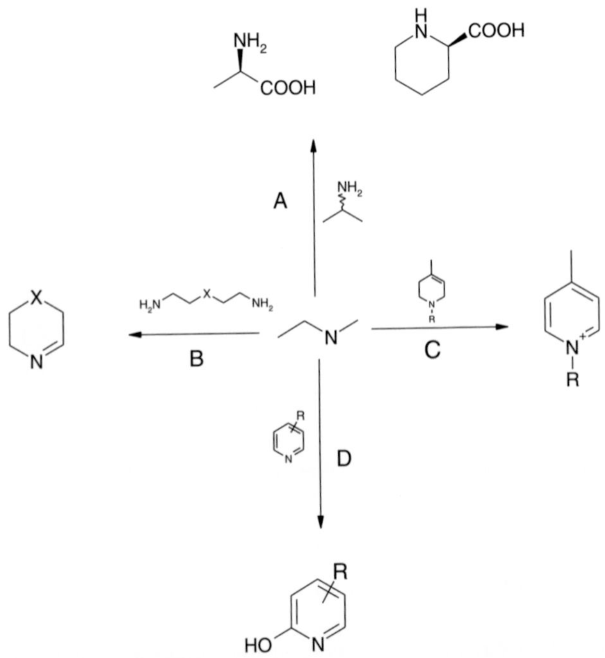

Figure 16.7-1.
Oxidations of C-N bonds with synthetic relevance. A: kinetic resolution, deracemization and stereoinversion (Sects. 16.7.2.1 and 16.7.3.1); B: preparation of aldehydes (and subsequent formation of imines) by oxidation of primary amines (Sect. 16.7.3.2); C: preparation of substituted pyridines (Sect. 16.7.3.2); D: hydroxylation of N-heteroaromatic compounds (Sect. 16.7.2.2).

16.7.2
Oxidations Catalyzed by Dehydrogenases

16.7.2.1
L-Alanine Dehydrogenase (L-Ala-DH, E.C. 1.4.1.1)

L-Alanine dehydrogenase (L-Ala-DH, E.C. 1.4.1.1) catalyzes the specific deaminative oxidation of L-alanine and thus can potentially be exploited for the resolution of racemic alanine (e. g. derived from the Strecker-synthesis). However, the oxidation of secondary alcohols and amines is thermodynamically unfavorable[1], so that the equilibrium of the reversible dehydrogenase reaction is on the substrate side. Therefore, an additional thermodynamic driving force has to be introduced into the system in order to drive the desired reaction towards completion. Moiroux and coworkers recently introduced such a system (Fig. 16.7-2)[2–5].

The general philosophy of their approach is the utilization of electrical power to remove the dehydrogenase products NADH and pyruvate (which is *in situ* transformed into the corresponding imine), thus driving the equilibrium reaction towards completion. The electrochemical oxidation and reduction reactions produce NAD⁺ and racemic alanine, respectively, as substrates for the dehydrogenase reaction. Using this procedure, not only a racemate resolution (with maximum 50% yield) but a deracemization (100% yield) is achieved. The overall rate-limiting step is the slow, non-enzymatic formation of the imine. Consequently, the process is very slow (at best, the complete conversion of a 10 mM solution of L-alanine required 140 h).

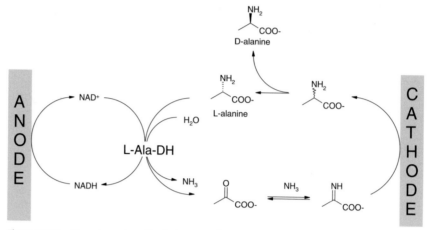

Figure 16.7-2. Stereoinversion of L-alanine to D-alanine catalyzed by L-alanine dehydrogenase (L-Ala-DH) in an electrochemical reactor.

16.7.2.2
Nicotinic Acid Dehydrogenase (Hydroxylase) (E.C. 1.5.1.13)

The membrane-bound molybdoenzyme[6] nicotinic acid dehydrogenase catalyzes the first step in the microbial degradation of nicotinic acid by inserting a hydroxyl function α to the nitrogen atom (Fig. 16.7-3). A possible mechanism for this reaction is given in Fig. 16.7-4[7].

The inserted hydroxyl function originates from water, which was confirmed by $H_2^{18}O$ experiments[6, 8]. While nicotinic acid dehydrogenase does not accept NAD$^+$ as electron acceptor, artificial mediators such as benzyl viologene and 2,3,5-triphenyltetrazolium dyes can replace NADP$^+$[9]. Various bacterial strains have been reported to convert a broad range of nicotinic acid derivatives (Table 16.7-1)[10, 12].

An industrial process (according to the first entry in Table 16.7-1) was set up by

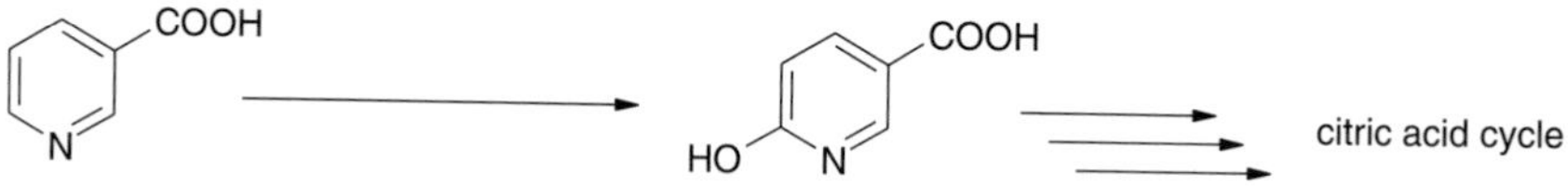

Figure 16.7-3. Microbial mineralization of nicotinic acid.

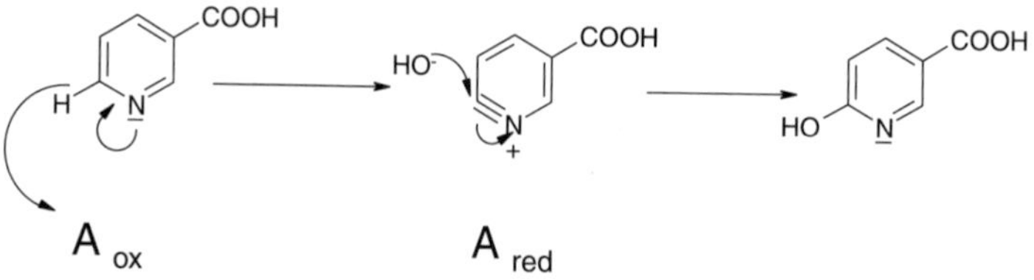

Figure 16.7-4. Proposed mechanism for enzymatic hydroxylation of nicotinic acid (A = acceptor). The reaction scheme is based on the so-called arine mechanism.

Table 16.7-1. Microbial α-hydroxylation of substituted pyridines.

Reactions catalyzed by whole cells	Final product concentration [g L^{-1}]	Enzymes and reference
Achromobacter xylosoxidans	74	Dehydrogenase[1]
Pseudomonas fluorescens	191	Dehydrogenase[2]
Agrobacterium sp.	301	Dehydrogenase[3]
Proteobacteria	6.4	Dehydrogenase[4]
Alcaligenes faecalis DSM6269	98	Dehydrogenase[5]
Alcaligenes sp. UK21	NR[a]	Dehydrogenase and decarboxylase[6]
Rhizobium sp. LA17	NR[a]	Dehydrogenase[7]
Comamonas testosteroni	45	Dehydrogenase[8]
	40	Nitrilase and Dehydrogenase[9]
Alcaligenes faecalis	55	Nitrilase and Dehydrogenase[5]
Agrobacterium sp.	40	Nitrilase and Dehydrogenase[10]
Rhodococcus erythropolis	8	Dehydrogenase[11]

a NR: not reported.

1 H. Kulla, *Chimia* **1991**, *45*, 81–85.
2 T. Nagasawa, B. Hurh, T. Yamane, *Biosci. Biotech. Biochem.* **1994**, *58*, 665–668.
3 B. Hurh, M. Ohshima, T. Yamane, T. Nasagawa, *J. Ferm. Bioeng.* **1994**, *77*, 382–385.
4 M. Ueda, R. Sashida, *J. Mol. Cat. B: Enzymatic* **1998**, *4*, 199–204.
5 A. Kiener, R. Glockler, K. Heinzmann, *J. Chem. Soc. Perkin Trans. I* **1993**, 1201–1202.
6 T. Yoshida, A. Uchida, T. Nagasawa. "Regiospecific ..."; Annu. Meet. Soc. Biosci. Bioeng., 1998, Japan.
7 T. Yoshida, T. Nagasawa, *Biosci. Biotech. Biochem.* **2000**, *89*, 111–118.
8 M. Yasuda, T. Sakamoto, R. Sashida, M. Ueda, Y. Morimoto, *Biosci. Biotech. Biochem.* **1995**, *59*, 572.
9 A. Kiener. USP5266469 (1993).
10 M. Wieser, K. Heinzmann, A. Kiener, *Appl. Microbiol. Biotechnol.* **1997**, *48*, 174.
11 A. Kiener, Y. van Gameren, M. Bokel.; USP 5,284,767, 1994.

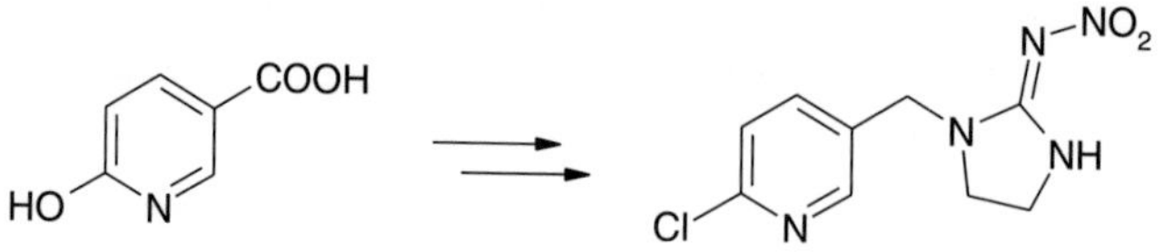

Figure 16.7-5. 6-Hydroxynicotinic acid as synthon for the pesticide Imidachloprid.

Lonza AG, Switzerland. 6-Hydroxynicotinic acid is precipitated from the fermentation broth as magnesium salt in the so-called pseudocrystal process, thus enabling not only easy downstream processing but also continuous fermentation[13]. 6-Hydroxynicotinic acid is the key building block in the synthesis of Imidachloprid (Fig. 16.7-5), an effective pesticide against hemipterans and other sucking insects[10, 11].

16.7.3
Oxidations Catalyzed by Oxidases

16.7.3.1
Amino Acid Oxidases

Among the enzymes catalyzing oxidations of carbon nitrogen bonds, the amino acid oxidases (AAO, E.C. 1.4.3.x) are the most interesting for synthetic applications. Compared to some specific amino acid oxidases such as aspartate oxidase or glutamate oxidase, the two D- and L-amino acid oxidases (E. C. 1.4.3.2 for L-AAO and E.C. 1.4.3.3 for D-AAO) are advantageous on account of their broad substrate

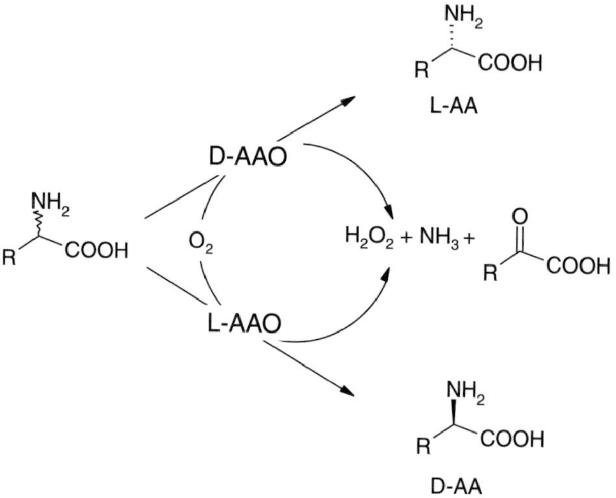

Figure 16.7-6. Resolution of racemic amino acids (AA) catalyzed by (D)- and (L)-specific amino acid oxidases (AAO).

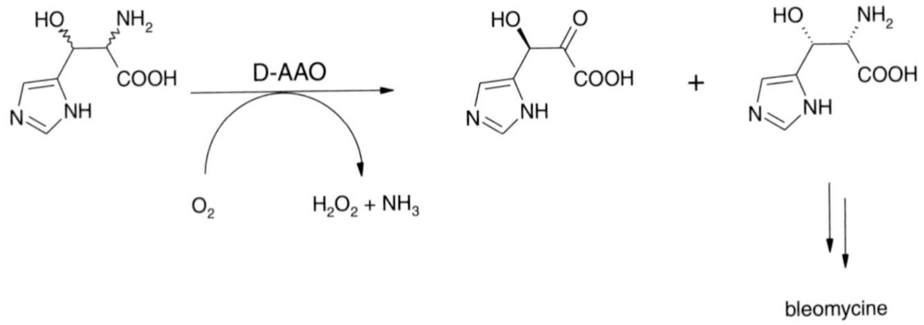

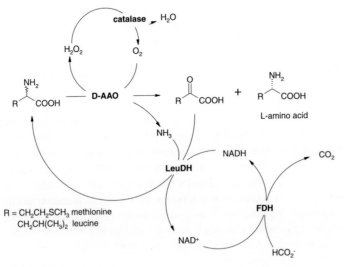

Figure 16.7-7. Resolution of D,L-*erythro*-β-hydroxyhistidine as the enantiospecific step in bleomycine synthesis.

Figure 16.7-8. Enzymatic deracemization of amino acids catalyzed by D-amino acid oxidase (D-AAO). Leucine dehydrogenase (LeuDH) transforms the oxidation product of the undesired amino acid enantiomer *in situ* into the racemic amino acid. Regeneration of NADH is performed by formate dehydrogenase (FHD).

spectrum and their strict stereospecificity[14, 15]. Therefore, AAOs are most commonly used for the resolution or deracemization of racemic amino acid mixtures (Fig. 16.7-6).

The approach outlined in Fig. 16.7-6 was used for example to remove traces of D-methionine from 99% pure L-methionine[16, 17] or to transform racemic phenylalanine quantitatively into D-phenylalanine and phenylpyruvic acid[18]. Coimmobilization with catalase on a solid matrix (Eupergit®) resulted in largely increased D-AAO stability. In an enzyme-membrane-reactor, space-time-yields as high as 90 g L^{-1} d^{-1} were reached. In another example, a racemic mixture of D,L-erythro-β-hydroxyhistidine was converted into the ketoacid and L-erythro-β-hydroxyhistidine[19]. The

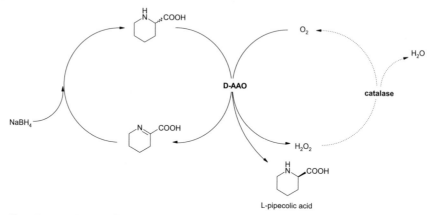

Figure 16.7-9. One-pot chemo-enzymatic deracemisation of D,L-pipecolic acid catalyzed by D-amino acid oxidase (D-AAO). Utilization of catalase was not reported.

latter compound is a key intermediate in the synthesis of the anti-tumor agent bleomycine (Fig. 16.7-7)

Simple racemate resolutions have a maximal yield of 50% for the desired compound. Furthermore, additional (potentially laborious) separation steps are necessary. As a consequence, alternative processes that involve the stereoinversion of the undesired enantiomer are gaining increasing interest [20]. One approach for these so-called deracemization processes is to reconvert the oxidation product either enzymatically (Fig. 16.7-8) or chemically (Fig. 16.7-9) to the racemic substrate.

The enzymatic variant of this concept was reported for the deracemization of D,L-methionine or D,L-leucine (Fig. 16.7-8) [17]. Soda and coworkers developed a chemo-enzymatic racemization procedure utilizing boron hydrides for non-enantioselective reduction of the undesired D-AAO product (Fig. 16.7-9) [21, 22]. Using the same procedure, the authors achieved conversion of D-proline into L-proline [21]. Furthermore, D,L-lactate and 2-hydroxy butyric acid were deracemized by utilizing L-lactate oxidase [23].

The D-AAO catalyzed oxidative deamination of cephalosporin C found industrial application (Hoechst Marion Roussel, Germany) as the first step in the so-called 7-aminocephalosporanic acid (7-ACA) process (Fig. 16.7-10) [24–26].

Using this process, this application of heavy metals and chlorinated hydrocarbons can be avoided, and the volumes of waste-gas as well as of mother liquors are drastically reduced [26].

16.7.3.2
Amine Oxidases

16.7.3.2.1 **Monoamine Oxidase (MAO, E.C. 1.4.3.4)**
The flavoenzymes monoamine oxidase A and B (MAO-A, MAO-B) [27] catalyze the oxidative deamination of various primary and secondary amines and the oxidation of tertiary amines. Their physiological role, as the various synonyms such as epineph-

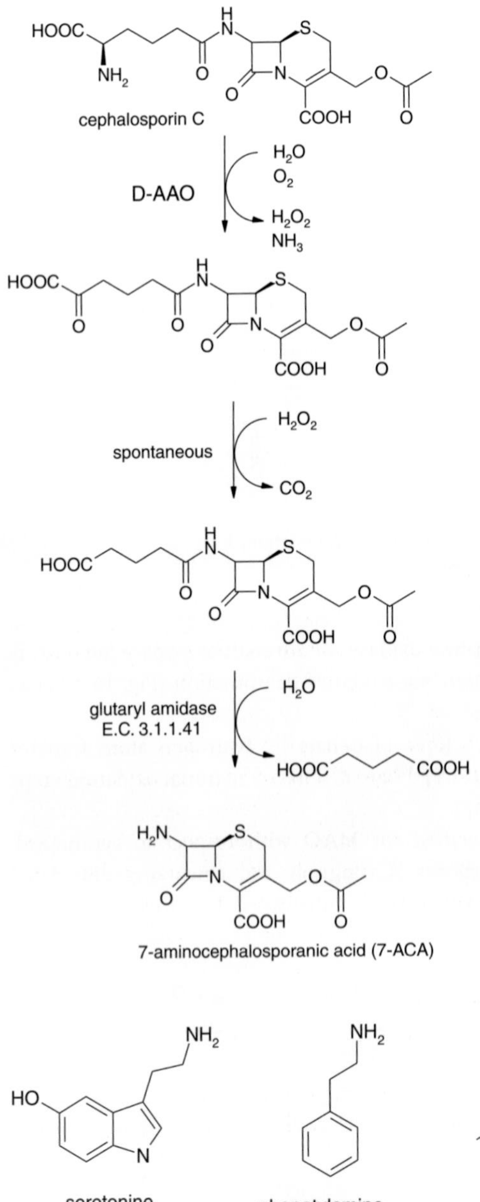

Figure 16.7-10. Enzymatic reaction sequence for the production of 7-ACA from cephalosporin C.

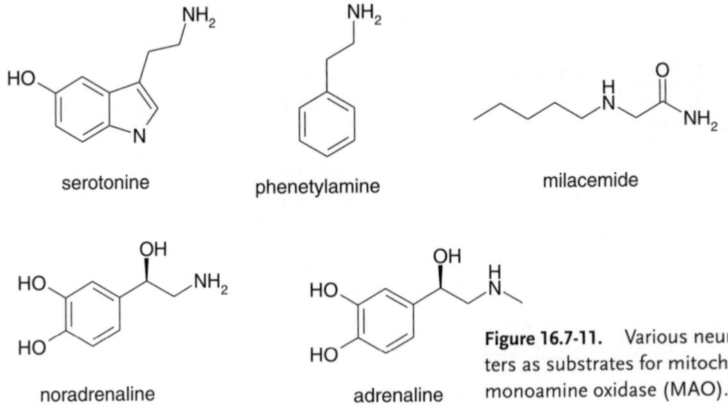

cephalosporin C

D-AAO

H_2O
O_2

H_2O_2
NH_3

spontaneous

H_2O_2

CO_2

glutaryl amidase
E.C. 3.1.1.41

H_2O

7-aminocephalosporanic acid (7-ACA)

serotonine

phenetylamine

milacemide

noradrenaline

adrenaline

Figure 16.7-11. Various neurotransmitters as substrates for mitochondrial monoamine oxidase (MAO).

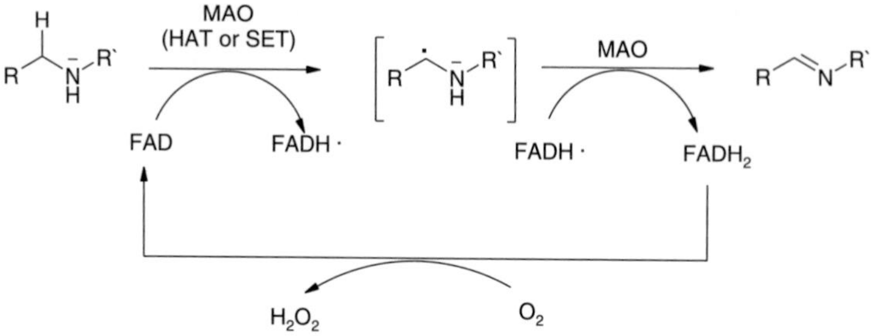

Figure 16.7-12. Proposed mechanisms for the oxidation of primary and secondary amines by monoamine oxidase (MAO).

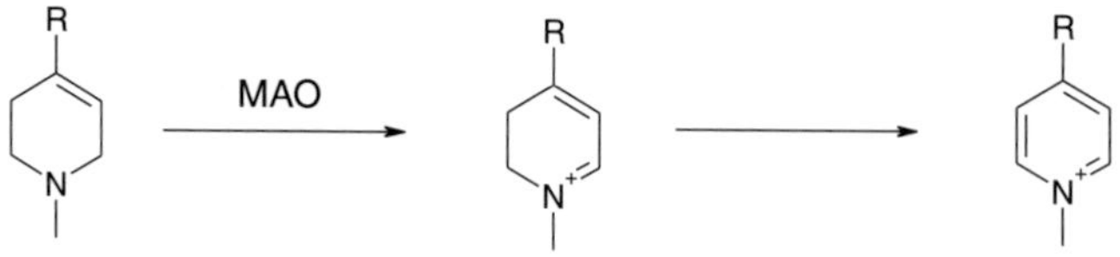

Figure 16.7-13. Oxidation of 1-methyl-4-aryl(heteroaryl)-1,2,3,6-tetrahydro-pyridines catalyzed by monoamine oxidase (MAO).

rine oxidase, serotonin oxidase, tyramine oxidase, or adrenaline oxidase suggest, is the transformation of neurotransmitters via oxidative deamination (Fig. 16.7-11) as well as the detoxification of xenobiotics [28–30].

The mechanism of MAO is still a topic of debate [31]; hydrogen atom transfer (HAT) [32] or single electron transfer (SET) [33] are discussed as initial oxidation steps in the overall mechanism (Fig. 16.7-12).

Various substrates have been specified for MAO with respect to synthetical, mechanistical and biochemical purposes. Castagnoli and coworkers elucidated structural requirements of MAO-B with various substituted 1-methyl-1,2,3,6-tetra-

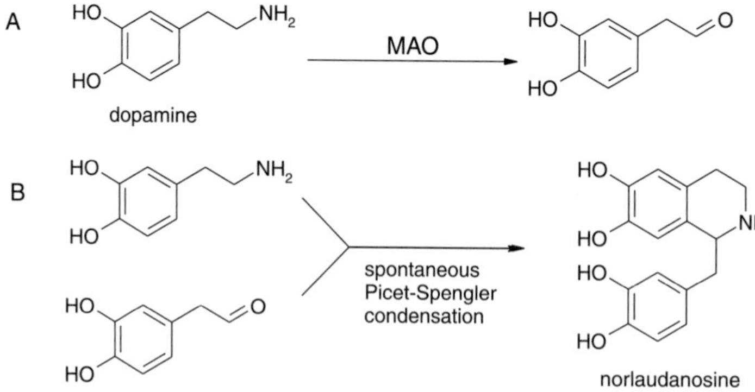

Figure 16.7-14. Preparation of norlaudanosine initiated by the oxidation of dopamine by monoamine oxidase (MAO) (A). The oxidation product reacts spontaneously in a Picet-Spengler condensation with unreacted dopamine (B).

Table 16.7-2. Oxidation of various amines catalyzed by monoamine oxidase in *n*-octane (0.5% v/v water) [12].

Substrate	Product	Yield [%]
		99
		99
		92
		69
		14

12 J. C. G. Woo, X. Wang, R. B. Silverman, *J. Org. Chem.* **1995**, *60*, 6235–6236.

hydropyridines to produce dihydropyridines that are further oxidized to pyridinium structures (Fig. 16.7-13) [31, 34].

MAO was used *in vivo* and *in vitro* as a catalyst for the production of norlaudanosine from dopamine (Fig. 16.7-14) [35]. Norlaudanosine is an important synthon for benzylisoquinoline alkaloids, providing the upper isoquinoline portion of the morphinan skeleton. *In vitro* and *in vivo* yields were in the range of 20%.

MAO-B was also tested in low water content organic media such as ether, tetrachloromethane, octane, benzene and cyclohexane. Under optimized conditions quantitative conversions of various substrates were achieved (Table 16.7-2) [36].

16.7.3.2.2 Diamine Oxidase (E. C. 1.4.3.6)

The copper-containing amine oxidases (copper amine oxidases, diamine oxidases) possess either a topaquinone or a 6-hydroxydopamine cofactor (Fig. 16.7-15), generally integrated in the oxidase primary structure. Tyrosine residues of the enzyme backbone in the active site are discussed as precursors for the prosthetic group [37].

As the name suggests, diamine oxidase catalyzes the oxidative deamination of diamines. Preferably α,ω-diamines such as putrescine (1,4-diaminobutane) or cadaverine (1,5-diaminopentane) (the names already suggest their smell), but also various derivatives are readily converted. Quite often cyclic imines are obtained via internal nucleophilic attack by the unreacted amino function (Fig. 16.7-16) [38–40].

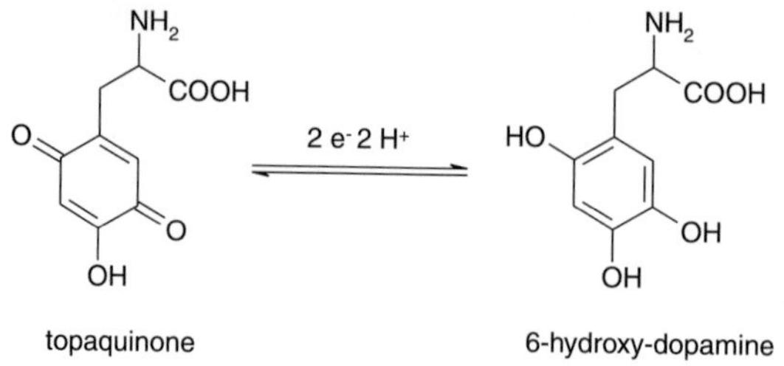

Figure 16.7-15. Topaquinone and 6-hydroxy-dopamine as prosthetic groups of diamine oxidases.

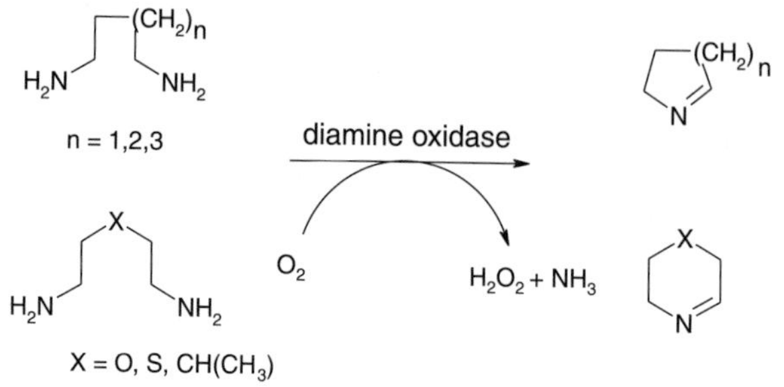

topaquinone

6-hydroxy-dopamine

Figure 16.7-16. Application of diamine oxidase in the synthesis of different azaheterocycles.

$n = 1,2,3$

diamine oxidase

O_2

$H_2O_2 + NH_3$

$X = O, S, CH(CH_3)$

In the presence of suitable nucleophiles (such as benzoyl acetic acid) the primary imines can be spontaneously further modified *in situ*. A convenient approach to obtain phenacyl-derivatives, building blocks in the synthesis of certain alkaloids, was reported [38]. In some cases, diamine oxidases exhibit activities complementary to monoamine oxidases. For example vanillylamine is far more efficiently converted into vanillin by a diamine oxidase from *Aspergillus niger* than by the monoamine oxidase from *E. coli* [11].

Even enantioselective oxidations of some alkyl-, benzyl-, or phenylethyl- (arylethyl-) amines were reported with diamine oxidase from pea settlings [41]. Porcine kidney diamine oxidase was used for the oxidative transformation of *Nitraria* alkaloids such as nazlinin [42].

For the conversion of poorly water-soluble amines (and to avoid product inhibition), diamine oxidase can also be applied in non-aqueous media [43].

16.7.4
Oxidations Catalyzed by Transaminases

Transaminases are generally not considered to be enzymes catalyzing redox reactions, which is obvious considering the meaning of the E. C. code for transferases (E. C. 2.6.1.x = transferring amino groups). Nevertheless, the exchange of an amino functionality between an amino acid and an α-keto acid implies the oxidation of the amino acid. Transaminases are described elsewhere in this book (Chapter 12).

References

1 L. G. Lee, G. M. Whitesides, *J. Am. Chem. Soc.* **1985**, *107*, 6999–7008.

2 J. M. Laval, J. Moiroux, C. Bourdillon, *Biotech. Bioeng.* **1991**, *38*, 788–796.

3 A. Anne, C. Bourdillon, S. Daninos, J. Moiroux, *Biotech. Bioeng.* **1999**, *64*, 101–107.

4 A.-E. Biade, C. Bourdillon, J.-M. Laval, G. Mairesse, J. Moiroux, *J. Am. Chem. Soc.* **1992**, *114*, 893–897.

5 A. Fassouane, J.-M. Laval, J. Moiroux, C. Bourdillon, *Biotech. Bioeng.* **1990**, *35*, 935–939.

6 M. Nagel, J. R. Andreesen, *Arch. Microbiol.* **1990**, *154*, 605–613.

7 G. Wittig, *Angew. Chem.* **1957**, *69*, 245–251.

8 D. E. Hughes, *Biochem. J.* **1955**, *60*, 303–310.

9 J. S. Holcenberg, E. R. Stadtman, *J. Biol. Chem.* **1969**, *244*, 1194–1203.

10 M. Petersen, A. Kiener, *Green Chem.* **1999**, 99–106.

11 T. Yoshida, T. Nagasawa, *Biosci. Biotech. Biochem.* **2000**, *89*, 111–118.

12 A. Tinschert, A. Tschech, K. Heinzmann, A. Kiener, *Appl. Microbiol. Biotech.* **2000**, *53*, 185–195.

13 P. Lehky, H. Kulla, S. Mischler. Verfahren zur Herstellung von 6-Hydroxynikotinsäure; EP 015 2948 A2: Lonza, Switzerland, 1995.

14 V. W. Rodwell, *Methods Enzmol.* **1971**, *17B*, 174–188.

15 P. Wikström, E. Szwajcer, P. Brodelius, K. Nilsson, K. Mosbach, *Biotechnol. Lett.* **1982**, *4*, 153–158.

16 K. Parkin, H. O. Hultin, *Biotech. Bioeng.* **1979**, 939–953.

17 N. Nakajima, D. Conrad, H. Sumi, N. Esaki, C. Wandrey, K. Soda, *Ferment. Bioeng.* **1990**, *70*, 322–325.

18 R. Fernandez-Lafuente, V. Rodriguez, J. Guisan, *Enz. Microb. Tech.* **1998**, *23*.

19 S. M. Hecht, K. M. Rupprecht, P. M. Jacobs, *J. Am. Chem. Soc.* **1979**, *101*, 3982–3983.

20 W. Kroutil, K. Faber, *Tetrahedron: Asym.* **1998**, *9*, 2901–2913.

21 J. W. Huh, K. Yokoigawa, N. Esaki, K. Soda, *J. Ferment. Bioeng.* **1992**, *74*, 189–190.

22 J. W. Huh, K. Yokoigawa, N. Esaki, K. Soda, *Biosci. Biotech. Biochem.* **1992**, *56*, 2081–2082.

23 K. Soda, T. Oikawa, K. Yokoigawa, *J. Mol. Cat. B: Enzymatic* **2001**, *11*, 149–153.

24 J. Verweij, E. D. Vroom, *Rec. Trav. Chim. Pays-Bas* **1993**, *112*, 66–81.

25 F. Alfani, M. Cantarella, A. Gallifuoco, *Biocat. Biotransf.* **1998**, *16*, 395–409.

26 A. Liese, K. Seelbach, C. Wandrey. In *Industrial Biotransformations*, Wiley-VCH, Weinheim, 2000, pp. 129–130 and 225–230.

27 R. B. Silverman, *Biochem. Soc. Trans.* **1991**, *19*, 201–206.

28 W. Weyler, Y.-P. P. Hsu, X. O. Breakefield, *Pharmacol. Ther.* **1990**, *47*, 391–417.

29 J. F. Powell, *Biochem. Soc. Trans.* **1991**, *19*, 199–214.

30 P. Dostert, M. Strolin Benedetti, K. F. Tipton, *Med. Res. Rev.* **1989**, *9*, 45–89.

31 S. K. Nimkar, S. Mabic, A. H. Anderson, S. L. Palmer, T. H. Graham, M. de Jonge, L. Hazelwood, S. J. Hislop, N. Castagnoli, *J. Med. Chem.* **1999**, *42*, 1828–1835.

32 A. Anderson, S. Kuttab, N. J. Castagnoli, *Biochem.* **1996**, *35*, 3335–3340.

33 B. Y. Zhong, R. B. Silverman, *J. Am. Chem. Soc.* **1997**, *119*, 6690–6691.

34 J. Yu, N. Castagnoli, *Bioinorg. Med. Chem.* **1999**, *7*.

35 L. K. Hoover, M. Moo-Young, R. L. Legge, *Biotech. Bioeng.* **1991**, *38*, 1029–1033.

36 J. C. G. Woo, X. Wang, R. B. Silverman, *J. Org. Chem.* **1995**, *60*, 6235–6236.

37 N. K. Williams, J. P. Klinman, *J. Mol. Cat. B: Enzymatic* **2000**, *8*, 95–101.

38 J. E. Cragg, R. B. Herbert, M. M. Kgaphola, *Tetrahedron Lett.* **1990**, *31*, 6907–6910.

39 A. M. Equi, A. M. Brown, A. Copper, S. K. Ner, A. B. Watson, D. J. Robins, *Tetrahedron* **1991**, *47*.

40 E. Santaniello, A. Manzocchi, P. A. Biondi, C. Secchi, T. Simonic, *J. Chem. Soc., Chem. Commun.* **1984**, 803–804.

41 A. R. Battersby, J. Staunton, M. C. Summers, *J. Chem. Soc., Chem. Commun.* **1974**, *465*, 548–549.

42 E. Cheng, J. Botzem, M. J. Wanner, B. E. Burm, G.-J. Koomen, *Tetrahedron* **1996**, *52*, 5725–6732.

43 J. A. Chaplin, C. L. Budde, Y. L. Khmelnitsky, *J. Mol. Cat. B: Enzymatic* **2001**, *13*, 69–75.

16.8
Oxidation at Sulfur

Karl-Heinz van Pee

16.8.1
Enzymes Oxidizing at Sulfur and their Sources

The oxidation at sulfur is catalyzed by a number of different enzymes produced by a variety of organisms. They have been isolated from a fungus[1], soybean[2], rat, pig and rabbit liver[3–5], horseradish[6], bacteria[7–9], milk[10], and human white blood cells[11].

The enzymes catalyzing oxidation reactions at sulfur belong to two different classes of enzymes: monooxygenases, including cytochrome P-450 monooxygenases and FAD-containing monooxygenases, and heme-containing peroxidases (Figs. 16.8-1 and 16.8-2, Table 16.8-1).

Some of these enzymes such as chloroperoxidase from *Caldariomyces fumago*, horseradish peroxidase, lactoperoxidase from bovine milk, and myeloperoxidase from human white blood cells are commercially available.

Others such as pig liver microsomal FAD-containing monooxygenase have to be isolated from tissue with very low yields[4] or like hydrocarbon monooxygenase from *Pseudomonas oleovorans*[12–13] require several protein components and cofactors, substantially limiting the use of these enzymes for the production of oxidized sulfur compounds.

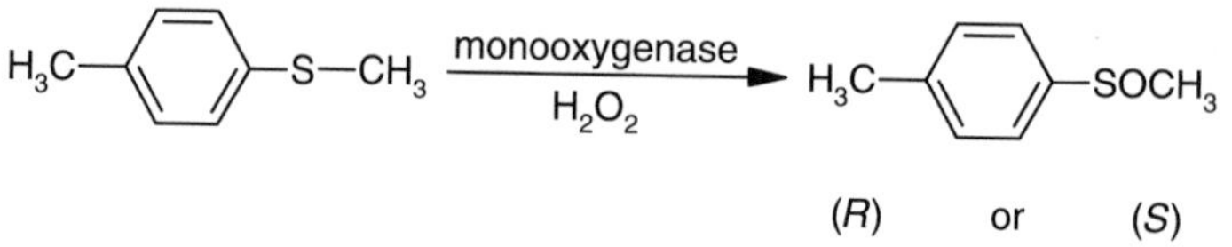

Figure 16.8-1. Oxidation of methyl *p*-tolyl sulfide to methyl *p*-tolyl sulfoxide by a monooxygenase. The product can either be of the *R*- or *S*-configuration depending on the monooxygenase used.

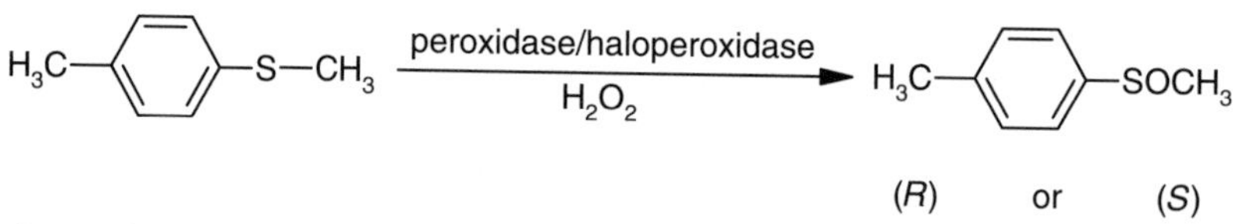

Figure 16.8-2. Oxidation of methyl *p*-tolyl sulfide to methyl *p*-tolyl sulfoxide by a peroxidase or haloperoxidase in the presence of hydrogen peroxide. The product can either be predominantly of the *R*- or *S*-configuration depending on the peroxidase or haloperoxidase used.

Table 16.8-1. Classification of enzymes oxidizing at sulfur and their sources.

Enzyme class	Source	Reference
Monooxygenases	pig liver microsomes	14
	rat liver microsomes	3
	rabbit liver microsomes	16
	bovine adrenals	17
	Pseudomonas oleovorans	12, 13
	Acinetobacter sp.	15
Peroxidases	soybean	2
	horseradish	23
Haloperoxidases	*Caldariomyces fumago*	1
	Ascophyllum nodosum	27
	Corallina officinalis	28
	human white blood cells	30
	bovine milk	25

16.8.2
Oxidation of Sulfides

16.8.2.1
Oxidation of Sulfides by Monooxygenases and by Whole Organsims

Fujimori et al.[14] used pig liver microsomal FAD-containing monooxygenase and phenobarbital-induced rabbit liver microsomal cytochrome P-450 to catalyze the oxidation of unsymmetrical sulfides to the corresponding optically active sulfoxides with varying degrees of enantiomeric excess (12–96%). Comparison of the oxygenation of racemic 2-methyl-2,3-dihydrobenzo[b] thiophene showed that the enantiotopic, diastereotopic, and enantiomeric differentiating abilities of the FAD-containing monooxygenase are higher than those of the cytochrome P-450 monooxygenase. They found that the oxygenation with the FAD-containing monooxygenase is sterically much more highly controlled than that with cytochrome P-450. Whereas higher ee-values are observed in the oxygenation of smaller sulfides with the FAD-containing monooxygenase, the oxygenation of large sulfides by the cytochrome P-450 monooxygenase results in higher ee values than those of sulfides bearing small substituents.

Hydrocarbon monooxygenase from *Pseudomonas oleovorans*[7, 8, 12] also catalyzes the stereoselective sulfoxidation of methyl thioether substrates[13] with up to 80% ee. The products obtained with this enzyme are probably of the *R*-configiration.

The (*S*)-(–)-sulfoxide is predominantly produced (82% *S*, 18% *R*) from *p*-tolyl ethyl sulfide when cyclohexanone monooxygenase from *Acinetobacter* sp. NCIB 9871[9] was used, whereas the the FAD-containing monooxygenase from hog liver microsomes oxidizes *p*-tolyl ethyl sulfide to yield the (*R*)-(+)-sulfoxide enantiomer as the major product (95% *R*, 5% *S*)[15].

The enzymatic oxidation of various diaryl, dialkyl, and aryl alkyl sulfides by cytochrome P-450 from rabbit liver resulted predominantly in the formation of the sulfoxides with the *R*-configuration[16].

The *S*-(-) configuration was predominantly obtained when two cytochrome P-450 isoenzymes from rat liver were used for the oxidation of *p*-tolyl ethyl sulfides[3].

Oxidation of phenyl 2-aminoethyl sulfide by dopamine β-hydroxylase from bovine adrenals in the presence of ascorbate as the electron donor resulted in the formation of phenyl 2-aminoethyl sulfoxide. The product was probably of the *S*-configuration[17].

Holland et al.[18] obtained the (*R*)-sulfoxides from various *para*-substituted phenyl 3-chloropropyl and phenyl 3-hydroxypropyl sulfides by biotransformation with the fungus *Mortierella isabellina* with an enantiomeric excess of 82–88%. The (*S*)-sulfoxides were produced using the fungus *Helminthosporium* sp. and the bacterium *Acinetobacter calcoaceticus* with ee values of > 95% and 94%, respectively.

16.8.2.2
Oxidation of Sulfides by Peroxidases and Haloperoxidases

A number of peroxidases were investigated for their use in oxidizing organic sulfides. *p*-Substituted thioanisols were oxidized by partially purified soybean sulfoxidase using 13(*S*)-hydroperoxylinoleic acid as the peroxide. Methyl *p*-tolyl sulfide gave the (*S*)-sulfoxide with about 90% ee[2].

The sulfoxidation of organic sulfides by chloroperoxidase from *Caldariomyces fumago* was investigated by different groups[19–26]. Colonna et al.[20] compared the oxidation of sulfides by this enzymes with that catalyzed by horseradish peroxidase. Chloroperoxidase catalyzed the formation of sulfoxides with *tert*-butyl and other peroxides with an *R* absolute configuration in up to 92% ee, whereas horseradish peroxidase gave racemic products. When sterically hindered oxidants such as cumyl hydroperoxides and chloroperoxidase were used, racemic or almost racemic products were obtained. *tert*-Butyl hydroperoxide also had the advantage of giving higher yields and higher ee.

Using vanadium bromoperoxidases from marine algae the (*S*)- or (*R*)-sulfoxides can be obtained from methyl phenyl sulfide derivatives, respectively, depending on the source of the enzyme. While bromoperoxidase from *Ascophyllum nodosum* produces the (*R*)-sulfoxide with 91% ee[27], the (*S*)-enantiomer is obtained with bromoperoxidases from *Corallina officinalis* and *C. pilulifera*[28].

When investigating the substrate selectivity using a series of aryl, alkyl, dialkyl, and heterocyclic sulfides, it was found that *p*-substitution led to higher enatioselectivity and higher chemical yields with respect to *o*-substitution[20]. A similar influence of the *p*-substitution was found for sulfoxidation catalyzed by bromoperoxidase from the marine alga *Ascophyllum nodosum*[27].

Benzyl methylsulfide, thioanisol, and thiobenzamide were oxidized by chloroperoxidase, lactoperoxidase, and horseradish peroxidase to the respective sulfoxides. Whereas lactoperoxidase and horseradish peroxidase had low activities towards benzyl methylsulfide, thiobenzamide was efficiently oxidized by lactoperoxidase. Chloroperoxidase had high activity in halide-independent reactions towards all three substrates[25]. This enzyme was also used for the asymmetric sulfoxidation of a series of cyclic sulfides. In all cases the (*R*)-sulfoxides were obtained. In the case of

Table 16.8-2. Products and absolute configuration obtained in the oxidation of various sulfides by different enzymes.

Sulfide	Predominant configuration of sulfoxide obtained	Enzyme or organism	Reference
Methyl phenyl	R	chloroperoxidase	19–21
	R	vanadium bromoperoxidase	27
	S	vanadium bromoperoxidase	28
Methyl p-tolyl	S	soybean hydroperoxide-dependent oxygenase	2
Ethyl p-tolyl	S	rat liver cytochrome P-450	3
		cyclohexanone monooxygenase	15
	R	FAD-containing monooxygenase	15
Methyl alkyl	R	alkane monooxygenase	13
Diaryl, dialkyl, aryl alkyl	R	rabbit liver cytochrome P-450	16
Phenyl 2-aminoethyl	S	dopamine β-hydroxylase	17
Phenyl 3-chloropropyl	R	Mortierella isabellina	18
Phenyl 3-chloropropyl	S	Helminthosporium sp.	18
2,3-Dihydrobenzo[b]thiophene	R	chloroperoxidase	29

2,3-dihydrobenzo[*b*]thiophene the yield was 99.5% with an ee of 99%[29]. Table 16.8-2 shows some examples of sulfides oxidized to sulfoxides by different enzymes and the absolute configuration of the products. When using peroxidases, care has to be taken, as the peroxidase-catalyzed oxidation is in competition with the spontaneous oxidation of the sulfides by the oxidant.

Depending on the enzyme used for oxidation of organic sulfides, sulfoxides with *S*- or *R*-configuration can be obtained with high ee, whereas at present there is only one chemical oxidation method which leads to high ee in alkyl aryl sulfoxides. This method uses chiral titanium complexes and cumene hydroperoxide for the oxidation of organic sulfides[26].

References

1 D. R. Morris, L. P. Hager, *J. Biol. Chem.* **1966**, *241*, 1763–1768.

2 E. Blee, F. Schuber, *Biochemistry* **1989**, *28*, 4962–4967.

3 D. J. Waxman, D. R. Light, C. Walsh, *Biochemistry* **1982**, *21*, 2499–2507.

4 D. M. Ziegler, L. L. Poulsen, *Methods Enzymol.* **1978**, *Vol. 52*, 142–151.

5 Y. Imai, R. Sato, *Biochem. Biophys. Res. Commun.* **1973**, *60*, 8–14.

6 L. M. Shannon, E. Kay, J. Y. Lew, *J. Biol. Chem.* **1966**, *241*, 2166–2171.

7 A. G. Katopodis, K. Wimalasena, J. Lee, S. W. May, *J. Am. Chem. Soc.* **1984**, *106*, 7928–7935.

8 S. W. May, L. G. Lee, A. G. Katopodis, J. Y. Kuo, K. Wimalasena, J. R. Thowsen, *Biochemistry* **1984**, *23*, 2187–2192.

9 N. A. Donoghue, D. B. Norris, P. W. Trudgill, *Eur. J. Biochem.* **1976**, *63*, 175–192.

10 C. Dumontet, B. Rousset, *J. Biol. Chem.* **1983**, *258*, 14166–14172.

11 J. Schultz, *Reticuloendothel. Syst.* **1980**, *2*, 231–254.

12 S. W. May, A. G. Katopodis, *Enzyme Microb. Technol.* **1986**, *8*, 17–21.

13 A. G. Katopodis, H. A. Smith, Jr., S. W. May, *J. Am. Chem. Soc.* **1988**, *110*, 897–899.

14 K. Fujimori, T. Matsuura, A. Mikami, Y. Watanabe, S. Oae, T. Iyanagi, *J. Chem. Soc. Perkin Trans. 1* **1990**, 1435–1440.

15 D. R. Light, D. J. Waxman, C. Walsh, *Biochemistry* **1982**, *21*, 2490–2498.

16 T. Takata, M. Yamazaki, K. Fujimori, Y. H. Kim, T. Iyanagi, S. Oae, *Bull. Chem. Soc. Jpn.* **1983**, *56*, 2300–2310.

17 S. W. May, R. S. Phillips, *J. Am. Chem. Soc.* **1980**, *102*, 5983–5984.

18 H. L. Holland, J.-X. Gu, A. Kerridge, A. Willetts, *Biocat. Biotrans.* **1999**, *17*, 305–317.

19 S. Colonna, N. Gaggero, A. Manfredi, L. Casella, M. Gullotti, *J. Chem. Soc., Chem. Commun.* **1988**, 1451–1452.

20 S. Colonna, N. Gaggero, A. Manfredi, L. Casella, M. Gullotti, G. Carrea, P. Pasta, *Biochemistry* **1990**, *29*, 10465–1048.

21 S. Colonna, N. Gaggero, L. Casella, G. Carrea, P. Pasta, *Tetrahedron: Asymmetry* **1992**, *3*, 95–106.

22 L. Casella, S. Colonna, G. Carrea, *Biochemistry* **1992**, *31*, 9451–9459.

23 S. Kobayashi, M. Nakano, T. Goto, T. Kimura, A. P. Schaap, *Biochem. Biophys. Res. Commun.* **1986**, *135*, 166–171.

24 S. Kobayashi, M. Nakano, T. Kimura, A. P. Schaap, *Biochemistry* **1987**, *26*, 5019–5022.

25 D. R. Doerge, *Arch. Biochem. Biophys.* **1986**, *244*, 678–685.

26 D. R. Doerge, N. M. Cooray, M. E. Brewster, *Biochemistry* **1991**, *30*, 8960–8964.

27 H. B. ten Brink, H. L. Holland, H. E. Shoemaker, H. van Lingen, R. Wever, *Tetrahedron: Asymmetry* **1999**, *10*, 4563–4572.

28 M. A. Andersson, S. G. Allenmark, *Tetrahedron* **1998**, *54*, 15293–15304.

29 S. G. Allenmark, M. A. Andersson, *Chirality* **1998**, *10*, 246–252.

30 M.-F. Tsan, *J. Cell.Physiol.* **1982**, *III*, 49–54.

16.9
Halogenation

Karl-Heinz van Pee

16.9.1
Classification of Halogenating Enzymes and their Reaction Mechanisms

16.9.1.1
Haloperoxidases and Perhydrolases

The only type of halogenating enzymes known until 1997 were peroxidases and perhydrolases which catalyze the formation of carbon halogen bonds using halide ions, hydrogen peroxide and an organic substrate activated for electrophilic attack.

According to the halide ions they can utilize they are arranged into three groups: iodoperoxidases, bromoperoxidases and chloroperoxidases. Iodoperoxidases catalyze the formation of carbon-iodine bonds, whereas bromoperoxidases catalyze iodination and bromination reactions, and chloroperoxidases catalyze the iodination, bromination, and chlorination of organic substrates. As haloperoxidases are oxidoreductases using hydrogen peroxidase as the oxidant for the oxidation of halide ions producing hypohalogenic acids, the existence of fluoroperoxidases can be ruled out. The overall reactions catalyzed by haloperoxidases and perhydrolases are shown in Fig. 16.9-1. All haloperoxidases isolated until 1984 were heme-containing enzymes[1]. The first non-heme haloperoxidase was isolated by Vilter[2]. Instead of heme, vanadium is responsible for the halogenating activity of this algal enzyme[3, 4]. Non-heme and non-metal "haloperoxidases" were isolated from bacteria[5–9], however, elucidation of the three-dimensional structure and the reaction

1) $X^- + H^+$ $\xrightarrow[\text{b) vanadium haloperoxidase}]{\text{a) heme-haloperoxidase}}$ HOX

2) R-H + HOX $\longrightarrow$ R-X + H_2O

1) $CH_3COOH + H_2O_2$ $\xrightarrow{\text{c) perhydrolase}}$ CH_3COOOH

2) $CH_3COOOH + X^-$ $\longrightarrow$ $CH_3COOH + HOX$

3) HOX + R-H $\longrightarrow$ R-X + H_2O

$X^- = Cl^-, Br^-, I^-$

Figure 16.9-1. Overall reaction catalyzed by (a) heme-[51] and (b) vanadium-containing haloperoxidases[51] and (c) perhydrolases[20].

mechansim of this type of halogenase showed that they are not real haloperoxidases. They are actually perhydrolases which produce hypohalogenic acids via the oxidation of halide ions by enzymatically formed peracetic acid[10–12]. Thus, in addition to grouping the haloperoxidases according to the range of halide ions oxidized, they can be classified according to their prosthetic group into heme type and non-heme type haloperoxidases[13].

The heme type haloperoxidases are inactivated during the halogenation reaction, because the heme group of these enzymes is attacked by the hypohalous acids produced by the enzymes[1]. Thus, heme type haloperoxidases have the disadvantage that the reaction velocity slows down considerably during the course of the reaction[14]. With non-heme type haloperoxidases this does not seem to be the case. They are not inactivated during the halogenation reaction and are very stable under reaction conditions[14]. However, the disadvantage of inactivation is partly compensated for by the fact that some of the heme type haloperoxidases have much higher specific activities than non-heme type haloperoxidases.

Some of the non-heme haloperoxidases are very stable with respect to organic solvents[15] which is of great importance when the substrates that are to be halogenated are not very soluble in water. In these cases water missible organic solvents can be added to the reaction mixture or a two phase-system can be used.

16.9.1.2
FADH$_2$-dependent Halogenases

In 1997 the existence of a novel class of halogenating enzymes was reported[16]. These halogenases showed no relationship to any of the known haloperoxidases[17, 18] and did not require hydrogen peroxide for halogenating activity. Initially these new halogenases were thought to require NADH[16], but more detailed studies showed that they actually require FADH$_2$[19, 20] which is produced by NADH-dependent flavin reductases. Figure 16.9-2 shows the hypothetical reaction mechanism of FADH$_2$-dependent halogenases.

16.9.2
Sources and Production of Enzymes

16.9.2.1
FADH$_2$-dependent Halogenases

Although FADH$_2$-dependent halogenases seem to be present in many bacteria producing halometabolites[19, 21, 22], only one example of this new class of halogenases has been isolated to homogeneity until now. This enzyme, tryptophan 7-halogenase, is produced by several *Pseudomonas* strains producing the antibiotic pyrrolnitrin such as *Pseudomonas fluorescens* and *P. aureofaciens* and by *Myxococcus fulvus*[23]. Monodechloroaminopyrrolnitrin-3-halogenase, another FADH$_2$-dependent halogenase from pyrrolnitrin-producing *Pseudomonas* strains has so far only been purified partially[16, 20].

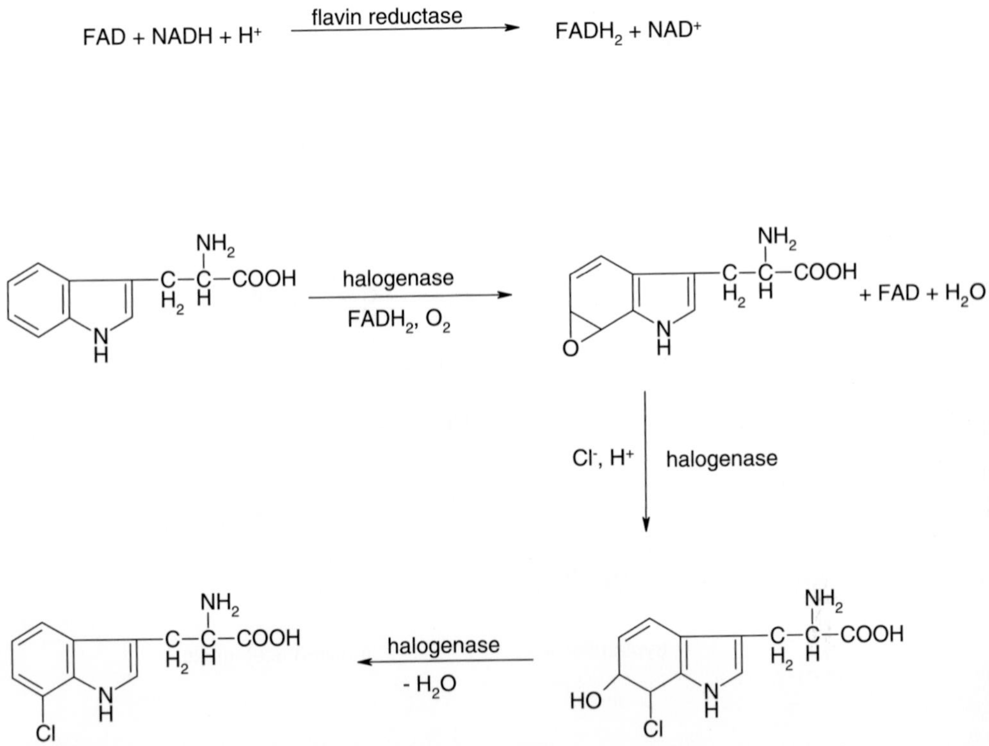

$$FAD + NADH + H^+ \xrightarrow{\text{flavin reductase}} FADH_2 + NAD^+$$

Figure 16.9-2. Hypothetical reaction mechanism of $FADH_2$-dependent tryptophan 7-halogenase as an example of $FADH_2$-dependent halogenases[20].

From biosynthetic and hybridization studies it is known that $FADH_2$-dependent halogenases are involved in the biosynthesis of many halometabolites produced by bacteria[19, 21, 22] and it can be expected that other $FADH_2$-dependent halogenases will be purified and characterized in the near future.

16.9.2.2
Haloperoxidases and Perhydrolases

Iodoperoxidases such as horseradish peroxidase[24] and thyroid peroxidase[25] can be isolated from many different organisms. Bromoperoxidases have been obtained in a pure form from mammals (lactoperoxidase)[26], sea urchin (ovoperoxidase)[27], marine algae[28–30], lichen[31], fungi (lignin peroxidase)[32] and bacteria[33, 34]. Chloroperoxidases have been found in mammals (myeloperoxidase[35] and eosinophil peroxidase[36]), a marine worm[37], and fungi[38–40]. Several perhydrolases have been isolated from bacteria[5–9, 41].

Chloroperoxidase can be produced in batch culture at concentrations of 280 mg L^{-1}[42] and 20 mg of lactoperoxidase can be isolated from 1 L of bovine milk[26]. The sources for these two enzymes, bovine milk and culture broth of *Caldariomyces fumago*, are easily obtained. Chloroperoxidase can also be obtained in larger quantities from the fungus *Curvularia inaequalis*[40]. Thus, a number of different haloperoxidases from various sources are available in quantities necessary for the enzymatic halogenation of organic compounds.

a)

tryptophan

tryptamine   indole-3-acetonitrile

3-methylindole   5-methylindole

b)

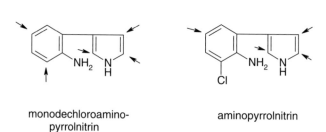

monodechloroamino-   aminopyrrolnitrin
pyrrolnitrin

Figure 16.9-3. Substrates accepted by tryptophan 7-halogenase: (a) indole derivatives, (b) phenylpyrrole derivatives; the positions of chlorination are indicated by arrows [43].

16.9.3
Substrates for Halogenating Enzymes and Reaction Products

16.9.3.1
Halogenation of Aromatic Compounds

The recently detected $FADH_2$-dependent halogenases are substrate specific. Trypto-phan 7-halogenase catalyzes the chlorination and bromination of D- and L-trypto-phan to 7-chloro- or 7-bromotryptophan, respectively[20]. This enzyme also accepts a number of other indole derivatives such as tryptamine, indole-3-acetonitrile, 3-me-thylindole and 5-methylindole as substrates (Fig. 16.9-3a)[43]. In addition to indoles, aminophenylpyrrole derivatives are also chlorinated by tryptophan 7-halogenase (Fig. 16.9-3b)[43].

Monodechloroaminopyrrolnitrin 3-halogenase catalyzes the regioselective chlor-ination of the aminophenylpyrrole derivative monodechloroaminopyrrolnitrin to aminopyrrolnitrin[16], however, nothing is known about the substrate specificity of this enzyme.

In contrast to $FADH_2$-dependent halogenases, haloperoxidases have no substrate specificity. The enzymatic iodination, bromination, and chlorination of a number of different aromatic compounds by haloperoxidases have been reported in the last few years. All aromatic substrates halogenated successfully by haloperoxidases are aromatic compounds activated for electrophilic substitution (Table 16.9-1).

Phenols and phenol ethers are very good substrates for haloperoxidases. The first aromatic substrate to be used in enzymatic iodination was tyrosine. This substrate was iodinated using chloroperoxidase from *Caldariomyces fumago*[44] and thyroid peroxidase[45]. Horseradish peroxidase and lactoperoxidase have been used to lable proteins with radioactive isotopes of iodide[1] and bromoperoxidase from *Penicillus capitatus* has been employed to lable human serum albumin with the radioactive isotope of bromine[46].

Phenolsulfonephthalein (Phenol Red) is brominated to 3,3',5,5'-tetrabromophe-nolsulfonephthalein (Bromophenol Blue) by many haloperoxidases[1, 15]. This reac-tion has been used for the detection of halogenating enzymes by different groups[9, 47].

Corbett et al.[48] obtained 2,6-dibromo-4-chloroaniline or 2,4,6-trichloroaniline when they incubated 4-chloroaniline with chloroperoxidase in the presence of hydrogen peroxide and bromide or chloride, respectively.

Several obscurolides, secondary metabolites produced by *Streptomyces viridochro-mogenes* T7[49], were brominated using perhydrolase from *Streptomyces aureofaciens* Tü24[50]. The obscurolides were monobrominated in the 2-position and dibromi-nated in the 2,4-positions of the aromatic ring system of the obscurolides (Fig. 16.9-4). In the case of dibromination, the hydroxymethyl group was replaced by bromine. No bromination of the olefinic double bond could be detected.

A number of aromatic heterocyclic compounds have been halogenated by different haloperoxidases.

Franssen et al.[51] used chloroperoxidase from *Caldariomyces fumago* to chlorinate

Table 16.9-1. FADH$_2$-dependent halogenases, haloperoxidases and perhydrolases used for biotransformation of aromatic compounds and their sources.

Enzyme (type)	Source	Substrate (halide)	Reference
Tryptophan 7-halogenase (FADH$_2$-dependent)	pyrrolnitrin-producing Pseudomonads	indole derivatives (Cl$^-$, Br$^-$)	20, 43
		phenylpyrroles (Cl$^-$, Br$^-$)	43
Monodechloroaminopyrrolnitrin 3-halogenase (FADH$_2$-dependent)	pyrrolnitrin-producing Pseudomonads	monodechloroamino-pyrrolnitrin (Cl$^-$, Br$^-$)	20
Chloroperoxidase (heme)	Caldariomyces fumago	phenol ether (Cl$^-$, Br$^-$)	67–68
		phenols (Cl$^-$, Br$^-$, I$^-$)	44
		anilines (Cl$^-$, Br$^-$)	48
		pyrazoles, pyridines (Cl$^-$)	51
		nucleic bases (Cl$^-$, Br$^-$, I$^-$)	14
Lactoperoxidase (heme)	bovine milk	phenols (I$^-$)	72
		estrone (I$^-$)	73
Thyroid peroxidase (heme)	thyroid glands	phenols (I$^-$)	45
Chloroperoxidase (heme-flavin)	Notomastus lobatus	phenols (Cl$^-$, Br$^-$)	37
Bromoperoxidase (non-heme)	Asophyllum nodosum	phenols (Br$^-$, I$^-$)	74
		phenol red (Br$^-$)	15
Bromoperoxidase (non-heme)	Corallina pilulifera	nucleic bases (Br$^-$, I$^-$)	14
		phenols (Br$^-$)	75
Perhydrolase (non-heme, non-metal)	Streptomyces aureofaciens	nikkomycin (Br$^-$)	52
		phenylpyrroles (Cl$^-$, Br$^-$)	56
		obscurolide (Br$^-$)	50
		phenol red (Br$^-$)	9
Perhydrolase (non-heme, non-metal)	Pseudomonas pyrrocinia	indole (Cl$^-$, Br$^-$)	29
		phenylpyrroles (Cl$^-$, Br$^-$)	7, 54, 56

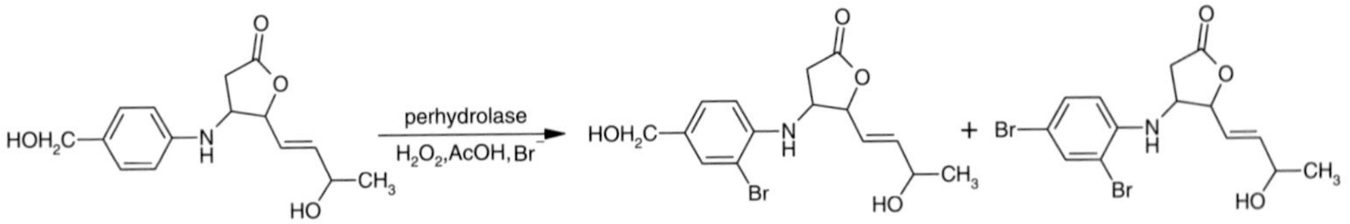

Figure 16.9-4. Bromination of obscurolide A₃ by perhydrolase from *Streptomyces aureofaciens* Tü24 [50].

pyrazole, 1-methylpyrazole and 3-methylpyrazole to their corresponding 4-chloroderivatives. The same enzyme was used to produce 5,7-dibromo-8-hydroxyquinoline from 8-hydroxyquinoline. 2-Aminopyridine was regiospecifically chlorinated to 2-amino-3-chloropyridine.

The chlorination, bromination, and iodination of various nitrogen-containing heterocycles catalyzed by chloroperoxidase from *Caldariomyces fumago* and bromoperoxidase from *Corallina pilulifera* were compared by Itoh et al. [14].

The nucleoside antibiotic nikkomycin Z was brominated using the perhydrolase from *Streptomyces aureofaciens* Tü24 [52]. Bromination occurred at the 6-position and at the 4,6-positions of the pyridine system of nikkomycin Z.

The antifungal antibiotic pyrrolnitrin [3-chloro-4-(2-nitro-3-chlorophenyl)pyrrole] was brominated at the 2-position of the pyrrole moiety by bromoperoxidase from *Streptomyces phaeochromogenes* [53]. Pyrrolnitrin was chlorinated at the 2-position and at the 2,5-positions of the pyrrole system by perhydrolases from *Pseudomonas pyrrocinia* and *Streptomyces aureofaciens*. The corresponding bromo-derivatives were also obtained with these enzymes [54].

Another phenylpyrrole compound, 2-(3,5-dibromo-2-methoxyphenyl)pyrrole was brominated to 2-bromo-, 2,3-dibromo-, 3,4-dibromo-, 2,3,4-tribromo-5-(3,5-dibromo-2-methoxyphenyl)pyrrole by perhydrolase from *Streptomyces aureofaciens* Tü24 [55]. When the same substrate was chlorinated using the perhydrolases from *Pseudomonas pyrrocinia* and *Streptomyces aureofaciens* Tü24, 2-chloro-, 3-chloro-, 4-chloro-, 2,3-dichloro-, 2,4-dichloro, and 3,4-dichloro-5-(3,5-dibromo-2-methoxyphenyl)pyrrole could be isolated (Fig. 16.9-5) [56].

16.9.3.2
Halogenation of Aliphatic Compounds

Haloperoxidases catalyze the halogenation of a wide range of alkene substrates. Ethylene was iodinated, brominated, and chlorinated to the corresponding 2-haloethanol by chloroperoxidase from *Caldariomyces fumago*. Using the same enzyme and propylene as the substrate the 1-halo-2-propanols and the 2-halo-1-propanols were obtained. 1,3-Butadiene was converted into 1-bromo-3-butene-2-ol, 2-bromo-3-butene-1-ol and 1,4-dibromo-2,3-butanediol by lactoperoxidase (Fig. 16.9-6). Bromination of allene by chloroperoxidase from *Caldariomyces fumago* resulted in 2-bromo-2-propen-1-ol [57]. Propylene, allyl chloride and allyl alcohol were halogenated to yield halohydrins and dihalogenated products [58]. When several halide ions were present in the reaction mixture, heterogeneous dihalides were obtained [59]. The chlorination

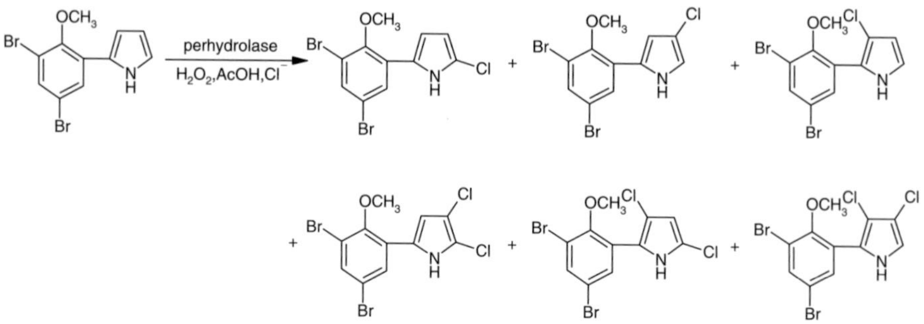

Figure 16.9-5. Chlorination of 2-(3,5-dibromo-2-methoxyphenyl)pyrrole by perhydrolase from *Streptomyces aureofaciens* Tü24 [56].

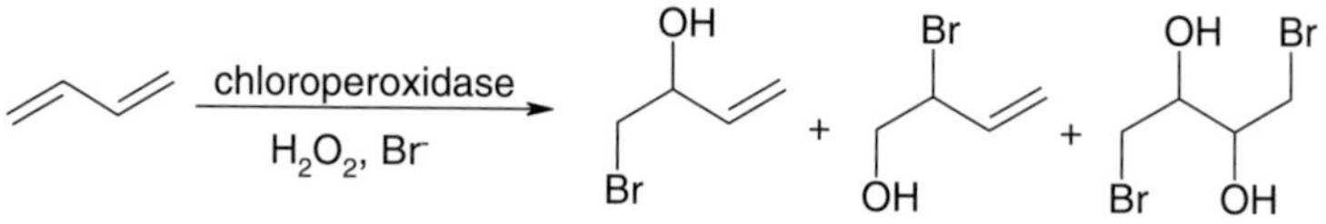

Figure 16.9-6. Bromination of 1,3-butadiene by lactoperoxidase from bovine milk [57].

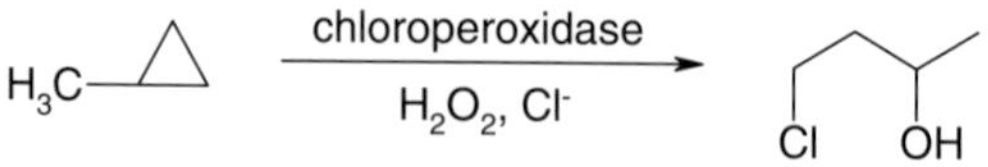

Figure 16.9-7. Chlorination of methyl cyclopropane by chloroperoxidase from *Caldariomyces fumago* [61].

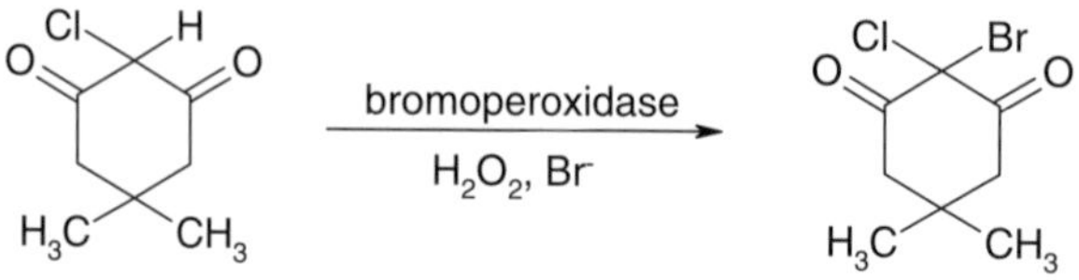

Figure 16.9-8. Bromination of monochlorodimedone, the substrate used for the search for halopereoxidases [1].

of propenylphosphonic acid resulted in the formation of 1-chloro-2-hydroxypropyl-phosphonic acid [60]. Phenyl acetylene was brominated to α-bromoacetophenone and α-dibromoacetophenone by lactoperoxidase. Chloroperoxidase from *Caldariomyces fumago* was used to chlorinate methyl cyclopropane to 4-chloro-2-hydroxy-butane (Fig. 16.9-7) [61].

Monochlorodimedone, the substrate used for the detection and isolation of haloperoxidases and perhydrolases (Fig. 16.9-8), and other β-diketones such as barbituric acid [62] is brominated at the 2-position by all known haloperoxidases and perhydrolases. Oxooctanoic acid and other β-ketoacids form mono- and dihalogen-ated ketones and carbon dioxide [63]. When β-alanine and taurine were used as substrates for myeloperoxidase the corresponding N-chloroamines could be de-tected [64–65].

As can be seen from the number of substrates halogenated by the different haloperoxidases and perhydrolases, these enzymes show no substrate specificity. Examples of aliphatic substrates halogenated by haloperoxidases and perhydrolases are shown in Table 16.9-2.

16.9.4
Regioselectivity and Stereospecificity of Enzymatic Halogenation Reactions

16.9.4.1
FADH$_2$-dependent Halogenases

FADH$_2$-dependent tryptophan 7-halogenase shows regioselctivity which is dependent on the substrate used. With tryptophan, the enzyme is highly regioselective and catalyzes only halogenation at position 7 of the indole ring. However, with other indole derivatives halogenation occurs at positions 2 and 3 of the indole ring (Fig. 16.9-3a). Chlorination of aminophenylpyrrole derivatives by tryptophan 7-halogenase also proceeds with relaxed regioselectivity (Fig. 16-3b) [43].

Nothing is know about the regioselectivity of monodechloroaminopyrrolnitrin 3-halogenase and other FADH$_2$-dependent halogenases, but biosynthetic investigations suggest that many of these halogenases catalyze halogenation reactions regioselectively [19, 66].

So far no investigations on the stereospecificity of this type of halogenating enzymes have been reported.

Haloperoxidases show very poor regioselectivity. There are only very few reports on regioselective reactions catalyzed by haloperoxidases. Franssen et al. [62] reported the regioselective chlorination of 2-aminopyridine to 2-amino-3-chloroaminopyridine by chloroperoxidase from *Caldariomyces fumago*. When anisole was brominated using chloroperoxidase from *Caldariomyces fumago* Walter and Ballschmitter [67] found a *para*-preference for the bromination reaction with a *para : ortho* ratio of 16 compared with 9 for the normal electrophilic bromination. The *para : ortho* ratio for the chlorination of anisole with the same enzyme obtained by Brown and Hager [68] was 1.9. This discrepancy could be due to the different reaction conditions used. Walter and Ballschmitter [67] used a 50 times higher anisole concentration and only about half the amount of chloroperoxidase compared with Brown and Hager [68]. If one takes into consideration that anisole could reach the active site of the enzyme and is present at a high concentration, a relatively large part of the substrate could be chlorinated at the active site with a certain orientation and only a smaller part would be chlorinated by enzymatically produced hypochlorous acid. This effect could be amplified by smaller amounts of the enzyme and thus lower concentrations of hypohalous acid produced. However, this would mean that halogenation occurring at the active site showed a higher degree of regioselectivity, even without a specific binding site for the organic substrate.

Similar results were obtained by Itoh et al. [14, 69] for the halogenation of different substrates using bromoperoxidase from *Corallina pilulifera* and chloroperoxidase from *Caldariomyces fumago*.

In addition to poor regioselectivity, haloperoxidases also show poor stereospeci-

Table 16.9-2. Haloperoxidases used for biotransformations of aliphatic compounds and their sources.

Haloperoxidase (type)	Source	Substrate (halide)	Reference
Chloroperoxidase (heme)	*Caldariomyces fumago*	alkenes (Cl⁻, Br⁻, I⁻)	57–60, 70, 75
		9(11)-dehydroprogesterone (Br⁻)	76
		2-hydroxymethylene testosterone (Br⁻)	77
		2-hydroxymethylene-17β-hydroxy- androstan-3-one (Br⁻)	78
		glycals (Cl⁻, Br⁻, I⁻)	71
		alkynes (Cl⁻, Br⁻, I⁻)	61
		cyclopropanes (Cl⁻, Br⁻)	61
Lactoperoxidase (heme)	bovine milk	β-diketones (Cl⁻, Br⁻)	44, 70, 79–81
		alkenes (Br⁻, I⁻)	57, 59, 82–84
		alkynes (Br⁻)	61
Myeloperoxidase (heme)	mammals	cyclopropanes (Br⁻)	61
		alanine, taurine (Cl⁻)	65
Bromoperoxidase (heme)	*Penicillus capitatus*	β-alanine (Cl⁻)	64
	Bonnemasoinia hamifera	α-amino acids, peptides (Br⁻)	85
Bromoperoxidase (non-heme)	*Ascophyllum nodosum*	barbituric acid (Br⁻)	62

ficity. Kollonitsch et al.[60] obtained optically inactive erythro-dimethyl 1-chloro-2-hydroxypropylphosphonate from *trans*-propenylphosphonic acid using chloroperoxidase from *Caldariomyces fumago*. Ramakrishnan et al.[70] investigated the bromination of racemic 2-*exo*-methylbicyclo-[2.2.1]hept-5-ene-2-*endo*-carboxylic acid to the δ-lactone and racemic bicyclo-[3.2.0]hept-2-en-6-one to the 2-*exo*-bromo-3-*endo*-hydroxybromohydrin. The products were obtained in racemic form. 2-Methyl-4-propylcyclopentane-1,3-dione was chlorinated to 2-chloro-2-methyl-4-propylcyclopentane-1,3-dione. Here the product was obtained as a 40 : 60 ratio of the racemic diastereomers. From these findings they concluded that active site chlorination by chloroperoxidase from *Caldariomyces fumago* proceeds without appreciable stereoselectivity.

On the other hand, Liu and Wong[71] described the stereoselective bromohydrations of D-galactal and L-fucal to 2-bromo-2-deoxy-D-galactose (β/α = 3) and 2-bromo-2-deoxy-L-fucose (β/α = 2), respectively. They also obtained the corresponding chlorinated products, however, in much lower yields.

16.9.5
Comparison of Chemical with Enzymatic Halogenation

$NADH_2$-dependent tryptophan 7-halogenase catalyzes the incoporation of a chloride atom into the indole ring at a position were direct chemical chlorination is not possible. The structures of metabolites containing halogenated indole rings suggest that similar halogenases exist which catalyze halogenation reactions at positions 2–7 of the indole ring. These enzymes are certainly very promising candidates as tools in organic synthesis, especially as they catalyze the incorporation of the halide atoms as nucleophiles, which allows regioselective and possibly stereoselective halogenation reactions.

Enzymatic halogenation catalyzed by haloperoxidases and perhydrolases involves the oxidation of halide ions to a halonium ion species which leads to the formation of hypohalous acids (Fig. 16.9-1). The products obtained by enzymatic halogenation with these enzymes are the same as the products obtained by chemical electrophilic halogenation with hypohalous acids. The differences in the *para* : *ortho* ratios in the halogenation of some aromatic compounds could be due to a mixture of halogenation at or near the active site and in solution.

The major advantage of enzymatic halogenation using haloperoxidases and perhydrolases is that the enzymes have a very low substrate specificity and that no free halogen is needed which makes halogenation catalyzed by these enzymes less hazardous than chemical halogenation.

Some of the non-heme haloperoxidases and perhydrolases are very stable, even against organic solvents, and easy to use as they do not need any cofactors. However, care has to be taken not to use too high concentrations of hydrogen peroxide, as this could lead to oxidation of the substrate.

References

1 S. L. Neidleman, J. Geigert, *Biohalogenation: Principles, Basic Roles and Applications*, Ellis Horwood Ltd., Chichester, UK. **1986**.

2 H. Vilter, *Le Jol. Bot. Mar.*, **1983**, *26*, 451–455.

3 H. Vilter, *Phytochemistry*, **1984**, *23*, 387–1390.

4 R. Wever, H. Plat, E. De Boer, *Biochim. Biophys. Acta* **1985**, *830*, 181–186.

5 K.-H. van Pee, G. Sury, F. Lingens, *Biol. Chem. Hoppe-Seyler* **1987**, *368*, 1225–1232.

6 B. E. Krenn, H. Plat, R. Wever, *Biochim. Biophys Acta* **1988**, *952*, 255–260.

7 W. Wiesner, K.-H. van Pee, F. Lingens, *J. Biol. Chem.* **1988**, *263*, 13725–13732.

8 R. Zeiner, K.-H. van P,e, F. Lingens, *J. Gen. Microbiol.* **1988**, *134*, 3141–3149.

9 M. Weng, O. Pfeifer, S. Krauss, F. Lingens, K.-H. van Pée, *J. Gen. Microbiol.* **1991**, *137*, 2539–2546.

10 H. J. Hecht, H. Sobek, T. Haag, O. Pfeifer, K.-H. van Pée, *Nature Struct. Biol.* **1994**, *1*, 532–537.

11 M. Picard, J. Gross, E. Lübbert, S. Tölzer, S. Krauss, K.-H. van Pée, A. Berkessel, *Angew. Chem. Int. Ed. Engl.* **1997**, *36*, 1196–1199.

12 B. Hofmann, S. Tölzer, I. Pelletier, J. Alten-buchner, K.-H. van Pée, H. J. Hecht, *J. Mol. Biol.* **1998**, *279*, 889–900.

13 N. Itoh, Y. Izumi, H. Yamada, *J. Biol. Chem.* **1986**, *261*, 5194–5200.

14 N. Itoh, Y. Izumi, H. Yamada, *Biochemistry* **1987**, *26*, 282–289.

15 E. De Boer, H. Plat, M. G. M. Tromp, R. Wever, M. C. R. Franssen, H. C. van der Plas, E. M. Meijer, H. E. Schoemaker, *Biotechnol. Bioeng.* **1987**, *30*, 607–610.

16 K. Hohaus, A. Altmann, W. Burd, I. Fischer, P. E. Hammer, D. S. Hill, J. M. Ligon, K.-H. van Pée, *Angew. Chem. Int. Ed. Engl.* **1997**, *36*, 2012–2013.

17 P. E. Hammer, D. S. Hill, S. T. Lam, K.-H. van Pée, J. M. Ligon, *Appl. Environ. Microbiol.* **1997**, *63*, 2147–2154.

18 S. Kirner, P. E. Hammer, D. S. Hill, A. Alt-mann, I. Fischer, L. J. Weislo, M. Lanahan, K.-H. van Pée, J. M. Ligon, *J. Bacteriol.* **1998**, *180*, 1939–1943.

19 K.-H. van Pée, S. Keller, T. Wage, I. Wy-nands, H. Schnerr, S. Zehner, *Biol. Chem.* **2000**, *381*, 1–5.

20 S. Keller, T. Wage, K. Hohaus, E. Eichhorn, K.-H. van Pée, *Angew. Chem. Int. Ed. Engl.* **2000**, *39*, 2300–2302.

21 S. Pelzer, R. Süßmuth, D. Heckmann, J. Recktenwald, P. Huber, G. Jung, W. Wohlle-ben, *Antimicrob. Agents Chemother.* **1999**, *43*, 1565–1573.

22 B. Nowak-Thompson, N. Chaney, J. S. Wing, S. J. Gould, J. E. Loper, *J. Bacteriol.* **1999**, *181*, 2166–2174.

23 P. E. Hammer, W. Burd, D. S. Hill, J. M. Ligon, K.-H. van Pée, *FEMS Microbiol. Lett.* **1999**, *180*, 39–44.

24 L. M. Shannon, E. Kay, J. Y. Lew, *J. Biol. Chem.*, **1966**, *241*, 2166–2172.

25 N. M. Alexander, *J. Biol. Chem.* **1959**, *234*, 1530–1533.

26 C. Dumontet, B. Rousset, *J. Biol. Chem.* **1983**, *258*, 14166–14172.

27 T. Deits, M. Farrance, E. S. Kay, L. Medill, E. E. Turner, P. J. Weidman, B. M. Shapiro, *J. Biol. Chem.* **1984**, *259*, 13525–13533.

28 D. G. Baden, M. D. Corbett, *Biochem. J.* **1980**, *187*, 205–211.

29 J. A. Manthey, L. P. Hager, *J. Biol. Chem.* **1981**, *256*, 11232–11238.

30 N. Itoh, Y. Izumi, H. Yamada, *Biochem. Biophys Res. Commun.* **1985**, *131*, 428–435.

31 H. Plat, B. E. Krenn, R. Wever, *Biochem. J.* **1987**, *248*, 1123–1131.

32 V. Renagathan, K. Miki, M. H. Gold. *Biochemistry* **1987**, *26*, 5127–5132.

33 K.-H. van Pée, F. Lingens, *J. Gen. Microbiol.* **1985**, *131*, 1911–1916.

34 M. Knoch, K.-H. van Pée, L. C. Vining, F. Lingens, *J. Gen. Microbiol.* **1989**, *135*, 2493–2502.

35 J. Schultz, *Reticuloendothel. Syst.* **1980**, *2*, 231–254.

36 J. T. Archer, G. Air, M. Jackas, D. B. Morell, *Biochim. Biophys. Acta* **1965**, *99*, 96–101.

37 Y. P. Chen, D. E. Lincoln, S. A. Woodin, C. R. Lovell, *J. Biol. Chem.* **1991**, *266*, 23909–23915.

38 D. R. Morris, L. P. Hager, *J. Biol. Chem.* **1966**, *241*, 1763–1768.

39 T.-N. E. Liu, T. M'Timkulu, J. Geigert, B. Wolf, S. L. Neidleman, D. Silva, J. C. Hunter-Cevera, *Biochem. Biophys. Res. Commun.* **1987**, *142*, 329–333.

40 B. H. Simons, P. Barnett, E. G. M. Vollen-

broek, H. L. Dekker, A. O. Muijsers, A. Messerschmidt, R. Wever, *Eur. J. Biochem.* **1995**, *229*, 566–574.

41 W. Wiesner, K.-H. van Pée, F. Lingens, *FEBS Lett.*, **1986**, *209*, 321–324.

42 R. D. Carmichael, M. A. Pickard, *Appl. Environ. Microbiol.* **1989**, *55*, 17–20.

43 K.-H. van Pée, M. Hölzer, in *Tryptophan, Serotonin and Melatonin – Basic Aspects and Application. Advances in Tryptophan Research* (Eds: G. Huether, W. Koch, T. J. Simat, H. Steinhardt, Plenum Press, New York **1999**.

44 L. P. Hager, D. R. Morris, F. S. Brown, H. Eberwein, *J. Biol. Chem.* **1966**, *241*, 1769–1777.

45 J. Nunez, *Methods Enzymol.* **1984**, *107*, 476–488.

46 J. A. Manthey, L. P. Hager, K. D. McElvany, *Methods Enzymol.* **1984**, *107*, 439–445.

47 J. C. Hunter-Cevera, L. Sotos, *Microb. Ecol.* **1986**, *12*, 121–127.

48 M. D. Corbett, B. R. Chipko, A. O. Batchelor, *Biochem. J.* **1980**, *187*, 893–903.

49 H. Hoff, H. Drautz, H.-P. Fiedler, H. Zähner, J. E. Schultz, W. Keller-Schierlein, S. Philipps, M. Ritzau, A. Zeek, *J. Antibiot.* **1992**, *45*, 1096–11107.

50 F. Thiermann, G. Bongs, K.-H. van Pée, D. Braun, H.-J. Cullmann, H.-P. Fiedler, H. Zähner, *10. DECHEMA-Jahrestagung der Biotechnologen, Karlsruhe, Germany,* **1992**.

51 M. C. R. Franssen, H. G. van Boven, H. C. van der Plas, *J. Heterocycl. Chem.* **1987**, *24*, 1313–1316.

52 H. Decker, U. Pfefferle, C. Bormann, H. Zähner, H.-P. Fiedler, K.-H. van Pée, M. Rieck, W. A. König, *J. Antibiot.* **1991**, *44*, 626–634.

53 K.-H. van Pée, F. Lingens, *FEBS Lett.* **1984**, *173*, 5–8.

54 G. Bongs, K.-H. van Pée, *Enzyme Microb. Technol.* **1994**, *16*, 53–60.

55 H. Laatsch, H. Pudleiner, B. Pelizaeus, K.-H. van Pée, *Liebigs Ann. Chem.* **1994**, 65–71.

56 V. N. Burd, K.-H. van Pée, F. Lingens, A. Voskoboev, *Bioorg. Khim* **1992**, *18*, 1002–1006.

57 J. Geigert, S. L. Neidleman, D. J. Dalietos, S. K. DeWitt, *Appl. Environ. Microbiol.* **1983**, *45*, 366–374.

58 J. Geigert, S. L. Neidleman, D. J. Dalietos, S. K. DeWitt, *Appl. Environ. Microbiol.* **1983**, *45*, 1575–1581.

59 S. L. Neidleman, J. Geigert, *Trends Biotechnol.* **1983**, *1*, 1–5.

60 J. Kollonitsch, S. Marburg, L. M. Perkins, *J. Am. Chem. Soc.* **1970**, *92*, 4489–4490.

61 J. Geigert, S. L. Neidleman, D. J. Dalietos, *J. Biol. Chem.* **1983**, *258*, 2273–2277.

62 M. C. R. Franssen, J. D. Jansma, H. C. van der Plas, E. de Boer, R. Wever, *Bioorg. Chem.* **1988**, *16*, 352–363.

63 R. F. Theiler, J. F. Siuda, L. P. Hager, *Science* **1978**, *202*, 1094–1096.

64 R. J. Selvaraj, J. M. Zgliczynski, B. B. Paul, A. J. Sbarra, *J. Infect. Dis.* **1978**, *137*, 481–485.

65 M. B. Grisham, M. M. Jefferson, D. F. Metton, E. L. Thomas, *J. Biol. Chem.* **1984**, *259*, 10404–10413.

66 K.-H. van Pée, *Annu. Rev. Microbiol.* **1996**, *50*, 375–399.

67 B. Walter, K. Ballschmitter, *Chemosphere* **1991**, *22*, 557–567.

68 F. S. Brown, L. P. Hager, *J. Am. Chem. Soc.* **1967**, *89*, 719–720.

69 N. Itoh, A. K. M. Q. Hasan, Y. Izumi, H. Yamada, *Eur. J. Biochem.* **1988**, *172*, 477–484.

70 K. Ramakrishnan, M. E. Oppenhuizen, S. Saunders, J. Fisher, *Biochemistry* **1983**, *22*, 3271–3277.

71 K. K.-C. Liu, C.-H. Wong, *J. Org. Chem.* **1992**, *40*, 3748–3750.

72 R. E. Huber, L. A. Edwards, T. J. Carne, *J. Biol. Chem.* **1989**, *264*, 1381–1386.

73 B. Matkovics, Z. Rakonczay, S. E. Rajki, L. Balaspiri, *Steroidologia* **1971**, *2*, 77–79.

74 H. Vilter, *Le Jol. Bot. Mar.* **1983**, *26*, 429–435.

75 H. Yamada, N. Itoh, S. Murakami, Y. Izumi, *Agric. Biol. Chem.* **1985**, *49*, 2961–2967.

76 S. L. Neidleman, S. D. Levine, *Tetrahedron Lett.* **1968**, *46*, 4057–4059.

77 S. L. Neidleman, M. A. Oberc, *J. Bacteriol.* **1968**, *95*, 2424–2425.

78 S. D. Levine, S. L. Neidleman, M. Oberc, *Tetrahedron* **1968**, *24*, 2979–2984.

79 R. P. Martyn, S. C. Branzer, G. T. Sperl, *Bios* **1981**, *52*, 8–12.

80 S. L. Neidleman, P. A. Diassi, B. Junta, R. M. Palmere, S. C. Pan, *Tetrahedron Lett.* **1966**, *44*, 5337–5342.

81 M. C. R. Franssen, H. C. van der Plas, *Bioorg. Chem.* **1987**, *15*, 59–70.

82 J. M. Boeynaems, D. Reagan, W. C. Hubbard, *Lipids* **1981**, *16*, 246–249.

83 J. M. Boeynaems, J. T. Watson, J. A. Oates, W. C. Hubbard, *Lipids* **1981**, *16*, 323–327.

84 J. Turk, W. R. Henderson, S. L. Klebanoff, W. C. Hubbard, *Biochim. Biophys. Acta* **1983**, *751*, 189–200.

85 M. Nieder, L. P. Hager, *Arch. Biochem. Biophys.* **1985**, *240*, 121–127.

17
Isomerizations

Nobuyoshi Esaki, T. Kurihara and K. Soda

17.1
Introduction

Isomerases catalyze the isomerization of substrates, and are classified into five groups as follows:

Racemases and epimerases (E. C. class 5.1)
They are defined as enzymes that catalyze the isomerization of a substrate through stereochemical reverse rearrangement of a substituent bound to a chiral center (usually a chiral carbon) in the substrate molecule. Racemases act on molecules containing only the asymmetric center concerned in the reaction. Epimerases act on substrates containing one or more asymmetric centers in addition to the reactive chiral center.

cis-trans-Isomerases (E. C. class 5.2)
They catalyze the interconversion of *cis-trans* geometrical isomers.

Sugar isomerases, tautomerases, Δ-isomerases, etc. (E. C. class 5.3)
Sugar isomerases catalyze the interconversion between aldose and ketose. Tautomerases catalyze a keto-enol tautomerization. Δ-Isomerases catalyze the shift of a double bond. The reactions catalyzed by these enzymes proceed through intramolecular oxidation and reduction.

Mutases (E. C. class 5.4)
They catalyze the transfer of a substituent to produce a structural isomer.

Cycloisomerases (E. C. class 5.5)
They catalyze the ring formation through an intramolecular lyase reaction.

Isomerizations catalyzed by most of these enzymes proceed through 1,1-, 1,2-, or 1,3-hydrogen shifts (Table 17-1), while mutases catalyze exchange of a hydrogen

Table 17-1. Enzyme-catalyzed isomerizations classified as hydrogen shifts.

Type	Examples	Category
1,1-Shifts	$R_2 \overset{R_1}{\underset{R_3}{-}}H \rightleftharpoons H \overset{R_1}{\underset{R_3}{-}}R_2$	Epimerases, Racemases
1,2-Shifts	$\overset{H}{\underset{H-C-OH}{C=O}} \rightleftharpoons \overset{H-C-OH}{\underset{C=O}{}}$	Aldose-ketose isomerases
1,3-Shifts	$\overset{R_1 \quad H}{\underset{R_2}{C}} \overset{}{\underset{C=C}{}}\overset{R_3}{\underset{R_4}{}} \rightleftharpoons \overset{R_1 \; H \; R_3}{\underset{R_2 \quad C \quad R_4}{C=C}}$	Allylic isomerizations

atom with particular functional groups such as amino, hydroxy, and α-amino-α-carboxymethyl groups attached at neighboring carbon atoms of the substrates through homolytic cleavage.

Here we describe enzymological properties of representative racemases, epimerases, and isomerases, and their application to production of various optically active compounds.

17.2
Racemizations and Epimerizations

Since the discovery of enzymatic racemization of lactate by lactic acid bacteria[1, 2], *Clostridium acetobutyricum*[3], and *Cl. butyricum*[4], a variety of racemases and epimerases have been demonstrated, and they are classified into the four groups as follows:

- Amino acid racemases and epimerases catalyzing racemization and epimerization at the chiral center containing an NH_2 or NH group (E. C. class 5.1.1);
- Mandelate racemase, lactate racemase, and others acting at the chiral center containing an OH group (E. C. class 5.1.2);
- Various carbohydrate epimerases such as UDP-D-glucose-4'-epimerase (E. C. class 5.1.3);
- Methylmalonyl CoA epimerase and some others, in whose substrates a CH_3 group is bound to the chiral centers (E. C. class 5.1.99).

Racemases and epimerases have been used for production of various optically active compounds from cheaply-available racemic substrates by combination of enzymes that act specifically on one of the isomers of the racemates to catalyze hydrolysis, oxidation, reduction, elimination, replacement, and other reactions. The racemases and epimerases used act exclusively on the substrates, but not on the products of the reaction. Thus, total conversion of the racemic substrates into the desired optically-active compounds is achieved. Here we describe enzymological characteristics of the representative racemases and epimerases, and their application to production of optically active compounds.

17.2.1
Pyridoxal 5'-phosphate-dependent Amino Acid Racemases and Epimerases

17.2.1.1
Alanine Racemase (E. C. 5.1.1.1)

Alanine racemase is a bacterial enzyme that catalyzes racemization of L- and D-alanine, and requires pyridoxal 5'-phosphate (PLP) as a cofactor. The enzyme plays an important role in the bacterial growth by providing D-alanine, a central molecule in the peptidoglycan assembly and cross-linking, and has been purified from various sources[5–16]. The enzyme has been used for the production of stereospecifically deuterated NADH and various D-amino acids by combination of L-alanine dehydrogenase (E. C. 1.4.1.1), D-amino acid aminotransferase (E. C. 2.6.1.21), and formate dehydrogenase (E. C. 1.2.1.2)[17, 18].

17.2.1.1.1 Gene Cloning and Primary Structure
Two distinct alanine racemase genes were cloned from the *Salmonella typhimurium* chromosome. One mapped at minute 37 on the chromosome is termed *dad*B, and the other mapped at minute 91 is termed the *alr* gene[19]. The *dad*B alanine racemase is formed inducibly and functions in the catabolism of L-alanine: the *alr* enzyme is synthesized constitutively, and functions in the anabolic assembly of peptidoglycan[19]. Alanine racemase genes were also cloned from *Bacillus stearothermophilus*[10], *Bacillus subtilis*[20], *Bacillus psychrosaccharolyticus*[14], and *Aquifex pyrophilus*[15]. Two distinct alanine racemase genes were assigned in the genome sequences of *Escherichia coli*[12], *B. subtilis*, *Pseudomonas aeruginosa*, and *Vibrio cholerae*, but only a single one occurs in the other bacterial genomes whose complete nucleotide sequences were determined as shown at internet sites such as http://www.geno me.ad.jp/kegg/catalog/org_list.html.

Uo et al.[16] have found that fission yeast, *Schizosaccharomyces pombe*, has also the alanine racemase gene, which is involved in the catabolism of D-alanine in *S. pombe* in the same manner as *dad*B of *S. typhimurium*. The yeast enzyme only shows any high degree of similarity to the alanine racemases of γ-proteobacteria (gram-negative phylum). Therefore, the gene of *S. pombe* has possibly been acquired from γ-proteobacteria through some events of horizontal gene transfer such as conjugation: *S. pombe* is known to be a recipient of the genes from *E. coli* through direct conjugation.

D-Alanine occurs in various natural compounds produced by fungus. For example, cyclosporin A contains D-alanine as a component and is produced by *Tolypocladium niveum*[12]. Alanine racemase is involved in the biosynthesis of D-alanine in this fungus.

17.2.1.1.2 Stability

The native *dadB* and *alr* racemases from *Salmonella typhimurium* are readily inactivated by digestion with α-chymotrypsin, trypsin, and subtilisin[22]. However, the *Bacillus stearothermophilus* enzyme is stable even after fragmentation into two pieces[23, 24]. *A. pyrophilus*, a hyperthermophilic bacterium, produces extremely stable alanine racemase[15]. It maintains catalytic activity in the presence of organic solvents as well. On the other hand, *Bacillus psychrosaccharolyticus*, a psychrophyilic bacterium, produces a thermo-labile enzyme[14]. However, it shows high catalytic activity at low temperatures, such as at 0 °C. Similar cold activity and thermal instability was found in the enzyme from a psychrophile isolated from raw milk, *Pseudomonas fluorescens*[11].

17.2.1.1.3 Reaction Mechanism

Reaction of alanine racemase proceeds through the steps shown in Fig. 17-1. PLP bound with the active-site lysyl residue (A) reacts with a substrate to form an external Schiff base (B) through transaldimination. The subsequent α-hydrogen abstraction

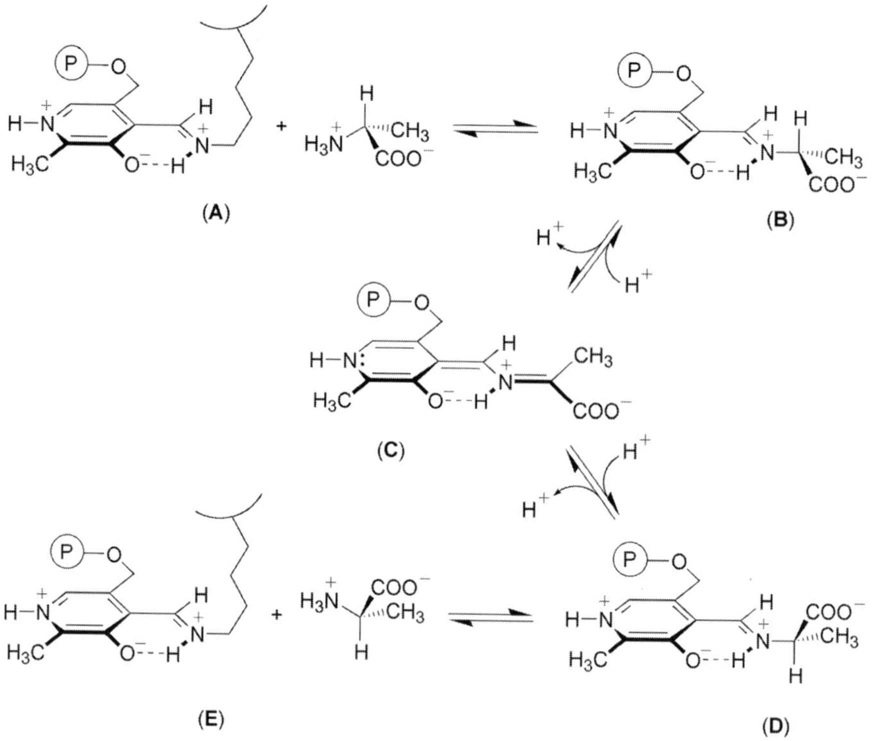

Figure 17-1. Mechanism of the alanine racemase reaction. A, An internal aldimine of PLP with a lysyl residue; B, an external aldimine of PLP with D-alanine; C, a quinonoid intermediate formed after removal of a hydrogen from alanyl external aldimines B or D; D, an external aldimine of PLP with L-alanine. Reprinted from Watanabe et al.[33].

results in the formation of a resonance-stable deprotonated intermediate (C). If reprotonation occurs at the α-carbon of the substrate moiety on the opposite face of the planar intermediate (C), then an antipodal aldimine (D) is formed. The ε-amino group of the lysine residue is substituted for the isomerized amino acid through transaldimination, and the internal aldimine (A) is regenerated. According to Dunathan [25], the C^α-H - bond to be broken is positioned perpendicularly to the plane of the conjugated π-system of the external Schiff base intermediate, in order to achieve maximum orbital overlap with the π electron system of the complex, resulting in a substantial rate enhancement for the cleavage of that bond.

The racemization reaction proceeds via either a one-base [26] or two-base [27] mechanism. The one base mechanism is characterized by the retention of the substrate-derived proton in the product (internal return) [26]. By this criterion, reactions catalyzed by α-amino-ε-caprolactam racemase [28] and amino acid racemase with low substrate specificity (E.C. 5.1.1.10) [26] have been considered to proceed through the one-base mechanism. However, such internal returns were not observed in the reactions catalyzed by alanine racemases from *E. coli* [26], *Bacillus stearothermophilus* [29], and *Salmonella typhimurium* (*dadB* and *alr*) [29]. The internal return is not expected to occur in the reactions of two-base mechanism. In fact, kinetic analyses [30] indicated that the alanine racemase reaction proceeds through a two-base mechanism: proton donors and proton acceptors are situated on both sides of the planar intermediate (Fig. 17-1, C) and accomplish removal and return of the α-hydrogen of the substrate amino acid. X-ray crystallographic studies [31, 32] suggested that Lys 39 and Tyr 265 of alanine racemase from *B. stearothermophilus* serve as the bases (Fig. 17-2). Watanabe *et al.* [33] showed that the lysyl residue binding PLP in the racemase (Lys 39) acts as the base catalyst specific to the D-enantiomer of alanine. The crystal structure of the enzyme complex with R-1-aminoethylphosphonic acid [32], a tight-bind inhibitor of the enzyme [34], demonstrated that the phenolic oxygen of Tyr 265 is appropriately aligned for proton abstraction from an L-isomer in the active site of the structure: Tyr 265 is the second base specifically acting on the L-alanyl-PLP aldimine.

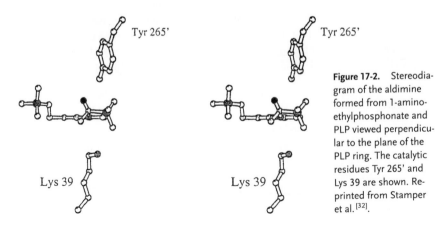

Tyr 265'

Tyr 265'

Lys 39

Lys 39

Figure 17-2. Stereodiagram of the aldimine formed from 1-aminoethylphosphonate and PLP viewed perpendicular to the plane of the PLP ring. The catalytic residues Tyr 265' and Lys 39 are shown. Reprinted from Stamper et al. [32].

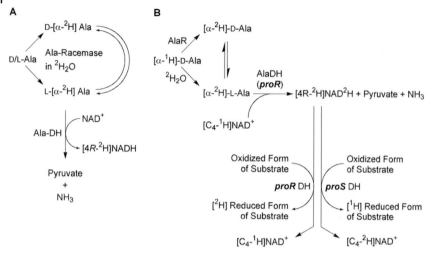

Figure 17-3. A, Preparation of [4S-^{2}H]-NADH by coupling of alanine racemase and L-alanine dehydrogenase. B, *In situ* determination of stereospecificity of H-transfer by ^{1}H-NMR. AlaR, AlaDH, and DH represent alanine racemase, L-alanine dehydrogenase, and dehydrogenase, respectively.

17.2.1.1.4 Production of Stereospecifically Deuterated NADH

NAD-linked dehydrogenases show either pro-S or pro-R-stereospecificity for hydrogen removal from the C4 position of the nicotinamide moiety of the reduced coenzymes. The stereospecificity of hydrogen transfer is examined by means of stereospecifically C4-deuterated NADH, which is prepared enzymatically from NAD$^+$ and deuterated substrates by tedious procedures.

Esaki et al. developed a simple method to produce the stereospecifically deuterated NADH by an NAD$^+$-dependent dehydrogenase by combination with amino acid racemase[35]. L-Alanine dehydrogenase transfers deuterium of [2-^{2}H]-L-alanine to NAD$^+$ to produce [4R-^{2}H]-NADH[36]. Alanine racemase catalyzes the C2-deuteration of D and L-alanine in ^{2}H$_2$O[10], and [4R-^{2}H]-NADH was produced from D-alanine and NAD$^+$ by coupling of the reactions catalyzed by alanine racemase and L-alanine dehydrogenase in ^{2}H$_2$O (Fig. 17-3A). Furthermore, this finding led to development of a simple procedure for the *in situ* analysis of stereospecificity of hydrogen transfer of NADH by an NAD-dependent dehydrogenase by means of ^{1}H-NMR (Fig. 17-3B)[35].

17.2.1.1.5 Production of D-Amino Acids

Considerable attention has been paid to multi-enzyme reaction systems as a means to the stereospecific production of L-amino acids[37]. Wichmann et al. have developed a continuously operated membrane reactor for production of L-leucine from α-ketoisocaproate[38]. The system is also applicable to the production of several other aliphatic L-amino acids such as L-valine, L-*tert*-leucine and [^{15}N]-L-leucine. The

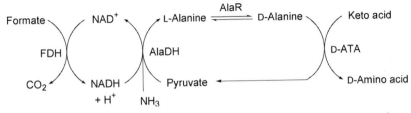

Figure 17-4. Synthesis of D-amino acids from α-keto acid, formate, NAD⁺, D-alanine, and ammonia by coupling of L-alanine dehydrogenase (AlaDH), formate dehydrogenase (FDH), alanine racemase (AlaR), and D-amino acid aminotransferase (D-ATA).

process has been successfully scaled-up for industrial production of these L-amino acids[39]. A similar system has been developed for production of L-phenylalanine[40, 41] and L-β-chloroalanine[40–43]. However, little attention has been paid to the stereospecific production of D-amino acids by means of multi-enzyme reaction systems, although D-amino acids have been paid considerable attention[44]. For example, substantial amounts of D-serine, D-aspartate and other D-amino acids occur in mammalian brain[45–47], and ¹³N-labeled D-amino acids are expected to be useful for the study of their metabolism in brain[48].

A simple procedure was established for the synthesis of various D-amino acids by means of four types of thermostable enzymes: alanine racemase, D-amino acid aminotransferase[49, 50], L-alanine dehydrogenase[5], and formate dehydrogenase (Fig. 17-4)[17]. The commercial preparation of formate dehydrogenase from *Candida boidinii* used by Wichmann et al.[38] is not sufficiently stable. However, Galkin et al.[52] cloned and expressed the gene of thermostable formate dehydrogenase in *E. coli*.

D-Phenylalanine and D-tyrosine, which are the poor substrates for D-amino acid aminotransferase, were synthesized in an optical purity of essentially 100%, but with yields of lower than 50%. However, the yields were increased by addition of excess amounts of the D-amino acid aminotransferase (Table 17-2)[17]. Selenium is an essential micronutrient for mammals, fish and several bacteria, although it is toxic at a high concentration[52, 53]. D-Selenomethionine was produced in an 80% yield based on 2-oxo-4-methylselenobutyrate[54]. Norvaline, valine, and α-aminobutyrate were also produced with high yields. However, α-aminobutyrate was synthesized as a racemic mixture. D-Norvaline was obtained at an enantiomeric excess of only 30%. The low optical purity is probably due to the action of L-alanine dehydrogenase: α-ketobutyrate and α-ketovalerate are reduced by L-alanine dehydrogenase at rates of 79 and 6.6% relative to that of pyruvate, respectively[10]. Moreover, alanine racemase also racemizes α-aminobutyrate and norvaline, though very slowly[55]. Thus, this method is not applicable to the stereospecific production of D-α-aminobutyrate and D-norvaline. The preparations of D-valine, D-methionine and D-norleucine also suffered contamination by the antipodes at concentrations of 4, 3, and 1%, respectively, due to the action of L-alanine dehydrogenase on the α-keto analogs of these amino acids[51]. However, D-glutamate, D-phenylalanine and D-tyrosine were efficiently produced in the system. The final concentration of D-glutamate produced

Table 17-2. Synthesis of D-amino acids from α-keto acids by combination of four purified enzymes: alanine racemase, L-alanine dehydrogenase, formate dehydrogenase, and D-amino acid aminotransferase.

Substrate	Product	Yield (%)[a]	ee (%)
α-Ketoglutarate	D-glutamate	98	100
α-Ketoisocaproate	D-leucine	80	>99
α-Ketocaproate	D-norleucine	82	98
α-Keto-γ-thiomethylbutyrate	D-methionine	95	94
α-Ketoisovalerate	D-valine	90	92
α-Ketovalerate	D-norvaline	92	30
α-Ketobutyrate	α-aminobutyrate	93	0
Phenylpyruvate	D-phenylalanine	72[b]	100
Hydroxyphenylpyruvate	D-tyrosine	70[b]	100

a The yields were determined after an 8 h incubation.
b The amount of D-amino acid aminotransferase used (30 units) was 10-fold higher than that in other systems

was only around 0.3 M, limited because of the equilibrium of the D-amino acid aminotransferase reaction. The method is most suitable for stereospecific conversion of α-keto acids into the corresponding D-amino acids, in particular labeled compounds, for example with ^{13}N by means of ^{13}N-NH$_3$.

The industrial use of the above-mentioned systems depends predominantly on the cost of the enzymes, although the intact cells of microorganisms containing the enzymes can be used as catalysts in order to decrease costs[56]. In most cases, however, additional genetic improvements through metabolic engineering are required, thereby new functional combinations are made by the rational transfer of pathways from one organism to another[57]. The transfer of the ethanol pathway from *Zymomonas mobilis* to other enteric bacteria represents an example of this approach[58]. In the above-mentioned system, various D-amino acids can be produced from the corresponding α-keto acids, if four functional genes are introduced into one microorganism. The simultaneous expression of all enzymes in a single cell

Table 17-3. Synthesis of D-amino acids from α-keto acids by *E. coli* cells harboring pFADA which codes for four enzyme genes: alanine racemase, L-alanine dehydrogenase, formate dehydrogenase, and D-amino acid aminotransferase.

Substrate	Product	Yield (%)[a]	ee (%)
α-Ketoglutarate	D-glutamate	85	100
α-Ketoisocaproate	D-leucine	76	>99[b]
α-Ketocaproate	D-norleucine	70	88
α-Keto-γ-thiomethylbutyrate	D-methionine	80	90
α-Ketoisovalerate	D-valine	85	92
α-Ketovalerate	D-norvaline	90	35
α-Ketobutyrate	α-aminobutyrate	95	0
Phenylpyruvate	D-phenylalanine	15	ND[c]
Hydroxyphenylpyruvate	D-tyrosine	5	ND

a The yields were determined after a 12 h incubation
b The optical purity determined by HPLC is >99.9%
c ND, not determined.

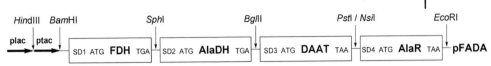

Figure 17-5. Construction of the plasmid used for the production of D-amino acids by expression in *E. coli* cells; formate dehydrogenase (FDH), L-alanine dehydrogenase (AlaDH), alanine racemase (AlaR), and D-amino acid aminotransferase (DAAT).

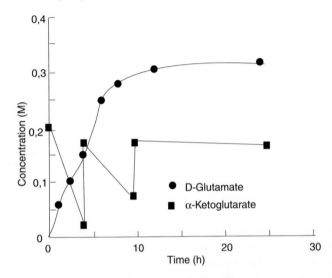

Figure 17-6. Time course for the production of D-glutamate with *E. coli* cells containing pFADA. α-Ketoglutarate was added after 4 and 10 h of incubation at a final concentration of approximately 0.2 M.

provides additional benefit for industrial applications: the intracellular pool of NAD^+ (supplied by the cell itself) could be used for NADH regeneration without any additional supplies.

Galkin et al.[18] constructed plasmids containing, in addition to the thermostable formate dehydrogenase gene, all three genes required for the synthesis of D-amino acids (Fig. 17-5). D-Enantiomers of glutamate and leucine were produced at high optical purity and high conversion rates with the recombinant *E. coli* cells harboring the plasmid for coding of the four heterologous genes (Table 17-3). α-Keto acids, particularly branched-chain and long-chain α-keto acids, are toxic, inhibiting the growth of *E. coli* when added at concentrations of only 15–30 mM. Therefore, Galkin et al. used the resting cells of the recombinant *E. coli* instead of growing ones. Moreover, the isolation of products in the resting-cell system is much easier than when using growth media containing complex ingredients such as yeast extracts. The final concentration of D-glutamate produced was around 0.3 M (Fig. 17-6).

17.2.1.2
Amino Acid Racemase with Low Substrate Specificity (E. C. 5.1.1.10)

An amino acid racemase which shows very broad substrate specificity was discovered in *Pseudomonas striata* (= *Ps. putida*), purified, and characterized[59]. The enzyme catalyzes racemization of various amino acids except aromatic and acidic amino acids. A similar enzyme also occurs in *Aeromonas punctata*[60]. Arginine racemase, which also shows a broad substrate specificity, has been demonstrated in *Pseudomonas graveolens* (= *Pseudomonas taetrolens*)[61]. These amino acid racemases do not act on threonine, valine and their analogs, whose β-methylene group is substituted. Recently, Lim et al.[62] found, in *Ps. putida* ATCC 17 642, a new amino acid racemase catalyzing not only racemization of various amino acids but also epimerization of D- and L-threonine by stereoconversion at the α-position: it catalyzes epimerization of L-to D-allo- and also of D- to L-*allo*-threonine.

Amino acid racemase with low substrate specificity catalyzes racemization of leucine and various other amino acids, which are also α-deuterated in ^2H_2O during their racemization[63]. Therefore, [4S-^{2}H]-NADH was produced in the same manner as described above with the racemase and L-leucine dehydrogenase (E.C. 1.4.1.9), which is pro-*S* specific[35].

Amino acid racemase with low substrate specificity of *Ps. putida* ATCC 17 642 does not racemize aromatic and acidic amino acids. However, phenylalanine and phenylglycine undergo α-hydrogen exchange with deuterium from the solvent when incubated with the racemase in ^2H_2O. Lim et al.[64] found that each enantiomer of both α-deuterated phenylalanine and phenylglycine are produced stereospecifically with retention of the C2 configuration. This α-hydrogen exchange reaction is applicable to the production of α-deuterated phenylalanine and phenylglycine.

Makiguchi and coworkers established a method to synthesize L-tryptophan from D,L-serine and indole by means of tryptophan synthase (E.C. 4.2.1.20) from *E. coli* and the amino acid racemase with low substrate specificity of *Ps. striata* (= *Ps. putida*)[65]. Both D,L-serine and indole are cheaply available by chemical synthesis. Tryptophan synthase catalyzes the β-replacement reaction of L-serine with indole to produce L-tryptophan, and the amino acid racemase with low substrate specificity converts unreacted D-serine into L-serine. Because the racemase does not act on tryptophan, almost all D,L-serine is converted into optically pure L-tryptophan. Makiguchi et al.[65] succeeded in producing L-tryptophan in a 200 L reactor using intact cells of *E. coli* and *Ps. putida*[65]. Under the optimal conditions established, 110 g L^{-1} of L-tryptophan was formed in molar yields of 91 and 100 % for added D,L-serine and indole, respectively, after 24 h of incubation with intermittent indole feeding. Continuous production of L-tryptophan was also achieved using immobilized cells of *E. coli* and *Ps. putida*. The maximum concentration of L-tryptophan formed was 5.2 g L^{-1} (99 % molar yield for indole).

S-Adenosyl-L-methionine is the important methyl donor in biological transmethylation to form S-adenosyl-L-homocysteine, which is hydrolyzed to adenosine and homocysteine by S-adenosyl-L-homocysteine hydrolase (E.C. 3.3.1.1) *in vivo*. However, equilibrium of the S-adenosyl-L-homocysteine hydrolase reaction favors the direction toward synthesis of S-adenosyl-L-homocysteine. Shimizu et al. developed a simple and efficient method for the high yield preparation of S-adenosyl-L-homocysteine with S-adenosyl-L-homocysteine hydrolase of *Alcaligenes faecalis*, in which the cellular content of S-adenosyl-L-homocysteine hydrolase was about 2.5 % of the total soluble protein. S-Adenosyl-L-homocysteine was produced at a concentration of about 80 g L^{-1} with a yield of nearly 100 %[66]. However, when racemic

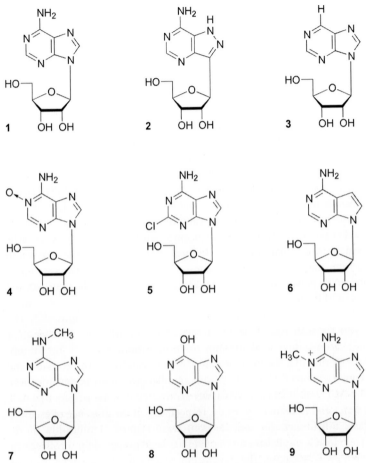

Figure 17-7. Structures of adenosine and related nucleosides which serve as substrates for *S*-adenosyl-L-homocysteine hydrolase. 1, Adenosine; 2, formycin A; 3, neburalin; 4, adenosine N^1-oxide; 5, 2-chloroadenosine; 6, tubercidine; 7, N^6-methyladenosine; 8, inosine; 9, 1-methyladenosine.

homocysteine was used, the D-enantiomer remained unreacted. When *Ps. striata* (= *Ps. putida*) cells were used as the catalyst, D-homocysteine was converted into *S*-adenosyl-L-homocysteine: the amino acid racemase with low substrate specificity acts on homocysteine, but not on *S*-adenosylhomocysteine. *A. faecalis* is better than *Ps. striata* in showing higher *S*-adenosyl-L-homocysteine hydrolase and lower adenosine deaminase activities than those of *Ps. striata*. Therefore, a mixture of both bacterial cells was used to produce 70 g L⁻¹ of *S*-adenosyl-L-homocysteine from D,L-homocysteine and adenosine with a molar yield of nearly 100%[66]. *S*-Adenosyl-L-homocysteine hydrolase acts on various adenosine analogs, and the corresponding *S*-nucleotidyl-L-homocysteines (Fig. 17-7) were synthesized from the analogs and D,L-homocysteine by means of both bacterial cells[67].

17.2.1.3
α-Amino-ε-caprolactam Racemase

α-Amino-ε-caprolactam (ACL) is a chiral heterocyclic compound synthesized from cyclohexene, which is a by-product in the industrial production of nylon. Fuku-mura[68–70] established an enzymatic method to produce L-lysine from D,L-ACL. The process is composed of two enzyme reactions: the selective hydrolysis of L-ACL to L-lysine, and the racemization of ACL (Fig. 17-8). The L-ACL-hydrolyzing enzyme (α-amino-ε-caprolactam hydrolase (E. C. class 3.5.2) is distributed in the cells of *Cryptococcus laurentii* and other yeasts[68–70], and its synthesis is induced by D,L-ACL. The enzyme purified to homogeneity from a cell extract of *C. laurentii* has a molecular weight of about 185 000, and is activated by $MnCl_2$ and $MgCl_2$[71]. L-ACL is the only substrate of the hydrolase: D-ACL and ε-caprolactam are not hydrolyzed.

   ACL racemase has been found in the cells of *Achromobacter obae* and other bacteria[72], and is a unique enzyme among racemases in acting exclusively on cyclic amides derived from α,ω-diamino acids. Ahmed et al.[73] purified the enzyme to homogeneity from the cell extract of *A. obae*, and characterized it. The enzyme is composed of a single polypeptide chain whose molecular weight is about 50 000, and contains 1 mol of PLP per mol of enzyme as a coenzyme. In addition to both isomers of ACL, D- and L-α-amino-δ-valerolactam also serve as effective substrates[74]. The enzyme catalyzes the exchange of the α-hydrogen of the substrate with deuterium or tritium during racemization in deuterium oxide or tritium oxide[28]. By tritium-incorporation experiments, the enzyme was shown to catalyze both inversion and retention of configuration of the substrate with a similar probability in each turnover. When [α-^{2}H]-D-ACL and unlabeled D-ACL were converted into the L-isomer by ACL racemase in water and in deuterium oxide, respectively, in the presence of excess L-ACL hydrolase, α-hydrogen (or α-deuterium) was retained significantly in the product[28]. Therefore, a single base mechanism has been proposed for the racemization catalyzed by ACL racemase. The ACL racemase gene has been cloned from the chromosomal DNA of *A. obae*, and its complete nucleotide sequence determined, which revealed that the enzyme consists of 435 amino acids and that its molecular weight is 45 568[75].

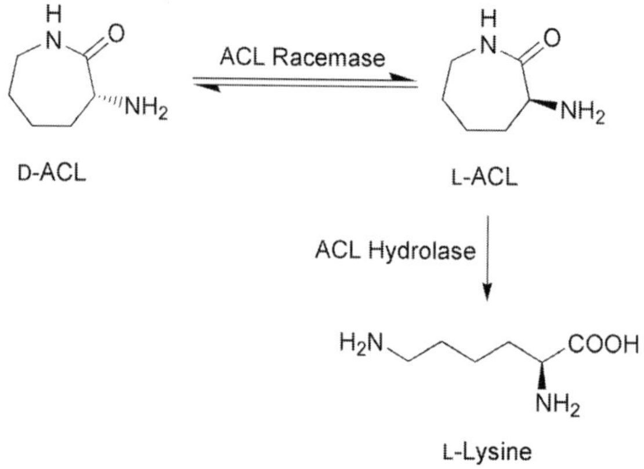

Figure 17-8. Total conversion of racemic ACL into L-lysine by coupling of ACL racemase and ACL hydrolase reactions.

17.2.2
Cofactor-independent Racemases and Epimerases Acting on Amino Acids

17.2.2.1
Glutamate Racemase (E. C. 5.1.1.3)

D-Glutamate as well as D-alanine is an important component of the peptidoglycan of bacterial cell walls[76], and is produced by glutamate racemase[77, 78]. Lactic acid bacteria show high activity of the enzyme[79], and glutamate racemase was first purified from *Pediococcus pentosaceus*[80, 81].

17.2.2.1.1 Gene Cloning
Nakajima et al. cloned the glutamate racemase gene of *Pediococcus pentosaceus*[80]. Glutamate racemase genes have been also cloned from various other sources: *Lactobacillus fermenti*[82, 83], *Lactobacillus brevis*[84], *E. coli*[85], *Bacillus pumilus*[86], *Aquifex pyrophilus*[87], and *Bacillus subtilis*[88, 89].

17.2.2.1.2 Enzymological Properties
The glutamate racemase gene from *Pediococcus pentosaceus* was over-expressed in the recombinant cells, but formed an inclusion body[81]. However, the enzyme was solubilized with 6 M urea, renatured by dialysis to remove urea, and purified to homogeneity with a high overall yield[81]. The amount of enzyme produced by the clone cells corresponded to about 38% of the total insoluble proteins. However, the glutamate racemase gene was solubilized *in vivo* in an active form when it was co-expressed with the gene of chaperonin GroESL[90]. Choi et al. isolated the active enzyme and purified it effectively[81]. The enzyme is composed of a subunit with a molecular mass of about 29 kDa. The enzyme acts specifically on glutamate with K_M values of 14 and 10 mM for D- and L-glutamates, respectively. None of other amino acids occurring in proteins including aspartate, asparagine, and glutamine are racemized. Other glutamate analogs (homocysteate, α-aminoadipate, glutamate γ-methyl ester, N-acetylglutamate, α-hydroxyglutarate, and cysteine sulfinate) are also inert. However, L-homocysteine sulfinate, a γ-sulfinate analogue of glutamate, is racemized at a rate of about 10% of that of L-glutamate. Amino acid racemases generally require PLP as a cofactor, but glutamate racemase is dependent on neither PLP nor on any other cofactor[80, 91]. Proline racemase (E. C. 5.1.1.4)[92], diaminopimelate epimerase (E. C. 5.1.1.7)[93] and hydroxyproline epimerase (E. C. 5.1.1.8)[94] also require no coenzyme.

Glutamate racemase from *E. coli* is unique because it is activated about 100 fold in the presence of UDP-N-acetylmuramoyl-L-alanine (UDP-MurNAc-L-Ala), the precursor of peptidoglycan[95]. UDP-MurNAc-L-Ala is ligated to D-glutamate, a product of the glutamate racemase reaction, by the catalysis of UDP-N-acetylmuramoyl-L-alanyl-D-glutamate synthetase (E. C. 6.3.2.9). Thus, the activation of the *E. coli* glutamate racemase by UDP-MurNAc-L-Ala has a physiological importance in the regulation of peptidoglycan biosynthesis[95]. In contrast, glutamate racemases of Gram-positive bacteria such as *Lactobacillus fermenti*, *Lactobacillus brevis*, *Bacillus pumilus* are not activated by UDP-MurNAc-L-Ala, though these enzymes show about 30% sequence similarities to the *E. coli* enzyme. The predominant difference between the *E. coli* enzyme and the glutamate racemases of the Gram-positive bacteria is that the former has a 21-amino acid extension at the *N*-terminus as compared with the latter enzymes: the *N*-terminal region is responsible for the activation[95].

Glutamate racemase produced in cell extracts of *Bacillus subtilis*, an abundant producer of poly-γ-glutamate, is a monomer with a molecular mass of about 30 kDa containing no cofactor[88]. It almost exclusively catalyzes racemization of glutamate and is mainly concerned in D-glutamate synthesis for poly-γ-glutamate production. *B. subtilis* produces another isozyme of glutamate racemase encoded by the *YrpC* gene[89]. Ashiuchi et al. cloned both enzyme genes and compared their enzymological properties[88, 89]. Enzymological properties of YrpC, such as the substrate specificity and optimum pH, are similar to those of the other glutamate racemase (Glr). The thermostability of YrpC, however, is considerably lower than that of Glr. In addition, YrpC shows higher affinity and lower catalytic efficiency for L-glutamate than Glr[89].

17.2.2.1.3 Structure and Mechanism

Glutamate racemase contains one essential cysteine residue per mol of enzyme, whose chemical modification results in complete inactivation[91]. Choi et al. determined the amount of tritium incorporated into the substrate and product enantiomer during incubation with the enzyme in tritium water, and found that tritium is exclusively incorporated into the product enantiomer regardless of the configuration of the substrate used[91]. This is compatible with a model in which two different bases participate in abstraction and return of α-hydrogen of the substrate. One of the two bases involved in catalysis is suggested to be the essential cysteine residue: a thiolate from one of the cysteines abstracts the α-proton, and the other cysteine thiol delivers a proton to the opposite face of the resulting carbanionic intermediate[91].

Kim et al.[87] cloned the glutamate racemase gene from *Aquifex pyrophilus*, a hyperthermophilic bacterium, and expressed it in *E. coli*. The enzyme shows strong thermostability in the presence of phosphate ion, and it retains more than half of its original activity after incubation at 85 °C for 90 min. Hwang et al.[96] crystallized the glutamate racemase of *A. pyrophilus* and determined the tertiary structure of the enzyme by X-ray crystallography. The enzyme is composed of two identical subunits, and each monomer consists of two α/β fold domains. Hwang et al. has also proposed a mechanism in which two cysteine residues are involved in the catalysis (Fig. 17-9)[96].

Glavas and Tanner replaced the two cysteine residues, Cys 73 and Cys 184, by serine, and analyzed the reactions catalyzed by the mutant enzymes: the elimination of water from a substrate analog, *N*-hydroxyglutamate, through a one-base requiring reaction[97]. The C73S mutant was a much poorer catalyst than the wild-type enzyme

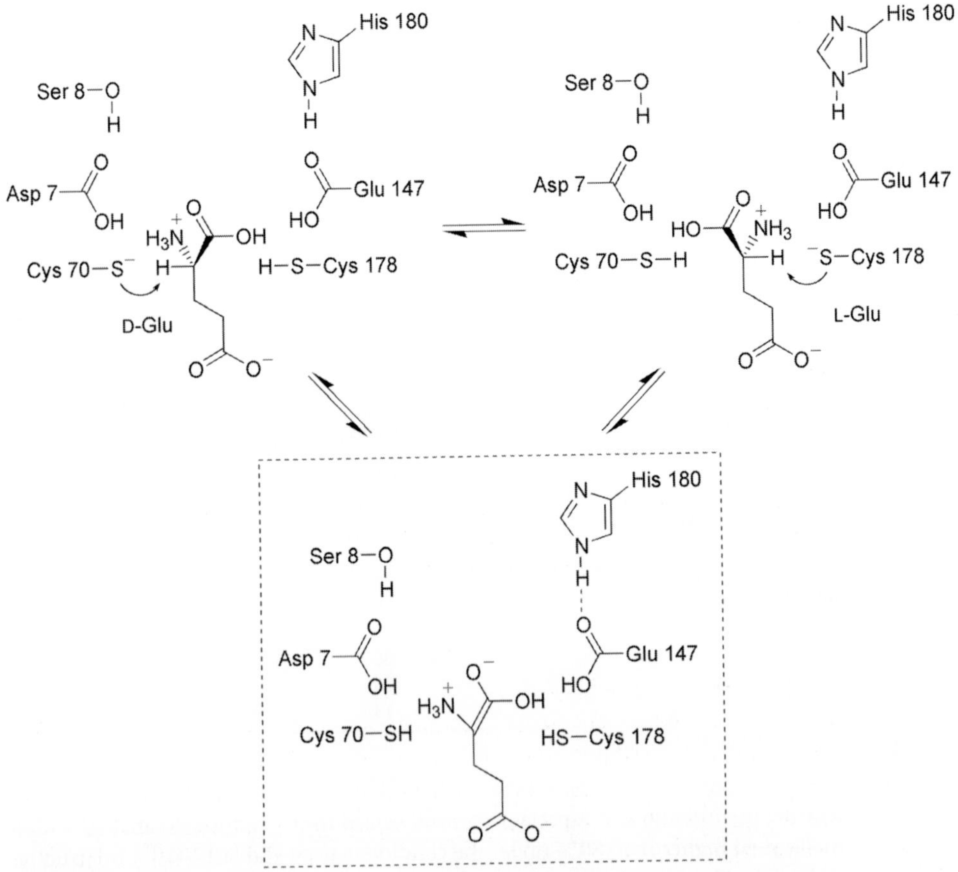

Figure 17-9. Mechanism of glutamate racemase reaction. Cys 70 and Cys178 serve as the bases to abstract an α-proton from the substrate, and a carbanion intermediate is formed. Alternatively, the racemization may proceed through a concerted mechanism. Reprinted from Hwang et al.[95].

toward D-N-hydroxyglutamate, whereas the C184S mutant was better than the wild-type. When L-N-hydroxyglutamate was used as a substrate, C73S was better but C184S was poorer than the wild-type. Thus, Glavas and Tanner concluded that Cys73 is responsible for the deprotonation of D-glutamate and Cys 184 is responsible for the deprotonation of L-glutamate[97].

17.2.2.1.4 Synthesis of D-Amino Acids with Glutamate Racemase

Nakajima et al.[98] have developed an efficient method for the synthesis of various D-amino acids from the corresponding α-keto acids and ammonia by coupling of four enzyme reactions catalyzed by D-amino acid aminotransferase[99], glutamate race-mase[79, 91], glutamate dehydrogenase and formate dehydrogenase (Fig. 17-10). Various D-amino acids are produced by this method. Under the optimum conditions established by Nakajima et al.[98], D-enantiomers of valine, alanine, α-aminobutyrate,

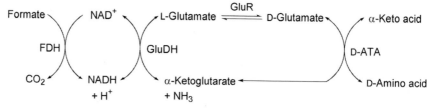

Figure 17-10. Enzymatic synthesis of D-amino acids by combination of glutamate racemase, glutamate dehydrogenase, D-amino acid aminotransferase and formate dehydrogenase reactions.

Table 17-4. Production of various D-amino acids by means of four purified enzymes: glutamate racemase, D-amino acid aminotransferase, glutamate dehydrogenase, and formate dehydrogenase[a].

D-Amino acids	Molar yield (%)
D-Valine	100
D-Alanine	100
D-α-Aminobutyrate	100
D-Aspartate	100
D-Leucine	84
D-Methionine	80
D-Serine	50
D-Histidine	36
D-Phenylalanine	28
D-Tyrosine	13

a Reprinted from N. Nakajima et al. [198].

leucine, methionine and aspartate are synthesized from their α-keto analogs with a molar yield higher than 80% under the conditions used (Table 17-4) [98]. D-Histidine and a few other D-amino acids, which are poor substrates of D-amino acid amino-transferase[99], are produced in a yield lower than 40% under the same conditions. However, Bae et al.[100] established an efficient method for production of D-phenylalanine and D-tyrosine by feeding α-keto acid intermittently in order to keep its concentration at less than 50 mM, above which the productivity decreased greatly (Fig. 17-11). By running the multi-enzyme system for 35 h, 48 g L^{-1} of D-phenyl-alanine and 60 g L^{-1} of D-tyrosine were produced with 100% of optical purity from the equimolar amounts of phenylpyruvate and hydroxyphenylpyruvate, respectively. An enzyme-membrane reactor system containing polyethyleneglycol-NAD$^+$ devel-oped by Wandrey and associates[101] is probably applicable to this system. The production level of D-amino acids are mainly dependent on the stability of glutamate racemase. Therefore, thermostable glutamate racemases produced by *A. pyrophi-lus*[87] and *B. subtilis*[88] are probably useful as catalyst of this multi-enzyme system.

Yagasaki et al.[102] developed a new method for the synthesis of D-glutamate from L-glutamate by means of *E. coli* recombinant cells harboring a plasmid containing glutamate racemase gene from *L. brevis* ATCC 8287. L-Glutamate was first racemized to D,L-glutamate at pH 8.5, and L-glutamate was then decarboxylated at pH 4.2 by glutamate decarboxylase, which was inherently produced by the *E. coli* host cells.

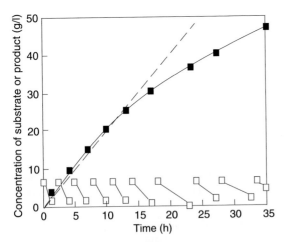

Figure 17-11. Production of D-phenylalanine by successive feeding of phenylpyruvate. Phenylpyruvate (□) was added intermittently. The dotted line indicates the expected productivity of D-phenylalanine (■) on the basis of the initial production rate. Reprinted from Bae et al.[100].

Starting from 100 g L^{-1} of L-glutamate, they obtained 50 g L^{-1} of D-glutamate in a 15 h reaction. D-Glutamate can be produced successively from L-glutamate with *L. brevis* ATCC 8287 cells because this strain produces both glutamate racemase and glutamate decarboxylase simultaneously. Thus, 50 g L^{-1} of optically pure D-glutamate was produced from 100 g L^{-1} of L-glutamate[103]. Oikawa et al.[104] replaced glutamate decarboxylase by glutamate oxidase because the oxidase has optimum pH values similar to that of glutamate racemase. They developed a bioreactor consisting of two columns sequentially connected and containing immobilized glutamate racemase from *B. subtilis* and L-glutamate oxidase from *Streptomyces* sp. X119–6: L-glutamate was racemized by the glutamate racemase column, and then L-glutamate was oxidized by the L-glutamate oxidase column. D-Glutamate was produced in about 90% of the theoretical yield[104].

17.2.2.2
Aspartate Racemase (E. C. 5.1.1.13)

D-Aspartate occurs in the peptidoglycan layer of bacterial cell walls, and is produced from L-aspartate through an aspartate racemase (E. C. 5.1.1.13) reaction[105]. The enzyme has been demonstrated as being present in various *Lactobacillus* and *Streptococcus* strains[106] such as *Lactobacillus fermenti*[105] and *Streptococcus faecalis*[107]. Recently, archaea such as *Desulfurococcus* strain SY[108] and *Thermococcus* strains[109] were shown to produce aspartate racemase. It is interesting to note that various other archaea such as *Pyrobaculum islandicum*, *Methanosarcina barkeri* and *Halobacterium salinarium* produce D-amino acids, although their function is not yet known[110].

Okada et al. purified the enzyme to homogeneity from the cell extract of *S. thermophilus*, the specific activity of the crude extract of which was elevated 3400-fold[106]. The gene encoding aspartate racemase was cloned from *S. thermophilus*, and overexpressed in *E. coli*[111]. The amount of the enzyme produced reached about 20% of the total soluble proteins of the *E. coli* clone cells. Thus, the enzyme was efficiently purified to homogeneity from the clone cells[111]. The enzyme is a homodimer of a subunit with a molecular weight of about 28000. In addition to aspartate, cysteate and cysteine sulfinate are the only substrates of the enzyme: they are racemized at a rate of 88 and 51%, respectively, of that of L-aspartate[112]. The presence of the acidic group at the β-carbon is essential; none of asparagine, cysteine, serine, and alanine are the substrates. Both isomers of glutamate are also inert. The K_M values for L- and D-aspartate are 35 and 8.7 mM, respectively.

Aspartate racemase requires no cofactors and contains an essential cysteine residue in the same manner as glutamate racemase[80]. When L- or D-aspartate was incubated with aspartate racemase in tritiated water, tritium was incorporated preferentially into the product enantiomer. This is consistent with the results of glutamate racemase as described above[91].

Yamauchi et al.[112] concluded that aspartate racemase also uses two bases to remove and return the α-proton of the substrate. Aspartate racemase contains three cysteine residues: Cys 84, Cys 190 and Cys 197, and only Cys 84 is essential for the enzyme activity. The alkylation of one cysteine residue/dimer with 2-nitro-5-thiocyanobenzoic acid results in a complete loss of activity. Therefore, the enzyme shows a half-of-the-sites-reactivity[112]. Yamauchi et al.[112] suggested that the enzyme has a composite active site formed at the interface of two identical subunits in the same manner as proposed for proline racemase[92].

Kumagai and coworkers[113] developed an enzymatic procedure to produce D-alanine from fumarate by means of aspartase (E.C. 4.3.1.1), aspartate racemase, and D-amino acid aminotransferase (Fig. 17-12). Aspartase catalyzes conversion of fumarate into L-aspartate, which is racemized to form D-aspartate. D-Amino acid aminotransferase catalyzes transamination between D-aspartate and pyruvate to produce D-alanine and oxalacetate. This 2-oxo acid is easily decarboxylated spontaneously to form pyruvate in the presence of metals. Thus, the transamination proceeds exclusively toward the direction of D-alanine synthesis, and total conversion of fumarate into D-alanine was achieved.

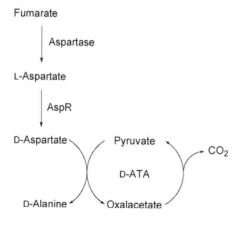

Figure 17-12. Enzymatic production of D-alanine by combination of aspartase, aspartate racemase, and D-amino acid aminotransferase reactions.

17.2.2.3
Diaminopimelate Epimerase (E. C. 5.1.1.7)

meso-α,ε-Diaminopimelate is the direct precursor of L-lysine, and is an essential component of the cell wall peptidoglycans in Gram negative bacteria. *meso*-α,ε-Diaminopimelate is formed from L-α,ε-diaminopimelate by diaminopimelate epimerase. The enzyme gene (*dap*F) was mapped at 85 min on the *E. coli* chromosome[114]. Richaud et al. isolated an *E. coli* mutant lacking diaminopimelate epimerase activity by insertional mutagenesis, and showed that the mutant does not require *meso*-α,ε-diaminopimelate in a minimal medium[114]. Thus, *meso*-α,ε-diaminopimelate epimerase encoded by the *dap*F gene is not essential for *E. coli*, but *meso*-α,ε-diaminopimelate still occurs in the mutant cells. Richaud et al. proposed that *E. coli* has another enzyme with diaminopimelate epimerase activity[114].

The diaminopimelate epimerase gene (*dap*F) was cloned from *E. coli*[114], and the amino acid sequence of the enzyme was deduced from the nucleotide sequence[115]. The enzyme was purified to homogeneity from the wild-type *E. coli* cells[93] and the recombinant *E. coli* cells carrying a plasmid coding for *dap*F gene[116]. The enzyme is composed of two identical subunits with a molecular weight of about 32 000. The enzyme is independent of PLP or of any other cofactors. The enzyme shows a V_{max} value of 132 µmol min^{-1} per mg of protein, and a K_M value of 0.24 mM for L-α,ε-diaminopimelate. The thiol group of Cys 73 of the enzyme is specifically labeled by a mechanism-based inactivator, 2-(4-amino-4-carboxybutyl)-2-aziridine carboxylic acid. Higgins et al. discovered an interesting similarity in amino acid sequences around the catalytically essential cysteine residue of proline racemase[92], hydroxyproline epimerase[94], and diaminopimelate epimerase (Cys 73), and proposed that PLP-independent racemases/epimerases derive from a common evolutionary origin[116]. However, no significant similarity in the entire amino acid sequence was found between diaminopimelate epimerase and glutamate racemase, and also between diaminopimelate epimerase and aspartate racemase.

Cirilli et al.[117] cloned the gene of diaminopimelate epimerase from *Haemophilus influenzae*, and purified and crystallized the enzyme. The enzyme is monomeric and has a unique protein fold, in which the amino terminal and carboxyl terminal halves of the molecule fold into structurally homologous and superimposable domains (Fig. 17-13). Cys 73 of the amino terminal domain is found in the disulfide linkage, at the domain interface, with Cys 217 of the carboxy terminal domain[117]. Thus, it is most conceivable that these two cysteine residues stay in reduced form in the active enzyme and function as the acid and base in the mechanism. Koo and Blanchard[118] explored a number of kinetic and isotope approaches to clarify the mechanism of the enzyme. However, which of the two cysteine residues is responsible for proton abstraction from the two enantiomeric Cα-H bonds is not yet known.

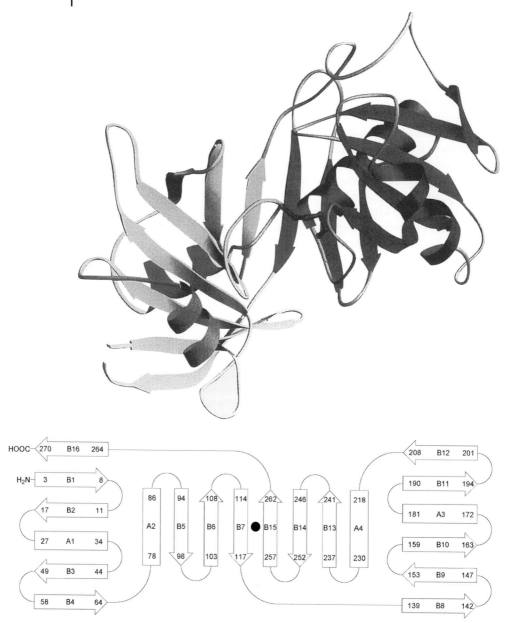

Figure 17-13. Top: Ribbon diagram of diamino-pimelate epimerase from *Haemophilus influenzae*. The disulfide bridge between Cys 73 and Cys 217 connects domain I (residues 1–117 and 263–274) and domain II (residues 118–262). Bottom: Topology of the secondary structural elements of diaminopimelate epimerase. The position of pseudo-2-fold symmetry axis is indicated by the black dot between β-strands B7 and B8. Reprinted from M. Cirilli et al.[117].

17.2.2.4
Proline Racemase (E. C. 5.1.1.4)

Proline racemase occurs in *Clostridium sticklandii*, which produces δ-aminovalerate from L-proline. Proline racemase and D-proline reductase are responsible for the conversion: L-proline is racemized by proline racemase to form D-proline, which is converted into δ-aminovalerate by D-proline reductase (E. C. 1.4.4.1).

Rudnick and Abeles purified proline racemase to 95 % homogeneity from *Clostridium sticklandii*, and characterized it[92]. The enzyme is composed of two identical subunits with a molecular weight of about 38 000, and is independent of any cofactors or metals. Most amino acid racemases require pyridoxal 5'-phosphate, which labilizes the bond between the α-hydrogen and the chiral center by aldimine formation with the α-amino group of the substrate. However, PLP is not involved in the reaction of proline racemase acting on an α-imino acid. The enzyme also acts on 2-hydroxy-L-proline and 2-allo-hydroxy-D-proline although slowly: they are epimerized at a rate of 2 and 5 % of the rate of L-proline racemization, respectively. L-Proline and D-proline showed K_M values of 2.9 and 2.5 mM, respectively[119].

Pyrrole-2-carboxylate is a competitive inhibitor of proline racemase, and stoichiometrically binds with the enzyme (1 mol per dimer). Thiol groups of the enzyme are alkylated by iodoacetate at a stoichiometry of 1 mol of cysteine residue per mol subunit. However, the enzyme is inactivated completely by modification of only one cysteine residue per dimer. Thus, Rudnick and Abeles proposed a reaction scheme in which the active site is located at the interface of two identical subunits, each of which furnishes one of the two active site thiol groups positioning appropriately at the composit active site: a thiolate anion derived from one thiol group abstracts the α-proton from the substrate, and another thiol group protonates the intermediate derived from the substrate from the opposite face[92]. They proposed occurrence of two forms of free proline racemase: one binds with D-proline and the other binds with L-proline. According to their proposed mechanism, the product enantiomer is released much faster than the release of the substrate-derived proton. The proton release also proceeds much faster than the interconversion of the two forms of the enzyme. Knowles and coworkers defined the energetics and delineated the complete free energy profile for the proline racemase reaction[119–125].

Yagasaki and Ozaki[126] developed a method for production of D-proline from L-proline using the recombinant proline racemase of *Clostridium sticklandii*. L-Proline was degraded by *Candida* sp. PRD-234, and optically pure D-proline was obtained.

17.2.3
Other Racemases and Epimerases Acting on Amino Acid Derivatives

17.2.3.1
2-Amino-Δ2-thiazoline-4-carboxylate Racemase

Sano et al.[127] have found several bacterial strains that are capable of producing L-cysteine from D,L-2-amino-2-thiazoline-4-carboxylate (ATC), an intermediate in the

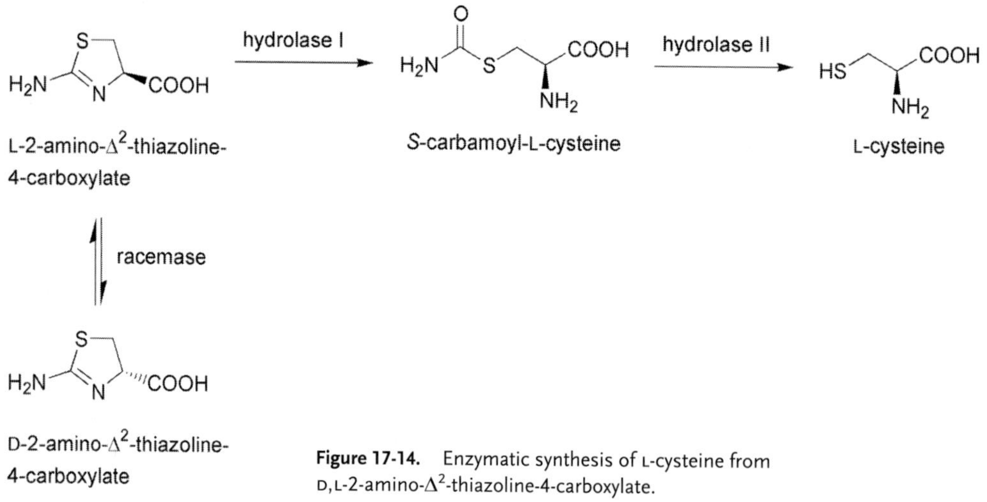

L-2-amino-Δ²-thiazoline-
4-carboxylate

S-carbamoyl-L-cysteine

L-cysteine

racemase

D-2-amino-Δ²-thiazoline-
4-carboxylate

Figure 17-14. Enzymatic synthesis of L-cysteine from
D,L-2-amino-Δ²-thiazoline-4-carboxylate.

chemical synthesis of D,L-cysteine. These include several *Pseudomonas* species isolated from soil and other strains belonging to different genera such as *E. coli*, *Bacillus brevis*, and *Micrococcus sodenensis*[127]. Three enzymes are probably involved in this pathway: L-ATC hydrolase, S-carbamoyl-L-cysteine hydrolase and ATC racemase (Fig. 17-14). *Pseudomonas thiazolinophilum* isolated from soil was shown to have the highest activity of the enzymes that produce L-cysteine from D,L-ATC. The enzymes are inducibly formed in the bacterial cells by addition of D,L-ATC to the growth medium.

Degradation of L-cysteine by cysteine desulfhydrase or other PLP enzymes present in the cells was successfully prevented by addition of hydroxylamine or semicarbazide to the incubation mixture. A mutant strain of *Ps. thiazolinophilum* lacking cysteine desulfhydrase was isolated and used to produce L-cysteine from D,L-ATC in a molar yield of 95% and at a product concentration of 31.4 g L^{-1}[128]. *Pseudomonas desmolytica* AJ 3872, one of the L-cysteine producers isolated was found to lack the ability to convert D-ATC into L-cysteine: it is an ATC racemase-deficient strain[129]. However, little is known about the enzymological properties and function of the racemase.

Among the three enzymes participating in L-cysteine production, L-ATC hydrolase was found to be the least stable[130]. However, the stability of L-ATC hydrolase was sharply enhanced as water activity decreased from 0.93 to 0.80. In the absence of sorbitol, the stability of L-ATC hydrolase increased in proportion to ionic strength. Thus, Ryu et al. succeeded in enhancing the half life of L-ATC hydrolase by 10-fold to 20-fold in sorbitol-salt mixtures[130].

17.2.3.2
Hydantoin Racemase

5-Substituted hydantoin derivatives have been used as precursors for D- and L-amino acids in chemical synthesis. However, they are hydrolyzed enantioselectively by the enzymes named hydantoinases: some act specifically on D-5-substituted hydantoins, and others on the L-isomers. N-Carbamoyl amino acids formed are also hydrolyzed enantiospecifically by N-carbamoyl amino acid amidohydrolases to produce D- or L-amino acids (Fig. 17-15). Since the Kanegafuchi Chemical Industry, Japan, commercialized an enzymatic procedure for the production of D-p-hydroxyphenylglycine, which is a building block for the semisynthetic β-lactam antibiotic amoxycillin, various processes for amino acid production by means of hydantoinases have been developed[131-133].

Subsequent to the discovery that hydantoin is hydrolyzed by extracts of mammalian livers[134] and plant seeds[135], various microorganisms have been shown to utilize D- and L-5-substituted hydantoins as a sole carbon or nitrogen source by means of D- as well as L-specific hydantoinases inducibly formed[131-133].

Distribution of D-hydantoinase in microorganisms has been shown by Yamada and coworkers[136]. The enzyme is identical to dihydropyrimidinase (E.C. 3.5.2.2), and is widely distributed in bacteria, in particular in *Klebsiella*, *Corynebacterium*, *Agrobacterium*, *Pseudomonas*, and *Bacillus*, and also in actinomycetes such as *Streptomyces* and *Actinoplanes*. The enzyme activity occurs also in eukaryotes: yeasts, molds, plants and mammals. *Pseudomonas putida* was found to be the best strain, which produced D-hydantoinase most abundantly and inducibly by addition of 5-methylhydantoin. Most of D-hydantoinase producers form N-carbamoyl D-amino acids from the corresponding 5-substituted hydantoins. Accordingly, to obtain free D-amino acids, N-carbamoyl amino acids need to be isolated and hydrolyzed chemically or enzymatically. However, a few bacterial strains produce N-carbamoyl D-amino acid amidohydrolase in addition to D-hydantoinase. Thus, optically pure D-amino acids were produced from D-hydantoins with these bacterial cells. Olivieri et al.[137] found that *Agrobacterium tumefaciens* cells grown on uracil as a sole nitrogen source catalyze the complete conversion of racemic hydantoins into D-amino acids. Hartley et al.[138] obtained a mutant strain which expresses both the hydantoinase and N-carbamoylamino acid amidohydrolase in the absence of an inducer. In contrast, other bacterial strains belonging to the genera of *Flavobacterium*[139], Arthro-

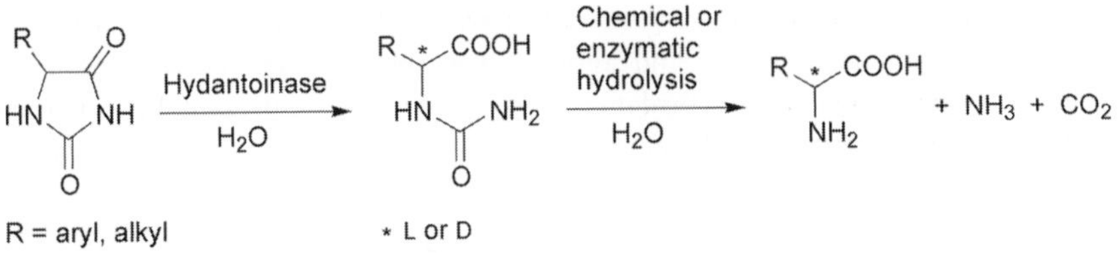

Figure 17-15. Enzymatic synthesis of D- or L-amino acids from 5-substituted D,L-hydantoins through N-carbamoyl-D- or L-amino acids.

bacter[140], *Pseudomonas*[141, 142], and *Bacillus*[143–145] convert whole racemic 5-substituted hydantoins into the corresponding L-amino acids. In these bacteria, 5-substituted hydantoins are hydrolyzed by L-hydantoinase to form *N*-carbamoyl L-amino acids, which are hydrolyzed further to L-amino acids by *N*-carbamoyl L-amino acid amidohydrolase in the same manner as described above except that the enzymes involved show opposite stereospecificity. 5-Mono-substituted hydantoins can racemize spontaneously under weakly alkaline conditions, and this chemical racemization participates at least partly in the total conversion of the racemic hydantoins into free L- or D-amino acids. However, if chemical racemization proceeds only slowly[146], a hydantoin racemase was suggested to occur and participate in the total conversion[146, 147].

Watabe et al.[148] isolated a plasmid which is responsible for the conversion of 5-substituted hydantoins into the corresponding L-amino acids from a soil bacterium, *Pseudomonas* sp. NS 671, which is able to convert racemic 5-substituted hydantoins into the corresponding L-amino acids. The genes involved in the conversion were cloned from the *Pseudomonas* plasmid into *E. coli*, and functions of four genes were identified and named *hyu*A, *hyu*B, *hyu*C and *hyu*E. Both *hyu*A and *hyu*B are required for the conversion of D- and L-5-substituted hydantoins into the corresponding *N*-carbamoyl-D- and *N*-carbamoyl-L-amino acids, respectively, although the individual reactions catalyzed by the gene products have not yet been identified. *Hyu*C codes for an *N*-carbamoyl-L-amino acid amidohydrolase, while *hyu*E is a hydantoin racemase gene[149]. Significant nucleotide sequence similarity was found between *hyu*A and *hyu*C (43%), and also between *hyu*B and *hyu*C (46%). Watabe et al. suggested that these genes have evolved from a common ancestor by gene duplication[148]. However, no proteins registered in NBRF and SWISS protein data bases showed similarity with the deduced amino acid sequences of the four genes.

Wagner and associates purified hydantoin racemase from *Arthrobacter aurescens* DSM 3747 and characterized it[133]. Watabe et al.[149] also purified the enzyme from *E. coli* clone cells harboring a plasmid coding for the enzyme gene derived from *Pseudomonas* sp. NS 671. The *Pseudomonas* enzyme is a hexamer composed of a subunit with a molecular weight of about 32 000, which is consistent with the value deduced from the amino acid sequence. The D- and L-isomers of 5-(2-methylthioethyl)hydantoin and 5-isobutyrylhydantoin are racemized effectively. D-5-(2-Methylthioethyl)hydantoin is racemized at a V_{max} value (79 µmol min^{-1} mg^{-1}) which is about 2.5 times higher than that for the L-isomer. Wiese et al.[150] cloned the hydantoin racemase gene from *Arthrobacter aurescens* DSM 3747 and purified the enzyme to homogeneity. The *Arthrobacter* enzyme has a molecular mass of 25.1 kDa[150] and acts on aromatic and aliphatic hydantoin derivatives such as 5-indolylmethylhydantoin, 5-benzylhydantoin, 5-(*p*-hydroxybenzyl)hydantoin, 5-(2-methylthioethyl)hydantoin, and 5-isobutylhydantoin[133], although hydantoins with arylalkyl side chains are preferred substrates[159]. Free amino acids, amino acid esters and amides are inert, but the enzyme suffers from inhibition by aliphatic substrates such as L-5-methylthioethylhydantoin. The hydrogen at the chiral center of a substrate, D-5-indolylmethylenehydantoin, is exchanged with solvent deuterium during racemization[150]. Pietzsch et al.[152] established a method for the synthesis of optically pure D-3-trimethylsilylalanine from D,L-5-trimethylsilylmethylhydantoin in 88% yield and 95% enantiomeric excess with whole resting cells of *Agrobacterium* sp. IP I 671, immobilized in a Ca-alginate matrix. On the other hand, L-3-trimethylsilylalanine was also prepared from the racemic substrate by enantiomer-specific hydrolysis of the L-form in the presence of L-N-carbamoylase from *Arthrobacter aurescens* DSM 3747[152].

Watabe et al. found that the *Pseudomonas* enzyme is inactivated by a substrate, L-5-methylhydantoin, during racemization[151]. However, the enzyme was not affected by the D-isomer. Both enantiomers of 5-isopropylhydantoin inactivated the enzyme to the same extent. Interestingly, divalent sulfur-containing compounds such as methionine, cysteine, glutathione, and biotin protected the enzyme effectively from inactivation. *E. coli* cells expressing the racemase are capable of racemizing all of these hydantoin derivatives: the enzyme is protected from inactivation by divalent sulfur compounds occurring in the cells. Watabe et al. concluded that the protective effect by the divalent sulfur-compounds is not due to their reducing activity[151]. Both *Pseudomonas*[151] and *Arthrobacter*[133] enzymes are inhibited strongly by Cu^{2+}. The *Arthrobacter* enzyme is completely inhibited by $HgCl_2$ and iodoacetamide, and stimulated by addition of dithiothreitol[150]. Therefore, the enzyme may contain essential cysteine residues, which are possibly modified by some activated intermediate derived from the particular substrates leading to the enzyme inactivation.

E. coli cells carrying a plasmid coding for *hyuA*, *hyuB*, *hyuC*, and *hyuE* convert only D-5-(2-methylthioethyl)hydantoin into L-methionine. On the other hand, *E. coli* cells harboring a plasmid coding for only *hyuA*, *hyuB*, and *hyuC* first convert the L-hydantoin, then the D-isomer is hydrolyzed slowly when the L-isomer is depleted. Therefore, Watabe et al. believe that D-5-(2-methylthioethyl)hydantoin is only converted into L-methionine in the presence of the hydantoin racemase[151]. The mechanism of stereospecific conversion of D,L-5-substituted hydantoins to the corresponding L-amino acids by *Pseudomonas* sp. strain NS 671 has been clarified by Ishikawa et al.[153]: D,L-5-substituted hydantoins are converted exclusively into the L-forms of the corresponding N-carbamoylamino acids by the hydantoinase in combination with hydantoin racemase, and then the N-carbamoyl-L-amino acids are converted into L-amino acids by N-carbamoyl-L-amino acid amidohydrolase (Fig. 17-16).

By directed evolution May et al.[154] succeeded in inverting the enantioselectivity of D-hydantoinase from *Arthrobacter* sp. DSM 9771 into an L-selective enzyme. The improved hydantoinase also acquired a five-fold increase in activity. The recombinant *E. coli* cells expressing three heterologous genes (i.e. the evolved L-hydantoinase, L-N-carbamoylase, and hydantoin racemase) were found to produce 91 mM L-methionine from 100 mM D-5-(2-methylthioethyl)hydantoin in less than 2 h[154].

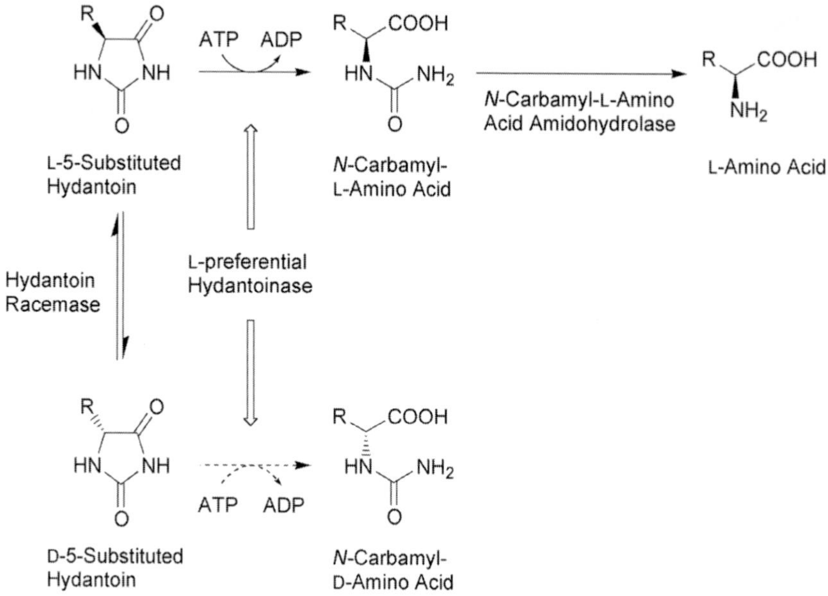

Figure 17-16. Stereospecific conversion of D,L-5-substituted hydantoins into the corresponding L-amino acids by *Pseudomonas* sp. NS 671. Reprinted from Ishikawa et al. [153].

17.2.3.3
N-Acylamino Acid Racemase

L-Aminoacylases (E. C. 3.5.1.14) catalyze the hydrolysis of the amide bond of various *N*-acyl-L-amino acids, such as *N*-acetyl-, *N*-chloroacetyl- and *N*-propionyl-L-amino acids [155], and is widely distributed in animals [155–157], plants [158, 159], and micro-organisms [160, 161]. Greenstein [155] first studied the reactivity of pig kidney enzyme, and showed its application to the optical resolution of racemic amino acids. Chibata et al. [162] found that L-aminoacylase is produced abundantly by fungal species belonging to the genera *Aspergillus* and *Penicillium*. L-Aminoacylases were purified from pig kidney and *A. oryzae*, and their reaction mechanism and physiological function were studied [160, 163–165]. Cho et al. [166] showed that various thermophilic *Bacillus* strains produce thermostable L-aminoacylase, and purified it to homogeneity from *Bacillus thermoglucosidius* DSM 2542, which produces the enzyme most abundantly. L-Aminoacylases of pig kidney, *Aspergillus oryzae* and *B. thermoglucosidius* share many features with each other: they contain Zn^{2+} as a prosthetic metal, are strongly activated by Co^{2+}, and have a pH optimum in the range of 8.0–8.5.

Sugie and Suzuki [167] demonstrated the occurrence of D-aminoacylase, which specifically hydrolyzes the amide bond of *N*-acyl-D-amino acids, in actinomycetes, and applied the enzyme to the production of D-phenylglycine. Recently, a new D-aminopeptidase was found in *Alcaligenes denitrificans*, and shown to act on various *N*-acyl-D-amino acids including *N*-acetyl-D-methionine [168, 169].

N-Acylamino acids are usually racemized much more readily than the corresponding free amino acids. Therefore, by combination of chemical racemization and enantioselective hydrolysis of *N*-acylamino acids, racemates of *N*-acylamino acids can be fully converted into the desired enantiomer of the free amino acids according to the stereospecificity of the aminoacylases used. For example, L-tryptophan is produced industrially by combination of chemical racemization of *N*-acetyltryptophan and enantiospecific hydrolysis of its L-isomer with the *Aspergillus* L-aminoacylase, which shows high reactivity towards *N*-acyl derivatives of aromatic L-amino acids. When *N*-acetyl-D,L-tryptophan is incubated with the fungal enzyme, *N*-acetyl-L-tryptophan is selectively hydrolyzed to L-tryptophan, which is then crystallized from the solution. *N*-Acetyl-D-tryptophan in the mother liquor is racemized with acetic anhydride, and the racemate is again used as a starting material. In principle, D- and L-amino acids can be produced from their corresponding *N*-acyl derivatives in the same manner, provided that *N*-acyl derivatives of the desired amino acids serve as the substrates of the available aminoacylases, and are racemized chemically without any major loss by decomposition. However, the chemical racemization can be achieved only under extreme conditions in order for the aminoacylases to be inactivated, and the enzymes are usually required to be saved for the subsequent cycles for reasons of economy. Therefore, the antipode of the substrate is separated from the enzyme and preferably from the product in order to avoid its possible racemization. Tosa et al. have developed a continuous method to produce L-tryptophan, which is now utilized in industry, by means of the *Aspergillus* L-aminoacylase immobilized on DEAE-Sephadex[170].

Takahashi and Hatano of Takeda Chemical Industries, Japan, succeeded in finding a racemase that acts on *N*-acylamino acids, but not on the corresponding free amino acids, and named it acylamino acid racemase[171]. They have established a method of producing optically active α-amino acids from the corresponding D,L-*N*-acylamino acids by means of the acylamino acid racemase and aminoacylases.

Acylamino acid racemase occurs widely in various actinomycete strains belonging to the genera of *Streptomyces*, *Actinomadura*, *Actinomyces*, *Jensenia*, and *Amycolatopsis*[172]. The enzyme was purified to homogeneity from *Streptomyces atratus* Y-53, which shows the highest enzyme activity among the strains tested[173]. The enzyme is composed of 6 subunits with identical molecular masses (about 41 000), and shows a molecular mass of 244 000 in the native state. Tokuyama and Hatano[174] purified thermostable *N*-acylamino acid racemase from *Amycolatopsis* sp. TS-1-60 and purified it to homogeneity. The molecular masses of the native enzyme and the subunit are 300 000 and 40 000, respectively. The enzyme is stable at 55 °C for 30 min. The enzyme catalyzes the racemization of *N*-acylamino acids such as *N*-acetyl-L- or D-methionine, *N*-acetyl-L-valine, *N*-acetyl-L-tyrosine and *N*-chloroacetyl-L-valine (Table 17-5). In addition, the enzyme also catalyzes racemization of dipeptide L-alanyl-L-methionine. By contrast, *N*-alkylamino acids and methyl and ethyl esters of *N*-acetyl-D- and L-methionine are not racemized. The apparent K_M values for *N*-acetyl-L-methionine and *N*-acetyl-D-methionine are 18.5 mM and 11.3 mM, respectively. The enzyme activity is markedly enhanced by the addition of divalent metal ions such as Co^{2+}, Mn^{2+} and Fe^{2+} and inhibited by addition of EDTA and *p*-

Table 17-5. Substrate specificity of acylamino acid racemase[a].

Substrate	Relative activity
N-Acetyl-D-methionine	100
N-Acetyl-L-methionine	100
N-Formyl-D-methionine	40
N-Formyl-L-methionine	63
N-Acetyl-D-alanine	33
N-Acetyl-L-alanine	21
N-Benzoyl-D-alanine	14
N-Acetyl-D-leucine	37
N-Acetyl-L-leucine	74
N-Acetyl-D-phenylalanine	64
N-Acetyl-L-phenylalanine	84
N-Chloroacetyl-D-phenylalanine	90
N-Chloroacetyl-L-phenylalanine	112
N-Acetyl-D-tryptophan	10
N-Acetyl-L-tryptophan	8
N-Acetyl-D-valine	35
N-Acetyl-L-valine	19
N-Chloroacetyl-D-valine	80
N-Chloroacetyl-L-valine	105
N-Acetyl-D-allo isoleucine	33

a Inert: D- and L-methionine, D- and L-alanine, D- and L-leucine, D-and L-phenylalanine,
D- and L-tryptophan, D- and L-valine.

chloromercuribenzoate. The gene of N-acylamino acid racemase was cloned from *Amycolatopsis* sp. TS-1–60[175], and overexpressed in *E. coli* host cells with T7 promoter[176]. The gene codes for a protein of 368 amino acids with a molecular mass of 39411 Da. Palmer et al.[177] found that N-acylamino acid racemase of *Amycolaptosis* sp. TS-1–60 is similar to an unidentified protein encoded by the *Bacillus subtilis* genome. N-Acylamino acid racemase efficiently catalyzes an *O*-succinylbenzoate synthase reaction, which is responsible for menaquinone biosynthesis.

Tokuyama et al.[172] found that most of acylamino acid racemase-producing strains produce not only acylamino acid racemase but also aminoacylases; one of either D- or L-aminoacylase or both of them. Moreover, acylamino acid racemase shows the optimum pH at around 8.0, which is close to that of aminoacylases. Therefore, N-acylamino acid can be converted as a whole into L- or D-amino acids in one step by means of microbial cells of appropriate strains producing either L- or D-aminoacylase in addition to acylamino acid racemase.

17.2.3.4
Isopenicillin N Epimerase

Isopenicillin N is a precursor of penicillin, and synthesized from δ-(L-aminoadipoyl)-L-cysteinyl-D-valine by isopenicillin N synthetase[178]. Isopenicillin N is then converted into penicillin N by isopenicillin N epimerase. Penicillin N is ring-expanded to deacetoxycepharosporin C by penicillin N expandase. The latter compound is

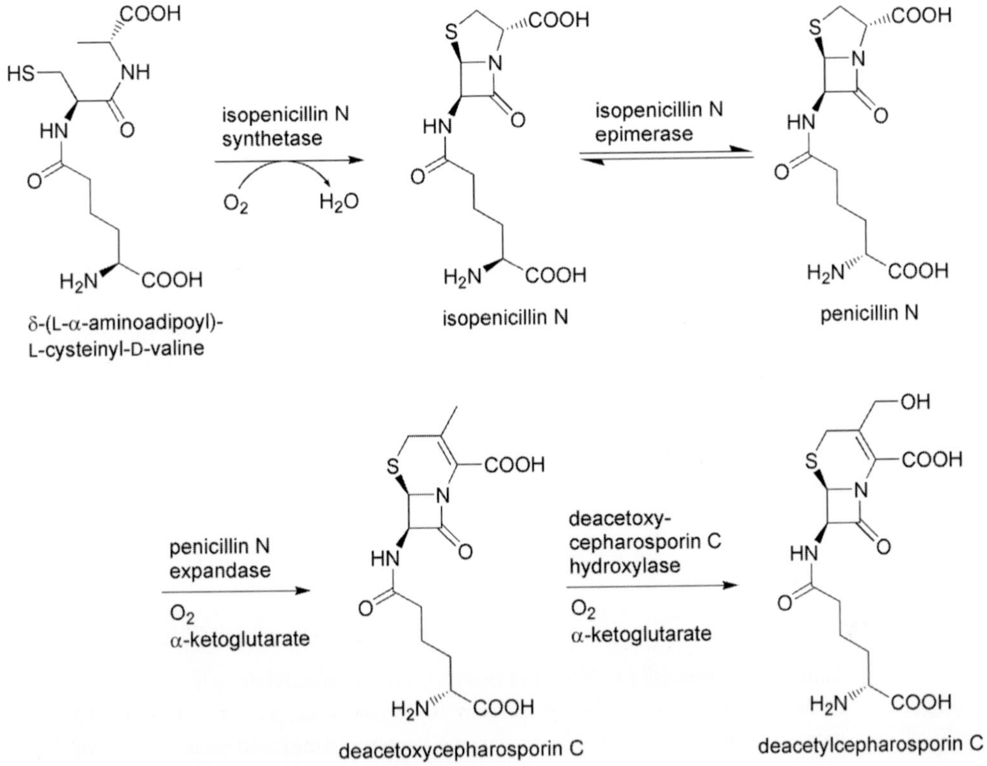

Figure 17-17. Biosynthetic pathway for cepharosporin C.

hydroxylated to form deacetylcepharosporin C by deacetoxycepharosporin C hydroxylase. These reactions proceed sequentially in the biosynthesis of cepharosporin C in *Streptomyces clavuligerus*, a producer of various β-lactam antibiotics[179, 180] (Fig. 17-17). However, in *Cepharosporium acremonium*, conversion of penicillin N into deacetoxycepharosporin C is catalyzed by a bifunctional enzyme, penicillin N expandase/deacetoxycepharosporin C hydroxylase in *Cepharosporium acremonium*[181].

Isopenicillin N epimerase activity, demonstrated in the extract of *Cepharosporium acremonium* protoplasts was found to be very unstable[182]. Usui and Yu[183], however, succeeded in purifying the enzyme to homogeneity after development of a simple assay procedure of the enzyme. They studied its enzymological properties[183]. The enzyme has a monomeric structure with a molecular mass of 47 000. The enzyme contains 1 mol of PLP per mol of protein. The enzyme shows a V_{max} value of 3.93 μmol min^{-1} per mg and a K_M of 0.30 mM for isopenicillin N, whereas it shows a V_{max} of 9.47 μmol min^{-1} per mg and a K_M of 0.78 mM for penicillin N. The K_{eq} value for the conversion between isopenicillin N and penicillin N is 1.09, which is in good agreement with the theoretical value. In addition to isopenicillin N and penicillin N, deacetoxycepharosporin C was epimerized only slowly: the rate relative to isopenicillin N is about 1 %. However, the following penicillin derivatives are inert: deacetylcepharosporin C, ceparosporin C, δ-(L-α-aminoadiopoyl)-L-cysteinyl-D-valine, L-α-aminoadipate, and D-α-aminoadipate. The enzyme is inhibited strongly by thiol reagents such as *p*-chloromercuribenzoate [183].

17.2.4
Racemization and Epimerization at Hydroxyl Carbons

Various epimerases acting on carbohydrate derivatives and acyl-CoA derivatives were demonstrated, purified, and characterized as reviewed previously [184]. Lactate racemase (E. C. 5.1.2.1) is the first racemase to he discovered [1-4]. The mechanism of lactate racemase reaction was studied with the enzyme preparations partially purified from *Clostridium butyricum* [185]. Hiyama et al. [186] highly purified the enzyme from *Lactobacillus sake*, but little is known about its enzymological properties. In contrast, mandelate racemase (E. C. 5.1.2.2) is the enzyme best characterized among various racemases and epimerases: its tertiary structure and functional groups that participate directly in catalysis has been clarified.

17.2.4.1
Mandelate Racemase (E. C. 5.1.2.2)

Mandelate racemase catalyzes the racemization of mandelate, which is the first step of the mandelate assimilation pathway in *Pseudomonas putida*. Although the mandelate pathway occurs widely in various bacteria, fungi and yeasts, most of them utilize one enantiomer or the other of mandelate in a benzoate-forming pathway. A few strains such as *Acinetobacter calcoaceticus* [187] and *Aspergillus nigar* [188] are capable of using both enantiomers with two complementary dehydrogenases with different stereospecificities. However, a single strain of *Pseudomonas putida* producing mandelate racemase can utilize both enantiomers [189].

In *Pseudomonas putida*, D-mandelate is converted into L-mandelate by mandelate racemase, then oxidized to benzoylformate by mandelate dehydrogenase (Fig. 17-18). Benzoylformate decarboxylase is the second enzyme of the pathway and catalyzes decarboxylation of benzoylformate to form benzaldehyde, which is oxidized to benzoate by NAD- and NADP-linked benzaldehyde dehydrogenases. The genes encoding these five enzymes constitute an operon that is induced by either enantiomer of mandelate [190]. Stecher et al. [191] established large-scale production of mandelate racemase by *Pseudomonas putida* ATCC12633 by optimization of enzyme induction: both glucose and mandelate were added to the culture right from the start as the carbon source. Thus, about 300-fold enhancement in the enzyme production was achieved. Strauss et al. [192] showed that immobilized mandelate racemase is an efficient biocatalyst used for repeated batch reactions to produce (*R*)-mandelate from (*S*)-mandelate under mild conditions.

Kenyon and coworkers purified mandelate racemase to homogeneity, and characterized it [189]. Divalent metal ions such as Mg^{2+}, Mn^{2+}, Co^{2+}, and Ni^{2+} were required for the catalysis. In addition to mandelate, *p*-hydroxymandelate and *p*-(bromome-

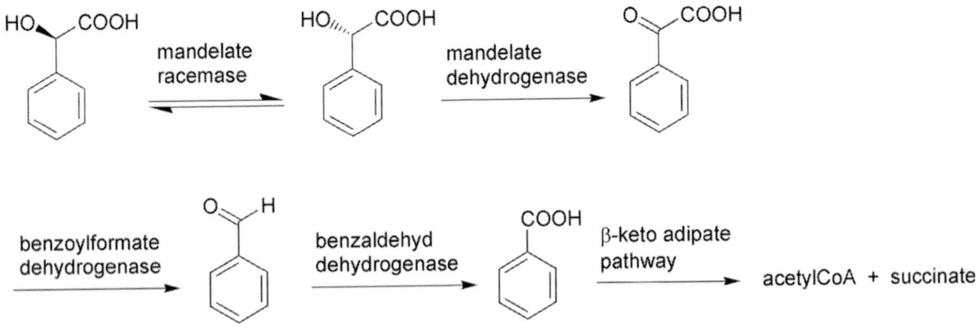

Figure 17-18. Mandelate assimilation pathway in *Pseudomonas putida*.

thyl)mandelate serve as the substrates. *p*-(Bromomethyl)mandelate is decomposed to *p*-(methyl)benzoylformate and bromide by action of the enzyme. The K_M values for D- and L-mandelate are 0.23 and 0.26 mM, respectively.

Ransom et al.[193] cloned the gene for mandelate racemase from *Pseudomonas putida* in *Pseudomonas aeruginosa* on the basis of the inability of the latter strain to grow on D-mandelate as a sole carbon source. The amino acid sequence was deduced from the nucleotide sequence, and the predicted molecular mass of the enzyme was 38 750[193]. The enzyme is composed of eight identical subunits. The crystal structure of mandelate racemase has been solved and refined at 2.5 Å resolution[194]. The secondary, tertiary and quaternary structures of mandelate racemase are quite similar to those of muconate lactonizing enzyme[195, 196]. Mandelate racemase is composed of two major structural domains and a small C-terminal domain. The N-terminal domain has an α + β structure, and the central domain has an α/β-barrel topology. The C-terminal domain consists of an L-shaped loop.

Divalent metal ions, which are essential catalytically, are ligated by three distal carboxyl groups of Asp 195, Glu 221, and Glu 247, all of which occur at the central domain[194]. The active site location was determined by analysis of a complex between mandelate racemase and *p*-iodomandelate, whose iodine atom has high electron density and contributes greatly to the analysis. The active site of the enzyme is located between the two major domains. The ionizable groups of Lys 166 and His 297 are located at the positions interacting with the chiral center of the substrate (Fig. 17-19). Neidhart et al.[194] proposed that they participate in general acid/base catalysis: Lys 166 abstracts the α-proton of L-mandelate, and His 297 abstracts the α-proton from D-mandelate. Landro et al.[197] then replaced His 297 by asparagine, analyzed the crystal structure of the H297N mutant enzyme at 2.2 Å resolution, and studied the mechanism of catalysis of the mutant enzyme. Although the mutant enzyme has no mandelate racemase activity, it catalyzes the stereospecific elimination of bromide from *p*-(bromomethyl)-L-mandelate at a rate equivalent to that catalyzed by the wild-type enzyme. Moreover, the mutant enzyme catalyzes exchange of the α-hydrogen of L- but not D-mandelate with deuterium in deuterium oxide at a rate 3.3 times less than that of the wild-type enzyme. Thus, Landro et al.[197, 198] concluded that the mandelate racemase reaction proceeds through a two-base

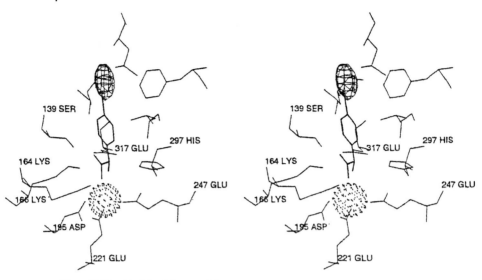

Figure 17-19. Models of the mandelate racemase active site with complexed substrate, *p*-iodomandelate. Reprinted from Neidhart et al.[194].

mechanism in which Lys 166 abstracts the α-proton from L-mandelate and His 297 abstracts the α-proton from D-mandelate (Fig. 17-20). In fact, the X-ray crystal studies of mandelate racemase inactivated by (*R*)-α-phenylglycidate revealed that the ε-amino group of Lys 166 is covalently bound to the distal carbon of the epoxide ring[199]. K166R mutant enzyme catalyzes the stereospecific elimination of bromide ion from *p*-(bromomethyl)mandelate to form *p*-(methyl)benzoylformate at a rate similar to that catalyzed by the wild-type enzyme[200], while H297N acts stereospecifically on (*S*)-*p*-(bromomethyl)mandelate[201]. This is compatible with the mechanism that Lys 166 and His 297 participate as the (*S*)- and (*R*)-specific catalyst, respectively. Bearne and Wolfenden[202] proposed that the complementary nature of the structures of mandelate racemase and its substrate is optimized in the transition state otherwise the general acid-general base catalysis will not become an efficient mode of catalysis.

17.3
Isomerizations

We describe here the enzymological characteristics and application of isomerases, especially D-xylose (glucose) isomerase, phosphoglucose isomerase, triose phosphate isomerase, L-rhamnose isomerase, L-fucose isomerase, maleate *cis-trans* isomerase, and unsaturated fatty acid *cis-trans* isomerase. *N*-Acetyl-D-glucosamine 2-epimerase is not an isomerase, but for convenience we will also describe the characteristics and use of the enzyme because this section deals with sugar-metabolizing enzymes.

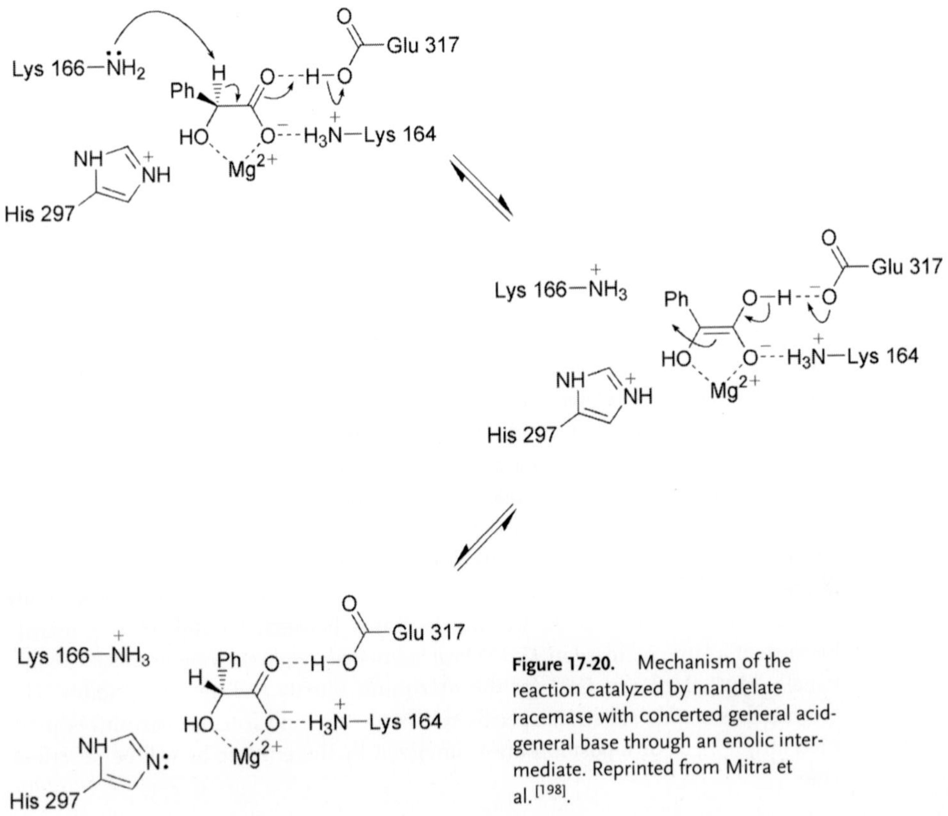

Figure 17-20. Mechanism of the reaction catalyzed by mandelate racemase with concerted general acid-general base through an enolic intermediate. Reprinted from Mitra et al.[198].

17.3.1
D-Xylose (Glucose) Isomerase (E. C. 5.3.1.5)

D-Xylose isomerase catalyzes the interconversion between D-xylose and D-xylulose (Fig. 17-21). Since this enzyme acts on D-glucose to produce D-fructose, it is often referred to as glucose isomerase (Fig. 17-21). The isomerization of glucose to fructose by this enzyme is a very important process for the industrial production of high fructose corn syrup. This enzyme is also applicable to the synthesis of many aldoses and ketoses because of its wide substrate specificity. The enzyme gene has been cloned from various microorganisms, and the enzyme has been overexpressed, purified, and characterized. Their three dimensional structures have also been determined [203–206].

17.3.1.1
Properties

Xylose isomerases have been purified from various microorganisms, such as *Lactobacillus brevis*, *Streptomyces* sp., *Bacillus stearothermophilus*, and *Actinoplanes*

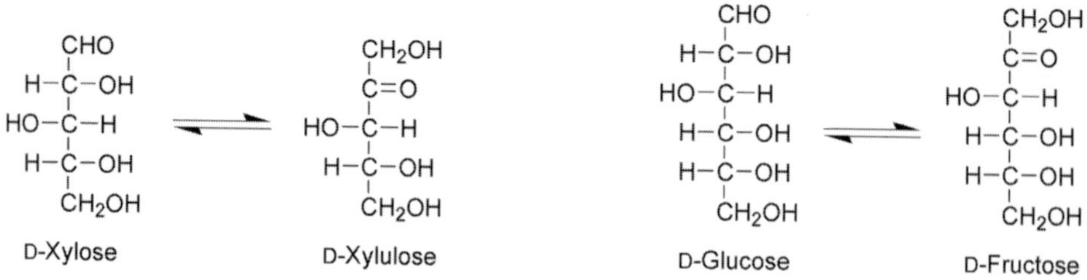

CHO   CH₂OH   CHO   CH₂OH

```
   CHO              CH2OH              CHO              CH2OH
    |                 |                 |                 |
 H-C-OH             C=O            H-C-OH             C=O
    |                 |                 |                 |
HO-C-H      ⇌    HO-C-H         HO-C-H             HO-C-H
    |                 |                 |                 |
 H-C-OH          H-C-OH         H-C-OH      ⇌     H-C-OH
    |                 |                 |                 |
  CH2OH            CH2OH          H-C-OH             H-C-OH
                                      |                 |
                                    CH2OH             CH2OH
```

D-Xylose   D-Xylulose   D-Glucose   D-Fructose

Figure 17-21.   Reactions catalyzed by D-xylose isomerase.

missouriensis[207–210]. They consist of four identical subunits whose molecular mass are in the range 42 000–51 000. The optimum pH usually ranges from 7.0 to 9.0. The cDNA for barley (*Hordeum vulgare*) enzyme gene has been cloned, and the recombinant enzyme characterized[211]. It is unique because it is a dimer composed of a subunit with a molecular mass of 53 620, which is much larger than those of microbial enzymes. Thermostable xylose isomerases were purified and characterized from many thermophilic bacteria[204, 205, 212–222]. The enzyme isolated from *Thermotoga neapolitana* is extremely thermostable, with the optimal activity being above 95 °C[216]. The catalytic efficiency (k_{cat}/K_M) of the enzyme is essentially constant between 60 and 90 °C, and decreases between 90 and 98 °C primarily because of a large increase in K_M. Xylose isomerase requires divalent metal cations, usually Mg^{2+}, Mn^{2+}, or Co^{2+} for the maximum activity and thermal stability. The enzyme has a wide substrate specificity[223]: glucose and fructose derivatives modified at the 3-, 5- or 6-position are isomerized by the enzyme as will be described later.

17.3.1.2
Reaction Mechanism

The reaction mechanism of xylose isomerase was proposed based on X-ray crystallography[224] and molecular mechanical and molecular orbital studies[225].

The α-pyranose form of the substrate binds to the active site of the enzyme, and the reaction is initiated by ring-opening involving hydrogen transfer from the first hydroxyl group to O5 (Fig. 17-22). After extension of the substrate, a water molecule abstracts the proton from the hydroxyl group at O2 of xylose and transfers it to Asp 257 in the second step. The following hydride shift causes isomerization. The O1 atom of the ketose is negatively charged and most probably abstracts a proton from Asp 257. The stable cyclic conformation is then formed.

This hydride shift reaction mechanism is quite different from the base-catalyzed enolization mechanism proposed for phospho sugar isomerases such as triosephosphate isomerase which generally do not require a metal ion for activity[226].

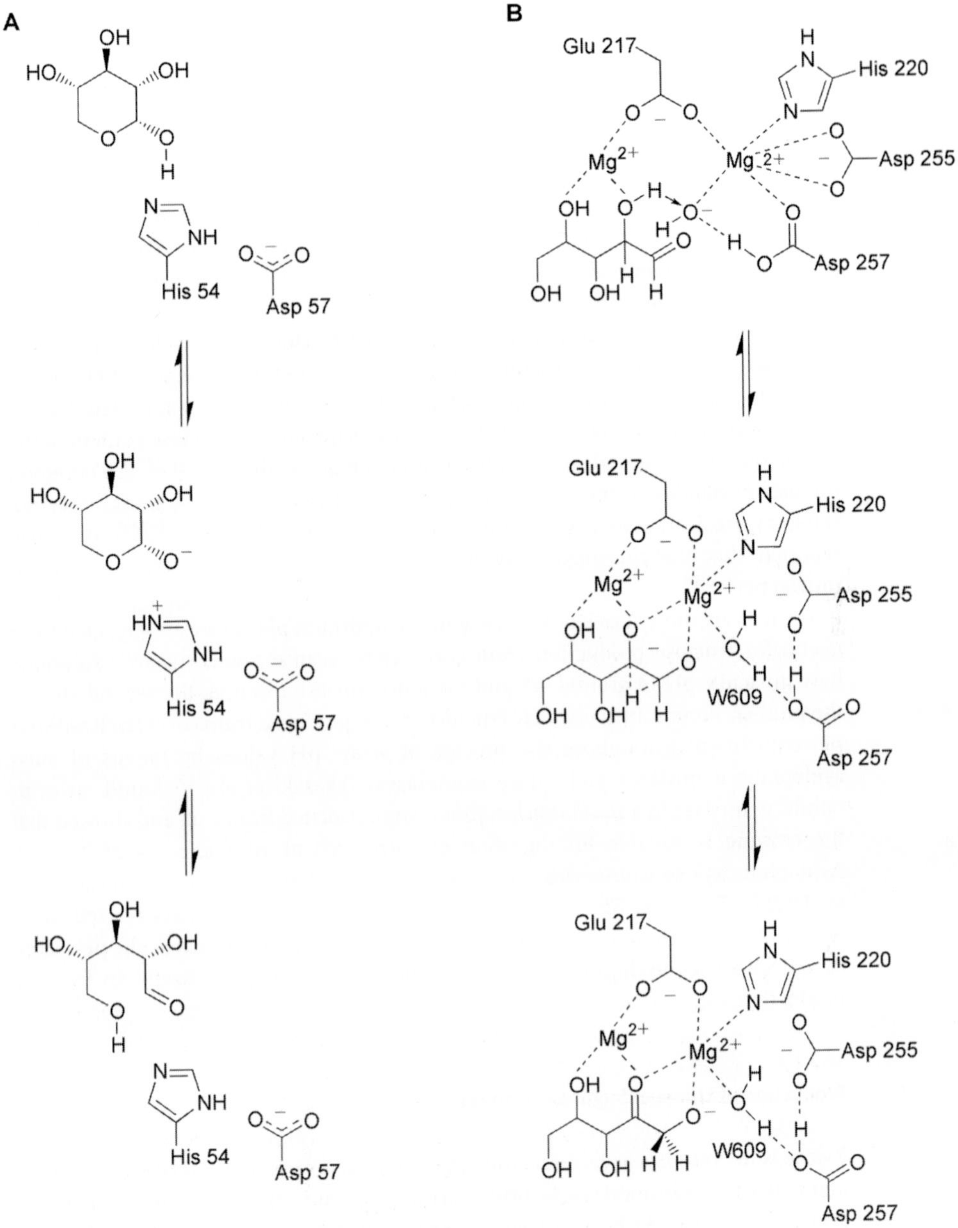

Figure 17-22. Reaction mechanism for xylose-xylulose conversion by D-xylose isomerase through ring opening (A) and hydride shift (B). Reprinted from Fuxreiter et al. [225].

17.3.1.3
Production of Fructose

Xylose isomerase derived from various microorganisms, such as *Actinoplanes missouriensis*, *Streptomyces griseofuscus*, *Flavobacterium arborescens*, *Streptomyces phaechromogenes*, *Bacillus coagulans*, *Streptomyces murinus*, *Streptomyces rubiginosus*, and *Streptomyces olivochromogenes*, is utilized in the annual conversion of 3 million tons of glucose into fructose for use as high fructose corn syrup. The enzyme is immobilized by glutaraldehyde cross-linking or adsorption on an insoluble resin for the fixed bed isomerization process [227].

The isomerization is reversible, and the final fructose content depends on the reaction temperature. The reaction is usually carried out in the region of 60–65 °C. However, a higher temperature gives a higher fructose content. It is reported that the degree of conversion is raised from 42%, which is the normal fructose content of the syrup, to 55% by isomerization with xylose isomerase at about 95 °C [227]. Therefore, the thermostability of the enzyme is an important issue. Recently, several thermostable xylose isomerases were found and characterized [204, 205, 212–222]. It is also reported that the thermostability of the enzyme is enhanced by site-directed mutagenesis [228].

α-Amylases and xylose isomerases with low optimum pH values are expected to be useful for fructose production from cornstarch because raw cornstarch solutions have an acidic pH of around 4.5 and the glucoamylase reaction, the second step in the process, prefers an acidic pH. Fructose can be produced from cornstarch without pH adjustment throughout the process at acidic pH values by means of such acidophilic α-amylases and xylose isomerases. Takasaki et al. [229] found an acidophilic α-amylase in a *Bacillus licheniformis* strain isolated from soil, and showed that the enzyme is suitable for digestion of cornstarch at an acidic pH of 4.5–5.0. Acidophilic xylose isomerases have been demonstrated in *Thermoanaerobacterium* sp. JW/SL-YS [217] and *Streptomyces* sp. SK [221], and purified and characterized. Both of these have optimum pH values around 6.5, but are highly active at acidic pHs such as 5.0. Since they are highly thermostable, they are expected to be useful for fructose production.

17.3.1.4
Production of Unusual Sugar Derivatives

Xylose isomerase has a wide substrate specificity, and 3-, 5-, or 6- substituted glucose and fructose are isomerized by this enzyme. Since this enzyme requires the 4-OH group for hexoses to be substrates, phosphoglucose isomerase instead of xylose isomerase is used for the synthesis of 4-substituted fructose as described below.

17.3.1.4.1 **Preparation of Glucose Derivatives Modified at Position 3 or 6**
Bock and coworkers [230] showed that D-glucose derivatives bearing modifications at the C3 or C6 position are converted by xylose isomerase from *Streptomyces* sp.

Figure 17-23. Conversion by xylose isomerase of (2R,3R)-configured aldotetrose modified at C5 into open-chain 2-ketoses (A), and L-erythrose into L-erythrulose (B). Reprinted from Ebner and Stütz[232].

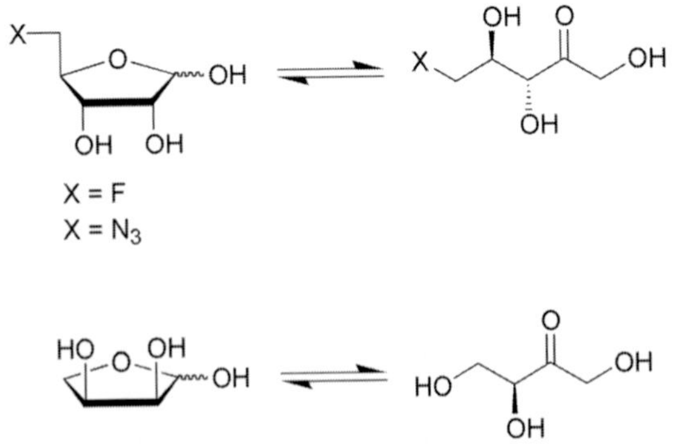

However, epimers of D-glucose are inert as substrates of the enzyme: D-mannose, D-allose, and D-galactose. Various 5-modified D-glucofuranoses are quantitatively converted into the corresponding D-fructopyranoses with the enzyme[231]. Ebner and Stütz[232] showed that various (2R,3R)-configured aldofuranoses such as D-erythrose and C5-modified D-ribose derivatives serve as substrates of the enzyme: D-erythrose is quantitatively converted into D-glycero-tetrulose, with D-ribofuranoses being the corresponding open-chain 2-ketoses (Fig. 17-23). L-Erythrose, the enantiomer of D-erythrose, is also isomerized quantitatively by the enzyme to L-erythrulose (L-*glycero*-tetrulose) (Fig. 17-23). Fructose bisphosphate aldolase catalyzes a stereospecific aldol condensation between dihydroxyacetone phosphate and a number of aldehydes to form hexoketose 1-phosphates, the phosphate groups of which are removed by hydrolysis. The resultant hexoketoses are converted stereospecifically into hexoaldose derivatives by xylose isomerase. Thus, unusual hexoaldose derivatives such as 3-deoxy-D-glucose, 6-deoxy-D-glucose, 6-*O*-methyl-D-glucose and 6-deoxy-6-fluoro-D-glucose were prepared by this method[223, 233].

17.3.1.4.2 Preparation of Fructose and Sorbose Derivatives Modified at Position 5

Xylose isomerase converts a wide range of D-glucose as well as L-idose derivatives modified at position 5 into the corresponding ketose. 5-Deoxy-5-fluoro-D-xylulose and a variety of 5,6-dimodified open-chain analogs of D-fructose, namely the 5,6-diazido-5,6-dideoxy, 6-azido-5,6-dideoxy, 6-azido-5,6-dideoxy-5-fluoro, 5,6-dideoxy-5-fluoro, 5,6-dideoxy-6-fluoro and 5,6-dideoxy-5,6-difluoro derivatives were prepared with glucose isomerase (Fig. 17-24)[234, 235].

17.3.1.4.3 Preparation of Sucrose Derivatives with Modified Fructose Moieties

Xylose isomerase is also used for the synthesis of modified sucroses, which is important in the study of the topographical aspects of the binding of sucrose to a sucrose carrier protein[236]. 6-Deoxy- and 6-deoxy-6-fluoroglucose chemically synthesized are isomerized to the corresponding 6-substituted fructose by xylose isomerase. The resultant substrates are subsequently condensed with UDP-glucose by sucrose synthase. Although the equilibrium of the first step lies towards the glucose

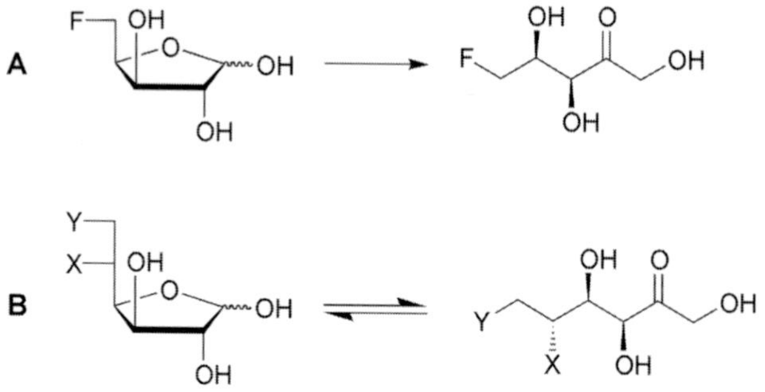

Figure 17-24. Production of 5-deoxy-5-fluoro-D-xylulose and 5,6-dimo-
dified open-chain analogs of D-fructose with xylose isomerase.
Reprinted from Hadwiger et al. [235].

derivatives, this problem is overcome by coupling the isomerization reaction with
the sucrose formation, which is irreversible. The second reaction completely drives
the isomerization reaction almost to completion. Incubation of 6-deoxy- or 6-deoxy-
6-fluoroglucose and UDP-glucose with both the xylose isomerase and sucrose
synthase afforded 6'-deoxy- and 6'-deoxy-6'-fluorosucrose in 73 and 53% isolated
yield, respectively.

17.3.2
Phosphoglucose Isomerase (E. C. 5.3.1.9)

Phosphoglucose isomerase catalyzes the interconversion of glucose 6-phosphate and
fructose 6-phosphate. This enzyme is involved in the gluconeogenesis, glycolytic
pathway, and pentose phosphate cycle. Since thermostable enzymes are generally
useful for industrial application, thermostable phosphoglucose isomerase was
purified from *Bacillus stearothermophilus* [237] and *Bacillus caldotenax* [238]. *B. stearothermophilus* produces two isozymes of phosphoglucose isomerase, and they were
overexpressed in *E. coli*, purified to homogeneity, crystallized [239], and the X-ray
structure of the enzyme was determined [240, 241]. The structure of the rabbit muscle
enzyme complexed with a competitive inhibitor D-gluconate 6-phosphate was also
determined by X-ray crystallography [242, 243]. The enzyme is a dimer with two α/β-
sandwich domains in each subunit. Lys 518 and His 388 are located at the active
center and are probably involved in the catalytic mechanism. Since gluconate
6-phosphate occurs predominantly in its cyclic form, phosphoglucose isomerase
probably catalyze the opening of the hexose ring to give initially its straight chain
form with Lys 518 and His 388. Then the enzyme undergoes isomerization of the

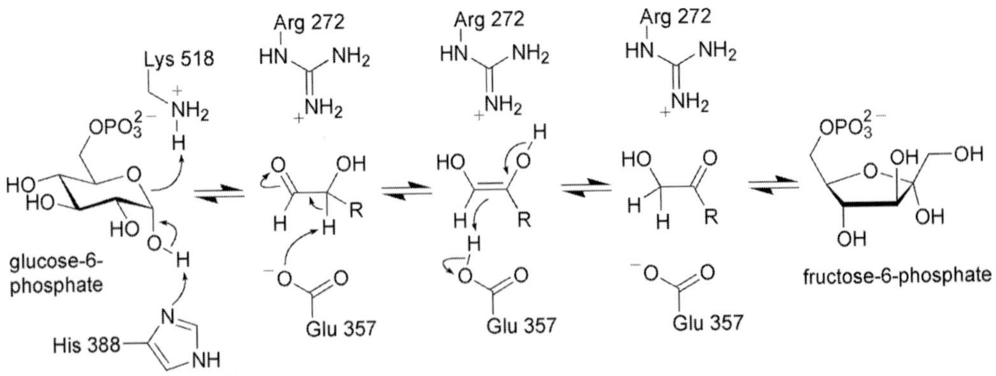

Figure 17-25. Mechanism of phosphoglucose isomerase reaction. His 388 and Glu 216 catalyze the ring opening. The side-chain of Glu357 abstracts a proton from the C2 position of the open chain form of the substrate, and the *cis*-enediol is formed. Then, a proton is transferred from the protonated Glu 357 to the C1 position of the intermediate. Reprinted from Jeffery et al. [242].

substrate through formation of a *cis*-enediol intermediate with the double bond between C1 and C2 (Fig. 17-25). Glu 357 transfers the proton from the C2 of glucose 6-phosphate to its C1 position. The side chain of Arg 272 stabilizes the negative charge of the intermediate (Fig. 17-25).

Xylose isomerase requires the 4-OH group for glucose derivatives to be substrates [230]. On the other hand, phosphoglucose isomerase can act on 4-substituted phosphoglucose. Therefore the latter enzyme is applicable to the preparation of glucose or fructose derivatives modified at position 4. For example, 4-deoxy-4-fluorofructose was prepared from 4-deoxy-4-fluoroglucose with phosphoglucose isomerase because xylose isomerase cannot isomerize 4-deoxy-4-fluoroglucose [236]. 4-Deoxy-4-fluorofructose was then converted into 4'-deoxy-4'-fluorosucrose, which is useful for the analysis of the interaction between sucrose and a sucrose carrier protein, with fructose-6-phosphate kinase [236].

Fructose 1,6-bisphosphate has attracted attention due to its important applications in the field of medicine, and is produced from glucose in three step by enzymatic reactions catalyzed by glucokinase, phosphoglucose isomerase, and phosphofructokinase. ATP is regenerated by acetate kinase (Fig. 17-26). Ishikawa and coworkers established an efficient method for production of fructose 1,6-bisphosphate in a

$$\text{Glucose} + \text{ATP} \xrightarrow{\text{GK}} \text{Glucose-6-phosphate (G6P)} + \text{ADP}$$

$$\text{G6P} \underset{\text{PGI}}{\rightleftharpoons} \text{Fructose-6-phosphate (F6P)}$$

$$\text{F6P} + \text{ATP} \xrightarrow{\text{PFK}} \text{FDP} + \text{ADP}$$

$$\text{ADP} + \text{Acetyl phosphate} \xrightarrow{\text{AK}} \text{ATP} + \text{Acetic acid}$$

Figure 17-26. Synthesis of fructose 1,6-bisphosphate from glucose by combination of glucokinase (GK), phosphoglucose isomerase (PGI), phosphofructokinase (PFK), and acetate kinase (AK) reactions.

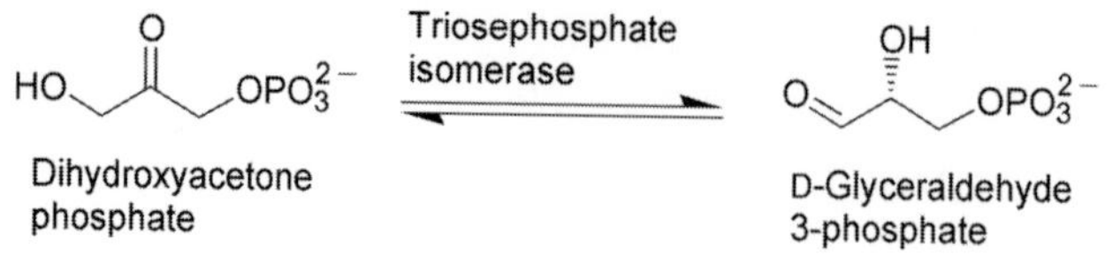

Figure 17-27. Reaction catalyzed by triosephosphate isomerase.

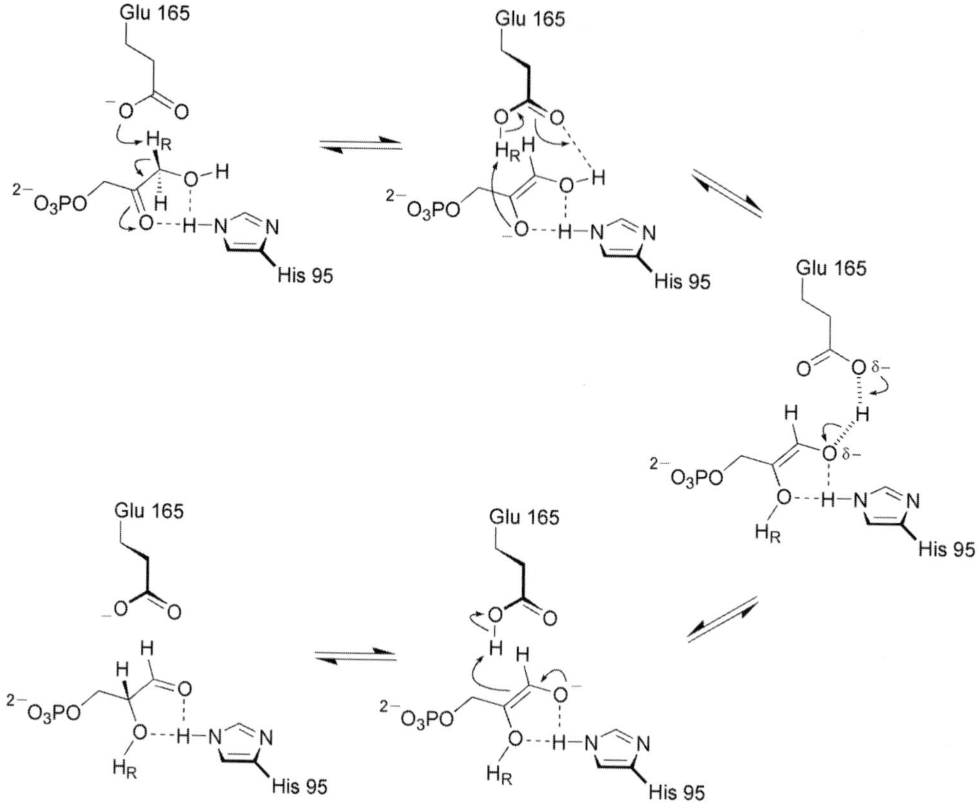

Figure 17-28. Triosephosphate isomerase reaction through a *cis*-enediol intermediate. The *pro-R* proton is removed from C1 of dihydroxyacetone phosphate by the side chain of Glu 165, and the carbonyl group of the substrate is polarized by the side chain of His 95. Reprinted from Harris et al. [249].

batch reactor system using the purified enzymes [244] and the crude extract of *Bacillus stearothermophilus* cells [245]. The yield of fructose 1,6-bisphosphate depended on the activity of glucokinase in the reactor [246].

17.3.3
Triosephosphate Isomerase (E. C. 5.3.1.1)

Triosephosphate isomerase is involved in the glycolytic pathway, and catalyzes the interconversion of dihydroxyacetone phosphate and D-glyceraldehyde phosphate (Fig. 17-27). The refined three-dimensional structures of chicken, yeast, and trypano-

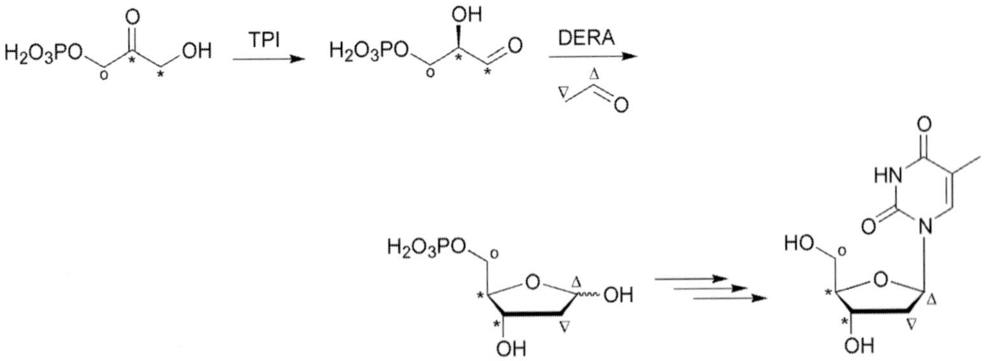

Figure 17-29. Synthesis of $[3',4'-{}^{13}C_2]$-thymidine from $[2',3'-{}^{13}C_2]$-dihydroxyacetone phosphate with triosephosphate isomerase (TPI) and D-2-deoxyribose-5-phosphate (DHAP). Asterisks indicate the positions selectively labeled with ^{13}C. Other positions that can be isotopically substituted are marked with °, △, and ▽. Reprinted from Ouwerkerk et al. [251].

somal enzymes have been elucidated [247]. The reaction is thought to proceed through a *cis*-enediol intermediate with Glu 165 and His 95 as acid and base catalysts (Fig. 17-28) [248, 249]. The side chain of Glu 165 removes the *pro-R* proton from the C1 of dihydroxyacetone phosphate, and that of neutral His 95 polarizes the carbonyl group of the substrate. Fructose 1,6-bisphosphate, a precursor molecule for sugar synthesis, can be prepared from dihydroxyacetone phosphate with this enzyme and aldolase [250]. Triosephosphate isomerase has been used for various other purposes. For example, $[3',4'-{}^{13}C_2]$-thymidine has been prepared from $[{}^{13}C_2]$-acetic acid through $[2',3'-{}^{13}C_2]$-dihydroxyacetone phosphate and D-$[3',4'-{}^{13}C_2]$-2-deoxyribose-5-phosphate with triosephosphate isomerase and D-2-deoxyribose-5-phosphate aldolase (E.C. 4.2.1.2) (Fig. 17-29) [251].

17.3.4
L-Rhamnose Isomerase (E.C. 5.3.1.14)

L-Rhamnose is an important component of bacterial cell walls, and is metabolized in *E. coli* through a pathway similar to that of glucose 6-phosphate in glycolysis. Rhamnose isomerase catalyzes the first reaction in the pathway to produce L-rhamnulose from L-rhamnose (Fig. 17-30). The enzyme gene was cloned from *E. coli* and overexpressed [252], and the enzyme was purified and characterized [252].

Rhamnose isomerase is composed of four identical subunits with a molecular mass of about 47 kDa. It has the maximum activity around 7.6, and requires Mn^{2+} to provide the highest activity. The enzyme shows no significant sequence similarity to any other ketol isomerases including xylose isomerase. However, rhamnose isomerase was found, by X-ray crystallography, to be most similar to xylose isomerase [252]. The monomer of rhamnose isomerase is composed of $(\beta/\alpha)_8$-barrels, and the structure and arrangement of the barrel are very similar to those of xylose isomerase. However, each of them has an additional α-helical domain, which is involved in subunit assembly and differs from each other only in its structure. The

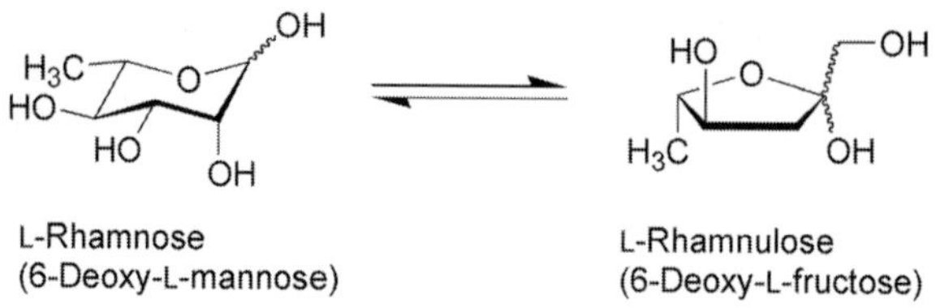

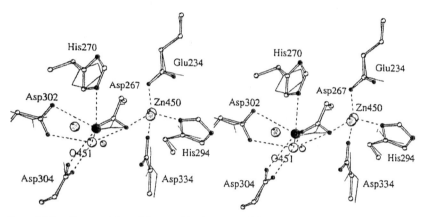

L-Rhamnose
(6-Deoxy-L-mannose)

L-Rhamnulose
(6-Deoxy-L-fructose)

Figure 17-30. Reaction catalyzed by rhamnose isomerase. Since both substrate and product occur in cyclic forms, L-rhamnose isomerase catalyzes ring opening before isomerization. Reprinted from Korndorfer et al. [252].

Figure 17-31. Superposition of the metal binding sites of rhamnose isomerase (residues named and drawn with thick bonds) and zylose isomerase (thin bonds). Reprinted from Korndorfer et al. [252].

residues surrounding the catalytic Mn^{2+} site (Asp 302, Asp 304 and His 270) are conserved in the two structures (Fig. 17-31). Therefore, the reaction catalyzed by rhamnose isomerase is thought to proceed through a metal-mediated hydride-shift mechanism in the same manner as xylose isomerase [252].

Bhuiyan et al. [253] immobilized L-rhamnose isomerase from *Pseudomonas* sp. LL172 on chitopearl beads, and used it to produce L-mannose from L-fructose. The immobilized enzyme was found to be stable: it retained about 90% of the initial activity after five repeated batch reactions. The concentration of L-mannose relative to L-fructose was about 3:7 at equilibrium. D-Allose was also produced from D-psicose with the immobilized L-rhamnose isomerase. Since D-psicose is readily produced from D-fructose with D-tagatose 3-epimerase, D-allose can be produced from D-fructose by combination of the two enzymes immobilized on chitopearl beads. Bhuiyan et al. [254] found that the reaction progresses steadily until 40% of the D-psicose is converted into D-allose. The immobilized D-tagatose 3-epimerase was also stable even after repeated uses, and D-allose was produced efficiently in the system.

17.3.5
L-Fucose Isomerase (E. C. 5.3.1.3)

Fucosylated oligosaccharides are important components of glycoproteins and glyco-lipids which are useful for cancer diagnosis and immunotyping. Therefore, efficient production methods for L-fucose and its analogs would be useful.

L-Fucose isomerase acts on D-arabinose, which was known as D-arabinose iso-merase in earlier literatures. L-Fucose is metabolized through a pathway similar to that of D-glucose in glycolysis, and L-fucose isomerase corresponds to glucose 6-phosphate isomerase. However, none of the aldose-ketose isomerases including glucose 6-phosphate isomerase shows sequence similarity to L-fucose isomerase. L-Fucose isomerase shares the common characteristics with other aldose-ketose isomerases acting on unphosphorylated substrates: the requirement of metal ions such as Mn^{2+} for L-fucose isomerase. Aldose-ketose isomerases acting on phos-phorylated substrates generally require no metal ions with the exception of phospho-mannose isomerase (E. C. 5.3.1.8) which requires Zn^{2+} for its activity.

Seemann and Schulz[255] determined the three-dimensional structure of L-fucose isomerase from *E. coli*, a hexamer from a subunit with a molecular mass of 64 976 Da. The enzyme shows no structural similarity to any other aldose-ketose isomerases analyzed thus far. However, Seemann and Schulz, on the basis of the tertiary structure, suggested that the L-fucose isomerase reaction proceeds through an ene-diol intermediate[255].

Fessner et al.[256] developed an efficient method for the synthesis of L-fucose analogs modified at the nonpolar terminus by means of L-fucose isomerase and L-fuculose 1-phosphate aldolase from *E. coli*. Various L-fucose analogs bearing linear or branched aliphatic side chains were prepared in about 30% overall yield with hydroxyaldehyde precursors and dihydroxyacetone phosphate as the starting materi-als (Fig. 17-32).

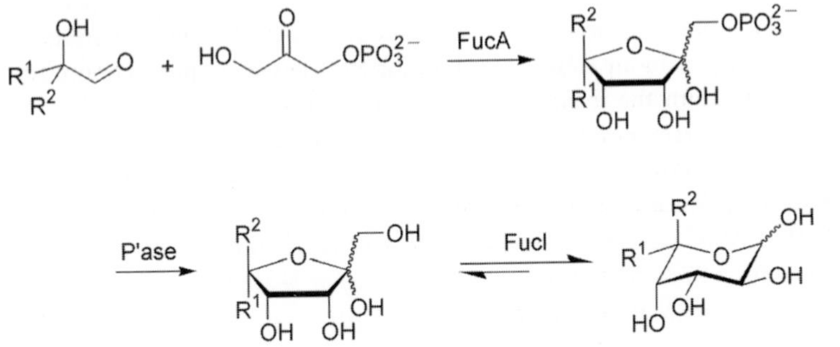

R^1	R^2
CH_3	H
CH_2-CH_3	H
$CH=CH_2$	H
$C\equiv CH$	H
CH_3	CH_3
CF_3	H

Figure 17-32. Enzymatic synthesis of L-fucose analogs with L-fucose 1-phosphate aldolase (FucA), phosphatase (P'ase), and L-fucose iso-merase (FucI). Reprinted from Fessner et al.[256].

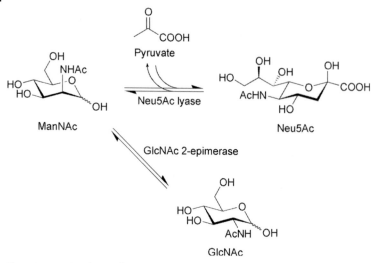

Figure 17-33. Synthesis of *N*-acetylneuraminate (Neu5Ac) from *N*-acetyl-
D-glucosamine (GlcNAc) and pyruvate through *N*-acetyl-D-mannosamine (ManNAc)
with *N*-acetylneuraminate and *N*-acetyl-D-glucosamine 2-epimerase. Reprinted from
Maru et al. [259].

17.3.6
N-Acetyl-D-glucosamine 2-Epimerase

N-Acetylneuraminate is a sialic acid with various biological functions that is widely
distributed in animals. It has been prepared only from natural resources such as
colominic acid, edible birds nests, milk or eggs. Alternatively, it has been prepared
enzymatically from *N*-acetyl-D-mannosamine and pyruvate with *N*-acetylneurami-
nate lyase as the catalyst [257, 258]. However, *N*-acetyl-D-mannosamine is expensive,
and the method is not suitable for large-scale production of *N*-acetylneuraminate.
Maru et al. [259] developed an elegant method for the enzymatic production of *N*-
acetylneuraminate from the inexpensive *N*-acetyl-D-glucosamine and pyruvate by
means of *N*-acetylneuraminate lyase and *N*-acyl-D-glucosamine 2-epimerase, whose
genes were cloned from *E. coli* [260] and pig kidney [261], respectively (Fig. 17-33).
Simultaneous use of these enzymes and feeding of appropriate amounts of pyruvate
to the reaction mixture enabled production of *N*-acetylneuraminate from *N*-acetyl-D-
glucosamine with a 77 % conversion rate, and 29 kg of *N*-acetylneuraminate were
obtained from 27 kg of *N*-acetyl-D-glucosamine.

17.3.7
Maleate *cis-trans* Isomerase (E. C. 5.2.1.1)

Maleate *cis-trans* isomerase catalyzes the conversion of maleate into fumarate. This
enzyme is applicable to the production of L-aspartate by coupling with the aspartase
reaction as shown in Fig. 17-34 [262, 263]. First, maleate is isomerized to fumarate by

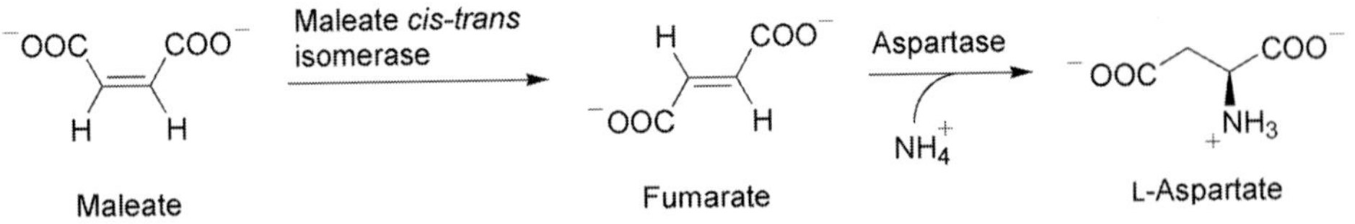

Figure 17-34. Synthesis of L-aspartate using maleate cis-trans isomerase and aspartase.

cis-trans isomerase, and then the fumarate formed is aminated to L-aspartate by aspartase. In this procedure, the resting cells of *Alcaligenes faecalis* containing both enzymes can be used as a catalyst. Thermostable maleate *cis-trans* isomerase was purified from *Bacillus stearothermophilus* MI-102 and characterized, and the enzyme gene was cloned and sequenced[264]. Two cysteine residues, Cys 80 and Cys 198, among the three conserved cysteines were found by site-directed mutagenesis studies to be catalytically important, although their catalytic roles are not yet known.

17.3.8
Unsaturated Fatty Acid *cis-trans* Isomerase

trans-Unsaturated fatty acids occur in membrane phospholipids of some bacterial genera such as *Pseudomonas* and *Vibrio*[265]. They are produced by *cis-trans* isomerase from *cis*-unsaturated fatty acids in response to environmental stresses such as elevated temperatures, increased salt concentrations, and the presence of organic solvents such as toluene[266–269]. The structural gene for the *cis-trans* isomerase was cloned from *Pseudomonas putida* P8[270]. The *E. coli* recombinant cells carrying the gene were shown to produce *trans*-unsaturated fatty acids in response to the organic solvent, although *E. coli* has no inherent ability to produce these fatty acids[270].

Okuyama et al.[271] purified the *cis-trans* isomerase from *Pseudomonas* sp. E-3 and characterized the enzyme catalyzing *cis-trans* isomerization toward 9-hexadecenoate. It catalyzes the *cis-to-trans* conversion of a double bond of *cis*-mono-unsaturated fatty acids with carbon chain lengths of 14, 15, 16, and 17 at positions 9, 10, or 11, but not at 6 or 7: the enzyme shows a strict specificity for both the position of the double bond and the chain length of the fatty acid. A similar enzyme was also discovered by Witholt and coworkers, which was purified from the periplasmic fraction of *Pseudomonas oleovorans*[272]. Not only 9-*cis*-hexadecenoate but also 11-*cis*-octadecenoate were found to serve as substrates of the enzyme. Moreover, the enzyme acted only on free unsaturated fatty acids and not on esterified fatty acids in contrast to the enzyme from *Pseudomonas* sp. E-3. Therefore, the *Pseudomonas oleovorans* enzyme differs from the enzyme of *Pseudomonas* sp. E-3 in substrate specificity, although both are monomeric enzymes with a molecular mass of about 80 kDa. The *cis-trans* isomerases are expected to be useful for biotransformation of unsaturated fatty acids.

17.4
Conclusion

Total conversion of racemic starting materials into a particular stereoisomer of a desired compound is very useful in the chemical industry. Half or more of the starting materials can be saved and steps for the laborious separation of the products from the starting material remaining reduced. Thus, racemases and epimerases are very useful in the chemical industry, when their reactions are coupled with some stereospecific reactions. Isomerases are also powerful catalysts for the production of particular enantiomers or diastereomers of interest from cheaply-available starting materials especially in the field of carbohydrate chemistry. Various new racemases and isomerases useful for industrial applications will no doubt be discovered from microorganisms at some point. However, established and well-known enzymes can be remodeled in order to expand their uses by various protein engineering technologies such as directed evolution. A good example for this is L-specific hydantoinase derived from D-specific hydantoinase[154]. The engineered enzymes can be incorporated into metabolic engineering studies in order to develop powerful microbial cells.

References

1 H. Katagiri, K. Kitahara, *J. Agr. Chem. Soc. Jpn.* **1936**, *12*, 844.

2 H. Katagiri, K. Kitahara, *Biochem. J.* **1937**, *31*, 909.

3 H. Katagiri, K. Kitahara, *J. Agr. Chem. Soc. Jpn.* **1936**, *12*, 1217.

4 E. L. Tatum, W. H. Peterson, E. B. Fred, *Biochem. J.* **1936**, *30*, 1892.

5 G. Rosso, K. Takashima, E. Adams, *Biochem. Biophys. Res. Commun.* **1969**, *34*, 134.

6 K. Yonaha, T. Yorifuji, T. Yamamoto, K. Soda, *J. Ferment. Technol.* **1975**, *53*, 579.

7 N. Esaki, C. T. Walsh, *Biochemistry* **1986**, *25*, 3261.

8 S. A. Wasserman, E. Daub, P. Grisafi, D. Botstein, C. T. Walsh, *Biochemistry* **1984**, *25*, 5182.

9 B. Badet, C. T. Walsh, *Biochemistry* **1985**, *24*, 1333.

10 K. Inagaki, K. Tanizawa, B. Badet, C. T. Walsh, H. Tanaka, K. Soda, *Biochemistry* **1986**, *25*, 3268.

11 K. Yokoigawa, H. Kawai, K. Endo, Y. Lim, N. Esaki, K. Soda, *Biosci. Biotechnol. Biochem.* **1993**, *57*, 93.

12 K. Hoffmann, E. Schneider-Scherzer, H. Kleinkauf, R. Zocher, *J. Biol. Chem.* **1994**, *269*, 12710.

13 T. Seow, K. Inagaki, T. Tamura, K. Soda, H. Tanaka, *Biosci. Biotechnol. Biochem.* **1998**, *62*, 242.

14 Y. Okubo, K. Yokoigawa, N. Esaki, K. Soda, H. Kawai, *Biochem. Biophys. Res. Commun.* **1999**, *256*, 333.

15 S. Kim, Y. Gyu, *J. Biochem. Mol. Biol.* **2000**, *33*, 82.

16 T. Uo, T. Yoshimura, N, Tanaka, K. Takegawa, N. Esaki, *J. Bacteriol.* **2001**, *183*, 2226.

17 A. Galkin, L. Kulakova, H. Yamamoto, K. Tanizawa, H. Tanaka, N. Esaki, K. Soda, *J. Ferment. Bioeng.* **1997**, *83*, 299.

18 A. Galkin, L. Kulakova, T. Yoshimura, K. Soda, N. Esaki, *Appl. Environ. Microbiol.* **1997**, *63*, 4651.

19 C. T. Walsh, *J. Biol. Chem.* **1989**, *264*, 2393.

20 E. Ferrari, D. J. Henner, M. Y. Yang, *Biotechnology* **1985**, *3*, 1003.

21 J. Wild, M. Lobocka, W. Walczak, T. Klopotowski, *Mol. Gen. Genet.* **1985**, *198*, 315.

22 N. G. Galakatos, C. T. Walsh, *Biochemistry* **1989**, *28*, 8167.

23 H. Toyama, K. Tanizawa, M. Wakayama, Q. Lee, T. Yoshimura, N. Esaki, K. Soda, *Agric. Biol. Chem.* **1991**, *55*, 2881.

24 H. Toyama, K. Tanizawa, T. Yoshimura, S.

Asano, H. -H. Lim, N. Esaki, K. Soda, *J. Biol. Chem.* **1991**, *266*, 13634.

25 H. C. Dunathan, *Proc. Natl. Acad. Sci. USA* **1966**, *55*, 713.

26 S.-J. Shen, H. G. Floss, H. Kumagai, H. Yamada, N. Esaki, K. Soda, S. A. Wasserman, C. T. Walsh, *J. Chem. Soc. Chem. Commun.* **1983**, 82.

27 G. J. Cardinale, R. H. Abeles, *Biochemistry* **1968**, *7*, 3970.

28 S. A. Ahmed, N. Esaki, H. Tanaka, K. Soda, *Biochemistry* **1986**, *25*, 385.

29 W. S. Faraci, C. T. Walsh, *Biochemistry* **1988**, *27*, 3267.

30 S. Sawada, Y. Tanaka, S. Hayashi, M. Ryu, T. Hasegawa, Y. Yamamoto, N. Esaki, K. Soda, S. Takahashi, *Biosci. Biotechnol. Biochem.* **1994**, *58*, 807.

31 J. P. Shaw, G. P. Petsko, D. Ringe, *Biochemistry* **1997**, *36*, 1329.

32 C. G. Stamper, A. A. Morollo, D. Ringe, *Biochemistry* **1998**, *37*, 10438.

33 A. Watanabe, Y. Kurokawa, T. Yoshimura, T. Kurihara, K. Soda, N. Esaki, *J. Biol. Chem.* **1999**, *274*, 4189.

34 B. Badet, K. Inagaki, K. Soda, C. T. Walsh, *Biochemistry* **1986**, *25*, 3275.

35 N. Esaki, H. Shimoi, N.Nakajima, T. Ohshima, H. Tanaka, K.Soda, *J. Biol. Chem.* **1989**, *264*, 9750.

36 T. Ohshima, K. Soda, *Eur. J. Biochem.* **1979**, *100*, 29.

37 K. Yonaha, K. Soda, *Biochem. Engin. Biotechnol.* **1986**, *33*, 95.

38 R. Wichmann, C. Wandrey, A. F. Buckmann, M. -R. Kula, *Biotechnol. Bioeng.* **1981**, *23*, 2789; A. S. Bommarius, M. Schwarm, K. Stingl, M. Kottenhahn, K. Huthmacher, K. Drauz, *Tetrahedron: Asymmetry*, **1995**, *6*, 2851.

39 W. Hummel, M.-R. Kula, *Eur. J. Biochem.* **1989**, *184*, 1.

40 Y. Asano, A. Nakazawa, *Agric. Biol. Chem.* **1987**, *51*, 2035.

41 Y. Asano, A. Yamada, K. Kato, Y. Yamaguchi, K. Hibino, K. Kondo, *J. Org. Chem.* **1990**, *55*, 5567.

42 Y. Kato, Y. Fukumoto, Y. Asano, *Appl. Microbiol. Biotechnol.* **1993**, *39*, 301.

43 T. Ohshima, C. Wandrey, M.-R. Kula, K. Soda, *Biotechnol. Bioeng.* **1985**, *27*, 1616.

44 K. Soda, K. Yonaha in: *Biotechnology 7a* (Eds.: H.-J. Rehm, G. Reed), VCH Verlagsgesellschaft, Weinheim, **1987**, 616.

45 A. Hashimoto, S. Kumashiro, T. Nishikawa, T. Oka, K. Takahashi, T. Mito, S. Takashima, N. Doi, Y. Mizutani, T. Yamazaki, *J. Neurochem.* **1993**, *61*, 348.

46 A. Hashimoto, T. Nishikawa, T. Oka, K. Takahashi, *J. Neurochem.* **1993**, *60*, 783.

47 T. Matsui, M. Sekiguchi, A. Hashimoto, U. Tomita, T. Nishikawa, K. Wada, *J. Neurochem.* **1995**, *65*, 454.

48 S. Filc-DeRicco, A. S. Gelbard, A. J. Cooper, K. C. Rosenspire, E. Nieves, *Cancer Res.* **1990**, *50*, 4839.

49 K. Yonaha, H. Misono, T. Yamamoto, and K. Soda, *J. Biol. Chem.* **1975**, *250*, 6983.

50 K. Tanizawa, Y. Masu, S. Asano, H. Tanaka, K. Soda, *J. Biol. Chem.* **1989**, *264*, 2445.

51 A. Galkin, L. Kulakova, V. Tishkov, N. Esaki, K. Soda, *Appl. Microbiol. Biotechnol.* **1995**, *44*, 479.

52 M. L. Scott in: *Organic Selenium Compounds: Their Chemistry and Biology* (Eds.: D. L. Klayman, W. H. H. Gunther), John Wiley & Sons, New York, **1973**, 629.

53 A. Shrift in: *Organic Selenium Compounds: Their Chemistry and Biology* (Eds.: D. L. Klayman, W. H. H. Gunther), John Wiley & Sons, New York, **1973**, 763.

54 N. Esaki, H. Shimoi, H. Tanaka, K. Soda, *Biotechnol. Bioeng.* **1989**, *34*, 1231.

55 Y. Sakamoto, S. Nagata, N. Esaki, H. Tanaka, K. Soda, *J. Ferment. Bioeng.* **1990**, *69*, 154.

56 S. M. Roberts, N. J. Turner, A. J. Willetts, M. K. Turner in: *Introduction to Biocatalysis Using Enzymes and Micro-organisms*, Cambridge University Press, New York, **1995**, 34.

57 J. E. Bailey, *Science* **1991**, *252*, 1668.

58 L. O. Ingram, F. Alterthum, K. Ohta, D. S. Beall in: *Developments in Industrial Microbiology* (*J. Indust. Microbiol.*, Suppl. No. 5), **1990**, *31*, pp. 21–30.

59 K. Soda, T. Osumi, *Methods Enzymol.* **1971**, *17 B*, 629.

60 K. Inagaki, K. Tanizawa, H. Tanaka, K. Soda, *Agric. Biol. Chem.* **1987**, *51*, 173.

61 T. Yorifuji, K. Ogata, K. Soda, *J. Biol. Chem.* **1971**, *246*, 5085.

62 Y. -H. Lim, K. Yokoigawa, N. Esaki, K. Soda, *J. Bacteriol.* **1993**, *175*, 4213.

63 K. Reynolds, J. Martin, S.-J. Shen, N. Esaki, K. Soda, H. G. Floss, *J. Basic Microbiol.* **1991**, *31*, 177.

64 Y. Lim, T, Yoshimura, K. Soda, N. Esaki, *J. Ferment. Bioeng.* **1998**, *86*, 400.

65 N. Makiguchi, N. Fukuhara, M. Shimada, Y. Asai, T. Nakamura, K. Soda in: *Biochemistry of Vitamin B6* (Eds.: T. Korpela, P. Christen), Birkheuser, Basel, **1987**, 457.

66 S. Shimizu, H. Yamada, *Trends Biotechnol.* **1984**, *2*, 137.

67 S. Shimizu, S. Shiozaki, T. Ohshiro, H. Yamada, *Agric. Biol. Chem.* **1984**, *48*, 1383.

68 T. Fukumura, *Agric. Biol. Chem.* **1976**, *40*, 1687.

69 T. Fukumura, *Agric. Biol. Chem.* **1976**, *40*, 1695.

70 T. Fukumura, *Agric. Biol. Chem.* **1976**, *41*, 1327.

71 T. Fukumura, G. Talbot, H. Misono, Y. Teramura, K. Kato, K. Soda, *FEBS Lett.* **1978**, *89*, 298.

72 T. Fukumura, *Agric. Biol. Chem.* **1976**, *41*, 1321.

73 S. A. Ahmed, N. Esaki, H. Tanaka, K. Soda, *Agric. Biol. Chem.* **1983**, *47*, 1887.

74 S. A. Ahmed, N. Esaki, H. Tanaka, K. Soda, *Agric. Biol. Chem.* **1983**, *47*, 1149.

75 N. Naoko, W. Oshihara, A. Yanai in: *Biochemistry of Vitamin B6* (Eds.: T. Korpela, P. Christen) Birkhauser, Basel, **1987**, 449.

76 H. R. Perkins, *Bacteriol. Rev.* **1963**, *27*, 18.

77 M. Tanaka, Y. Kato, S. Kinoshita, *Biochem. Biophys. Res. Commun.* **1961**, *4*, 114.

78 W. F. Diven, *Biochim. Biophys. Acta* **1969**, *191*, 702.

79 N. Nakajima, K. Tanizawa, H. Tanaka, K. Soda, *Agric. Biol. Chem.* **1988**, *52*, 3099.

80 N. Nakajima, K. Tanizawa, H. Tanaka, K. Soda, *Agric. Biol. Chem.* **1986**, *50*, 2823.

81 S. -Y. Choi, N. Esaki, T. Yoshimura, K. Soda, *Protein Express. Purif.* **1992**, *2*, 90.

82 K. A. Gallo, J. R. Knowles, *Biochemistry* **1993**, *32*, 3981.

83 M. Mansur, J. L. Garcia, J. M. Guisan, E. Garcia-Calvo, *Biotechnol. Lett.* **1998**, *20*, 57.

84 M. Yagasaki, K. Iwata, S. Ishino, M. Azuma, A. Ozaki, *Biosci. Biotechnol. Biochem.* **1995**, *59*, 610.

85 T. Yoshimura, M. Ashiuchi, N. Esaki, C. Kobatake, S. Choi, K. Soda, *J. Biol. Chem.* **1993**, *268*, 24242.

86 L. Liu, T. Yoshimura, K. Endo, N. Esaki, K. Soda, *J. Biochem.* **1997**, *121*, 1155.

87 S. Kim, I. Choi, S. Kim, Y. Yu, *Extremophiles* **1999**, *3*, 175.

88 M. Ashiuchi, K. Tani, K. Soda, H. Misono, *J. Biochem.* **1998**, *123*, 1156.

89 M. Ashiuchi, K. Soda, H. Misono, *Biosci. Biotechnol. Biochem.* **1999**, *63*, 792.

90 M. Ashiuchi, T. Yoshimura, T. Kitamura, Y. Kawata, J. Nagai, S. Gorlatov, N. Esaki, K. Soda, *J. Biochem.* **1995**, *117*, 495.

91 S.-Y. Choi, N. Esaki, T. Yoshimura, K. Soda, *J. Biochem.* **1992**, *112*, 139.

92 G. Rudnick, R. H. Abeles, *Biochemistry* **1975**, *14*, 4515.

93 J. S. Wiseman, J. S. Nichols, *J. Biol. Chem.* **1984**, *259*, 8907.

94 S. J. Rawaswamy, *J. Biol. Chem.* **1984**, *259*, 249.

95 H. T. Ho, P. J. Falk, K. M. Ervin, B. S. Krishnan, L. F. Discotto, T. J. Dougherty, M. J. Pucci, *Biochemistry* **1995**, *34*, 2464.

96 Y. Hwang, S. Cho, S. Kim, H. Sung, Y. Yu, Y. Cho, *Nat. Struct. Biol.* **1999**, *6*, 422.

97 S. Glavas, M. Tanner, *Biochemistry* **1999**, *38*, 4106.

98 N. Nakajima, K. Tanizawa, H. Tanaka, K. Soda, *J. Biotechnol.* **1986**, *8*, 243.

99 K. Tanizawa, Y. Masu, S. Asano, H. Tanaka, K. Soda, *J. Biol. Chem.* **1989**, *264*, 2445.

100 H. Bae, S. Lee, S. Hong, M. Kwak, N. Esaki, K. Soda, M. Sung, *J. Mol. Catal. B: Enzymatic* **1999**, *6*, 241.

101 C. Wandrey, R. Wichmann, A. S. Jandel in: *Enzyme Engineering* (Eds.: I. Chibata, S. Fukui, L. B. Wingard, Jr.), Vol. 6, Plenum Press, New York, **1982**, 61.

102 M. Yagasaki, M. Azuma, S. Ishino, A. Ozaki, *J. Ferment. Bioeng.* **1995**, *79*, 70.

103 M. Yagasaki, A. Ozaki, Y. Hashimoto, *Biosci. Biotechnol. Biochem.* **1993**, *57*, 1499.

104 T. Oikawa, M. Watanabe, H. Makiura, H. Kusakabe, K. Yamade, K. Soda, *Biosci. Biotechnol. Biochem.* **1999**, *63*, 2168.

105 M. M. Johnston, W. F. Diven, *J. Biol. Chem.* **1969**, *244*, 5414.

106 H. Okada, M. Yohda, Y. Giga-Hama, Y. Ueno, S. Ohdo, H. Kumagai, *Biochim. Biophys. Acta* **1991**, *1078*, 377.

107 H. C. Lamont, W. L. Staudenbauer, J. L. Strominger, *J. Biol. Chem.* **1972**, *247*, 5103.

108 M. Yohda, I. Endo, Y. Abe, T. Ohta, T. Iida, T. Maruyama, Y. Kagawa, *J. Biol. Chem.* **1996**, *271*, 22017.

109 M. Matsumoto, H. Homma, Z. Long, K. Imai, T. Iida, T. Maruyama, Y. Aikawa, I. Endo, M. Yohda, *J. Bacteriol.* **1999**, *181*, 6560.

110 Y. Nagata, K. Tanaka, I. Iida, Y. Kera, R. Ya-mada, Y. Nakajima, T. Fujiwara, Y. Fuku-mori, T. Yamanaka, Y. Koga, S. Tsuji, K. Kawaguchi-Nagata, *Biochim. Biophys. Acta Protein Struct. Mol. Biol.* **1999**, *1435*, 160.

111 M. Yohda, H. Okada, H. Kumagai, *Biochim. Biophys. Acta* **1991**, *1089*, 234.

112 T. Yamauchi, S. -Y. Choi, H. Okada, M. Yohda, H. Kumagai, N. Esaki, K. Soda, *J. Biol. Chem.* **1992**, *267*, 18 361.

113 H. Kumagai in: *Tanpakushitsu Kagaku*, Hir-okawa Shoten, Tokyo, **2002**, in the press.

114 C. Richaud, W. Higgins, D. Mengin-Le-creulx, P. Stragier, *J. Bacteriol.* **1987**, *169*, 1454.

115 C. Richaud, C. Printz, *Nucleic Acids Res.* **1988**, *16*, 10 367.

116 W. Higgins, C. Tardif, C. Richaud, M. A. Krivanek, A. Cardin, *Eur. J. Biochem.* **1989**, *186*, 137.

117 M. Cirilli, R. Zheng, G. Scapin, J. S. Blan-chard, *Biochemistry* **1998**, *37*, 16 452.

118 C. W. Koo, J. S. Blanchard-John, *Biochem-istry* **1999**, *38*, 4416.

119 L. M. Fisher, W. J. Albery, J. R. Knowles, *Bio-chemistry* **1986**, *25*, 2529.

120 L. M. Fisher, W. J. Albery, J. R. Knowles, *Bio-chemistry* **1986**, *25*, 2538.

121 L. M. Fisher, J. G. Belasco, T. W. Bruice, W. J. Albery, J. R. Knowles, *Biochemistry* **1986**, *25*, 2543.

122 J. G. Belasco, W. J. Albery, J. R. Knowles, *Biochemistry* **1986**, *25*, 2552.

123 J. G. Belasco, T. W. Bruice, W. J. Albery, J. R. Knowles, *Biochemistry* **1986**, *25*, 2558.

124 J. G. Belasco, T. W. Bruice, L. M. Fisher, W. J. Albery, J. R. Knowles, *Biochemistry* **1986**, *25*, 2564.

125 W. J. Albery, J. R. Knowles, *Biochemistry* **1986**, *25*, 2572.

126 M. Yagasaki, A. Ozaki, *J. Mol. Catal. B: Enzymatic* **1998**, *4*, 1.

127 K. Sano, K. Yokozeki, K. Tamura, N. Yasuda, I. Noda, K. Mitsugi, *Appl. Environ. Microbiol.* **1977**, *34*, 806.

128 K. Sano, K. Mitsugi, *Agric. Biol. Chem.* **1978**, *42*, 2315.

129 K. Sano in: *Biochemistry of Vitamin B6* (Eds.: T. Korpela, P. Christen), Birkheuser, Basel, **1987**, 453.

130 O. Ryu, W. Oh, S. Yoo, C. Shin, *Biotechnol. Lett.* **1995**, *17*, 275.

131 C. Syldatk, A. Leufer, R. Muller, H. Hoke in: *Advaces in Biochemical Engineering/Biotech-nology*, (Ed.: A. Fiechter), Vol. 41, Springer Verlag, Berlin, **1990**, 29.

132 C. Syldatk, R. Muller, M. Siemann, K. Krohn, F. Wagner in: Biocatalytic *Production of Amino Acids and Derivatives* (Eds.: J. D. Rozzell, F. Wagner), Hanser, Munich, **1992**, 75.

133 C. Syldatk, R. Müller, M. Pietzsch, F. Wagner in: *Biocatalytic Production of Amino Acids and Derivatives* (Eds.: J. D. Roz-zell, F. Wagner), Hanser, Munich, **1992**, 129.

134 F. Bernheim, M. L. C. Bernheim, *J. Biol. Chem.* **1946**, *163*, 683.

135 F. Bernheim, *Fed. Proc.*, **1947**, 6, 238.

136 H. Yamada, S. Takahashi, Y. Kii, H. Kuma-gai, *J. Ferment. Technol.* **1978**, *56*, 484.

137 R. Olivieri, E. Fascetti, L. Angelini, L. De-gen, *Biotechnol. Bioeng.* **1983**, *23*, 2173.

138 C. Hartley, S. Kirchmann, S. G. Burton, R. A. Dorrington, *Biotechnol. Lett.* **1998**, *20*, 707.

139 Y. Nishida, K. Nakamachi, K. Nabe, T. Tosa, *Enzyme Microb. Technol.* **1987**, *9*, 721.

140 C. Syldatk, D. Cotoras, G. Dombach, C. Gross, H. Kallwass, F. Wagner, *Biotechnol. Lett.* **1987**, *9*, 25.

141 K. Yokozeki, S. Nakamori, S. Yamanaka, C. Eguchi, K. Mitsugi, F. Yoshinaga, *Agric. Biol. Chem.* **1987**, *51*, 715.

142 K. Yokozeki, K. Kubota, *Agric. Biol. Chem.* **1987**, *51*, 721.

143 R. Tsugawa, S. Okumura, T. Ito, N. Katsuya, *Agric. Biol. Chem.* **1966**, *30*, 27.

144 H. Yamada, K. Oishi, K. Aida, T. Uemura, *Nippon Nogeikagaku Kaishi* **1969**, *43*, 528.

145 A. Yamashiro, K. Yokozeki, H. Kano, K. Kubota, *Agric. Biol. Chem.* **1988**, *52*, 2851.

146 M. Battilotti, U. Barberini, *J. Mol. Cat.* **1988**, *43*, 343.

147 J. Knabe, W. Wumm, *Arch. Pharm.* **1980**, *313*, 538.

148 K. Watabe, T. Ishikawa, Y. Mukohara, H. Nakamura, *J. Bacteriol.* **1992**, *174*, 962.

149 K. Watabe, T. Ishikawa, Y. Mukohara, H. Nakamura, *J. Bacteriol.* **1992**, *174*, 3461.

150 A. Wiese, M. Pietzsch, C. Syldatk, R. Mattes, J. Altenbuchner, *J. Biotechnol.* **2000**, *80*, 217.

151 K. Watabe, T. Ishikawa, Y. Mukohara, H. Nakamura, *J. Bacteriol.* **1992**, *174*, 7989.

152 M. Pietzsch, T. Waniek, R. Smith, S. Brato-vanov, S. Bienz, Stefan, C. Syldatk, *Mon-atsh. Chem.* **2000**, *131*, 645.

153 T. Ishikawa, K. Watabe, Y. Mukohara, H. Nakamura, *Biosci. Biotechnol. Biochem.* **1997**, *61*, 185.

154 O. May, P. T. Nguyen, F. H. Arnold, *Nature Biotechnol.* **2000**, *18*, 317.

155 J. P. Greenstein, *Methods Enzymol.* **1957**, *3*, 554.

156 Y. Endo, *Biochim. Biophys. Acta* **1976**, *438*, 532.

157 Y. Endo, *Biochim. Biophys. Acta* **1976**, *628*, 13.

158 L. C. Lugay, J. N. Aronson, *Biochim. Biophys. Acta* **1969**, *191*, 397.

159 W. S. Pierpoint, *Phytochemistry* **1973**, *12*, 2359.

160 I. Gentzen, H. -G. Löffler, F. Schneider, *Z. Naturforsch.*, **1980**, *35c*, 544.

161 M. Kikuchi, I. Koshiyama, D. Fukushima, *Biochim. Biophys. Acta* **1983**, *744*, 180.

162 I. Chibata, T. Ishikawa, S. Yamada, *Bull. Agric. Chem. Soc. Jpn.* **1957**, *21*, 300.

163 W. Kordel, F. Schneider, *Biochim. Biophys. Acta* **1976**, *445*, 446.

164 I. V. Galaev, V. K. Svedas, *Biochim. Biophys. Acta* **1982**, *701*, 389.

165 K. H. Röhm, R. L. V. Etten, *Eur. J. Biochem.* **1986**, *160*, 327.

166 H. -Y. Cho, K. Tanizawa, H. Tanaka, K. Soda, *Agric. Biol. Chem.* **1987**, *51*, 2793.

167 M. Sugie, H. Suzuki, *Agric. Biol. Chem.* **1980**, *44*, 1089.

168 Y.-C. Tsai, C.-P. Tseng, K.-M. Hsiao, L.-Y. Chen, *Appl. Environ. Microbiol.* **1988**, *54*, 984.

169 M. Moriguchi, K. Ideta, *Appl. Environ. Microbiol.* **1988**, *54*, 2767.

170 T. Tosa, T. Mori, N. Fuse, I. Chibata, *Enzymologia* **1966**, *31*, 214.

171 K. Takahashi, K. Hatano, *European Patent Application* **1989**, 030 4021A2.

172 S. Tokuyama, K. Hatano, T. Takahashi, *Biosci. Biotechnol. Biochem.* **1994**, *58*, 24.

173 S. Tokuyama, H. Miya, K. Hatano, T. Takahashi, *Appl. Microbiol. Biotechnol.* **1994**, *40*, 835.

174 S. Tokuyama, K. Hatano, *Appl. Microbiol. Biotechnol.* **1995**, *42*, 853.

175 S. Tokuyama, K. Hatano, *Appl. Microbiol. Biotechnol.* **1995**, *42*, 884.

176 S. Tokuyama, K. Hatano, *Appl. Microbiol. Biotechnol.* **1996**, *44*, 774.

177 D. R. J. Palmer, J. B. Garrett, V. Sharma, R. Meganathan, P. C. Babbitt, J. A. Gerlt, *Biochemistry* **1999**, *38*, 4252.

178 Y. Sawada, J. E. Baldwin, P. D. Singh, N. A. Solomon, A. L. Demain, *Antimicrob. Agents Chemother.* **1980**, *18*, 465.

179 S. E. Jensen, D. W. S. Westlake, S. Wolfe, *J. Antibiot.* **1982**, *35*, 483.

180 S. E. Jensen, D. W. S. Westlake, R. J. Bowers, S. Wolfe, *J. Antibiot.* **1982**, *35*, 1351.

181 J. E. Dotzlaf, W. -K. Yeh, *J. Bacteriol.* **1987**, *169*, 1611.

182 J. E. Baldwin, J. W. Keeping, P. D. Singh, C. A. Vallejo, *Biochem. J.* **1981**, *194*, 645.

183 S. Usui, C. -A. Yu, *Biochim. Biophys. Acta* **1989**, *999*, 78.

184 E. Adams, *Adv. Enzymol. Relat. Areas Mol. Biol.* **1976**, *44*, 69.

185 J. S. Pepple, D. Dennis, *Biochim. Biophys. Acta* **1976**, *429*, 1036.

186 T. Hiyama, S. Fukui, K. Kitahara, *J. Biochem.* **1968**, *64*, 99.

187 G. D. Hageman, E. Y. Rosenberg, G. L. Kenyon, *Biochemistry* **1970**, *9*, 4029.

188 M. Jamaluddin, P. V. Rao Subba, C. S. Vaidyanathan, *J. Bacteriol.* **1970**, *101*, 786.

189 G. L. Kenyon, G. D. Hageman, *Adv. Enzymol. Relat. Areas Mol. Biol.* **1979**, *50*, 325.

190 M. L. Wheelis, R. Y. Stanier, *Genetics* **1970**, *66*, 245.

191 H. Stecher, U. Felfer, K. Faber, *J. Biotechnol.* **1997**, *56*, 33.

192 U. T. Strauss, A. Kandelbauer, K. Faber, *Biotechnol. Lett.* **2000**, *22*, 515.

193 S. C. Ransom, J. A. Gehlt, V. M. Powers, G. L. Kenyon, *Biochemistry* **1988**, *27*, 540.

194 D. J. Neidhart, P. L. Howell, G. A. Petsko, V. M. Powers, R. Li, G. L. Kenyon, J. A. Gerlt, *Biochemistry* **1991**, *30*, 9264.

195 D. J. Neidhart, G. L. Kenyon, J. A. Gerlt, G. A. Petsko, *Nature* **1990**, *347*, 692.

196 P. C. Babbitt, G. T. Mrachko, M. S. Hasson, G. W. Huisman, R. Kolter, D. Ringe, G. A. Petsko, G. L. Kenyon, J. A. Gerlt, *Science* **1995**, *267*, 1159.

197 J. A. Landro, A. T. Kallarakal, S. C. Ransom, J. A. Gerlt, J. W. Kozarich, D. J. Neidhart, G. L. Kenyon, *Biochemistry* **1991**, *30*, 927.

198 B. Mitra, A. T. Kallarakal, J. W. Kozarich, J. A. Gerlt, J. G. Clifton, G. A. Petsko, G. L. Kenyon, *Biochemistry* **1995**, *34*, 2777.

199 J. A. Landro, J. A. Gerlt, J. W. Kozarich, C. W. Koo, V. J. Shah, G. L. Kenyon, D. J. Neidhart, S. Fujita, G. A. Petsko, *Biochemistry* **1994**, *33*, 635.

200 A. T. Kallarakal, B. Mitra, J. W. Kozarich, J. A. Gerlt, G. L. Kenyon, J. G. Clifton, G. A.

Petsko, G. L. Kenyon, *Biochemistry* **1995**, *34*, 2788.

201 S. L. Schafer, W. C. Barrett, A. T. Kallarakal, B. Mitra, J. W. Kozarich, J. A. Gerlt, J. G. Clifton, G. A. Petsko, G. L. Kenyon, *Biochemistry* **1996**, *35*, 5662.

202 S. L. Bearne, R. Wolfenden, *Biochemistry* **1997**, *36*, 1646.

203 K. A. Briggs, W. E. Lancashire, B. S. Hartley, *EMBO J.* **1984**, *3*, 611.

204 K. Dekker, H. Yamagata, K. Sakaguchi, S. Udaka, *Agric. Biol. Chem.* **1991**, *55*, 221.

205 K. Dekker, A. Sugiura, H. Yamagata, K. Sakaguchi, S. Udaka, *Appl. Microbiol. Biotechnol.* **1992**, *36*, 727.

206 S. D. Feldmann, H. Sahm, G. A. Sprenger, *Mol. Gen. Genet.* **1992**, *234*, 201.

207 K. Yamanaka, *Biochim. Biophys. Acta* **1968**, *151*, 670.

208 Y. Takasaki, Y. Kosugi, A. Kanbayas, *Agric. Biol. Chem.* **1969**, *33*, 1527.

209 N. Muramatsu, Y. Nosoh, *Arch. Biochem. Biophys.* **1971**, *144*, 245.

210 C. S. Gong, L. F. Chen, G. T. Tsao, *Biotechnol. Bioeng.* **1980**, *22*, 833.

211 P. Kristo, R. Saarelainen, R. Fagerstrom, S. Aho, M. Korhola, *Eur. J. Biochem.* **1996**, *237*, 240.

212 S. H. Brown, C. Sjoholm, R. M. Kelly, *Biotechnol. Bioeng.* **1993**, *41*, 878.

213 C. Lee, J. G. Zeikus, *Biochem. J.* **1991**, *273*, 565.

214 K. Dekker, H. Yamagata, K. Sakaguchi, S. Udaka, *J. Bacteriol.* **1991**, *173*, 3078.

215 J. Chauthaiwale, M. Rao, *Appl. Environ. Microbiol.* **1994**, *60*, 4495.

216 C. Vieille, J. M. Hess, R. M. Kelly, J. G. Zeikus, *Appl. Environ. Microbiol.* **1995**, *61*, 1867.

217 S. Liu, J. Wiegel, F. C. Gherardini, *J. Bacteriol.* **1996**, *178*, 5938.

218 S. S. Deshmukh, V. Shankar, *Biotechnol. Appl. Biochem.* **1996**, *24*, 65.

219 C. J. Moes, I. S. Pretorius, W. H. Van-Zyl, *Biotechnol. Lett.* **1996**, *18*, 269.

220 B. C. Park, S. Koh, C. Chang, S. W. Shu, D. S. Lee, S. M. Byun, *Appl. Biochem. Biotechnol.* **1997**, *62*, 15.

221 B. K. Srih, S. Bejar, *Biotechnol. Lett.* **1998**, *20*, 553.

222 C. Chang, H. K. Song, B. C. Park, D. S. Lee, S. W. Suh, *Acta Crystallogr. Sect D Biol. Crystallogr.* **1999**, *55*, 294.

223 J. R. Durrwachter, H. M. Sweers, K. Nozaki, C. H. Wong, *Tetrahedron Lett.* **1986**, *27*, 1261.

224 C. A. Collyer, K. Henrick, D. M. Blow, *J. Mol. Biol.* **1990**, *212*, 211.

225 M. Fuxreiter, Ö. Farkas, G. Náray-Szabó, *Protein Eng.* **1995**, 925.

226 I. A. Rose, *Philos. Trans. R. Soc. London, Ser. B* **1981**, *293*, 131.

227 V. J. Jensen, S. Rugh, *Methods Enzymol.* **1987**, *136*, 356.

228 M. Meng, M. Bagdasarian, J. G. Zeikus, *Biotechnology* **1993**, *11*, 1157.

229 Y. Takasaki, S. Furutani, S. Hayashi, K. Imada, *J. Ferment. Bioeng.* **1994**, *77*, 94.

230 K. Bock, M. Meldal, B. Meyer, L. Wiebe, *Acta Chem. Scand., Ser. B* **1983**, *37*, 101.

231 A. Berger, A. de Raadt, G. Gradnig, M. Grasser, H. Löw, A. E. Stütz, *Tetrahedron Lett.* **1992**, *33*, 7125.

232 M. Ebner, A. E. Stütz, *Carbohydr. Res.* **1998**, *305*, 331.

233 J. R. Durrwachter, D. G. Drueckhammer, K. Nozaki, H. M. Sweers, C. H. Wong, *J. Am. Chem. Soc.* **1986**, *108*, 7812.

234 A. Berger, A. de Raadt, G. Gradnig, M. Grasser, H. Low, A. E. Stutz, *Tetrahedron Lett.* **1992**, *33*, 7125.

235 P. Hadwiger, P. Mayr, B. Nidetzky, A. E. Stutz, A. Tauss, *Tetrahedron: Asymmetry* **2000**, *11*, 607.

236 P. J. Card, W. D. Hitz, K. G. Ripp, *J. Am. Chem. Soc.* **1986**, *108*, 158.

237 N. Muramatsu, Y. Nosoh, *Arch. Biochem. Biophys.* **1971**, *144*, 245.

238 M. Takama, Y. Nosoh, *J. Biochem.* **1980**, *87*, 1821.

239 C. D. Hsiao, C. C. Chou, Y. Y. Hsiao, Y. J. Sun, M. Meng, *J. Structural Biol.* **1997**, *120*, 196.

240 Y. J. Sun, C. C. Chou, W. S. Chen, R. T. Wu, M. Meng, C. D. Hsiao, *Proc. Natl. Acad. Sci. U. S. A.* **1999**, *96*, 5412.

241 C. C. Chou, Y. J. Sun, M. Meng, C. D. Hsiao, *J. Biol. Chem.* **2000**, *275*, 23154.

242 C. J. Jeffery, B. J. Bahnson, W. Chien, D. Ringe, G. Petsko, *Biochemistry* **2000**, *39*, 955.

243 C. J. Jeffery, R. Hardré, L. Salmon, *Biochemistry* **2001**, *276*, 1560.

244 A. Widjaja, M. Shirishima, M. Yasuda, H. Ogino, H. Nakajima, H. Ishikawa, *J. Biosci. Bioeng.* **1999**, *87*, 611.

245 A. Widjaja, M. Yasuda, H. Ogino, H. Naka-

jima, H. Ishikawa, *J. Biosci. Bioeng.* **1999**, *87*, 693.

246 A. Widjaja, H. Ogino, M. Yasuda, K. Ishimi, H. Ishikawa, *J. Biosci. Bioeng.* **1999**, *88*, 640.

247 R. K. Wierenga, M. E. M. Noble, *J. Mol. Biol.* **1992**, *224*, 1115.

248 T. C. Alber, R. C. Davenport, G. K. Farber, D. A. Giammona, A. M. Glasfeld, W. D. Horrocks, M. Kanaoka, E. Lolis, G. A. Petsko, D. Ringe, G. Tiraby, *ACS Symp. Ser.* **1989**, *392*, 34.

249 T. K. Harris, R. N. Cole, F. I. Comer, A. S. Mildvan, *Biochemistry* **1998**, *37*, 16828.

250 C. H. Wong, G. M. Whitesides, *J. Org. Chem.* **1983**, *48*, 3199.

251 N. Ouwerkerk, J. H. van Boom, J. Lugtenburg, J. Raap, *Eur. J. Org. Chem.* **2000**, 861.

252 I. P. Korndorfer, W. D. FessnerB. W. Matthews, *J. Mol. Biol.* **2000**, *300*, 917.

253 S. H. Bhuiyan, Y. Itami, K. Izumori, *J. Ferment. Bioeng.* **1997**, *84*, 558.

254 S. H. Bhuiyan, Y. Itami, Y. Rokui, T. Katayama, K. Izumori, *J. Ferment. Bioeng.* **1998**, *85*, 539.

255 J. E. Seemann, G. E. Schulz, *J. Mol. Biol.* **1997**, *273*, 256.

256 W. D. Fessner, C. Gosse, G. Jaeschke, O. Eyrisch, *Eur. J. Org. Chem.* **2000**, 125.

257 C. Auge, S. David, C. Gautheron, *Tetrahedron Lett.* **1984**, *25*, 4663.

258 M.-J. Kim, W. J. Hennen, H. M. Sweers, C.-H. Wong, *J. Am. Chem. Soc.* **1992**, *114*, 10138.

259 I. Maru, J. Ohnishi, Y. Ohta, Y. Tsukada, *Carbohydr. Res.* **1998**, *306*, 575.

260 Y. Ohta, Y. Tsukada, T. Sugimori, K. Murata, A. Kimura, *Agric. Biol. Chem.* **1989**, *53*, 477.

261 I. Maru, Y. Ohta, K. Murata, Y. Tsukada, *J. Biol. Chem.* **1996**, *271*, 16294.

262 Y. Takamura, I. Kitamura, M. Iikura, K. Kono, A. Ozaki, *Agric. Biol. Chem.* **1966**, *30*, 338.

263 Y. Takamura, I. Kitamura, M. Iikura, K. Kono, A. Ozaki, *Agric. Biol. Chem.* **1966**, *30*, 345.

264 K. Hatakeyama, M. Goto, Y. Uchida, M. Kobayashi, M. Terasawa, H. Yukawa, *Biosci. Biotechnol. Biochem.* **2000**, *64*, 569.

265 H. Keweloh, H. J. Heipieper, *Lipids* **1996**, *31*, 129.

266 N. Morita, A. Shibahara, K. Yamamoto, K. Shinkai, G. Kajimoto, H. Okuyama, *J. Bacteriol.* **1993**, *175*, 916.

267 R. Diefenbach, H. Keweloh, *Arch. Microbiol.* **1994**, *162*, 120.

268 Q. Chen, D. B. Janssen, B. Witholt, *J. Bacteriol.* **1995**, *177*, 6894.

269 H. J. Heipieper, G. Meulenbeld, Q. van Oirschot, J. A. M. de Bont, *Appl. Environ. Microbiol.* **1996**, *62*, 2773.

270 R. Holtwick, F. Meinhardt, H. Keweloh, *Appl. Environ. Microbiol.* **1997**, *63*, 4292.

271 H. Okuyama, A. Ueno, D. Enari, N. Morita, T. Kusano, *Arch. Microbiol.* **1998**, *169*, 29.

272 V. Pedrotta, B. Witholt, *J. Bacteriol.* **1999**, *181*, 3256.

18
Introduction and Removal of Protecting Groups

Dieter Kadereit, Reinhard Reents, Duraiswamy A. Jeyaraj and Herbert Waldmann

18.1
Introduction

The proper introduction and removal of protecting groups is one of the most important and widely carried out synthetic transformation in preparative organic chemistry. In particular, in the highly selective construction of complex, polyfunctional molecules, e. g. oligonucleotides, oligosaccharides, peptides and conjugates thereof, and in the synthesis of alkaloids, macrolides, polyether antibiotics, prostaglandins and other natural products, regularly the problem arises that a given functional group has to be protected or deprotected selectively under the mildest conditions and in the presence of functionalities of similar reactivity, as well as in the presence of structures that are sensitive to acids, bases, oxidation and reduction. Numerous classical chemical methods have been developed for the manipulation of protecting groups [1-3]. Nevertheless, severe problems still remain caused by the need to introduce or remove selectively specific blocking functions which can not, or only with great difficulties, be solved by using classical chemical tools only. However, the arsenal of the available protecting group techniques has been substantially enriched by the application of biocatalysts. In addition to their stereodiscriminating properties, enzymes offer the opportunity to carry out highly chemo- and regioselective transformations. They often operate at neutral, weakly acidic or weakly basic pH values and in many cases combine a high selectivity for the reactions they catalyze and the structures they recognize with a broad substrate tolerance. Therefore, the application of these biocatalysts to effect the introduction and/or removal of suitable protecting groups offers viable alternatives to classical chemical methods [4-11].

18.2
Protection of Amino Groups [4–12]

18.2.1
N-Terminal Protection of Peptides

The selective protection and liberation of the α-amino function, the carboxy group and the various side chain functionalities of polyfunctional amino acids constitute some of the most fundamental problems in peptide chemistry. Consequently, numerous efficient protective functions based on chemical techniques have been developed to a high level of practicability. [1–3, 13, 14] However, since the mid-1970s, a systematic search for blocking groups being removable with a biocatalyst has been carried out [4–12]. In addition to the mild deprotection conditions they promise, protecting groups of this type are expected to be particularly useful for the construction and manipulation of larger peptide units, i.e. for transformations which, for solubility reasons, in general have to be carried out in aqueous systems. Also applications in the reprocessing of peptides obtained by recombinant DNA technology are foreseen (for an interesting appropriate example see Chapter 12.5).

Initial attempts to introduce an enzyme-labile amino protecting group involved the use of chymotrypsin for the removal of *N*-benzoylphenylalanine (Bz-Phe) from the tripeptide Bz-Phe-Leu-Leu-OH [15]. The desired dipeptide H-Leu-Leu-OH was obtained in 80% yield under mild conditions (pH 7.3, room temperature). Chymotrypsin, however, is an endopeptidase with a rather broad substrate tolerance, catalyzing the hydrolysis of peptide bonds on the carboxy groups of hydrophobic and of aromatic amino acid residues. Since such amino acids appear widely in peptides, and since no method is available to protect them against attack by the enzyme during the attempted deprotection, the use of chymotrypsin is problematic. Its use is therefore limited to special cases [16] in which no danger of competitive cleavage at undesired sites has to be feared. A protease of much narrower specificity is trypsin which catalyzes the hydrolysis of peptide bonds at the carboxylic group of lysine and arginine. These amino acids carry polar, chemically reactive side chain functional groups which can be protected by various techniques [13, 14]. The high specificity of trypsin together with the possibility of hiding the critical amino acids which function as primary points of tryptic cleavage allowed for the development of a broadly applicable system for the protection of the α-amino group of peptides [12, 17–19]. In several studies the application of trypsin-labile protecting groups, along with suitable blocking functions for the side chains of arginine and lysine were described [17–23]. Thus, for instance Z-Arg-OH served as the enzymatically removable protecting group in a stepwise synthesis of deamino-oxytocin 1 (Fig. 18-1) [18, 19].

Starting with a pentapeptide the amino acid chain was elongated with Z-Arg-protected amino acid *p*-nitrophenyl esters. The *N*-terminal Z-Arg protecting group was successively removed in moderate to high yield and without attack on the other peptide bonds by treatment with trypsin. Unfortunately, the preparation of the protected arginine *p*-nitrophenyl esters is difficult, thus preventing this method from becoming generally useful for the stepwise assembly of larger peptides. The trypsin-

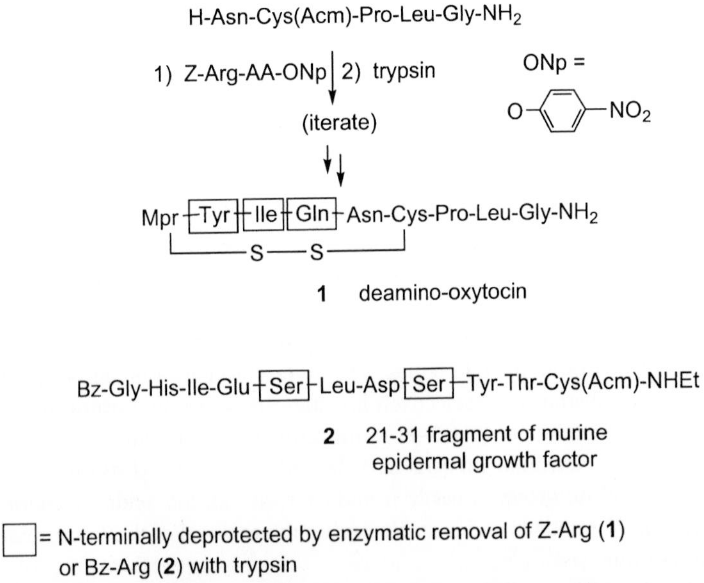

H-Asn-Cys(Acm)-Pro-Leu-Gly-NH₂

1) Z-Arg-AA-ONp | 2) trypsin   ONp =

(iterate)   O—⟨benzene⟩—NO₂

Mpr ⌐Tyr┼Ile┼Gln⌐Asn-Cys-Pro-Leu-Gly-NH₂
     └──S──S──┘

1   deamino-oxytocin

Bz-Gly-His-Ile-Glu⌐Ser⌐Leu-Asp⌐Ser⌐Tyr-Thr-Cys(Acm)-NHEt

2   21-31 fragment of murine
         epidermal growth factor

☐ = N-terminally deprotected by enzymatic removal of Z-Arg (**1**)
    or Bz-Arg (**2**) with trypsin

Figure 18-1. Construction of oligopeptides via removal of N-terminal arginine
residues with trypsin.

labile blocking groups have, however, proven to be very useful for the construction of oligo- and polypeptides via condensation of preformed peptide fragments. An illustrative example consists of a chemoenzymatic construction of the 21–31 fragment **2** of murine epidermal growth factor (Fig. 18-1). In the course of this synthesis the deblocking by trypsin was applied twice [16]. The enzyme first liberated the N-terminus of a tetrapeptide and subsequently of a heptapeptide. In a synthesis [24] of human β-lipotropin an Ac-Arg-residue was introduced by a solid-phase technique at the N-terminus of the 29 C-terminal amino acids of the desired polypeptide. After cleavage from the resin and protection of the side chain functionalities, the arginine moiety was removed with trypsin, leaving the peptide chain intact. Finally, coupling of this 61–89 fragment to a partially protected 1–60 segment, and subsequent deprotection delivered β-lipotropin. Further examples are found in syntheses of oxypressin [12], Met-enkephalin [25] and Glu⁴-oxytocin [12].

In addition to chymotrypsin and trypsin, the collagenase from *Clostridium histolyticum* has been proposed as a catalyst for the removal of N-terminally attached dummy amino acids from peptides [26]. The enzyme recognizes the tetrapeptides Pro-X-Gly-Pro and cleaves the X-Gly bond. The use of this biocatalyst permitted the construction of des-pyroglutamyl-[15-leucine]human little gastrin I by selective hydrolysis of the dipeptide Pz-Pro-Leu (Pz = 4-phenylazobenzyloxycarbonyl) from the N-terminus of the octadecapeptide Pz-Pro-Leu-Gly-Pro-Trp-Leu-(Glu)₅-Ala-Tyr-Gly-Trp-Leu-Asp-Phe-NH₂. Transformations of this type are analogous to the naturally occuring conversion of prohormones into hormones and may prove to be useful for the processing of peptide factors produced by recombinant DNA technology.

Despite the impressive syntheses that have been made possible using proteases, the use of these enzymes is always accompanied by the danger of a competitive (and sometimes unexpected and unforeseeable) cleavage of the peptide backbone at an undesired site. At a minimum, complex protecting group schemes may become necessary if the amino acid which serves as the recognition structure for the protease occurs several times in the peptide chain to be constructed. This disadvantage can be overcome if a biocatalyst devoid of peptidase activity is used for the liberation of the N-terminal amino group. This principle has been illustrated by the application of penicillin G acylase from *E. coli*[27–44] in industry for the large scale synthesis of semisynthetic penicillins and by using a phthalyl imidase from *Xanthobacter agilis*[45–47] (*vide infra*). Penicillin G acylase attacks phenylacetic acid (PhAc) amides and esters but does not hydrolyze peptide bonds. The acylase accepts a broad range of protected peptides as substrates and selectively liberates the N-terminal amino group under almost neutral conditions (pH 7–8, room temperature) leaving the amide bonds as well as the C-terminal methyl, allyl, benzyl and *tert*-butyl esters unaffected[28–35, 38]. The PhAc group is easily introduced into amino acids by chemical[48] or enzymatic[49] methods and is stable during the removal of the C-terminal protecting groups employed[29–32].

Recently, it has been shown that a phthalyl amidase isolated from *Xanthobacter agilis* is able to deprotect a variety of phthalimido substrates once the substrates are partially hydrolyzed to their monoacids (Fig. 18-2)[45–47]. The phthalyl group is commonly used for amine protection, because it completely blocks this functionality by double acylation[2, 3]. The enzymatic phthalyl removal proceeds via a two step process of weakly basic hydrolysis to yield the monoacid 4 and subsequent treatment with the phthalyl amidase (Fig. 18-2). Because the hydrolysis of the phthalimide 3 to the corresponding monoacid 4 can be catalyzed by imidases such as the rat liver imidase,[50] this procedure in particular represents a powerful alternative to the classical phthalyl deprotection which requires relatively drastic conditions and toxic reagents. However, the general applicability of the enzymatic phthalyl removal is yet to be investigated.

If the construction of PhAc- or phthalyl-peptides is carried out by chemical activation of the PhAc-amino acids, the application of the non-urethane blocking group results in ca. 6% racemization[29, 30]. However, this disadvantage can be overcome by forming the peptide bonds enzymatically, e.g. with trypsin[51], chymotrypsin[51] or carboxypeptidase Y[39, 51], or by using urethane-type protecting groups (*vide infra*). For such condensation reactions and the subsequent enzymatic removal of the PhAc group, a continuous process was developed which has the potential to be transferable to a larger scale[39].

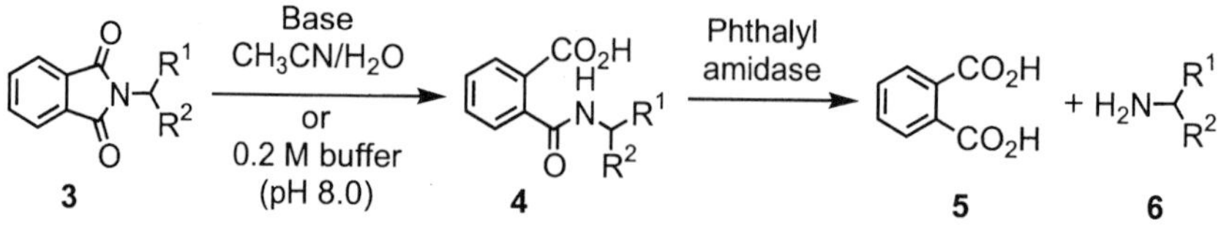

Figure 18-2. Enzymatic removal of the phthalyl group.

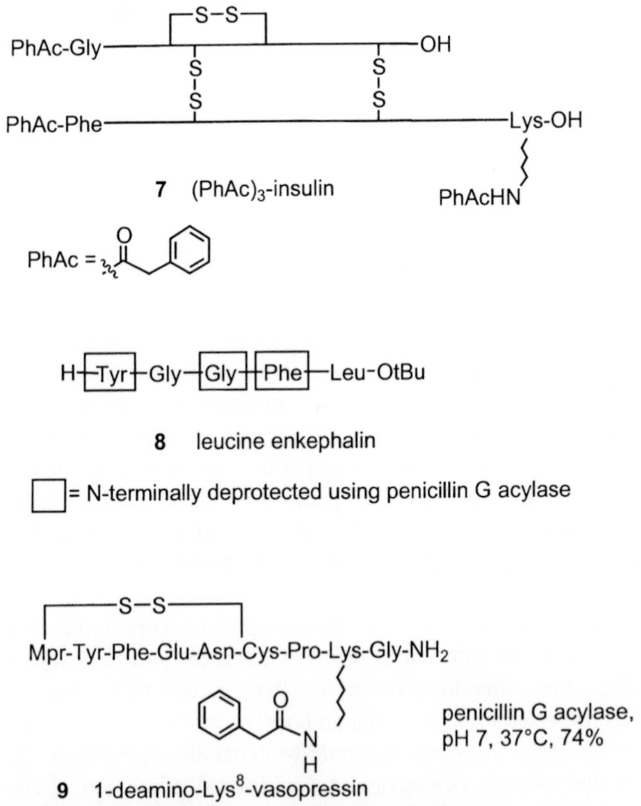

7 (PhAc)₃-insulin

$$PhAc = $$

H–Tyr–Gly–Gly–Phe–Leu-OtBu

8 leucine enkephalin

☐ = N-terminally deprotected using penicillin G acylase

penicillin G acylase,
pH 7, 37°C, 74%

9 1-deamino-Lys⁸-vasopressin

Figure 18-3. Application of the phenylacetamido (PhAc) group as an enzymatically removable amino protecting group.

The applicability of the penicillin acylase-catalyzed deprotection for the construction of larger peptides has been demonstrated by the complete deprotection of the porcine insulin derivative **7** carrying three PhAc groups[27], presumably at the N-terminal glycine of the A-chain, the N-terminal phenylalanine of the B-chain and the side chain of the lysine in position 29 of the B-chain (Fig. 18-3). The enzymatic hydrolysis proceeded to completeness and the peptide backbone was not attacked. A further interesting example is given by a recent biocatalyzed synthesis of leucine enkephalin *tert*-butyl ester **8**[38] in which all critical steps are performed by enzymes, two of them through the agency of penicillin G acylase: i) phenylacetates are introduced as N-terminal protecting groups of the amino acid esters by using penicillin G acylase, ii) the elongation of the peptide chain is carried out with papain or α-chymotrypsin, iii) the deprotection of the N-terminal amino group is achieved again by means of penicillin G acylase. These examples and also the application of this technique for aspartame syntheses[28, 40, 41], as well as the deprotection of glutathione derivatives[35] demonstrate that penicillin G acylase can be used advantageously for the N-terminal unmasking of peptides. In addition, the enzyme has been used for the liberation of the side chain functionalities of lysine and cysteine, as well as in β-lactam, nucleoside and carbohydrate chemistry (*vide infra*).

18.2.2
Enzyme-labile Urethane Protecting Groups

The enzyme-labile *N*-protecting functions described so far are simple acyl groups which typify the danger of razemization during chemical peptide syntheses. This problem can, in general, be overcome by the use of urethane blocking functions. However, so far only few examples of a biocatalytic removal of classical urethane protecting groups such as the Z- and Boc-group are known[52]. Apparently, the enzymatic attack on the urethane carbonyl group, which would initiate the cleavage process, is too inefficient to be useful for synthetic purposes. To overcome this problem, two different strategies were developed. Both concepts have in common the fact, that the enzyme-labile bond is no longer part of the urethane. However, the first approach includes the introduction of a spacer (the AcOZ- and PhAcOZ groups), while the second strategy relies on the cleavage of a glycosidic C - O-bond of a glycoside urethane by the respective biocatalyst, e.g. a glucosidase (the BGloc group).

Through the introduction of a spacer between the group which is recognized by the enzyme and the urethane, the substrate is kept at a distance from the enzyme during the reaction (Fig. 18-4). Therefore, any steric effects caused by the bulk of certain amino acids are expected to be minimal and, as the amino acid sequence does not influence the reactivity, this concept should be generally applicable to the synthesis of peptides and peptide conjugates. An additional advantage of the introduction of the spacer is the option to choose the group that is recognized by the enzyme and thus the enzyme itself.

This concept was first realized by using p-hydroxybenzyl alcohol as a spacer in the p-(acetoxy)-benzyloxycarbonyl (AcOZ) group which encorporates an acetic acid ester as the enzyme-labile bond (Fig. 18-4). Accordingly, the AcOZ group can be removed under conditions typical for acetyl ester hydrolysis, for instance by treatment with lipases or esterases[53–55]. As lipases display a broad specificity, other esters present in the substrate molecule might be hydrolyzed during the AcOZ removal. Thus, the p-(phenylacetyl)benzyloxycarbonyl (PhAcOZ) group was developed, which takes advantage of the high selectivity of penicillin G acylase for the phenylacetyl group (Fig. 18-4). The versatiliy of this enzyme-labile urethane protecting group was demonstrated by the synthesis of phosphorylated[56–60], glycosylated[56–60] and lipidated[61] peptides.

A second approach takes advantage of a characteristic property of glycosidases. It is well known that glycosidases hydrolyze their substrates by cleaving the glycosidic bond via nucleophilic attack at the anomeric carbon atom. Therefore, a carbohydrate-derived urethane protecting group would provide the desired enzyme-lability. In additional, such sugar derivatives have increased solubility in aqueous solutions, a necessary requirement for all biotransformations. This concept was successfully realized by using glucose and galactose as the carbohydrate component

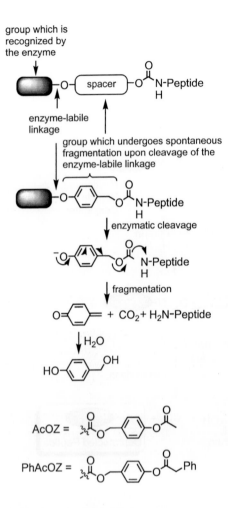

group which is recognized by the enzyme

enzyme-labile linkage

group which undergoes spontaneous fragmentation upon cleavage of the enzyme-labile linkage

enzymatic cleavage

fragmentation

Figure 18-4. Principle of the spacer-based protecting groups AcOZ and PhAcOZ.

(Fig. 18-5) [62, 63]. During the synthesis the carbohydrate hydroxy functions are blocked by either benzyl ethers in the tetra-*O*-benzyl-D-glucopyranosyloxycarbonyl (BGloc) group or acetyl groups in the tetra-*O*-acetyl-D-glucopyranosyloxycarbonyl (AGloc) or the tetra-*O*-acetyl-β-D-galactopyranosyloxycarbonyl (AGaloc) protecting groups. The removal of these carbohydrate-based protecting groups proceeds via a two step process by removing the hydroxy blocking function in a first step followed by treatment with a glucosidase (AGloc, BGloc) or galactosidase (AGaloc), respectively. In the case of the acetyl derivatives AGloc and AGaloc a sequential two step process as well as a one-pot procedure were developed for the deprotection reaction, allowing for a convenient deprotection protocol as demonstrated for dipeptide **11** (Fig. 18-5) [62].

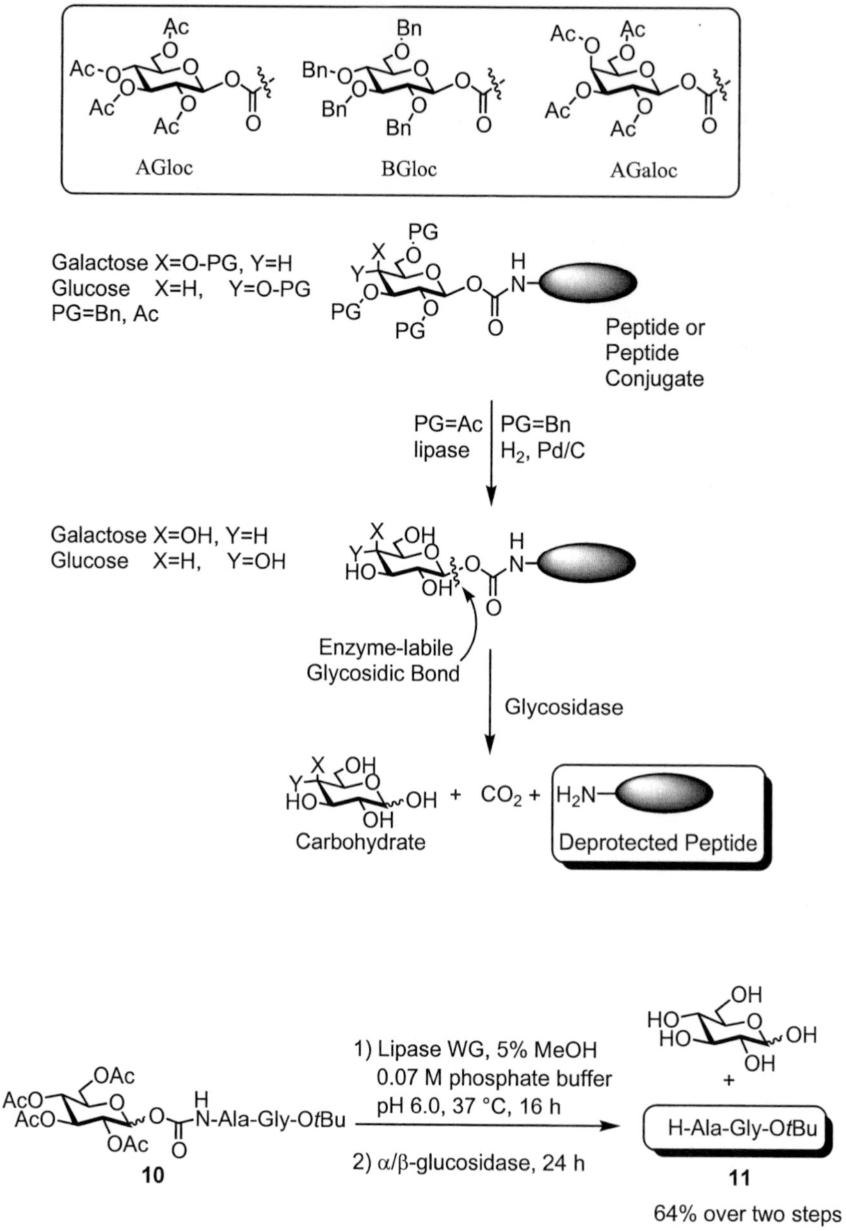

Figure 18-5. Carbohydrate-based urethane protecting groups.

18.2.3
Protection of the Side Chain Amino Group of Lysine

During chemical peptide syntheses and if trypsin is used for the construction of the peptide bonds or *N*-terminal deprotection, the side chain amino group of lysine generally has to be protected to prevent side reactions[13, 14]. This goal can be achieved enzymatically by applying the penicillin G acylase-catalyzed removal of the PhAc group (*vide supra*)[64]. Thus, the first application of the PhAc group in peptide chemistry was a synthesis of 1-deamino-Lys^8-vasopressin from the protected congener **9**, during which the lysine side chain was masked as the phenylacetamide (Fig. 18-3). After the peptide chain had been assembled and the disulfide bond was formed by oxidative cyclization, the PhAc group could be removed enzymatically in 74 % yield without side reaction. A further interesting example which demonstrates that this technique can be applied advantageously to the synthesis of even larger peptides is found in the complete deprotection of (PhAc)$_3$porcine insuline (*vide supra*, Fig. 18-3)[27] and modified insuline fragments[65]. Since penicillin acylase is commercially available and devoid of peptidase activity[66], this method appears to be generally useful for the construction of lysine-containing oligopeptides.

In addition to the PhAc group, pyroglutamyl amides (Glp) were proposed as enzymatically removable blocking functions for the lysine side chain[23]. Their removal was achieved with pyroglutamate aminopeptidase from calf liver. Thus, all *N*-protecting groups were split off from the protected RNAse 1–10 fragment Glp-Lys(Glp)-Glu-Thr-Ala-Ala-Ala-Lys(Glp)-Phe-Glu-Arg-OH and from a model dipeptide. The general usefulness of this method remains to be demonstrated, however.

18.2.4
Protection of Amino Groups in β-Lactam Chemistry

The enzymatic removal of acyl groups plays an important role in the industrial production of semisynthetic penicillins and cephalosporins. To this end, penicillin G **12** (R = CH$_2$-Ph) and penicillin V **12** (R = CH$_2$-O-Ph), or the respective cephalosporins are first deacylated by means of penicillin acylases (Fig. 18-6)[67, 68]. The 6-aminopenicillanic acid and the 7-aminocephalosporanic acid thus obtained are subsequently acylated by non-enzymatic or enzymatic methods to give the semisynthetic antibiotics **13**.

The manufacture of therapeutically important cephalosporins from penicillin G and V includes a chemical ring expansion of the thiazolidine ring to a dihydrothiazine. In the course of this sequence the amino group remains protected as phenylacetyl or phenoxyacetyl amide, which is finally removed using penicillin G or V acylase. Of particular importance is the choice of a suitable protecting function for the COOH group. It must be stable during the ring expansion but removable without damaging the ceph-3-em nucleus. As an alternative to chemical methods, the use of the phenylacetoxymethylene ester was suggested for this purpose[41, 69]. It is easily introduced and is stable during the construction of the cephalosporin framework (Fig. 18-6). Together with the phenylacetamide the ester can eventually be

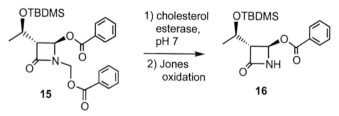

Figure 18-6. Enzymatic deprotection of amino- and carboxy groups in β-lactam chemistry.

removed in high yield from penicillin G and the cephalosporins **14** by penicillin G acylase. The formaldehyde formed in the deprotection is not harmful to the enzyme.

In a new approach to the well known versatile β-lactam building blocks, an enzymatic deprotection of an acylated methylol amide was applied with advantages (Fig. 18-6) [70]. Thus, the dibenzoate **15** was regioselectively saponified by cholesterol esterase at pH 7 giving rise to a monoacylated aminal. After Jones oxidation and subsequent loss of formaldehyde, the azetidinone **16** was obtained, which can be transformed into various enantiomerically pure penem and carbapenem building blocks.

As an alternative to the well established phenylacetyl group in β-lactam chemistry, recently a biocatalyzed procedure for the removal of phthalyl imide has been described (Fig. 18-2) [45, 71]. Its general usefulness remains to be demonstrated, however.

18.2.5
Protection of Amino Groups of Nucleobases

In general, the amino groups of the nucleobases adenine, guanine and cytosine in general must be protected during oligonucleotide synthesis to prevent undesired side reactions. To this end, they usually are converted into amides which are finally hydrolyzed under fairly basic conditions. If the amino functions are, however, masked as phenylacetamides, the protecting functions can be cleaved off by again employing penicillin G acylase (Fig. 18-7) [72–78]. The enzyme, for instance, se-lectively liberates the amino groups of the deoxynucleosides **17** without attacking the acetates in the carbohydrate parts and without damage to the acid-labile N-glycosidic bonds.

The biocatalyzed phenylacetyl removal can be carried out using both solubilized or immobilized substrates [77]. The latter methodology has been developed using controlled pore glass (CPG) as a solid support (Fig. 18-7).

18.3
Protection of Thiol Groups [4–6, 8, 12]

18.3.1
Protection of the Side Chain Thiol Group of Cysteine

The liberation of the β-mercapto group of cysteine was also achieved by means of the penicillin G acylase mediated hydrolysis of phenylacetamides [33–35]. To this end, the SH group was masked with the phenylacetamidomethyl (PhAcm) blocking function (Fig. 18-7). After penicillin acylase-catalyzed hydrolysis of the amide incorporated in the acylated thioaminal (see, e. g. **18**), a labile S-aminomethyl compound is formed which immediately liberates the desired thiol. This technique was for instance applied in a synthesis of glutathione which was isolated as the disulfide **19**. In a related glutathione synthesis the method was used for the simultaneous liberation of the SH- and the N-terminal amino function of glutamine [34, 35].

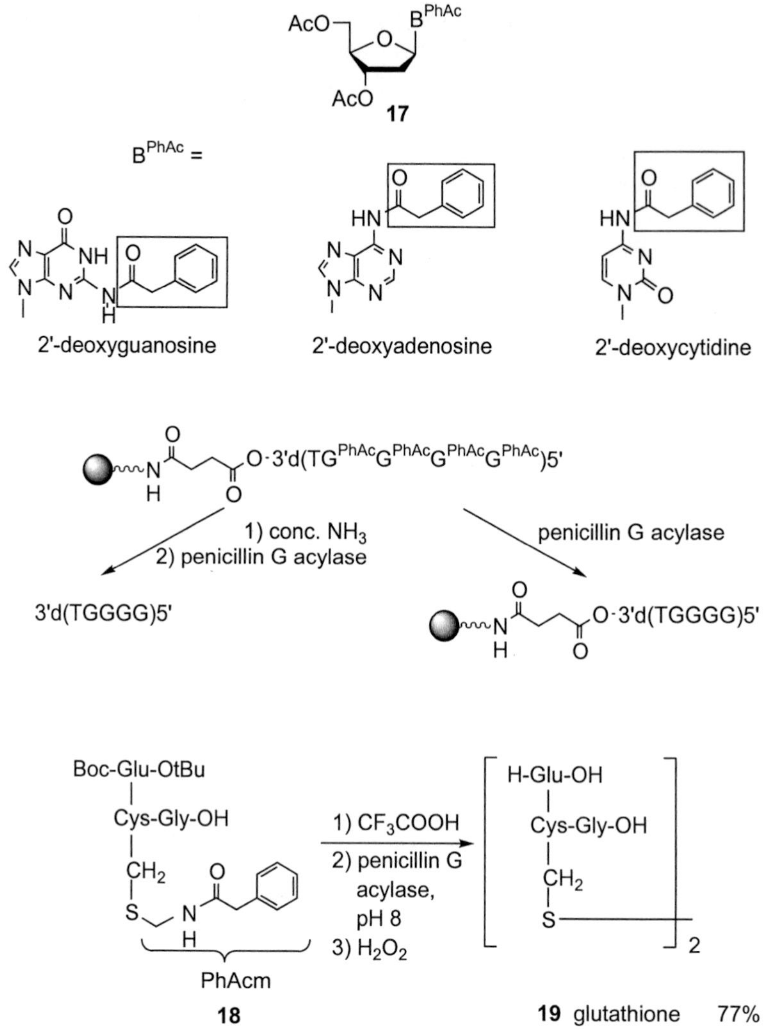

$B^{PhAc} =$

2'-deoxyguanosine   2'-deoxyadenosine   2'-deoxycytidine

Figure 18-7. Enzymatic deprotection of the amino groups of nucleobases and the mercapto group of cysteine by means of penicillin G acylase. The shaded balls represent controlled pore glass (CPG).

18.4
Protection of Carboxy Groups [4–9, 12, 79]

18.4.1
C-Terminal Protection of Peptides

As in the enzymatic liberation of the *N*-terminus of peptides, initial attempts to achieve an enzyme-catalyzed deprotection of the corresponding carboxyl groups concentrated on the use of the endopeptidases chymotrypsin[80–82], trypsin[81, 83, 84] and thermolysin[85], a protease obtained from *Bacillus thermoproteolyticus* which hydrolyzes peptide bonds on the amino side of hydrophobic amino acid residues (e. g. leucine, isoleucine, valine, phenylalanine). This latter biocatalyst enables the cleavage of the "supporting" tripeptide ester H-Leu-Gly-Gly-OEt from a protected undecapeptide to take place (pH 7, room temperature). The octapeptide thereby obtained was composed exclusively of hydrophilic amino acids. Owing to the broad substrate specificity of thermolysin and the resulting possibility of unspecific peptide hydrolysis this method can not be regarded as being generally applicable.

The exploitation of the esterase activities of chymotrypsin and trypsin opened routes to the hydrolysis of several peptide methyl, ethyl and *tert*-butyl esters at pH 6.4 to 8 and room temperature[80, 81]. The transformations are not only successful with peptides carrying the respective enzyme-specific amino acids at the *C*-terminus, but in several cases different amino acids were also tolerated at this position. However, severe drawbacks of this methodology are that numerous peptides are poor substrates or are not accepted at all. Moreover, a competitive cleavage of the peptide bonds occurs if the peptides contain trypsin- or chymotrypsin-labile sequences. Therefore, these proteases appear not to be generally useful for a safe *C*-terminal deprotection as well.

The disadvantages of using by the endopeptidases can be overcome by using carboxypeptidase Y from baker's yeast[25, 86, 87]. This serine-exopeptidase also has esterase activity and is characterized by quite different pH-optima for the peptidase and the esterase activity (pH >8.5). Even in the presence of various organic cosolvents the enzyme selectively removes the carboxy protecting groups from a variety of differently protected di- and oligopeptide methyl and ethyl esters[25, 87] without attacking the peptide bonds. An additional attractive feature is, that its esterase activity is restricted to α-esters, consequently β- and γ-esters of aspartic and glutamic acid, respectively, are not attacked. Carboxypeptidase Y was used advantageously for the stepwise *C*-terminal elongation of the peptide chain in aqueous solution employing a solubilizing poly(ethylene glycol) derived polymeric support as the *N*-terminal blocking group[86]. In a further remarkable synthesis which did not include the use of a polymeric *N*-protecting group, Met-enkephalin **20** was built up employing carboxypeptidase Y for *C*-terminal deprotection of intermediary generated peptide amides as well as for the formation of the peptide bonds (Fig. 18-8)[25].

The additional opportunity to hydrolyze selectively *C*-terminal peptide amides with carboxypeptidase Y is of particular interest if, as is demonstrated in the above mentioned example, enzymatic methods are applied to the formation of the peptide bonds, because amino acid amides are often the nucleophiles of choice in these biocatalyzed processes. For this purpose a peptide amidase from the flavedo of oranges shows very promising properties[88–90]. The enzyme is equipped with a broad substrate specificity and accepts Boc-, Trt-, Z- and Bz-protected and *N*-terminally unprotected peptide amides (Fig. 18-8). The *C*-terminal amides are saponified in high yields at pH 7.5 and 30 °C without affecting the *N*-terminal blocking groups or the peptide bonds. A noticeable advantage of this biocatalyst is

H—[Tyr]—Gly—Gly—[Phe]—Met–OH

20   methionine enkephalin

[ ] = C-terminally deprotected by enzymatic saponification of the peptide amide with carboxypeptidase Y;Tyr was N-terminally deprotected by removal or Bz-Arg with trypsin

$$\text{PG-peptide-NH}_2 \xrightarrow[\text{pH 7.5, 30°C}]{\substack{\text{amidase from the} \\ \text{flavedo of oranges}}} \text{PG-peptide-OH}$$

PG	peptide	conv. [%]
Bz	Tyr-Ser	100
Boc	Leu-Val	20
Trt	Gly-Leu-Val	100
Z	Gly-Gly-Leu	100

Figure 18-8. C-terminal deprotection of peptide amides by carboxypeptidase Y and an amidase from the flavedo of oranges.

that N-deprotected amino acid amides, in contrast to the respective peptide amides, do not belong to its substrates. They can, therefore, be used as nucleophiles in peptide syntheses catalyzed by this enzyme, i.e. the formation of the peptide bond together with the subsequent C-terminal deprotection is achieved in a single step.

A further possibility for the enzymatic removal of C-terminal blocking groups is opened up by the application of enzymes which generally display a high esterase/protease ratio. Such a biocatalyst is the alkaline protease from *Bacillus subtilis DY* which shows similarities to Subtilisin Carlsberg. For this enzyme the ratio of esterase to protease activity is >10^5. It selectively removes methyl, ethyl and benzyl esters from a variety of Trt-, Z- and Boc-protected di- and tripeptides and a pentapeptide at pH 8 and 37 °C (Fig. 18-9) [91].

The N-terminal urethanes and the peptide linkages are left intact. A further protease which fulfills the requirements for a successfull application in peptide chemistry is alcalase, a serine endopeptidase from *Bacillus licheniformis* whose major component is subtilisin A (Subtilisin Carlsberg) [92–94]. It can advantageously be employed with advantage to selectively saponify peptide methyl and benzyl esters (Fig. 18-9). In a solvent system consisting of 90 % *tert*-butanol and 10 % buffer (pH 8.2) even highly hydrophobic and in aqueous solution insoluble Fmoc peptides were accepted as substrates and deprotected at the C-terminus without any disturbing side reactions. A selective classical alkaline saponification of methyl esters would be impossible due to the base-sensitivity of the Fmoc group.

$$\text{PG-peptide-OR} \xrightarrow[\text{pH 8, 37°C}]{\substack{\text{alkaline protease from} \\ \textit{Bacillus subtilis} \text{ DY}}} \text{PG-peptide-OH}$$

PG	peptide	R	yield [%]
Z	Tyr(tBu)-Glu-Leu	Me	93
Boc	Leu-Glu-Val	Bzl	85
Trt	Ala-Glu-Asp-Leu-Glu	Bzl	80

$$\text{PG-peptide-OR} \xrightarrow[\substack{\text{90 vol\% tert-butanol,} \\ \text{10 vol\% buffer}}]{\text{alcalase, pH 8.2, 35°C}} \text{PG-peptide-OH}$$

PG	peptide	R	yield [%]
Fmoc	Ala-Val-Ile	Me	85
Fmoc	Asn-Phe	Bzl	90
Boc	Met-Leu-Phe	Me	80
Z	Met-Asp(OMe)-Phe	Me	90

Figure 18-9. *C*-terminal deprotection of peptide esters by the alkaline protease from *Bacillus subtilis* DY and alcalase.

A very promising and unusually stable biocatalyst is thermitase, a thermostable extracellular serine protease from the thermophilic microorganism *Thermoactino-myces vulgaris* whose esterase/protease ratio amounts to >1000 : 1. The enzyme shows a broad amino acid side chain specificity and cleaves methyl, ethyl, benzyl, methoxybenzyl and *tert*-butyl esters from a variety of Nps-, Boc-, Bpoc- and Z-protected di- and oligopeptides in high yields at pH 8 and 35–55 °C (Fig. 18-10) [33, 34, 95–97]. In addition, it is specific for the α-carboxy groups of Asp and Glu. To enhance the solubility of the substrates, furthermore, up to 50 vol% of organic cosolvents such as DMF and DMSO may be added which also serve to reduce the remaining peptidase activity to a negligible amount [34, 97].

In the discussion of the protease-catalyzed cleavage of the *N*-terminal protecting groups it has already been pointed out that the use of biocatalysts belonging to this class of enzymes in general, i.e. also for the *C*-terminal deblocking, may lead to an undesired hydrolysis of peptide bonds. In particular, this has to be expected if the respective ester or amide to be hydrolyzed turns out to be only a poor substrate, which is only attacked slowly, an experience not uncommon if unnatural substrates are subjected to enzyme mediated transformations. This undesired possibility would, however, be overcome if enzymes were used which were not able to split amides at all. This principle has been realized in the development of the heptyl

thermitase, pH 8, 55°C

PG-peptide-OR $\xrightarrow{\hspace{3cm}}$ PG-peptide-OH

10-60 vol% organic cosolvent

PG	peptide	R	yield [%]
Z	Leu-Val-Glu(tBu)-Ala	Me	92
Boc	Pro-Gly	Me	73
Bpoc	Tyr(tBu)-Glu-Leu	Me	55
Nps	Ser(Bzl)-His(Dnp)-Leu-Val-Glu(tBu)-Ala	Me	90

Figure 18-10. *C*-terminal deprotection of peptide esters by thermitase.

lipase from
Rhizopus niveus

PG-peptide-OR $\xrightarrow{\hspace{3cm}}$ PG-peptide-OH

pH 7, 37°C

21 R = $(CH_2)_6CH_3$
22 R = $(CH_2)_2Br$

PG	peptide	R	yield [%]
Boc	Ser-Thr	Hep	95
Z	Thr-Ala	Hep	85
Aloc	Met-Gly	Hep	90
Z	Ser-Phe	EtBr	84
Boc	Val-Ala	EtBr	95

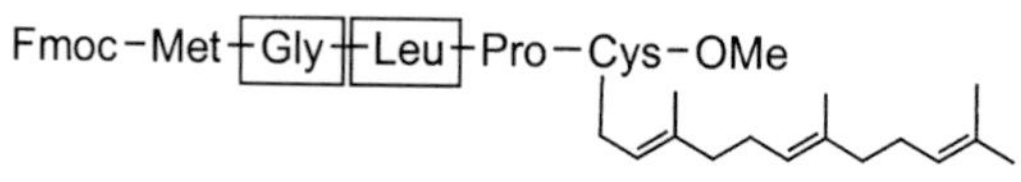

Fmoc – Met ┤Gly├┤Leu├ Pro – Cys – OMe

23 C-terminal pentapeptide of the N-Ras protein

☐ = C-terminally deprotected by employing lipase from *Rhizopus niveus*

Figure 18-11. *C*-terminal deprotection of peptide esters by lipase from *Rhizopus niveus*.

(Hep), [4–9, 31, 32, 98–100] the 2-bromoethyl (EtBr) [4–6, 31, 32, 101] and the *p*-nitrobenzyl (PNB) esters [102] as carboxy protecting groups for peptide synthesis which can be enzymatically removed by means of lipases or esterases, respectively (Fig. 18-11).

The Hep-esters proved to be chemically stable during the removal of the *N*-terminal Z-, Boc- and the Aloc-group from the dipeptides **21**. The selective removal of the Hep-esters was achieved by a lipase-catalyzed hydrolysis. From several enzymes investigated, a biocatalyst isolated from the fungus *Rhizopus niveus* was superior to the others with respect to substrate tolerance and reaction rate. The enzyme accepts a variety of Boc-, Z- and Aloc-protected dipeptide Hep-esters as substrates and hydrolyzes the ester functions in high yields at pH 7 and 37 °C without damaging the urethane protecting groups and the amide bonds (Fig. 18-11)[98, 99]. Z- and Boc-dipeptide-2-bromoethyl esters 22 are also attacked, at a comparable or in some cases even higher rate. In the presence of either one of the enzyme-labile protecting groups the N- and C-terminal amino acid can be varied considerably. With increasing steric bulk and lipophilicity of the amino acids, in particular the C-terminal one, the rate of the enzymatic reactions decreases. If the C-terminal amino acid is proline, the enzymatic reaction does not take place. The lipase-mediated deprotection of peptides was for instance successfully applied in the construction of the C-terminal pentapeptide methyl ester 23 of the N-Ras-protein, which is localized in the plasma membrane and which plays a vital role in cellular signal transduction (Fig. 18-11)[103].

The use of lipases for the removal of protecting groups from peptides in addition to the absence of protease activity has several advantages. Various enzymes belonging to this class and stemming from different natural sources (including mammals, bacteria, fungi and thermophilic organisms) are commercially available and fairly inexpensive. This variety provides the opportunity of replacing a chosen biocatalyst by a better one if a particular substrate is only attacked slowly (vide infra). The lipases are not specific for L-amino acids but also tolerate the presence of the D-enantiomer[104]. A noticeable feature is that, in contrast to proteases and esterases, they operate at the interface between water and organic solvents[105]. This is particularly important if longer peptides, which are composed of hydrophobic amino acids and/or carrying side chain protecting groups, and that do not dissolve well in the aqueous systems, have to be constructed.

The full capacity of the lipase mediated technique for C-terminal deprotection was demonstrated by the synthesis of complex base-labile phosphopeptides[44] and O-glycopeptides, which are sensitive to both acids and bases[106, 107]. To this end, e. g. the serine glycoside 24 was selectively deprotected at the C-terminus by lipase from the fungus Mucor javanicus (Fig. 18-12).

The carboxylic acid 25 liberated thereby was then coupled with an N-terminally deprotected glycodipeptide and after subsequent enzyme-mediated deprotection the glycotripeptide carboxylic acid 26 was obtained in high yield. This compound was finally condensed with a tripeptide to give the complex diglycohexapeptide 27, which carries the characteristic linkage region of a tumor-associated glycoprotein antigen found on the surface of human breast cancer cells. In the course of these enzymatic transformations, the N-terminal urethanes, the peptide bonds, the acid- and base-labile glycosidic linkages and the acetyl protecting groups, being sensitive to bases, were not attacked. In these cases lipase from Rhizopus niveus which was the enzyme of choice for simple peptides only attacked the substrates slowly, so that a different biocatalyst had to be used. This demonstrates the above mentioned advantage of being able to apply several catalytic proteins of comparable activity but different substrate tolerance for the solution of a given synthetic problem.

The viability and the wide applicability of the principle of using enzymes for the removal of individual protecting groups from complex multifunctional compounds such as lipo- and glycopeptides is furthermore proven by the finding that proteases can also be used for this purpose. Thus, by means of thermitase-catalysis the C-

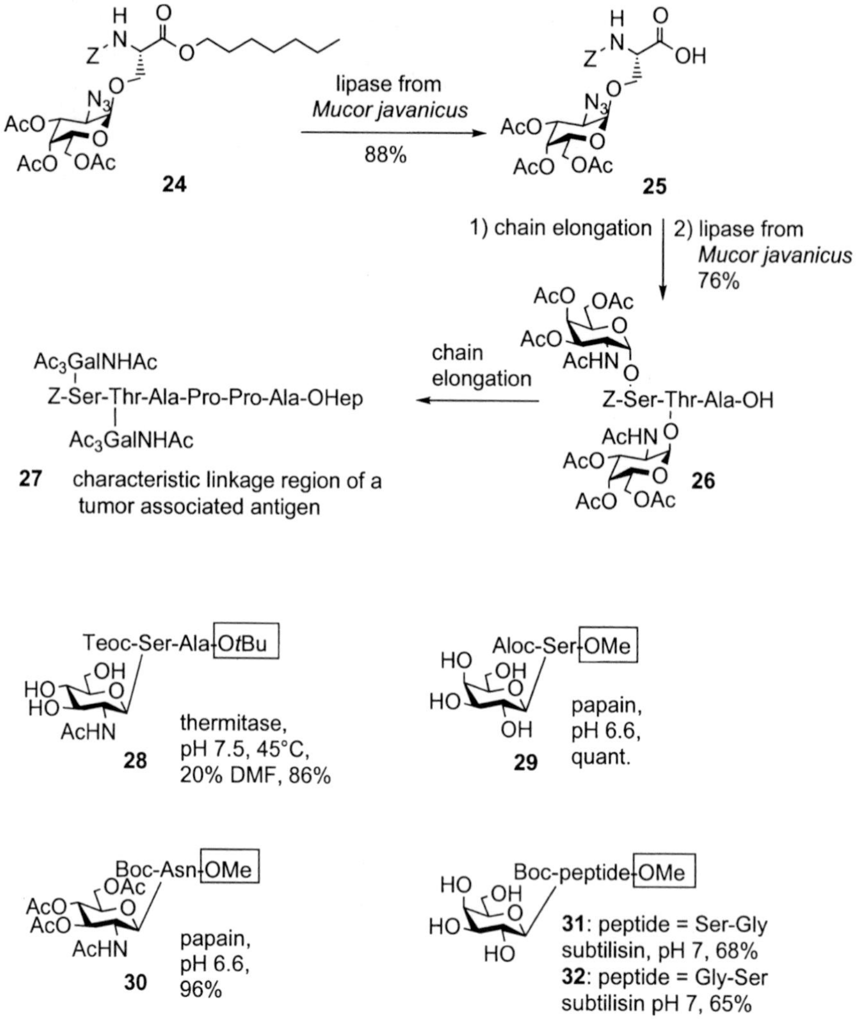

Figure 18-12. Construction of acid- and base labile glycopeptides via enzyme-mediated C-terminal deprotection.

terminal *tert*-butyl ester was removed from the glycopeptide **28** (Fig. 18-12)[34, 108]. In a different study, this enzyme was also used for the cleavage of methyl and *p*-nitrobenzyl esters[109]. From the serine glycoside **29**[110, 111] and from the asparagine conjugate **30**[112] the methyl esters could be cleaved off without disturbing side reactions by using papain as the biocatalyst. Similarly, the liberation of the *C*-terminal carboxy group of the glycosylated dipeptides **31** and **32** was achieved by means of subtilisin-catalyzed hydrolysis[113]. However, in these cases papain could not be used since this protease preferably cleaved the peptide bonds. This example again highlights the danger associated with the use of a protease for the removal of protecting groups from peptides.

A problem arising regularly in the enzymatic deprotection is the poor solubility of the fully blocked peptides in the required aqueous media, resulting in a limited accessibility of the substrates to the enzymes. To overcome this difficulty, in many cases solubilizing organic cosolvents are added, however, a more general and viable approach consists of the introduction of solubilizing protecting groups, e. g. in the enzyme-mediated formation of peptide bonds (see Chapter B 2.5) [114]. An enzymatically removable solubilizing ester protecting group could be found in the ethylene glycol derived esters such as the methoxyethyl (ME) esters [78, 115], and the methox-

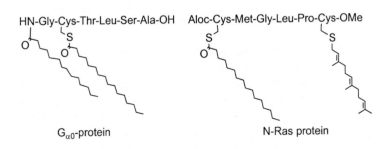

PG-peptide-O $\overbrace{}$ O $\}_n$ CH₃ $\xrightarrow[\text{pH 7, 37 °C}]{\text{lipase}}$ PG-peptide-OH

33

n=1: methoxyethyl (ME)
n=2: methoxyethoxyethyl (MEE)

Boc-peptide-O $\overbrace{}$ NMe₃⁺ Br⁻ $\xrightarrow[\text{pH 6.5, r.t.}]{\substack{\text{butyrylcholine} \\ \text{esterase from} \\ \text{horse serum}}}$ Boc-peptide-OH

34

Cho

H-Ser-Gly-Asp(OH)-OH

adenovirus 2 nucleoprotein

H-Thr-Gln-Thr-Ser-Ser-Ser-Gly-OH
OP(O)(OH)₂

serum response factor (SRF)

HN-Gly-Cys-Thr-Leu-Ser-Ala-OH

G$_{\alpha 0}$-protein

Aloc-Cys-Met-Gly-Leu-Pro-Cys-OMe

N-Ras protein

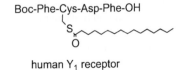

Boc-Phe-Cys-Asp-Phe-OH

human Y₁ receptor

Figure 18-13. Use of hydrophilic esters as solubilizing enzymatically removable protecting groups for the synthesis of characteristic protein fragments.

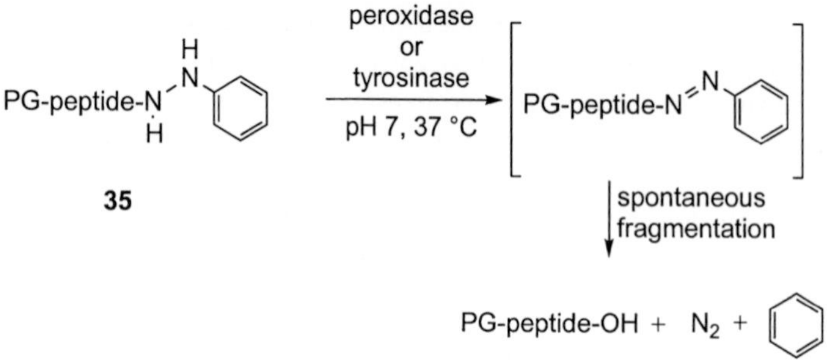

Figure 18-14. Phenylhydrazide as a carboxy protecting group.

yethoxyethyl (MEE) esters [78, 115–117] and in the choline esters (Fig. 18-13) [58, 59, 76, 78, 118–121]. The ME and MEE esters serve both as hydrophilic analogues of the heptyl esters discussed above and can therefore be removed by the same biocatalysts such as the lipase from *Mucor javanicus*. Their increased solubility in aqueous media has been used successfully in the synthesis of small peptides and peptide conjugates including glyco- [115–117] and nucleopeptides [78].

Similarly, the respective dipeptide choline esters **34** are readily soluble in purely aqueous media (i. e. without added cosolvent) and are converted into the corresponding carboxylic acids under the mildest conditions, and without side attack on the peptide bonds and the *N*-terminal urethanes, by means of the commercially available butyrylcholine esterase from horse serum. The increased hydrophilicity of peptide choline esters was used advantageously used for the synthesis of peptides and very sensitive peptide conjugates such as lipidated peptides [118–121], phosphorylated and glycosylated peptides [58, 59] and nucleopeptides (Fig. 18-13) [76, 78].

Recently, phenylhydrazide has been introduced as an enzyme-labile carboxy protecting group [122, 123]. This protecting group can be removed by mild enzymatic oxidation using a peroxidase [122, 123] or mushroom tyrosinase [124] (Fig. 18-14).

18.4.2
Protection of the Side Chain Groups of Glutamic and Aspartic Acid

The stepwise removal of arginine methyl ester by proteases has been investigated as a possibility for the enzymatic deprotection of the side chain carboxylate groups of the aminodicarboxylic acids aspartic acid (Asp) and glutamic acid (Glu). To this end, Z-Asp(ArgOMe)-NH$_2$ and Z-Glu(ArgOMe)-NH$_2$ were converted into Z-Asp(OH)-NH$_2$ and Z-Glu(OH)-NH$_2$ by subsequent treatment with trypsin, which hydrolyzes the arginine methyl esters, and with porcine pancreatic carboxypeptidase B, which splits off the arginines [125]. Since the second step is slow and requires high concentrations of the carboxypeptidase, this method can, most probably, not be applied routinely in peptide synthesis because it introduces too much of a danger of competitive side reactions.

However, enzymatic transformations have proved to be useful for the synthesis of selectively functionalized aspartic and glutamic acid derivatives. For instance, alcalase selectively hydrolyzes the α-benzyl esters of H-Asp(Bzl)-OBzl and H-Glu(Bzl)-OBzl in 82% and 85% yield, respectively, on a decagramm scale[126]. Similarly, aspartyl- and glutamylpeptides can be deprotected selectively at the C-terminus by this enzyme, however, in these cases an undesirable attack on the peptide bonds may occur[127]. In addition, Z-Asp(OAll)-OAll is converted into Z-Asp(OAll)-OH in quantitative yield by papain[128]. Also a lipase from *Candida cylindracea* is able to differentiate between the two carboxylic acid groups of glutamic acid. From the respective di-cyclopentyl ester it preferably (ratio 20 : 1) removes the γ-ester in 90% yield[129]. In addition, the enzyme thermitase and the alkaline protease from *Bacillus subtilis* (*vide supra*) also have great potential for the selective manipulation of dicarboxylic amino acids.

The examples given in Sections 18.2 to 18.4 demonstrate that the selective deprotection of peptides can be achieved advantageously by making use of enzymatic reactions. In the light of the increasing number of available biocatalysts it appears that in the near future a host of new and superior enzymatically removable blocking groups for the synthesis of peptides will be developed. However, these techniques will definitely not be used for the preparation of simple small peptides in the laboratory. Most probably they will be applied to the synthesis of sensitive polyfunctional compounds and long oligopeptides, the construction of which is cumbersome by standard chemical methods. Furthermore, they offer significant advantages if a technical process for the manufacturing of a given peptide has to be developed. Finally, together with the recently developed methods for the biocatalyzed formation of peptide bonds (see Chapter 12.5)[130], enzymatic protecting group techniques could prove to be the tools of choice for the construction of peptides in aqueous solution, the practical development of which has been tried for several decades[131,132].

18.5
Protection of Hydroxy Groups [4–9, 133–136]

Mono- and oligosaccharides, alkyl- and arylglycosides and various other glycoconjugates generally include a multitude of hydroxyl groups of comparable chemical reactivity. Also, the synthesis of oligonucleotides and nucleosides, ß-lactams, alkaloids, steroids and peptides often requires the selective protection of one or more alcoholic functions. Consequently, for the directed construction of polyhydroxy compounds these functional groups have to be manipulated selectively, in general making cumbersome protection and deprotection steps necessary. Although numerous chemical techniques are available to mask or to liberate hydroxyl groups, [1–3] the development of enzymatic methods for this purpose has been progressing steadily and appears to complement the arsenal of classical tools. In addition, the enzymatic protection of hydroxy goups (and vice versa of carboxy groups) in racemic compounds as well as their enzyme-catalyzed deprotection has been used extensively for the separation of enantiomeric alcohols and carboxylic acids (see Chapter 11).

18.5.1

Protection of Monosaccharides[133, 137]

The selective protection and deprotection of carbohydrates can be achieved with various classical chemical techniques[1–3, 138–140]. In addition, however, owing to the synthetic challenge the multifunctional carbohydrates pose, enzymatic techniques for the introduction of blocking groups into sugars and/or their subsequent removal offer further, different opportunities.

The enzymatic acylation of sugars in aqueous solution has been reported but gives low yields as the equilibrium for the reaction favors hydrolysis. However, enzymatic acylation in dry organic solvents has shown substantial success. While direct enzymatic esterification of alcohols with acids is often not practical, good to excellent yields have been obtained using transesterification techniques (Table 18-1). The displacement of the equilibrium toward products has been accomplished by using an excess of the acyl donor and by using activated, irreversible acyl donors such as trihaloethyl esters[141], enol esters[142], acid anhydrides or oxime esters[134, 136]. In particular, the enol esters have the advantage that the liberated enol tautomerizes to a ketone or an aldehyde, thereby shifting the equilibrium toward the desired products and consequently giving higher yields. This technology, however, is not restricted to carboxylic acid derivatives being the acyl donor. Organic carbonates[143], either activated as the vinyl[144] or, even better, as an oxime[145] derivative, allow for the enzyme-catalyzed synthesis of carbonates such as the methoxycarbonyl, the benzyloxycarbonyl (Z) and the allyloxycarbonyl (Aloc) carbonate. The last two examples can later be removed by non-enzymatic means.

The high polarity of sugars and their derivatives requires that polar solvents be used to dissolve them. Solvents found to be suitable include pyridine, DMSO, DMF and dimethylacetamide. However, these solvents also often inactivate enzymes, although some enzymes, for instance the lipases from the porcine pancreas (PPL), from *Candida antarctica* (CAL), from *Candida cylindracea* (CCL, later renamed *Candida rugosa*) and the lipase from *Pseudomonas cepacia* (PSL) as well as the proteases subtilisin and proleather, maintain their inherent acitvity[146]. A less polar solvent such as THF allows the use of a broader variety of lipases, but does not dissolve unmodified pyranoses. Nevertheless, it should be noted that even glucose suspended in THF has been successfully acylated by using lipase of *Candida antarctica*[147].

To remain active in an organic solvent, the enzyme must contain a small amount of water which is required for maintaining the correct protein structure. In the absence of this essential water, highly polar compounds such as carbohydrates form excessively tight enzyme-product complexes. This inhibits association and dissociation of substrates and products from the active site and thus slows down the reaction. Accordingly, the addition of drying agents such as zeolite CaA not only influences activity of the the biocatalyst but also its selectivity. For instance, the acylation of 1-*O*-methyl β-D-glycopyranoside **49** catalyzed by lipase SP 435 (an immobilized lipase from *Candida antarctica*) in ethyl butanote as the solvent and acyl donor led to acylation predominantly in the 6-position[148, 149]. If zeolite CaA was added, a mixture of 2,6- and 3,6-bisacylated pyranosides (95 : 5) was formed. In the presence of zeolite CaA and *tert*-butanol as a cosolvent, again monoacylation in the 6-position was observed.

Alternatively, precipitation of the enzyme from aqueous solution at its optimum pH prior to its use in an organic solvent has also been reported to increase the enzyme's activity greatly.

The results of enzymatic acylation of several pyranose and furanose sugars are shown in Table 18-1. Other lipophilic carbohydrate derivatives such as alkyl glycosides also display a higher solubility in less polar organic solvents, in which most lipases tend to be more stable than in polar solvents.

A further interesting finding is that heat stable lipases are capable of transferring long-chain fatty acids to the 6-hydroxy group of ethyl glucoside on a kilogram-scale, utilizing the molten fatty acids themselves as solvents [171]. On a somewhat smaller scale, the acylation of glucose has also been carried out using only a minute amount of solvent [172] or in supercritical CO_2 [173, 174].

The regioselectivity observed in the acylation of underivatized pyranoses in principle parallels that recorded for the classical chemical introduction of acyl groups into carbohydrates. However, if the 6-OH groups are protected first or deoxygenated, in the corresponding enzymatic reactions selectivities are observed which can not be realized with classical chemical methods. By careful choice of solvent and lipase, it is possible to modifiy selectively a number of C6 protected pyranoses at the secondary hydroxy groups (Table 18-2).

By combination of enzymatic with non-enzymatic protection group chemistry, carbohydrates can be selectively modified in the primary and secondary hydroxy positions. To demonstrate this versatility, the straightforward synthesis of differently mono-acylated glucose derivatives is described in Fig. 18-15. For instance, 6-*O*-butyrylated glucose **66a** (R = n-butanoyl; prepared enzymatically, see Table 18-1) is converted into the 3,6-dibutanoate **93** by lipase from *Chromobacterium viscosum* (CVL) or from *Aspergillus niger* (ANL). The 2,6-dibutanoate **94** can conveniently be built up with the lipase from porcine pancreas (PPL; Fig. 18-15) [164]. Similar observations were reported for n-octylglucoside, but for the corresponding galactose- and mannose 6-esters the selectivity was lower. In contrast, the chemical butyrylation of glucose derivative **66a** with the acid anhydride in pyridine gave a complex mixture of various diesters without any significant regiodiscrimination. The enzymatic approach was also used to convert the 6-*O*-tritylglucose **66b** (R = Trt) into the 3-butanoate **95** by a chemoenzymatic approach with lipase from *Chromobacterium viscosum* (CVL), and the 6-*tert*-butyl-diphenylsilylated glucose **66c** (R = TBDPS) could be acylated exclusively at the 2-position when employing lipase from *Candida cylindracea* (CCL) [164]. From the disubstituted glucoses obtained by the enzyme-catalyzed reactions, the protecting functions in the 6-position could be split off chemically or enzymatically, thus making the glucose esters **95** and **96** carrying a single acyl group in the 2- or the 3-position available in a convenient way (Fig. 18-15).

The monoacylated saccharides used in these studies dissolve in several organic solvents, of which tetrahydrofuran and methylenedichloride were found to be

Table 18-1. Selective acylation of the primary hydroxy group in monosaccharides.

Compound No.	Structure	Enzyme[a]	Solvent	Acyl Donor	Position	Yield (%)	Ref.
36		PPL	pyridine	$RCO_2CH_2CCl_3$	6	19–35	[141]
		CAL	dioxane	$ROCO_2N=CMe_2$	6	15–72	[145]
		CAL	THF	$RCO_2CH=CH_2$	6		[147]
		PSL	pyridine	$MeCO_2CH_2CCl_3$	6	79	[146]
		PSL	pyridine	$EtCO_2CH_2CCl_3$	6	29	[146]
		proleather	pyridine	$PhCO_2CH_2CCl_3$	6	33	[146]
		subtilisin	DMF	$PrCO_2CH_2CCl_3$	6	60	[150]
		subtilisin	pyridine	$PrCO_2CH_2CCl_3$	6	64	[150]
		optimase M-440	pyridine	Boc-Phe-OCH_2CF_3	6		[151]
37		PPL	pyridine	$MeCO_2CH_2CCl_3$	6	57	[141]
		PSL	pyridine	$RCO_2N=CMe_2$	6	70–85	[152]
		CAL	dioxane	$ROCO_2N=CMe_2$	6	43–68	[145]
38		PPL	pyridine	$MeCO_2CH_2CCl_3$	6	36	[141]
		CCL	benzene/pyridine 2:1	$MeCO_2CH=CH_2$	6		[142]
		PSL	pyridine	$RCO_2N=CMe_2$	6	65–80	[152]
		CAL	dioxane	$ROCO_2N=CMe_2$	6	44–53	[145]
		protease N	DMF	$MeCO_2C(Me)=CH_2$	6	40	[153]
39		CCL	benzene/pyridine 2:1	$MeCO_2C(Me)=CH_2$	6	73	[142]
		protease N	DMF	$MeCO_2C(Me)=CH_2$	6		[153]
		subtilisin 8399	DMF	$MeCO_2CH=CH_2$	6	92	[154]
		subtilisin BNP'	97 % DMF	Boc-Gly-OCH_2CN	6	65	[155]
40		CAL	pyridine	$RCO_2N=CMe_2$	6	45–83	[156]
		PSL	dioxane	$RCO_2N=CMe_2$	6	50–72	[156]

Table 18-1. (cont.).

Compound No. Structure	Enzyme[a]	Solvent	Acyl Donor	Position	Yield (%)	Ref.
41	CAL PSL	pyridine dioxane	$RCO_2N=CMe_2$ $RCO_2N=CMe_2$	6 6	57–81 47–62	[156] [156]
42	PSL	pyridine	$RCO_2N=CMe_2$	1	68–86	[152]
43	CAL CAL CAL	pyridine dioxane THF	$RCO_2N=CMe_2$ $RCO_2N=CMe_2$ Pr_2O	5 5 5	50–64 37–52	[152] [145] [157,158]
44	CAL CAL	pyridine dioxane	$RCO_2N=CMe_2$ $ROCO_2N=CMe_2$	5 5	45–70 38–49	[152] [145]
45	CAL	THF	Pr_2O	5		[157,158]
46	PPL	pyridine	$C_{11}H_{23}CO_2CH_2CCl_3$	5	40	[159]

Table 18-1. (cont.).

Compound No. Structure	Enzyme[a]	Solvent	Acyl Donor	Position	Yield (%)	Ref.
47a	CAL	acetone/pyridine 3:1	$C_{11}H_{23}CO_2H$	6	67	[160]
47b	CAL	tBuOH	$C_{11}H_{23}CO_2Et$	6	51	[161]
48	CCL	benzene/pyridine 2:1	$MeCO_2CH=CH_2$	6		[142]
	CAL	THF/pyridine (4:1)	$MeCO_2CH=CH_2$	3,6		[162]
	CAL	$PrCO_2Et/t$BuOH (1:1)	$PrCO_2Et$	6		[148,149]
49	CAL	$CH_2=CHCO_2Et/t$BuOH (1:1)	$CH_2=CHCO_2Et$	6		[149]
	CAL	THF/pyridine (4:1)	$MeCO_2CH=CH_2$	6		[162]
50	CAL	THF	$MeCO_2CH=CH_2$	6		[162]
	CAL	$PrCO_2Et/t$BuOH (1:1)	$PrCO_2Et$	6		[148,149]
	CAL	tBuOH	$Ph(CH_2)_3CO_2H$	6	52	[163]
	CVL	THF	$PrCO_2CH_2CCl_3$	6; 3,6 (1:1)		[164]
	ANL	THF	$PrCO_2CH_2CCl_3$	6; 3,6 (10:1)		[164]
51	CAL	$CH_2=CHCO_2Et$	$CH_2=CHCO_2Et$	6		[149]

Table 18-1. (cont.).

Compound No. Structure	Enzyme[a]	Solvent	Acyl Donor	Position	Yield (%)	Ref.
52	PPL CAL CAL	pyridine THF/pyridine (4:1) CH₂=CHCO₂Et/tBuOH (1:1)CH₂=CHCO₂Et	PrCO₂CH₂CF₃ MeCO₂CH=CH₂	6; 3,6 (3:1) 2,6	79	[165] [162] [148,149]
53	PSL	MeCO₂CH=CH₂/THF	MeCO₂CH=CH₂	6	93	[166]
54	CAL CAL	THF/pyridine (4:1) CH₂=CHCO₂Et/ tBuOH (1:1)CH₂=CHCO₂Et	MeCO₂CH=CH₂	6; 2,6; 3,6 (1:1.3:1.8) 6; 2,6; 3,6 (2:1:1)		[162] [148,149]
55	PSL	MeCO₂CH=CH₂/THF	MeCO₂CH=CH₂	6	90	[166]
56	PSL	MeCO₂CH=CH₂	MeCO₂CH=CH₂	6	75	[167]

Table 18-1. (cont.).

Compound No.	Structure	Enzyme[a]	Solvent	Acyl Donor	Position	Yield (%)	Ref.
57		PSL	$MeCO_2CH=CH_2$/THF	$MeCO_2CH=CH_2$	6	94	[166]
58		PPL	pyridine	$PrCO_2CH_2CF_3$	6	81	[165]
59		PPL	THF	$MeCO_2CH_2CF_3$	5	77	[168]
60		PPL	THF	$MeCO_2CH_2CF_3$	5	77	[168]
61		PPL	THF	$MeCO_2CH_2CF_3$	5	84	[168]

Table 18-1. (cont.).

Compound No. Structure	Enzyme[a]	Solvent	Acyl Donor	Position	Yield (%)	Ref.
62	PPL	THF	$MeCO_2CH_2CF_3$	5 3	39 17	[168]
63	CCL	EtOAc	$MeCO_2CH=CH_2$	6	90	[169]
64	CCL	EtOAc	$MeCO_2CH=CH_2$	6	93	[169]
65	CCL	THF	$MeCO_2CH=CH_2$	9	60	[170]

a Many enzymes were usually screened for activity, only the best results are listed. CAL: *Candida antarctica* lipase; CCL: lipase from *Candida cylindracea* (later renamed *Candida rugosa*); PPL: porcine pancreas lipase; PSL: *Pseudomonas cepacia* lipase.

Table 18-2. Selective acylation of secondary hydroxy groups in monosaccharides.

Compound No. Structure	Enzyme[a]	Solvent	Acyl Donor	Position	Yield (%)	Ref.
66 a: R=butyryl b: R=trityl c: R=TBDPS	ANL	THF	$PrCO_2CH_2CCl_3$	3 (66a)	80	[164]
	CVL	THF	$PrCO_2CH_2CCl_3$	3 (66a)		[164]
	PPL	THF	$PrCO_2CH_2CCl_3$	2 (66a)	51	[164]
	CVL	THF	$PrCO_2CH_2CCl_3$	3 (66b)	88	[164]
	PFL	$MeCO_2CH=CH_2$	$MeCO_2CH=CH_2$	2 (66b)		[175]
	CCL	CH_2Cl_2	$PrCO_2CH_2CCl_3$	2 (66c)	45	[164]
67	CVL	THF	$PrCO_2CH_2CCl_3$	2 3	20 31	[164]
68	CVL	THF	$PrCO_2CH_2CCl_3$	2 3	13 52	[164]
69	lipase from Mucor miehei	$MeCO_2CH=CH_2$	$MeCO_2CH=CH_2$	2		[175]
70	PFL lipase from fl2Mucor miehei	$MeCO_2CH=CH_2$ $MeCO_2CH=CH_2$	$MeCO_2CH=CH_2$ $MeCO_2CH=CH_2$	3 2		[175] [175]

Table 18-2. (cont.).

Compound No. Structure	Enzyme[a]	Solvent	Acyl Donor	Position	Yield (%)	Ref.
71 (Ph, HO, HO, R) **a:** R=OMe **b:** R=SEt **c:** R=OPh	PSL	RCO$_2$CH$_2$CF$_3$/THF	RCO$_2$CH$_2$CF$_3$	2 (71a)		[176]
	PSL	RCO$_2$CH=CH$_2$/THF	RCO$_2$CH=CH$_2$	2 (71a)	98	[177]
	PFL	MeCO$_2$CH=CH$_2$	MeCO$_2$CH=CH$_2$	2 (71a)	94	[178,179]
	PFL	MeCO$_2$CH=CH$_2$	MeCO$_2$CH=CH$_2$	2 (71a)	73	[180]
		tBuCO$_2$CH=CH$_2$/THF	tBuCO$_2$CH=CH$_2$	2 (71b)	76	
				2 (71c)		
72 (Ph, HO, HO, R, OH) **a:** R=OMe **b:** R=SEt	PSL	RCO$_2$CH$_2$CF$_3$/THF	RCO$_2$CH$_2$CF$_3$	3 (72a)		[176]
		RCO$_2$CH=CH$_2$/THF	RCO$_2$CH=CH$_2$	3 (72a)		
	PSL	MeCO$_2$CH=CH$_2$	MeCO$_2$CH=CH$_2$	3 (72a)	86	[177]
	PFL	MeCO$_2$CH=CH$_2$	MeCO$_2$CH=CH$_2$	3 (72a)	86	[178,179]
	PFL	MeCO$_2$CH=CH$_2$	MeCO$_2$CH=CH$_2$	3 (72b)	86	[180,181]
73 (Me, HO, HO, OH, OMe)	PPL	THF/pyridine (4:1)	PrCO$_2$CH$_2$CF$_3$	2	93	[182]
74 (HO, OR, HO, OH, OMe) **a:** R=butyryl **b:** R=trityl **c:** R=benzyl	PPL	THF/pyridine (4:1)	PrCO$_2$CH$_2$CF$_3$	2 (74a)	84	[165,182]
	PFL	THF/pyridine (4:1)	PrCO$_2$CH$_2$CF$_3$	2 (74a)	81	[165]
	CCL	CH$_2$Cl$_2$/pyridine (4:1)	PrCO$_2$CH$_2$CF$_3$	2 (74a)	80	[165]
	PFL	MeCO$_2$CH=CH$_2$	MeCO$_2$CH=CH$_2$	2 (74b)		[175]
	PFL	MeCO$_2$CH=CH$_2$	MeCO$_2$CH=CH$_2$	2 (74c)		[175]
75 (HO, OTrt, HO, OMe, OH)	PFL	MeCO$_2$CH=CH$_2$	MeCO$_2$CH=CH$_2$	3		[175]

Table 18-2. (cont.).

Compound No. Structure	Enzyme[a]	Solvent	Acyl Donor	Position	Yield (%)	Ref.
76 **a:** R=OAll **b:** R=SEt	PSL	MeCO$_2$CH=CH$_2$	MeCO$_2$CH=CH$_2$	3 (76a)	91	[177]
	PFL	MeCO$_2$CH=CH$_2$/THF	MeCO$_2$CH=CH$_2$	3 (76b)	10	[180]
77	PSL	MeCO$_2$CH=CH$_2$	MeCO$_2$CH=CH$_2$	2	90	[177]
78	PFL	MeCO$_2$CH=CH$_2$	MeCO$_2$CH=CH$_2$	3		[175]
79	PSL	RCO$_2$CH=CH$_2$	RCO$_2$CH=CH$_2$	3	92	[177]
80	PPL	THF	PrCO$_2$CH$_2$CF$_3$	4	65	[165]
	PFL	THF	PrCO$_2$CH$_2$CF$_3$	4	68	[165]

Table 18-2. (cont.).

Compound No.	Structure	Enzyme[a]	Solvent	Acyl Donor	Position	Yield (%)	Ref.
81		PPL	THF/PrCO$_2$CH$_2$CF$_3$ (4:1)	PrCO$_2$CH$_2$CF$_3$	4	70	[183]
82		PSL	dioxane	RCO$_2$N=CMe$_2$	3	54–67	[156]
84		PSL	dioxane	RCO$_2$N=CMe$_2$	3	48–56	[156]
84		PPL PFL	THF/pyridine (4:1) THF/pyridine (4:1)	PrCO$_2$CH$_2$CF$_3$ PrCO$_2$CH$_2$CF$_3$	2 2	78 84	[184] [184]
85		PFL	THF	PrCO$_2$CH$_2$CF$_3$	2	40	[184]

Table 18-2. (cont.).

Compound No.	Structure	Enzyme[a]	Solvent	Acyl Donor	Position	Yield (%)	Ref.
86		PSL	MeCN	$MeCO_2CH=CH_2$	3,4	85	[185]
87		PSL	Hexane	$MeCO_2CH=CH_2$	2,4 3,4	70 28	[185]
88		CAL CAL PSL CAL	dioxane dioxane $MeCO_2CH=CH_2$ $PrCO_2Et/tBuOH$	$RCO_2N=CMe_2$ $MeOCO_2N=CMe_2$ $MeCO_2CH=CH_2$ $PrCO_2Et$	4 4 4 4; diester	70–72 42	[156] [156] [186,187] [188]
89		CAL	$PrCO_2Et$	$PrCO_2Et$	4		[188]
90		PSL	$MeCO_2CH=CH_2$	$MeCO_2CH=CH_2$	4		[187,189]

Table 18-2. (cont.).

Compound No. Structure	Enzyme[a]	Solvent	Acyl Donor	Position	Yield (%)	Ref.
91 a: R=acetyl b: R=benzoyl c: R=PhAc	PFL	DME	$RCO_2CH=CH_2$	3	84–93	[190]
92	HLL RJL	benzene benzene	$PrCO_2CH_2CCl_3$ $PrCO_2CH_2CCl_3$	2 3	66 79	[191] [191]

a Many enzymes were normally screened for activity, only the best results are listed. ANL: *Aspergillus niger* lipase; CAL: *Candida antarctica* lipase; CCL: lipase from *Candida cylindracea* (later renamed *Candida rugosa*; CRL); CVL: *Chromobacterium viscosum* lipase; HLL: *Humicula lanuginosa* lipase; PFL: *Pseudomonas fluorescens* (later renamed *Pseudomonas cepacia*) lipase; PPL: porcine pancreas lipase; PSL: *Pseudomonas cepacia* lipase; RJL: *Rhizopus japonicus* lipase.

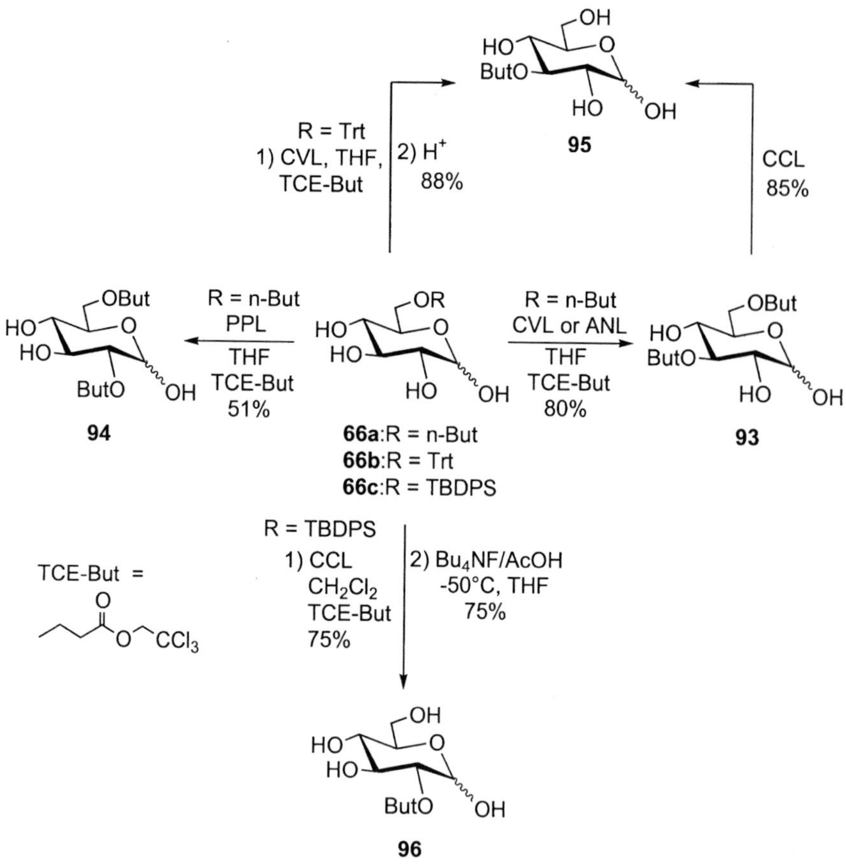

Figure 18-15. Selective enzymatic introduction of protecting groups into partially acylated hexoses.

particularly suitable for the enzymatic reactions. This was also observed in the lipase-mediated acylation of the methyl glycosides of both D- and L-fucose and -rhamnose, respectively[184]. Using lipase from *Pseudomonas fluorescence* (PFL), both D-carbohydrates were converted into the 2-monobutanoates with high regioselectivity. The naturally occurring L-enantiomers of these 6-deoxysugars, however, were esterified preferably at the 4-hydroxy groups. These results contrast favorably with chemical derivatizations, since the 4-hydroxy groups of the 6-deoxy-L-carbohydrates have only slight reactivity toward chemical acylating reagents. In addition, methyl-L-fucoside can be converted into the 3-butanoate with lipase from *Candida cylindracea*. The introduction of an acyl-substituent into the 6-positions of the D-fucoside and the L-rhamnoside does not influence the regioselectivity of the enzymatic acylation[165].

Finally, it should be mentioned, that some attempts were made to differentiate between the hydroxy groups of fructose by enzymatic methods, however, with lipases as well as with subtilisin, only mixtures of the 1- and 6-isomers were obtained [141, 150, 192]. Regioselectively monosubstituted fructoses can, however, be obtained by an enzymatic approach from sucrose (*vide infra*).

18.5.2
Deprotection of Monosaccharides [133, 137]

Initial attempts to apply lipases for the enzymatic removal of acyl groups from glucose pentaacetate only resulted in low levels of selectivity [193, 194]. However, later on lipase from porcine pancreas (PPL) [168] was found to hydrolyze exclusively the anomeric acetate from peracetylated pyranoses while the esterase from *Rhodosporium toruloides* (RTE) [195] releases the primary hydroxy group in preferance (Table 18-4). On the other hand, if the anomeric center is derivatized as a methyl glycoside, the regioselective enzymatic liberation of the 6-OH group becomes feasible with a number of hydrolytic enzymes [168, 195–199]. Thus, from methyl α-D-glucose tetra-octanoate **97a** and the corresponding tetrapentanoate **97b**, lipase from *Candida cylindracea* (CCL) removes only the primary ester group in yields of ca. 75 %. Similarly, the α-D-galactoside **103**, as well as the corresponding mannoside **104b** and the 2-acetamido-2-deoxy-mannoside **105** were converted into the 6-deprotected pyranosides in 29–50 % yield (Table 18-3), but the 2-acetamido-2-deoxy-glucoside was only a poor substrate. In the latter cases the regioselectivity was less pronounced and the 4,6-dideoxy derivatives were also formed in ca. 20 % yield. In addition to this class of compounds, lipases also accept hexopyranosides carrying several different functionalities (e.g. acetals [197], enol ethers [169, 200] and, in particular, 1,6-anhydropyranoses as substrates (Tables 18-3 and 18-4). In all cases the reaction conditions are so mild that the acid sensitive structures of these compounds remain unaffected. Particularly remarkable is the regioselectivity displayed by lipase from *Pseudomonas cepacia* (PSL) in the deprotection of the glycal **131** [169, 200]. The biocatalyst exclusively attacks the 3-acetate and leaves the primary ester intact. The enzymatic deprotection strategy can also be used to synthesize carbohydrates carrying a single acyl group in selected positions. Thus, 3,6-dibutyryl glucose **93** (prepared by enzymatic acylation of glucose) was converted into the 3-butanoate **95** by lipase mediated hydrolysis of the 6-ester (Fig. 8-15) [164]. The principles and the enzymes mentioned above which allow the regio- and chemoselective protection and deprotection of the various pyranoses to be carried out were also successfully applied to the enzymatic manipulation of acyl groups in furanoses. Of particular interest in this context is the finding that the five-membered rings can also be handled by the biocatalysts with a pronounced regioselectivity, although furanoses can adopt more flexible conformations with similar energies in solution.

The cleavage of the primary acetyl groups from the furanosides **106–111** could be carried out in high yields with lipase from *Candida cylindracea* (Table 18-3) [168]. For the 2-deoxy-α-D-ribofuranoside and the α- and the β-xylo-compounds the hydrolysis was less selective. From the peracetylated furanoses **125** and **126** the anomeric acyl group was removed with total selectivity by means of lipase from *Aspergillus niger* (Table 18-4).

1,6-Anhydropyranoses serve as convenient starting materials for various synthetic

Table 18-3. Selective deacylation of primary hydroxy groups in monosaccharides.

Compound No. Structure	Enzyme[a]	Solvent	Position	Yield (%)	Ref.
97 **a:** R=octanoyl **b:** R=pentanoyl **c:** R=acetyl	CCL	0.1 M phosphate buffer	6	78 (97a)	[168]
	CCL	0.1 M phosphate buffer	6	75 (97b)	[168]
	CCL	0.1 M phosphate buffer	6	90 (97b)	[196]
	CCL	0.1 M phosphate buffer, Bu_2O (10%)	6	(97c)	[197]
	PEG-modified CCL	Cl_3CCH_3	6	27 (97c)	[198]
			4,6	48 (97c)	
	CRL	0.1 M phosphate buffer	6	91 (97c)	[199]
	RTE	citrate buffer	6	77 (97c)	[195]
98 R=octanoyl	CCL	0.1 M phosphate buffer	6	77	[168]
99	PPL	0.1 M phosphate buffer, acetone 10:1	6	90	[166]
100	PPL	0.1 M phosphate buffer, acetone 10:1	6	82	[166]
101	PPL	0.1 M phosphate buffer, acetone 10:1	6	75	[166]

Table 18-3. (cont.).

Compound No. Structure	Enzyme[a]	Solvent	Position	Yield (%)	Ref.
102 (ButO–, HO–, ButO–, OH, OH)	CCL	0.1 M Tris·HCl	6	85	[164]
103 (RO–, OR, RO–, RO, OMe) a: R=acetyl b: R=pentanoyl	RTE	citrate buffer	6	85 (103a)	[195]
	CCL	0.1 M phosphate buffer	6	29 (103b)	[168]
104 (RO–, RO, RO–, RO, OMe) a: R=acetyl b: R=pentanoyl	CRL	0.1 M phosphate buffer	6	94 (104a)	[199]
	RTE	citrate buffer	6	70 (104a)	[195]
	CCL	0.1 M phosphate buffer	6	33 (104b)	[168]
105 (PentO–, NHAc, PentO, PentO, OMe)	CCL	0.1 M phosphate buffer	6	50	[168]
106 (AcO–, OMe, AcO, OAc)	CCL	0.1 M phosphate buffer, 10% DMF	6	85	[168]

Table 18-3. (cont.).

Compound No. Structure	Enzyme[a]	Solvent	Position	Yield (%)	Ref.
107	CCL	0.1 M phosphate buffer, 10% DMF	5	96	[168]
108	CCL	0.1 M phosphate buffer, 10% DMF	5	98	[168]
109	CCL	0.1 M phosphate buffer, 10% DMF	5 3	50 30	[168]
110	CCL	0.1 M phosphate buffer, 10% DMF	5 3	40 50	[168]
111	CCL	0.1 M phosphate buffer, 10% DMF	5	63	[168]

a Many enzymes were normally screened for activity, only the best results are listed. ANL: Aspergillus niger lipase; CCL: lipase from Candida cylindracea (later renamed Candida rugosa; CRL); PPL: porcine pancreas lipase; RTE: Rhodosporium toruloides esterase.

Table 18-4. Selective deacylation of secondary hydroxy groups in monosaccharides.

Compound No. Structure	Enzyme[a]	Solvent	Position	Yield (%)	Ref.
112	CCL	0.1 M phosphate buffer	4,6	73	[168]
	RTE	citrate buffer	6	54	[195]
	PPL	0.05 M phosphate buffer, 10% DMF	1	70	[168]
	PFL	phosphate buffer, MeCN (7:3)	1	80	[201]
	CCL	phosphate buffer, MeCN (7:3)	4	50	[201]
	CCL	phosphate buffer, MeCN (7:3)	6	75	[201]
113	PPL	0.05 M phosphate buffer, 10% DMF	1		[168]
	CAL	butanone	1	95	[202]
114	PPL	0.05 M phosphate buffer, 10% DMF	1	96	[168]
115	ANL	phosphate buffer, MeCN (10:1)	1,4	41	[203]
	RTE	citrate buffer	6	80	[204]
116	PPL	0.05 M phosphate buffer, 10% DMF	1	75	[168]
	RTE	citrate buffer	6	67	[195]
117	ANL	0.1 M phosphate buffer, 10% acetone	2	58	[205]

Table 18-4. (cont.).

Compound No. Structure	Enzyme[a]	Solvent	Position	Yield (%)	Ref.
118	*hog kidney* acylase	phosphate buffer, DMF (10:1)	2	93	[206]
	Aspergillus niger pecti-nase	phosphate buffer, DMF (10:1)	3	27	[206]
	ANL	phosphate buffer, DMF (10:1)	4	11	[206]
119	PPL	0.05 M phosphate buffer, 10% DMF	1	95	[168]
	RTE	citrate buffer	6	88	[195]
120	ANL	0.1 M phosphate buffer, 10% acetone	3	61	[205]
121	PPL	0.05 M phosphate buffer, 10% DMF	1	88	[168]
122	PSL	tAmyl-OH	4	84	[207]
	PEG-modified CCL	Cl₃CCH₃	4	82	[198]
123	PPL	0.05 M phosphate buffer, 10% DMF	1	54	[168]

Table 18-4. (cont.).

Compound No.	Structure	Enzyme[a]	Solvent	Position	Yield (%)	Ref.
124		PPL	0.05 M phosphate buffer, 10% DMF	1	71	[168]
125		ANL	0.1 M phosphate buffer, 10% DMF	1	63	[168]
126		ANL	0.1 M phosphate buffer, 10% DMF	1	50	[168]
127	a: R=acetyl b: R=butyryl	RJL	0.1 M phosphate buffer	2	47 (127a)	[208]
				4	15 (127a)	[208]
		WGL	0.1 M phosphate buffer	3	67 (127a)	[208]
		PLE	0.1 M phosphate buffer	4	69 (127a)	[208]
		PPL	0.05 M citrate-phosphate buffer	4	42 (127a)	[209]
		CVL	0.1 M phosphate buffer	4	91 (127b)	[210]
		CCL	0.1 M phosphate buffer	2,4	77 (127b)	[210]
128		CCL	0.1 M phosphate buffer	4	85–90	[211]
		alcalase	0.1 M phosphate buffer	2	82	[211]

Table 18-4. (cont.).

Compound No. Structure	Enzyme[a]	Solvent	Position	Yield (%)	Ref.
129	CCL	0.1 M phosphate buffer	2		[210]
	CCL	0.1 M phosphate buffer	2	90	[212]
	PPL	0.1 M phosphate buffer	2	16	[212]
			4	19	
			2,4	65	
130	WGL	phosphate buffer, DMF (10:1)	2	60	[206]
131	PSL	0.25 M phosphate buffer	3	90	[169]
	acetyl esterase from the flavedo of oranges	0.15 M NaCl buffer	3,4	24	[74,213]
			3,4,6	22	
132	PGA	0.1 M phosphate buffer	3	80–85	[190]

a Many enzymes were normally screened for activity, only the best results are listed. ANL: *Aspergillus niger* lipase; CAL: *Candida antarctica* lipase; CCL: lipase from *Candida cylindracea* (later renamed *Candida rugosa*; CRL); PGA: penicillin-G-acylase; PLE: porcine liver esterase; PPL: porcine pancreas lipase; PSL: *Pseudomonas cepacia* lipase; RJL: *Rhizopus japonicus* lipase; RTE: *Rhodosporium toruloides* esterase; WGL: wheat germ lipase.

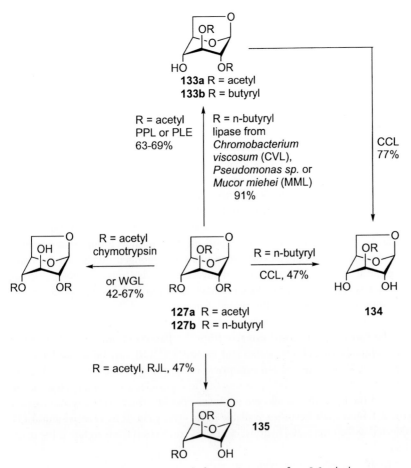

Figure 18-16. Selective enzymatic removal of protecting groups from 1,6-anhydropyranoses.

purposes in carbohydrate chemistry. Therefore, the directed manipulation of their hydroxy groups is of particular interest. Each of the three OH-groups in 1,6-anhydroglucopyranose can be liberated selectively making use of enzymatic reactions (Fig. 18-16, Table 18-4) [208–210, 212]. Thus, the 4-protecting group was split off from the triacetate **127a** using lipase from porcine pancreas (PPL) [209] or pig liver esterase (PLE) [208, 209]. The acetate in the 3-position could be attacked preferentially using chymotrypsin [209] or lipase from wheat germ (WGL) [208], and the 3,4-diacetate **135** was obtained by hydrolysis with lipase from *Rhizopus javanicus* (RJL) [208]. In each case, however, other derivatives were formed as undesired by products. High yields could be obtained from the tri-n-butanoate **127b**. It was converted into the 2,3-dibutanoate **133b** in 91% yield by means of several lipases, but the enzyme from *Candida cylindracea* (CCL) removed two acyl groups successively to yield the monobutanoate **134**. Similarly, the analogous 3-azido-1,6-anhydropyranose **128** is regioselectively deacylated at O2 and O4 by means of lipase OF from *Candida cylindracea* and alcalase, respectively[211]. Of particular importance is the stereochemistry at C4 of the bicyclic substrates. If the alcohol at this position is equatorial, as for instance in the corresponding 1,6-anhydrogalactopyranose **129** and the analogous lactone **130**, several enzymes act only in a random fashion or not at all[210]. However, the acyl group in the 2-position seems to be preferred (Table 18-4). The results obtained from these studies indicate that the reactivity of acyl protecting groups in 1,6-anhydropyranoses toward hydrolysis by lipases decreases in the order $C4_{ax} > C2_{ax} > C3_{ax} \gg C4_{eq}$.

The above mentioned investigations revealed that the lipase-mediated hydrolysis proceeds at higher reaction rate and, in many cases with better selectivity, if butanoates or pentanoates are employed as substrates instead of acetates. However, the use of enzymatic deacylations is by no means restricted to simple alkanoates. An illustrative and impressive example is found in the hydrolysis of generally base-stable carbohydrate pivaloylates using an esterase from rabbit serum (ERS) [214–217]. For instance, the biocatalyst selectively splits off the 6-pivaloyl group from α-methyl 3,4,6-tripivaloyl-2-acetamido-2-deoxy-glucoside. On prolonged incubation the complete removal of pivaloylates from carbohydrates is also possible. Of particular significance is, that the enzyme does not have to be purified, but that crude serum preparations are sufficient for the preparative purposes. A further enzyme which allows the chemo- and regioselective unmasking of different carbohydrate derivatives to be carried out is acetyl esterase from the flavedo of oranges, a biocatalyst which preferably hydrolyzes acetic acid esters[32, 218]. It can be applied for the synthesis of selectively deacylated pyranoses. Thus, from pentaacetylglucose **112** the 2,3,4,6-tetraacetate is obtained by means of the regioselective saponification of the 1-acetate. If the hydrolysis is allowed to proceed further, the 6-acetate is also cleaved and the 2,3,4-triacetate becomes available in ca. 40 % yield. If tri-*O*-acetyl-glucal **131** is subjected to the enzymatic hydrolysis, at 40 % conversion the 6-acetate is the main product.

By introducing acyl groups which are specifically recognized by certain enzymes into carbohydrates, not only the regioselectivity but also the chemoselectivity of the biocatalysts can be exploited. This can, for instance, be achieved by the selective saponification of phenylacetates catalyzed by penicillin G acylase[30–32]. The enzyme liberates the 2-OH group of 1,3,4,6-tetraacetyl-2-phenylacetyl glucose without affecting the acetic acid esters. In this case, moreover, an ester of a secondary hydroxy function is chemoselectively hydrolyzed in the presence of the chemically more reactive acetates at the 6-position and at the anomeric center. This approach was also adopted for the enzymatic deprotection of the glucal **132**. Thus, its 3-OH group was liberated without cleaving the acetates that were present[190].

18.5.3

Di- and Oligosaccharides [137]

For enzymatic protecting group manipulations on di- and oligosaccharides in particular the use of subtilisin together with dimethylformamide as the solvent is advantageous. As has already been pointed out, the use of DMF is often critical, since its dissolving ability is high enough to solubilize even highly polar polyhydroxy compounds (comparable experiments with pyridine as the solvent generally failed)[141]. Only a few reports about the successful use of other solvents such as pyridine[219] or *tert*-butanol[220] have been published.

Subtilisin accepts several disaccharides as substrates and transfers butyric acid from ethyl or trichloroethyl butanoate to the primary 6'-hydroxy functions of the nonreducing monosaccharide of the β-(1–3)-linked cellobiose **136** and the respective maltobiose (Fig. 18-17)[150, 220]. For lactose the regioselectivity was less pronounced, however, methyl and benzyl β-D-lactoside **137** were converted into the 6'-butanoates in 71–73% yield[221]. Rutinose in which the primary hydroxy group of the glucose moiety is blocked (see also **149**, Fig. 18-19), is selectively substituted in the 3-position[222]. In addition, higher maltooligomers could also be acylated in the 6-position of the terminal nonreducing carbohydrate. For instance, 6''-O-butyrylmaltotriose was isolated in 29% yield, but also the corresponding tetra-, penta-, hexa- and heptamer were substrates for the biocatalyst. These enzymatic esterifications open a route to discriminating between the primary hydroxy groups in di- and oligosaccharides in a convenient and straightforward way. Classical chemical one step methods of comparable selectivity are not available for this purpose[139, 140], and multistep sequences usually have to be carried out if the selective protection of a specific primary hydroxy group in a di- or oligosaccharide is desired.

Owing to its great commercial importance as a renewable resource, sucrose **138** has been subjected to several enzymatic hydroxy group manipulations. This nonreducing disaccharide turned out to be a substrate for subtilisin also[150]. In contrast to chemical acylations in which the most reactive OH-groups are found in the 6- and the 6'-position, the enzyme selectively transfers various acyl functions to the 1'-alcohol (Fig. 18-17)[150, 192, 223]. This acylation was usually carried out in DMF as a solvent, but the use of anhydrous pyridine gave similar results[219]. The monoacylated disaccharides **139** thereby obtained could then be further transformed enzymatically. On the one hand, with the lipase from *Chromobacterium viscosum* (CVL) the free primary 6-OH group was acylated in 31% yield. On the other hand, the 1'-esters **139** are substrates for yeast α-glucosidase which hydrolyzes the glycosidic bond and thus makes the 1-O-acylfructoses **140**, potentially useful as chiral synthons, available[192]. Alternatively, the 6'-OH-group in sucrose **138** can be selectively acylated, if the carbohydrate is converted into the 2,1':4,6-bisacetal prior to the treatment with a lipase (Novozym™ 435)[224].

On considering hydrolysis, several enzymes were investigated[225–229]. Depending on the biocatalyst used, acetyl groups from different positions of octaacetyl sucrose **141** could be removed selectively in useful yields. For instance, alcalase and protease N preferably attack the acetate on O1'[226, 230], the lipase from *Candida cylindracea* preferably liberates the OH-group on C4' of the furanoid ring[225, 230] and wheat germ lipase preferentially liberates the 1'-, 4'- and 6'-OH-groups (Fig. 18-17)[223, 231].

The deacylation of the octaacetates of cellobiose, lactose, maltose and melibiose with *Aspergillus niger* lipase leads to the formation of the respective carbohydrate heptaacetates with a free anomeric OH-group at C1 in high yield[230, 232]. With

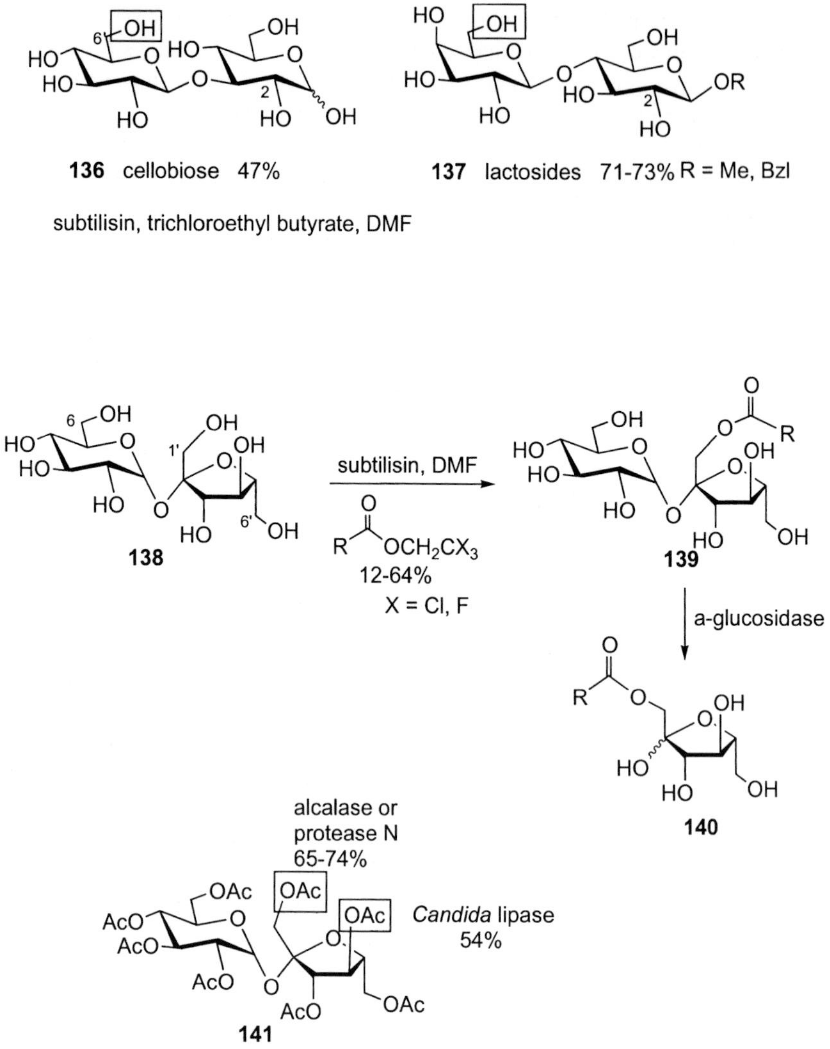
136 cellobiose 47%
137 lactosides 71-73% R = Me, Bzl
subtilisin, trichloroethyl butyrate, DMF
138
subtilisin, DMF
12-64%
X = Cl, F
139
a-glucosidase
140
alcalase or
protease N
65-74%
Candida lipase
54%
141

Figure 18-17. Selective enzymatic protection and deprotection of disaccharides.

prolonged reaction times, the acetates at C1 and C2 are hydrolyzed from cellobiose and lactose octaacetate in 51% or 42% yield, respectively.

18.5.4
Nucleosides [135, 233]

The directed protection of nucleoside functional groups is a fundamental problem in nucleoside and nucleotide chemistry. Although several chemical methods are available for the regioselective acylation of the nucleoside carbohydrates, enzymatic methods offer significant advantages with respect to yield, regioselectivity and the number of synthetic steps which have to be carried out.

Earlier studies focussed on the use of the dihydrocinnamoyl group as an enzyme-labile nucleoside protecting function which can be removed through the agency of α-chymotrypsin[234, 235]. Although the enzyme shows an interesting tendency to attack preferably the 5'-position, this technique was not exploited further. Highly re-giodiscriminating biocatalyzed acyl transfer reactions to the carbohydrate parts of various nucleosides could be carried out again employing the protease subtilisin together with dimethylformamide as solvent. In particular, a mutant of this enzyme, obtained via site specific mutations appears to display advantageous properties. It transfers the acetyl group from isopropenyl acetate to the primary hydroxy functions of various purine and pyrimidine nucleosides and 2'-deoxynucleosides **142** in high yields (Fig. 18-18)[236]. Commercially available subtilisin (protease N from Amano) provided the same compounds with identical yields and selectivities, however, five times more enzyme is required for this purpose. In addition, in the transfer of butyric acid from trichloroethyl butanoate to adenosine and uridine, carried out earlier[150], this biocatalyst showed inferior properties with respect to regioselectivity and yields.

The selective introduction of protecting groups into the hydroxy functions of different nucleosides can also be achieved by means of lipases. Thus, unprotected pyrimidine and purine 2'-deoxynucleosides **143** (X = H) are selectively converted into the 3'-O-acylated derivatives **144** in 64–82% yield making use of lipase from *Pseudomonas cepacia* (PSL) and employing oxime carbonates as acyl donors (Fig. 18-18)[237–239]. Similarly, by applying oxime esters or acid anhydrides, different ester functions can be selectively introduced into the 3'-position of nucleotides by using the lipases from *Candida cylindracea* (CCL), porcine pancreas (PPL) or *Pseudomonas cepacia* (PSL)[240–244]. If lipase from *Candida antarctica* (CAL) is used, however, the esters and carbonates are predominantly generated at the primary 5'-OH group of (deoxy)nucleotides[238, 239, 241, 242, 244–247]. Furthermore, in the case of ribonucleo-tides, complete regioselectivity can be achieved by using the same methodology[241]. The regioselectivity of the CAL-catalyzed alkoxycarbonylation is profoundly influenced significantly by the structure of the starting oxime carbonate[248]. In the alkoxycarbonylation of thymidine the use of the phenyl derivative leads to almost exclusive formation of the 5' carbonate, while the corresponding allyl carbonate is introduced without any regioselectivity.

An investigation of the enzyme-catalyzed acylation of α-, xylo-, anhydro-, and arabino-nucleosides showed that in these cases the primary 5'-hydroxy group can be selectively acylated using lipase from *Candida antarctica* (CAL)[249–251]. A selective derivatization of the 3'-OH-group, however, was unsuccessful.

When acylations of nucleosides with acid anhydrides in the presence of lipase from *Pseudomonas fluorescence* (PFL) in DMF or DMSO as the solvent first pro-ceeded, the regioselectivity was unsatisfactory[252]. However, this lipase together with subtilisin can be utilized to effect highly specific deacylations of various pyrimidine nucleosides **145** (Fig. 18-18)[253]. Thus, lipase from *Pseudomonas fluorescence* (PFL) preferably attacks the hexanoyl group on the secondary hydroxy function of the *N*-

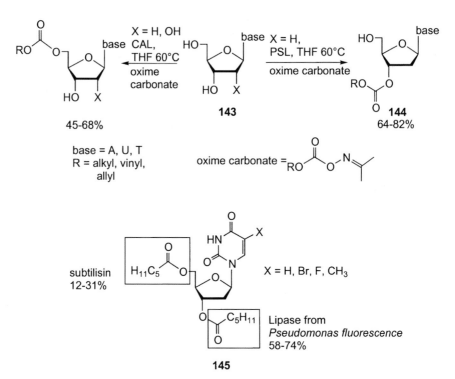

R	H	OH	H	OH	H	OH
base	H₃C (thymine)	(uracil)		(cytosine)		(adenine)
yield [%]	quant.	90	80	80	80	65

Figure 18-18. Selective enzymatic protection and deprotection of the carbohydrate parts of nucleosides.

glycosides, giving rise to the 5-esters in good yields. On the other hand, subtilisin gives rise to the 3-esters with moderate results. It should be noted, however, that in both cases from considerable to large amounts (6–71%) of the completely deprotected nucleosides were also formed. Subtilisin in phosphate buffer also selectively hydrolyzes the 5'-acetate of purine and pyrimidine triacetylated esters to give the corresponding 2',3'-diacetylribonucleosides in 40–92 % yield [254]. A similar preference was observed for the lipase from porcine pancreas, but with poorer selectivity and a slower reaction rate. This enzyme, however, deacetylated the deoxynucleoside 3',5'-di-*O*-acetylthymidine at the 5'-position in almost quantitative yield [255]. In contrast, if lipase from *Candida cylindracea* (CCL) was used in the catalysis, the 3'-ester of this diacetate was preferentially hydrolyzed [255].

Using acetyl esterase of the flavedo of oranges, bisacylated purine deoxynucleotides can be selectively deprotected at the 3'-hydroxy group in 31–40 % yield [74]. Interestingly, by introducing a phenylacetyl group for amino protection in the purine moiety the regioselectivity of the acetyl removal is reversed. Now the primary acetate is hydrolyzed by acetyl esterase in 22–52 % yield.

In addition, the complete hydrolysis of an anomeric mixture of peracetylated 2'-deoxynucleosides by wheat germ lipase or porcine liver esterase has been used to synthesize the pure β-anomer of e. g. thymidine, this being the only completely deprotected product [256]. The alcoholysis peractylated uridines catalyzed by *Candida antarctica* lipase leads to the formation of the completely deprotected nucleotide [257]. Although this reaction can be stopped after removal of the first acetyl group, no regioselectivity was observed for the formation of di-*O*-acetyluridine.

18.5.5
Further Aglycon Glycosides

In addition to nucleosides, several other naturally occurring carbohydrate derivatives can be selectively protected/deprotected by means of enzymatic techniques. For instance, salicin **146**, a wood component that contains a primary hydroxy group located in a glucose moiety and a second one in a benzylic position, was butyrylated exclusively at the 6-OH of the monosaccharide in 35% yield by applying subtilisin and trichloroethyl butanoate in DMF (Fig. 18-19) [150]. Under the same conditions, in riboflavin (vitamin B₂) **147** only the primary alcohol was esterified in 25% yield [150], and colchicoside **148a** as well as a thio analog **148b** were converted into the 6'-butanoates by treatment with trichloroethyl butanoate in pyridine in the presence of subtilisin [258]. The corresponding 6'-acetates of **148a,b** were obtained by treatment with vinyl acetate in the presence of *Candida antarctica* lipase as the biocatalyst (Fig. 18-19) [162]. Similarly, the carbohydrate parts of flavonoid disaccharides were regioselectively functionalized. Thus, for instance in the disaccharide rutin **149** and the related hesperidin only the 3"-OH group of the glucose moiety was esterified upon treatment with trifluoroethyl butanoate and subtilisin in 53% yield (Fig. 18-19) [222]. In the presence of lipase from *Candida antarctica*, however, both the 3"- and the 4'''-positions were acetylated [162]. If only the glucose moiety is present in the molecule, as in the related isoquercitrin **150**, the regioselectivty in the subtilisin-

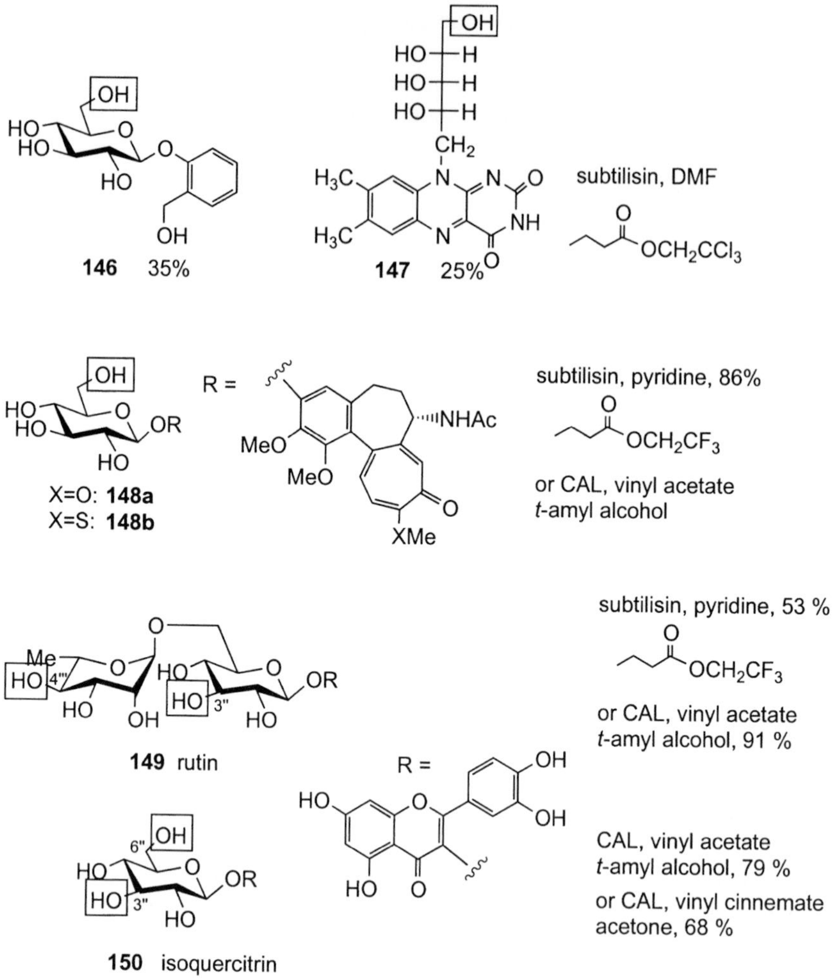

146 35%

147 25%

subtilisin, DMF

X=O: **148a**
X=S: **148b**

subtilisin, pyridine, 86%

or CAL, vinyl acetate
t-amyl alcohol

149 rutin

subtilisin, pyridine, 53 %

or CAL, vinyl acetate
t-amyl alcohol, 91 %

R =

CAL, vinyl acetate
t-amyl alcohol, 79 %

or CAL, vinyl cinnemate
acetone, 68 %

150 isoquercitrin

Figure 18-19. Selective enzymatic acylation of aglycon glycosides.

catalyzed reaction was less pronounced[259]. However, in the presence of lipase from *Candida antarctica* the 3",6"-bisacylated product is formed if vinyl acetate is used as the acyl donor[162]. Interestingly, by using vinyl cinnamate as the acyl donor, this biocatalyst only acylates the primary 6"-hydroxy group[260]. Naringine **151** was converted into the 6-glucosyl ester in the presence of subtilisin (Fig. 18-20). In all cases the rhamnose and the phenolic hydroxyls remained unattacked (for the protection of phenolic hydroxy groups in flavonoids see Sect. 18.5.8).

The steroidal glucoside ginsensoside Rg$_1$ **152** can be selectively monoacylated in high yields at the 6'-position using *Candida antarctica* lipase as the biocatalyst[261, 262]. In this case, similar results were obtained with different acyl donors such as vinyl acetate, dibenzyl malonate and bis(trichloethyl) malonate (Fig. 18-20).

Two impressive examples of selective enzymatic deacylations of complex sub-

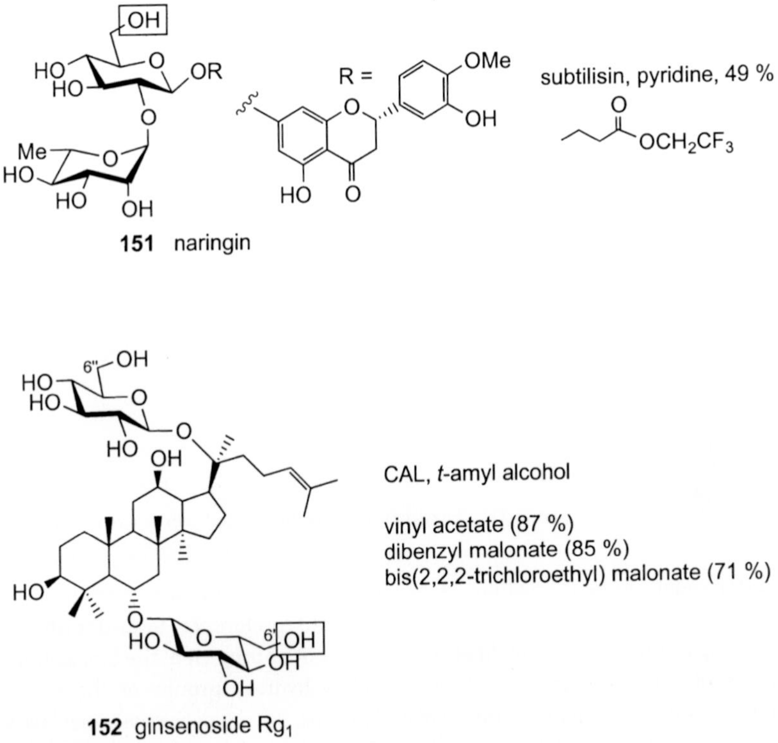

151 naringin

subtilisin, pyridine, 49 %

152 ginsenoside Rg₁

CAL, *t*-amyl alcohol

vinyl acetate (87 %)
dibenzyl malonate (85 %)
bis(2,2,2-trichloroethyl) malonate (71 %)

Figure 18-20. Selective enzymatic acylation of aglycon glycosides.

strates consist in the removal of all acetates from the peracetylated ß-ᴅ-glucopyr-
anosyl ester **153** of abscisinic acid[263] and of the gibberellinic acid derivative **154**[264],
containing one glucose tetraacetate glycosidically bound and a second one attached
as an ester (Fig. 18-21). In both cases the removal of the acetyl groups by chemical
methods in particular was complicated by an undesired cleavage of the ester linkages
to the glucoses. However, the four acetyl groups present in **153** could be hydrolyzed
chemoselectively by means of helicase, an enzyme occurring in the seeds of
Helianthus annus, whereby the unprotected glucose ester was formed in 82 % yield
without destroying the ester bond between abscisinic acid and glucose. Similarly, the
biocatalyst removed all acetates from **154**. In this case the yield reached only 8 %, it
should, however, be kept in mind that ten acetic acid esters had to be cleaved in the
enzymatic process and that the aglycon is rather complex.

In conclusion, the various enzyme-mediated protecting group manipulations
carried out on numerous carbohydrate derivatives indicate that biocatalysts can be
used advantageously in the protecting group chemistry of carbohydrates. In particular, subtilisin and several lipases from different sources (from porcine pancreas,
from *Candida cylindracea, Aspergillus niger, Chromobacterium viscosum, Mucor javanicus, Pseudomonas fluorescence* and from wheat germ) allow the chemo- and
regioselective acylation and deprotection of various saccharides, the structures of

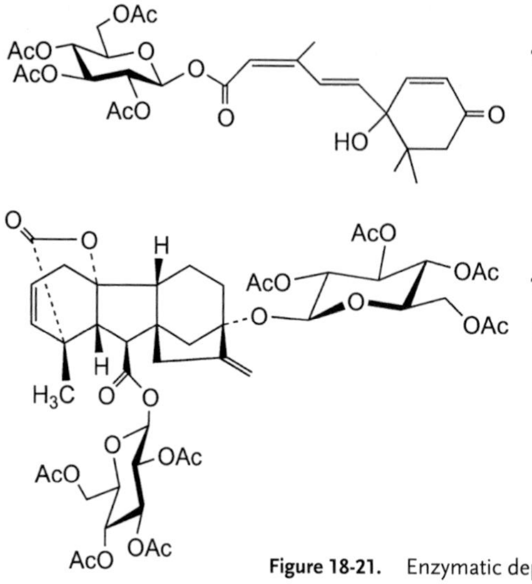

Figure 18-21. Enzymatic deprotection of complex glucosyl esters.

which differ widely, to be carried out. A general principle that emerges from these studies is that the enzymes exhibit a predominant preference toward primary hydroxy groups. If these functionalities are not present or protected, the biocatalysts are capable of selectively manipulating secondary hydroxy groups or the esters thereof. In the introduction and removal of acyl groups, the regioselectivity displayed by the enzymes often parallels the findings recorded for classical chemical transformations, although it is significantly higher in many cases. Furthermore, in several cases regioselectivities were observed in the biocatalyzed processes which can not or only slightly be achieved by means of chemical methods. Finally, it should be realized that subtilisin and the lipases are capable of introducing specific acyl groups into the carbohydrates which can later be removed selectively by different enzymatic or chemical methods.

18.5.6
Polyhydroxylated Alkaloids

The plant alkaloid castanospermine **155** and the related piperidine alkaloid 1-deoxynojirimicin **160**, like several other polyhydroxylated octahydroindolizidines, piperidines and pyrrolidines, are potent glycosidase inhibitors. These nitrogen bases are of considerable interest for the study of biosynthetic processes and, in addition, castanospermine and some of its derivatives may be of clinical value as antineoplastic agents and as drugs in the treatment of AIDS.

In the light of the analogy between the structures of these alkaloids and glucose, some of the above mentioned enzymatic methods for the selective functionalization of carbohydrates were applied to prepare several acyl derivatives of **155** and **160**. Thus, subtilisin transfers the acyl moieties from several activated esters to the 1-OH group of the bicyclic base in moderate to high yields (Fig. 18-22) [265, 266]. Again,

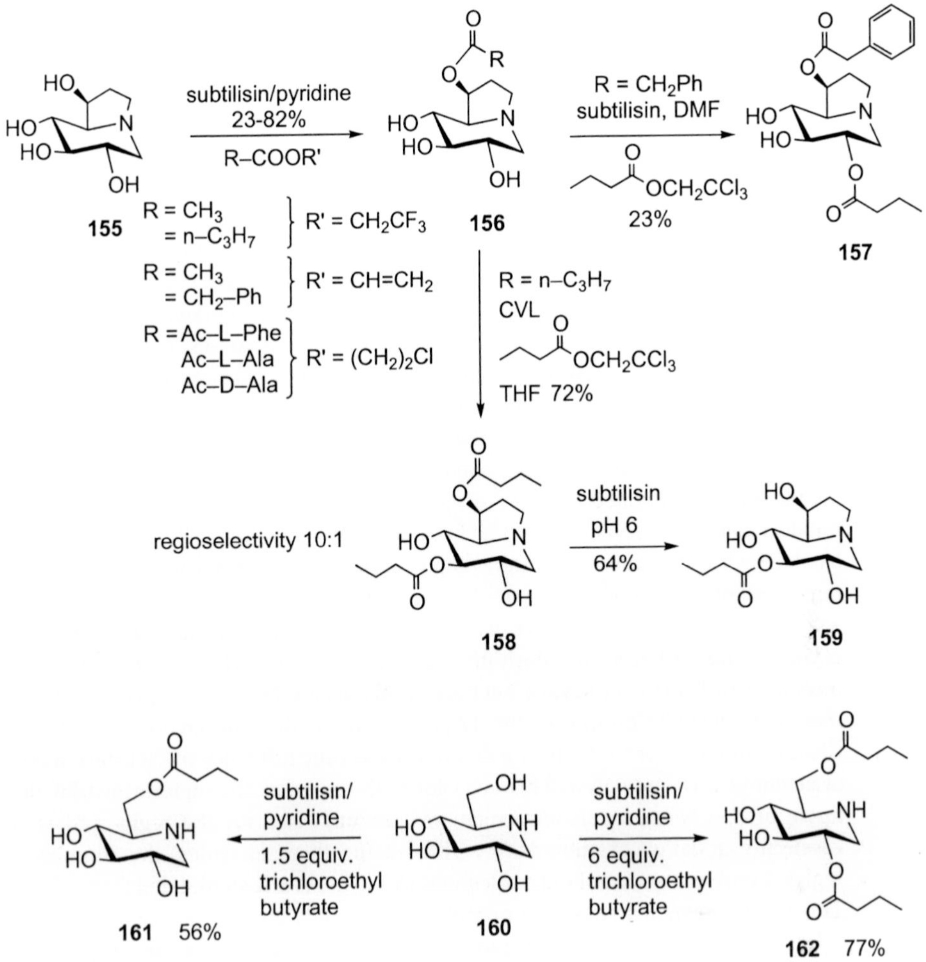

Figure 18-22. Selective enzymatic protection of polyhydroxylated alkaloids.

pyridine had to be used as the solvent for the polyhydroxy compound. The monoesters **156** obtained by this technique, like the monoesters of hexoses could subsequently be dissolved in THF and were further acylated by means of different enzymes, e. g. to the 6-butanoate **157** and the 1,7-dibutanoate **158**. Finally, the 1-ester was removed from **158** by subtilisin in aqueous solution to deliver the 7-butanoate **159** in 64 % yield.

In contrast to castanospermine, 1-deoxynojirimicine **160** contains a primary hydroxy group as well as a much more nucleophilic amino function. If a small excess of trifluoroethyl butanoate is employed, subtilisin converts this alkaloid preferably into the 6-monoester **161** (Fig. 18-22) [266]. However, with 6 equiv. of the acylating agent, the 2,6-diester **162** is formed in 77 % yield. This diester **162** may be subsequently deacylated regioselectively at the 6-position by means of several different enzymes.

It should be noted that under the conditions of the enzymatic acylation the amino group is not derivatized, an observation which has also been made in related cases [266, 267], e.g. *N*-terminally deprotected serine-peptides.

18.5.7
Steroids

Enzymatic acyl transfer reactions are also practical processes for the acylation of hydroxy groups in steroids. The lipase from *Chromobacterium viscosum* (CVL) for instance selectively transfers butyric acid from trifluoroethyl butanoate to equatorial (ß) C3-alcoholic functions that are present in a variety of sterols, e.g. **163** and the respective 5,6-didehydro compound (Fig. 18-23) [268]. Axially oriented alcohols at C3 and secondary alcohols at C17 or in the sterol side chains are not derivatized. In addition to the equatorial alcohols, the compounds being accepted as substrates by the lipase must have the A/B-ring fusion in the *trans* configuration. In the B-ring a double bond is tolerated, in the A-ring, however, it is not. Similarly, lipase from *Candida antarctica* acylates the 3-hydroxy group in steroids such as **163** and its 5,6-didehydro derivative [269]. Interestingly, acylation in this position is preferred regardless of the orientation of the hydroxy group. For instance, treatment of **164** with vinyl acetate in the presence of *Candida antarctica* lipase leads to the formation of corresponding 3-acetylated derivative in 82 % yield. In contrast, subtilisin does not recognize the hydroxy group at C3 of the steroid nucleus, but rather transfers the acyl moiety to alcoholic groups in the 17-position or in the side chains (Fig. 18-23). Changes in the A- or in the B-ring do not dramatically influence the selective mode of action of this biocatalyst. This behavior is the same as that determined for the lipase of *Pseudomonas cepacia*, which was recently used for the regio- and stereoselective acylation of steroids [270]. Thus, using these enzymes, the completely regioselective protection of either alcoholic group in several steroid diols is possible. This feature opened a route to a new chemoenzymatic process for the oxidation of selected positions of the steroid framework via an enzymatic protection/oxidation/ deprotection sequence. Chemoenzymatic approaches of this type are expected to provide attractive alternatives to the currently utilized enzymatic oxidation of steroids by hydroxysteroid dehydrogenases.

A further biocatalyst comes into play when bile acids serve as starting materials, e.g. deoxycholic acid methyl ester **165** [271]. The *cis*-configuration of the A/B-ring fusion prevents the application of lipase from *Chromobacterium viscosum* (CVL) and the aliphatic chain hinders the esterification of the C12α hydroxy group by subtilisin. The lipase from *Candida cylindracea* (CCL) has proved to be the most suitable enzyme for the enzymatic acylation of bile acids. In hydrophobic solvents, i.e. hexane, toluene, butyl ether, benzene, etc. (except acetone) and employing trichloroethyl butanoate as the acyl donor, the 3α-*O*-butanoyldeoxycholic acid methyl ester **166** is formed in 80 % yield without any by-products, suggesting that the enzyme is ineffective towards 12α-OH. In addition, the 7α-OH and the 7β-OH, present in **167** and **168** are not esterified by the enzyme. In both cases, the 3-butanoate is also formed (Fig. 18-23).

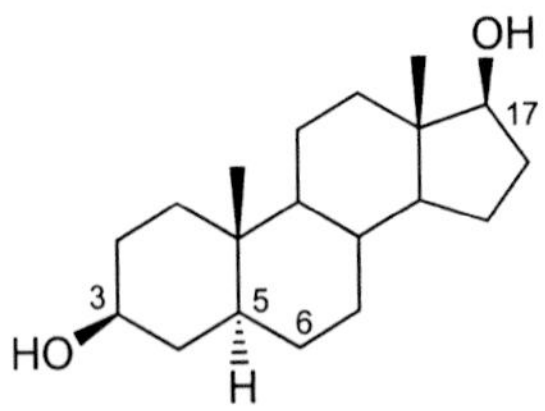

163

CVL: 3-monoburyrate 83%
CAL: 3-monoburyrate
subtilisin: 17-monobutyrate 60%

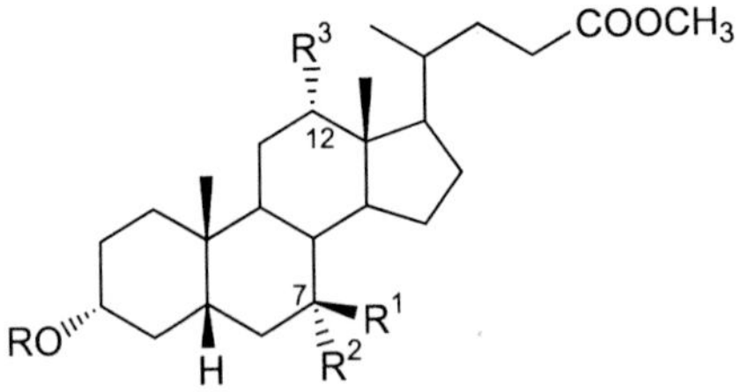

CCL, trichloroethyl butyrate, hydrophobic solvent
164 R = R^1 = R^2 = OH, R^3 = OH
165 R = R^1 = R^2 = H, R^3 = OH
166 R = But, R^1= R^2 = H, R^3 = OH   80%
167 R = But, R^1 = H, R^2 = OH, R^3 = H
168 R = But, R^1 = OH, R^2 = H, R^3 = H

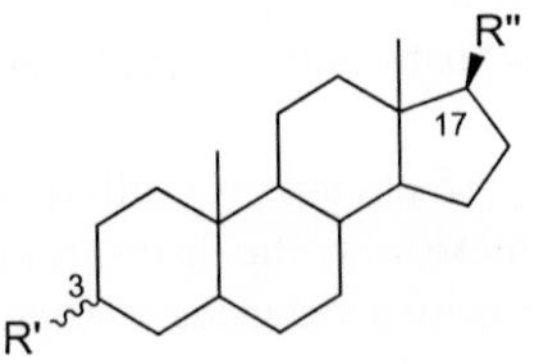

CCL:
169 R'= 3α-OAc, R"= 17β-OAc no reaction
170 R'= 3β-OAc, R"= 17β-OAc
→ 3β-OH, R" = 17β-OAc 79%

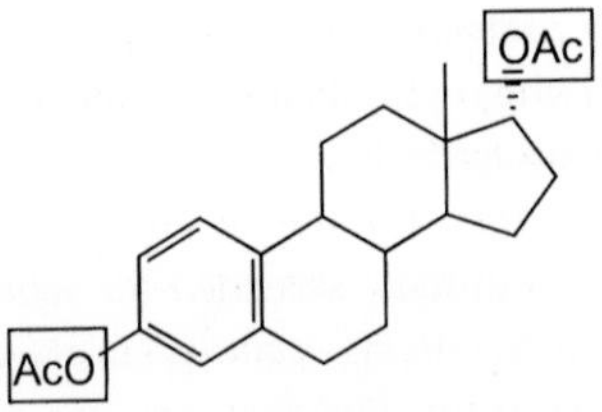

171

CCL:
3,17α-dihydroxyestradiol   60%
3-hydroxy–17α-acetoxyestradiol   25%

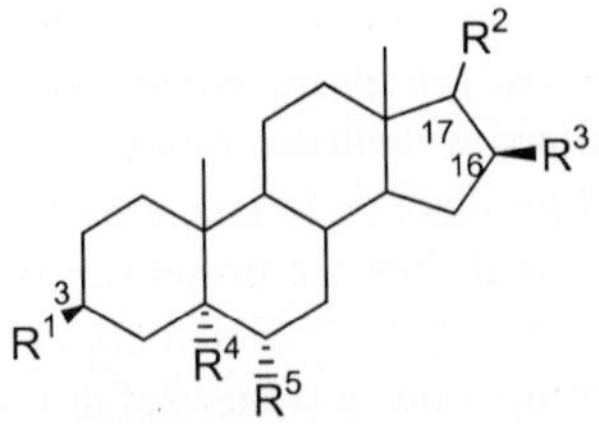

172 R^1=R^3=OAc, R^2=(O), R^4=R^5=H
173 R^1=R^5=OAc, R^2=(O), R^3=R^4=H
174 R^1=R^2=R^3=OAc, R^4=R^5=H

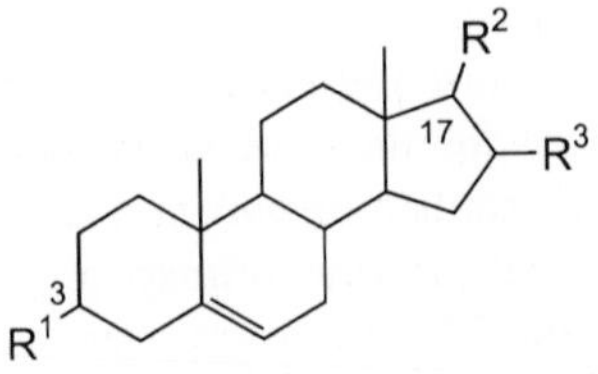

175 R^1= R^2=OAc, R^3=(O)
176 R^1=OAc, R^2 =C(O)CH$_3$, R^3=H

Figure 18-23.   Selective enzymatic protection of steroids.

Saponification of steroid esters can also be steered with *Candida cylindracea* lipase (CCL) [272, 273]. This process occurs in the presence of octanol in organic solvents and is characterized by a pronounced stereospecificity and regioselectivity. Thus, the 3α-esters of 3α,17β-diacetoxy steroid **169** resisted liberation, whereas the 3ß-isomer **170** is transformed into the corresponding alcohol in 79% yield. The 17α-acetate of 3,17α-diacetoxy estradiol **171** is also saponified, but at a slower rate than the 3-acetate (Fig. 18-23). In the case of the androstane derivatives **172** and **175** different selectivities of *Candida antarctica* lipase (CAL) and CCL were observed[273]. Thus, the alcoholysis of **172** in the presence of CAL afforded the C3 deprotected product in 75% yield whereas CCL led to the removal of the acetate at C16 in 66% yield. Treatment of **173**, **174** and **176** with CCL led to the cleavage of the C3 acetate in 79%, 87% and 83% yield, respectively[273, 274].

18.5.8
Phenolic Hydroxy Groups

Polyphenolic compounds occur widely distributed in nature and may possess a variety of interesting biological properties, e.g. antibiotic, antiviral and antitumor activity. The synthesis and further elaboration of these compounds often requires the selective protection or deprotection of specific phenolic hydroxy groups. To achieve this goal, the methods highlighted above for the various aliphatic polyols can also be applied successfully.

For example, for the the enzyme-catalyzed acetylation of phenols six different lipases was initially screened for activity[275, 276]. Out of these, only the lipase from *Chromobacterium viscosum* (CVL) showed significant activity. In a subsequent study, the lipase from *Pseudomonas cepacia* (PSL) turned out to be a more efficient biocatalyst, which was succesfully used for the regioselective acylation of various aromatic dihydroxycarbonyl compounds[277], and (+)-catechin[278]. Thus, by using PSL as the biocatalyst the dihydroxy aldehydes and ketones **177**, **178** and related compounds were selectively acetylated in conversions ranging from 20 to 97% using vinyl acetate as the acyl donor (Fig. 18-24)[277]. (+)-Catechin **179** was also subjected to irreversible acyl transfer conditions. In this case, both the 5- and 7-monoacetates were obtained in 40% and 32% yield, respectively[278]. Interestingly, the inability of the lipase from *Aspergillus niger* to acylate aromatic hydroxy groups has consequently been used for the selective acylation of primary aliphatic hydroxy functions in molecules containing both aromatic and aliphatic OH-groups[279]. In fact, even PSL preferentially acylates primary aliphatic hydroxy groups if they are present in the compound[280].

In the deprotection of peracetylated polyphenolic compounds a somewhat different scheme has emerged. In this area, a broader spectrum of lipases has been used successfully. For example, the pentaacetyl derivative of catechine **179** was treated with PSL under alcoholysis conditions (THF, n-butanol) to give the 3,3',4'-trisacetate in 50% yield after 12 hours[278]. On longer exposure to the biocatalyst, the 3-monoacetyl derivative was isolated in 95% yield.

Thus, the coumarine **180**, the chromanone **181**, the chalcone **182**, the flavanone **183** as well as several flavones, e.g. **183** and **185** were regioselectively deacylated by employing different lipases in organic solvents (Fig. 18-24). Porcine pancreatic lipase (PPL) predominantly attacks one of the phenolic acetates present in **180-183** with

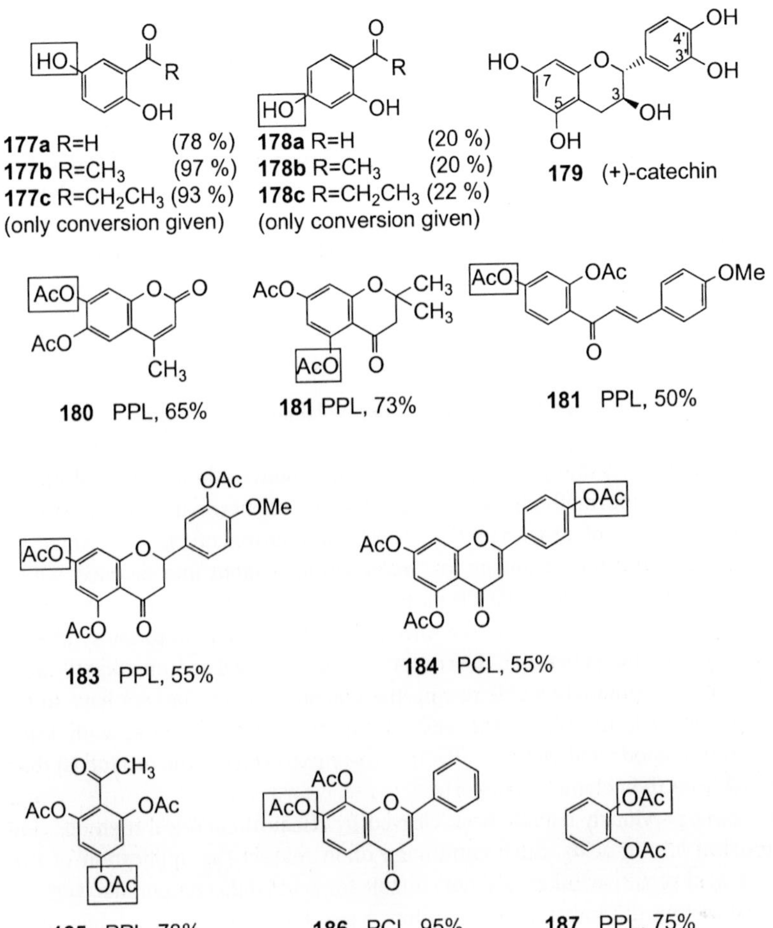

177a R=H   (78 %)
177b R=CH₃   (97 %)
177c R=CH₂CH₃ (93 %)
(only conversion given)

178a R=H   (20 %)
178b R=CH₃   (20 %)
178c R=CH₂CH₃ (22 %)
(only conversion given)

179 (+)-catechin

180 PPL, 65%

181 PPL, 73%

181 PPL, 50%

183 PPL, 55%

184 PCL, 55%

185 PPL, 78%

186 PCL, 95%

187 PPL, 75%

Figure 18-24. Selective enzymatic protection and deprotection of polyphenolic compounds.

good to high regioselectivity and produces the respective selectively protected compounds available in good yields [281–283]. The flavone acetates **184** and **186** can be partially deacylated with high regioselectivity by transesterification using lipase from *Pseudomonas cepacia* (PSL) and n-butanol in THF. [284,285] However, in other cases the positional specificity displayed by the enzyme was less pronounced. This technique has allowed for an efficient construction of a selectively O-methylated flavonoid [284].

In addition, aryl alkyl ketones which are important starting materials for the synthesis of polyphenolic natural products may be manipulated selectively by making use of an enzymatic saponification [283, 285–287]. In general, in these cases the sterically better accessible ester groups are cleaved, as for instance in **185** [285]. All of these examples have in common the fact that a carbonyl group is either directly or vinylogously attached to the aryl moiety. Without such a function present in the molecule, the biocatalysts failed to differentiate the ester groups or completely deacylated the substrates. However, by using the lipases from porcine pancreas (PPL) or *Candida cylindracea* (CCL) immobilized on microemulsion-based gels it was possible to monodeacylate resorcinol and related diesters such as **187** in high yields[158]. Alternatively, by using *tert*-butyl methyl ether saturated with water as the solvent, it was possible to monodeacetylate diacetoxynaphthalenes selectively[288]. The influence of the solvent was exemplified by charging the solvent system to acetone/buffer: under such conditions only completely deacylated products were obtained.

18.6
Biocatalysis in Polymer Supported Synthesis: Enzyme-labile Linker Groups

Combination chemistry and parallel synthesis of compound libraries on polymeric supports are efficient methods for the generation of new substances with a predetermined profile of properties[289–291]. The anchoring of one reactant to a polymeric support has the advantage that an excess of reagent may be used, while purification is kept manageable. This is particularly important if the reaction is to be carried out with several reactants in the same reaction vessel. Solid phase synthesis involves the use of linkers between the compounds to be varied combinationally and the solid supports which are stable during the reactions. These linkers have to be cleavable as desired, usually at the end of the synthetic sequence, with high selectivity and in good yield, without affecting the structure(s) of the product(s) that are released from the polymeric supports.

Linkers have previously usually been cleaved by classical chemical methods, for instance using strong acids. Such conditions often restrict the application of the linkers, i.e. acid-sensitive linkers are not suitable for acid-labile compounds, such as carbohydrates. Specific linkers have therefore been developed for acid-labile compounds, such as silylether linkages, thioether linkages[292], and ester linkages[293]. Although such linkers may be cleaved in the presence of acid-labile groups, they have the disadvantage that they are themselves quite labile to common chemical reagents that one might want to employ on the solid phase. For example, esters and silylethers are unstable to bases and thioethers are unstable in the presence of oxidants, such as *m*-chloroperbenzoic acid, and to electrophilic reagents, such as alkylating agents.

In principle linker groups are polymer-enlarged versions of blocking functions used in regular solution phase chemistry. Therefore, enzymatic transformations that may be employed for the removal of protecting groups in solution in principle may also open up alternative opportunities for releasing compounds from polymeric supports. The linkers developed so far can be divided into exo- and endo-linkers (Fig. 18-25) cleavable by exo- endo-enzymes, respectively, as proposed by Flitsch et al. [294].

Exo-linkers are composed of three units: (i) a group providing the site for enzyme catalyzed hydrolysis (R^1); (ii) a site for attachment of the target molecule (R^3); and (iii) a site for attachment to a further optional spacer (R^2).

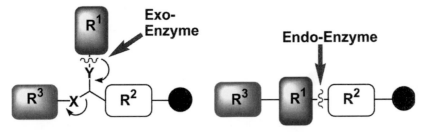

Exo-linkers cleaved by exo-enzymes   Endo-linkers cleaved by endo-enzymes

R^1: group providing the site for enzyme catalyzed hydrolysis,
R^2: optional intermediate linked to a solid support,
R^3: residue to be synthesized and varied in the course of a
     synthesis on the support,
     X: O, N(H), N(R''), C(O)O, S, C(O)N(H) or C(O)N(R''),
     R'' is a noninterfering substituent, Y: O or NH.

Figure 18-25. Graphical representation of exo- and endo-linkers.

Endo-linkers are linkers in which the target molecule (R^3), the group, which provides a site for enzyme catalyzed hydrolysis (R^1) and a further optional spacer (R^2) are attached to the polymeric support in a linear arrangement. By means of enzyme mediated dissection, the target molecule, in many cases tagged with the functional group recognized by the enzyme, is released.

Examples of endo-cleavable linkers have been reported (Table 18-5). However, in many cases the product is tagged with part of the linker. For instance, the endo-peptidase chymotrypsin cleaves endo-linkers towards the middle of a peptide-chain or "internally". Not only does this limit the methodology to a very small number of enzymes, but it may also restrict the structure of molecules that can be generated. For instance, this method will typically (but not necessarily, see Figs. 18-26 and 18-27) generate compounds containing *C*-terminal aromatic amino acids, which are necessary for recognition by chymotrypsin. By contrast, exo-linkers do not restrict the structure of the reactant and can be cleaved by more readily available exo-enzymes, which act at the end of a chain or "externally" (Table 18-5). Furthermore exo-cleavable linkers yield untagged products upon cleavage from the solid support.

18.6.1
Endo-linkers

For a better overview, examples of endo-linkers and the enzymes used for the cleavage of the product from the solid phase which have been described in the literature so far are given in Table 18-5.

Wong and coworkers[295] introduced a silica-based solid support with a specific enzymatically cleavable linker for the synthesis of glycopeptides and oligosaccharides. They found that styrene- and sugar-based polymers tend to swell which leads to a low coupling yield. Their choice of solid support is aminopropyl silica based on the

Table 18-5. Examples of endo-linkers and the appropriate cleavage enzymes.

Linker	Enzyme	Examples	Ref.
	α-Chymotrypsin	Glycopeptide synthesis	[296]
	α-Chymotrypsin	Oligosaccharide synthesis	[297–298]
	Ceramide glycanase	Oligosaccharide synthesis	[299]
	Phosphodiesterase	Peptide synthesis	[301]

facts that: (a) it is compatible with both aqueous and organic solvents, (b) it has a large surface area accessible to biomolecules, and (c) it has sufficient density of functional groups.

A hexaglycine spacer was attached to the solid support to give a substitution of 0.2 mmol g^{-1} of dry silica and the excess amino groups were then capped using acetic anhydride. In the next step a selectively cleavable, α-chymotrypsin sensitive, phenylalanine ester **189** was implemented for the release of the products from the solid support under mild conditions. Then it was transformed to **190** followed by reactions with glycosyl transferases to yield **191**. Finally, the desired glycopeptide was cleaved from the solid support in high yield by treatment of **191** with α-chymotrypsin (Fig. 18-26).

Nishimura and coworkers [296–297] described a novel method for the enzymatic synthesis of oligosaccharide derivatives employing an α-chymotrypsin sensitive linker. The synthesis of the water soluble GlcNAc-polymer **197**, sensitive to α-chymotrypsin, is shown in Fig. 18-27. Oxazoline derivative **193** was coupled with 6-(N-benzyloxycarbonyl-L-phenylalanyl)-amino-hexanol-1 (**194**) followed by N-deprotection of the phenylalanine and subsequent condensation with 6-acrylamido caproic acid **195**. De-O-acetylation gave the polymerizable GlcNAc derivative **196**. Finally, co-polymerization of acrylamide and monomer **196** in the presence of ammoniumpersulphate (APS) and N,N,N',N'-tetramethyl ethylene diamine (TMEDA) gave the

Figure 18-26. Glycopeptide synthesis and α-chymotrypsin catalyzed release from the solid support.

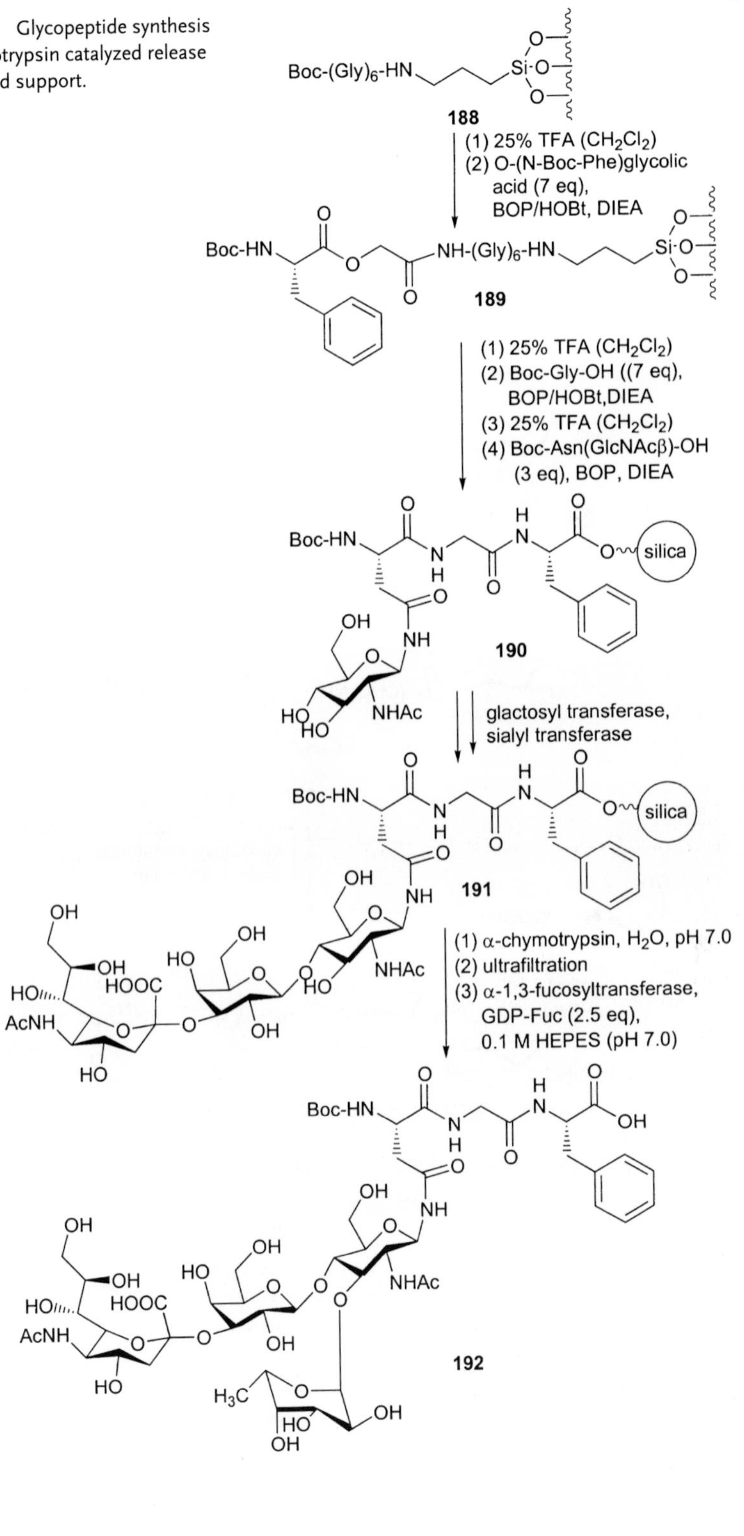

Figure 18-27. Oligosaccharide synthesis and α-chymotrypsin catalyzed release from the solid support.

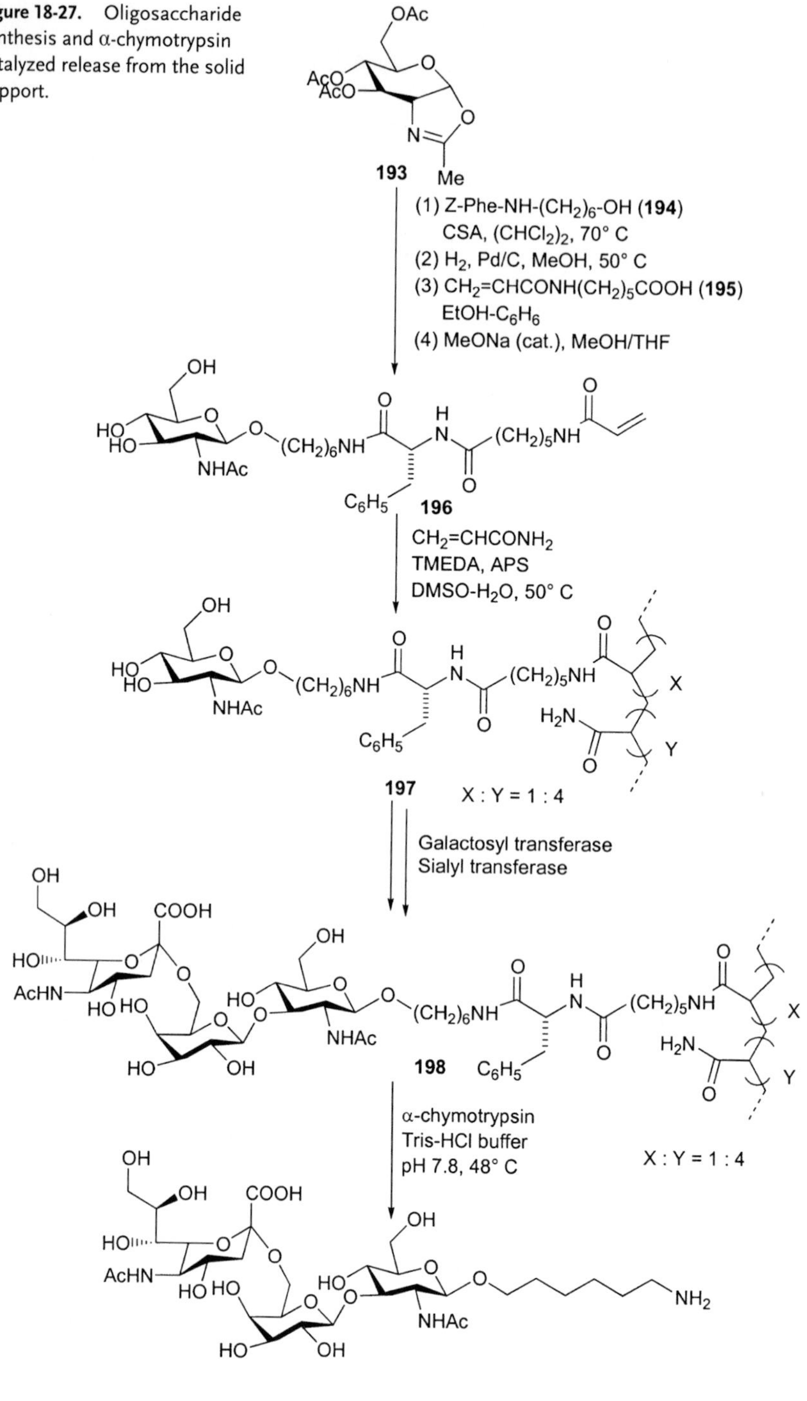

193

(1) Z-Phe-NH-(CH$_2$)$_6$-OH (**194**)
   CSA, (CHCl$_2$)$_2$, 70° C
(2) H$_2$, Pd/C, MeOH, 50° C
(3) CH$_2$=CHCONH(CH$_2$)$_5$COOH (**195**)
   EtOH-C$_6$H$_6$
(4) MeONa (cat.), MeOH/THF

196

CH$_2$=CHCONH$_2$
TMEDA, APS
DMSO-H$_2$O, 50° C

197   X : Y = 1 : 4

Galactosyl transferase
Sialyl transferase

198   C$_6$H$_5$

X : Y = 1 : 4

α-chymotrypsin
Tris-HCl buffer
pH 7.8, 48° C

199

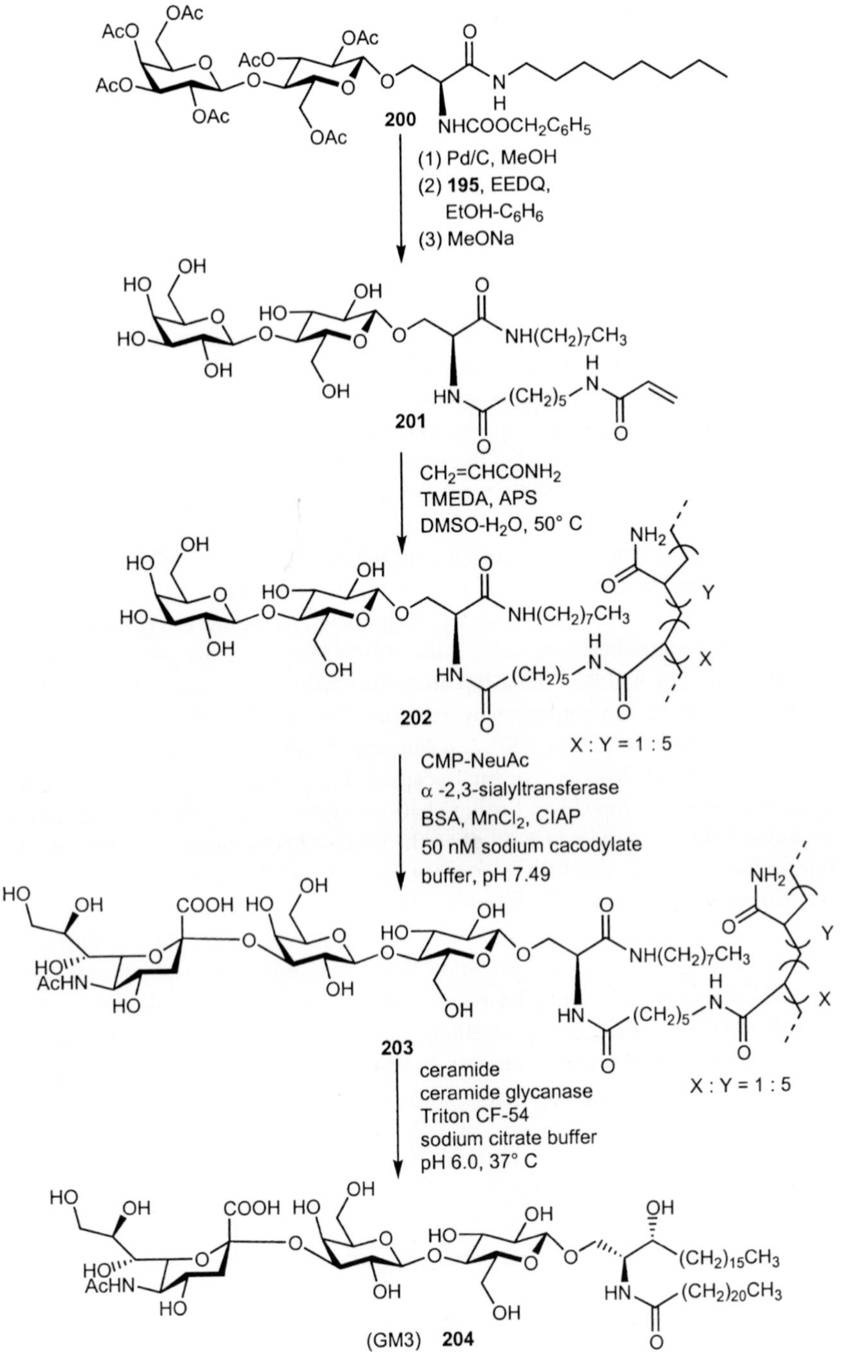

Figure 18-28. Ceramide glycanase mediated release by transglycosylation.

polymer **197** in high yield. The polymer **197** was then subjected to galactosylation and subsequent sialylation with the corresponding glycosyl transferases to yield **198**. The final product **199** was cleaved from the water-soluble support by treatment with α-chymotrypsin at 40 °C for 24 h in 72% overall yield from **197**.

Nishimura and Yamada [298] introduced a water-soluble polymeric support having a linker recognized by ceramide glycanase for a synthesis of ganglioside GM3 (**204**). Synthesis of the polymerizable lactose derivative **201** with a ceramide glycanase sensitive linker is shown in Fig. 18-28. The lactosyl ceramide (LacCer) mimetic glycopolymer **202** is obtained from the monomeric precursor **201** by co-polymerization with acrylamide.

This solid support **202** was converted into the intermediate product **203** by sialylation using βGal1→3/4GlcNAc α-2,3-sialyltransferase. Finally, the polymeric support was cleaved by transglycosylation with leech ceramide glycanase in the presence of excess ceramide as the acceptor to give the desired product **204** in high yield (Fig. 18-28). An advantage of the water-soluble polymer is that the transformation can be monitored by NMR spectroscopy during the enzymatic glycosylation steps.

Arrays of up to 1000 peptide nucleic acid (PNA) oligomers of different sequence were synthesized by Jensen et al. on polymer membranes (Fig. 18-29) [299]. The PNA chain was linked to the peptide spacer glutamic acid-(γ-*tert*-butyl ester)-(ε-aminohexanoic acid)-(ε-aminohexanoic acid) (Glu[OtBu]-εAhx-εAhx) via an enzymatically cleavable Glu-Lys handle. The Glu[OtBu]-εAhx-εAhx spacer was coupled to the amino-functionalized membrane by standard Fmoc-Chemistry. Then the membranes were mounted in an ASP 222 Automated SPOT Robot and a grid of the desired format was dispensed at each position. The free amino groups outside the spotted areas were capped and further chain elongation was performed with Fmoc-protected PNA monomers to synthesize the desired PNA oligomers. After completion of the synthesis, the PNA oligomers were cleaved from the solid support by incubation with bovine trypsin solution in ammonium bicarbonate at 37 °C for 3 h.

One of the very first papers concerning endo-linkers was published by Elmore et al. (Fig. 18-30) [300]. They described a new linker containing a phosphodiester group for solid phase peptide synthesis using a Pepsyn K (polyacrylamide) resin. After completion of coupling and deprotection cycles, the phosphodiester **207** was cleaved with a phosphodiesterase. In this way β-casomorphin, Leu-enkephalin and a col-

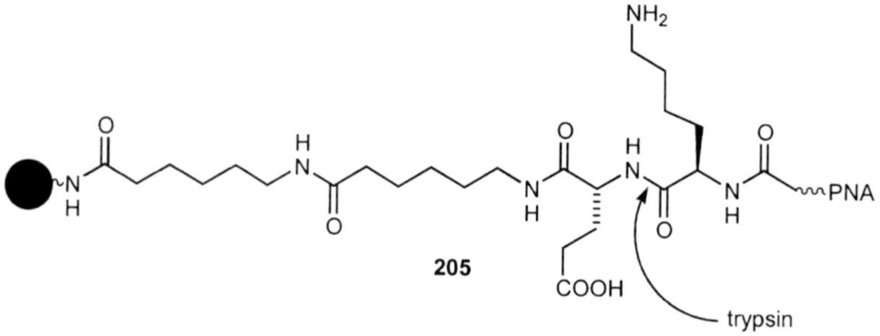

Figure 18-29. Trypsin mediated cleavage of a peptide bond in PNA oligomer synthesis.

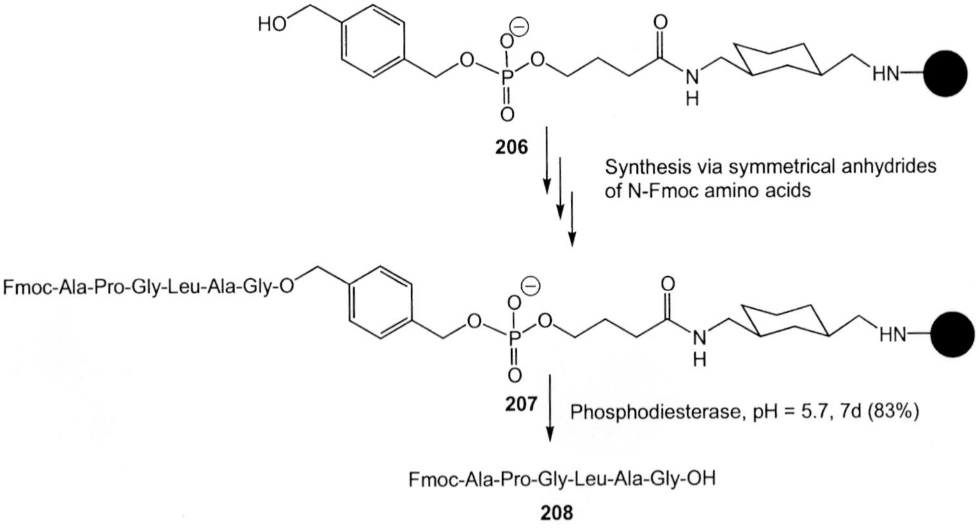

Figure 18-30. Synthesis of a collagenase substrate on a phosphodiesterase-scissile linker.

lagenase substrate were synthesised in high yields. In the context of enzymatic cleavage of linkers on polymeric supports particular attention was paid to the general question of whether enzymatic transformations on resins are viable and high yielding. An in-depth treatment of this problem is beyond the scope of this review. However, a few examples for the application of biocatalyzed transformations on solid supports will serve to illustrate that such transformations can indeed be employed advantageously for various purposes.

Meldal et al. described the proteolytic cleavage of the alanine-tyrosine bond in a resin-bound decapeptide by treatment with the 27 kDa protease subtilisin BNP' to demonstrate the accessibility of the interior of the newly designed SPOCC-resin[301] to enzymes[302].

Furthermore, enzymatic hydrolysis of model isopeptides N^ε-oligo(L-methionyl)-L-lysine from Bio-beads[303] by pepsin, chymotrypsin, cathepsin C (dipeptidyl peptidase IV) and intestinal aminopeptidase N was investigated using high-performance liquid chromatography to identify and quantify the hydrolysis products[304].

Larsen et al. reported the enzymatic cleavage of a desB30 insulin B-chain from a presequence (Lys(Boc))$_6$. This spacer shifts the conformation of the growing peptide chain from a β-structure to a random coil conformation and reduces peptide-chain aggregation, which otherwise causes serious synthetic problems. Novasyn KA-[305] was employed as a solid support, but unfortunately, no information about the enzyme used was reported[306].

Barany et al. were the first to exploit the different enzyme accessibilities of surface and interior areas of a given bead and the resulting differentiated bead was used to synthesize a peptide library on the surface and the code for this on the interior simultaneously[307]. This clever strategy is illustrated in Fig. 18-31. Selective cleavage of short N^α-protected peptide substrates with chymotrypsin from the surface area of

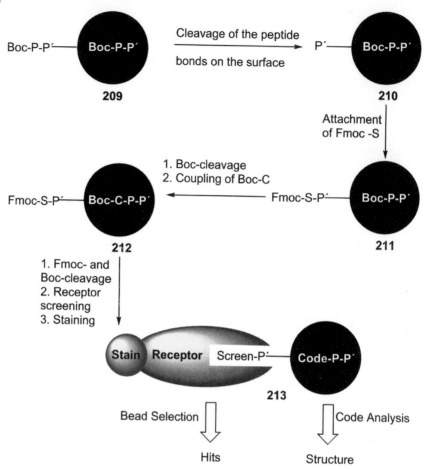

Figure 18-31. Peptide encoded combinatorial peptide libraries via enzyme-mediated spatial segregation. P-P': substrate with a scissile bond between P and P'; S: terminal residue of the screening structure, C: terminal residue of the coding structure.

TentaGel-AM-beads **209** leaves the majority of the peptide attachment sites in the interior uncleaved to afford **210** ("shaving" methology). The first residue is attached using orthogonal FMOC-chemistry to provide **211**. Coding is done by using standard BOC-chemistry on the interior of the bead to yield **212**. Repetition of this process furnishes a surface peptide, which is encoded internally (**213**).

This generation of two structures on the same bead allowed the investigation of the synthesized peptide library (1×10^5 members) with different receptors (anti-β-endorphin antibody, streptavidin and thrombin). After the staining procedure had been carried out, the beads that showed a color were selected for sequencing and the coding peptides present within the bead were used to deduce the binding structures. This screening led to the discovery of a new thrombin ligand, which binds with an affinity one order of magnitude higher than the natural motif.

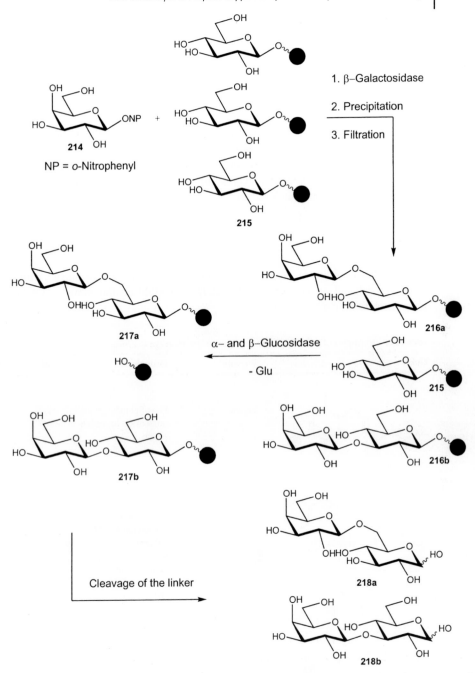

Figure 18-32. General strategy for the liquid-phase synthesis of disaccharides using glycosidases.

Fernandez-Mayoralas and coworkers[308] used the high substrate specifity of enzymes in their synthesis of galactose-glucose-disaccharides (218) on an MPEG-support[309]. After galactosylation of glucose immobilized on the soluble support (215) using β-galactosidase, the unreacted monosaccharide glucose was removed by the combined use of α- and β-glucosidases to obtain only MPEG-bound disaccharides (216, Fig. 18-32). Finally the disaccharides 218 obtained were released from the support by ethanolysis.

Schmitz and Reetz described the solid phase enzymatic synthesis of oligonucleotides on Kieselguhr-PDMA-resins via T4 RNA ligase. Concomitantly, they found that RNase A selectively cleaves the last bound nucleotide at the ribose sugar leaving a 3',5'- diphosphorylated oligomer behind on the resin, but application in actual synthesis has not yet been undertaken[310].

18.6.2
Exo-linkers

An exo-linker according to Fig. 18-25 must contain an enzyme labile group R^1, which is recognized and attacked by the biocatalyst. Possible combinations could be: phenylacetamide/penicillin amidase, ester/esterase, monosaccharide/glycosidase,

Table 18-6. Examples of exo-linkers and the appropriate cleavage enzymes.

Linker	Enzyme	Examples	Ref.
	Lipase	Picet-Spengler reaction, nucleoside immobilization	[313]
	Penicillin acylase	Palladium cat. C-C-couplings, Mitsunobu- and Diels-Alder reactions, 1,3-dipolar cycloadditions	[318, 319]
	Penicillin acylase	Immobilization of alcohols (e.g. Fmoc protected serine methyl ester, glycosides) and amines (e.g. phenyl-alanine)	[312, 313]

phosphate/phosphatase, sulfate/sulfatase and peptides/peptidases[311]. Up till now only the following systems have been worked out (Table 18-6).

In independent and simultaneous investigations Flitsch and coworkers [311, 312] and Waldmann and coworkers[313, 314] developed a selectively cleavable exo-linker, which can be cleaved with penicillin G acylase, a commercially available and widely used enzyme[77].

Penicillin acylase catalyzes the hydrolysis of phenylacetamides and has been used in peptide synthesis for the cleavage of protecting groups[6, 315]. In linker **219** developed by Flitsch and coworkers[311, 312] (Fig. 18-33) -XR represents the alcohol or amine group of the target molecule. Hydrolysis of the phenylacetamide moiety generates the hemiaminal **221** which readily fragments in an aqueous medium and thereby releases the desired products, RXH. The thioethyl group present in the anchor group of **219** was activated by treatment with *N*-iodosuccinimide (NIS) followed by displacement with a variety of alcohols (**223–225**). To prove the possible application of this linker in solid phase carbohydrate synthesis protected glycosides **226** and **227** were coupled to linker **219** and released enzymatically. Flitsch et al. also described the immobilization and enzymatic cleavage on a variety of amines[311]. Nevertheless, the application of this enzyme-labile linker group in multi-step syntheses on the solid phase and subsequent enzyme-initiated release from the polymeric support has not been described yet.

Waldmann and coworkers described designed exo-linker **228**[313, 314]. The anchor group comprises a 4-acyloxy-3-carboxybenzyloxy group, which is recognized and attacked by the biocatalyst, so that a spontaneously fragmenting intermediate is generated, thereby releasing the desired compound (Fig. 18-34)[53, 54, 57]. The linker **228** is attached as an amide to the solid phase. Cleavage of the acyl group by a lipase generated a phenolate **229**, which fragments to give a quinone methide **230** and releases the product **231**. The quinone methide remains on the solid phase and is trapped by water or an additional nucleophile.

Following on from this cleavage principle, amines (bound as urethanes), alcohols (bound as carbonates), and carboxylic acids (bound as esters) can be detached from the polymeric carrier. The substrate specificity of the enzyme guarantees that only the intended ester is cleaved. TentaGelS-NH$_2$ was chosen as the polymeric support, i.e. a polystyrene resin equipped with terminally NH$_2$-functionalized oligoethylene-glycol units. It has a polar surface and swells in aqueous solutions allowing the biocatalyst access to the polymer matrix[316].

The applicability of the enzyme-labile anchor group was demonstrated by the synthesis of tetrahydro-β-carbolins **237** employing the Pictet-Spengler reaction (Figure 18-35). The benzylic alcohol group of the linker **232** was first esterified with Boc-L-tryptophan, and after its *N*-terminal deprotection the support-bound trypto-phan **233** was reacted with aliphatic and aromatic aldehydes to give imines **234**, which cyclized immediately in reasonable to high yields to the tetrahydro-β-carbolins **235**. Lipase RB 001–05 selectively attacked the acetate incorporated into the linker and generated the corresponding phenolate **236**, which then fragmented sponta-neously. Following these multistep transformations the desired tetrahydro-β-carbo-lins **237** were obtained in 70–80 % yield.

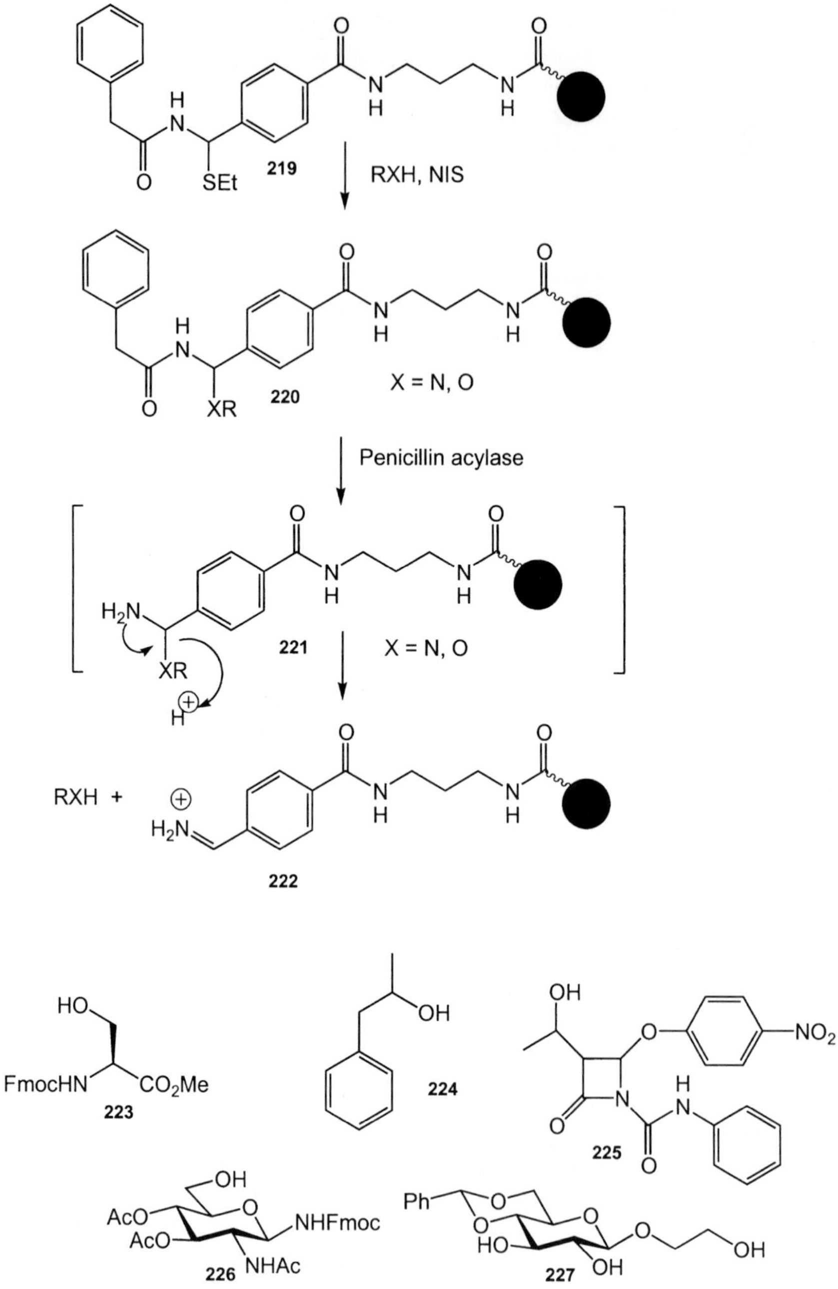

Figure 18-33. Loading and cleavage of a penicillin acylase scissile linker.

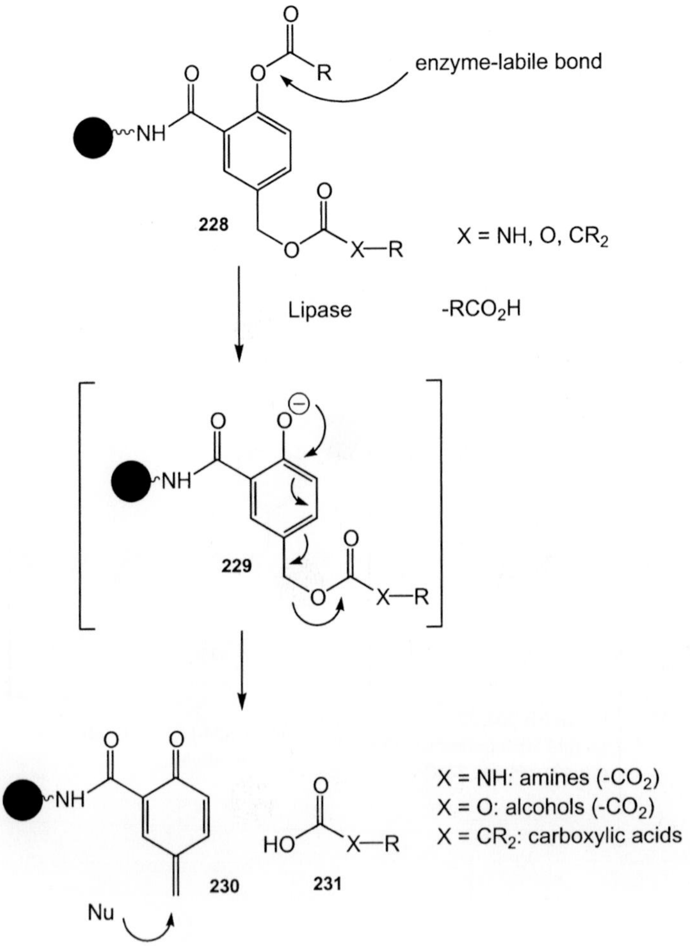

Figure 18-34. Principle for the development of the enzyme-labile 4-acyloxy-benzyloxy linker group.

Waldmann and coworkers developed a second exo-linker following a new approach [317, 318] which makes use of a safety-catch linker. It is based on the enzymatic cleavage of a functional group embodied in the linker. In this way an intermediate is generated, which subsequently cyclizes intramolecularly according to the principle of assisted removal [2, 319–322] and thereby releases the desired target compounds (Fig. 18-36). The linker group is immobilized as a urethane on the amino-functionalized carrier **238**. It facilitates the attachment of a variety of molecules such as alkyl halides, alcohols or amines bound as carboxylic acid esters and amides. According to the safety-catch principle, the separation of the desired products proceeds in a two-step process. First, penicillin G acylase hydrolyzes the phenylacetamide with complete chemo- and regioselectivity, under exceptionally mild conditions (pH 7.0, room temperature or 37 °C) [30, 72, 74]. Then the activated intermediate generated,

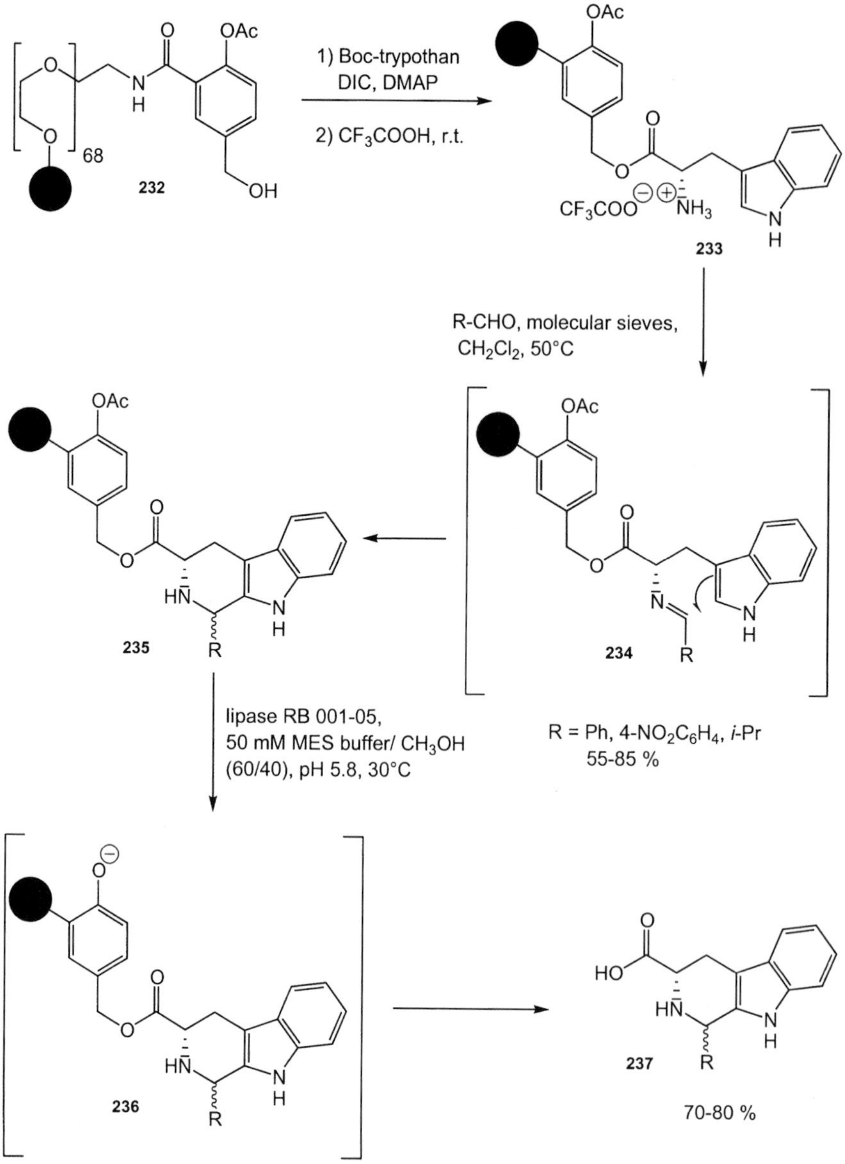

Figure 18-35. Solid phase synthesis of tetrahydro-β-carbolins and subsequent detachment by enzyme initiated fragmentation of the anchor group.

i.e. benzylamine **239**, cyclizes to polymer-bound lactam **240** and releases the desired target molecule **241**.

POE 6000 was used as the polymeric support, a soluble polyethyleneglycol derivative functionalized at both termini with an amino group and with an average molecular mass of 6000 Da[323–324]. After completion of the homogeneous reactions

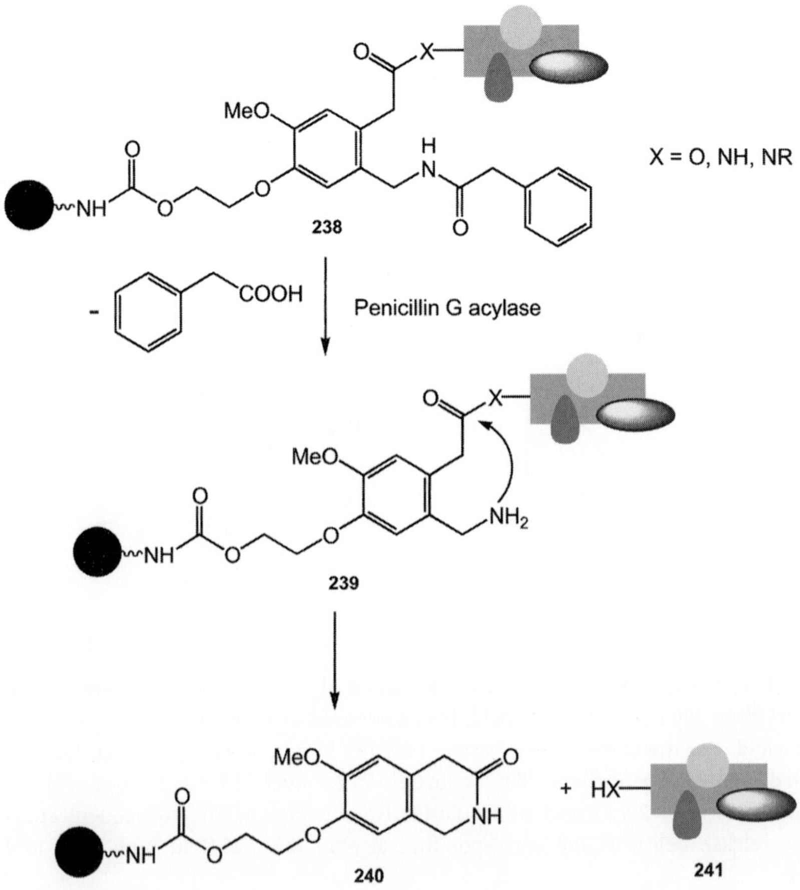

Figure 18-36. Principle of the enzyme-labile safety catch linker.

it can be precipitated, filtered off, and washed with diethyl ether, thereby facilitating the separation of surplus reagents and the side products. Furthermore it allows for NMR spectroscopic monitoring of the reactions[325]. Most importantly, it is soluble in aqueous solutions, thereby allowing efficient access of the enzyme to the polymer-fixed linker group.

The suitability of the polymer-linker conjugate was examined for a variety of transformations, in particular Pd⁰-catalyzed reactions. For instance, the polymer-bound aryl iodide **242** was transformed quantitatively in a Heck reaction to a cinnamic acid ester **243** and to biphenyl **245** in a Suzuki reaction. It gave an alkine **244** in a Sonogashira reaction (Fig. 18-37). The desired benzyl alcohols **246–248** were released by incubation of the corresponding polymer conjugates **243–245** with penicillin G acylase at pH 7 and 37 °C in high yields and isolated with a purity of >95 % by simple extraction with diethyl ether.

Furthermore, the applicability in a Mitsunobu esterification reaction and a Diels-Alder reaction was proven (Fig. 18-38). The polymer-bound benzyl alcohol **249** was

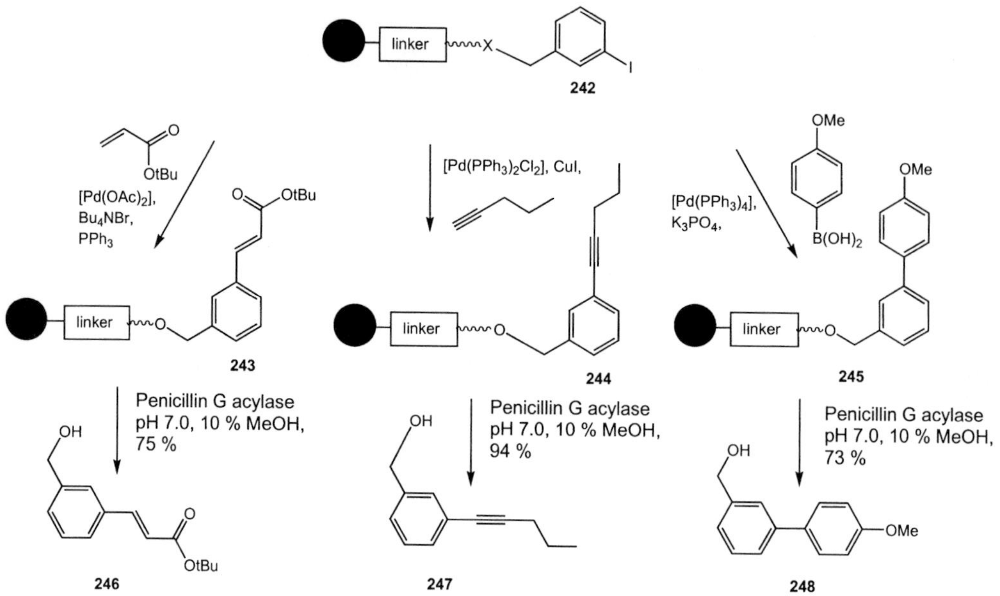

Figure 18-37. Pd^0-catalyzed reactions on enzyme labile linker-conjugates.

reacted with 4-acetamidophenol in the presence of the Mitsunobu reagent to give phenyl ether **250** in quantitative yield. It was released from the polymeric support in high yield. For the Diels-Alder reaction, polymer-bound acrylic acid ester **252** was treated with cyclopentadiene. The cycloaddition product **253** was formed with an endo/exo ratio of 2.5 : 1 and with quantitative conversion. The subsequent enzymatic release delivered the corresponding alcohol (**251, 254**) in high yield and purity.

18.7
Outlook

During recent decades substantial progress was achieved in the development of enzymatic protecting group techniques. In particular, it was demonstrated that these methods offer viable alternatives to classical chemical approaches. Not only do the biocatalyzed transformations complement the arsenal of chemically removable protecting groups, but in many cases they additionally offer the opportunity to carry out useful functional group interconversions with selectivities which can not or only barely be matched by chemical techniques. However, the overwhelming majority of the investigations carried out in this area has restricted themselves to the study of the protection and deprotection of model compounds. Complex synthetic schemes were nearly always avoided. Whereas this appears to be particularly true in the area of carbohydrates, noticeable examples which demonstrate the capacity of these bio-catalyzed processes were recorded in peptide and peptide conjugate chemistry, i. e. in

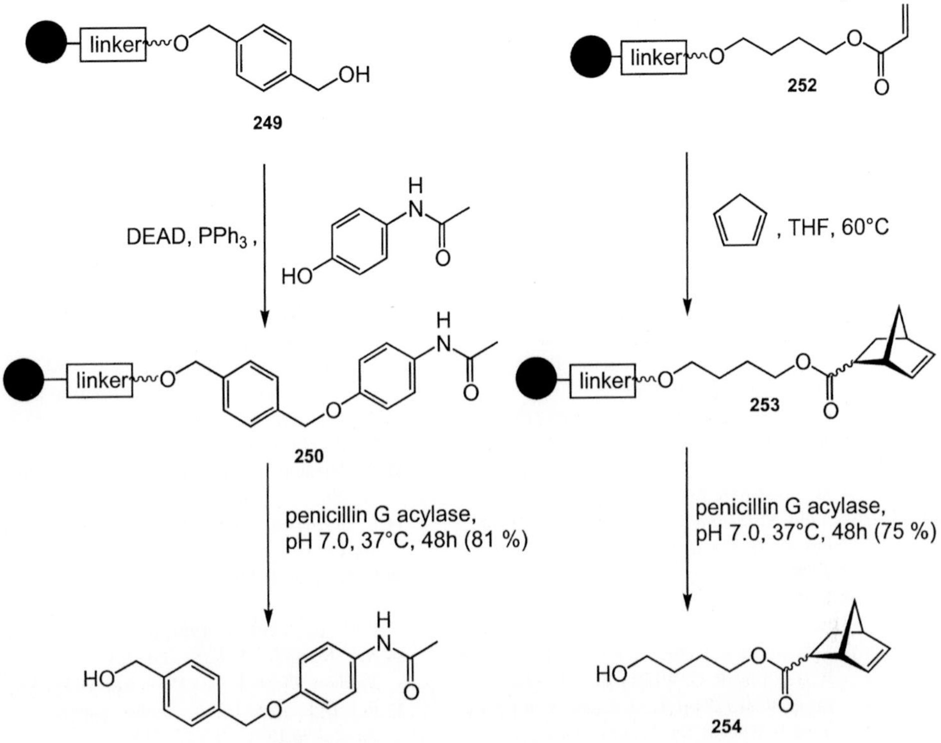

Figure 18-38. Mitsunobu and Diels-Alder reaction on enzyme labile linker-conjugates.

the synthesis of lipo-, glyco and nucleopeptides. The data and observations highlighted above, however, provide a solid basis for the application of biocatalysts in the handling of protecting group problems in complex multistep syntheses.

On the other hand, the use of biocatalysts in protecting group chemistry in the sense of a general method deserves and is certainly awaiting further intensive development. Numerous applications of the known enzymes appear to be possible in all areas of preparative chemistry. In addition, the use of catalytic proteins which have not yet been applied to carry out protecting group manipulations and of biocatalysts unknown today or which will be developed in the future, e.g. by evolutionary approaches, will create new opportunities for improved organic syntheses.

References

1 H. Kunz, H. Waldmann in: (Eds.: B. M. Trost, I. Flemming, E. Winterfeldt), Comprehensive Organic Synthesis Vol. 6, Pergamon Press, Oxford **1991**, pp. 631–701.

2 T. W. Greene, P. G. M. Wuts, *Protective Groups in Organic Synthesis*, Wiley and Sons, New York, **1999**.

3 P. J. Kocienski, *Protecting Groups*, Georg Thieme Verlag, Stuttgart, **1994**.

4 H. Waldmann, *Kontakte (Darmstadt)* **1991**, 2, 33–54.

5 A. Reidel, H. Waldmann, *J. Prakt. Chem.* **1993**, *335*, 109.

6 H. Waldmann, D. Sebastian, *Chem. Rev.* **1994**, *94*, 911–937.

7 T. Kappes, H. Waldmann, *Liebigs Ann. Chem.* **1997**, 803–813.

8 T. Pathak, H. Waldmann, *Curr. Opin. Chem. Biol.* **1998**, *2*, 112–120.

9 B. Sauerbrei, T. Kappes, H. Waldmann, *Top. Curr. Chem.* **1997**, *186*, 65–86.

10 D. Kadereit, J. Kuhlmann, H. Waldmann, *ChemBioChem* **2000**, *1*, 144–169.

11 D. Kadereit, H. Waldmann, *Monatsh. Chem.* **2000**, *131*, 571–584.

12 J. D. Glass in: *The Peptides* (Eds.: S. Udenfried, J. Meienhofer) Academic Press, San Diego, **1987**, pp. 167–184.

13 E. Wünsch, *Methoden Org. Chem. (Houben-Weyl), Vol. XV/I and II*, Thieme, Stuttgart, **1974**.

14 M. Goodman, A. Felix, L. Moroder, C. Toniolo, *Methods of Organic Chemistry (Houben-Weyl), Vol. E22a-c*, Thieme, Stuttgart, **2000**.

15 R. W. Holley, *J. Am. Chem. Soc.* **1955**, *77*, 2552–2553.

16 F. Widmer, S. Bayne, G. Houen, B. A. Moss, R. D. Rigby, R. G. Whittaker, J. T. Johansen in: *Peptides 1984* (Ed.: U. Ragnarsson) Almquist & Wiksell, Stockholm, **1985**, pp. 193–200.

17 J. D. Glass, C. Meyers, I. L. Schwartz, R. Walter in: (Ed.: V. Wolman) *Proc. 18th Europ. Pept. Symp.*, Wiley and Sons, New York, **1974**, pp. 141–151.

18 C. Meyers, J. D. Glass, *Proc. Natl. Acad. Sci. USA* **1975**, *72*, 2193–2196.

19 C. Meyers, J. D. Glass in: *Peptides: Chemistry, Structure and Biology* (Eds.: R. Walter, J. Meienhofer), Ann Arbor Science Publishers, Ann Arbor, **1975**, pp. 325–331.

20 L. Patthy, E. L. Smith, *J. Biol. Chem.* **1975**, *250*, 557–564.

21 J. D. Glass, M. Pelzig, C. S. Pande, *Int. J. Pept. Prot. Res.* **1979**, *13*, 28–34.

22 C. S. Pande, M. Pelzig, J. D. Glass, *Proc. Natl. Acad. Sci. USA* **1980**, *77*, 895–899.

23 J. D. Glass, C. S. Pande in: *Peptides: Structure and Function* (Eds.: V. Hruby, D. H. Rich), Pierce Chem. Co., Rockford/Ill **1983**, pp. 203–206.

24 J. Blake, C. H. Li, *Proc. Natl. Acad. Sci. USA* **1983**, *80*, 1556–1559.

25 F. Widmer, K. Breddam, J. T. Johansen in: *Peptides 1980* (Ed.: K. Brunfeldt), Scriptor, Kopenhagen, **1981**, pp. 46–55.

26 E. Wünsch, L. Moroder, W. Göhring, P. Thamm, R. Scharf, *Hoppe-Seyler's Z. Physiol. Chem.* **1981**, *362*, 1285–1287.

27 Q.-C. Wang, J. Fei, D.-F. Cui, S.-G. Zhu, L.-G. Xu, *Biopolymers* **1986**, *25*, 109–114.

28 C. Fuganti, P. Grasselli, P. Casati, *Tetrahedron Lett.* **1986**, *27*, 3191–3194.

29 H. Waldmann, *Tetrahedron Lett.* **1988**, *29*, 1131–1134.

30 H. Waldmann, *Liebigs Ann. Chem.* **1988**, 1175–1180.

31 H. Waldmann, P. Braun, H. Kunz, *Biomed. Biochim. Acta* **1991**, *50*, 243–248.

32 H. Waldmann, A. Heuser, P. Braun, M. Schulz, H. Kunz in: *Microbial Reagents in Organic Synthesis* (Ed.: S. Servi), Kluwer, Dordrecht, **1992**, pp. 113–122.

33 P. Hermann, *Wiss. Z. Univ. Halle* **1987**, *36*, 17–29.

34 P. Hermann, *Biomed. Biochim. Acta* **1991**, *50*, 19–31.

35 G. Greiner, P. Hermann in: *Peptides 1990* (Eds.: E. Giralt, P. Malon), Escom Science Publishers, Leiden, **1991**, pp. 277–278.

36 A. L. Margolin, V. K. Svedas, I. V. Berezin, *Biochim. Biophys. Acta* **1980**, *616*, 283–289.

37 R. J. Didziapetris, V. K. Svedas, *Biomed. Biochim. Acta* **1991**, *50*, 237–242.

38 R. Didziapetris, B. Drabnig, V. Schellenberger, H.-D. Jakubke, V. Svedas, *FEBS Lett.* **1991**, *287*, 31–33.

39 H. J. Schütz, C. Wandrey, W. Leuchtenberger, *Abstracts of the Ninth Engineering Foundation Conference on Enzyme Engineering*, New York, **1987**, pp. 1–11.

40 I. Stoineva, B. Galunsky, V. Lozanov, I. Ivanov, D. Petkov, *Tetrahedron* **1992**, *48*, 1115–1122.

41 E. Baldaro, C. Fuganti, S. Servi, A. Tagliani, M. Terreni in: *Microbial Reagents in Organic Synthesis* (Ed.: S. Servi), Kluwer, Dordrecht, **1992**, pp. 175–188.

42 H. Waldmann, A. Heuser, S. Schulze, *Tetrahedron Lett.* **1996**, *37*, 8725–8728.

43 V. K. Svedas, A. I. Beltser, *Ann. N. Y. Acad. Sci.* **1998**, *864*, 524–527.

44 D. Sebastian, A. Heuser, S. Schulze, H. Waldmann, *Synthesis* **1997**, 1098–1110.

45 B. S. Briggs, A. J. Kreuzman, C. Whitesitt, W.-K. Yeh, M. Zmijewski, *J. Mol. Catal. B: Enzym.* **1996**, *2*, 53–69.

46 C. A. Costello, A. J. Kreuzmann, M. J. Zmijewski, *Tetrahedron Lett.* **1996**, *37*, 7469–7472.

47 C. A. Costello, A. J. Kreuzman, M. J. Zmijewski, *Tetrahedron Lett.* **1997**, *38*, 1.

48 T. Suyama, T. Tagoda, S. Kanao, *Yakugaku Zasshi* **1965**, *85*, 279.

49 A. Pessina, P. Lüthi, P. L. Luisi, J. Prenosil, Y. Zhang, *Helv. Chim. Acta* **1988**, *71*, 631–641.

50 Y.-S. Yang, S. Ramaswamy, W. B. Jakoby, *J. Biol. Chem.* **1993**, *268*, 10870–10875.

51 F. Widmer, M. Ohno, M. Smith, N. Nelson, C. B. Anfinsen in: *Peptides 1982* (Eds.: P. Blaha, P. Malon), de Gruyter, Berlin, **1983**, pp. 375–379.

52 E. Matsumura, T. Shin, S. Murao, M. Sakaguchi, T. Kawano, *Agric. Biol. Chem.* **1985**, *49*, 3643–3645.

53 H. Waldmann, M. Schelhaas, E. Nägele, J. Kuhlmann, A. Wittinghofer, H. Schroeder, J. R. Silvius, *Angew. Chem., Int. Ed. Engl.* **1997**, *36*, 2238–2241.

54 H. Waldmann, E. Nägele, *Angew. Chem., Int. Ed. Engl.* **1995**, *34*, 2259–2262.

55 E. Nägele, M. Schelhaas, N. Kuder, H. Waldmann, *J. Am. Chem. Soc.* **1998**, *120*, 6889–6902.

56 T. Pohl, H. Waldmann, *Angew. Chem., Int. Ed. Engl.* **1996**, *35*, 1720–1723.

57 T. Pohl, H. Waldmann, *J. Am. Chem. Soc.* **1997**, *119*, 6702–6710.

58 J. Sander, H. Waldmann, *Angew. Chem., Int. Ed. Engl.* **1999**, *38*, 1247–1250.

59 J. Sander, H. Waldmann, *Chem. Eur. J.* **2000**, *6*, 1564–1577.

60 T. Kappes, H. Waldmann, *Perkin Trans. I* **2000**, 1449–1453.

61 R. Machauer, H. Waldmann, *Angew. Chem. Int. Ed. Engl.* **2000**, *39*, 1449–1453.

62 A. G. Gum, T. Kappes-Roth, H. Waldmann, *Chem. Eur. J.* **2000**, *6*, 3714–3721.

63 T. Kappes, H. Waldmann, *Carbohydrate Res.* **1998**, *305*, 341–349.

64 F. Brtnik, T. Barth, K. Jost, *Coll. Czech. Chem. Commun.* **1981**, *46*, 1983–1989.

65 I. Svoboda, D. Brandenburg, T. Barth, H.-G. Gattner, J. Jiracek, J. Velek, I. Blaha, K. Ubik, V. Kasicka, J. Pospisek, P. Hrbas, *Biol. Chem. Hoppe-Seyler* **1994**, *375*, 373–378.

66 J. Jiracek, T. Barth, J. Velek, I. Blaha, J. Pospisek, I. Svoboda, *Collect. Czech. Chem. Commun.* **1992**, *57*, 2187–2191.

67 T. A. Savidge in: *Biotechnology of Industrial Antibiotics* (Ed.: E. J. Vandamme) ' Marcel Dekker, New York, **1984**, p. 171.

68 J. G. Shewale, B. S. Deshpande, V. K. Sudhakaran, S. S. Ambedkar, *Process Biochem. Int.* **1990**, 97–103.

69 E. Baldaro, D. Faiardi, C. Fuganti, P. Grasselli, A. Lazzarini, *Tetrahedron Lett.* **1988**, *29*, 4623–4624.

70 E. Hungerbühler, M. Biollaz, I. Ernest, J. Kalvoda, M. Lang, P. Schneider, G. Sedelmeier, in: *New Aspects of Organic Chemistry I* (Eds.: Z. Yoshida, T. Shiba, Y. Oshiro), VCH, Weinheim **1989**, pp. 419–451.

71 T. D. Black, B. S. Briggs, R. Evans, W. L. Muth, S. Vangala, M. J. Zmijewski, *Biotechn. Lett.* **1996**, *18*, 875–880.

72 M. A. Dineva, B. Galunsky, V. Kasche, D. D. Petkov, *Bioorg. Med. Chem. Lett.* **1993**, *3*, 2781–2784.

73 M. A. Dineva, D. D. Petkov, *Nucleosides Nucleotides* **1996**, *15*, 1459–1467.

74 H. Waldmann, A. Heuser, A. Reidel, *Synlett* **1994**, 65–67.

75 H. Waldmann, S. Gabold, *J. Chem. Soc., Chem. Commun.* **1997**, 1861–1862.

76 V. Jungmann, H. Waldmann, *Tetrahedron Lett.* **1998**, *39*, 1139–1142.

77 H. Waldmann, A. Reidel, *Angew. Chem., Int. Ed. Engl.* **1997**, *36*, 647–649.

78 S. Flohr, V. Jungmann, H. Waldmann, *Chem. Eur. J.* **1999**, *5*, 669–681.

79 M. Schultz, H. Kunz, *Exs* **1995**, *73*, 201–228.

80 E. Walton, J. O. Rodin, C. H. Stammer, F. W. Holly, *J. Org. Chem.* **1962**, *27*, 2255–2257.

81 G. Kloss, E. Schröder, *Hoppe-Seyler's Z. Physiol. Chem.* **1964**, *336*, 248–256.

82 N. Xaus, P. Clapes, E. Bardaji, J. L. Torres, X. Jorba, J. Mata, G. Valencia, *Biotechnol. Lett.* **1989**, *11*, 393–396.

83 C. F. Hayward, R. E. Offord in: *Peptides 1969* (Ed.: E. Scoffone), North-Holland Publ., Amsterdam, **1971**, pp. 116–120.

84 G. M. Anantharamaiah, R. W. Roeske in: *Peptides: Synthesis, Structure, Function* (Eds.: D. H. Rich, E. Gross), Pierce Chemical Co., Rockford/Ill, **1982**, pp. 45–47.

85 M. Ohno, C. B. Anfinsen, *J. Am. Chem. Soc.* **1970**, *92*, 4098–4102.

86 G. Royer, G. M. Anantharamaiah, *J. Am. Chem. Soc.* **1979**, *101*, 3394–3396.

87 G. P. Royer, H. Y. Hsiao, G. M. Anantharamaiah, *Biochimie* **1980**, *62*, 537–541.

88 D. Steinke, M.-R. Kula, *Angew. Chem., Int. Ed. Engl.* **1990**, *29*, 1139–1141.

89 D. Steinke, M.-R. Kula, *Biomed. Biochim. Acta* **1991**, *50*, 143–148.

90 D. Kammermeier-Steinke, A. Schwarz, C. Wandrey, M. R. Kula, *Enzyme Microb. Technol.* **1993**, *15*, 764–769.

91 B. Aleksiev, P. Schamlian, G. Widenov, S. Stoev, S. Zachariev, E. Golovinsky, *Hoppe-Seyler's Z. Physiol. Chem.* **1981**, *362*, 1323–1329.

92 S.-T. Chen, S.-C. Hsiao, C.-H. Chang, K.-T. Wang, *Synth. Commun.* **1992**, *22*, 391–398.

93 S.-T. Chen, S.-Y. Chen, S.-C. Hsiao, K.-T. Wang, *Biomed. Biochim. Acta* **1991**, *50*, 181–186.

94 S.-T. Chen, S.-H. Wu, K.-T. Wang, *Int. J. Pept. Prot. Res.* **1991**, *37*, 347–350.

95 P. Hermann, L. Salewski in: *Peptides 1982* (Eds.: K. Blaha, P. Malon), de Gruyter, Berlin, **1983**, pp. 399–402.

96 P. Hermann, H. Baumann, C. Herrnstadt, D. Glanz, *Amino Acids* **1992**, *3*, 105–118.

97 S. Dudek, S. Friebe, P. Hermann, *J. Chromatogr.* **1990**, *520*, 333–338.

98 P. Braun, H. Waldmann, W. Vogt, H. Kunz, *Synlett* **1990**, 105–107.

99 P. Braun, H. Waldmann, W. Vogt, H. Kunz, *Liebigs Ann. Chem.* **1991**, 165–170.

100 D. Sebastian, H. Waldmann, *Tetrahedron Lett.* **1997**, *38*, 2927–2930.

101 H. Waldmann, H. Kunz, P. Braun, unpublished results.

102 J. Zock, C. Cantwell, J. Swartling, R. Hodges, T. Pohl, K. Sutton, P. R. Jr., D. McGilvray, S. Queener, *Gene* **1994**, *151*, 37–43.

103 H. Waldmann, P. Stöber, unpublished results.

104 A. Margolin, A. Klibanov, *J. Am. Chem. Soc.* **1987**, *109*, 3802–3804.

105 C.-S. Chen, C. J. Sih, *Angew. Chem., Int. Ed. Engl.* **1989**, *28*, 695–708.

106 P. Braun, H. Waldmann, H. Kunz, *Synlett* **1992**, 39–40.

107 P. Braun, H. Waldmann, H. Kunz, *Bioorg. Med. Chem.* **1993**, *1*, 197–207.

108 M. Schulz, P. Hermann, H. Kunz, *Synlett* **1992**, 37–38.

109 S. Reissmann, G. Greiner, *Int. J. Pept. Protein Res.* **1992**, *40*, 110–113.

110 D. Cantacuzene, S. Attal, S. Bay, *Bioorg. Med. Chem. Lett.* **1991**, 197–200.

111 D. Cantacuzene, S. Attal, S. Bay, *Biomed. Biochim. Acta* **1991**, *50*, 231–236.

112 H. Ishii, K. Unabashi, Y. Mimura, Y. Inoue, *Bull. Chem. Soc. Jpn.* **1990**, *63*, 3042–3043.

113 S. Attal, S. Bay, D. Cantacuzene, *Tetrahedron* **1992**, *48*, 9251–9260.

114 A. Fischer, A. Schwarz, C. Wandrey, A. Bommarius, G. Knaup, K. Drauz, *Biomed. Biochim. Acta* **1991**, *50*, 169–174.

115 M. Gewehr, H. Kunz, *Synthesis* **1997**, 1499–1511.

116 J. Eberling, P. Braun, D. Kowalczyk, M. Schultz, H. Kunz, *J. Org. Chem.* **1996**, *61*, 2638–2646.

117 H. Kunz, D. Kowalczyk, P. Braun, G. Braum, *Angew. Chem. Int. Ed. Engl.* **1994**, *33*, 336–339.

118 M. Schelhaas, S. Glomsda, M. Haensler, H.-D. Jakubke, H. Waldmann, *Angew. Chem., Int. Ed. Engl.* **1996**, *35*, 106–109.

119 M. Schelhaas, E. Nägele, N. Kuder, B. Bader, J. Kuhlmann, A. Wittinghofer, H. Waldmann, *Chem. Eur. J.* **1999**, *5*, 1239–1252.

120 A. Cotté, B. Bader, J. Kuhlmann, H. Waldmann, *Chem. Eur. J.* **1999**, *5*, 922–936.

121 K. Kuhn, H. Waldmann, *Tetrahedron Lett.* **1999**, *40*, 6369–6372.

122 A. N. Semenov, I. V. Lomonosova, V. I. Berezin, M. I. Titov, *Biotechnol. Bioeng.* **1993**, *42*, 1137–1141.

123 A. N. Semenov, I. V. Lomonosova, V. I. Berezin, M. I. Titov, *Bioorg. Khim.* **1991**, *17*, 1074–1076.

124 G. H. Müller, H. Waldmann, *Tetrahedron Lett.* **1999**, *40*, 3549–3552.

125 J. Glass, M. Pelzig, *Proc. Natl. Acad. Sci. USA* **1977**, *74*, 2739–2741.

126 S.-T. Chen, K.-T. Wang, *Synthesis* **1987**, 581–582.

127 S.-T. Chen, K.-T. Wang, *J. Chem. Res. (S)* **1987**, 308–309.

128 N. Xaus, P. Clapés, E. Bardaji, J. L. Torres, X. Jorba, J. Mata, G. Valencia, *Tetrahedron* **1989**, *45*, 7421–7426.

129 S.-H. Wu, F.-Y. Chu, C.-H. Chang, K.-T. Wang, *Tetrahedron Lett.* **1991**, *32*, 3529–3530.

130 H.-D. Jakubke, P. Kuhl, A. Könnecke, *Angew. Chem. Int. Ed. Engl.* **1985**, *24*, 85–93.

131 T. Wieland, *Acta Chim. Hung. (Budapest)* **1965**, *44*, 5–9.

132 K. v. d. Bruch, H. Kunz, *Angew. Chem. Int. Ed. Engl.* **1990**, *29*, 1457–1460.

133 D. G. Drueckhammer, W. J. Hennen, R. L. Pederson, C. F. Barbas, III, C. M. Gau-

theron, T. Krach, C. H. Wong, *Synthesis* **1991**, 499–525.

134 K. Faber, S. Riva, *Synthesis* **1992**, 895–910.

135 A. K. Prasad, J. Wengel, *Nucleosides Nucleotides* **1996**, *15*, 1347–1359.

136 V. Gotor, *Biocatal. Biotransform.* **2000**, *18*, 87–103.

137 N. B. Bashir, S. J. Phythian, A. J. Reason, S. M. Roberts, *J. Chem. Soc., Perkin Trans. 1* **1995**, 2203–2222.

138 A. H. Haines, *Adv. Carbohydr. Chem. Biochem.* **1976**, *33*, 11–109.

139 A. H. Haines, *Adv. Carbohydr. Chem. Biochem.* **1981**, *39*, 13–70.

140 J. Stanek, *Top. Curr. Chem.* **1990**, *154*, 209–256.

141 M. Therisod, A. M. Klibanov, *J. Am. Chem. Soc.* **1986**, *108*, 5638–5640.

142 Y. F. Wang, J. J. Lalonde, M. Momongan, D. E. Bergbreiter, C. H. Wong, *J. Am. Chem. Soc.* **1988**, *110*, 7200–7205.

143 D. Pioch, P. Lozano, J. Graille, I. M. F. Cirad, *Biotechnol. Lett.* **1991**, *13*, 633–636.

144 M. Pozo, R. Pulido, V. Gotor, *Tetrahedron* **1992**, *48*, 6477–6484.

145 R. Pulido, V. Gotor, *J. Chem. Soc., Perkin Trans. 1* **1993**, 589–592.

146 T. Watanabe, R. Matsue, Y. Honda, M. Kuwahara, *Carbohydr. Res.* **1995**, *275*, 215–220.

147 B. Haase, G. Machmuller, M. P. Schneider, *Schriftenr. "Nachwachsende Rohst."* **1998**, *10*, 218–224.

148 A. T. J. W. de Goede, M. van Oosterom, M. P. J. van Deurzen, R. A. Sheldon, H. van Bekkum, F. van Rantwijk, *Stud. Surf. Sci. Catal.* **1993**, *78*, 513–520.

149 A. T. J. W. de Goede, W. Benckhuijsen, F. van Rantwijk, L. Maat, H. van Bekkum, *Recl. Trav. Chim. Pays-Bas* **1993**, *112*, 567–572.

150 S. Riva, J. Chopineau, A. P. G. Kieboom, A. M. Klibanov, *J. Am. Chem. Soc.* **1988**, *110*, 584–589.

151 O.-J. Park, H. G. Park, J.-W. Yang, *Biotechnol. Lett.* **1996**, *18*, 473–478.

152 R. Pulido, F. Lopez Ortiz, V. Gotor, *J. Chem. Soc., Perkin Trans. 1* **1992**, 2891–2898.

153 M. J. Kim, W. J. Hennen, H. M. Sweers, C. H. Wong, *J. Am. Chem. Soc.* **1988**, *110*, 6481–6486.

154 J. L. C. Liu, G. J. Shen, Y. Ichikawa, J. F. Rutan, G. Zapata, W. F. Vann, C. H. Wong, *J. Am. Chem. Soc.* **1992**, *114*, 3901–3910.

155 W. Fitz, C.-H. Wong, *J. Org. Chem.* **1994**, *59*, 8279–8280.

156 R. Pulido, V. Gotor, *Carbohydr. Res.* **1994**, *252*, 55–68.

157 A. K. Prasad, M. D. Soerensen, V. S. Parmar, J. Wengel, *Tetrahedron Lett.* **1995**, *36*, 6163–6166.

158 V. S. Parmar, K. S. Bisht, H. N. Pati, N. K. Sharma, A. Kumar, N. Kumar, S. Malhotra, A. Singh, A. K. Prasad, J. Wengel, *Pure Appl. Chem.* **1996**, *68*, 1309–1314.

159 D. J. Pocalyko, A. J. Carchi, B. Harirchian, *J. Carbohydr. Chem.* **1995**, *14*, 265–270.

160 S. M. Andersen, I. Lundt, J. Marcussen, S. Yu, *Carbohydr. Res.* **1999**, *320*, 250–256.

161 M. Woudenberg-van-Oosterom, F. Van Rantwijk, R. A. Sheldon, *Fett/Lipid* **1996**, *98*, 390–393.

162 B. Danieli, M. Luisetti, G. Sampognaro, G. Carrea, S. Riva, *J. Mol. Catal. B: Enzym.* **1997**, *3*, 193–201.

163 R. T. Otto, U. T. Bornscheuer, C. Syldatk, R. D. Schmid, *Biotechnol. Lett.* **1998**, *20*, 437–440.

164 M. Therisod, A. M. Klibanov, *J. Am. Chem. Soc.* **1987**, *109*, 3977–3981.

165 D. Colombo, F. Ronchetti, L. Toma, *Tetrahedron* **1991**, *47*, 103–110.

166 P. L. Barili, G. Catelani, F. D'Andrea, E. Mastrorilli, *J. Carbohydr. Chem.* **1997**, *16*, 1001–1010.

167 X. Zhang, T. Kamiya, N. Otsubo, H. Ishida, M. Kiso, *J. Carbohydr. Chem.* **1999**, *18*, 225–239.

168 W. J. Hennen, H. M. Sweers, Y. F. Wang, C. H. Wong, *J. Org. Chem.* **1988**, *53*, 4939–4945.

169 E. W. Holla, *Angew. Chem., Int. Ed. Engl.* **1989**, *28*, 220–221.

170 A. Bianco, C. Melchioni, G. Ortaggi, P. Romagnoli, M. Brufani, *J. Mol. Catal. B: Enzym.* **1997**, *3*, 209–212.

171 K. Adelhorst, F. Björkling, S. E. Godtfredsen, O. Kirk, *Synthesis* **1990**, 112–115.

172 L. Q. Cao, U. T. Bornscheuer, R. D. Schmid, *Fett/Lipid* **1996**, *98*, 332–335.

173 P. Pasta, G. Mazzola, G. Carrea, S. Riva, *Biotechnol. Lett.* **1989**, *11*, 643–648.

174 C. Tsitsimpikou, H. Stamatis, V. Sereti, H. Daflos, F. N. Kolisis, *J. Chem. Technol. Biotechnol.* **1998**, *71*, 309–314.

175 D. A. MacManus, E. N. Vulfson, *Carbohydr. Res.* **1995**, *279*, 281–291.

176 L. Panza, S. Brasca, S. Riva, G. Russo, *Tetrahedron: Asymmetry* **1993**, *4*, 931–932.

177 L. Panza, M. Luisetti, E. Crociati, S. Riva, *J. Carbohydr. Chem.* **1993**, *12*, 125–130.

178 M. J. Chinn, G. Iacazio, D. G. Spackman, N. J. Turner, S. M. Roberts, *J. Chem. Soc., Perkin Trans. 1* **1992**, 661–662.

179 M. J. Chinn, G. Iacazio, D. G. Spackman, N. J. Turner, S. M. Roberts, *J. Chem. Soc., Perkin Trans. 1* **1992**, 2045.

180 J. J. Gridley, A. J. Hacking, H. M. I. Osborn, D. G. Spackman, *Tetrahedron* **1998**, *54*, 14925–14946.

181 J. J. Gridley, A. J. Hacking, H. M. I. Osborn, D. G. Spackman, *Synlett* **1997**, 1397–1399.

182 D. Colombo, F. Ronchetti, A. Scala, L. Toma, *J. Carbohydr. Chem.* **1992**, *11*, 89–94.

183 P. Ciuffreda, F. Ronchetti, L. Toma, *J. Carbohydr. Chem.* **1990**, *9*, 125–129.

184 P. Ciuffreda, D. Colombo, F. Ronchetti, L. Toma, *J. Org. Chem.* **1990**, *55*, 4187–4190.

185 R. Lopez, E. Montero, F. Sanchez, J. Canada, A. Fernandez-Mayoralas, *J. Org. Chem.* **1994**, *59*, 7027–7032.

186 C. Chon, A. Heisler, N. Junot, F. Levayer, C. Rabiller, *Tetrahedron: Asymmetry* **1993**, *4*, 2441–2444.

187 N. Boissiere-Junot, C. Tellier, C. Rabiller, *J. Carbohydr. Chem.* **1998**, *17*, 99-115.

188 M. Woudenberg-van Oosterom, C. Vitry, J. M. A. Baas, F. van Rantwijk, R. A. Sheldon, *J. Carbohydr. Chem.* **1995**, *14*, 237–246.

189 N. Junot, J. C. Meslin, C. Rabiller, *Tetrahedron: Asymmetry* **1995**, *6*, 1387–1392.

190 E. W. Holla, *J. Carbohydr. Chem.* **1990**, *9*, 113–119.

191 F. Nicotra, S. Riva, F. Secundo, L. Zucchelli, *Tetrahedron Lett.* **1989**, *30*, 1703–1704.

192 G. Carrea, S. Riva, F. Secundo, B. Danieli, *J. Chem. Soc., Perkin Trans. 1* **1989**, 1057–1061.

193 J.-F. Shaw, A. M. Klibanov, *Biotechnol. Bioeng.* **1987**, *29*, 648–651.

194 A. L. Fink, G. W. Hay, *Can. J. Biochem.* **1969**, *47*, 353–359.

195 T. Horrobin, C. H. Tran, D. Crout, *J. Chem. Soc., Perkin Trans. I* **1998**, 1069–1080.

196 H. M. Sweers, C. H. Wong, *J. Am. Chem. Soc.* **1986**, *108*, 6421–6422.

197 M. Kloosterman, E. W. J. Mosmuller, H. E. Schoemaker, E. M. Meijer, *Tetrahedron Lett.* **1987**, *28*, 2989–2992.

198 Y. Kodera, K. Sakurai, Y. Satoh, T. Uemura, Y. Kaneda, H. Nishimura, M. Hiroto, A. Matsushima, Y. Inada, *Biotechnol. Lett.* **1998**, *20*, 177–180.

199 K.-F. Hsiao, F.-L. Yang, S.-H. Wu, K.-T. Wang, *Biotechnol. Lett.* **1995**, *17*, 963–968.

200 T. Matsui, Y. Kita, Y. Matsushita, M. Nakayama, *Chem. Express* **1992**, *7*, 45–48.

201 A. Bastida, R. Fernández-Lafuente, G. Fernández-Lorente, J. M. Guisán, G. Pagani, M. Terreni, *Bioorg. Med. Chem. Lett.* **1999**, *9*.

202 O. Kirk, M. W. Christensen, F. Beck, T. Damhus, *Biocatal. Biotransform.* **1995**, *12*, 91–97.

203 L. Gardossi, R. Khan, P. A. Konowicz, L. Gropen, B. S. Paulsen, *J. Mol. Cat. B: Enzym.* **1999**, *6*, 89–94.

204 D. Chaplin, D. H. G. Crout, S. Bornemann, D. W. Hutchinson, R. Khan, *J. Chem. Soc., Perkin Trans. 1* **1992**, 235–237.

205 K. F. Hsiao, S. H. Wu, K. T. Wang, *Bioorg. Med. Chem. Lett.* **1993**, *3*, 2125–2128.

206 C. Vogel, S. Kramer, A. J. Ott, *Liebigs Ann./ Recl.* **1997**, 1425–1428.

207 R. Lopez, C. Perez, A. Fernandez-Mayoralas, S. Conde, *J. Carbohydr. Chem.* **1993**, *12*, 165–171.

208 R. Csuk, B. I. Glänzer, *Z. Naturforsch., B: Chem. Sci.* **1988**, *43*, 1355–1357.

209 J. Zemek, S. Kucar, D. Anderle, *Collect. Czech. Chem. Commun.* **1987**, *52*, 2347–2352.

210 M. Kloosterman, M. P. De Nijs, J. G. J. Weijnen, H. E. Schoemaker, E. M. Meijer, *J. Carbohydr. Chem.* **1989**, *8*, 333–341.

211 E. W. Holla, V. Sinnwell, W. Klaffke, *Synlett* **1992**, 413–414.

212 A. Ballesteros, M. Bernabe, C. Cruzado, M. Martin-Lomas, C. Otero, *Tetrahedron* **1989**, *45*, 7077–7082.

213 H. Waldmann, A. Heuser, *Bioorg. Med. Chem.* **1994**, *2*, 477–482.

214 S. Tomic, J. Tomasic, L. Sesartic, B. Ladesic, *Carbohydr. Res.* **1987**, *161*, 150–155.

215 S. Tomic, D. Ljevacovic, J. Tomasic, *Carbohydr. Res.* **1989**, *188*, 222–227.

216 S. Tomic, A. Trescec, D. Ljevakovic, J. Tomasic, *Carbohydr. Res.* **1991**, *210*, 191–198.

217 D. Ljevacovic, S. Tomic, J. Tomasic, *Carbohydr. Res.* **1992**, *230*, 107–115.

218 H. Waldmann, A. Heuser, P. Braun, H. Kunz, *Ind. J. Chem.* **1992**, *31B*, 799.

219 J. O. Rich, B. A. Bedell, J. S. Dordick, *Biotechn. Bioeng.* **1995**, *45*, 426–434.

220 M. Woudenberg-van Oosterom, F. van

Rantwijk, R. A. Sheldon, *Biotechnol. Bioeng.* **1996**, *49*, 328–333.

221 S. Cai, S. Hakomori, T. Toyokuni, *J. Org. Chem.* **1992**, *57*, 3431–3437.

222 B. Danieli, P. D. Bellis, G. Carrea, S. Riva, *Helv. Chim. Acta* **1990**, *73*, 1837–1844.

223 M. A. Cruces, C. Otero, M. Bernabe, M. Martin-Lomas, A. Ballesteros, *Ann. N. Y. Acad. Sci.* **1992**, *672*, 436–443.

224 D. B. Sarney, M. J. Barnard, D. A. MacManus, E. N. Vulson, *J. Am. Oil Chem. Soc.* **1996**, *73*, 1481–1487.

225 M. Kloosterman, J. G. J. Weijnen, N. K. De Vries, J. Mentech, I. Caron, G. Descotes, H. E. Schoemaker, E. M. Meijer, *J. Carbohydr. Chem.* **1989**, *8*, 693–704.

226 K.-Y. Chang, S.-H. Wu, K. T. Wang, *Carbohydr. Res.* **1991**, *222*, 121–129.

227 K.-Y. Chang, S.-H. Wu, K. T. Wang, *J. Carbohydr. Chem.* **1991**, *10*, 251–261.

228 G.-T. Ong, S.-H. Wu, K. T. Wang, *Bioorg. Med. Chem. Lett.* **1992**, *2*, 161–164.

229 G. T. Ong, S. H. Wu, K. T. Wang, *Bioorg. Med. Chem. Lett.* **1992**, *2*, 631.

230 D. C. Palmer, F. Terradas, *Tetrahedron Lett.* **1994**, *35*, 1673–1676.

231 S. Bornemann, J. M. Cassells, J. S. Dordick, A. J. Hacking, *Biocatalysis* **1992**, *7*, 1–12.

232 G.-T. Ong, K.-Y. Chang, S.-H. Wu, K.-T. Wang, *Carbohydr. Res.* **1994**, *265*, 311–318.

233 M. Ferrero, V. Gotor, *Monatsh. Chem.* **2000**, *131*, 585–616.

234 H. S. Sachdev, N. A. Starkovsky, *Tetrahedron Lett.* **1969**, *9*, 733–736.

235 A. Taunton-Rigby, *J. Org. Chem.* **1973**, *38*, 977–985.

236 C.-H. Wong, S.-T. Chen, W.-J. Hennen, J. A. Bibbs, Y.-F. Wang, J. L.-C. Liu, M. W. Pantoliano, M. Whitlow, P. N. Bryan, *J. Am. Chem. Soc.* **1990**, *112*, 945–953.

237 F. Moris, V. Gotor, *J. Org. Chem.* **1992**, *57*, 2490–2492.

238 L. F. Garcia-Alles, F. Moris, V. Gotor, *Tetrahedron Lett.* **1993**, *34*, 6337–6338.

239 L. F. Garcia-Alles, V. Gotor, *Tetrahedron* **1995**, *51*, 307–316.

240 V. Gotor, F. Moris, *Synthesis* **1992**, 626–628.

241 F. Moris, V. Gotor, *J. Org. Chem.* **1993**, *58*, 653–660.

242 F. Moris, V. Gotor, *Tetrahedron* **1994**, *50*, 6927–6934.

243 R. V. Nair, M. M. Salunkhe, *Synth. Commun.* **2000**, *30*, 3115–3120.

244 S. Ozaki, K. Yamashita, T. Konishi, T. Maekawa, M. Eshima, A. Uemura, L. Ling, *Nucleosides Nucleotides* **1995**, *14*, 401–404.

245 F. Moris, V. Gotor, *Tetrahedron* **1992**, *48*, 9869–9876.

246 L. F. Garcia-Alles, J. Magdalena, V. Gotor, *J. Org. Chem.* **1996**, *61*, 6980–6986.

247 J. Magdalena, S. Fernandez, M. Ferrero, V. Gotor, *Tetrahedron Lett.* **1999**, *40*, 1787–1790.

248 L. F. Garcia-Alles, V. Gotor, *J. Mol. Catal. B: Enzym.* **1999**, *6*, 407–410.

249 F. Moris, V. Gotor, *Tetrahedron* **1993**, *49*, 10089–10098.

250 V. Gotor, F. Moris, L. F. Garcia-Alles, *Biocatalysis* **1994**, *10*, 295–305.

251 M. Mahmoudian, J. Eaddy, M. Dawson, *Biotechnol. Appl. Biochem.* **1999**, *29*, 229–233.

252 A. Uemura, K. Nozaki, J.-I. Yamashita, M. Yasumoto, *Tetrahedron Lett.* **1989**, *30*, 3817–3818.

253 A. Uemura, K. Nozaki, J.-I. Yamashita, M. Yasumoto, *Tetrahedron Lett.* **1989**, *30*, 3819–3820.

254 H. K. Singh, G. L. Cote, R. S. Sikorski, *Tetrahedron Lett.* **1993**, *34*, 5201–5204.

255 D. H. G. Crout, A. M. Dachs, S. E. Glover, D. W. Hutchinson, *Biocatalysis* **1990**, *4*, 177–183.

256 D. L. Damkjaer, M. Petersen, J. Wengel, *Nucleosides Nucleotides* **1994**, *13*, 1801–1807.

257 L. E. Iglesias, M. A. Zinni, M. Gallo, A. M. Iribarren, *Biotechnol. Lett.* **2000**, *22*, 361–365.

258 B. Danieli, P. D. Bellis, G. Carrea, S. Riva, *Gazz. Chim. Ital.* **1991**, *121*, 123–125.

259 B. Danieli, P. D. Bellis, G. Carrea, S. Riva, *Heterocycles* **1989**, *29*, 2061–2064.

260 N. Nakajima, K. Ishihara, T. Itoh, T. Furuya, H. Hamada, *J. Biosci. Bioeng.* **1999**, *87*, 105–107.

261 B. Danieli, S. Riva, *Pure & Appl. Chem.* **1994**, *66*, 2215–2218.

262 B. Danieli, M. Luisetti, S. Riva, A. Bertinotti, E. Ragg, L. Scaglioni, E. Bombardelli, *J. Org. Chem.* **1995**, *60*, 3637–3642.

263 H. Lehmann, O. Miersch, H. R. Schüttle, *Z. Chem.* **1975**, *15*, 443.

264 G. Schneider, O. Miersch, H.-W. Liebisch, *Tetrahedron Lett.* **1977**, 405–406.

265 D. L. Delinck, A. L. Margolin, *Tetrahedron Lett.* **1990**, *31*, 3093–3096.

266 A. L. Margolin, D. L. Delinck, M. R. Whalon, *J. Am. Chem. Soc.* **1990**, *112*, 2849–2854.

267 L. Gardossi, D. Bianchi, A. M. Klibanov, *J. Am. Chem. Soc.* **1991**, *113*, 6328–6329.

268 S. Riva, A. M. Klibanov, *J. Am. Chem. Soc.* **1988**, *110*, 3291–3295.

269 A. Bertinotti, G. Carrea, G. Ottolina, S. Riva, *Tetrahedron* **1994**, *50*, 13165–13172.

270 E. Santaniello, P. Ferraboschi, S. Reza-Elahi, *Monatsh. Chem.* **2000**, *131*, 617–622.

271 S. Riva, R. Bovara, G. Ottolina, F. Secundo, G. Carrea, *J. Org. Chem.* **1989**, *54*, 3161–3164.

272 V. C. O. Njar, E. Caspi, *Tetrahedron Lett.* **1987**, *28*, 6549–6552.

273 A. Baldessari, A. C. Bruttomesso, E. C. Gros, *Helv. Chim. Acta* **1996**, *79*, 999–1004.

274 A. Baldessari, M. S. Maier, E. G. Gros, *Tetrahedron Lett.* **1995**, *36*, 4349–4352.

275 G. Nicolosi, M. Piattelli, C. Sanfilippo, *Tetrahedron* **1992**, *48*, 2477–2482.

276 D. Lambusta, G. Nicolosi, M. Piattelli, C. Sanfilippo, *Indian J. Chem., Sect. B* **1993**, *32B*, 58–60.

277 G. Nicolosi, M. Piattelli, C. Sanfilippo, *Tetrahedron* **1993**, *49*, 3143–3148.

278 D. Lambusta, G. Nicolosi, A. Patti, M. Piatelli, *Synthesis* **1993**, 1155–1158.

279 K.-F. Hsiao, F.-L. Yang, S.-H. Wu, K.-T. Wang, *Biotechnol. Lett.* **1996**, *18*, 1277–1282.

280 P. Allevi, P. Ciuffreda, A. Longo, M. Anastasia, *Tetrahedron: Asymmetry* **1998**, *9*, 2915–2924.

281 V. S. Parmar, A. K. Prasad, N. K. Sharma, K. S. Bisht, R. Sinha, P. Taneja, *Pure Appl. Chem.* **1992**, *64*, 1135–1139.

282 V. S. Parmar, A. K. Prasad, N. K. Sharma, S. K. Singh, H. N. Pati, S. Gupta, *Tetrahedron* **1992**, *31*, 6495–6498.

283 K. S. Bisht, O. D. Tyagi, A. K. Prasad, N. K. Sharma, S. Gupta, V. S. Parmar, *Bioorg. Med. Chem.* **1994**, *2*, 1015–1020.

284 M. Natoli, G. Nicolisi, M. Piattelli, *Tetrahedron Lett.* **1990**, *31*, 7371–7374.

285 M. Natoli, G. Nicolisi, M. Piattelli, *J. Org. Chem.* **1992**, *57*, 5776–5778.

286 V. S. Parmar, H. N. Pati, A. Azim, R. Kumar, Himanshu, K. S. Bisht, A. K. Prasad, W. Errington, *Bioorg. Med. Chem.* **1998**, *6*, 109–118.

287 A. K. Prasad, H. N. Pati, A. Azim, S. Trikha, Poonam, *Bioorg. Med. Chem.* **1999**, *7*, 1973–1977.

288 P. Ciuffreda, S. Casati, E. Santaniello, *Tetrahedron* **2000**, *56*, 317–321.

289 F. Balkenhohl, C. von dem Bussche Hünnefeld, A. Lansky, C. Zechel, *Angew. Chem. Int. Ed. Engl.* **1996**, *35*, 2289–2337.

290 J. S. Fruchtel, G. Jung, *Angew. Chem., Int. Ed. Engl.* **1996**, *35*, 17–42.

291 L. A. Thompson, J. A. Ellman, *Chem. Rev.* **1996**, *96*, 555–600.

292 L. Yan, C. M. Taylor, R. Goodnow, D. Kahne, *J. Am. Chem. Soc.* **1994**, *116*, 6953–6954.

293 R. L. Halhomb, H. M. Huang, C. H. Wong, *J. Am. Chem. Soc.* **1994**, *116*, 11315–11322.

294 This nomenclature was introduced by S. L. Flitsch et al. See references 312 and N. J. Turner, *Curr. Org. Chem.* **1997**, *1*, 21–36.

295 M. Schuster, P. Wang, J. C. Paulson, C. H. Wong, *J. Am. Chem. Soc.* **1994**, *116*, 1135–1136.

296 K. Yamada, S. I. Nishimura, *Tetrahedron Lett.* **1995**, *36*, 9493–9496.

297 K. Yamada, E. Fujita, S. I. Nishimura, *Carbohydr. Res.* **1997**, *305*, 443–461.

298 S. Nishimura, K. Yamada, *J. Am. Chem. Soc.* **1997**, *119*, 10555–10556.

299 J. Weiler, H. Gausepohl, N. Hauser, O. N. Jensen, J. D. Hoheisel, *Nucleic Acids Res.* **1997**, *25*, 2792–2799.

300 D. T. Elmore, D. J. S. Guthrie, A. D. Wallace, S. R. E. Bates, *J. Chem. Soc., Chem. Commun.* **1992**, 1033–1034.

301 SPOCC resin is based on the cross-linking of long-chain poly(ethylene glycol) (PEG) terminally substituted with oxetane by cationic ring-opening polymerization.

302 J. Rademann, M. Grotli, M. Meldal, K. Bock, *J. Am. Chem. Soc.* **1999**, *121*, 5459–5466.

303 Bio-beads consists of 1% cross-linked polystyrene with 1.25 mmol chloromethyl substitution per g of dry resin, respectively benzhydrylamine polymer 1% cross-linked polystyrene with 0.24 mmol NH_2 per g of dry resin, Bio-Rad Laboratories (Richmond, CA, USA).

304 H. Gaertner, A. Puigserver, *Eur. J. Biochem.* **1984**, *145*, 257–263.

305 Novasyn KA constits of kieselguhr supported dimethylacrylamide functionalized with sarcosine methylester.

306 B. D. Larsen, A. Holm, *J. Pept. Res.* **1998**, *52*, 470–476.

307 J. Vagner, G. Barany, K. S. Lam, V. Krchnak, N. F. Sepetov, J. A. Ostrem, P. Strop, M. Lebl, *Proc. Nat. Acad. Sci. U. S. A.* **1996**, *93*, 8194–8199.

308 G. Corrales, A. Fernandez-Mayoralas, E. Garcia-Junceda, Y. Rodriguez, *Biocatal. Biotransform.* **2000**, *18*, 271–281.

309 MPEG consists of a poly(ethylene glycol) 5000 monomethyl ester, see: J. J. Krepinsky, Advances in polymer-supported solution synthesis of oligosaccharides, In *Modern Methods in Carbohydrate Synthesis* (Eds. S. A. Khan and R. A. O'Neill), Harwood Academic Publishers, The Netherlands, **1996**, 194–224.

310 C. Schmitz, M. T. Reetz, *Org. Lett.* **1999**, *1*, 1729–1731.

311 Flitsch, S. L., Lahja, S., and Turner, N. J. Solid phase preparation and enzymic and non-enzymic bond cleavage of sugars and glycopeptides, PCT Int. Appl. **1997**, patent EP 960 5535, CAN 127:81736.

312 G. Böhm, J. Dowden, D. C. Rice, I. Burgess, J. F. Pilard, B. Guilbert, A. Haxton, R. C. Hunter, N. J. Turner, S. L. Flitsch, *Tetrahedron Lett.* **1998**, *39*, 3819–3822.

313 B. Sauerbrei, V. Jungmann, H. Waldmann, *Angew. Chem., Int. Ed. Engl.* **1998**, *37*, 1143–1146.

314 H. Waldmann, B. Sauerbrei, U. Grether, Enzyme cleavable linker for solid phase synthesis, (BASF A.-G., Germany), Ger. Offen. **1998**, CAN 128:114573.

315 D. Kadereit, H. Waldmann, *Chem. Rev.*, **2001**, *101*, 3367–3396.

316 W. Rapp in: *Combinatorial Peptide and Non-Peptide Libraries* (Ed.: G. Jung), VCH, Weinheim, **1996**, p. 425.

317 U. Grether, H. Waldmann, *Chem. Eur. J.* **2001**, *7*, 959–971.

318 U. Grether, H. Waldmann, *Angew. Chem., Int. Ed. Engl.* **2000**, *39*, 1629–1632.

319 B. F. Cain, *J. Org. Chem.* **1976**, *41*, 2029–2031.

320 I. D. Entwistle, *Tetrahedron Lett.* **1994**, *35*, 4103–4106.

321 F. Cubain, *Rev. Roum. Chim.* **1973**, *18*, 449–461.

322 G. Just, G. Rosebery, *Synth.Commun.* **1973**, *3*, 447–451.

323 D. J. Gravert, K. D. Janda, *Chem. Rev.* **1997**, *97*, 489–509.

324 E. Bayer, M. Mutter, *Nature* **1972**, *237*, 512–513.

325 J. M. Dust, Z. H. Fang, J. M. Harris, *Macromolecules* **1990**, *23*, 3742–3746.

19

Replacing Chemical Steps by Biotransformations: Industrial Application and Processes Using Biocatalysis

Andreas Liese

'*Bacteria are capable of bringing about chemical reactions of amazing variety and sublety in an extremely short time ... Many bacteria are of very great importance to industry where they perform tasks which would take much time and trouble by ordinary chemical methods.*'

Sir Cyril Hinshelwood, 1956 [1]

19.1
Introduction

Starting with big promises and many expectations in the seventies biocatalytic processes have left the status of a lab curiosity together with many prejudices far behind and are now established on an industrial scale [2]. Product examples range from amino acids, sugars, chiral alcohols and amines, and highly functionalized building blocks for pharmaceuticals to bulk chemicals such as acrylamide or propane-1,3-diol. When speaking about biotechnological processes one has to distinguish between fermentation processes and biotransformations. In a fermentation process the desired product is synthesized from nutrients and trace elements by either microorganisms (bacteria, yeasts, fungi) or higher cells such as mammalian or plant cells. The phrase "biotransformation" or "biocatalysis" is commonly used to describe a one-step or multi-step transformation of a precursor to the desired product using whole cells and/or (partly) purified enzymes. Whole cell processes are often used for redox reactions using the metabolism of the living cell for cofactor regeneration. In some cases the cell is used as a compartment containing the enzymes in a confinement allowing easier separation of the entire biocatalyst using centrifugation or microfiltration. If one has to deal with membrane-bound enzymes, whole cell biocatalysts are to be preferred.

Numerous authors have given overviews over biotransformations used in industry [3–11]. A very recent monograph summarizes almost 100 processes including many details on reaction conditions, screening of the biocatalyst or the product application [2]. The use of biocatalysis from the viewpoint of a chemist in the laboratory is also summarized in several books. Recent ones are [12–14].

In this contribution we shall focus on those examples where the biocatalytic step has distinct advantages over the corresponding chemical method, or even has replaced or is about to replace other methods. The reasons may be better regio-, stereo- or chemoselectivity, better product purity or simplified downstream processing. Often the incorporation of biocatalytic steps reduces the amount or toxicity of waste.

19.2
Types and Handling of Biocatalysts

This chapter tries to give a brief introduction to the types of biocatalysts, their requirements and methods of handling them. A more detailed treatment can be found in other chapters of this book or in the literature [15–17].

The biocatalyst may be a whole cell or a partly purified enzyme. In the first case the cell may be regarded as a mini-reactor with all necessary cofactors and enzymes to catalyze multiple steps concentrated in one cell. In the second case the main catalytically active species is isolated and purified.

For the whole cell systems either prokaryotic cells such as *Escherichia coli* or eukaryotic cells such as *Saccharomyces cerevisiae* or *Zymomonas mobilis* are used. Prokaryotic cells do not posses a nucleus. The nuclear material is contained in the cytoplasm of the cell. Therefore introduction and processing of foreign DNA to obtain a genetically engineered strain is simple. They are relatively small in size (0.2–10 μm) and exist mostly as single cells. Eukaryotic cells are higher microorganisms and have a true nucleus separated by a nuclear membrane. They are larger in size (5–30 μm) and sometimes form more complex structures. For both types the bioreactor has to fulfill certain requirements. An adequate supply of nutrients as well as oxygen into the bioreactor has to be assured. Parameters such as pH, oxygen, feed rate and temperature in the bioreactor must be kept within certain limits in order to guarantee optimum growth and/or metabolic activity of the cells. Especially when recombinant microorganisms are employed genetic stability during cultivation has to be observed carefully. Substrates, products and/or solvents required may be toxic for the cells and may therefore have to be added in low amounts to secure a low stationary concentration. The example of a ketone reduction with whole cells of *Zygosaccharomyces rouxii* shows one possible solution. The toxic substrate is adsorbed on XAD-7 resin (80 g/L resin, resulting in a concentration of 40 g/L reaction volume), and the resin is added to the fermentation broth. The equilibrium concentration in the aqueous phase is approximately 2 g/L. The product is adsorbed on the resin as well, thus providing integrated downstream processing [18–19].

When purified enzymes are used, basically the same requirements have to be met. The purification may cause additional costs, but contrary to a biochemical characterization it is not necessary to purify the protein to homogeneity. On the contrary, the remaining protein content in a partly purified enzyme may increase its stability. The only requirement is to have a functional pure enzyme, meaning that activities catalyzing undesired side reactions have to be absent. This is the major advantage of purified enzymes over whole cell processes: side reactions may be easily avoided, and substrates that are toxic for the cell or which may not be able to enter the cell can be converted. For enzymes the thermal deactivation or deactivation by interphases (liquid-liquid, liquid-gas) may be limiting.

For industrial biotransformations, catalyst recovery and reuse are major issues. This may be desirable either for reasons of downstream processing or for repeated use in order to reduce the specific catalyst costs per kg of product produced. A very simple method is the use of membrane filtration. Because of the increasing number of membranes from different materials (polymers, metal or ceramics) this is an attractive alternative. Whereas for whole cells microfiltration or centrifugation can be applied, for the recovery of soluble enzymes ultrafiltration membranes have to be used [20–22]. Often immobilization on a support is chosen to increase the catalyst's stability as well as to facilitate its recovery. The main advantages of immobilization are:

- easy separation,
- often increased stability,
- use of fixed or fluidized bed reactors (for continuous processes).

Disadvantages are:

- loss of absolute activity due to the immobilization process,
- mass transport limitations,

There is no general or best method of immobilization; the protocol has to be developed individually for each catalyst.

The most common methods for the immobilization are entrapment in matrices such as alginate beads, cross-linking, and covalent or adsorptive binding to a carrier. A very recent method is the development of cross-linked enzyme crystals (CLECS) (see also Chapter 6) [23]. A survey of different immobilization methods can be found in in [24, 25].

19.3
Examples

The examples presented here are taken from [2]. Only those biotransformations were chosen where a classical chemical step was replaced. The enzymes involved are mainly from the groups of oxidoreductases (E.C. class 1) and hydrolases (E.C. class 3). There are a few examples of lyases (E.C. class 4) and one example of an isomerase (E.C. class 5). The processes involving oxidoreductases mainly use whole cells because of the problem of cofactor regeneration. The examples are sorted in the order of the main classes of the Enzyme Commission (E.C.). The big letter E denotes the biotransformation in the syntheses schemes.

19.3.1
Reduction Reactions Catalyzed by Oxidoreductases (E.C. 1)

19.3.1.1
Ketone Reduction Using Whole Cells of *Neurospora crassa* (E.C. 1.1.1.1) [26–29]

The key step in the synthesis of Trusopt®, which is a topically active treatment for glaucoma, is the enantioselective reduction of 5,6-dihydro-6-methyl-4H-thieno[2,3b] thiopyran-4-one-7,7-dioxide (Fig. 19-1).

The biological route overcomes the problem of incomplete inversion of the *cis*-alcohol in the chemical synthesis (Fig. 19-2). The reaction is carried out below pH 5 to prevent epimerization of the (6S)-methyl ketosulfone in aqueous media.

The (R)-3-hydroxy-butyrate (Fig. 19-1), which is responsible for the stereochemistry of the methyl group in the sulfone ring, can be produced by depolymerization of biopolymers, e.g. Biopol from Zeneca. This is a natural polyester produced by some microorganisms as a storage compound.

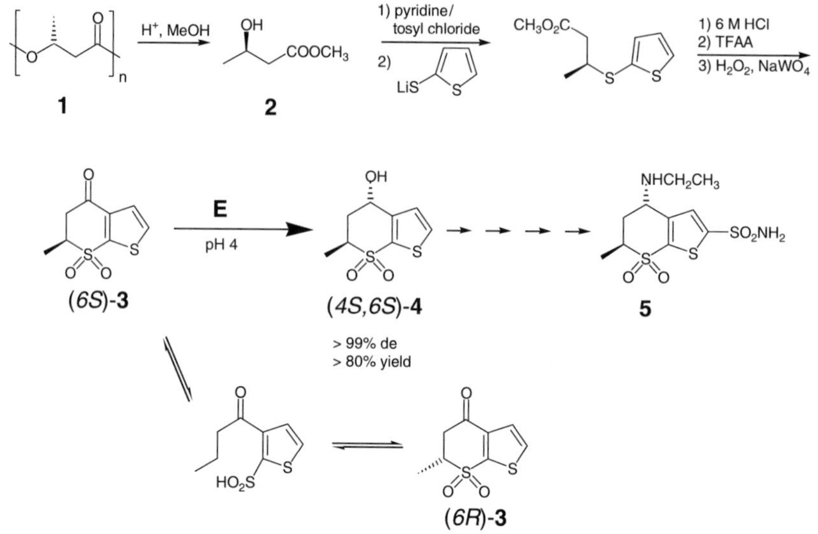

1 = biologically derived homopolymer
2 = (R)-3-hydroxy-methyl butyrate
3 = 5,6-dihydro-6-methyl-4H-thieno[2,3b]thiopyran-4-one-7,7-dioxide
4 = 5,6-dihydro-4-hydroxy-6-methyl-4H-thieno[2,3b]thiopyran-7,7-dioxide
5 = trusopt, MK-0507
E = alcohol dehydrogenase, whole cells from *Neurosporia crassa*

Figure 19-1. Synthesis of Trusopt® (5) via enzymatic ketone reduction (Astra-Zeneca).

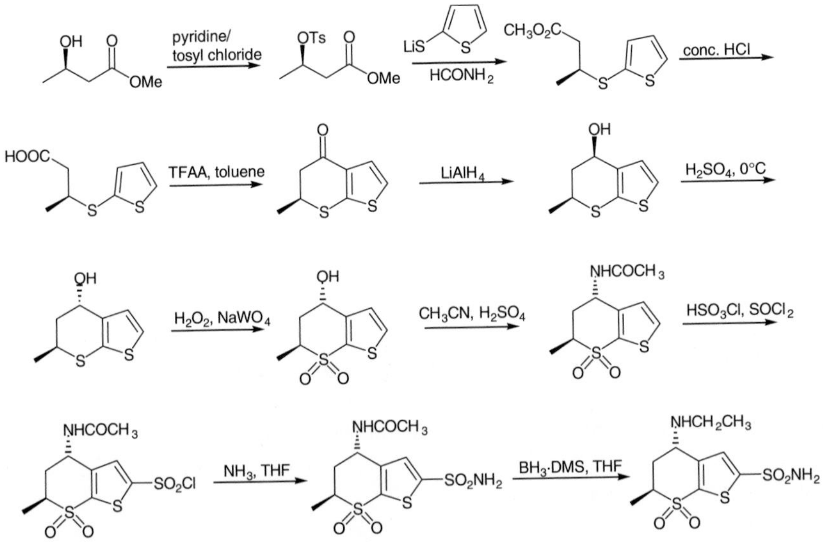

Figure 19-2. Chemical synthesis of Trusopt®.

19.3.1.2
Ketoester Reduction Using Cell Extract of *Acinetobacter calcoaceticus* (E.C. 1.1.1.1) [30–32]

The biotransformation (Fig. 19-3) is an alternative to the chemical synthesis via the chlorohydrine and selective hydrolysis of the acyloxy group (Fig. 19-4). After final fractional distillation this synthesis has an overall yield of 41%. The biotransformation has a yield of 92%. The diketoester can be obtained as shown in Fig. 19-5.

6-Benzyloxy-(3R,5S)-dihydroxy-hexanoic acid ethyl ester is a key chiral intermediate for anticholesterol drugs that act by inhibition of hydroxy methyl glutaryl coenzyme A (HMG CoA) reductase.

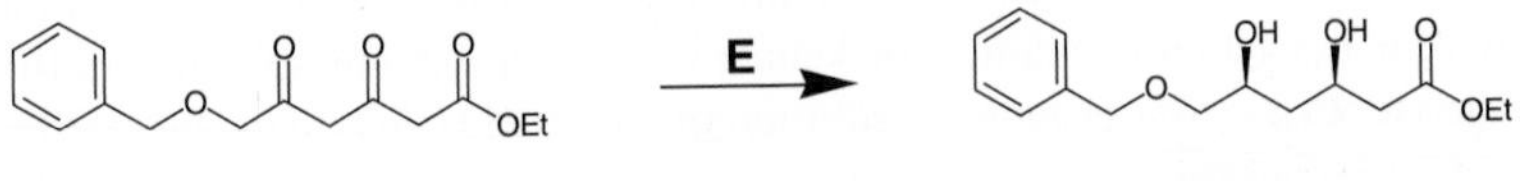

1 = 6-benzyloxy-3,5-dioxo-hexanoic acid ethyl ester
2 = 6-benzyloxy-3,5-dihydroxy-hexanoic acid ethyl ester
E = alcohol dehydrogenase from *Acinetobacter calcoaceticus*

Figure 19-3. Synthesis of key intermediate of anticholesterol drugs (Bristol-Myers Squibb).

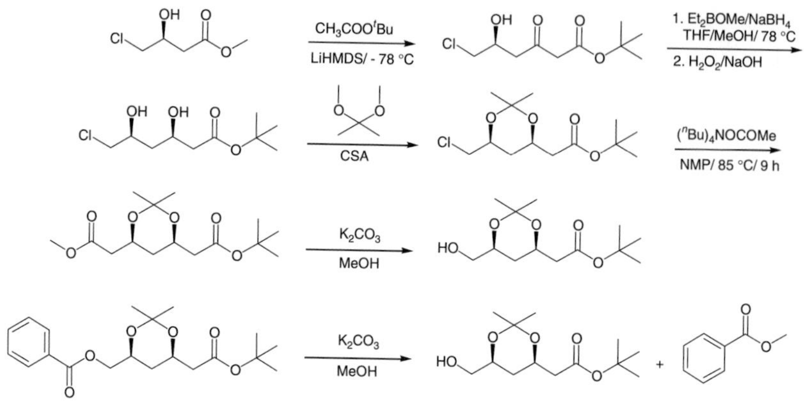

Figure 19-4. Chemical synthesis of key intermediate of anticholesterol drugs.

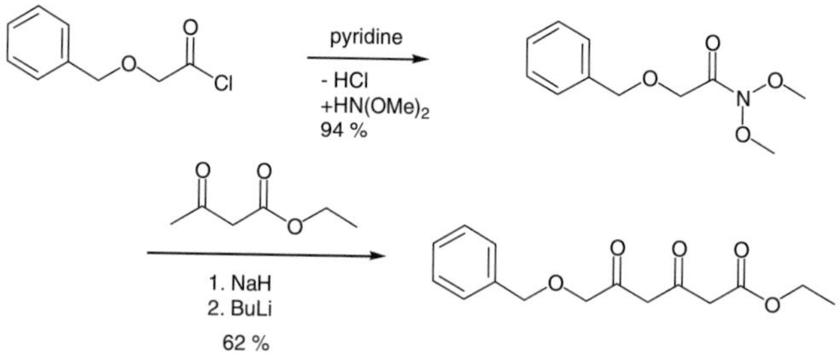

Figure 19-5. Synthesis of starting material 6-benzyloxy-3,5-dihydroxy-hexanoic acid ethyl ester.

19.3.1.3

Enantioselective Reduction with Whole Cells of *Candida sorbophila* (E. C. 1.1.X.X) [33–35]

Here the biotransformation (Fig. 19-6) is preferred over the chemical reduction with commercially available asymmetric catalysts (BH$_3$– or noble-metal-based), since with the chemocatalysts the desired high enantiomeric excess ($ee > 98\%$, 99.8% after purification) is not achievable. Since the ketone has only a very low solubility in the aqueous phase, 1 kg ketone is added as solution in 4 L 0.9 M H$_2$SO$_4$ to the bioreactor. The bioreduction is essentially carried out in a two-phase system, consisting of the aqueous phase and small droplets made up of substrate and product. The downstream processing consists of multiple extraction steps with methyl ethyl ketone and precipitation induced by pH titration of the pyridine functional group (pK_a = 4.66) with NaOH. The (R)-amino alcohol is an important intermediate for the synthesis of β-3-agonists that can be used for obesity therapy and to decrease the level of associated type II diabetes, coronary artery disease and hypertension.

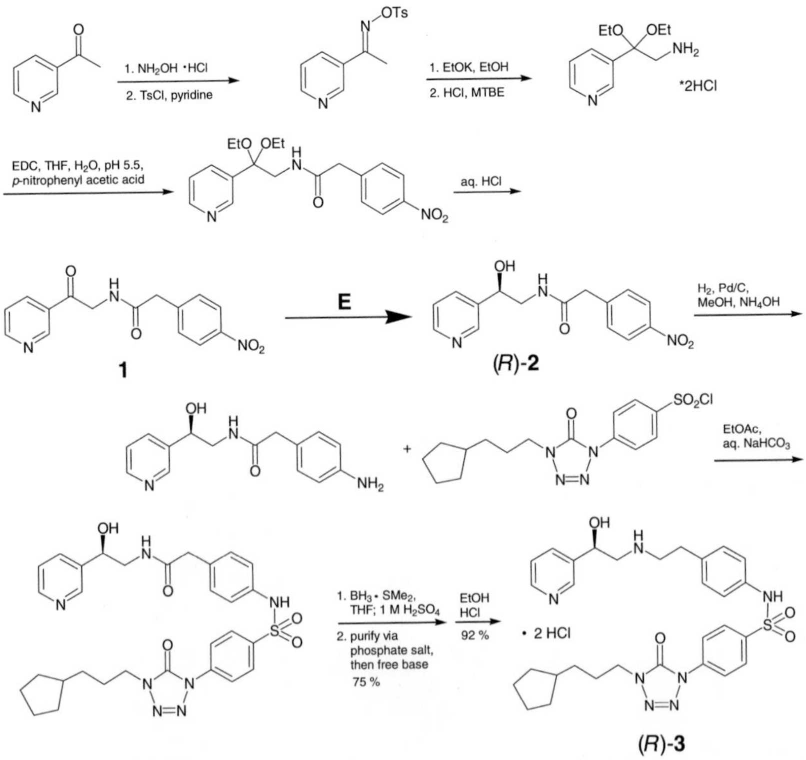

Figure 19-6. Synthesis of key intermediate of β-3-agonist (Merck Research Laboratories).

1 = 2-(4-nitro-phenyl)-*N*-(2-oxo-2-pyridin-3-ethyl)-acetamide
2 = (*R*)-*N*-(2-hydroxy-2-pyridin-3-yl-ethyl)-2-(4-nitro-phenyl)-acetamide
3 = β-3-agonist
E = dehydrogenase, whole cells of *Candida sorbophila*

19.3.2
Oxidation Reactions Catalyzed by Oxidoreductases (E.C. 1)

19.3.2.1
Alcohol Oxidation Using Whole Cells of *Gluconobacter suboxydans* (E.C. 1.1.99.21) [36–38]

In 1923 the bacterium *Acinetobacter suboxydans* was isolated and, starting in 1930, was used for the industrial oxidation of L-sorbitol to L-sorbose in the Reichstein-Grüssner synthesis of vitamin C [39]. Bayer uses the same type of reaction, but instead of *Acinetobacter* the bacterium *Gluconobacter suboxydans* is used in the oxidation of *N*-protected 6-amino-L-sorbitol to the corresponding 6-amino-L-sorbose, which is an intermediate in miglitol production (Fig. 19-7). 1-Desoxynojirimycin is produced by chemical intramolecular reductive amination of 6-amino-L-sorbose. In contrast, the

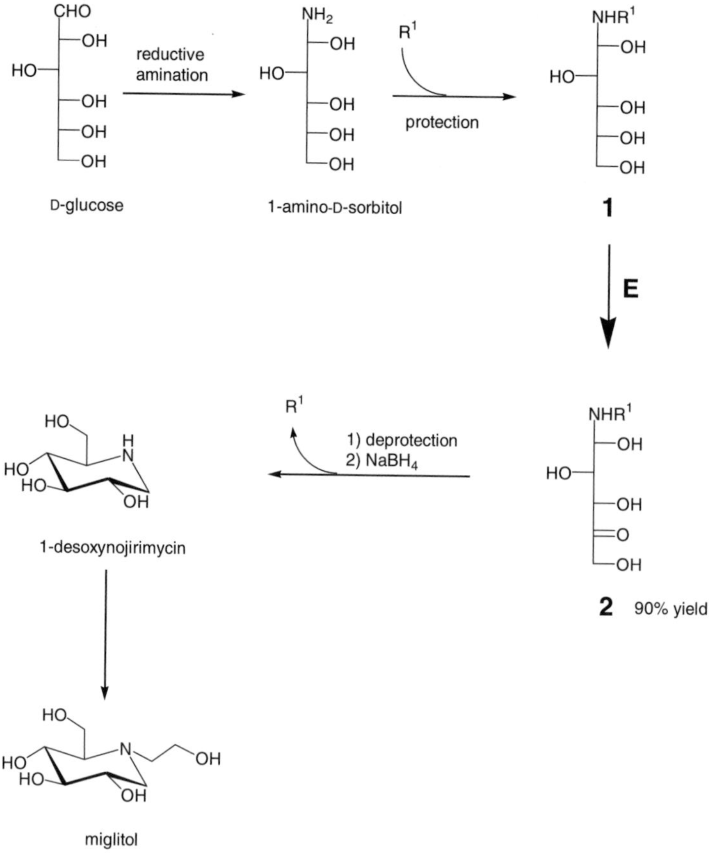

Figure 19-7. Synthesis of key intermediate for miglitol (Bayer).

1 = 1-amino-D-sorbitol (N-protected)
2 = 6-amino-L-sorbose (N-protected)
E = D-sorbitol dehydrogenase, whole cells of *Gluconobacter oxidans*

published chemical synthesis of 1-desoxynojirimycin and its derivatives requires multiple steps and laborious protecting-group chemistry. Miglitol and derivatives thereof are pharmaceuticals for the treatment of carbohydrate metabolism disorders (e.g. diabetes mellitus).

19.3.2.2
Oxidative Deamination Catalyzed by Immobilized D-Amino Acid Oxidase from *Trigonopsis variabilis* (E.C. 1.4.3.3) [40–42]

This oxidative deamination catalyzed by immobilized enzymes is part of the 7-aminocephalosporanic acid (7-ACA) process. Ketoadipinyl-7-aminocephalosporanic acid decarboxylates *in situ* in the presence of H_2O_2, which is formed by the

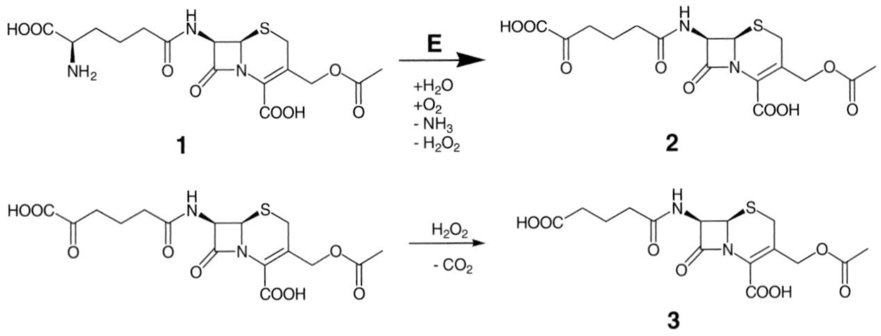

1 = cephalosporin C
2 = α-ketoadipinyl-7-aminocephalosporanic acid
3 = glutaryl-7-aminocephalosporanic acid (7-ACA)
E = D-aminoacid oxidase, immobilized enzyme from *Trigonopsis variablis*

Figure 19-8. Synthesis of glutaryl-7-aminocephalosporanic acid (7-ACA) (Hoechst-Marion-Roussel).

biotransformation step yielding glutaryl-7-ACA. The reaction solution is directly transferred to the 7-ACA production (see Sect. 19.3.4.2 for details and a comparison with the chemical synthesis).

19.3.2.3
Kinetic Resolution by Oxidation of Primary Alcohols Catalyzed by Whole Cells from *Rhodococcus erythropolis* (E. C. 1.X.X.X) [43–45]

(R)-Isopropylideneglycerol is a useful C_3-synthon in the synthesis of (S)-β-blockers, e. g. (S)-metoprolol. Also, (R)-isopropylideneglyceric acid may be used as the starting material for the synthesis of biologically active products. The resolution is carried out by selective microbial oxidation of the (S)-enantiomer (Fig. 19-9). The chemical synthesis of (R)-isopropylideneglycerol starts either from unnatural L-mannitol or from L-ascorbic acid (Fig. 19-10). In comparison to the biotransformation, here stoichiometric quantities of lead tetra acetate are needed.

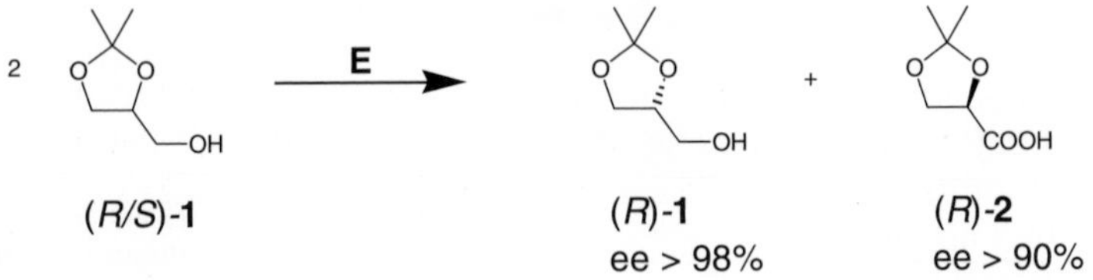

1 = isopropylideneglycerol
2 = isopropylideneglyceric acid
E = oxidase, whole cells from *Rhodococcus erythropolis*

Figure 19-9. Synthesis of (R)-isopropylideneglycerol and (R)-isopropylideneglyceric acid (International BioSynthetics).

From L-mannitol:

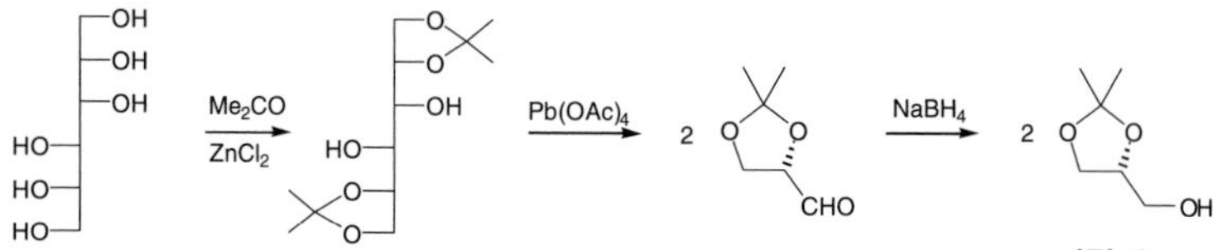

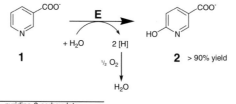

(R)-1

From L-ascorbic acid:

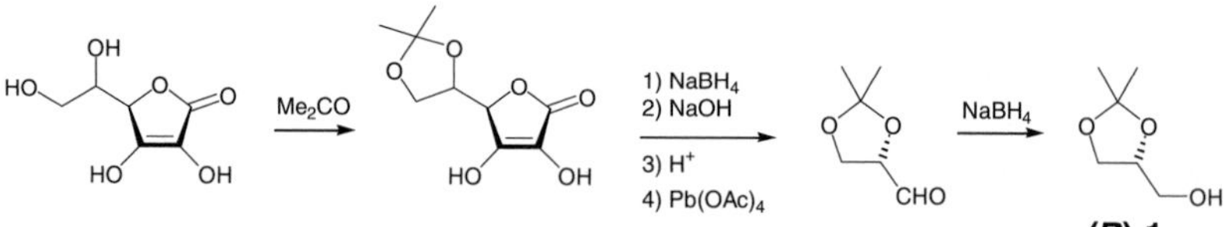

1) NaBH4
2) NaOH
3) H$^+$
4) Pb(OAc)4

(R)-1

Figure 19-10. Chemical synthesis of (R)-isopropylideneglycerol.

19.3.2.4
Hydroxylation of Nicotinic Acid (Niacin) Catalyzed by Whole Cells of *Achromobacter xylosoxidans* (E. C. 1.5.1.13) [46–48]

6-Hydroxynicotinate is a versatile building block used chiefly in the synthesis of modern insecticides. The 6-hydroxynicotinate-producing strain (Fig. 19-11) was found by accident, when in the mother liquor of a niacin-producing chemical plant precipitated white crystals of 6-hydroxynicotinate were found. The second enzyme of the nicotinic acid pathway, the decarboxylating 6-hydroxynicotinate hydroxylase becomes strongly inhibited at niacin concentrations higher than 1%, whereas the operation of niacin hydroxylase is unaffected. In contrast to the biotransformation, the chemical synthesis of 6-substituted nicotinic acids is difficult and expensive because of the necessity for the separation of by-products that are produced by non-regioselective hydroxylations.

E

+ H$_2$O   2 [H]

1

$\frac{1}{2}$ O$_2$

H$_2$O

2 > 90% yield

Figure 19-11. Synthesis of 6-hydroxynicotinate (Lonza).

1 = niacin = nicotinic acid = pyridine-3-carboxylate
2 = 6-hydroxynicotinate = 6-hydroxy-pyridine-3-carboxylate
E = nicotinic acid hydrolase, whole cells from *Achromobacter xylosoxidans*

19.3.2.5
Reduction of Hydrogen Peroxide Concentration by Catalase (E.C. 1.11.1.6) [49]

During oxidative coupling to dinitrodibenzyl (DNDB), hydrogen peroxide is formed as a by-product. It is not possible to decompose H$_2$O$_2$ by adding heavy-metal catalysts

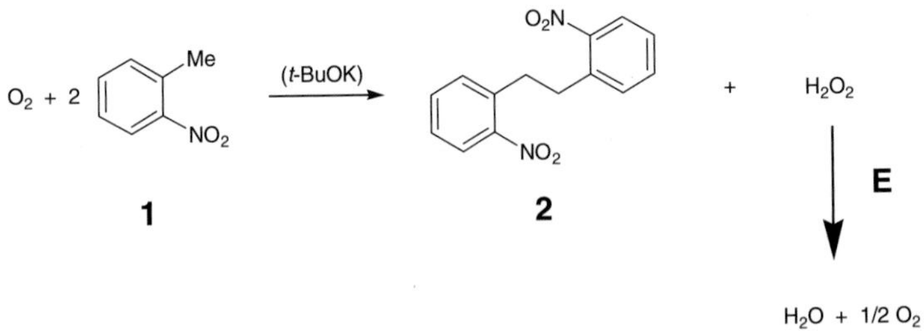

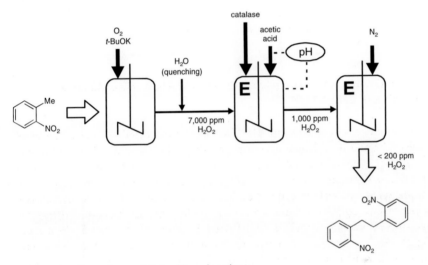

1 = nitrotoluene
2 = dinitrodibenzyl (DNDB)
E = catalase, enzyme from microbial source

Figure 19-12. Degradation of hydrogen peroxide (Novartis).

Figure 19-13. Flow scheme of dinitrodibenzyl synthesis.

because only an incomplete conversion is reached. Additionally, subsequent process steps with DNDB are problematic because of contamination with heavy-metal catalyst. The biotransformation is the only relevant method of decomposing the undesired side product H_2O_2 (Fig. 19-12). The enzyme of choice is catalase derived from a microbial source, which has advantages compared to beef catalase since the activity remains constant over a broad pH range from 6.0 to 9.0, temperatures up to 50 °C are tolerated, and salt concentrations up to 25% do not affect the enzyme stability. The reaction is carried out in a cascade of three continuously operated stirred tank reactors (Fig. 19-13). The H_2O_2 concentration is reduced from 7000 ppm to < 200 ppm in the product solution. The third vessel is aerated with nitrogen to degas the product solution. The dinitrodibenzyl is used as a pharmaceutical intermediate.

19.3.3
Hydrolytic Cleavage and Formation of C-O Bonds by Hydrolases (E.C. 3)

19.3.3.1
Kinetic Resolution of Glycidic Acid Methyl Ester by Lipase from *Serratia marcescens* (E.C. 3.1.1.3)[50–53]

Trans-(2R,3S)-(4-methoxyphenyl)glycidic acid methyl ester is an intermediate in the synthesis of diltiazem, a coronary vasodilator and a calcium channel blocker with

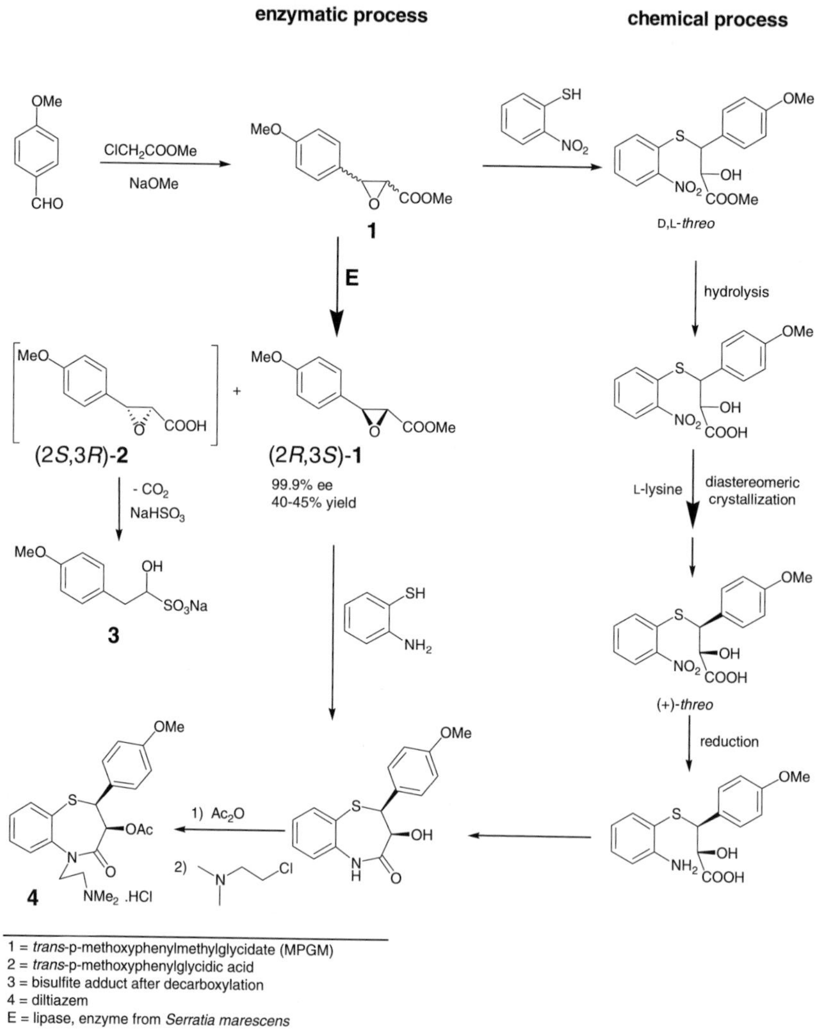

1 = *trans*-p-methoxyphenylmethylglycidate (MPGM)
2 = *trans*-p-methoxyphenylglycidic acid
3 = bisulfite adduct after decarboxylation
4 = diltiazem
E = lipase, enzyme from *Serratia marescens*

Figure 19-14. Comparison of chemical and biocatalytical route to diltiazem (Tanabe Seiyaku Co., Ltd.).

antianginal and antihypersensitive activity. It is produced worldwide in excess of > 100 t a^{-1}.

In comparison to the chemical route, only 5 steps (instead of 9) are necessary with the biotransformation (Fig. 19-14). The kinetic resolution is carried out in an earlier step with a lower molecular weight compound during the synthesis, resulting in a reduction of waste. By redesigning the synthesis route using a biotransformation, the manufacturing costs of diltiazem were decreased to two thirds of those of the original process including a chemical resolution[54].

The lipase from *Serratia marcescens* has a high enantioselectivity ($E = 135$) for the (2R,3S)-(4-methoxyphenyl)glycidic acid methyl ester, which acts as a competitive inhibitor. The formed acid (hydrolyzed (+)-methoxyphenylglycidate) is unstable and decarboxylates to give 4-methoxyphenylacetaldehyde; this aldehyde strongly inhibits and deactivates the enzyme. It is removed by transfer to the aqueous phase by formation of a water-soluble adduct with sodium hydrogen sulfite added to the aqueous phase. The bisulfite acts also as a buffer to maintain constant pH during synthesis.

The enantioselective hydrolysis is carried out in an organic-aqueous two-phase reactor (toluene/water), where the phase contact is established by a hydrophilic hollow-fiber membrane (polyacrylonitrile). The lipase is immobilized onto a spongy layer by pressurized adsorption. The productivity is about 40 kg *trans*-(2R,3S)-(4-methoxyphenyl)glycidic acid methyl ester m^{-2} a^{-1}. This process has been operated since 1993.

19.3.3.2
Kinetic Resolution of Diester by Protease Subtilisin Carlsberg from *Bacillus* sp. (E.C. 3.4.21.62) [55, 56]

(R)-(2-Methylpropyl)-butanedioic acid 4-ethyl ester is used as a chiral building block for potential collagenase inhibitors (e. g. Ro 31–9790) in the treatment of osteoarthritis. The diester is reacted as a 20 % emulsion in 30 mM aqueous NaHCO$_3$ using Protease® L 660 or Alcalase® 2.5 L (9 % each, with respect to the racemic diester) (Fig. 19-15). The unconverted (S)-diester can be extracted in a solvent such as toluene and racemized by heating the anhydrous extract with catalytic amounts of sodium ethanolate. The resulting racemic diester can be recycled, thus improving the overall

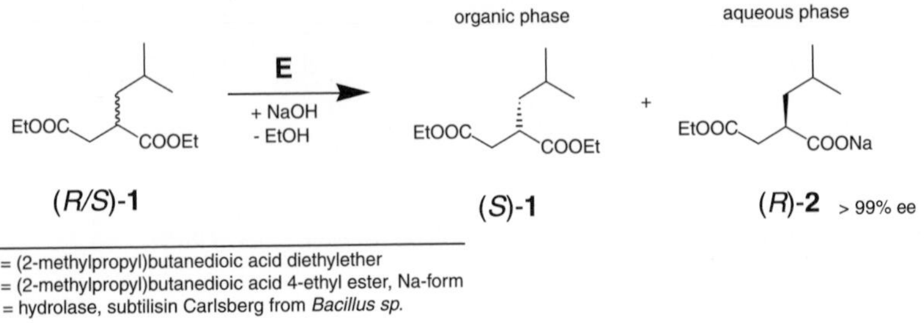

$$\text{(R/S)-1} \quad\quad \text{(S)-1} \quad\quad \text{(R)-2} \;\; {}_{> 99\% \text{ ee}}$$

1 = (2-methylpropyl)butanedioic acid diethylether
2 = (2-methylpropyl)butanedioic acid 4-ethyl ester, Na-form
E = hydrolase, subtilisin Carlsberg from *Bacillus sp.*

Figure 19-15. Synthesis of (R)-(2-methylpropyl)butanedioic acid diethylether (Hoffmann La-Roche).

yield from 45% to 87%. The reaction was repeatedly carried out on a 200 kg scale with respect to the racemic diester. The enzyme is highly stereoselective even at high substrate concentrations (20%). The chemoenzymatic route starting from the cheap bulk agents maleic anhydride and isobutylene replaced the existing chemical research synthesis for bulk amounts (Fig. 19-16).

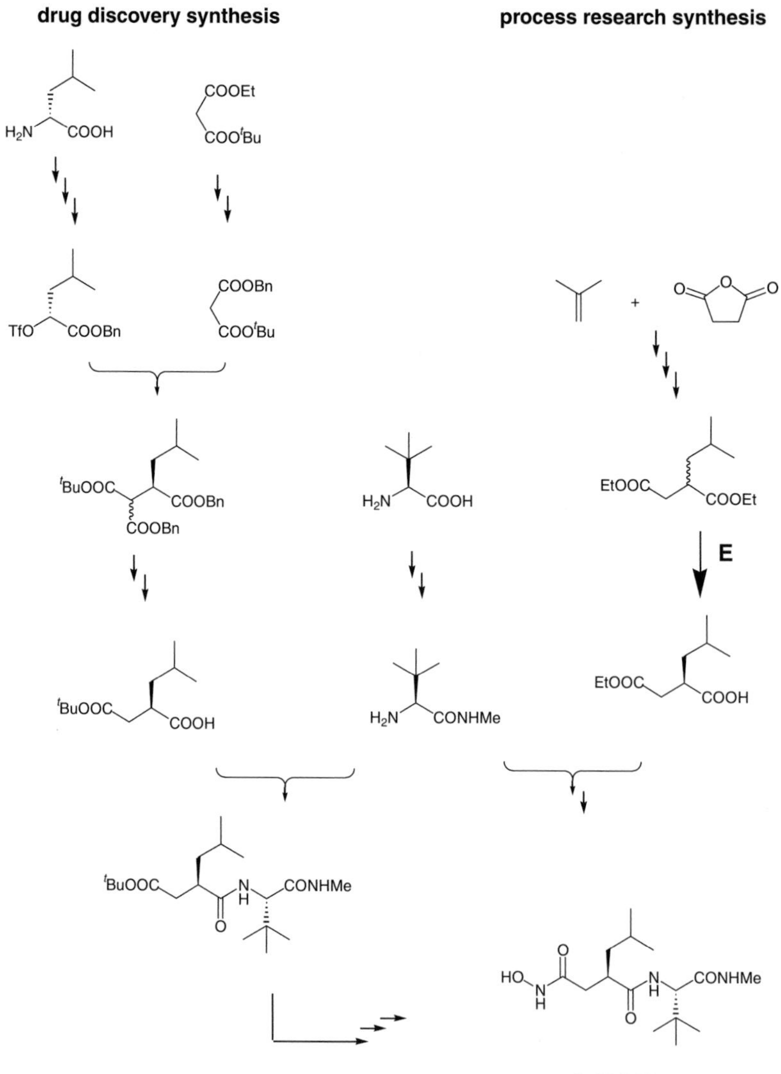

Figure 19-16. Comparison of drug discovery and process research route of Ro 31-9790.

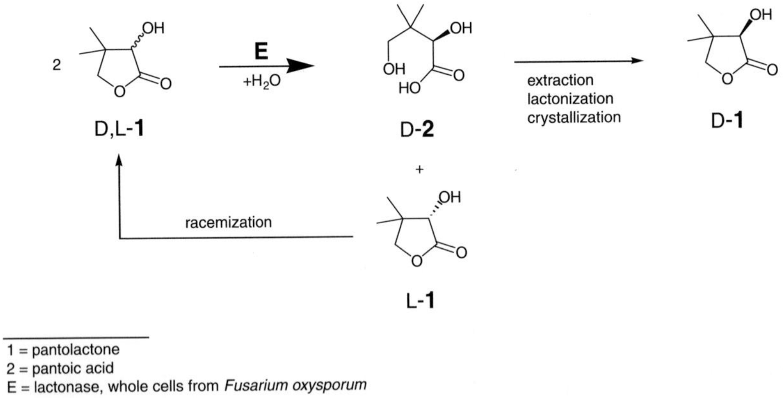

1 = pantolactone
2 = pantoic acid
E = lactonase, whole cells from *Fusarium oxysporum*

Figure 19-17. Synthesis of D-pantolactone (Fuji Chemical Industries).

19.3.3.3
Kinetic Resolution of Pantolactones and Derivatives thereof by a Lactonase from *Fusarium oxysporum* (E. C. 3.1.1.25) [57]

Pantenoic acid is used as a vitamine B_2 complex. D- and L-pantolactone are used as chiral intermediates in chemical synthesis. The enantioselective hydrolysis is carried out in the aqueous phase with a substrate concentration of 2.69 M = 350 g L^{-1} (Fig. 19-17). For the synthesis whole cells are immobilized in calcium alginate beads and used in a fixed bed reactor. The immobilized cells retain more than 90 % of their initial activity after 180 days of continuous use. At the end of the reaction L-pantolactone is extracted and reracemized to D,L-pantolactone, which is recycled to the reactor. The D-pantenoic acid is chemically lactonized to D-pantolactone and extracted. By applying cells from *Brevibacterium protophormia* the L-lactone is available. The biotransformation eliminates several steps that are necessary in the chemical resolution process (Fig. 19-18).

19.3.3.4
Hydrolysis of Starch to Glucose by Action of Two Enzymes: α-Amylase (E. C. 3.2.1.1) and Amyloglucosidase (E. C. 3.2.1.3) [58–60]

The process is part of the production of high fructose corn syrup. After several improvements, this process (Fig. 19-19) provides an effective way for an important, low-cost sugar substitute derived from grain. At various stages enzymes are applied in this process [61, 62]. The corn kernels are softened to separate oil, fiber and proteins by centrifugation. The enzymatic steps are cascaded to yield the source product for the invertase process after liquefaction in continuous cookers, debranching and filtration (Fig. 19-20). Since starches from different natural sources have different compositions, the procedure is not unique. The process ends, if all starch is completely broken down to limit the amount of oligomers of glucose and dextrins. Additionally, recombination of molecules has to be prevented. The thermostable

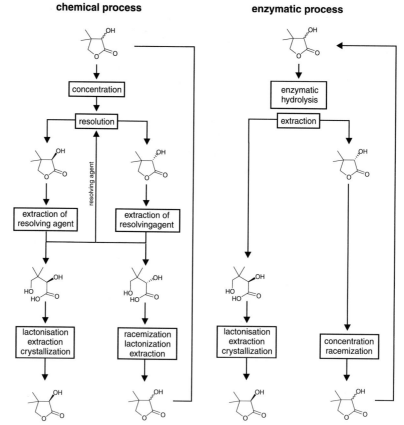

chemical process　　　　　　　　　　**enzymatic process**

Figure 19-18. Comparison of chemical and biocatalytic route for the enantioselective synthesis of pantolactone.

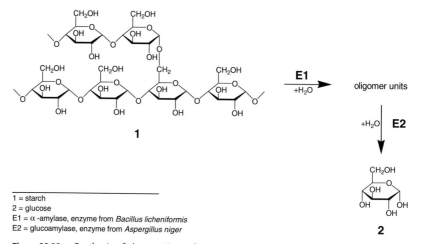

1 = starch
2 = glucose
E1 = α -amylase, enzyme from *Bacillus licheniformis*
E2 = glucoamylase, enzyme from *Aspergillus niger*

Figure 19-19. Synthesis of glucose (Several companies).

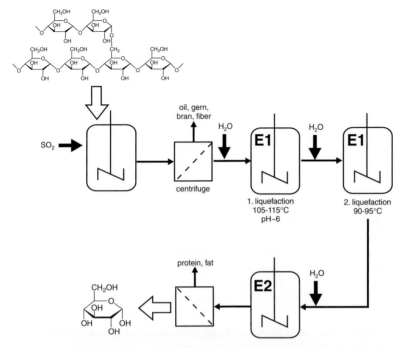

Figure 19-20. Flow scheme for the hydrolysis of starch to glucose.

enzyme can be used up to 115 °C. The enzymes need Ca^{2+} ions for stabilization and activation. Since several substances in corn can complex cations, the cation concentration is increased requiring a further product purification, i. e. making it necessary to refine the product. There is no alternative industrial chemical process for starch liquefaction. The worldwide production is about 10^7 t a^{-1}.

19.3.4
Formation or Hydrolytic Cleavage of C-N Bonds by Hydrolases (E. C. 3)

19.3.4.1
Enantioselective Acylation of Racemic Amines Catalyzed by Lipase from *Burkholderia plantarii* (E. C. 3.1.1.3) [63–65]

The lipase catalyzes the kinetic resolution of racemic amines, e. g. 1-phenyl-ethylamine (Fig. 19-21) [11]. Products are intermediates for pharmaceuticals and pesticides. They can also be used as chiral synthons in asymmetric synthesis. As acylating agent ethylmethoxyacetate is used, because the reaction rate is more than 100 times faster than that with butyl acetate. Probably an enhanced carbonyl activity induced by the electronegative α-substituents accounts for the activating effect of the methoxy group. The lipase is immobilized on polyacrylate. The lowered activity caused by use of in organic solvent (*tert*-methylbutylether = MTBE) can be increased

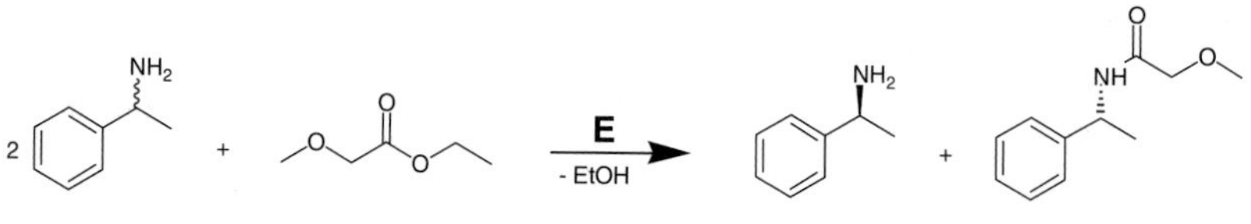

(R,S)-**1** **2** (S)-**1** (R)-**3**

\> 99% ee \> 93% ee

1 = 1-phenylethylamine
2 = ethylmethoxyacetate
3 = phenylethylmethoxyamide
E = lipase, enzyme from *Burkholderia plantarii*

Figure 19-21. Kinetic resolution of phenylethylamine (BASF).

(about 1000 times and more) by freeze drying a solution of the lipase together with fatty acids (e.g. oleic acid). Because of the use of MTBE a high starting material concentration of 1.65 M 1-phenylethylamine can be established. The enantioselectivity is greater than 500. The (R)-phenylethylmethoxyamide can easily be hydrolyzed to the (R)-phenylethylamine. The unconverted (S)-enantiomer can be racemized using a palladium catalyst.

19.3.4.2
7-Aminocephalosporanic Acid Formation by Amide Hydrolysis Catalyzed by Glutaryl Amidase (E. C. 3.1.1.41) [66–69]

The second step of the 7-aminocephalosporanic acid (7-ACA) process is the deamidation of glutaryl-7-ACA (Fig. 19-22), the first step is described in Sect. 19.3.2.2. 7-ACA is an intermediate for semi-synthetic cephalosporins. Hoechst Marion Roussel uses the glutaryl amidase immobilized on a spherical carrier. Toyo Jozo and Asahi Chemical immobilize the glutaryl amidase on porous styrene anion exchange resin with subsequent cross-linking with 1% glutardialdehyde. The catalyst is applied in a fixed bed reactor in a repetitive batch mode (70 cycles). Here, an enzymatic process has replaced an existing chemical process for environmental reasons (Fig. 19-23):

In the first step, the zinc salt of cephalosporin C is produced, followed by the protection of the functional groups (NH_2 and COOH) with trimethylchlorosilane.

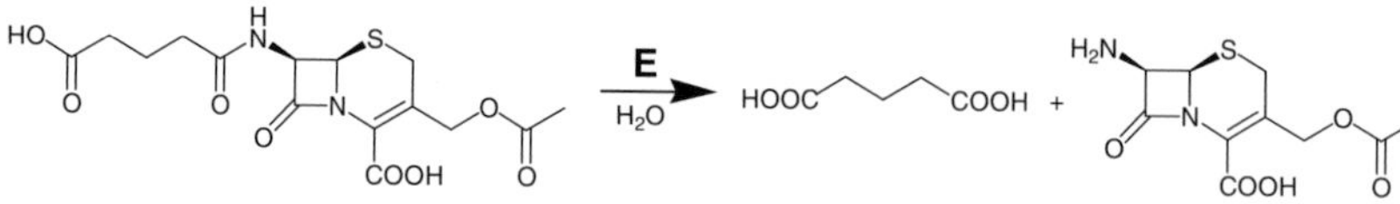

1 **2**

1 = glutaryl-7-aminocephalosporanic acid
2 = 7-aminocephalosporanic acid (7-ACA)
E = glutaryl amidase, enzyme from *Escherichia coli*

Figure 19-22. Synthesis of 7-aminocephalosporanic acid (7-ACA)
(Asahi Chemical, Hoechst Marion Roussel, Toyo Jozo).

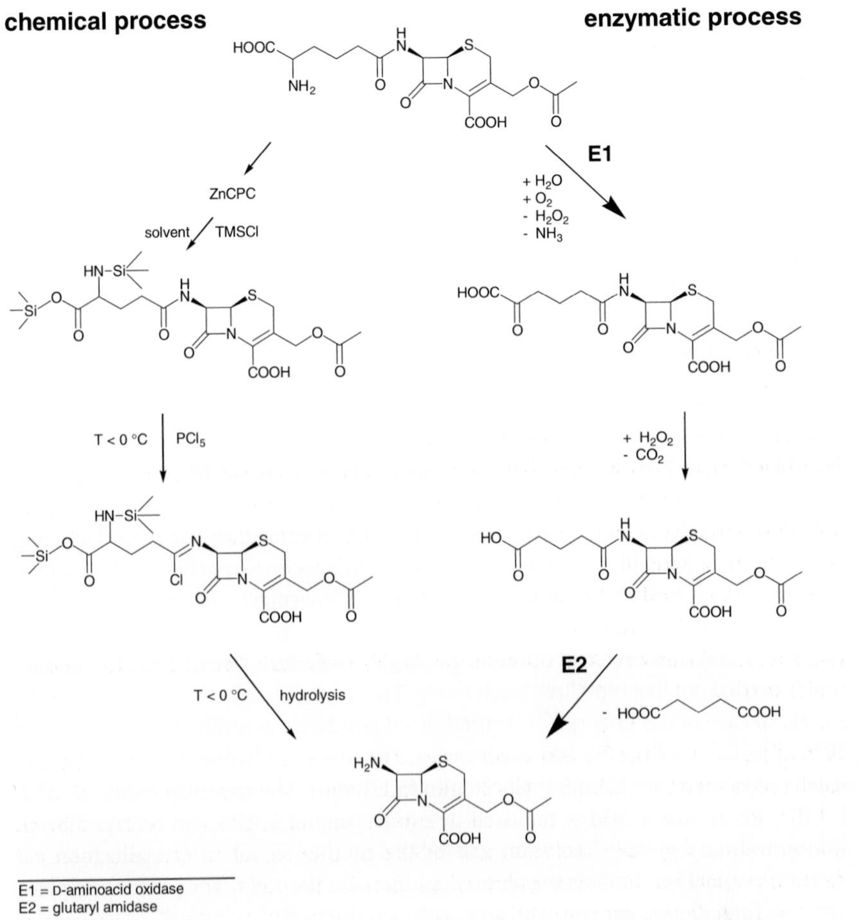
chemical process
enzymatic process
ZnCPC
solvent
TMSCl
T < 0 °C
PCl5
T < 0 °C
hydrolysis
E1
+ H2O
+ O2
- H2O2
- NH3
+ H2O2
- CO2
E2
- HOOC COOH
E1 = D-aminoacid oxidase
E2 = glutaryl amidase

Figure 19-23. Comparison of chemical and biocatalytical route for the synthesis of 7-ACA.

The imide chloride is synthesized in the subsequent step at 0 °C with phosphorous pentachloride. Hydrolysis of this imide chloride yields 7-ACA. By replacement of this synthesis with the biotransformation, the use of heavy-metal salts ($ZnCl_2$) and chlorinated hydrocarbons as well as precautions for highly flammable compounds can be circumvented. The off-gas quantities were reduced from 7.5 to 1.0 kg. Mother liquors requiring incineration were reduced from 29 to 0.3 t. Residual zinc that was recovered as $Zn(NH_4)PO_4$ is reduced from 1.8 to 0 t. The absolute costs of environmental protection are reduced by 90% per tonne of 7-ACA. Asahi Chemical and Toyo Jozo have produced 7-ACA since 1973 with a capacity of 90 t a^{-1} and Hoechst Marion Roussel since 1996 with a capacity of 200 t a^{-1}.

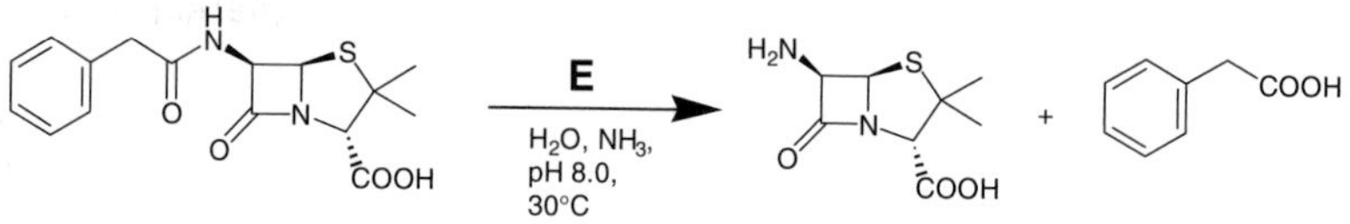

1 = penicillin-G
2 = 6-amino penicillanic acid (6-APA)
3 = phenylacetic acid
E = penicillin amidase, enzyme from *Escherichia coli*

Figure 19-24. Synthesis of 6-amino penicillanic acid (multiple companies).

19.3.4.3
Penicillin G Hydrolysis by Penicillin Amidase from *Escherichia coli* (E.C. 3.5.1.11) [68–71]

6-Amino penicillanic acid (6-APA) is used as the intermediate for manufacturing semi-synthetic penicillins. Companies applying this technology (Fig. 19-24) include Unifar, Turkey; Asahi Chemicals, Japan; Fujisawa Pharmaceutical Co., Japan; Gist-Brocades/DSM, The Netherlands; Novo-Nordisk, Sweden; Pfizer, USA. The enzyme is isolated and immobilized, often on Eupergit®C (Röhm, Germany). The production is carried out in a repetitive batch mode. The immobilized enzyme is retained by sieves. In case of the Eupergit®C immobilized amidase the residual activity is about 50% of initial activity after 800 batch cycles. Therefore the hydrolysis time after 800 batch cycles increases from initially 60 min to 120 min. The space-time yield is 445 g L^{-1} d^{-1}. Phenylacetic acid is removed by extraction and 6-APA can be crystallized. Concentrating the "split" solution and/or the mother liquor of crystallization via vacuum evaporation or reverse osmosis can increase the yield. The production plant operates for 300 days per year with an average production of 12.8 batch cycles per day (production campaigns of 800 cycles per campaign). Asahi Chemical utilizes a penicillin amidase from *Bacillus megaterium* that is immobilized on aminated porous polyacrylonitrile fibers. The production is carried out in a recirculation reactor consisting of 18 parallel columns with immobilized enzyme. Each column has a volume of 30 L. The circulation of the reaction solution is established with a flow rate of 6 000 L h^{-1}. One cycle time takes 3 h. The lifetime of each column is 360 cycles. Purification of 6-APA is done by isoelectric precipitation at pH 4.2 with subsequent filtration and washing with methanol.

7-Amino deacetoxy cephalosporanic acid (7-ADCA) is also produced by the same technology.

Several chemical steps are replaced by a single enzyme reaction (Fig. 19-25). Organic solvents, the use of low temperature (– 40 °C) and the need for absolutely anhydrous conditions, which used to make the process difficult and expensive, are no longer necessary in the enzymatic process.

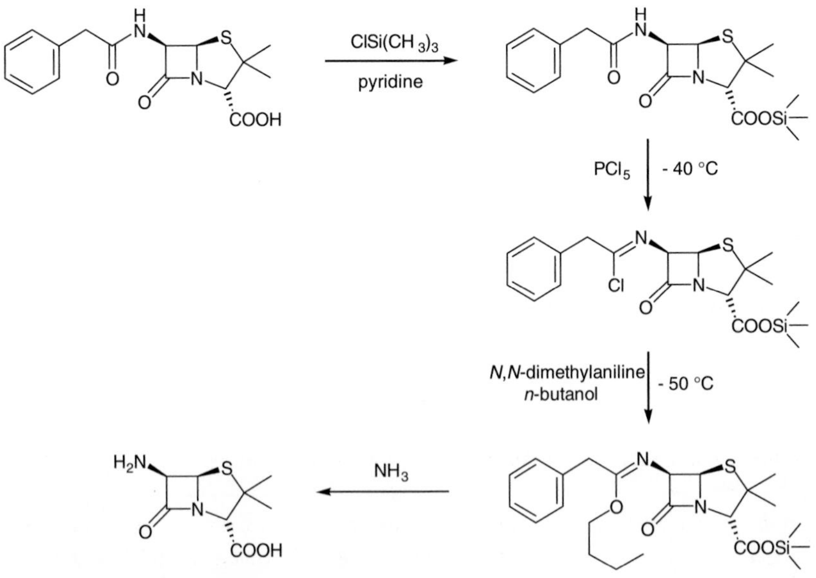

Figure 19-25. Chemical process for 6-APA.

19.3.4.4
Kinetic Resolution of α-Amino Acid Amides Catalyzed by Aminopeptidase from Pseudomonas putida (E.C. 3.4.1.11) [72–75]

Enantiomerically pure α-H-amino acids are intermediates in the synthesis of antibiotics used for parenteral nutrition and for food and feed additives (see also Chapter 12.2). Examples are D-phenylglycine and 4-hydroxyphenylalanine for semi-synthetic β-lactam antibiotics and L-phenylalanine for the peptidic sweetener aspartame. DSM used this process to produce also L-homophenylalanine, a potential precursor molecule for several ACE-inhibitors.

The α-amino amides as substrates for this enantiospecific, biocatalytic amide hydrolysis can be readily obtained from the appropriate aldehydes via the Strecker synthesis (Fig. 19-26).

As whole cell catalyst, *Pseudomonas putida*, which accepts a wide range of substrates, is applied. Subsequent to the biotransformation, benzaldehyde is added, resulting in precipitation of the D-amide Schiff base, which can be easily isolated by filtration. An acidification step leads to the D-amino acid. The L-amino acid can be reused after racemization so that a theoretical yield of 100% D-amino acid is possible.

The same process can be used for the synthesis of 100% of L-amino acids by racemizing the Schiff base of the D-amide in a short time using small amounts of base in organic solvents.

Using *in vivo* protein engineering not only mutant strains of *Pseudomonas putida*

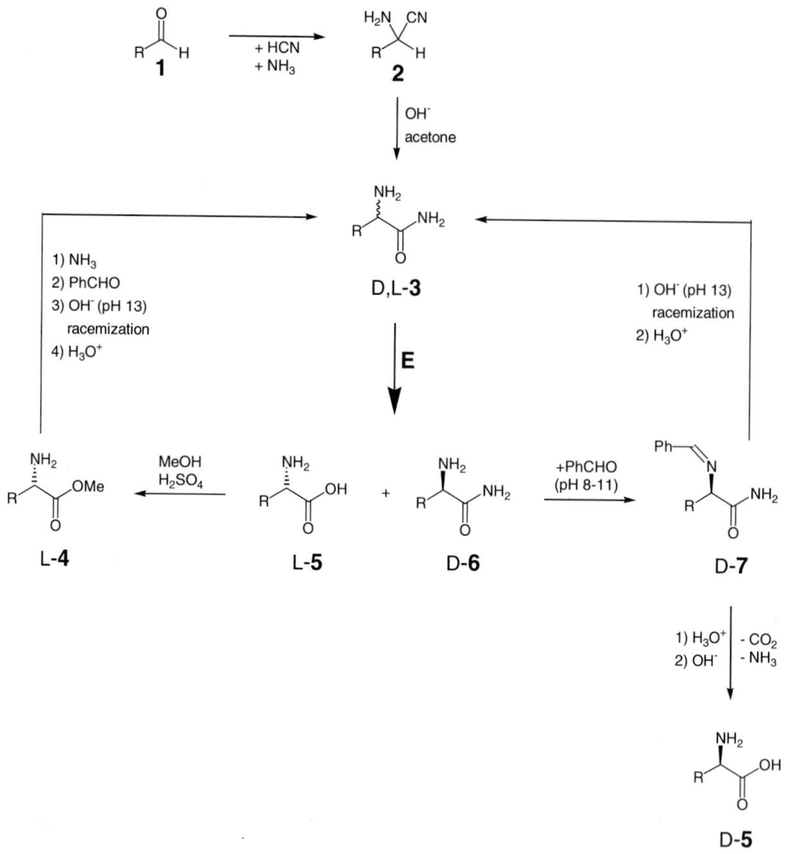

1 = aldehyde
2 = amino nitrile
3 = α-amino acid amide
4 = α-amino acid methyl ester
5 = α-amino acid
6 = α-amino acid amide
7 = schiff base of α-amino acid amide
E = aminopeptidase, whole cells from *Pseudomonas putida*

Figure 19-26. Production of L-and D-α-amino acids by kinetic resolution of α-amino acids amides (DSM).

exhibiting L-amidase and also D-amidase but also amino acid amide racemase activities were obtained. Using these mutants a convenient synthesis of α-H-amino acids with 100 % yield would be possible with one cell system. It is noteworthy that only α-H-substrates can be used. By screening, a new biocatalyst of the strain *Mycobacterium neoaurum* was found, which is capable of converting α-substituted amino acid amides.

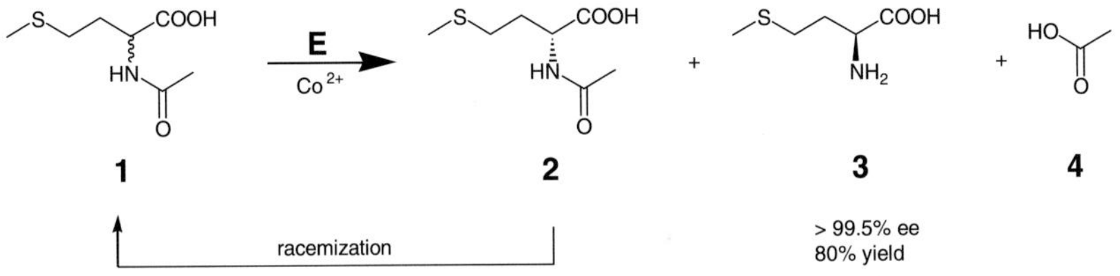

1 = *N*-acetyl-D,L-methionine
2 = *N*-acetyl-D-methionine
3 = L-methionine
4 = acetic acid
E = aminoacylase, enzyme from *Aspergillus niger*

Figure 19-27. Biocatalytical production of L-methionine by kinetic resolution (Degussa).

19.3.4.5
Production of L-Methionine by Kinetic Resolution with Aminoacylase of *Aspergillus oryzae* (E. C. 3.5.1.14) [76–79]

The *N*-acetyl-D,L-amino acid precursors are conveniently accessible through either acetylation of D,L-amino acids with acetyl chloride or acetic anhydride in a Schotten-Baumann reaction or via amidocarbonylation [80]. For the acylase reaction, Co^{2+} as metal effector is added to yield an increased operational stability of the enzyme. The unconverted acetyl-D-methionine is racemized by acetic anhydride in alkali, and the racemic acetyl-D,L-methionine is reused. The racemization can also be carried out in a molten bath or by an acetyl amino acid racemase. Product recovery of L-methionine is achieved by crystallization, because L-methionine is much less soluble than the acetyl substrate. The production is carried out in a continuously operated stirred tank reactor. A polyamide ultrafiltration membrane with a cutoff of 10 kDa retains the enzyme, thus decoupling the residence times of catalyst and reactants. L-methionine is produced with an *ee* > 99.5 % and a yield of 80 % with a capacity of > 300 t a^{-1}. At Degussa, several proteinogenic and non-proteinogenic amino acids are produced in the same way e. g. L-alanine, L-phenylalanine, α-amino butyric acid, L-valine, L-norvaline and L-homophenylalanine.

19.3.4.6
Production of D-*p*-Hydroxyphenyl Glycine by Dynamic Resolution with Hydantoinase from *Bacillus brevis* (E. C. 3.5.2.2) [8, 81–83]

D-*p*-Hydroxyphenyl glycine is a key raw material for the semisynthetic penicillins such as ampicillin and amoxycillin. It is also used in photographic developers. Racemic hydantoins are synthesized starting from phenol derivatives, glyoxylic acid and urea via the Mannich condensation (Fig. 19-28). The D-specific hydantoinase is applied as immobilized whole cells in a batch reactor. The unreacted L-hydantoins are readily racemized under the alkaline conditions (pH 8) of enzymatic hydrolysis, yielding quantitative conversion. This process enables the stereospecific preparation of various amino acids, such as L-tryptophane, L-phenylalanine, D-valine, D-alanine

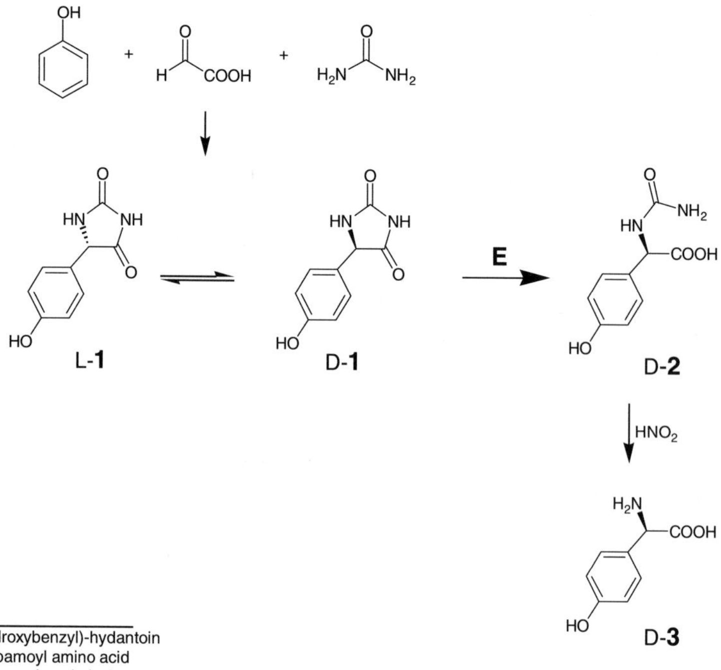

1 = 5-(p-hydroxybenzyl)-hydantoin
2 = D-*N*-carbamoyl amino acid
3 = D-4-hydroxyphenyl glycine
E = D-hydantoinase, whole cells from *Bacillus brevis*

Figure 19-28. Synthesis of D-amino acids (Kanegafuchi).

and D-methionine. Instead of chemical treatment with sodium nitrite, a carbamoylase (EC 3.5.1.77) can also be applied to remove the carbamoyl group. Several other companies have developed patented processes to produce D-hydroxyphenylglycine (Ajinomoto, DSM, SNAM-Progetti, Recordati and others).

Here the biotransformation competes with the classical chemical route (Fig. 19-29), which employs bromocamphorsulfonic acid (Br-CAS) as the resolving agent. In both routes phenol is used as raw material since *p*-hydroxybenzaldehyde is too expensive. The hydantoinase process for phenylglycines does not necessarily need an extra racemization step since the hydantoin is racemized *in situ* at an alkaline pH. Because of the dynamic resolution in the case of this biotransformation, higher yields are reached.

19.3.4.7

Dynamic Resolution of α-Amino-ε-caprolactam by the Action of Lactamase (E. C. 3.5.2.11) and Racemase (E. C. 5.1.1.15) [84, 85]

Again a dynamic resolution is carried out, but this time the racemization is introduced by an enzyme, a racemase from *Achromobacter obae* (Fig. 19-30). The lactamase and racemase are applied as whole cells and are fortunately active at the same pH, so that they can be used in one reactor. Reaction conditions enabling

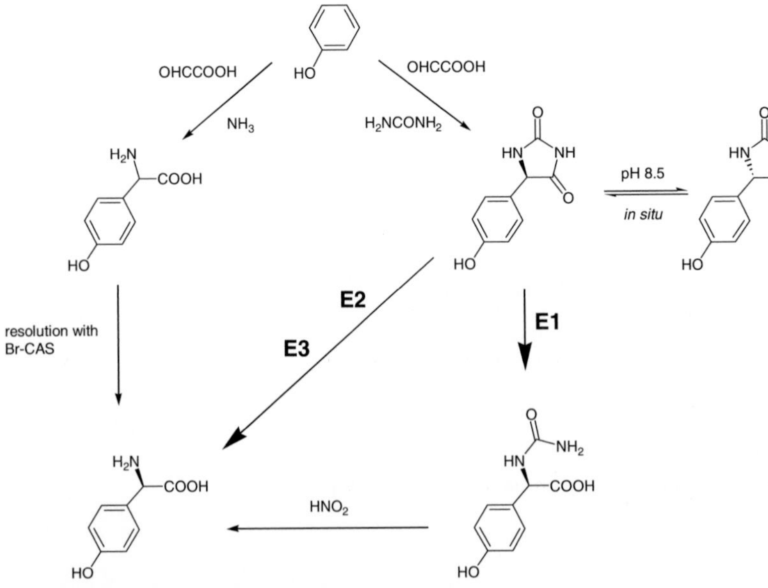

Figure 19-29. Comparison of chemical and biocatalytical route for the synthesis of D-amino acids (Kanegafuchi).

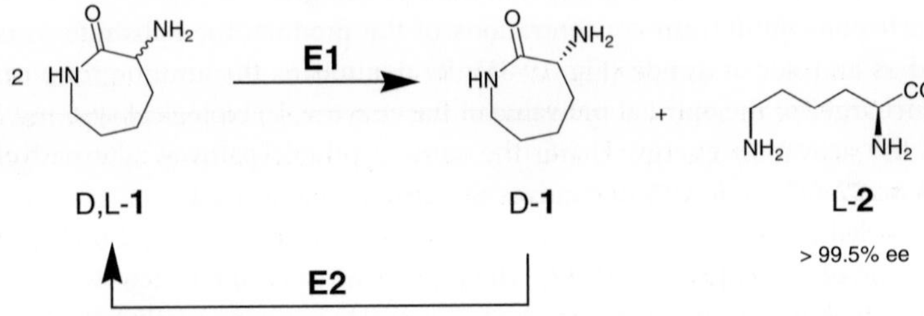

E1 = D-hydantoinase, whole cells from *Bacillus brevis*
E2/E3 = D-hydantoinase/*N*-carbamoyl-D-amino acid hydrolase, whole cells, strain *Pseudomonas* sp. contains both enzymes

1 = α-amino-ε-caprolactam (ACL)
2 = lysine
E1= L-aminolactam-hydrolase, whole cells from *Cryptococcus laurentii*
E2= amino-lactam-racemase, whole cells from *Achromobacter obae*

Figure 19-30. Synthesis of L-lysine (Toray Industries).

chemical racemization would reduce the enzyme stability. L-Lysine was produced with an *ee* of 99.5 % at a capacity of 4 000 t a^{-1}. This process has been totally replaced by highly effective fermentation methods.

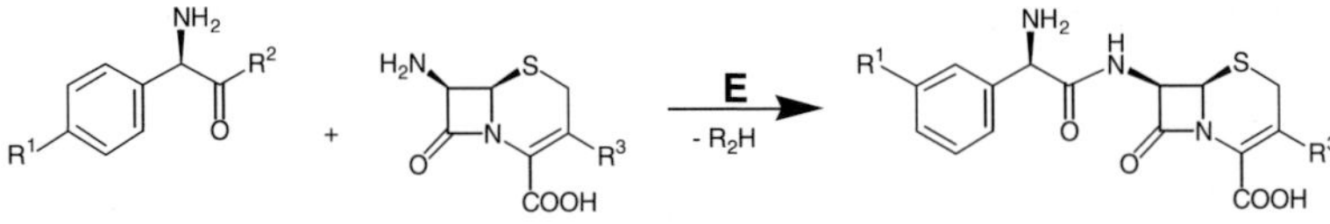

1a = phenylglycineamide (R^1=H, R^2=NH$_2$) = PGA
1b = phenylglycinemethylester (R^1=H, R^2=OMe) = PGM
1c = hydroxyphenylglycineamide (R^1=OH, R^2=NH$_2$) = HPGA
1d = hydroxyphenylglycinemethylester (R^1=OH, R^2=OMe) = HPGM
2a = 7-aminodeacetoxycephalosporanic acid (R^3=Me) = 7-ADCA
2b = 7-aminodeacetoxymethyl-3-chlorocephalosporanic acid (R^3=Cl) = 7-ACCA
3a = cefaclor (R^1=H, R^3=Cl)
3b = cephalexin (R^1=H, R^3=Me)
3c = cefadroxil (R^1=OH, R^3=Me)
E = penicillin acylase

Figure 19-31. Synthesis of β-lactam antibiotics (Chemferm).

19.3.4.8
Synthesis of β-Lactam Antibiotics Catalyzed by Penicillin Acylase (E.C. 3.5.1.11) [86–89]

The penicillin acylases do not accept charged amino groups. Therefore phenyl-glycine itself cannot be used at a pH value at which the carboxyl function is uncharged, because the amino group will then be charged.

To reach non-equilibrium concentrations of the product, the substrate must be activated as an ester or amide (Fig. 19-31). By this means the amino group can be partly uncharged at the optimal pH value of the enzyme. In biological systems, ATP delivers the activation energy. Using the same synthetic pathway alternatively to 7-ADCA and 7-ACCA, 6-APA derivatives can also be synthesized.

The established chemical synthesis started from benzaldehyde and included the fermentation of penicillin (Fig. 19-32). The process consists of ten steps with a waste stream of 30–40 kg waste per kg product. The waste contains methylene chloride, other solvents, silylating agents and many products from side-chain protection and acylating promoters. In comparison, the chemoenzymatic route needs only six steps including three biocatalytic ones. The biotransformations E1 and E2 in Fig. 19-32 can be found in Sect. 19.3.4.3 and 19.3.4.4.

19.3.4.9
Synthesis of Azetidinone β-Lactam Derivatives Catalyzed by Penicillin Acylase (E.C. 3.5.1.11) [90, 91]

It was thought that the Pen G amidase would exhibit only a limited substrate spectrum, since it does not hydrolyze the phenoxyacetyl side chain of penicillin V. Nevertheless, Eli Lilly shows that the Pen G amidase acylates the amino function of *cis*-3-amino-azetidinone with the methyl ester of phenoxyacetic acid (Fig. 19-33). The

chemical process

benzaldehyde   *Penicillium*

↓ Strecker synthesis   ↓ fermentation

D,L-phenylglycine   penicillin G

↓ classical resolution   ↓ ring enlargement

D-(-)-phenylglycine   cephalosporin

↓ protection   ↓ deacylation

Dane salt   7-ADCA

↓ activation   ↓ protection

mixed anhydride   protected 7-ADCA

↓ coupling deprotecting

cefalexin

enzymatic process

benzaldehyde   *Penicillium*

↓ fermentation

↓ Strecker synthesis   penicillin G

D,L-phenylglycine-amide ester   cephalosporin

E1 ↓ deacylation

E2 ↓ kinetic resolution   7-ADCA

↓ protection

D-(-)-phenylglycine-amide/ ester   protected 7-ADCA

E3 ↓ coupling

cefalexin

E1 = penicillin amidase
E2 = amino peptidase
E3 = penicillin acylase

Figure 19-32. Comparison of the chemical and biocatalytical synthesis of cefalexin.

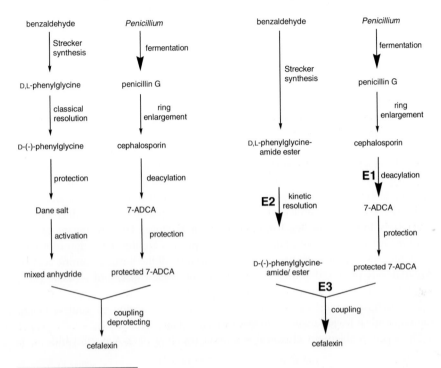

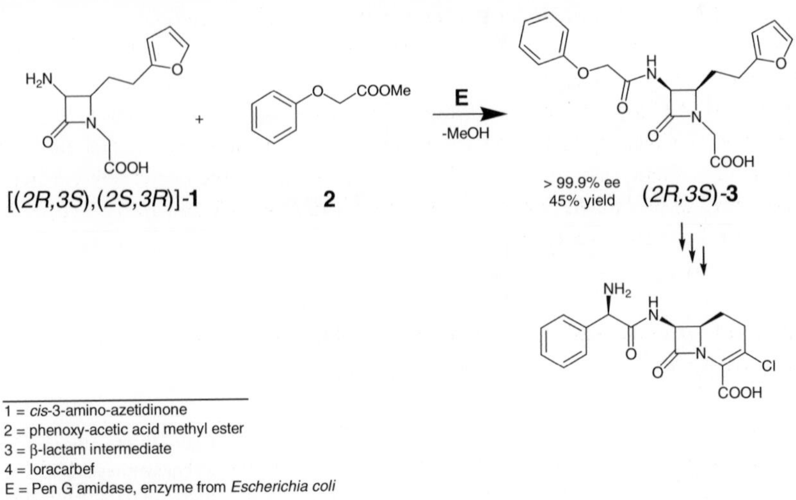

$[(2R,3S),(2S,3R)]$-**1**   **2**   **E** / -MeOH   > 99.9% ee 45% yield $(2R,3S)$-**3**

1 = *cis*-3-amino-azetidinone
2 = phenoxy-acetic acid methyl ester
3 = β-lactam intermediate
4 = loracarbef
E = Pen G amidase, enzyme from *Escherichia coli*

Figure 19-33. Synthesis of azetidione β-lactam derivatives (Eli Lilly).

acylation occurs using methyl phenylacetate (MPA) or methyl phenoxyacetate (MPOA) as the acylating agents. The penicillin amidase is immobilized on Eupergit (Roehm GmbH, Germany).

The chemical resolution of the racemic azetidinone is only low yielding. The (2*R*,3*S*)-azetidinone is a key intermediate in the synthesis of the carbacephalosporin antibiotic loracarbef.

19.3.4.10
Enantioselective Synthesis of an Aspartame Precursor with Thermolysin from *Bacillus proteolicus* (E.C. 3.4.24.27) [92, 93]

Since the reaction (Fig. 19-34) is limited by the equilibrium the products have to be removed from the reaction mixture to reach high yields. Therefore an excess of racemic phenylalanine methylester (which is inert to the reaction) is added. The carboxylic anion of the protected aspartame forms a poorly soluble adduct with D-Phe-OCH$_3$ that precipitates from the reaction mixture. The precipitate can be removed easily by filtration. Final steps of the process are the separation of D-Phe-ester, removal of protecting groups and racemization of the formed L-amino acid. α-Aspartame is produced with > 99.9% and a worldwide capacity of ~ 10,000 t a^{-1}, ~ 2,500 t a^{-1} by enzymatic coupling.

The bacterial strain was found in the Rokko Hot Spring in central Japan. Consequently it is very stable up to temperatures of 60 °C.

The main problem in chemical synthesis coupling of Z-Asp anhydride with

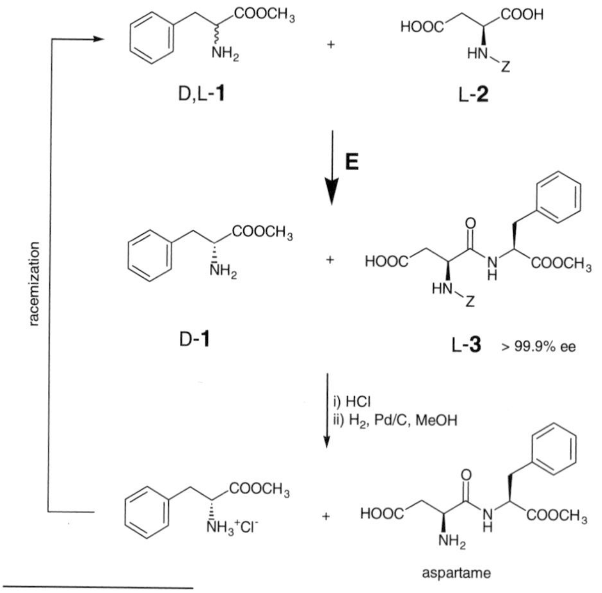

1 = phenylalanine methylester
2 = aspartic acid (protected)
3 = α-aspartame (protected)
E = thermolysin, enzyme from *Bacillus proteolicus*

Figure 19-34. Biocatalytical synthesis of aspartame (HSC, Holland Sweetener Company).

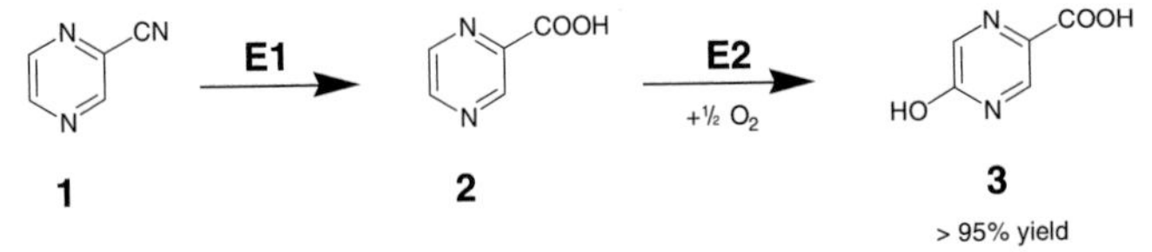

1 = 2-cyanopyrazine
2 = pyrazine-2-carboxylic acid
3 = 5-hydroxypyrazine-2-carboxylic acid
E1/E2 = nitrilase/hydroxylase, whole cells, strain *Agrobacterium* sp. contains both enzymes

Figure 19-35. Biocatalytical synthesis of 5-hydroxypyrazine-2-carboxylic acid (Lonza).

L-Phe-OCH$_3$ is the by-product formation of β-aspartame. This isomer is of bitter taste and has to be completely removed from the α-isomer. The advantages of the enzymatic route are: (i) No β-isomer is produced, (ii) the enzyme is completely stereoselective, so that racemic mixtures of the substrate or the appropiate enantiomer of the amino acid can be used, (iii) no racemization occurs during synthesis and (iv) the reaction takes place in aqueous media under mild conditions.

19.3.4.11
Hydrolysis of Heterocyclic Nitrile by Nitrilase from *Agrobacterium* sp. (E. C. 3.5.5.1) [94–96]

5-Hydroxypyrazine-2-carboxylic acid (Fig. 19-35) is a versatile building block in the synthesis of new antituberculous agents, e.g. 5-chloro-pyrazine-2-carboxylic acid esters. The regioselective hydroxylation of pyrazine-2-carboxylic acid is catalyzed by a hydroxylase (E2, E. C. 1.5.1.13). This second enzyme is also in the applied suspended whole cells from *Agrobacterium* sp. The biomass is separated by ultrafiltration (cutoff 10 kDa) after the biotransformation. 5-Hydroxypyrazine-2-carboxylic acid is precipitated from the permeate by acidification with sulfuric acid to pH 2.5.

In contrast to the biotransformation, the chemical synthesis of 5-substituted pyrazine-2-carboxylic acid leads to a mixture of 5- and 6-substituted pyrazine-carboxylic acids and requires multiple steps.

19.3.5
Formation of C-O Bonds by Lyases

19.3.5.1
Synthesis of Carnitine Catalyzed by Carnitine Dehydratase in Whole Cells (E. C. 4.2.1.89) [8, 46, 97–99]

L-Carnitine is used in infant health, sport and geriatric nutrition. The biotransformation is catalyzed by carnitine dehydratase in whole cells (Fig. 19-36). (*R*)-carnitine is produced with > 99.5 % conversion of butyrobetaine and > 99.5 % *ee*. The mutant strain has blocked the L-carnitine dehydrogenase and excretes the accumulated product. The purified enzyme could not be used for the biotransformation because of its high instability. Apart from usual batch fermentations, continuous production

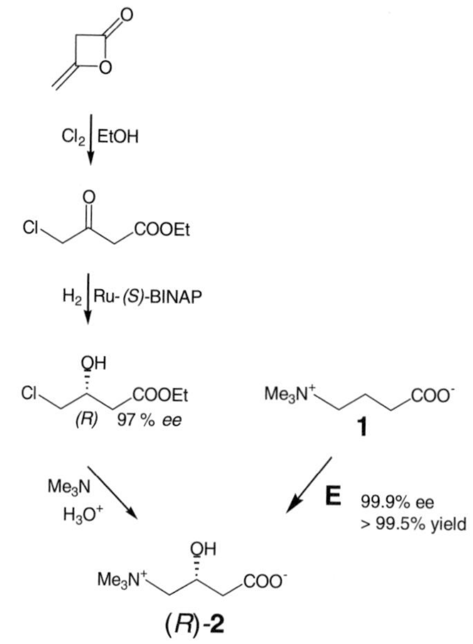

1 = 4-butyro betaine
2 = carnitine
E = carnitine dehydratase, whole cells from *Escherichia coli*

Figure 19-36. Comparison of chemical and biocatalytical synthesis of carnitine (Lonza).

is also feasible since the cells go into a "maintenance state" with high metabolic activity and low growth rate. The cells can be recycled after separation from the fermentation broth by filtration. A chemical resolution process with L-tartaric acid that was developed at Lonza was no longer competitive with the biotechnological route. A more attractive chemical route would be the Ru-BINAP catalyzed asymmetric hydrogenation of 4-chloroacetoacetate (Fig. 19-36). Here an *ee* of 97% is yielded.

19.3.6
Formation of C-N Bonds by Lyases (E. C. 4)

19.3.6.1
Synthesis of L-Dopa Catalyzed by Tyrosine Phenol Lyase from *Erwinia herbicola* (E. C. 4.1.99.2) [100–103]

The product is applied for the treatment of Parkinsonism that is caused by a lack of L-dopamine and its receptors in the brain. L-Dopamine is synthesized in organisms by decarboxylation of L-3,4-dihydroxyphenylalanine (L-dopa). Since L-dopamine cannot pass the blood-brain barrier L-dopa is applied in combination with dopadecarboxylase-inhibitors to avoid formation of L-dopamine outside the brain. Ajinomoto produces L-dopa by this lyase-biotransformation with suspended whole cells in a fed batch reactor on a scale of 250 t a^{-1}. Much earlier, Monsanto has successfully scaled up the chemical synthesis of L-dopa (Fig. 19-38).

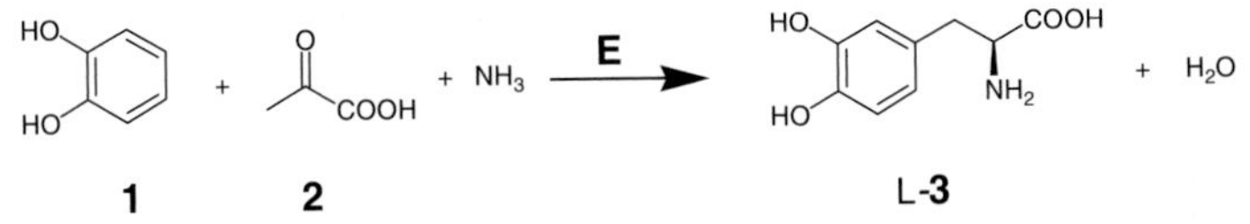

1 = catechol
2 = pyruvic acid
3 = dopa
E = tyrosine phenol lyase, whole cells from *Erwinia herbicola*

Figure 19-37. Biocatalytical synthesis of L-dopa (Ajinomoto).

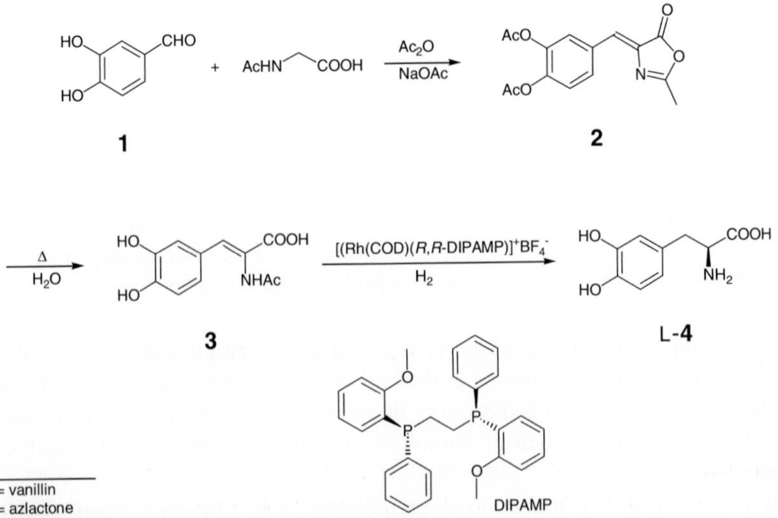

1 = vanillin
2 = azlactone
3 = Z-enamide
3 = dopa

Figure 19-38. Chemical synthesis of L-dopa (Monsanto).

The enantioselective hydrogenation of 3,4-dihydroxy-*N*-acetylamino cinnamic acid is catalyzed by the cationic Rh-biphosphine complex DIPAMP, in which the enantioselectivity is introduced by the chiral phosphine[104, 105]. The hydrogenation proceeds quantitatively with 94 % *ee*. The optically pure L-dopa is separated from the catalyst by crystallization.

19.3.6.2
Synthesis of 5-Cyano Valeramide by Nitrile Hydratase from *Pseudomonas chlororaphis* B23 (E. C. 4.2.1.84) [106, 107]

5-Cyanovaleramide is used as intermediate for the synthesis of the DuPont herbicide azafenidine (Fig. 19-39). The whole cells from *Pseudomonas chlororaphis* are immobilized in calcium alginate beads. The biotransformation itself is catalyzed by a nitrile hydratase that converts a nitrile into the corresponding amide by addition of water. Nitrile hydratases belonging to the enzyme class of lyases (E. C. 4) are not be

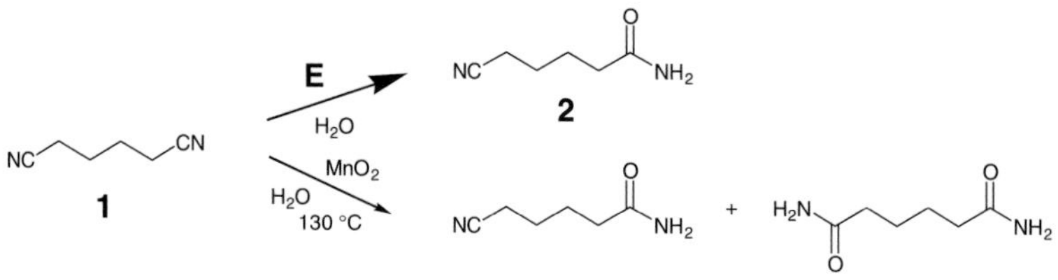

1 = adiponitrile
2 = 5-cyano-valeramide
E = nitrile hydratase, whole cells from *Pseudomonas chlororaphis*

Figure 19-39. Comparison of chemical and biocatalytical synthesis of 5-cyano-valeramide (DuPont).

confused with the nitrilases belonging to the class of hydrolases (E.C. 3) that hydrolyze nitriles to the corresponding carbon acids. For strain selection it was important that the cells did not show any amidase activity that would further hydrolyze the amide to the carboxylic acid. The biotransformation is carried out in a two-phase system with pure adiponitrile forming the organic phase. A reaction temperature of 5 °C is chosen, since the solubility of the by-product adipodiamide is only 37–42 mM in 1–1.5 M 5-cyanovaleramide. A batch reactor is preferred over a fixed-bed reactor, because of the lower selectivity to 5-cyanovaleramide that was observed and the possibility of precipitation of adipodiamide and plugging of the column. Excess water is removed at the end of the reaction by distillation. The by-product adipodiamide is precipitated by dissolution of the resulting oil in methanol at > 65 °C. The raw product solution is directly transferred to the herbicide synthesis.

By this method 13.6 tonnes have been produced in fifty-eight repetitive batch cycles with 97 % conversion and 96 % selectivity. This biotransformation was chosen over the chemical transformation because of the higher conversion and selectivity, production of more product per catalyst weight (3 150 kg per kg dry cell weight), and less waste. The catalyst consumption is 0.006 kg per kg product.

19.3.6.3
Synthesis of the Commodity Chemical Acrylamide Catalyzed by Nitrile Hydratase from *Rhodococcus rodochrous* (E.C. 4.2.1.84) [108–112]

Acrylamide (Fig. 19-40) is an important commodity monomer used in coagulators, soil conditioners and stock additives for paper treatment and paper sizing, and for adhesives, paints and petroleum recovering agents.

Since acrylonitrile is the most poisonous of the nitriles, screening for micro-organisms was conducted with low molecular weight nitriles instead.

Acrylamide is unstable and polymerizes easily; therefore the process is carried out at a low temperature (5 °C). Although the cells, which are immobilized on poly-acrylamide gel, and the contained enzyme are very stable towards acrylonitrile, the starting material has to be fed continuously to the reaction mixture because of inhibition effects at higher concentrations. The biotransformation is started with an

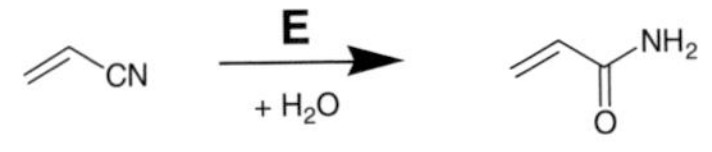

Figure 19-40. Biocatalytical synthesis of acrylamide (Nitto Chemical Industry).

1 = acrylonitrile
2 = acrylamide
E = nitrile hydratase, whole cells from *Rhodococcus erythropolis*

acrylonitrile concentration of 0.11 M and is stopped at an acrylamide concentration of 5.6 M. The process is operated at a capacity of 30 000 t a^{-1}.

This nitrile hydratase acts also on other nitriles with yields of 100%. The most impressive example is the conversion of 3-cyanopyridine to nicotinamide. The product concentration is about 1 465 g L^{-1}. This conversion (1.17 g L^{-1} dry cell mass) can be named "pseudocrystal enzymation", since at the start of the reaction the educt is solid and with ongoing reaction it is solubilized.

The chemical synthesis uses copper salt as catalyst for the hydration of acrylonitrile and has several disadvantages:

- The rate of acrylamide formation is lower than that of acrylic acid formation.
- The double bond of the starting material and the product causes the formation of by-products such as ethylene, cyanohydrin and nitrilotrispropionamide.
- Polymerization occurs.
- Copper needs to be separated from the product (an extra step in the chemical synthesis).

The biotransformation has the advantages that no recovering of unreacted nitrile is necessary since the conversion is 100% and no copper catalyst removal is needed. This is also the first case of a biocatalytic conversion of a bulk fiber monomer.

19.3.6.4
Synthesis of Nicotinamide Catalyzed by Nitrile Hydratase from *Rhodococcus rodochrous* (E. C. 4.2.1.84) [48, 113]

Nicotinamide (vitamin B3) is used as a vitamin supplement for food and animal feed. It is the same strain that is also used in the industrial production of acrylamide (see Sect. 19.3.6.3). The biotransformation is carried out on a scale of 3000 t a^{-1} (Fig. 19-41).

In contrast to the chemical alkaline hydrolysis of 3-cyanopyridine with 4% by-product of nicotinic acid (96% yield) the biotransformation works with absolute selectivity and no acid or base is required. The biotransformation (a continuous process) is operated at low temperature and atmospheric pressure. In contrast to the old synthesis route of nicotinamide at Lonza, the new one is environmentally friendly and safe. There is only one organic solvent used throughout the whole process in four highly selective continuous and catalytic reactions. The process water, NH$_3$ and H$_2$ are recycled.

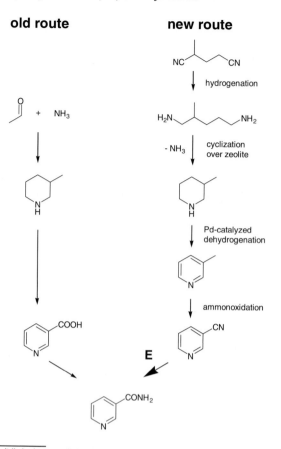

E = nitrile hydratase, whole cells from *Rhodococcus erythropolis*

Figure 19-41. Comparison of chemical and biocatalytical synthesis of nicotinamide (Lonza).

19.3.7
Epimerase

19.3.7.1
Epimerization of Glucosamine Catalyzed by Epimerase from *E. coli* (E.C. 5.1.3.8) [114–116]

N-Acetyl-D-mannosamine serves as the *in situ* generated substrate for the synthesis of N-acetylneuraminic acid. Since N-acetyl-D-mannosamine is quite expensive it is synthesized from N-acetyl-D-glucosamine by epimerization at C_2. This biotransformation is integrated into the production of N-acetylneuraminic acid (Neu5Ac).

By application of N-acylglucosamine 2-epimerase it is possible to start with the inexpensive N-acetyl-D-glucosamine instead of N-acetyl-D-mannosamine (Fig. 19-42). The epimerase is used for the *in situ* synthesis of N-acetyl-D-mannosamine

1 = N-acetyl-D-glucosamine
2 = N-acetyl-D-mannosamine
E = GlcNAc 2-epimerase, enzyme from *Escherichia coli*

Figure 19-42. Biocatalytical epimerization of glucosamine to mannosamine (Marukin Shoyu).

(ManNAc). Since the equilibrium is on the side of the starting material, the reaction is driven by the subsequent biotransformation of ManNAc together with pyruvate to Neu5Ac.

The *N*-acylglucosamine 2-epimerase is cloned from porcine kidney, transformed and overexpressed in *Escherichia coli*. To reach maximal activitiy, ATP and Mg^{2+} need to be added. Since the whole synthesis is reversible, high GlcNAc concentrations are used.

The chemical epimerization of GlcNAc is used by Glaxo. The equilibrium of the chemical epimerization is on side of *N*-acetyl-D-glucosamine (GlcNAc:ManNAc = 4:1). After neutralization and addition of isopropanol GlcNAc precipitates. In the remaining solution a ratio of GlcNAc:ManNAc = 1:1 is reached. After evaporation to dryness and extraction with methanol the ratio of GlcNAc:ManNAc is shifted to 1:4.

19.4
Some Misconceptions about Industrial Biotransformations

There are a lot of prejudices against biotransformations. The major ones are:

- Biocatalysts are too expensive.
- Biocatalysts only work under mild conditions.

The first prejudice that biocatalysts are too expensive is only partly true. If the cost per mol or per unit weight is calculated they certainly are expensive. For example, penicillin amidase costs $ 10 000/kg on a bulk scale. On the other hand the cost contribution of penicillin amidase in the "splitting" of penicillin G is only $ 1/kg of product[117]. In the case of L-aspartic acid production the cost contribution of aspartase is even lower, $ 0.1/kg. This demonstrates that it is not the absolute catalyst cost but the cost contribution of the catalyst to the final product cost that has to be considered and compared. This is also true for chemical catalysts; e. g., the bulk price of BINAP is $ 40 000/kg[117]. Important parameters influencing the cost contribution are the total turnover number (mol product/mol catalyst) and the turnover frequency (mol product/mol catalyst and unit time).

The second prejudice, that biocatalysts only work in an aqueous phase with low concentrations of starting material is also only partly true. The natural environment

Table 19-1. Highest concentrations applied in industrial biotransformations.

EC	enzyme	substrate	concentration	medium
1.1.99.21	D-Sorbitol dehydrogenase	1-Amino-D-sorbitol (N-protected)	1.00 M	Aqueous
4.2.1.2	Fumarase	Fumaric acid	1.00 M	Aqueous
3.4.21.62	Subtilisin	Phenylalanine isopropylester	1.20 M	Aqueous
3.1.1.3	Lipase	1-Phenylethylamine	1.65 M	MTBE
3.5.2.6	β-Lactamase	γ-Lactam	1.83 M	Aqueous
4.2.1.84	Nitrile hydratase	Adiponitrile	2.01 M	Aqueous/organic
4.3.1.1	L-Aspartase	Fumaric acid	2.00 M	Aqueous
4.1.1.12	Aspartate β-decarboxylase	Aspartic acid	2.50 M	Aqueous
3.1.1.25	Lactonase	Pantolactone	2.69 M	Aqueous
3.1.1.3	Lipase	Palmitic acid	3.10 M	2-Propanol
3.1.1.3	Lipase	Cyclopentenylester	4.16 M	Aqueous/organic
4.2.1.84	Nitrile hydratase	Acrylonitrile	5.60 M (product)	Aqueous
4.3.1.5	L-Phenylalanine ammonia-lyase	*trans*-Cinnamic acid	9.31 M (NH₃)	Aqueous

is in general the aqueous phase and ambient temperature. But the examples described above demonstrate that biocatalysts can be also applied in emulsions or even pure organic solvents (Table 19-1). Here, moreover, very high concentrations are reached, e.g. in the case of acrylamide up to 5.6 M.

19.5
Outlook

Despite the progress biocatalysis has made in the last few years its potential is still increasing. By improved screening methods new catalysts will be detected and made available in large amounts by cloning and overexpression. Directed evolution will be used to improve properties such as stability or selectivity[118, 119]. Metabolic engineering will be used to analyze and remove bottlenecks in the metabolism or to create novel biocatalysts[120].

References

1 *The New Scientist*, 1ˢᵗ issue, **22 Nov. 1956**.
2 A. Liese, K. Seelbach, C. Wandrey, *Industrial Biotransformations*; Wiley-VCH, Weinheim, **2000**.
3 U. Bornscheuer, Industrial Biotransformations, in H. Rehm, G. Reed, A. Pühler, P. Stadler (eds.), *Biotechnology Series* Vol. 8b, Wiley-VCH, Weinheim, **2000**, 277–294.
4 J. Peters in *Biotechnology* 2ⁿᵈ edn., Vol 8a, H. J. Rehm, G. Reed (eds.), Wiley-VCH, Weinheim, **1998**, 391–474.
5 A. Liese, M. Villela Filho, Production of fine chemicals using biocatalysis, *Curr. Op. Biotech.* **1999**, *10 (6)*, 595–603.

6 R. N. Patel, *Stereoselective Biocatalysis*, Marcel Dekker, New York, **2000**.

7 A. Tanaka, T. Tosa and T. Kobayashi (eds.), *Industrial Application of Immobilized Biocatalysts*, Marcel Dekker, New York, **1993**.

8 R. A. Sheldon, *Chirotechnology: Industrial Synthesis of Optically Active Compounds*, Marcel Decker, New York, **1993**.

9 A. N. Collins, G. N. Sheldrake, J. Crosby (eds.), *Chirality in Industry*, Wiley, Chichester, **1992**.

10 U. T. Bornscheuer, *Enzymes in Lipid Modification*, Wiley-VCH, Weinheim, **2000**.

11 A. Schmid, J. S. Dordick, B. Hauer, A. Kiener, M. Wubbolts, B. Witholt, Industrial biocatalysis today and tomorrow, *Nature* **2001**, *409*, 258–268.

12 K. Faber, *Biotransformations in Organic Chemistry*, Springer, Berlin, **2000**.

13 H. Griengl, *Biocatalysis*, Springer, Berlin, **2000**.

14 U. T. Bornscheuer, R. J. Kazlauskas, *Hydrolases in Organic Synthesis*, Wiely-VCH, Weinheim, **1999**.

15 J. E. Bailey, D. F. Ollis, *Biochemical Engineering Fundamentals*, McGraw-Hill, New York, **1986**.

16 H.-J. Rehm, *Biotechnology*, Wiley-VCH, Weinheim, **2000**.

17 D. Weuster-Botz, *Die Rolle der Reaktionstechnik in der mikrobiellen Verfahrensentwicklung*, Forschungszentrum Jülich, Jülich, **1999**.

18 B. A. Anderson, M. M. Hansen, A. R. Harkness, C. L. Henry, J. T. Vicenzi, M. J. Zmijewski, Application of a practical biocatalytic reduction to an enantioselective synthesis of the 5*H*-2,3-benzodiazepine LY200164, *J. Am. Chem. Soc.* **1995**, *117*, 12358–12359.

19 J. T. Vicenzi, M. J. Zmijewski, M. R. Reinhard, B. E. Landen, W. L. Muth, P. G. Marler, Large-scale stereoselective enzymatic ketone reduction with in situ product removal via polymeric adsorbent resins, *Enzyme Microb. Technol.* **1997**, *20*, 494–499.

20 U. Kragl, *Immobilized Enzymes and Membrane Reactors*, in T. Godfrey, S. West (eds.), *Industrial Enzymology*, Macmillan Press, London, **1996**, 275–283.

21 R. D. Noble, S. A. Stern, *Membrane Separations Technology. Principles and Applications*, Elsevier, Amsterdam, **1995**.

22 M. Mulder, *Basic Principles of Membrane Technology*, Kluwer Academic Publishers, Dortrecht, **1996**.

23 T. Zelinski, H. Waldmann, Cross-linked enzyme crystals (CLECs): Efficient and stable biocatalysts for preparative organic chemistry, *Angew. Chem. Int. Ed. Engl.* **1996**, *7*, 722–725.

24 W. Keim, B. Drießen-Hölscher, *Heterogenization of Complexes and Enzymes* in G. Ertl, H. Knözinger, J. Weitkamp (eds.), *Handbook of Heterogeneous Catalysis*, Wiley-VCH, Weinheim **1997**, 231–240.

25 P. Rasor, in D. E. De Vos, I. F. J Vankelecom, P. A. Jacobs (eds.) *Chiral Catalyst Immobilization and Recycling*, Wiley-VCH, **2000**, 96–104.

26 A. J. Blacker, R. A. Holt, *Development of a Multistage Chemical and Biological Process to an Optically Active Intermediate for an Antiglaucoma Drug*, in: A. N. Collins, G. N. Sheldrake and J. Crosby (eds.), *Chirality In Industry II*, John Wiley & Sons, New York **1997**, 246–261.

27 T. J. Blacklock, P. Sohar, J. W. Butcher, T. Lamanec, E. J. J. Grabowski, An enantioselective synthesis of the topically-active carbonic anhydrase inhibitor MK-0507: 5,6-dihydro-(*S*)-4-(ethylamino)-(*S*)-6-methyl-4H-thieno[2,3-*b*]thiopyran-2-sulfonamide 7,7-dioxide hydrochloride, *J. Org. Chem.* **1993**, *58*, 1672–1679.

28 R. A. Holt, Microbial asymmetric reduction in the synthesis of a drug intermediate, *Chimica Oggi* **1996**, *9*, 17-20.

29 R. A. Holt and S. R. Rigby, *Process for microbial reduction producing 4(S)-hydroxy-6(S)methyl-thienopyran derivatives*, Zeneca Limited, **1996** US 5580764.

30 R. N. Patel, C. G. McNamee, A. Banerjee, L. J. Szarka, *Stereoselective microbial or enzymatic reduction of 3,5-dioxo esters to 3-hydroxy-5-oxo, 3-oxo-5-hydroxy, and 3,5-dihydroxy esters*, E. R. Squibb & Sons, Inc., EP 056 9998A2 **1993**.

31 R. N. Patel, A. Banerjee, C. G. McNamee, D. Brzozowski, R. L. Hanson, L. J. Szarka, Enantioselective microbial reduction of 3,5-dioxo-6-(benzyloxy) hexanoic acid ethyl ester, *Enzyme Microb. Technol.* **1993**, *15*, 1014–1021.

32 S. Y. Sit, R. A. Parker, I. Motoc, W. Han, N. Balasubramanian, Synthesis, biological profile, and quantitative structure-activity relationship of a series of novel 3-hydroxy-3-methylglutaryl coenzyme A reductase inhibitors, *J. Med. Chem.* **1990**, *33 (11)*, 2982–2999.

33 M. Chartrain, J. Chung, C. N. Roberge, *N-(R)-(2-hydroxy-2-pyridine-3-yl-ethyl)-2-(4-nitro-phenyl)-acetamide*, Merck & Co., Inc., US 584 6791, **1998**.

34 J. Chung, G. Ho, M. Chartrain, C. Roberge, D. Zhao, J. Leazer, R. Farr, M. Robbins, K. Emerson, D. Mathre, J. McNamara, D. Hughes, E. Grabowski, P. Reider, Practical chemoenzymatic synthesis of a pyridylethanolamino beta-3 adrenergic receptor antagonist – stereospecific beta-3 agonist used in obesity therapy, produced by yeast-mediated asymmetric reduction of a ketone, *Tetrahedron Lett.* **1999**, *40 (37)*, 6739–6743.

35 M. Chartrain, C. Roberge, J. Chung, J. McNamara, D. Zhao, R. Olewinski, R. Hunt, P. Salmon, D. Roush, S. Yamazaki, T. Wang, E. Grabowski, B. Buckland, R. Greasham, Asymmetric bioreduction of (2-(4-nitro-phenyl)-N-(2-oxo-2-pyridin-3-yl-ethyl)-acetamide) to its corresponding (R) alcohol [(R)-N-(2-hydroxy-2-pyridin-3-yl-ethyl)-2-(4-nitro-phenyl)-acetamide] by using *Candida sorbophila* MY 1833, *Enz. Microb. Technol.* **1999**, *25*, 489–496.

36 G. Kinast, M. Schedel, *Verfahren zur Herstellung von 6-Amino-6-desoxy-L-sorbose*, Bayer AG, DE 283 4122 A1, **1978**.

37 G. Kinast, M. Schedel, *Herstellung von N-substituierten Derivaten des 1-Desoxynojirimycins*, Bayer AG, DE 285 3573 A1, **1978**.

38 G. Kinast, M. Schedel, Vierstufige 1-Desoxynojirimycin-Synthese mit einer Biotransformation als zentralem Reaktionsschritt, *Angew. Chem.* **1981**, *93*, 799–800.

39 T. Reichstein, H. Grüssner, Eine ergiebige Synthese der L-Ascorbinsäure (Vitamin C) *Helv. Chim. Acta* **1934**, *17*, 311-328.

40 K. Matsumoto, Production of 6-APA, 7-ACA, and 7-ADCA by immobilized penicillin and cephalosporin amidases, in A. Tanaka, T. Tosa, T. Kobayashi, (eds.), *Industrial Application of Immobilized Biocatalysts*, Marcel Dekker, New York **1993**, 67–88.

41 T. Tanaka, T. Tosa, T. Kobayashi, *Industrial application of immobilized biocatalysts*, Marcel Dekker Inc., New York, **1993**.

42 J. Verweij, E. de Vroom, Industrial transformations of penicillins and cephalosporins, *Rec. Trav. Chim. Pays-Bas* **1993**, *112 (2)*, 66–81.

43 J. Crosby, Synthesis of optically active compounds: a large scale perspective, *Tetrahedron* **1991**, *47*, 4789–4846.

44 M. A. Bertola, H. S. Koger, G. T. Phillips, A. F. Marx, V. P. Claassen, *A process for the preparation of R and S-2,2-R1,R2–1,3-dioxolane-4-methanol*, Gist Brocades NV and Shell Int. Research, EU 244 912, **1987**.

45 G. Hirth, W. Walther, Synthesis of the (R)- and (S)-glycerol acetonides. Determination of the optical purity, *Helv. Chim. Acta* **1985**, *68*, 1863.

46 H. G. Kulla, Enzymatic hydroxylations in industrial application, *Chimia* **1991**, *45*, 81–85.

47 P. Lehky, H. Kulla, S. Mischler, *Verfahren zur Herstellung von 6-Hydroxynikotinsäure*, Lonza AG, EP 015 2948 A2, **1995**.

48 M. Petersen, A. Kiener, Biocatalysis – Preparation and functionalization of N-heterocycles, *Green Chem.* **1999**, *2*, 99–106.

49 U. Onken, E. Schmidt, T. Weissenrieder, Enzymatic H_2O_2 decomposition in a three-phase suspension, Ciba-Geigy; *International conference on biotechnology for industrial production of fine chemicals*, 93rd event of the EFB; Zermatt, Schweiz, **29. 09. 1996**.

50 J. L. López, S. L. Matson, Liquid-liquid extractive membrane reactors, *Bioproc. Technol.* **1991**, *11*, 27–66.

51 S. L. Matson, *Method and apparatus for catalyst containment in multiphase membrane reactor systems*, PCT WO 87/02381, PCT US 86/02089, **1987**.

52 H. Matsumae, M. Furui, T. Shibatani, T. Tosa, Production of optically active 3-phenylglycidic acid ester by the lipase from *Serratia marcescens* in a hollow-fiber membrane reactor, *J. Ferment. Bioeng.* **1994**, *78*, 59–63.

53 H. Matsumae, T. Shibatani, Purification and characterization of the lipase from *Serratia marcescens* Sr41 8000 responsible for asymmetric hydrolysis of 3-phenylglycidic acid esters, *J. Ferment. Bioeng.* **1994**, *77*, 152–158.

54 T. Shibatani, K. Omori, H. Akatsuka, E. Kawai, H. Matsumae, Enzymatic resolution of diltiazem intermediate by *Serratia marcescens* lipase: molecular mechanism of lipase secretion and its industrial application *J. Mol. Cat. B.: Enz.* **2000**, *10*, 141–149.

55 B. Wirz, T. Weisbrod, H. Estermann, Enzymatic reactions in process research – The

importance of parameter optimization and workup, *Chimica Oggi.* **1995**, *14*, 37–41.

56 S. Doswald, H. Estermann, E. Kupfer, H. Stadler, W. Walther, T. Weisbrod, B. Wirz, W. Wostl, Large scale preparation of chiral building blocks for the P3 site of renin inhibitors, *Bioorg. Med. Chem.* **1994**, *2*, 403–410.

57 S. Shimizu, J. Ogawa, M. Kataoka, M. Kobayashi, *Screening of Novel Microbial Enzymes for the Production of Biologically and Chemically Useful Compounds*, in T. Schepper (ed.) *New Enzymes for Organic Synthesis*, Springer, New York, **1997**, 45–88.

58 J. Holm, I. Bjoerck, S. Ostrowska, A. C. Eliasson, N. G. Asp, K. Larsson, I. Lundquist, Digestibility of amylose-lipid complexes in vitro and in-vivo, *Stärke* **1983**, *35*, 294–297.

59 K. Kainuma, Applied glycoscience-past, present and future, *Food Ingredients J. Jpn.* **1998**, *178*, 4–10.

60 J. J. M. Labout, Conversion of liquefied starch into glucose using a novel glucoamylase system, *Stärke* **1985**, *37*, 157–161.

61 K. Kainuma, Applied glycoscience-past, present and future, *Foods Ingredients J. Jpn.* **1998**, *178*, 4–10.

62 J. Holm, I. Bjoerck, A. C. Eliasson, Digestibility of amylose-lipid complexes in vitro and in-vivo, *Prog. Biotechnol.* **1985**, *1*, 89–92.

63 F. Balkenhohl, B. Hauer, W. Lander, U. Schnell, U. Pressler, H. R. Staudemaier, *Lipase katalysierte Aylierung von Alkoholen mit Diketenen*, BASF AG, DE 432 9293, **1995**.

64 F. Balkenhohl, K. Ditrich, B. Hauer, W. Lander, Optisch aktive Amine durch Lipasekatalysierte Methoxyacetylierung, *J. prakt. Chem.* **1997**, *339*, 381–384.

65 M. T. Reetz, K. Schimossek, Lipase-catalyzed dynamic kinetic resolution of chiral amines: use of palladium as the racemization catalyst, *Chimia* **1996**, *50*, 668.

66 W. Aretz, Hoechst Marion Roussel, personal information **1998**.

67 C. Christ, *Biochemical Production of 7-Aminocephalosporanic Acid*, in: *Ullmann's Encyclopedia of Industrial Chemistry*, Vol. B8 (H.-J. Arpe, ed.), VCH Verlagsgesellschaft, Weinheim, **1995**, 240–241.

68 J. Verweij, E. de Vroom, Industrial transformations of penicillins and cephalosporins, *Rec. Trav. Chim. Pays-Bas* **1993**, *112 (2)*, 66–81.

69 K. Matsumoto, Production of 6-APA, 7-ACA, and 7-ADCA by immobilized penicillin and cephalosporin amidases, in A. Tanaka, T. Tosa, T. Kobayashi (eds.), *Industrial Application of Immobilized Biocatalysts*, Marcel Dekker, New York, **1993**, 67–88.

70 D. Krämer, C. Boller, personal information, **1998**.

71 J. Tramper, Chemical versus biochemical conversion: when and how to use biocatalysts, *Biotechnol. Bioeng.* **1996**, *52*, 290–295.

72 J. Kamphuis, E. M. Meijer, W. H. J. Boesten, T. Sonke, W. J. J. van den Tweel, H. E. Schoemaker, New developments in the synthesis of natural and unnatural amino acids, in: *Enzyme Engineering XI*, Vol. 672 (D. S. Clark, D. A. Estell, eds.), Ann. N. Y. Acad. Sci. **1992**, 510-527.

73 H. E. Schoemaker, W. H. J. Boesten, B. Kaptein, H. F. M. Hermes, T. Sonke, Q. B. Broxterman, W. J. J. van den Tweel, J. Kamphuis, Chemo-enzymatic synthesis of amino acids and derivatives, *Pure Appl. Chem.* **1992**, *64*, 1171–1175.

74 W. J. J. van den Tweel, T. J. G. M. van Dooren, P. H. de Jonge, B. Kaptein, A. L. L. Duchateau, J. Kamphuis, *Ochrobacterium anthropi* NCIMB 40 321: a new biocatalyst with broad-spectrum L-specific amidase activity, *Appl. Microbiol. Biotechnol.* **1993**, *39*, 296–300.

75 J. Kamphuis, H. F. M. Hermes, J. A. M. van Balken, H. E. Schoemaker, W. H. J. Boesten E. M. Meijer, Chemo-enzynatic of enantiomerically pure alpha-H and alpha-alkyl alpha-amino acids and derivates, in G. Lubec, Rosenthal, G. A. (eds.) *Amino acids: Chemistry, Biology, Medicine*, ESCOM Science Pupl., Leiden, **1990**, 119–125.

76 A. S. Bommarius, K. Drauz, H. Klenk, C. Wandrey, Operational stability of enzymes – acylase-catalyzed resolution of N-acetyl amino acids to enantiomerically pure L-amino acids, *Ann. N. Y. Acad. Sci.* **1992**, *672*, 126–136.

77 W. Leuchtenberger, M. Karrenbauer, U. Plöcker, Scale-up of an enzyme membrane reactor process for the manufacture of L-enantiomeric compounds, Enzyme Engineering 7, *Ann. N. Y. Acad. Sci.* **1984**, *434*, 78.

78 C. Wandrey, E. Flaschel, Process development and economic aspects in Enzyme Engineering. Acylase L-methionine system. In T. K. Ghose, A. Fiechter, N. Blakebrough (eds.), *Advances in Biochemical Engineering 12*, Springer-Verlag, Berlin, **1979**, 147-218.

79 C. Wandrey, R. Wichmann, W. Leuchtenberger, M. R. Kula, *Process for the continuous enzymatic change of water soluble α-ketocarboxylic acids into the corresponding amino acids*, Degussa AG, US 4 304 858, **1981**.

80 M. Beller, M. Eckert, W. Moradi, First amidocarbonylation with nitriles for the synthesis of N-aacyl amino acids, *Synlett*, **1999**, 108.

81 M. Ikemi, Industrial chemicals: enzymic transformation by recombinant microbes, *Bioprocess Technology* **1994**, 19, 797–813.

82 G. Schmidt-Kastner, P. Egerer, Amino acids and peptides. In: *Biotechnology, Vol. 6a* (Kieslich, K., ed.) Verlag Chemie, Weinheim, **1984**,387–419.

83 J. Crosby, Synthesis of optically active compounds: a large scale perspective, *Tetrahedron* **1991**, 47, 4789–4846.

84 B. Atkinson, F. Mavituna, *Biochemical Engineering and Biotechnology Handbook*, Stockton Press, New York, **1991**

85 G. Schmidt-Kastner, P. Egerer, Amino acids and peptides. in Kieslich, K. (ed.) *Biotechnology*, Vol. 6a, Verlag Chemie, Weinheim, **1984**, pp. 387–419.

86 A. Bruggink, Biocatalysis and process integration in the synthesis of semi-synthetic antibiotics, *CHIMIA* **1996**, 50, 431–432.

87 A. Bruggink, E. C. Roos, E. de Vroom, Penicillin acylase in the industrial production of β-lactam antibiotics, *Org. Proc. Res. Dev.* **1998**, 2, 128–133.

88 K. Clausen, *Method for the prepartion of certain β-lactam antibiotics*, Gist-Brocardes N. V., US 5,470,717, **1995**.

89 O. Hernandez-Justiz, R. Fernandez-Lafuente, J. M. Terrini, Guisan, Use of aqueous two-phase systems for in situ extraction of water soluble antibiotics during their synthesis by enzymes immobilized on porous supports, *Biotech. Bioeng.* **1998**, 59, 1, 73–79.

90 A. Zaks, D. R. Dodds, Application of biocatalysis and biotransformations to the synthesis of pharmaceuticals, *Drug Discovery Today* **1997**, 2 (12), 513–530.

91 M. J. Zmijewski, B. S. Briggs, A. R. Thompson, I. G. Wright, Enantioselective acylation

of a beta-lactam intermediate in the synthesis of Loracarbef using penicillin G amidase, *Tetrahedron Lett.* **1991**, 32 (13), 1621–1622.

92 K. Oyama, The industrial production of aspartame. in A. N. Collins, G. N. Sheldrake, J. Crosby (eds.), *Chirality in Industry*, John Wiley & Sons Ltd., New York, **1992**, 237–247.

93 T. Harada, Y. Shinnanyo-shi, S. Irino, Y. Kunisawa, K. Oyama, *Improved enzymatic coupling reaction of N-protected-L-aspartic acid and phenylalanine methyl ester*, Holland Sweetener Company, The Netherlands, EP 0768384, **1996**.

94 A. Kiener, *Mikrobiologisches Verfahren zur Herstellung von 5-Hydroxy-2-pyrazincarbonsäure*, Lonza AG, EP 578137 A1, **1994**.

95 A. Kiener, J.-P. Roduit, A. Tschech, A. Tinschert, K. Heinzmann, Regiospecific enzymatic hydroxylations of pyrazinecarboxylic acid and a practical synthesis of 5-chloropyrazine-2-carboxylic acid, *Synlett* **1994**, 10, 814–816.

96 M. Wieser, K. Heinzmann, A. Kiener, Bioconversion of 2-cyanopyrazine to 5-hydroxypyrazine-2-carbolic acid with *Agrobacterium sp.* DSM 6336, *Appl. Microbiol. Biotechnol.* **1997**, 48, 174–180.

97 M. Kitamura, T. Ohkuma, T. Takaya, R. Noyori, A practical asymmetric synthesis of carnitine, *Tetrahedron Lett.* **1988**, 29, 1555–1556.

98 J. Macy, H. Kulla, G. Gottschalk, H_2-dependent anaerobic growth of *Escherichia coli* on L-malate: succinate formation, *J. Bacteriol.* **1976**, 125, 423–428.

99 Th. P. Zimmermann, K. T. Robins, J. Werlen, F. W. Hoeks, Bio-transformation in the production of L-carnitine, in Collins, A. N., Sheldrake, G. N., Crosby, J. (eds.), *Chirality in Industry*, John Wiley and Sons Ltd, New York, **1997**, 287–305.

100 T. Tsuchida, Y. Nishimoto, T. Kotani, K. Iiizumi, *Production of L-3,4-dihydroxyphenylalanine*, Ajinimoto Co., Ltd., JP 5123177A, **1993**.

101 D. J. Ager, *Handbook of Chiral Chemicals*, Marcel Dekker, New York, **1999**.

102 A. Yamamoto, K. Yokozeki, K. Kubota, *Production of aromatic amino acids*, Ajinomoto Co., Ltd., JP 1010995A, **1989**.

103 H. Yamada, Screening of novel enzymes for the production of useful compounds, in K.

Kieslich, C. P. van der Beek, J. A. M. de Bont, W. J. J. van den Tweel (eds.), *New Frontiers in Screening for Microbial Biocatalysis, Studies in Organic Chemistry 53*, Elsevier, Amsterdam, **1998**, 13–17.

104 B. Cornils, W. A. Herrmann, R. Schlögl, C.-H. Wong, *Catalysis from A to Z*, Wiley-VCH, Weinheim, **2000**.

105 W. Knowles, M. Sabacky, B. Vineyard, D. Weinkauff, Asymmetric hydrogenation with a complex of rhodium and a chiral bisphosphine, *J. Am. Chem. Soc.* **1975**, *97*, 2567–2568.

106 E. C. Hann, A. Eisenberg, S. K. Fager, N. E. Perkins, F. G. Gallagher, S. M. Cooper, J. E. Gavagan, B. Stieglitz, S. M. Hennesey, R. DiCosimo, 5-Cyanovaleramide production using immobilized *Pseudomonas chlororaphis* B23, *Bioorg. Med. Chem.* **1999**, *7*, 2239–2245.

107 H. Yamada, K. Ryuno, T. Nagasawa, K. Enomoto, I. Watanabe, Optimum culture conditions for production by Pseudomonas chlororaphis B23 of nitrile hydratase, *Agric. Biol. Chem.* **1986**, *50*, 2859–2865.

108 T. Nagasawa, H. Shimizu, H. Yamada, The superiority of the third-generation catalyst, *Rhodococcus rhodochrous* J1 nitrile hydratase, for industrial production of acrylamide, *Appl. Microb. Biotechnol.* **1993**, *40*, 189–195.

109 H. Shimizu, C. Fujita, T. Endo, I. Watanabe, *Process for preparing glycine from glycinonitrile*, Nitto Chemical Industry Co., Ltd., US 5238827, **1993**.

110 H. Shimizu, J. Ogawa, M. Kataoka, M. Kobayashi, Screening of novel microbial enzymes for the production of biologically and chemically usesful commpounds, in T. K. Ghose, A. Fiechter, N. Blakebrough, N. (eds.), *New Enzymes for Organic Synthesis; Adv. Biochem. Eng. Biotechnol.* **1997**, *58*, 56–59.

111 H. Yamada, M. Kobayashi, Nitrile hydratase and its application to industrial production of acrylamide, *Biosci. Biotech. Biochem.* **1996**, *60 (9)*, 1391–1400.

112 H. Yamada, Y. Tani, *Process for biological preparation of amides*, Nitto Chemical Industry Co., Ltd., US 4637982, **1987**.

113 J. Heveling, Catalysis at Lonza: From metallic glasses to fine chemicals, *Chimia* **1996**, *50*, 114–118.

114 U. Kragl, D. Gygax, O. Ghisalba, C. Wandrey, Enzymatic process for preparing N-acetylneuraminic acid, *Angew. Chem. Int. Ed. Engl.* **1991**, *30*, 827–828.

115 I. Maru, J. Ohnishi, Y. Ohta, Y. Tsukada, Simple and large-scale production of N-acetylneuraminic acid from N-acetyl-D-glucosamine and pyruvate using N-acyl-D-glucosamine 2-epimerase and N-acetylneuraminate lyase, *Carb. Res.* **1998**, *306*, 575–578.

116 I. Maru, Y. Ohta, K. Murata, Y. Tsukada, Molecular cloning and indentification of N-acyl-D-glucosamine 2-epimerase from procine kidney as a renin-binding protein, *Biol. Chem.* **1996**, *271*, 16294–16299.

117 J. D. Rozzell, Commercial scale biocatalysis: Myths and realities, *Bioorg. Med. Chem.* **1999**, *7*, 2253–2261.

118 F. H. Arnold, Combinatorial and computational challenges for biocatalyst design, *Nature*, **2001**, *409*, 253–257.

119 U. Bornscheuer, M. Pohl, Improved biocatalysts by directed evolution and rational-protein design, *Curr. Opin. Chem. Biol.*, **2000**, *5*, 137–142

120 M. Chartrain, P. Salmon, D. Robinson, B. Buckland, Metabolic engineering and directed evolution for the production of pharmaceuticals, *Curr. Opin. Biotechnol.* **2000**, *11*, 209–214.

20
Tabular Survey of Commercially Available Enzymes

Peter Rasor

Enzymes are catalysts. Nature has designed them to perform specific tasks necessary for the survival of the organism producing the enzyme. The organic chemist tends to name enzymes "biocatalysts" which means nothing more than catalysts of *bio*logical origin.

These biocatalysts bring some confusion to the well-structured world of organic chemistry:

- the names are unfamiliar,
- each enzyme has a variety of names which are all used making it as difficult as distinguishing characters in a Russian novel (e. g. Penicillin G-amidase and Penicillin acylase),
- when it comes to microoogranisms or plants, the origin of enzymes is described in Latin (type face italic),
- in order to add to the confusion, the names of microorganisms may change over time, for example *Pseudomonas cepacia* is now *Burkholderia cepacia*, *Candida cylindracea* is *Candida rugosa*,
- even mammalian sources can be described differently – esterase from *hog* liver or *pig* liver, but lipase from *porcine* pancreas (type face not italic).

For identifying synonyms or finding out the correct name of an enzyme, the Enzyme Nomenclature Database (EC database) can be searched or downloaded under http://www.expasy.ch/enzyme/.

If the chemist is still not confused and has mastered this hurdle, the manufacturers or suppliers introduce brand names for marketing reasons, and may even change names once in a while. Additionally, not every supplier gives full information on the origin of the biocatalyst and may use old names of microorganisms while other suppliers already use new names.

Furthermore, the same biocatalyst by description may behave differently in a specific reaction: for example, lipase from *Candida rugosa* from Amano (Lipase AY) differs from Lipase MY or OF from Meito Sangyo with respect to activity and stereoselectivity because it consists of a number of catalytically active species which differ depending on the production strain used and thus, on the manufacturer.

Table 20-1. Abbreviation of most commonly used biocatalysts

Abbreviation	Lipase from	Abbreviation	Lipase from
ANL	*Aspergillus niger*	PcamL	*Penicillium camembertii*
BCL (PCL)	*Burkholderia cepacia* (formerly *Pseudomonas cepacia*)	PFL	*Pseudomonas fluorescens*
CAL	*Candida antarctica*	PfragiL	*Pseudomonas fragi*
CAL-A	*Candida antarctica*, type A	PPL	*Porcine pancreas*
CAL-B	*Candida antarctica*, type B	ProqL	*Penicillium roquefortii*
CLL	*Candida lipolytics*	PSL	*Pseudomonas* sp.
CRL (CCL)	*Candida rugosa* (formerly *C. cylindracea*)	RML (MML)	*Rhizomucor miehei* (formerly *Mucor miehei*)
CVL	*Chromobacterium viscosum* (identical to *Pseudomonas glumae*)	ROL	Rhizopus oryzae (other names: RNL – *Rhizopus niveus*, RDL – *Rhizopus delemar*, RJL – *Rhizopus javanicus*)
GCL	*Geotrichum candidum*	TLL (HLL)	*Thermomyces lanuginosa* (formerly *Humicola lanuginosa*)

Abbreviation	Esterase from	Abbreviation	Alcohol dehydrogenase from
PLE	Pig liver	YADH	Yeast
		TBADH	*Thermoanaerobium brockii*
		HLADH	Horse liver

Since the full enzyme name according to the EC nomenclature is rather long, the most commonly used enzymes have gotten abbreviations. For esterases and lipases there are certain rules: in most cases, the first (two or three) letters characterize the source, the last the type of enzyme (E for esterase, L for lipases) (see Table 20-1). Alcohol dehydrogenases are treated similarly.

All this may explain why many publications give only incomplete information on the exact type of enzyme used in the work described and why many references to enzymes are simply wrong. The author strongly recommends to provide at least the following information:

Parameter	Example 1	Example 2
Name of the product	Lipase Type XIII	CHIRAZYME L-2, lyo
Description (if the name of the product is a brand name or non-descriptive)	Lipase from *Pseudomonas* sp.	Lipase from *Candida antarctica*, type B
Formulation	Powder	Powder
Manufacturer	Sigma	Roche Diagnostics

Table 20-2. Available screening sets

Enzyme type	Company
Alcohol dehydrogenases	ThermoGen, BioCatalytics
Esterases & lipases	Altus, Fluka, Roche, ThermoGen
Nitrilases	BioCatalytics
Proteases	Altus
Transaminases (aminotransferases)	BioCatalytics

In the laboratory protocol, lot. no. and activity (incl. assay no. or assay conditions) must be recorded as well in order to track variation in results because of lot to lot inconsistency.

Every development of a biocatalytic reaction starts with a screening for the most appropriate enzyme. Some companies offer screening sets (or kits) containing the most commonly used enzymes (Table 20-2). Some Sets are single use (Altus, ThermoGen) while others contain enough material to perform depending on the scale 5–20 experiments (BioCatalytics, Fluka, Roche). These sets may include enzymes available on industrial scale or on research scale only.

The following companies offer screening set/kits for quick enzyme selection (Table 20-2). While some companies include only industrial scale enzymes, others contain enzymes only available at lab quantities. Diversa Co. offers an enzyme subscription program for lipases, esterases, nitrilases, cellulases, glycosidases, phosphatases, and transaminases (aminotransferases).

Some enzymes of Novozymes A/S (formerly Novo Nordisk A/S) were widely distributed on an experimental stage (SP nnn). Table 20-3 lists the most important

Table 20-3. List of experimental enzymes by Novozymes, current products names and suppliers

Old name	Characterization	Current brand name	Availability
SP 361	Immobilized enzyme mixture		discontinued
SP 409	from Rhodococcus sp. containing nitrilase, nitril hydratase, esterase, epoxide hydrolase and amidase activity		
SP 382	Immobilized lipase from Candida antarctica, containing type A & B		discontinued
SP 435	Immobilized lipase from Candida antarctica, type B, rec. in Aspergillus oryzae	Novozym 435 CHIRAZYME L-2, Carrier 2	Novo-Nordisk Roche Diagnostics
SP 523	Lipase powder from Thermomyces lanuginosus (formerly Humicola lanuginosa)	CHIRAZYME L-8, lyo	Roche Diagnostics
SP 524	Lipase powder from Rhizomucor miehei, rec. in Aspergillus oryzae	CHIRAZYME L-9, lyo	Roche Diagnostics
SP 525	Lipase powder from Candida antarctica, type B, rec. in Aspergillus oryzae	CHIRAZYME L-2, lyo	Roche Diagnostics
SP 526	Lipase from Candida antarctica, type A, rec. in Aspergillus oryzae	CHIRAZYME L-5, lyo	Roche Diagnostics

Table 20-4. Enzyme producers/suppliers and brief characterization

Company	Adress	Tel./Fax/Email/WWW	Focus/Characterization[1]
Altus	Altus Biologics Inc. 625 Putnam Avenue Cambridge, MA 02139–4807 USA	Tel.: +1 (617) 299-2900 Fax: +1 (617) 299–2999 Email: info@altus.com http://www.altus.com	Manufacturer of stabilized enzymes for use in industrial, biocatalytical, diagnostic, and medicinal applications. No enzyme production itself. Biocatalytical process development.
Amano	Amano Pharmaceutical Co., Ltd. 2–7, 1-chome, Nishiki Naka-ku, Nagoya, 460–8630 Japan	Tel.: +81 (52) 211-3032 Fax: +81 (52) 211–3054 http://www.amano-enzyme.co.jp	Specialty enzyme producer for industrial, biocatalytical, diagnostic, and medicinal applications.
Asahi Chemical Industry Co.	Diagnostics Division Hibiya-Mitsui Building 1–2 Yurakucho 1-chome, Chiyoda-ku Tokyo 100–8440 Japan	Tel.: +81 (3) 3259-5776 Fax: +81 (3) 3259–5741 Email: shindan@ml.asahi-kasei.co.jp http://www.asahi-kasei.co.jp	Speciality enzyme producer for diagnostic and medicinal applications.
Biocatalysts	Biocatalysts Ltd Main Avenue, Treforest Industrial Estate Pontypridd, Wales, CF37 5UD United Kingdom	Tel.: +44 (0) 1443843712 Fax: +44 (0) 1443841214 Email: sales@biocats.com http://www.biocatalysts.com	Producer and distributor of enzymes for use in industrial and diagnostic applications.
BioCatalytics	BioCatalytics Inc. 39 Congress Street, Suite 303 Pasadena, CA 91105–3022 USA	Tel.: +1 (626) 229–0588 Fax: +1 (626) 535–9465 Email: info@biocatalytics.com http://www.biocatalytics.com	Biocatalytical process development. Experimental enzymes for biocatalysis. Limited production capacity. Distributor for Roche in USA and Canada.
Biozyme Laboratories International Ltd.	USA and Canada: 9939 Hibert Street Suite 101 San Diego, CA 92131–1029 USA	Tel.: +1 (858) 549-4484 or (800) 423-8199 Fax: (858) 549–0138 Email: bioinfo@biozyme.com	Speciality enzyme producer for diagnostic and medicinal applications
	All other countries: Biozyme Laboratories Ltd. Unit 6, Gilchrist Thomas Estate Blaenavon, South Wales, NP4 9RL United Kingdom	Tel.: (+44) 1495790678 Fax: (+44) 1495791780 Email: info@biozyme.co.uk http://www.biozyme.com/	
Calbiochem. Co., CN Biosciences	Calbiochem-Novabiochem Corporation 10394 Pacific Center Court San Diego, CA 92121 Mailing Address: P.O. Box 12087 La Jolla, CA 92039–2087 USA	Tel.: +1 (858) 4509600 or (800) 8543417 Fax: +1 (858) 4533552 Email: orders@calbiochem.com. technical@calbiochem.com http://www.calbiochem.com http://www.cnbi.com	Supplier of enzymes and biochemicals on research scale. Focus on life science, not biocatalysis.

Table 20-4. (cont.).

Company	Adress	Tel./Fax/Email/WWW	Focus/Characterization[1]
Diversa	Diversa Corporation 4955 Directors Place San Diego, CA 92121–1609 USA	Tel.: +1 (858) 526–5000 Fax: +1 (858) 526–5551 Email: information@diversa.com http://www.diversa.com	Discovery and development of industrial enzymes. No general biocatalyst portfolio.
DSM Gist-Brocades	DSM Food Specialties P.O. Box 1 2600 MA Delft The Netherlands	Tel.: +31 (15) 279 3474 Fax: +31 (15) 279 3540 http://www.dsm.nl/dfs/	Enzyme producer for industrial applications (feed & food).
Fluka	see Sigma-Aldrich Fluka		
Genencor	Genencor International, Inc. 200 Meridian Centre Blvd. Rochester, NY 14618–3916 USA	Tel.: +1 (716) 256-5200 Fax: +1 (716) 256–6952 Email: ysmith@genencor.com http://www.genencor.com	Enzyme producer for industrial applications.
Jülich Enzyme Products	Juelich Enzyme Products GmbH Karl-Heinz-Beckurts-Str. 13 D-52428 Jülich Germany	Tel.: +49 (2461) 348188 Fax: +49 (2461) 348186 E-mail: juelichep@aol.com http://www.juelich-enzyme.com	Experimental enzymes for biocatalysis. Limited enzyme production capacity.
Lee Scientific	Lee Scientific Inc. 2924 Mary Ave. St. Louis, MO 63144 USA	Tel.: +1 (314) 968-1091 Fax: +1 (314) 968–9851 Email: burtonlee@leescientific.com http://www.leescientific.com/	Specialty enzyme producer. Focus on life science and diagnostics. Some bio-catalysts.
Meito Sangyo Co. Ltd.	Fine Chemicals Dept. Meito Sangyo Co. Ltd. Sankeido Bldg., 4–3-15, Muromachi, Nihonbashi Chuo-ku, Tokyo 103–0022 Japan	Tel.: +81 (3) 3242-1795 Fax: +81 (3) 3242–1792 Email: jdt02625@nifty.ne.jp	Producer and distributor of enzymes for use in industrial and diagnostic applications.
Novozyme A/S	Europe, Middle East & Africa: Novozymes France S.A. Immeuble Challenge 92 79, Avenue Frantois Arago 92017 Nanterre Cedex, France	http://www.novozymes.com Tel.: +33 146 140746 Fax: +33 146 140766	Largest enzyme producer for industrial applications. Distribution agreement with Roche for chiral organic synthesis market.
	Latin America: Novozymes Latin America Limited Rua professor Francisco Ribeiro 683 CEP 83707–660 – Araucaria – Parana Brazil	Tel.: +55 416411000 Fax: +55 416431443	

Table 20-4.   (cont.).

Company	Adress	Tel./Fax/Email/WWW	Focus/Characterization[1]
	USA: Novozymes North America Inc. 77 Perry Chapel Church Road Franklinton, N.C. 27525 Postal Address: State Road 1003 P.O. BOX 576 Franklinton, NC 27525	Tel.: +1 91 94 94 30 00 Fax: +1 91 94 94 34 50	
	Asia Pacific, Hong Kong: Novozymes Asia Pacific Regional Office 7/F Chinachem Century Tower 178 Gloucester Road, Wanchai	Tel.: +852 25 19 33 80 Fax: +852 28 77 06 59	
Recordati S.p.A.	Via Matteo Civitali, 1 20148 Milan Italy	Tel.: +39 (02) 48 78 71 http://www.recordati.it	Manufacturer of industrial enzymes for beta-lactam antibiotics.
Roche Diagnostics	Roche Diagnostics GmbH Roche Molecular Bio-chemicals Sandhofer Str. 116 68298 Mannheim Germany	Tel.: +49 (621) 759 85 93 Fax: +49 (621) 759 89 86 Email: ute.hill@roche.com http://indbio.roche.com	Speciality enzyme producer for industrial, biocatalytical, diagnostic, and medicinal applications. Broad range of enzymes.
	USA & Canada: Refer to BioCatalytics Inc.		
Seravac	Seravac USA, Inc. 13220 Evening Creek Drive San Diego, CA 92128 USA	Tel.: +1 (858) 679–40 50 or (800) 679–40 50 Fax: (858) 679–14 38 Email: enzymes@seravac.com http://www.seravac.com	Speciality enzyme producer for diagnostic and medicinal applications.
Sigma-Aldrich Fluka (SAF)	Sigma Co. 3050 Spruce Street St. Louis, MO 63103 Mail: P.O. Box 14508 St. Louis, MO 63178 USA	Tel.: (314) 771–57 65 Fax: (314) 771–57 57 Email: sigma@sial.com http://www.sigma-aldrich.com	Manufacturer and distributor of enzymes and bio-chemicals on research scale. Very broad range of enzymes (Sigma and Fluka). Limited range of biocatalysts at Aldrich. Within the group, Fluka has the focus on biocatalysts on research scale. Production of selected enzymes up to medium scale.
	Fluka Chemical LLC. Industriestrasse 25 CH-9471 Buchs Mail: P.O. Box 260 CH-9471 Buchs Switzerland	Tel.: +41 (81) 755 28 28 Fax: +41 (81) 756 54 49 EMail: fluka@sial.com http://www.sigma-aldrich.com	
ThermoGen	ThermoGen, Inc. 2501 Davey Road Woolridge, IL 60517 USA	Tel.: +1 (630) 783-46 00 Fax: +1 (630) 783–49 09 info@thermogen.com http://www.thermogen.com	Enzyme discovery. Limited enzyme production capacity. Biocatalytical process development.

Table 20-4. (cont.).

Company	Adress	Tel./Fax/Email/WWW	Focus/Characterization[1]
Toyobo Co. Ltd.	Toyobo Co. Ltd. Biochemical Operations Department 17–9 Nihonbashi Koami-cho Chuo-ku Tokyo 103–8530 Japan	Tel.: +81 (3) 3660–4819 Fax: +81 (3) 3660–4951 EMail: toshiro–kikuchi@bio.toyobo.co.jp http://www.toyobo.co.jp/e/	Speciality enzyme producer for diagnostic and medicinal applications.
Unitica Ltd.	Medical Products Division Unitika Ltd. 4–1-3, Kyutaro-machi, Chuo-ku, Osaka 541–8566 Japan	Tel.: +81(6) 6281–5021 Fax: +81 (6) 6281–5256 Email : medical@unitika.co.jp http://www.unitika.co.jp/ home–e.htm	Specialty enzyme producer for diagnostic and medicinal applications.
Wako Pure Chemicals Industries, Ltd.	1–2, Doshomachi 3-Chome, Chuo-Ku, Osaka 540–8605 Japan	Tel.: +81 (6) 6203-3741 Fax: +81 (6) 6222–1203 http://search.wako-chem.co.jp	Manufacturer and distributor of enzymes and biochemicals on research scale. Focus on life science, not biocatalysis.
Worthington Biochemical	Worthington Biochemical Corp. 730 Vassar Ave Lakewood, NJ 08701	Tel.: +1 (732) 942–1660 Fax: +1 (732) 942–9270 http://www.worthington-biochem.com/	Manufacturer and distributor of enzymes and biochemicals. Focus on life science and diagnostics.

1 Industrial applications include detergents, feed and food, pulp & paper, etc.

enzymes and gives the current brand names, wherever possible. Some enzymes have been discontinued at Novozymes but replacements are available from Roche Diagnostics (CHIRAZYME product line).

Catalytic antibodies are not yet widely available. Aldrich is offering two aldolase monoclonal antibodies.

The major enzyme producers and/or suppliers are listed and briefly characterized in Table 20-4. The author is aware that the list of enzyme producers is not complete.

The author has made the attempt to list enzymes that are commercially available (Table 20-5) and thus can be used in biocatalysis. He knows that the list is incomplete and therefore, the reader should not rely solely on this list but rather check the suppliers listed in Table 20-4. Enzyme manufacturers also update their product portfolio continuously, so this list probably needs updating before the book is even in print.

A special word is necessary with respect to the Sigma-Aldrich-Fluka conglomerate: Fluka has taken the lead in biocatalysis, while Sigma serves mostly the life science market. Especially since the Sigma catalog is a book in itself, only enzymes from Fluka are listed. The reader should be aware that the majority of enzymes is available from Sigma as well, and with respect to enzymes not typically used in biocatalysis, the portfolio may be even greater.

Explanations to Table 20-5:
The table is sorted by the **EC number.** In most cases the number is given in the

respective chapter and can be used to find the enzyme in the table. If the EC no. is not known, at least the general reaction of the enzyme class is given according to the EC nomenclature.

Underneath the *EC name, synonyms* are given. The **general reaction** according EC nomenclature is denoted too. Afterwards, the **product (enzyme) names** are listed, one entry for each **manufacturer** per enzyme. If the product is sold under a **brand name**, this name is listed too. In one enzyme class, the entries are sorted by *origin*.

The *availability* is characterized in three categories: lab, pilot and industrial scale. It refers to the scale with respect to biocatalytical reactions. The author recognized that this categorization is somewhat arbitrary and in some cases may not be correct because the actual production scale is not generally known. Hopefully, though, it will prove to be useful as a rough guide.

Enzyme producers are devoted to certain markets like food & feed, detergents, diagnostics or research. Large enzyme producers such as Novozymes, Genencor or DSM Gist-brocades are categorized as "industrial", specialty enzyme producers like Amano, Asahi or Roche Diagnostics serve various markets and thus, scale varies from pilot to industrial. Since the enzyme demand for diagnostics is much lower than for biocatalysis, typical diagnostic enzymes are labeled as "pilot" although the manufacturing process is certainly standardized and therefore, could be call "industrial" as well. Companies serving the life sciences market (e.g. Sigma-Aldrich Fluka, Roche Diagnostics) have manufacturing capacities from small ("lab") scale to medium scale (here termed as "pilot"). It should also be recognized that the term "pilot scale" in a context other than this table has a different meaning when comparing for example Sigma, Roche Diagnostics, and Novozymes.

Table 20.5. Commercially available enzymes.

Oxidoreductases. Acting on the CH-OH group of donors. With NAD(+) or NADP(+) as acceptor.	1.1.1.-
Alcohol Dehydrogenase Screening Kit; Origin: microorganism, rec. in E. coli ThemoGen: ThermoCat Alcohol Dehydrogenase Kits	Lab
Ketoreductase, broad-range; Origin: microorganism, rec. in E. coli BioCatalytics: KRED-1001	Lab
Ketoreductase, broad-range; Origin: microorganism, rec. in E. coli BioCatalytics: KRED-1002	Lab
Ketoreductase, broad-range; Origin: microorganism, rec. in E. coli BioCatalytics: KRED-1003	Lab
Ketoreductase, broad-range; Origin: microorganism, rec. in E. coli BioCatalytics: KRED-1004	Lab
Ketoreductase, broad-range; Origin: microorganism, rec. in E. coli BioCatalytics: KRED-1005	Lab
Ketoreductase, broad-range; Origin: microorganism, rec. in E. coli BioCatalytics: KRED-1006	Lab
Ketoreductase, broad-range; Origin: microorganism, rec. in E. coli BioCatalytics: KRED-1007	Lab
Ketoreductase, broad-range; Origin: microorganism, rec. in E. coli BioCatalytics: KRED-1008	Lab
Cholesterol Dehydrogenase; Origin: Nocardia sp. Amano: Amano 5 [CHDH-5]	Pilot
7-Hydroxysteroid Dehydrogenase; Origin: Pseudomonas sp. Asahi	Pilot

Alcohol dehydrogenase. Aldehyde reductase.	1.1.1.1 An alcohol + NAD(+) = an aldehyde or ketone + NADH.
Alcohol Dehydrogenase; Origin: Candida parapsilosis Jülich Enzyme Products	Lab
Alcohol Dehydrogenase; Origin: horse liver Fluka	Lab
Alcohol Dehydrogenase; Origin: microorganisms Biocatalysts: Sec ADH 300	Lab
Alcohol Dehydrogenase; Origin: Rhodococcus erythropolis Jülich Enzyme Products	Lab
Alcohol Dehydrogenase; Origin: yeast Biozyme	Pilot
Alcohol Dehydrogenase; Origin: yeast Fluka	Pilot
Alcohol Dehydrogenase; Origin: yeast Roche Diagnostics: Alcohol Dehydrogenase (YADH), lyo.	Pilot
Alcohol Dehydrogenase Origin: yeast Roche Diagnostics: Alcohol Dehydrogenase (YADH), susp.	Pilot
Alcohol Dehydrogenase; Origin: Zymomonas mobilis Unitika	Pilot

Alcohol dehydrogenase (NADP+). Aldehyde reductase (NADPH).	1.1.1.2 An alcohol + NADP(+) = an aldehyde + NADPH.
Alcohol Dehydrogenase; Origin: Lactobacillus kefir Fluka	Pilot

Table 20.5. (cont.).

Alcohol Dehydrogenase; Origin: Lactobacillus kefir Jülich Enzyme Products	Lab
Alcohol Dehydrogenase; Origin: Thermoanaerobium brockii Fluka	Pilot

Acetoin dehydrogenase. Diacetyl reductase.	**1.1.1.5** Acetoin + NAD(+) = diacetyl + NADH.
Acetoin Dehydrogenase; Origin: Lactobacillus kefir Fluka	Lab
Diketone Reductase; Origin: Lactobacillus kefir Jülich Enzyme Products	Lab

Glycerol dehydrogenase.	**1.1.1.6** Glycerol + NAD(+) = glycerone + NADH.
Glycerol Dehydrogenase; Origin: Bacillus megaterium Asahi	Pilot
Glycerol Dehydrogenase; Origin: Geotrichum candidum Fluka	Lab
Glycerol Dehydrogenase; Origin: Klebsiella pneumoniae (formerly Enterobacter aerogenes) Roche Diagnostics: Glycerol Dehydrogenase	Lab
Glycerol Dehydrogenase (GlDH); Origin: microorganisms Unitika	Pilot

Glycerol-3-phosphate dehydrogenase (NAD+).	**1.1.1.8** Sn-glycerol 3-phosphate + NAD(+) = glycerone phosphate + NADH.
Glycerol-3-phosphate Dehydrogenase; Origin: rabbit muscle Fluka	Lab

L-iditol 2-dehydrogenase. Polyol dehydrogenase. Sorbitol dehydrogenase.	**1.1.1.14** L-iditol + NAD(+) = L-sorbose + NADH.
Sorbitol Dehydrogenase (SorDH); Origin: microorganisms Unitika	Pilot
Sorbitol Dehydrogenase; Origin: sheep liver Fluka	Lab
Sorbitol Dehydrogenase; Origin: sheep liver Roche Diagnostics: Sorbitol Dehydrogenase (SDH)	Lab

L-lactate dehydrogenase. L-lactic acid dehydrogenase. L-lactic dehydrogenase.	**1.1.1.27** (S)-lactate + NAD(+) = pyruvate + NADH.
L-Lactate Dehydrogenase; Origin: beef heart Biozyme	Pilot
L-Lactate Dehydrogenase; Origin: bovine heart Fluka	Lab
L(+)-Lactate Dehydrogenase; Origin: hog muscle Roche Diagnostics: L(+)-Lactate Dehydrogenase (L-LDH)	Pilot
L-Lactate dehydrogenase; Origin: pig heart Biozyme	Pilot
L(+)-Lactate Dehydrogenase; Origin: pig muscle Roche Diagnostics: L(+)-Lactate Dehydrogenase (L-LDH)	Pilot
L-Lactate dehydrogenase; Origin: pig muscle Biozyme	Pilot

Table 20.5. (cont.).

L-Lactate dehydrogenase; Origin: rabbit muscle
Biozyme

Pilot

L-Lactate Dehydrogenase; Origin: rabbit muscle
Fluka

Lab

Lactate Dehydrogenase; Origin: Staphylococcus sp.
Amano: Amano 3 [LDH-3]

Pilot

D-lactate dehydrogenase.   **1.1.1.28**
D-lactic acid dehydrogenase. D-lactic dehydrogenase.   (R)-lactate + NAD(+) = pyruvate + NADH.

D-Lactate Dehydrogenase; Origin: Lactobacillus leichmanii
Fluka

Lab

D(-)-Lactate Dehydrogenase; Origin: Lactobacillus leichmannii
Roche Diagnostics: D(-)-Lactate Dehydrogenase (D-LDH)

Pilot

D-Lactate Dehydrogenase; Origin: microorganisms
Toyobo

Pilot

D-Lactate Dehydrogenase; Origin: microorganisms
Unitika: D-Lactate Dehydrogenase (D-LDH)

Pilot

3-hydroxybutyrate dehydrogenase.   **1.1.1.30**
D-beta-hydroxybutyrate dehydrogenase.   (R)-3-hydroxybutanoate + NAD(+) = acetoacetate + NADH.

3-Hydroxybutyrate Dehydrogenase
Asahi

Pilot

D-3-Hydroxybutyrate Dehydrogenase; Origin: Pseudomonas sp.
Toyobo

Pilot

3-Hydroxybutyrate Dehydrogenase; Origin: Rhodobacter sphaeroides (formerly Rhodopseudomonas sphaeroides)
Roche Diagnostics: 3-Hydroxybutyrate Dehydrogenase (3-HBDH), Grade II

Lab

3-Hydroxybutyrate Dehydrogenase; Origin: Rhodopseudomonas spheroides
Fluka

Lab

Malate dehydrogenase.   **1.1.1.37**
Malic dehydrogenase.   (S)-malate + NAD(+) = oxaloacetate + NADH.

Malate Dehydrogenase; Origin: microorganisms
Toyobo

Pilot

Malate Dehydrogenase; Origin: microorganisms
Unitika: Malate Dehydrogenase (MDH)

Pilot

Malate dehydrogenase; Origin: pig heart
Biozyme

Pilot

Malate Dehydrogenase; Origin: porcine heart
Fluka

Lab

Malate Dehydrogenase; Origin: Thermus sp.
Amano: Amano 3 [MDH-3]

Pilot

Isocitrate dehydrogenase (NADP+).   **1.1.1.42**
Oxalosuccinate decarboxylase. IDH.   Isocitrate + NADP(+) = 2-oxoglutarate + CO(2) + NADPH.

Isocitrate Dehydrogenase; Origin: porcine heart
Fluka

Lab

Isocitrate Dehydrogenase; Origin: porcine heart
Fluka

Lab

Table 20.5.   (cont.).

Phosphogluconate dehydrogenase (decarboxylating).   1.1.1.44

Phosphogluconic acid dehydrogenase. 6-phosphogluconic dehydrogenase. 6-phosphogluconic carboxylase. 6PGD.

6-phospho-D-gluconate + NADP(+) = D-ribulose 5-phosphate + CO_2 +NADPH.

6-Phosphogluconate Dehydrogenase (6PGDH); Origin: Thermoactinomces intermedius	
Unitika	Pilot
6-Phosphogluconic Dehydrogenase; Origin: Torula yeast	
Fluka	Lab
6-Phosphogluconic Dehydrogenase; Origin: yeast	
Fluka	Lab

Glucose 1-dehydrogenase.   1.1.1.47

Beta-D-glucose + NAD(P)(+) = D-glucono-1,5-lactone + NAD(P)H.

Glucose Dehydrogenase; Origin: Bacillus megaterium	
Fluka	Pilot
Glucose Dehydrogenase; Origin: Bacillus sp.	
Amano: Amano 2 [GLUCDH-2]	Pilot
Glucose Dehydrogenase; Origin: Cryptococcus uniguttulatus	
Asahi	Pilot
Glucose Dehydrogenase; Origin: microorganisms	
Toyobo	Pilot

Glucose-6-phosphate 1-dehydrogenase.   1.1.1.49

G6PD.

D-glucose 6-phosphate + NADP(+) = D-glucono-1,5-lactone 6-phosphate +NADPH.

Glucose-6-Phosphate Dehydrogenase	
Asahi	Pilot
Glucose-6-Phosphate Dehydrogenase (G6PDH); Origin: Bacillus stearothermophilus	
Unitika	Pilot
Glucose-6-phosphate Dehydrogenase; Origin: baker's yeast	
Fluka	Pilot
Glucose-6-phosphate dehydrogenase; Origin: Leuconostoc mesenteroides	
Biozyme	Pilot
Glucose-6-phosphate Dehydrogenase; Origin: Leuconostoc mesenteroides	
Fluka	Pilot
Glucose-6-Phosphate Dehydrogenase; Origin: Leuconostoc mesenteroides	
Toyobo	Pilot
Glucose-6-phosphate Dehydrogenase; Origin: Leuconostoc mesenteroides	
Roche Diagnostics: Glucose-6-phosphate Dehydrogenase (G6P-DH), susp.	Pilot
Glucose-6-phosphate Dehydrogenase; Origin: Leuconostoc mesenteroides, rec. in E. coli	
Roche Diagnostics: Glucose-6-phosphate Dehydrogenase (G6P-DH), lyo.	Pilot
Glucose-6-phosphate Dehydrogenase; Origin: Torula yeast	
Fluka	Lab
Glucose-6-phosphate dehydrogenase; Origin: yeast	
Biozyme	Pilot
Glucose-6-phosphate Dehydrogenase; Origin: yeast	
Fluka	Pilot
Glucose-6-phosphate Dehydrogenase; Origin: yeast	
Fluka	Pilot
Glucose-6-phosphate Dehydrogenase; Origin: yeast	
Roche Diagnostics: Glucose-6-phosphate Dehydrogenase (G6P-DH), lyo.	Pilot

Table 20.5. (cont.).

Glucose-6-Phosphate Dehydrogenase (G6PDH); Origin: Zymomonas mobilis

Unitika   Pilot

3-alpha-hydroxysteroid dehydrogenase (B-specific).	**1.1.1.50**
Hydroxyprostaglandin dehydrogenase. 3-alpha-HSD.	Androsterone + NAD(P)(+) = 5-alpha-androstane-3,17-dione + NAD(P)H.

3-Hydroxysteroid Dehydrogenase

Asahi   Pilot

3-alpha-Hydroxysteroid Dehydrogenase (3alphaHSDH); Origin: microorganisms

Unitika   Pilot

3-alpha-Hydroxysteroid Dehydrogenase; Origin: Pseudomonas testosteroni

Fluka   Lab

Xanthine dehydrogenase.	**1.1.1.204**
Xanthine oxidoreductase.	Xanthine + NAD(+) + H(2)O = urate + NADH.

Xanthine Dehydrogenase

Asahi   Pilot

12-alpha-Hydroxysteroid Dehydrogenase PP; Origin: Clostridium spec.

Jülich Enzyme Products   Lab

12-alpha-Hydroxysteroid Dehydrogenase; Origin: microorganisms

Asahi   Pilot

Glucose oxidase.	**1.1.3.4**
Glucose oxyhydrase. Beta-D-glucose:oxygen 1-oxido-reductase. Glucose aerodehydrogenase. D-Glucose-1-oxidase.	Beta-D-glucose + O(2) = D-glucono-1,5-lactone + H(2)O(2).

Glucose Oxidase

Seravac   Industrial

Glucose oxidase; Origin: Aspergillus niger

Amano: Hyderase   Industrial

Glucose oxidase; Origin: Aspergillus niger

Amano: Hyderase L   Industrial

Glucose oxidase; Origin: Aspergillus niger

Biozyme   Pilot

Glucose Oxidase; Origin: Aspergillus niger

Fluka   Industrial

Glucose oxidase; Origin: Aspergillus niger

Novozymes: Gluzyme®   Industrial

Glucose Oxidase; Origin: Aspergillus niger overproducer

Roche Diagnostics: Glucose Oxidase (GOD)   Industrial

Glucose Oxidase; Origin: Aspergillus sp.

Amano: Amano 2 [GO-2]   Pilot

Glucose Oxidase; Origin: Aspergillus sp.

Amano: Amano LC [GOLC]   Pilot

Glucose Oxidase; Origin: Aspergillus sp.

Amano: Amano LD2 [GOLD-2]   Pilot

Glucose Oxidase; Origin: Aspergillus sp.

Toyobo   Pilot

Glucose Oxidase; Origin: microorganism, rec. in yeast

Roche Diagnostics: Glucose Oxidase (GOD)   Pilot

Glucose Oxidase; Origin: Penicillium sp.

Biocatalysts   Industrial

Table 20.5. (cont.).

Cholesterol oxidase. | **1.1.3.6**
Cholesterol-O2 oxidoreductase. | Cholesterol + O(2) = cholest-4-en-3-one + H(2)O(2).

Cholesterol Oxidase
Asahi | Pilot

Cholesterol Oxidase; Origin: Brevibacterium sterolicum, rec. in microorganism
Roche Diagnostics: Cholesterol Oxidase | Pilot

Cholesterol Oxidase; Origin: microorganisms
Amano: Amano 6 [CHO-6] | Pilot

Cholesterol Oxidase; Origin: microorganisms
Asahi | Pilot

Cholesterol Oxidase; Origin: microorganisms
Toyobo | Pilot

Cholesterol Oxidase; Origin: Nocardia erythropolis
Fluka | Pilot

Cholesterol Oxidase; Origin: Pseudomonas sp.
Amano: Amano 1 [CHO-1] | Pilot

Cholesterol Oxidase; Origin: Pseudomonas sp.
Amano: Amano 2 [CHO-2] | Pilot

Cholesterol Oxidase; Origin: Pseudomonas sp.
Fluka | Pilot

Cholesterol Oxidase; Origin: Streptomyces cinnamomeus
Asahi | Pilot

Galactose oxidase. | **1.1.3.9**
Beta-Galactose oxidase. | D-galactose + O(2) = D-galacto-hexodialose + H(2)O(2).

Galactose Dehydrogenase; Origin: Agrobacterium sp.
Biocatalysts | Pilot

Alcohol oxidase. | **1.1.3.13**
Methanol oxidase. AOX. | A primary alcohol + O(2) = an aldehyde + H(2)O(2).

Alcohol Oxidase; Origin: Candida sp.
Asahi | Pilot

Alcohol oxidase, broad-range; Origin: microorganism, rec. in E. coli
BioCatalytics: BRAO-1001 | Lab

Alcohol oxidase; Origin: Pichia pastoris
Biozyme | Pilot

Alcohol Oxidase; Origin: Pichia pastoris
Jülich Enzyme Products | Lab

Choline oxidase. | **1.1.3.17**
 | Choline + O(2) = betaine aldehyde + H(2)O(2).

Choline Oxidase; Origin: Alcaligenes sp.
Fluka | Pilot

Choline Oxidase; Origin: Arthrobacter globiformis
Asahi | Pilot

Glycerol-3-phosphate oxidase. | **1.1.3.21**
 | Sn-glycerol 3-phosphate + O(2) = glycerone phosphate + H(2)O(2).

L-Glycerophosphate Oxidase
Asahi | Pilot

Glycerol 3-phosphate Oxidase; Origin: Aerococcus viridans
Fluka | Lab

Table 20.5. (cont.).

L-Glycerophosphate Oxidase; Origin: Aerococcus viridans Asahi	Pilot
L-Glycerol-3-phosphate Oxidase; Origin: microorganism, rec. in E. coli Roche Diagnostics: L-Glycerol-3-phosphate Oxidase (GPO), stabilized	Pilot
L-alpha-Glycerophosphate Oxidase; Origin: microorganisms Toyobo	Pilot
L-alpha-Glycerophosphate Oxidase; Origin: Pediococcus sp. Toyobo	Pilot
L-alpha-Glycerophosphate Oxidase; Origin: Streptococcus sp. Amano: Amano 2 [GPO-2]	Pilot

Xanthine oxidase.

1.1.3.22

Xanthine oxidoreductase. Hypoxanthine oxidase.
Hypoxanthine-xanthine oxidase. Schardinger enzyme.

$$Xanthine + H_2O + O_2 = urate + H_2O_2.$$

Xanthine oxidase; Origin: buttermilk Biozyme	Pilot
Xanthine Oxidase; Origin: buttermilk Fluka	Pilot
Xanthine Oxidase; Origin: cow milk Roche Diagnostics: Xanthine Oxidase	Pilot

Fructose 5-dehydrogenase.

1.1.99.11

D-Fructose dehydrogenase.

$$D\text{-fructose} + acceptor = 5\text{-dehydro-}D\text{-fructose} + reduced\ acceptor.$$

D-Fructose Dehydrogenase; Origin: Gluconobacter sp. Toyobo	Pilot

Formate dehydrogenase.

1.2.1.2

$$Formate + NAD^+ = CO_2 + NADH.$$

Formate Dehydrogenase; Origin: Candida boidinii Fluka	Pilot
Formate Dehydrogenase; Origin: Candida boidinii Jülich Enzyme Products	Lab
Formate Dehydrogenase, rec.; Origin: Candida boidinii, overexpressed in E. coli Roche Diagnostics: Formate Dehydrogenase (FDH), rec.	Industrial
Formate Dehydrogenase rec.; Origin: E. coli Fluka	Lab
Formate Dehydrogenase; Origin: microorganisms Unitika: Formate Dehydrogenase (FDH)	Pilot
Formate Dehydrogenase; Origin: Pseudomonas sp. Fluka	Lab
Formate Dehydrogenase; Origin: Pseudomonas sp. Fluka	Lab
Formate Dehydrogenase; Origin: Xilaria digitata (formerly Candida biodinii) Roche Diagnostics: Formate Dehydrogenase (FDH)	Industrial
Formate Dehydrogenase; Origin: yeast Fluka	Pilot

Aldehyde dehydrogenase (NAD(P)+).

1.2.1.5

$$An\ aldehyde + NAD(P)^+ + H_2O = an\ acid + NAD(P)H.$$

Aldehyde Dehydrogenase; Origin: baker's yeast Fluka	Lab

Table 20.5. (cont.).

Aldehyde dehydrogenase; Origin: yeast
Biozyme
Pilot

Aldehyde Dehydrogenase; Origin: yeast
Roche Diagnostics: Aldehyde Dehydrogenase (AldDH)
Lab

Glyceraldehyde 3-phosphate dehydrogenase (phosphorylating).

1.2.1.12

NAD-dependent glyceraldehyde-3-phosphate dehydrogenase.
Triosephosphate dehydrogenase. GAPDH.

D-glyceraldehyde 3-phosphate + phosphate + NAD(+) = 3-phospho-D-glyceroyl phosphate + NADH.

Glyceraldehyde-3-Phosphate Dehydrogenase (GapDH); Origin: Bacillus stearothermophilus
Unitika
Pilot

Glyceraldehyde-3-phosphate dehydrogenase; Origin: rabbit muscle
Biozyme
Pilot

Glyceraldehyde-3-phosphate Dehydrogenase; Origin: rabbit muscle
Fluka
Lab

Formaldehyde dehydrogenase.

1.2.1.46

Formaldehyde + NAD(+) + H(2)O = formate + NADH.

Formaldehyde Dehydrogenase; Origin: Pseudomonas putida
Fluka
Lab

Formaldehyde Dehydrogenase; Origin: Pseudomonas sp.
Toyobo
Pilot

Pyruvate oxidase.

1.2.3.3

Pyruvic oxidase.

Pyruvate + phosphate + O(2) + H(2)O = acetyl phosphate + CO(2) +H(2)O(2).

Pyruvate Oxidase ; Origin: Aerococcus viridans
Asahi
Pilot

Pyruvate Oxidase; Origin: Lactobacillus plantarum, rec. E. coli
Roche Diagnostics: Pyruvate Oxidase (PyrOD)
Pilot

Bilirubin oxidase.

1.3.3.5

Bilirubin + O(2) = biliverdin + H(2)O.

Bilirubin Oxidase; Origin: Myrothecium sp.
Amano: Amano 2 [BO-2]
Pilot

Acyl-CoA oxidase.

1.3.3.6

Acyl-CoA + O(2) = trans-2,3-dehydroacyl-CoA + H(2)O(2).

Acyl-CoA Oxidase; Origin: Arthrobacter sp.
Asahi
Pilot

Acyl-CoA Oxidase; Origin: microorganisms
Amano: Amano 3 [ACO-3]
Pilot

Alanine dehydrogenase.

1.4.1.1

L-alanine + H(2)O + NAD(+) = pyruvate + NH(3) + NADH.

Alanine Dehydrogenase
Asahi
Pilot

L-Alanine Dehydrogenase; Origin: Bacillus cereus
Jülich Enzyme Products
Lab

Alanine Dehydrogenase; Origin: Bacillus stearothermophilus
Unitika: Alanine Dehydrogenase (AlaDH)
Pilot

L-Alanine Dehydrogenase; Origin: Bacillus subtilis
Fluka
Lab

Table 20.5. (cont.).

Glutamate dehydrogenase (NAD(P)+). Glutamic dehydrogenase.	L-glutamate + H(2)O + NAD(P)(+) = 2-oxoglutarate + NH(3) + NAD(P)H.	**1.4.1.3**
Glutamate dehydrogenase; Origin: beef liver Biozyme		Pilot
Glutamate Dehydrogenase; Origin: bovine liver Fluka		Pilot
L-Glutamate Dehydrogenase; Origin: bovine liver Roche Diagnostics: L-Glutamate Dehydrogenase (GlDH), lyo.		Pilot
Glutamate Dehydrogenase; Origin: microorganisms Toyobo		Pilot
Glutamate Dehydrogenase; Origin: Proteus sp. Toyobo		Pilot
Leucine dehydrogenase.	L-leucine + H(2)O + NAD(+) = 4-methyl-2-oxopentanoate + NH(3) + NADH.	**1.4.1.9**
Leucine Dehydrogenase; Origin: Bacillus cereus Biocatalysts		Pilot
Leucine Dehydrogenase; Origin: Bacillus sp. Toyobo		Pilot
Leucine Dehydrogenase; Origin: Bacillus stearothermophilus Unitika: Leucine Dehydrogenase (LeuDH)		Pilot
Phenylalanine dehydrogenase.	L-phenylalanine + H(2)O + NAD(+) = phenylpyruvate + NH(3) + NADH.	**1.4.1.20**
Phenylalanine Dehydrogenase; Origin: microorganisms Unitika: Phenylalanine Dehydrogenase (PheDH)		Pilot
Phenylalanine Dehydrogenase; Origin: Sporosarcina sp. Biocatalysts		Lab
D-amino acid oxidase.	A D-amino acid + H(2)O + O(2) = a 2-oxo acid + NH(3) + H(2)O(2).	**1.4.3.3**
D-Amino Acid Oxidase; Origin: hog kidney Fluka		Lab
D-Amino Acid Oxidase; Origin: hog kidney Fluka		Lab
D-Amino Acid oxidase; Origin: porcine kidney Biozyme		Pilot
D-Amino Acid Oxidase; Origin: Trigonopsis variabilis Recordati: DAAO Beads		Industrial
D-Amino Acid Oxidase, carrier-fixed; Origin: Trigonopsis variabilis Roche Diagnostics: D-Amino Acid Oxidase (D-AOD), carrier-fixed		Industrial
D-Amino acid Oxidase, immobilized; Origin: Trigonopsis variabilis Fluka		Industrial
Amine oxidase (flavin-containing). Monoamine oxidase. Tyramine oxidase. Tyraminase. Amine oxidase.	RCH(2)NH(2) + H(2)O + O(2) = RCHO + NH(3) + H(2)O(2).	**1.4.3.4**
Tyramine Oxidase; Origin: Arthrobacter sp. Asahi		Pilot

Table 20.5. (cont.).

Dihydrofolate reductase.

Tetrahydrofolate dehydrogenase.

1.5.1.3

5,6,7,8-tetrahydrofolate + NADP(+) = 7,8-dihydrofolate + NADPH.

Dihydrofolate Reductase; Origin: bovine liver

Fluka

Lab

Sarcosine oxidase.

1.5.3.1

Sarcosine + H(2)O + O(2) = glycine + formaldehyde + H(2)O(2).

Sarcosine Oxidase

Asahi

Pilot

Sarcosine Oxidase; Origin: microorganisms

Toyobo

Pilot

With other acceptors.

1.5.99.

Dimethylamine Dehydrogenase; Origin: Paracoccus spec.

Jülich Enzyme Products

Lab

Trimethylamine dehydrogenase.

TMADh.

1.5.99.7

Trimethylamine + H(2)O + acceptor = dimethylamine + formaldehyde +reduced acceptor.

Trimethylamine Dehydrogenase; Origin: Paracoccus spec.

Jülich Enzyme Products

Lab

Glutathione reductase (NADPH).

1.6.4.2

NADPH + oxidized glutathione = NADP(+) + 2 glutathione.

Glutathione Reductase; Origin: baker's yeast

Fluka

Lab

NADPH dehydrogenase.

NADPH diaphorase.

1.6.99.1

NADPH + acceptor = NADP(+) + reduced acceptor.

Diaphorase (NADPH); Origin: Bacillus megaterium

Asahi

Pilot

Diaphorase I; Origin: Bacillus stearothermophilus

Unitika

Pilot

Urate oxidase.

Uricase.

1.7.3.3

Urate + O(2) + H(2)O = 5-hydroxyisourate + H(2)O(2).

Uricase; Origin: Arthrobacter globiformis

Asahi

Pilot

Uricase; Origin: Bacillus fastidiosus

Fluka

Lab

Uricase; Origin: Bacillus sp.

Toyobo

Pilot

Uricase; Origin: pig liver

Biozyme

Pilot

Dihydrolipoamide dehydrogenase.

Lipoamide reductase (NADH). E3 component of alpha-ketoacid dehydrogenase complexes. Lipoyl dehydrogenase. Dihydrolipoyl dehydrogenase.

1.8.1.4.

Dihydrolipoamide + NAD(+) = lipoamide + NADH.

Diaphorase (NADH); Origin: Bacillus megaterium

Asahi

Pilot

Diaphorase II; Origin: Bacillus stearothermophilus

Unitika

Pilot

Table 20.5. (cont.).

Diaphorase; Origin: Clostridium kluyveri	
Fluka	Pilot

Laccase.	**1.10.3.2**
Urishiol oxidase.	4 benzenediol + O(2) = 4 benzosemiquinone + 2 H(2)O.

Laccase A; Origin: Agaricus bisporus	
Jülich Enzyme Products	Lab

Laccase C; Origin: Coriolus versicolor	
Jülich Enzyme Products	Lab

Laccase; Origin: rec. microorganism	
Novozymes: DeniLite™	Industrial

L-ascorbate oxidase.	**1.10.3.3**
Ascorbase.	2 L-ascorbate + O(2) = 2 dehydroascorbate + 2 H(2)O.

Ascorbate Oxidase	
Asahi	Pilot

Ascorbate Oxidase; Origin: Cucumber	
Amano: Amano 2 [ASO-2]	Pilot

Ascorbate oxidase; Origin: Cucurbita sp.	
Biozyme	Pilot

Ascorbate Oxidase; Origin: Cucurbity sp.	
Fluka	Pilot

Ascorbate Oxidase; Origin: microorganisms	
Amano: Amano 3 [ASO-3]	Pilot

Oxidoreductases.	**1.11.-.-**
Acting on a peroxide as acceptor (peroxidases).	

Bromoperoxidase; Origin: Corallina officinalis	
Fluka	Lab

Catalase.	**1.11.1.6**
	2 H(2)O(2) = O(2) + 2 H(2)O.

Catalase	
Biocatalysts: CATALASE	Industrial

Catalase	
Seravac	Industrial

Catalase; Origin: Aspergillus niger	
Amano: Catalase NL "Amano"	Industrial

Catalase; Origin: Aspergillus niger	
Biozyme	Pilot

Catalase; Origin: Aspergillus niger	
Fluka	Industrial

Catalase; Origin: Aspergillus niger	
Novozymes: Catazyme®	Industrial

Catalase; Origin: Aspergillus niger	
Roche Diagnostics: Catalase, technical grade	Industrial

Catalase; Origin: Aspergillus niger, rec.	
Novozymes: Terminox™ Ultra	Industrial

Catalase; Origin: beef liver	
Biozyme	Pilot

Catalase; Origin: beef liver	
Roche Diagnostics	Industrial

Table 20.5. (cont.).

Catalase; Origin: bovine liver
Fluka

Industrial

Catalase, immobilized on Eupergit C; Origin: bovine liver
Fluka

Lab

Catalase; Origin: Corynebacterium glutamicum
Roche Diagnostics

Industrial

Catalase; Origin: Micrococcus lysodeikticus
Fluka

Lab

Catalase; Origin: microorganisms
Fluka

Pilot

Catalase; Origin: microorganisms
Toyobo

Industrial

Peroxidase.	**1.11.1.7**
Myeloperoxidase.	Donor + H_2O_2 = oxidized donor + 2 H_2O.

Lactoperoxidase; Origin: bovine milk
Biozyme

Pilot

Lactoperoxidase; Origin: bovine milk
Fluka

Pilot

Peroxidase; Origin: Coprinus cinereus
Novozymes: Novozym 502

Industrial

Peroxidase; Origin: Coprinus cinereus
Novozymes: NS18010

Industrial

Peroxidase; Origin: horse radish
Fluka

Industrial

Peroxidase; Origin: horseradish
Amano: Amano 2 [PO-2]

Pilot

Peroxidase; Origin: horseradish
Amano: Amano 3 [PO-3]

Pilot

Peroxidase; Origin: horseradish
Biocatalysts

Industrial

Peroxidase; Origin: horseradish
Biozyme

Pilot

Peroxidase; Origin: horseradish
Roche Diagnostics: Peroxidase (POD), Grade I

Industrial

Peroxidase; Origin: horseradish
Roche Diagnostics: Peroxidase (POD), Grade II

Pilot

Peroxidase ; Origin: horseradish
Seravac

Industrial

Peroxidase; Origin: horseradish
Toyobo

Pilot

Glutathione peroxidase.	**1.11.1.9**
	2 glutathione + H_2O_2 = oxidized glutathione + 2 H_2O.

Glutathione Peroxidase; Origin: bovine erythrocytes
Fluka

Lab

Chloride peroxidase.	**1.11.1.10**
Chloroperoxidase.	2 RH + 2 chloride + H_2O_2 = 2 RCl + 2 H_2O.

Chloroperoxidase; Origin: Caldariomyces fumago
Fluka

Pilot

Table 20.5.   (cont.).

Chloroperoxidase; Origin: Leptoxyphium fumago Jülich Enzyme Products	Lab

Lipoxygenase. Lipoxidase. Carotene oxidase. Lipoperoxidase.	**1.13.11.12** Linoleate + O(2) = (9Z,11E)-(13S)-13-hydroperoxyoctadeca-9,11-dienoate.

Lipoxygenase II; Origin: pea, rec. in E. coli Biocatalysts	Pilot
Lipoxygenase III; Origin: pea, rec. in E. coli Biocatalysts	Pilot
Lipoxidase; Origin: soybean Fluka	Lab
Lipoxidase; Origin: soybeen Biozyme	Pilot

Lactate 2-monooxygenase. Lactate oxidative decarboxylase. Lactate oxidase. Lactate oxygenase.	**1.13.12.4** (S)-lactate + O(2) = acetate + CO(2) + H(2)O.

Lactate Oxidase Asahi	Pilot
Lactate Oxidase; Origin: Pediococcus sp. Fluka	Pilot

Oxidoreductases. **Acting on paired donors with incorporation of** **molecular oxygen.** **With NADH or NADPH as one donor, and** **incorporation of one atom of oxygen.**	**1.14.13.-**

2-Tridecanone Monooxygenase; Origin: Pseudomonas cepacia Fluka	Lab

Cyclopentanone monooxygenase.	**1.14.13.16** Cyclopentanone + NADPH + O(2) = 5-valerolactone + NADP(+) + H(2)O.

Cyclopentanone Monooxygenase; Origin: Pseudomonas sp. Fluka	Lab

Cyclohexanone monooxygenase. Cyclohexanone oxygenase.	**1.14.13.22** Cyclohexanone + NADPH + O(2) = 6-hexanolide + NADP(+) + H(2)O.

Cyclohexanone Monooxygenase; Origin: Acinetobacter sp. Fluka	Lab
Cyclohexanone Monooxygenase; Origin: E. coli overproducer Fluka	Lab
Cyclohexanone Monooxygenase; Origin: Nocardia globerula Fluka	Lab
Cyclohexanone Monooxygenase; Origin: Xanthobacter sp. Fluka	Lab

2-hydroxybiphenyl 3-monooxygenase.	**1.14.13.44** 2-hydroxybiphenyl + NADH + O(2) = 2,3-dihydroxybiphenyl + NAD(+) + H(2)O.

2-Hydroxybiphenylmonooxygenase; Origin: E. coli Fluka	Lab

Table 20.5. (cont.).

Camphor 5-monooxygenase.
1.14.15.1

Camphor 5-exo-methylene hydroxylase. Cytochrome p450-cam.
(+)-camphor + putidaredoxin + O(2) = (+)-exo-5-hydroxycamphor + oxidizedputidaredoxin + H(2)O.

(+)-Camphor Monooxygenase; Origin: Pseudomonas putida
Fluka
Lab

Monophenol monooxygenase.
1.14.18.1

Tyrosinase. Phenolase. Monophenol oxidase. Cresolase.
L-tyrosine + L-DOPA + O(2) = L-DOPA + DOPAquinone + H(2)O.

Tyrosinase; Origin: mushroom
Fluka
Lab

Progesterone monooxygenase.
1.14.99.4

Progesterone hydroxylase.
Progesterone + AH(2) + O(2) = testosterone acetate + A + H(2)O.

Progesterone Monooxygenase; Origin: Cylindrocapron radicicola
Fluka
Lab

Superoxide dismutase.
1.15.1.1

2 peroxide radical + 2 H(+) = O(2) + H(2)O(2).

Superoxide Dismutase; Origin: Bacillus stearothermophilus
Unitika: Superoxide Dismutase (SOD)
Pilot

Superoxide dismutase; Origin: bovine erythrocytes
Biozyme
Pilot

Superoxide Dismutase; Origin: bovine erythrocytes
Fluka
Pilot

Superoxide Dismutase; Origin: bovine erythrocytes
Roche Diagnostics: Superoxide Dismutase (SOD)
Lab

Superoxide Dismutase; Origin: bovine liver
Fluka
Pilot

Catechol O-methyltransferase.
2.1.1.6

S-adenosyl-L-methionine + catechol = S-adenosyl-L-homocysteine +guaiacol.

Brenzkatechin-O-methyl-Transferase
Fluka
Lab

Transketolase.
2.2.1.1

Glycoaldehyde transferase.
Sedoheptulose 7-phosphate + D-glyceraldehyde 3-phosphate = D-ribose 5-phosphate + D-xylulose 5-phosphate.

Transketolase; Origin: baker's yeast
Fluka
Lab

Transketolase ; Origin: E. coli
Fluka
Lab

Transketolase; Origin: E. coli K12 (rec.)
Jülich Enzyme Products
Lab

Transaldolase.
2.2.1.2

Dihydroxyacetone transferase. Glycerone transferase.
Sedoheptulose 7-phosphate + D-glyceraldehyde 3-phosphate = D-erythrose 4-phosphate + D-fructose 6-phosphate.

Transaldolase ; Origin: Candida utilis
Fluka
Lab

Transaldolase; Origin: E. coli K12 (rec.)
Jülich Enzyme Products
Lab

Table 20.5. (cont.).

Glucosamine-phosphate N-acetyltransferase.
Phosphoglucosamine transacetylase. Phosphoglucosamine acetylase.

2.3.1.4

Acetyl-CoA + D-glucosamine 6-phosphate = CoA + N-acetyl-D-glucosamine 6-phosphate.

Phosphotransacetylase; Origin: Bacillus stearothermophilus
Unitika: Phosphotransacetylase (PTA)

Lab

Carnitine O-acetyltransferase.
Carnitine acetylase.

2.3.1.7

Acetyl-CoA + carnitine = CoA + O-acetylcarnitine.

Carnitine Acetyltransferase; Origin: pigeon breast muscle
Fluka

Lab

Gamma-glutamyltransferase.
Gamma-glutamyltranspeptidase. Glutamyl transpeptidase.

2.3.2.2

(5-L-glutamyl)-peptide + an amino acid = peptide + 5-L-glutamyl-aminoacid.

---

gamma-Glutamyltransferase; Origin: beef kidney
Biozyme

Pilot

---

gamma-Glutamyl Transpeptidase; Origin: hog kidney
Fluka

Lab

Phosphorylase.
Muscle phosphorylase A and B. Amylophosphorylase. Polyphosphorylase.

2.4.1.1

{(1,4)-alpha-D-glucosyl}(N) + phosphate = {(1,4)-alpha-D-glucosyl}(N-1)+ alpha-D-glucose 1-phosphate.

Phosphorylase b; Origin: rabbit muscle
Fluka

Lab

Sucrose synthase.
UDP-glucose-fructose glucosyltransferase. Sucrose-UDP glucosyltransferase.

2.4.1.13

UDP-glucose + D-fructose = UDP + sucrose.

Sucrose Synthase; Origin: rice grains
Jülich Enzyme Products

Lab

1,4-alpha-glucan branching enzyme.
Glycogen branching enzyme. Amylo-(1,4 to 1,6)transglucosidase. Branching enzyme. Amylo-(1,4-1,6)-transglycosylase.

2.4.1.18

Formation of 1,6-glucosidic linkages of glycogen.

Transglucosidase; Origin: Aspergillus niger
Amano: Transglucosidase L "Amano"

Industrial

Lactose synthase.
UDP-galactose-glucose galactosyltransferase. N-acetyllactosamine synthase.

2.4.1.22

UDP-galactose + D-glucose = UDP + lactose.

---

beta-1,4-Galactosyl Transferase; Origin: bovine milk
Fluka

Pilot

Beta-N-acetylglucosaminyl-glycopeptide beta-1,4-galactosyltransferase.
Glycoprotein 4-beta-galactosyltransferase. Thyroid galactosyltransferase. UDP-galactose-glycoprotein galactosyltransferase.

2.4.1.38

UDP-galactose + N-acetyl-beta-D-glucosaminylglycopeptide = UDP +beta-D-galactosyl-1,4-N-acetyl-beta-D-glucosaminylglycopeptide.

Beta-1,4-Galactosyltransferase; Origin: Saccharomyces cerevisiae (rec.)
Jülich Enzyme Products

Lab

Table 20.5. (cont.).

N-acetyllactosaminide alpha-1,3-galactosyltransferase.	**2.4.1.151**
Galactosyltransferase.	UDP-galactose + beta-D-galactosyl-(1,4)-N-acetyl-D-glucosaminyl-R = UDP + alpha-D-galactosyl-(1,3)-beta-D-galactosyl-(1,4)-N-acetyl-D-glucos

alpha-1,3-Galactosyl-Transferase; Origin: E. coli, rec. Fluka	Lab
alpha-1,3-Galactosyltransferase; Origin: E. coli, rec. Fluka	Lab

Purine-nucleoside phosphorylase.	**2.4.2.1**
Inosine phosphorylase. PNPase.	Purine nucleoside + phosphate = purine + alpha-D-ribose1-phosphate.

Purine-Nucleoside phosphorylase Asahi	Pilot
Purine-Nucleoside Phosphorylase; Origin: microorganisms Toyobo	Pilot

Transferases. Transferring nitrogenous groups. Transaminases (aminotransferases).	**2.6.1.-**

Transaminase, branched-chain, L-specific; Origin: microorganism, rec. in E. coli BioCatalytics: AT-102	Lab
Transaminase, broad-range, D-specific; Origin: microorganism, rec. in E. coli BioCatalytics: AT-103	Lab
Transaminase, broad-range, L-specific; Origin: microorganism, rec. in E. coli BioCatalytics: AT-101	Lab

Aspartate aminotransferase.	**2.6.1.1**
Transaminase A. Glutamic--oxaloacetic transaminase. Glutamic--aspartic transaminase.	L-aspartate + 2-oxoglutarate = oxaloacetate + L-glutamate.

Glutamic-oxaloacetic transaminase; Origin: pig heart Biozyme	Pilot
Glutamate-Oxaloacetate Transaminase; Origin: pig heart (mitochondrial) Roche Diagnostics: Glutamate-Oxaloacetate Transaminase (GOT)	Pilot
Glutamic-Oxalacetic Transaminase; Origin: porcine heart Fluka	Pilot

Alanine aminotransferase.	**2.6.1.2**
Glutamic--pyruvic transaminase. Glutamic--alanine transaminase.	L-alanine + 2-oxoglutarate = pyruvate + L-glutamate.

Glutamate-Pyruvate Transaminase; Origin: pig heart Roche Diagnostics: Glutamate-Pyruvate Transaminase (GPT)	Pilot
Glutamic-pyruvic transaminase; Origin: pig heart Biozyme	Pilot
Glutamic-Pyruvic Transaminase; Origin: porcine heart Fluka	Pilot

Hexokinase.	**2.7.1.1**
Glucokinase. Hexokinase type IV.	ATP + D-hexose = ADP + D-hexose 6-phosphate.

Hexokinase Asahi	Pilot
Hesperidinase; Origin: Penicillium decumbens Amano: Hesperidinase "Amano" Conc.	Industrial

Table 20.5. (cont.).

Hexokinase; Origin: Saccharomyces sp. Toyobo	Pilot
Hexokinase; Origin: yeast Biozyme	Pilot
Hexokinase; Origin: yeast Fluka	Industrial
Hexokinase; Origin: yeast Fluka	Industrial

	2.7.1.2
Glucokinase. Glucose kinase.	ATP + D-glucose = ADP + D-glucose 6-phosphate.

Glucokinase; Origin: Bacillus stearothermophilus Unitika: Glucokinase (GlcK)	Pilot
Glucokinase; Origin: Zymomonas mobilis Unitika: Glucokinase (GlcK)	Pilot

	2.7.1.11
6-phosphofructokinase. Phosphohexokinase. Phosphofructokinase I.	ATP + D-fructose 6-phosphate = ADP + D-fructose 1,6-bisphosphate.

Phosphofructokinase; Origin: Bacillus stearothermophilus Unitika: Phosphofructokinase (PFK)	Pilot

	2.7.1.30
Glycerol kinase. Glycerokinase. ATP:glycerol 3-phosphotransferase.	ATP + glycerol = ADP + glycerol 3-phosphate.

Glycerol Kinase Asahi	Pilot
Glycerol Kinase; Origin: Arthrobacter sp. Amano: Amano 2 [GK-2]	Pilot
Glycerokinase; Origin: Bacillus stearothermophilus Roche Diagnostics: Glycerokinase, lyo.	Pilot
Glycerokinase; Origin: Bacillus stearothermophilus Roche Diagnostics: Glycerokinase, sol.	Pilot
Glycerol Kinase; Origin: E. coli Fluka	Pilot
Glycerol Kinase; Origin: microorganisms Toyobo	Pilot

	2.7.1.40
Pyruvate kinase. Phosphoenolpyruvate kinase. Phosphoenol transphosphorylase.	ATP + pyruvate = ADP + phosphoenolpyruvate.

Pyruvate Kinase; Origin: Bacillus stearothermophilus Unitika: Pyruvate Kinase (PK)	Pilot
Pyruvate Kinase; Origin: pig heart Biozyme	Pilot
Pyruvate Kinase; Origin: rabbit muscle Biozyme	Pilot
Pyruvate Kinase; Origin: rabbit muscle Fluka	Lab
Pyruvate Kinase; Origin: Zymomonas mobilis Unitika: Pyruvate Kinase (PK)	Pilot

Table 20.5. (cont.).

Streptomycin 6-kinase.   2.7.1.72
Streptidine kinase. Streptomycin 6-phosphotransferase. APH(6).

$$\text{ATP} + \text{streptomycin} = \text{ADP} + \text{streptomycin 6-phosphate.}$$

Streptokinase; Origin: Streptococcus hemolyticus
Fluka

Lab

Acetate kinase.   2.7.2.1
Acetokinase.

$$\text{ATP} + \text{acetate} = \text{ADP} + \text{acetyl phosphate.}$$

Acetate Kinase; Origin: Bacillus stearothermophilus
Unitika: Acetate Kinase (AK)

Pilot

Acetate Kinase; Origin: E. coli
Fluka

Lab

Phosphoglycerate kinase.   2.7.2.3

$$\text{ATP} + \text{3-phospho-D-glycerate} = \text{ADP} + \text{3-phospho-D-glyceroyl phosphate.}$$

Phosphoglycerate Kinase; Origin: Bacillus stearothermophilus
Unitika: Phosphoglycerate Kinase (PGK)

Pilot

Creatine kinase.   2.7.3.2

$$\text{ATP} + \text{creatine} = \text{ADP} + \text{phosphocreatine.}$$

Creatine Kinase; Origin: beef heart
Biozyme

Pilot

Creatine Kinase; Origin: pig heart
Biozyme

Pilot

Creatine Kinase; Origin: rabbit muscle
Biozyme

Pilot

Creatine Phosphokinase; Origin: rabbit muscle
Fluka

Lab

Adenylate kinase.   2.7.4.3
Myokinase. Adenylic kinase. Adenylokinase.

$$\text{ATP} + \text{AMP} = \text{ADP} + \text{ADP.}$$

Adenylate Kinase; Origin: Bacillus stearothermophilus
Unitika: Adenylate Kinase (AdK)

Pilot

Polyribonucleotide nucleotidyltransferase.   2.7.7.8
Polynucleotide phosphorylase.

$$\{\text{RNA}\}(\text{N+1}) + \text{phosphate} = \{\text{RNA}\}(\text{N}) + \text{a nucleoside diphosphate.}$$

Polynucleotide Phosphorylase; Origin: Bacillus stearothermophilus
Unitika: Polynucleotide Phosphorylase (PNPase)

Pilot

Acylneuraminate cytidylyltransferase.   2.7.7.43
CMP-N-acetylneuraminic acid synthetase. CMP-NeuNAc
synthetase. CMP-sialate pyrophosphorylase. CMP-sialate
diphosphorylase.

$$\text{CTP} + \text{N-acylneuraminate} = \text{diphosphate} + \text{CMP-N-acylneuraminate.}$$

CMP-Neu5Ac Synthetase; Origin: E. coli K 235/CS1
Jülich Enzyme Products

Lab

Hydrolases.   3.1.1.-
Acting on ester bonds.
Carboxylic ester hydrolases.

Lipases & Esterases Screening Set; Origin: mammalian sources and microorganisms
Roche Diagnostics: CHIRAZYME Lipases & Esterases, Screening Set Industrial Enzymes 2

Industrial

Table 20.5. (cont.).

Carboxylesterase.	3.1.1.1
Ali-esterase. B-esterase. Monobutyrase. Cocaine esterase.	A carboxylic ester + H_2O = an alcohol + a carboxylic anion.

Esterase basic kit Fluka	Lab
Esterase; Origin: Bacillus sp. Fluka	Lab
Esterase; Origin: Bacillus stearothermophilus Fluka	Lab
Esterase; Origin: Bacillus thermoglucosidasius Fluka	Lab
Esterase; Origin: Candida lipolytica Fluka	Lab
Esterase ; Origin: Candida rugosa Altus	Industrial
Esterase; Origin: hog liver Fluka	Industrial
Esterase, immobilized on Eupergit® C; Origin: hog liver Fluka	Pilot
Esterase Isoenzyme 1; Origin: hog liver Fluka	Pilot
Esterase; Origin: horse liver Fluka	Lab
Esterase Screening Kit ; Origin: microorganism, rec. in E. coli ThemoGen: QuickScreen Esterase Kits	Lab
Esterase; Origin: Mucor miehei Fluka	Lab
Esterase; Origin: pig liver Jülich Enzyme Products: Esterase PL	Lab
Esterase. immobilized; Origin: pig liver Roche Diagnostics: CHIRAZYME E-1, c.-f., lyo.	Industrial
Pig Liver Esterase; Origin: pig Liver Altus	Industrial
Pig Liver Esterase; Origin: pig liver Roche Diagnostics: PLE, technical grade, susp.	Industrial
Esterase; Origin: pig liver, fraction 1 Roche Diagnostics: CHIRAZYME E-1, lyo.	Industrial
Esterase; Origin: pig liver, fraction 2 Roche Diagnostics: CHIRAZYME E-2, lyo.	Industrial
Esterase; Origin: Rhizopus arrhizus Jülich Enzyme Products: Esterase EL9	Lab
Esterase; Origin: Rhodotorula pilimanae Jülich Enzyme Products: Esterase EL5	Lab
Esterase; Origin: Saccharomyces cerevisiae Fluka	Lab
Desacetyl-esterase; Origin: Therm.sp., rec. in E. coli. Recordati: Desa-REC	Industrial
Esterase; Origin: Thermoanaerobium brockii Fluka	Lab

Table 20.5. (cont.).

Triacylglycerol lipase. Lipase. Triglyceride lipase. Tributyrase.	3.1.1.3 Triacylglycerol + H(2)O = diacylglycerol + a fatty acid anion.
Lipase basic kit Fluka	Lab
Lipase extension kit Fluka	Lab
Monoglyceride Lipase Asahi	Pilot
Lipase; Origin: Achromobacter sp. Meito Sangyo: Lipase AL	Industrial
Lipase, immobilized; Origin: Achromobacter sp. Meito Sangyo: Lipase ALC/ALG	Industrial
Lipase; Origin: Alcaligenes sp. Altus	Industrial
Lipase; Origin: Alcaligenes sp. Meito Sangyo: Lipase PL	Industrial
Lipase; Origin: Alcaligenes sp. Meito Sangyo: Lipase QLL	Industrial
Lipase; Origin: Alcaligenes sp. Meito Sangyo: Lipase QLM	Industrial
Lipase, immobilized; Origin: Alcaligenes sp. Meito Sangyo: Lipase PLC/PLG	Industrial
Lipase, immobilized; Origin: Alcaligenes sp. Meito Sangyo: Lipase QLC/QLG	Industrial
Lipase; Origin: Alcaligines sp. Roche Diagnostics: CHIRAZYME L-10, lyo.	Industrial
Lipase; Origin: Aspergillus niger Altus	Industrial
Lipase; Origin: Aspergillus niger Amano: Lipase A "Amano" 6	Industrial
Lipase; Origin: Aspergillus niger Amano: Lipase AS	Industrial
Lipase; Origin: Aspergillus niger Amano: Lipase DS	Industrial
Lipase; Origin: Aspergillus niger Fluka	Industrial
Lipase; Origin: Aspergillus niger Fluka	Industrial
Lipase, immobilized in Sol-Gel-AK; Origin: Aspergillus niger Fluka	Lab
Lipase; Origin: Aspergillus oryzae Fluka	Industrial
Lipase; Origin: Burkholderia cepacia Meito Sangyo: Lipase SL	Industrial
Lipase ; Origin: Burkholderia sp. Fluka	Pilot
Lipase; Origin: Candida antarctica Fluka	Lab
Lipase A; Origin: Candida antarctica Fluka	Industrial

Table 20.5.   (cont.).

Lipase, immobilized; Origin: Candida antarctica Fluka	Industrial
Lipase, immobilized in Sol-Gel-AK; Origin: Candida antarctica Fluka	Pilot
Lipase, immobilized in Sol-Gel-AK on sintered glass; Origin: Candida antarctica Fluka	Lab
Lipase, type A; Origin: Candida antarctica Altus	Industrial
Lipase, type B; Origin: Candida antarctica Altus	Industrial
Lipase B; Origin: Candida antarctica, rec. Fluka	Industrial
Lipase; Origin: Candida antarctica, type A Roche Diagnostics: CHIRAZYME L-5, lyo.	Industrial
Lipase; Origin: Candida antarctica, type A Roche Diagnostics: CHIRAZYME L-5, sol.	Industrial
Lipase, immobilized; Origin: Candida antarctica, type A Roche Diagnostics: CHIRAZYME L-5, c.-f., lyo.	Industrial
Lipase; Origin: Candida antarctica, type A, rec. in Aspergillus oryzae Novozymes: NovoCor AD	Industrial
Lipase; Origin: Candida antarctica, type A, rec. in Aspergillus oryzae Novozymes: Novozym® 868	Industrial
Lipase; Origin: Candida antarctica, type B Roche Diagnostics: CHIRAZYME L-2, lyo.	Industrial
Lipase; Origin: Candida antarctica, type B Roche Diagnostics: CHIRAZYME L-2, sol.	Industrial
Lipase, immobilized; Origin: Candida antarctica, type B Roche Diagnostics: CHIRAZYME L-2, c.-f., C2, lyo. (Novozym 435)	Industrial
Lipase, immobilized; Origin: Candida antarctica, type B Roche Diagnostics: CHIRAZYME L-2, c.-f., C3, lyo.	Industrial
Lipase, immobilized; Origin: Candida antarctica, type B Roche Diagnostics: CHIRAZYME L-2, c.-f., lyo.	Industrial
Lipase; Origin: Candida antarctica, type B, rec. in Aspergillus oryzae Novozymes: Nocozym 525 L	Industrial
Lipase, immobilized; Origin: Candida antarctica, type B, rec. in Aspergillus oryzae Novozymes: Novozym® 435	Industrial
Lipase; Origin: Candida cylindracae Jülich Enzyme Products: Lipase LE11	Lab
Lipase; Origin: Candida cylindracea Biocatalysts	Industrial
Lipase; Origin: Candida cylindracea Fluka	Industrial
Lipase; Origin: Candida cylindracea Fluka	Industrial
Lipase; Origin: Candida cylindracea Meito Sangyo: Lipase MY	Industrial
Lipase; Origin: Candida cylindracea Meito Sangyo: Lipase OF	Industrial
Lipase; Origin: Candida cylindracea Meito Sangyo: Lipase OFL	Industrial

Table 20.5. (cont.).

Lipase, immobilized; Origin: Candida cylindracea Meito Sangyo: Lipase OFC/OFG	Industrial
Lipase, immobilized in Sol-Gel-AK; Origin: Candida cylindracea Fluka	Pilot
Lipase; Origin: Candida lipolytica Fluka	Pilot
Lipase; Origin: Candida lypolytica Altus	Industrial
Lipase; Origin: Candida rugosa Altus	Industrial
Lipase; Origin: Candida rugosa Altus: ChiroCLEC-CR (dry)	Industrial
Lipase; Origin: Candida rugosa Altus: ChiroCLEC-CR (slurry)	Industrial
Lipase; Origin: Candida rugosa Amano: Lipase AY "Amano" 30	Industrial
Lipase; Origin: Candida rugosa Amano: Lipase AYS	Industrial
Lipase; Origin: Candida rugosa (formerly C. cylindracea) Roche Diagnostics: CHIRAZYME L-3, lyo.	Industrial
Lipase, purified; Origin: Candida rugosa (formerly C. cylindracea) Roche Diagnostics: CHIRAZYME L-3, purified, lyo.	Pilot
Lipase, purified, immobilized; Origin: Candida rugosa (formerly C. cylindracea) Roche Diagnostics: CHIRAZYME L-3, purified, c.-f., C2, lyo.	Pilot
Lipase; Origin: Candida utilis Fluka	Lab
Lipase; Origin: Chromobacterium viscosum Altus	Industrial
Lipase; Origin: Cromobacterium viscosum Asahi	Pilot
Lipase; Origin: Geotrichum candidum Altus	Industrial
Lipase; Origin: hog pancreas Fluka	Lab
Lipase; Origin: hog pancreas Fluka	Industrial
Lipase, immobilized in Sol-Gel-AK; Origin: hog pancreas Fluka	Pilot
Pancreatin; Origin: hog pancreas Fluka	Industrial
Lipase; Origin: Humicola sp. Roche Diagnostics: CHIRAZYME L-8, sol.	Industrial
Lipase B, covalently linked to carrier; Origin: microorganisms Fluka	Pilot
Lipoprotein Lipase; Origin: microorganisms Amano: Amano 6 [LPL-6]	Pilot
Lipase; Origin: Mucor javanicus Altus	Industrial
Lipase; Origin: Mucor javanicus Amano: Lipase M "Amano" 10	Industrial

Table 20.5. (cont.).

Lipase; Origin: Mucor javanicus Fluka	Industrial
Lipase; Origin: Mucor meihei Altus	Industrial
Lipase; Origin: Mucor miehei Fluka	Lab
Lipase; Origin: Mucor miehei Roche Diagnostics: CHIRAZYME L-9, lyo.	Industrial
Lipase; Origin: Mucor miehei Roche Diagnostics: CHIRAZYME L-9, sol.	Industrial
Lipase, immobilized; Origin: Mucor miehei Fluka: Lipozyme®, immobilized	Industrial
Lipase, immobilized; Origin: Mucor miehei Roche Diagnostics: CHIRAZYME L-9, c.-f., C2, lyo.	Industrial
Lipase, immobilized; Origin: Mucor miehei Roche Diagnostics: CHIRAZYME L-9, c.-f., dry	Industrial
Lipase, immobilized in Sol-Gel-AK; Origin: Mucor miehei Fluka	Lab
Lipase, immobilized in Sol-Gel-AK on sintered glass; Origin: Mucor miehei Fluka	Lab
Lipase; Origin: Mucor miehei, rec. Fluka	Industrial
Lipase; Origin: Penicillium camembertii Amano: Lipase G "Amano" 50	Industrial
Lipase; Origin: Penicillium roqueforti Fluka	Pilot
Lipase; Origin: Penicillium roquefortii Amano: Lipase R	Industrial
Lipase; Origin: porcine pancreas Altus	Industrial
Lipase; Origin: porcine pancreas Roche Diagnostics: CHIRAZYME L-7, lyo.	Industrial
Lipase; Origin: porcine pancreas Roche Diagnostics: Lipase	Pilot
Lipase; Origin: Protein engineered in rec. Aspergillus Novozymes: Lipolase Ultra	Industrial
Lipase; Origin: Protein engineered in rec. Aspergillus Novozymes: LipoPrime™	Industrial
Lipase; Origin: Pseudomonas aeroginosa Altus	Industrial
Lipase; Origin: Pseudomonas cepacia Altus	Industrial
Lipase; Origin: Pseudomonas cepacia Altus: ChiroCLEC-PC (dry)	Industrial
Lipase; Origin: Pseudomonas cepacia Altus: ChiroCLEC-PC (slurry)	Industrial
Lipase; Origin: Pseudomonas cepacia Amano: Lipase PS	Industrial
Lipase; Origin: Pseudomonas cepacia Amano: Lipase PS-C	Industrial

Table 20.5.   (cont.).

Lipase; Origin: Pseudomonas cepacia Amano: Lipase PS-D	Industrial
Lipase; Origin: Pseudomonas cepacia Fluka	Industrial
Lipase; Origin: Pseudomonas cepacia Fluka	Industrial
Lipase, immobilized in Sol-Gel-AK; Origin: Pseudomonas cepacia Fluka	Pilot
Lipase, immobilized in Sol-Gel-AK on sintered glass; Origin: Pseudomonas cepacia Fluka	Lab
Lipase, immobilized on Ceramic particles; Origin: Pseudomonas cepacia Fluka	Lab
Lipase; Origin: Pseudomonas fluorescens Amano: Lipase AK	Industrial
Lipase; Origin: Pseudomonas fluorescens Fluka	Pilot
Lipase, immobilized in Sol-Gel-AK; Origin: Pseudomonas fluorescens Fluka	Pilot
Lipase, immobilized in Sol-Gel-AK on sintered glass; Origin: Pseudomonas fluorescens Fluka	Lab
Lipase, immobilized on Eupergit C; Origin: Pseudomonas fluorescens Fluka	Lab
Lipase; Origin: Pseudomonas sp. Roche Diagnostics: CHIRAZYME L-6, lyo.	Industrial
Lipase, immobilized; Origin: Pseudomonas sp. Toyobo	Industrial
Lipoprotein Lipase; Origin: Pseudomonas sp. Amano: Amano 3 [LPL-3]	Pilot
Lipoprotein Lipase; Origin: Pseudomonas sp. Toyobo	Pilot
Lipase; Origin: Pseudomonas stutzeri Meito Sangyo: Lipase TL	Industrial
Lipase; Origin: Rhizomucor miehei DSM Gist-brocades: Piccantase	Industrial
Lipase; Origin: Rhizomucor miehei Fluka	Industrial
Lipase; Origin: Rhizomucor miehei, rec. in Aspergillus oryzae Novozymes: Novozym® 388	Industrial
Lipase; Origin: Rhizomucor miehei, rec. in Aspergillus oryzae Novozymes: Palatase®	Industrial
Lipase, immobilized; Origin: Rhizomucor miehei, rec. in Aspergillus oryzae Novozymes: Lipozyme® RM IM	Industrial
Lipase; Origin: Rhizopus arrhizus Fluka	Industrial
Lipase; Origin: Rhizopus delemar Altus	Industrial
Lipase; Origin: Rhizopus delemar Fluka	Industrial
Lipase; Origin: Rhizopus niveus Altus	Industrial

Table 20.5. (cont.).

Lipase; Origin: Rhizopus niveus Amano: Newlase F	Industrial
Lipase; Origin: Rhizopus niveus Fluka	Pilot
Lipase; Origin: Rhizopus niveus Jülich Enzyme Products: Lipase LE9	Lab
Lipase; Origin: Rhizopus oryzae Altus	Industrial
Lipase; Origin: Rhizopus oryzae Amano: Lipase F-AP15	Industrial
Lipase; Origin: Rhizopus oryzae Amano: Lipase F-DS	Industrial
Lipase; Origin: Rhizopus sp. Meito Sangyo: Lipase UL	
Lipase; Origin: Thermomyces lanuginosa Altus	Industrial
Lipase; Origin: Thermomyces lanuginosa, rec. Aspergillus oryzae Novozymes: Novozym 677 BG	Industrial
Lipase; Origin: Thermomyces lanuginosa rec. in Aspergillus oryzae Novozymes: Greasex®	Industrial
Lipase; Origin: Thermomyces lanuginosa rec. in Aspergillus oryzae Novozymes: Lipolase®	Industrial
Lipase; Origin: Thermomyces lanuginosa rec. in Aspergillus oryzae Novozymes: Novozym® 27007	Industrial
Lipase; Origin: Thermomyces lanuginosa rec. in Aspergillus oryzae Novozymes: Novozym® 398	Industrial
Lipase; Origin: Thermomyces lanuginosa rec. in Aspergillus oryzae Novozymes: Novozym® 735	Industrial
Lipase; Origin: Thermomyces lanuginosa rec. in Aspergillus oryzae Novozymes: Novozym® 871	Industrial
Lipase; Origin: Thermomyces sp. (formerly Humicola sp.) Roche Diagnostics: CHIRAZYME L-8, lyo.	Industrial
Lipase; Origin: Thermus aquaticus Fluka	Lab
Lipase; Origin: Thermus flavus Fluka	Lab
Lipase; Origin: Thermus thermophilus Fluka	Lab
Lipase; Origin: wheat germ Fluka	Pilot

Phospholipase A2.   **3.1.1.4**

Phosphatidylcholine 2-acylhydrolase. Lecithinase A.   Phosphatidylcholine + H_2O = 1-acylglycerophosphocholine + a
Phosphatidase. Phosphatidolipase.   fattyacid anion.

Phospholipase A2; Origin: Agkistrodon halys Fluka	Lab
Phospholipase A2; Origin: bovine pancreas Fluka	Pilot
Phospholipase A2; Origin: hog pancreas Fluka	Pilot

Table 20.5.   (cont.).

Phospholipase A2; Origin: porcine pancreas
Biocatalysts
Pilot

Phospholipase A2; Origin: porcine pancreas
Novozymes: Lecitase®
Industrial

Phospholipase A2; Origin: Streptomyces violaceoruber
Asahi
Pilot

Acetylcholinesterase.   3.1.1.7
True cholinesterase. Choline esterase I. Cholinesterase.
Acetylcholine + H(2)O = choline + acetate.

Acetylcholinesterase; Origin: bovine erythocytes
Biozyme
Pilot

Acetylcholine Esterase; Origin: Electrophorus electricus
Fluka
Lab

Cholinesterase.   3.1.1.8
Pseudocholinesterase. Acylcholine acylhydrolase. Butyrylcholine
esterase. Non-specific cholinesterase.
An acylcholine + H(2)O = choline + a carboxylic acid anion.

Butyrylcholine esterase; Origin: horse serum
Biozyme
Pilot

Butyrylcholine Esterase; Origin: horse serum
Fluka
Lab

Pectinesterase.   3.1.1.11
Pectin methylesterase. Pectin demethoxylase. Pectin methoxylase.
Pectin + N H(2)O = N methanol + pectate.

Pectinesterase
Novozymes: Cellubrix®
Industrial

Pectinesterase
Novozymes: Novoclair™ FCE
Industrial

Pectinesterase
Novozymes: Pectinex BE
Industrial

Pectinesterase; Origin: Aspergillus niger
Amano: Pectinase P
Industrial

Pectinesterase; Origin: Aspergillus niger
Amano: Pectinase PL "Amano"
Industrial

Pectin Esterase; Origin: orange peel
Fluka
Industrial

Pectinesterase; Origin: rec. microorganism
Novozymes: Novoshape®
Industrial

Pectinesterase; Origin: rec. microorganism
Novozymes: Pectinex SMASH
Industrial

Sterol esterase.   3.1.1.13
Cholesterol esterase. Cholesterol ester synthase. Triterpenol
esterase.
A steryl ester + H(2)O = a sterol + a fatty acid.

Cholesterol Esterase
Asahi
Pilot

Cholesterol Esterase, lyo.; Origin: Candida rugosa (formerly C. cylindracea)
Roche Diagnostics: Cholesterol Esterase, lyo.
Pilot

Cholesterol Esterase, sol.; Origin: Candida rugosa (formerly C. cylindracea)
Roche Diagnostics: Cholesterol Esterase, sol.
Pilot

Cholesterol Esterase; Origin: hog pancreas
Fluka
Pilot

Table 20.5. (cont.).

Cholesterol esterase; Origin: pig pancreas Biozyme	Pilot
Cholesterol Esterase; Origin: Pseudomonas sp. Amano: Amano 2 [CHE-2]	Pilot
Cholesterol Esterase; Origin: Pseudomonas sp. Amano: Amano 3 [CHE-3]	Pilot
Cholesterol Esterase; Origin: Pseudomonas sp. Asahi	Pilot
Cholesterol Esterase; Origin: Pseudomonas sp. Toyobo	Pilot

Tannase. **3.1.1.20**

Digallate + H(2)O = 2 gallate.

Tannase; Origin: Aspergillus ficuum Jülich Enzyme Products	Lab

Lipoprotein lipase. **3.1.1.34**

Clearing factor lipase. Diglyceride lipase. Diacylglycerol lipase.   Triacylglycerol + H(2)O = diacylglycerol + a fatty acid anion.

Lipoprotein Lipase; Origin: Chromobacterium viscosum Fluka	Pilot
Lipoprotein Lipase; Origin: Pseudomonas fluorescens Fluka	Pilot
Lipoprotein Lipase; Origin: Pseudomonas sp. Fluka	Pilot

Alkaline phosphatase. **3.1.3.1**

Alkaline phosphomonoesterase. Phosphomonoesterase.   An orthophosphoric monoester + H(2)O = an alcohol + phosphate.
Glycerophosphatase.

Phosphatase, alkaline Seravac	Pilot
Phosphatase, alkaline; Origin: Bacillus sp. Biocatalysts	Pilot
Phosphatase alkaline; Origin: bovine intestinal mucosa Fluka	Industrial
Phosphatase alkaline; Origin: calf intestinal mucosa Fluka	Industrial
Phosphatase alkaline, immobilized on Agarose; Origin: calf intestinal mucosa Fluka	Industrial
Phosphatase alkaline, immobilized on Agarose; Origin: calf intestinal mucosa Fluka	Industrial
Phosphatase, alkaline; Origin: calf intestine Roche Diagnostics: Phosphatase, alkaline, EIA Grade	Industrial
Phosphatase, alkaline; Origin: calf intestine or kidney Biozyme	Pilot
Phosphatase, alkaline, highly active; Origin: calf intestine, rec. in Pichia pastoris Roche Diagnostics: Phosphatase, alkaline, EIA Grade, highly active	Industrial
Phosphatase, alkaline; Origin: E. coli Fluka	Lab
Phosphatase, alkaline ; Origin: Escherichia coli Asahi	Pilot
Phosphatase, alkaline; Origin: microorganisms Unitika	Pilot

Table 20.5. (cont.).

Acid phosphatase.   3.1.3.2

Acid phosphomonoesterase. Phosphomonoesterase. Glycerophosphatase.

An orthophosphoric monoester + $H(2)O$ = an alcohol + phosphate.

Phosphatase, acid; Origin: potato
Roche Diagnostics: Phosphatase, acid, grade II   Pilot

Phosphatase, acid; Origin: potatoes
Fluka   Pilot

3-phytase.   3.1.3.8

Phytase. Phytate 3-phosphatase. Myo-inositol-hexaphosphate 3-phosphohydrolase.

Myo-inositol hexakisphosphate + $H(2)O$ = D-myo-inositol1,2,4,5,6-pentakisphosphate + phosphate.

Phytase; Origin: Aspergillus niger
Amano   Industrial

Phytase; Origin: Peniophora lycii, rec. In Asp. oryzae
Novozymes: Bio-feed® Phytase   Industrial

Phospholipase C.   3.1.4.3

Lipophosphodiesterase I. Lecithinase C. Clostridium welchii alpha-toxin. Clostridium oedematiens beta- and gamma-toxins.

A phosphatidylcholine + $H(2)O$ = 1,2-diacylglycerol + cholinephosphate.

Phospholipase C; Origin: Bacillus cereus
Asahi   Pilot

Phospholipase C; Origin: Bacillus cereus
Fluka   Pilot

Phospholipase C; Origin: Clostridium perfringens
Fluka   Pilot

Phospholipase D.   3.1.4.4

Lipophosphodiesterase II. Lecithinase D. Choline phosphatase.

A phosphatidylcholine + $H(2)O$ = choline + a phosphatidate.

Glycerophosphorylcholine Phosphodiesterase; Origin: microorganisms
Asahi   Pilot

Phospholipase D; Origin: Streptomyces chromofuscus
Asahi   Pilot

Phospholipase D; Origin: Streptomyces chromofuscus
Fluka   Pilot

Phospholipase D; Origin: Streptomyces sp.
Asahi   Pilot

Sphingomyelin phosphodiesterase.   3.1.4.12

Acid sphingomyelinase. Neutral sphingomyelinase.

Sphingomyelin + $H(2)O$ = N-acylsphingosine + choline phosphate.

Sphingomyelinase; Origin: Streptomyces sp.
Asahi   Pilot

Deoxyribonuclease I.   3.1.21.1

Pancreatic DNase. DNase. Thymonuclease.

Endonucleolytic cleavage to 5'-phosphodinucleotide and5'-phosphooligonucleotide end-products.

Deoxyribonuclease I; Origin: bovine pancreas
Fluka   Pilot

Ribonuclease T1.   3.1.27.3

Guanyloribonuclease. Aspergillus oryzae ribonuclease. RNase N1. RNase N2.

Two-stage endonucleolytic cleavage to 3'-phosphomononucleotides and3'-phosphooligonucleotides ending in G-P with 2',3'-cyclic phosphateintermediates.

Ribonuclease T1; Origin: Aspergillus oryzae
Fluka   Lab

Table 20.5. (cont.).

Pancreatic ribonuclease.	**3.1.27.5**
RNase. RNase I. RNase A. Pancreatic RNase.	Endonucleolytic cleavage to 3'-phosphomononucleotides and 3'-phosphooligonucleotides ending in C-P or U-P with 2',3'-cyclicphosphate intermediates.

Ribonuclease; Origin: beef pancreas	
Biozyme	Pilot
Ribonuclease A; Origin: bovine pancreas	
Fluka	Pilot
Ribonuclease, immobilized on EupergitC; Origin: bovine pancreas	
Fluka	Lab

Aspergillus nuclease S1.	**3.1.30.1**
Endonuclease S1. Single-stranded-nucleate endonuclease. Deoxyribonuclease S1.	Endonucleolytic cleavage to 5'-phosphomononucleotide and 5'-phosphooligonucleotide end-products.

Nuclease; Origin: Penicillium citrinum	
Amano: Enzyme RP-1	Industrial
Nuclease P1; Origin: Penicillium citrinum	
Fluka	Lab

Micrococcal nuclease.	**3.1.31.1**
Micrococcal endonuclease.	Endonucleolytic cleavage to 3'-phosphomononucleotide and 3'-phosphooligonucleotide end-products.

Nuclease micrococcal; Origin: Staphylococcus aureus	
Fluka	Lab

Alpha-amylase.	**3.2.1.1**
1,4-alpha-D-glucan glucanohydrolase. Glycogenase.	Endohydrolysis of 1,4-alpha-glucosidic linkages in oligosaccharides and polysaccharides.

Amylase; Origin: Aspergillus niger	
Amano: Gluczyme NL4.2	Industrial
alpha-Amylase; Origin: Aspergillus oryzae	
Fluka	Industrial
Amylase; Origin: Aspergillus oryzae	
Amano: Amylase DS	Industrial
Amylase; Origin: Aspergillus oryzae	
Amano: Biozyme F10 SD	Industrial
Amylase; Origin: Aspergillus oryzae	
Amano: Biozyme S Conc.	Industrial
Taka-Diastase; Origin: Aspergillus oryzae	
Fluka	Pilot
alpha-Amylase; Origin: Bacillus amyloliquefaciens	
Fluka	Industrial
alpha-Amylase; Origin: Bacillus licheniformis	
Fluka	Industrial
alpha-Amylase; Origin: Bacillus subtilis	
Fluka	Industrial
Amylase; Origin: Bacillus subtilis	
Amano: Amylase A "Amano" Conc.	Industrial
alpha-Amylase; Origin: fungus	
Novozymes: Fungamyl®	Industrial
alpha-Amylase; Origin: hog pancreas	
Fluka	Lab

Table 20.5. (cont.).

Amylase; Origin: microbacterium Amano: AMT "Amano"	Industrial
Amylase; Origin: microorganisms Novozymes: Aquazym®	Industrial
Amylase; Origin: microorganisms Novozymes: BAN (Bacterial Amylase Novo)	Industrial
alpha-Amylase; Origin: rec. microorganism Novozymes: Duramyl™	Industrial
alpha-Amylase; Origin: rec. microorganism Novozymes: Liquozyme®	Industrial
alpha-Amylase; Origin: rec. microorganism Novozymes: Termamyl®	Industrial
alpha-Amylase; Origin: rec. microorganism Novozymes: Termamyl, Type LS	Industrial
alpha-Amylase; Origin: rec. microorganism Novozymes: Thermozyme™	Industrial
Amylase; Origin: Rhizopus niveus Amano: Gluczyme 12	Industrial

Beta-amylase. 1,4-alpha-D-glucan maltohydrolase. Saccharogen amylase. Glycogenase.	**3.2.1.2** Hydrolysis of 1,4-alpha-glucosidic linkages in polysaccharides so asto remove successive maltose units from the non-reducing ends of the chains.

beta-Amylase; Origin: barley Fluka	Pilot
beta-Amylase; Origin: sweet potato Fluka	Pilot

Glucan 1,4-alpha-glucosidase. Glucoamylase. 1,4-alpha-D-glucan glucohydrolase. Amyloglucosidase. Gamma-amylase.	**3.2.1.3** Hydrolysis of terminal 1,4-linked alpha-D-glucose residues successively from non-reducing ends of the chains with release ofbeta-D-glucose.

Amyloglucosidase; Origin: Aspergillus niger Fluka	Industrial
Amyloglucosidase; Origin: rec. microorganism Novozymes: AMG	Industrial

Cellulase. Endoglucanase. Endo-1,4-beta-glucanase. Carboxymethyl cellulase.	**3.2.1.4** Endohydrolysis of 1,4-beta-D-glucosidic linkages in cellulose.

Cellulase Novozymes: Novozym® 342	Industrial
Cellulase; Origin: Aspergillus niger Amano: Cellulase A "Amano" 3	Industrial
Cellulase; Origin: Aspergillus niger Amano: Cellulase DS	Industrial
Cellulase; Origin: Aspergillus niger Fluka	Industrial
Cellulase; Origin: fungus Novozymes: Celluzyme®	Industrial
Cellulase; Origin: Humicola insolens Fluka	Lab

Table 20.5. (cont.).

Cellulase; Origin: rec. microorganism Novozymes: Carezyme®	Industrial
Cellulase; Origin: rec. microorganism Novozymes: DeniMax®	Industrial
Cellulase; Origin: Trichoderma longibrachiatum Fluka	Industrial
Cellulase; Origin: Trichoderma reesei Fluka	Lab
Cellulase; Origin: Trichoderma viride Amano: Cellulase T "Amano" 4	Industrial
Cellulase; Origin: Trichoderma viride Jülich Enzyme Products	Lab

		3.2.1.6
Endo-1,3(4)-beta-glucanase. Endo-1,4-beta-glucanase. Endo-1,3-beta-glucanase. Laminarinase.	Endohydrolysis of 1,3- or 1,4-linkages in beta-D-glucans when the glucose residue whose reducing group is involved in the linkage to be hydrolysed is itself substituted at C-3.	

beta-glucanase Novozymes: Cereflo®	Industrial
beta-glucanase Novozymes: Finizym®	Industrial
beta-glucanase, heat-stable Novozymes: Ultraflo®	Industrial
beta-Glucanase; Origin: Aspergillus niger Fluka	Industrial
beta-Glucanase; Origin: Bacillus subtilis Fluka	Industrial

		3.2.1.7
Inulinase. Inulase.	Endohydrolysis of 2,1-beta-D-fructosidic linkages in inulin.	

Inulinase; Origin: Aspergillus niger Fluka	Lab

		3.2.1.8
Endo-1,4-beta-xylanase. 1,4-beta-D-xylan xylanohydrolase.	Endohydrolysis of 1,4-beta-D-xylosidic linkages in xylans.	

Xylanase Novozymes: Pulpzyme™ HC	Industrial
Xylanase; Origin: Aspergillus niger Amano: Hemicellulase "Amano" 90	Industrial
Xylanase; Origin: Aspergillus niger Amano: Hemicellulase "Amano" 90	Industrial
Xylanase; Origin: bacteria Fluka	Industrial
Xylanase; Origin: rec. microorganism Novozymes: Pentopan™ Mono	Industrial
Xylanase; Origin: rec. microorganism Novozymes: Shearzyme™	Industrial
Xylanase; Origin: Trichoderma viride Fluka	Lab

Table 20.5. (cont.).

Dextranase.	**3.2.1.11**
Alpha-1,6-glucan-6-glucanohydrolase.	Endohydrolysis of 1,6-alpha-D-glucosidic linkages in dextran.

Dextranase
Novozymes: Dextranase

Industrial

Dextranase; Origin: Chaetomium erraticum)
Amano: Dextranase L "Amano"

Industrial

Dextranase; Origin: Paecilomyces lilacinus
Fluka

Pilot

Chitinase.	**3.2.1.14**
Chitodextrinase. 1,4-beta-poly-N-acetylglucosaminidase. Poly-beta-glucosaminidase.	Hydrolysis of the 1,4-beta-linkages of N-acetyl-D-glucosamine polymers of chitin.

Chitinase; Origin: bean leaves
Jülich Enzyme Products: Chitinase BB

Lab

Chitinase; Origin: Streptomyces griseus
Fluka

Pilot

Chitinase; Origin: sugar beet
Fluka

Lab

Chitinase; Origin: sugar-beet leaves
Jülich Enzyme Products: Chitinase ZR

Lab

Polygalacturonase.	**3.2.1.15**
Pectin depolymerase. Pectinase.	Random hydrolysis of 1,4-alpha-D-galactosiduronic linkages in pectate and other galacturonans.

Pectinase; Origin: Aspergillus niger
Fluka

Industrial

Pectinase; Origin: mould
Fluka

Industrial

Pectinase; Origin: Rhizopus sp.
Fluka

Industrial

Lysozyme.	**3.2.1.17**
Muramidase.	Hydrolysis of the 1,4-beta-linkages between N-acetyl-D-glucosamine and N-acetylmuramic acid in peptidoglycan heteropolymers of the prokaryotes cell walls.

Lysozyme
Seravac

Pilot

Lysozyme; Origin: chicken egg white
Biozyme

Industrial

Lysozyme; Origin: hen egg white
Fluka

Industrial

Exo-alpha-sialidase.	**3.2.1.18**
Sialidase. Neuraminidase. N-acylneuraminate glycohydrolase. Alpha-neuraminidase.	Hydrolysis of alpha-(2->3)-, alpha-(2->6)-, alpha-(2->8)-glycosidic linkages of terminal sialic residues in oligosaccharides, glycoproteins, glycolipids, colominic acid and synthetic substrates.

Neuraminidase; Origin: Clostridium perfringens
Fluka

Pilot

Neuraminidase, immobilized on Agarose4B; Origin: Clostridium perfringens
Fluka

Pilot

Neuraminidase; Origin: microorganisms
Unitika

Pilot

Table 20.5. (cont.).

Neuraminidase; Origin: Streptococcus sp. Toyobo	Pilot
Neuraminidase; Origin: Vibrio cholerae Fluka	Lab

	3.2.1.20
Alpha-glucosidase. Maltase. Glucoinvertase. Glucosidosucrase. Maltase-glucoamylase.	Hydrolysis of terminal, non-reducing 1,4-linked D-glucose residues with release of D-glucose.

alpha-Glucosidase; Origin: Aspergillus niger Fluka	Industrial
alpha-Glucosidase; Origin: Bacillus stearothermophilus Unitika: alpha-Glucosidase (alpha-Glu)	Pilot
alpha-Glucosidase; Origin: microorganisms Toyobo	Pilot
alpha-Glucosidase; Origin: yeast Biozyme	Pilot
alpha-Glucosidase; Origin: yeast Fluka	Lab
alpha-Glucosidase; Origin: yeast overproducer Roche Diagnostics: alpha-Glucosidase (alpha-Gluc)	Industrial

	3.2.1.21
Beta-glucosidase. Gentobiase. Cellobiase. Amygdalase.	Hydrolysis of terminal, non-reducing beta-D-glucose residues with release of beta-D-glucose.

Beta-Glucosidase Seravac	Pilot
beta-Glucosidase; Origin: almonds Fluka	Pilot
beta-Glucosidase; Origin: sweet almond Toyobo	Pilot
beta-Glucosidase; Origin: sweet almonds Biozyme	Pilot

	3.2.1.22
Alpha-galactosidase. Melibiase.	Melibiose + H(2)O = galactose + glucose.

alpha-Galactosidase; Origin: rec. microorganism Novozymes: Alpha-gal™	Industrial

	3.2.1.23
Beta-galactosidase. Lactase.	Hydrolysis of terminal, non-reducing beta-D-galactose residues in beta-D-galactosides.

Lactase; Origin: Aspergillus oryzae Amano: Lactase 14-DS	Industrial
Lactase; Origin: Aspergillus oryzae Amano: Lactase F "Amano"	Industrial
beta-Galactosidase; Origin: E. coli Fluka	Pilot
beta-Galactosidase; Origin: E. coli overproducer Roche Diagnostics: beta-Galactosidase	Industrial
beta-Galactosidase; Origin: Escherichia coli Toyobo	Pilot
beta-Galactosidase; Origin: Kluviromyces lactis Recordati: beta-Galactosidase, Lattasi beads	Industrial

Table 20.5.   (cont.).

beta-Galactosidase; Origin: Kluyveromyces fragilis
Fluka
                                                                          Pilot

Lactase; Origin: Kluyveromyces lactis
Novozymes: Lactozym®
                                                                          Industrial

beta-Galactosidase; Origin: microorganisms
Unitika: beta-Galactosidase (beta-Gal)
                                                                          Pilot

Alpha-mannosidase.   3.2.1.24
Hydrolysis of terminal, non-reducing alpha-D-mannose residues in
alpha-D-mannosides.

alpha-Mannosidase
Seravac
                                                                          Pilot

Beta-fructofuranosidase.   3.2.1.26
Invertase. Saccharase.
Hydrolysis of terminal non-reducing beta-D-fructofuranoside
residues in beta-D-fructofuranosides.

Invertase; Origin: Saccharomyces cerevisiae
Fluka
                                                                          Industrial

Beta-glucuronidase.   3.2.1.31
A beta-D-glucuronoside + $H(2)O$ = an alcohol + D-glucuronate.

Beta-Glucuronidase
Seravac
                                                                          Pilot

beta-Glucuronidase; Origin: bovine liver
Fluka
                                                                          Lab

beta-Glucuronidase; Origin: E. coli
Fluka
                                                                          Lab

beta-Glucuronidase; Origin: E. coli K12
Fluka
                                                                          Lab

beta-Glucuronidase; Origin: Helix pomatia
Fluka
                                                                          Lab

Hyaluronoglucosaminidase.   3.2.1.35
Hyaluronidase.
Random hydrolysis of 1,4-linkages between
N-acetyl-beta-D-glucosamine and D-glucuronate residues in
hyaluronate.

Hyaluronidase
Seravac
                                                                          Pilot

Hyaluronidase; Origin: bovine testes
Fluka
                                                                          Lab

Hyaluronidase; Origin: ovine or bovine testes
Biozyme
                                                                          Pilot

Hyaluronidase; Origin: sheep testes
Fluka
                                                                          Lab

Hyaluronidase; Origin: Streptomyces hyalurolyticus
Fluka
                                                                          Lab

Hyaluronidase; Origin: Streptomyces sp.
Amano: Amano 1 [HY-1]
                                                                          Pilot

Hyaluronidase; Origin: Streptomyces sp.
Amano: Amano 3 [HY-3]
                                                                          Pilot

Table 20.5.   (cont.).

Glucan endo-1,3-beta-D-glucosidase.   3.2.1.39
(1->3)-beta-glucan endohydrolase. Endo-1,3-beta-glucanase.   Hydrolysis of 1,3-beta-D-glucosidic linkages in 1,3-beta-D-glucans.
Laminarinase.

beta-1,3-D-Glucanase; Origin: Helix pomatia
Fluka   Pilot

Alpha-dextrin endo-1,6-alpha-glucosidase.   3.2.1.41
Pullulanase. Pullulan 6-glucanohydrolase. Limit dextrinase.   Starch-debranching enzyme, hydrolyses (1-6)-alpha-glucosidic
Debranching enzyme.   linkages in pullulan and starch to form maltotriose.

Pullulanase; Origin: Bacillus sp.
Amano: Debranchingenzyme "Amano" 8   Industrial

Pullulanase; Origin: Bacillus sp.
Fluka   Industrial

Pullulanase; Origin: rec. microorganism
Novozymes: Promozyme®   Industrial

Beta-N-acetylhexosaminidase.   3.2.1.52
Beta-hexosaminidase. Hexosaminidase.   Hydrolysis of terminal non-reducing N-acetyl-D-hexosamine
N-acetyl-beta-glucosaminidase.   residues in N-acetyl-beta-D-hexosaminides.

beta-N-Acetylglucosaminidase; Origin: jack bean
Fluka   Lab

Agarase.   3.2.1.81
                                          Hydrolysis of 1,3-beta-D-galactosidic linkages in agarose, giving the
                                                      tetramer as the predominant product.

Agarase; Origin: Pseudomonas atlantica
Fluka   Lab

Thioglucosidase.   3.2.3.1
Myrosinase. Sinigrinase. Sinigrase.   A thioglucoside + H(2)O = a thiol + sugar.

Myrosinase; Origin: Senapis alba (white mustard seed)
Biocatalysts   Pilot

Epoxide hydrolase.   3.3.2.3
Epoxide hydratase. Arene-oxide hydratase.   An epoxide + H(2)O = a glycol.

Epoxide Hydrolase; Origin: Agrobacterium sp.
Fluka   Lab

Epoxide Hydrolase; Origin: Aspergillus niger
Fluka   Lab

Epoxide Hydrolase; Origin: Rhodococcus rhodochrous
Fluka   Lab

Epoxide Hydrolase; Origin: Rhodotorula glutinis
Fluka   Lab

Hydrolases.   3.4.-.-
Acting on peptide bonds (peptide hydrolases).

Protease; Origin: Aspergillus melleus
Amano: Protease DS   Industrial

Protease; Origin: Aspergillus melleus
Amano: Protease P "Amano" 6   Industrial

Protease; Origin: Aspergillus niger
Altus   Industrial

Protease; Origin: Aspergillus niger
Amano: Acid Protease A   Industrial

Table 20.5.   (cont.).

Protease; Origin: Aspergillus niger Amano: Acid Protease DS	Industrial
Protease; Origin: Aspergillus niger Jülich Enzyme Products	Lab
Protease ; Origin: Aspergillus oryzae Altus	Industrial
Protease; Origin: Aspergillus oryzae Altus	Industrial
Protease; Origin: Aspergillus oryzae Altus	Industrial
Protease; Origin: Aspergillus oryzae Amano: Protease A "Amano" 2G	Industrial
Protease; Origin: Aspergillus oryzae Amano: Protease A-DS	Industrial
Protease; Origin: Aspergillus oryzae Amano: Protease M "Amano"	Industrial
Protease; Origin: Aspergillus oryzae Jülich Enzyme Products	Lab
Proteinase 2A; Origin: Aspergillus oryzae Fluka	Industrial
Protease; Origin: Aspergillus sp. Altus	Industrial
Protease, neutral; Origin: Bacillus amyloliquefaciens Novozymes: Neutrase®	Industrial
Protease; Origin: Bacillus licheniformis Novozymes: Bio-Feed® Pro	Industrial
Protease; Origin: Bacillus licheniformis Novozymes: Novozym® FM	Industrial
Protease, Endopeptidase ; Origin: Bacillus licheniformis Novozymes: Alcalase®	Industrial
Proteinase; Origin: Bacillus licheniformis Fluka	Industrial
Protease; Origin: Bacillus sp. Altus	Industrial
Protease; Origin: Bacillus sp. Novozymes: NovoCor S	Industrial
Protease, alkaline; Origin: Bacillus sp. Novozymes: Esperase®	Industrial
Proteinase, neutral; Origin: Bacillus sp. Toyobo	Industrial
Endoproteinase; Origin: Bacillus sp., rec. Fluka	Pilot
Protease; Origin: Bacillus sp., rec. Novozymes: Novo-Pro™ D	Industrial
Protease; Origin: Bacillus sp., rec. Novozymes: Pyrase®	Industrial
Protease; Origin: Bacillus stearothermophilus Amano: Protease S "Amano"	Industrial
Protease; Origin: Bacillus subtilis Amano: Proleather FG-F	Industrial

Table 20.5. (cont.).

Protease; Origin: Bacillus subtilis Amano: Protease N "Amano"	Industrial
Protease; Origin: Bacillus subtilis Amano: Protease NL "Amano"	Industrial
Protease; Origin: Bacillus subtilis Jülich Enzyme Products	Lab
Proteinase; Origin: Bacillus subtilis Fluka	Industrial
Proteinase; Origin: Bacillus subtilis var. biotecus A Fluka	Industrial
Protease; Origin: Carica papaya L. Amano: Papain W-40	Industrial
Protease, Proline-Specific Endopeptidase; Origin: Flavobacterium sp. Toyobo	Pilot
Protease, neutral to acidic; Origin: Fungus Novozymes: NovoCor P	Industrial
Protease, alkaline; Origin: microorganism, rec. in Bacillus sp. Novozymes: Savinase®	Industrial
Protease; Origin: microorganisms DSM Gist-brocades: Fermizyme	Industrial
Protease; Origin: Penicillum sp. Altus	Industrial
Protease; Origin: Protein engineered in rec. Bacillus Novozymes: Kannase™	Industrial
Protease, alkaline; Origin: Protein engineered, rec. in Bacillus sp. Novozymes: Everlase®	Industrial
Protease; Origin: Rhizomucor miehei, rec. in Aspergillus oryzae DSM Gist-brocades: Optiren	Industrial
Protease, neutral to acidic; Origin: Rhizomucor sp. Novozymes: NovoCor® AB	Industrial
Protease; Origin: Rhizopus niveus Amano: Acid Protease	Industrial
Protease; Origin: Rhizopus oryzae Amano: Peptidase R	Industrial
Pronase; Origin: Streptomyces griseus Fluka	Industrial
Pronase; Origin: Streptomyces griseus Fluka	Industrial
Pronase nonspecific protease; Origin: Streptomyces griseus Roche Diagnostics: Pronase nonspecific protease	Industrial
Protease, alkalophilic; Origin: Streptomyces sp. Toyobo	Pilot

Leucyl aminopeptidase. Cytosol aminopeptidase. Leucine aminopeptidase. Peptidase S.	**3.4.11.1** Release of an N-terminal amino acid, Xaa-l-Xbb-, in which Xaa is preferably Leu, but may be other amino acids including Pro although not Arg or Lys, and Xbb may be Pro.
Leucine Aminopeptidase, cytosol; Origin: hog kidney Fluka	Lab
Leucine aminopeptidase; Origin: pig kidney Biozyme	Pilot

Table 20.5. (cont.).

Xaa-Pro dipeptidase.   3.4.13.9
X-Pro dipeptidase. Proline dipeptidase. Imidodipeptidase.   Hydrolysis of Xaa-I-Pro dipeptides; also acts on
Prolidase.   aminoacyl-hydroxyproline analogs. No action on Pro-I-Pro.

Prolidase; Origin: Lactococcus lactis
Fluka
                                                                                    Lab

Prolidase; Origin: pig kidney
Biozyme
                                                                                    Pilot

---

Dipeptidyl-peptidase I.   3.4.14.1
Cathepsin C. Cathepsin J. Dipeptidyl aminopeptidase I. Dipeptidyl   Release of an N-terminal dipeptide, Xaa-Xbb-I-Xcc, except when
transferase.   Xaa is Arg or Lys, or Xbb or Xcc is Pro.

Cathepsin C, sol.; Origin: bovine spleen
Roche Diagnostics: Cathepsin C, sol.
                                                                                    Industrial

---

Transferred entry: 3.4.16.5 and 3.4.16.6.   3.4.16.1

Carboxypeptidase Y; Origin: baker's yeast
Fluka
                                                                                    Lab

Carboxypeptidase Y; Origin: yeast
Roche Diagnostics: Carboxypeptidase Y, Sequencing Grade
                                                                                    Lab

---

Carboxypeptidase A.   3.4.17.1
Carboxypolypeptidase.
                                             Peptidyl-L-amino acid + H(2)O = peptide + L-amino acid.

Carboxypeptidase A
Seravac
                                                                                    Pilot

Carboxypeptidase A; Origin: bovine pancreas
Fluka
                                                                                    Lab

---

Membrane Pro-X carboxypeptidase.   3.4.17.16
Carboxypeptidase P. Microsomal carboxypeptidase.   Release of a C-terminal residue other than proline, by
                                                            preferentialcleavage of prolyl bond.

Carboxypeptidase P; Origin: Penicillium janthinellum
Fluka
                                                                                    Lab

---

Pyroglutamyl-peptidase I.   3.4.19.3
5-oxoprolyl-peptidase. Pyrrolidone-carboxylate peptidase.   5-oxoprolyl-peptide + H(2)O = 5-oxoproline + peptide.
Pyrrolidone carboxyl peptidase. Pyroglutamyl aminopeptidase.

Pyroglutamate Aminopeptidase; Origin: calf liver
Fluka
                                                                                    Pilot

---

Hydrolases.   3.4.21.-
Acting on peptide bonds (peptide hydrolases).
Serine endopeptidases.

Endoproteinase Pro-C; Origin: microorganism, rec. in E. coli
Fluka
                                                                                    Lab

---

Chymotrypsin.   3.4.21.1
Chymotrypsin A. Chymotrypsin B. Alpha-chymotrypsin.   Preferential cleavage: Tyr-I-Xaa, Trp-I-Xaa, Phe-I-Xaa, Leu-I-Xaa.

alpha-Chymotrypsin
Seravac
                                                                                    Pilot

alpha-Chymotrypsin; Origin: Bacillus licheniformis
Altus
                                                                                    Industrial

alpha-Chymotrypsin; Origin: bovine pancreas
Fluka
                                                                                    Pilot

Table 20.5. (cont.).

Trypsin.	**3.4.21.4**
Alpha- and beta-trypsin.	Preferential cleavage: Arg-l-Xaa, Lys-l-Xaa.

Trypsin	
Seravac	Pilot

Trypsin	
Seravac	Pilot

Trypsin; Origin: bovine pancreas	
Fluka	Industrial

Trypsin; Origin: pig Pancreas	
Biozyme	Industrial

Trypsin ; Origin: porcine pancreas	
Altus	Industrial

Trypsin; Origin: porcine pancreas	
Biocatalysts: Trypsin	Industrial

Trypsin; Origin: porcine pancreas	
Biocatalysts: Trypsin 250	Industrial

Trypsin; Origin: porcine pancreas	
Biocatalysts: Trypsin 250	Industrial

Trypsin; Origin: porcine pancreas	
Novozymes: Crystalline Porcine Trypsin	Industrial

Trypsin (Chrymotrypsin as minor constituent); Origin: porcine pancreas	
Novozymes: PTN (Pancreatic Trypsin Novo)	Industrial

Thrombin.	**3.4.21.5**
Fibrinogenase.	Preferential cleavage: Arg-l-Gly; activates fibrinogen to fibrin andreleases fibrinopeptide A and B.

Thrombin; Origin: bovine plasma	
Fluka	Pilot

Enteropeptidase.	**3.4.21.9**
Enterokinase.	Selective cleavage of 6-Lys-l-Ile-7 bond in trypsinogen.

Enteropeptidase	
Seravac	Pilot

Glutamyl endopeptidase.	**3.4.21.19**
Staphylococcal serine proteinase. V8 proteinase. Protease V8. Endoproteinase Glu-C.	Preferential cleavage: Asp-l-Xaa, Glu-l-Xaa.

Endoprotease Glu-C; Origin: Endophrins	
Altus	Pilot

Endoproteinase Glu-C; Origin: Staphylococcus aureus strain V8	
Fluka	Lab

Endoproteinase Glu-C; Origin: Staphylococcus aureus strain V8	
Roche Diagnostics: Endoproteinase Glu-C, Sequencing Grade	Lab

Pancreatic elastase.	**3.4.21.36**
Pancreatopeptidase E. Pancreatic elastase I.	Hydrolysis of proteins, including elastin. Preferential cleavage:Ala-l-Xaa.

Elastase; Origin: hog pancreas	
Fluka	Pilot

Elastase; Origin: pig pancreas	
Biozyme	

Table 20.5. (cont.).

Elastase; Origin: porcine pancreas
Altus

Industrial

Lysyl endopeptidase.   **3.4.21.50**
Achromobacter proteinase I. Lysyl bond specific proteinase. Preferential cleavage: Lys-l-Xaa, including Lys-l-Pro.

Endoproteinase Lys-C; Origin: Lysobacter enzymogenes
Fluka

Lab

Endoproteinase Lys-C; Origin: Lysobacter enzymogenes
Roche Diagnostics: Endoproteinase Lys-C, Sequencing Grade

Lab

Subtilisin.   **3.4.21.62**
Hydrolysis of proteins with broad specificity for peptide bonds, and a preference for a large uncharged residue in P1. Hydrolyses peptide amides.

Protease, Subtilisin Carlsberg; Origin: Bacillus licheniformis
Novozymes: Subtilisin A

Pilot

Subtilisin; Origin: Bacillus licheniformis
Altus: ChiroCLEC-BL (dry)

Industrial

Subtilisin; Origin: Bacillus licheniformis
Altus: ChiroCLEC-BL (slurry)

Industrial

Subtilisin; Origin: Bacillus licheniformis
Altus: PeptiCLEC-BL (dry)

Industrial

Subtilisin; Origin: Bacillus licheniformis
Altus: PeptiCLEC-BL (slurry)

Industrial

Subtilisin; Origin: Bacillus licheniformis
Fluka

Industrial

Subtilisin; Origin: Bacillus licheniformis
Roche Diagnostics: CHIRAZYME P-1, lyo.

Industrial

Subtilisin; Origin: Bacillus licheniformis
Roche Diagnostics: CHIRAZYME P-1, sol.

Industrial

Subtilisin Carlsberg; Origin: Bacillus licheniformis
Altus

Industrial

Proteinase K.   **3.4.21.64**
Endopeptidase K. Tritirachium alkaline proteinase. Tritirachium album proteinase K.   Hydrolysis of keratin and of other proteins, with subtilisin-like specificity. Hydrolyses peptides amides.

Proteinase N; Origin: Bacillus subtilis
Fluka

Industrial

Proteinase K; Origin: Tritirachium album
Fluka

Industrial

Proteinase K, immobilized on EupergitC; Origin: Tritirachium album
Fluka

Lab

Proteinase K, lyo.; Origin: Tritirachium album
Roche Diagnostics: Proteinase K, lyo.

Industrial

Papain.   **3.4.22.2**
Papaya peptidase I.   Hydrolysis of proteins with broad specificity for peptide bonds, with preference for a residue bearing a large hydrophobic sidechain at the P2 position. Does not accept Val at P1'.

Papain
Biocatalysts: PROMOD 144L

Industrial

Papain; Origin: Carica papaya
Fluka

Industrial

Table 20.5. (cont.).

Papain; Origin: Carica papaya Roche Diagnostics: Papain	Industrial
Papain, immobilized on Eupergit® C; Origin: Carica papaya Fluka	Lab
Papain ; Origin: Papaya Latex Altus	Industrial

	3.4.22.6
Chymopapain. Papaya proteinase II.	Specificity similar to that of papain.

Chymopapain; Origin: Papaya Latex Altus	Industrial

	3.4.22.8
Clostripain. Clostridiopeptidase B.	Preferential cleavage: Arg-l-Xaa, including Arg-l-Pro bond, but not Lys-l-Xaa.

Clostripain; Origin: Clostridium histolyticum Altus	Industrial
Clostripain; Origin: Clostridium histolyticum Fluka	Pilot
Endoproteinase Arg-C; Origin: Clostridium histolyticum Roche Diagnostics: Endoproteinase Arg-C, Sequencing Grade	Lab
Clostripain rec.; Origin: E. coli Fluka	Pilot
Endoproteinase Arg-C; Origin: mice (submaxillary glands) Fluka	Lab

	3.4.22.32
Stem bromelain. Bromelain.	Broad specificity for cleavage of proteins, but strong preference for Z-Arg-Arg-l-NHMec amongst small molecule substrates.

Bromelain; Origin: pinapple stem Altus	Industrial
Bromelain; Origin: pineapple stem Fluka	Pilot

	3.4.23.1
Pepsin A. Pepsin.	Preferential cleavage: hydrophobic, preferably aromatic, residues in P1 and P1' positions. Cleaves 1-Phe-l-Val-2, 4-Gln-l-His-5, 13-Glu-l-Ala-14, 14-Ala-l-Leu-15, 15-Leu-l-Tyr-16,

Pepsin; Origin: hog stomach Fluka	Industrial
Pepsin; Origin: pig stomach mucosa Biozyme	Pilot
Pepsin ; Origin: porcine pancreas Altus	Industrial

	3.4.23.4
Chymosin. Rennin.	Broad specificity similar to that of pepsin A. Clots milk by cleavageof a single bond in casein (kappa chain).

Chymosin; Origin: calf stomach Altus	Industrial

Table 20.5.   (cont.).

Microbial collagenase.
3.4.24.3

Clostridium histolyticum collagenase. Clostridiopeptidase A.
Collagenase A. Collagenase I.

Digestion of native collagen in the triple helical region at Xaa-l-Gly bonds. With synthetic peptides, a preference is shown for Gly at P3 and P1'; Pro and Ala at P2 and P2'; and hydroxyproline, Ala or Arg

Collagenase; Origin: Clostridium histolyticum
Fluka

Pilot

Collagenase; Origin: Clostridium histolyticum
Fluka

Pilot

Collagenase; Origin: Clostridium sp.
Amano: Amano 1 [CL-1]

Pilot

Collagenase; Origin: Clostridium sp.
Amano: Amano S [CL-S]

Pilot

Thermolysin.
3.4.24.27

Bacillus thermoproteolyticus neutral proteinase.

Preferential cleavage: Xaa-l-Leu > Xaa-l-Phe.

Thermolysin; Origin: Bacillus thermoproteolyticus
Altus: PeptiCLEC-TR (dry)

Industrial

Thermolysin; Origin: Bacillus thermoproteolyticus
Altus: PeptiCLEC-TR (slurry)

Industrial

Thermolysin; Origin: Bacillus thermoproteolyticus
Fluka

Pilot

Peptidyl-Asp metalloendopeptidase.
3.4.24.33

Endoproteinase Asp-N.

Cleavage of Xaa-l-Asp, Xaa-l-Glu and Xaa-l-cysteic acid bonds.

Endoproteinase Asp-N; Origin: Pseudomonas fragi (mutant)
Fluka

Lab

Endoproteinase Asp-N; Origin: Pseudomonas fragi (mutant)
Roche Diagnostics: Endoproteinase Asp-N, Sequencing Grade

Lab

Hydrolases.
Acting on carbon-nitrogen bonds, other than peptide bonds.
3.5.-.-

Peptide Amidase; Origin: Citrus sinensis
Fluka

Lab

Hydrolases.
Acting on carbon-nitrogen bonds, other than peptide bonds.
In linear amides.
3.5.1.-

Glutaryl Acylase, immobilized
Fluka

Industrial

Glutaryl acylase, carrier-fixed; Origin: E. coli overproducer
Roche Diagnostics: Glutaryl acylase, carrier-fixed (Gl-Ac)

Industrial

Glutaryl-7-ACA Acylase; Origin: microorganism, rec. in E. coli
Recordati: GAA Beads

Industrial

Asparaginase.
3.5.1.1

L-asparaginase. L-asparagine amidohydrolase.

L-asparagine + H_2O = L-aspartate + NH_3.

L-Asparaginase; Origin: E. coli
Fluka

Lab

Table 20.5. (cont.).

Glutaminase.	**3.5.1.2**
L-glutamine amidohydrolase.	L-glutamine + H_2O = L-glutamate + NH_3.

Glutaminase; Origin: Bacillus subtilis	Industrial
Amano: Glutaminase F "Amano" 100	

Amidase.	**3.5.1.4**
Acylamidase. Acylase.	A monocarboxylic acid amide + H_2O = a monocarboxylate + NH_3.

Amidase; Origin: Pseudomonas aeruginosa, rec. in E. coli	Lab
Fluka	

Urease.	**3.5.1.5**
	Urea + H_2O = CO_2 + 2 NH_3.

Urease; Origin: jack bean	Industrial
Fluka	
Urease; Origin: jack bean	Pilot
Biozyme	
Urease; Origin: jack bean	Industrial
Roche Diagnostics: Urease	
Urease ; Origin: jack bean	Industrial
Seravac	
Urease; Origin: jack bean	Pilot
Toyobo	
Urease, immobilized on Eupergit C; Origin: jack bean	Lab
Fluka	

Penicillin amidase.	**3.5.1.11**
Penicillin acylase.	Penicillin + H_2O = a fatty acid anion + 6-aminopenicillanate.

Penicillin Acylase; Origin: E. coli	Industrial
Altus	
Penicillin Acylase; Origin: E. coli	Industrial
Altus: ChiroCLEC-EC (dry)	
Penicillin Acylase; Origin: E. coli	Industrial
Altus: ChiroCLEC-EC (slurry)	
Penicillin Amidase; Origin: E. coli	Industrial
Fluka	
Penicillin Amidase, immobilized on Eupergit® C; Origin: E. coli	Industrial
Fluka	
Penicillin G Amidase, immobilized; Origin: E. coli	Industrial
Fluka	
Penicillin-G Acylase; Origin: E. coli	Industrial
Recordati: PGA beads, Standard	
Penicillin-G Acylase; Origin: E. coli	Industrial
Recordati: PGA beads, Superenzyme	
Penicillin G Amidase; Origin: E. coli overproducer	Industrial
Roche Diagnostics: Penicillin G Amidase (PGA-450)	

Aminoacylase.	**3.5.1.14**
Histozyme. Hippuricase. Benzamidase. Dehydropeptidase II.	An N-acyl-L-amino acid + H_2O = a fatty acid anion + an L-amino acid.

Acylase I; Origin: Aspergillus melleus	Industrial
Fluka	

Table 20.5. (cont.).

Acylase I, immobilized on Eupergit C; Origin: Aspergillus sp.
Fluka	Pilot

Aminoacylase; Origin: Aspergillus sp.
Amano: Acylase	Industrial

Acylase I; Origin: hog kidney
Fluka	Pilot

Acylase I; Origin: pig kidney
Biozyme	Pilot

Acylase; Origin: Streptomyces chartreusis
Fluka	Lab

Acylase; Origin: Streptomyces griseocarneus
Fluka	Lab

Acylase; Origin: Streptomyces hachijoensis
Fluka	Lab

Acylase; Origin: Streptomyces toyocaensis
Fluka	Lab

Acylase; Origin: Streptomyces zaomyceticus
Fluka	Lab

Dihydropyrimidinase. **3.5.2.2**
Hydantoinase. 5,6-dihydrouracil + H(2)O = 3-ureidopropionate.

Hydantoinase; Origin: Agrobacterium radiobacter
Recordati: Hyda-REC	Industrial

D-Hydantoinase; Origin: Azuki beans
Fluka	Lab

D-Hydantoinase 1, carrier-fixed; Origin: Bacillus thermoglucosidasius, rec. in E. coli
Roche Diagnostics: D-Hydantoinase 1, carrier-fixed	Industrial

D-Hydantoinase, recombinant, immobilized; Origin: E. coli
Fluka	Industrial

Beta-lactamase. **3.5.2.6**
Penicillinase. Cephalosporinase. A beta-lactam + H(2)O = a substituted beta-amino acid.

beta-Lactamase I; Origin: Bacillus cereus
Fluka	Lab

beta-Lactamase; Origin: Enterobacter cloacae
Fluka	Lab

Creatininase. **3.5.2.10**
Creatinine amidohydrolase. Creatinine + H(2)O = creatine.

Creatininase; Origin: microorganisms
Asahi	Pilot

Arginase. **3.5.3.1**
Arginine amidinase. Canavanase. L-arginine + H(2)O = L-ornithine + urea.

Arginase
Biozyme: Bovine liver	Pilot

L-Arginase; Origin: bovine liver
Fluka	Lab

Table 20.5. (cont.).

Creatinase. Creatine amidinohydrolase.	**3.5.3.3** Creatine + H(2)O = sarcosine + urea.

Creatinase; Origin: Bacillus sp. Asahi	Pilot

Creatinase; Origin: Flavobacterium sp. Fluka	Pilot

Hydrolases. **Acting on carbon-nitrogen bonds, other than peptide bonds.** **In cyclic amidines.**	**3.5.4.-**

Deaminase; Origin: Aspergillus melleus Amano: Deamizyme 50000	Industrial

Adenosine deaminase. Adenosine aminohydrolase.	**3.5.4.4** Adenosine + H(2)O = inosine + NH(3).

Adenosine Deaminase; Origin: calf intestinal mucosa Fluka	Pilot

Adenosine Deaminase; Origin: calf intestine Roche Diagnostics: Adenosine Deaminase	Industrial

Nitrilase.	**3.5.5.1** A nitrile + H(2)O = a carboxylate + NH(3).

Nitrilase, broad-range; Origin: microorganism, rec. in E. coli BioCatalytics: NIT-101	Lab

Nitrilase, broad-range; Origin: microorganism, rec. in E. coli BioCatalytics: NIT-102	Lab

Nitrilase, broad-range; Origin: microorganism, rec. in E. coli BioCatalytics: NIT-103	Lab

Inorganic pyrophosphatase. Inorganic diphosphatase. Pyrophosphate phosphohydrolase. Diphosphate phosphohydrolase.	**3.6.1.1** Diphosphate + H(2)O = 2 phosphate.

Pyrophosphatase, inorganic; Origin: baker's yeast Fluka	Lab

Pyrophosphatase, inorganic; Origin: E. coli Fluka	Lab

Pyruvate decarboxylase. Alpha-carboxylase. Pyruvic decarboxylase. Alpha-ketoacid carboxylase.	**4.1.1.1** A 2-oxo acid = an aldehyde + CO(2).

Pyruvate Decarboxylase; Origin: Zymomonas mobils, E. coli (rec.) Jülich Enzyme Products	Lab

Oxaloacetate decarboxylase. Oxalate beta-decarboxylase.	**4.1.1.3** Oxaloacetate = pyruvate + CO(2).

Oxalacetate Decarboxylase; Origin: Pseudomonas sp. Fluka	Lab

Oxaloacetate Decarboxylase; Origin: Pseudomonas sp. Asahi	Pilot

Table 20.5. (cont.).

Acetolactate decarboxylase.
4.1.1.5

(S)-2-hydroxy-2-methyl-3-oxobutanoate = (R)-2-acetoin + CO_2.

Acetolactate Decarboxylase; Origin: Bac.Brevis in rec, Bac. subtilis
Novozymes: Maturex®

Industrial

alpha-Acetolactate Decarboxylase; Origin: Bacillus subtilis
Fluka

Lab

Lysine decarboxylase.
4.1.1.18

L-lysine = cadaverine + CO_2.

L-Lysine Decarboxylase; Origin: Bacterium cadaveris
Fluka

Lab

Tyrosine decarboxylase.
4.1.1.25

L-tyrosine = tyramine + CO_2.

L-Tyrosine Decarboxylase; Origin: Streptococcus faecalis
Fluka

Lab

Phosphoenolpyruvate carboxylase.
4.1.1.31

Phosphate + oxaloacetate = H_2O + phosphoenolpyruvate + CO_2.

Phosphoenolpyruvate carboxylase; Origin: maize leaves
Biozyme

Pilot

Phosphoenolpyruvate Carboxylase; Origin: maize leaves
Fluka

Lab

Phosphoenolpyruvate Carboxylase; Origin: microorganisms
Toyobo

Pilot

Phenylalanine decarboxylase.
4.1.1.53

L-phenylalanine = phenethylamine + CO_2.

L-Phenylalanine Decarboxylase; Origin: Streptococcus faecalis
Fluka

Lab

Methionine decarboxylase.
4.1.1.57

L-methionine = 3-methylthiopropanamine + CO_2.

L-Methionine decarboxylase; Origin: Streptomyces sp.
Fluka

Lab

Deoxyribose-phosphate aldolase.
4.1.2.4

Phosphodeoxyriboaldolase. Deoxyriboaldolase.

2-deoxy-D-ribose 5-phosphate = D-glyceraldehyde 3-phosphate
+acetaldehyde.

2-Deoxyribose-5-phosphate aldolase; Origin: Lactobacillus plantarum
Fluka

Lab

Threonine aldolase.
4.1.2.5

L-threonine = glycine + acetaldehyde.

Threonin Aldolase ; Origin: Candida humicola
Fluka

Lab

Threonin Aldolase; Origin: Pseudomonas putida
Fluka

Lab

Mandelonitrile lyase.
4.1.2.10

Hydroxynitrile lyase. (R)-oxynitrilase.

Mandelonitrile = cyanide + benzaldehyde.

(R)-Oxynitrilase; Origin: bitter almonds (Prunus amygdalus)
Jülich Enzyme Products

Lab

Table 20.5. (cont.).

R-Oxynitrilase rec.; Origin: Pichia pastoris
Fluka   Pilot

Hydroxymandelonitrile lyase.   **4.1.2.11**
Hydroxynitrile lyase.   (S)-4-hydroxymandelonitrile = cyanide + 4-hydroxybenzaldehyde.

(S)-Oxynitrilase; Origin: Sorghum bicolor or S. vulgare
Jülich Enzyme Products   Lab

Fructose-bisphosphate aldolase.   **4.1.2.13**
Aldolase. Fructose-1,6-bisphosphate triosephosphate-lyase.   D-fructose 1,6-bisphosphate = glycerone phosphate +
                                                                           D-glyceraldehyde 3-phosphate.

Fructose-1,6-bisphosphate aldolase; Origin: Bacillus subtilis
Fluka   Lab

Aldolase; Origin: rabbit muscle
Fluka   Lab

Aldolase; Origin: rabbit muscle
Roche Diagnostics: Aldolase   Lab

Aldolase; Origin: Staphylococcus aureus
Fluka   Lab

Aldolase ; Origin: Staphylococcus carnosus
Fluka   Lab

Fructose 1,6-bisphosphate Aldolase; Origin: Staphylococcus carnosus
Jülich Enzyme Products   Lab

Aldolase; Origin: Thermus aquaticus
Fluka   Lab

2-dehydro-3-deoxyphosphogalactonate aldolase.   **4.1.2.21**
6-phospho-2-dehydro-3-deoxygalactonate aldolase.   2-dehydro-3-deoxy-D-galactonate 6-phosphate = pyruvate
6-phospho-2-keto-3-deoxygalactonate aldolase.   +D-glyceraldehyde 3-phosphate.
2-oxo-3-deoxygalactonate 6-phosphate aldolase.

6-Phospho-2-dehydro-3-deoxygalactonate aldolase
Fluka   Lab

Dihydroneopterin aldolase.   **4.1.2.25**
                       2-amino-4-hydroxy-6-(D-erythro-1,2,3-trihydroxypropyl)-7,8-dihydr
                                                                        opteridine =
                       2-amino-4-hydroxy-6-hydroxymethyl-7,8-dihydropteridine +

Dihydroneopterin Aldolase
Fluka   Lab

Hydroxynitrilase.   **4.1.2.39**
Hydroxynitrile lyase. Oxynitrilase.   2-hydroxyisobutyronitrile = cyanide + acetone.

Hydroxynitrile Lyase; Origin: Hevea brasiliensis, rec. Pichia sp.
Roche Diagnostics: Hydroxynitrile Lyase (HNL)   Industrial

N-acetylneuraminate lyase.   **4.1.3.3**
N-acetylneuraminic acid aldolase.   N-acetylneuraminate = N-acetyl-D-mannosamine + pyruvate.

N-Acetyl-neuraminic acid aldolase; Origin: E. coli
Fluka   Lab

Neuraminic Acid Aldolase; Origin: E. coli K12
Jülich Enzyme Products   Lab

N-Acetylneuraminic Acid Aldolase; Origin: microorganisms
Toyobo   Pilot

Table 20.5. (cont.).

N-Acetylneuraminic Acid Aldolase; Origin: microorganisms
Unitika: N-Acetylneuraminic Acid Aldolase (Nana-Ald) — Pilot

Citrate lyase. — 4.1.3.6
Citrase. Citratase. Citritase. Citridesmolase. — Citrate = acetate + oxaloacetate.

Citrate Lyase; Origin: Enterobacter aerogenes
Fluka — Lab

4-hydroxy-2-oxoglutarate aldolase. — 4.1.3.16
2-keto-4-hydroxyglutarate aldolase. 2-oxo-4-hydroxyglutarate — 4-hydroxy-2-oxoglutarate = pyruvate + glyoxylate.
aldolase. KHG-aldolase.

4-Hydroxy-2-oxoglutarate aldolase ; Origin: E.coli
Fluka — Lab

4-hydroxy-4-methyl-2-oxoglutarate aldolase. — 4.1.3.17
4-hydroxy-4-methyl-2-oxoglutarate = 2 pyruvate.

4-Hydroxy-4-methyl-2-oxoglutarate aldolase
Fluka — Lab

Tryptophanase. — 4.1.99.1
L-tryptophan indole-lyase. Tnase. — L-tryptophan + H(2)O = indole + pyruvate + NH(3).

Tryptophanase; Origin: microorganisms
Unitika: Tryptophanase (Trp) — Pilot

Tyrosine phenol-lyase. — 4.1.99.2
Beta-tyrosinase. — L-tyrosine + H(2)O = phenol + pyruvate + NH(3).

beta-Tyrosinase; Origin: microorganisms
Unitika: beta-Tyrosinase (Bty) — Pilot

Carbonate dehydratase. — 4.2.1.1
Carbonic dehydratase. Carbonic anhydrase. — H(2)CO(3) = CO(2) + H(2)O.

Carbonic anhydrase; Origin: Bbvine erythocytes
Biozyme — Pilot

Carbonic Anhydrase; Origin: bovine erythrocytes
Fluka — Lab

Carbonic Anhydrase Isozyme II; Origin: bovine erythrocytes
Fluka — Lab

Chondroitin ABC lyase. — 4.2.2.4
Chondroitinase. Chondroitin ABC eliminase. — Eliminative degradation of polysaccharides containing
1,4-beta-D-hexosaminyl and 1,3-beta-D-glucuronosyl or
1,3-alpha-L-iduronosyl linkages to disaccharides containing

Chondroitinase ABC; Origin: Proteus vulgaris
Fluka — Lab

Pectin lyase. — 4.2.2.10
Eliminative cleavage of pectin to give oligosaccharides with
terminal 4-deoxy-6-methyl-alpha-D-galact-4-enuronosyl groups.

Pectolyase (EC 3.2.1.15 & 4.2.2.10); Origin: Aspergillus japonicus
Fluka — Industrial

Phenylalanine ammonia-lyase. — 4.3.1.5
L-phenylalanine = trans-cinnamate + NH(3).

Phenylalanine Deaminase; Origin: Rhodotorula glutinis
Fluka — Lab

Table 20.5. (cont.).

L-3-cyanoalanine synthase.	**4.4.1.9**
	L-cysteine + cyanide = H(2)S + L-3-cyanoalanine.

beta-Cyanoalanine Synthase; Origin: microorganisms
Unitika: beta-Cyanoalanine Synthase (Bcs)
Pilot

Alanine racemase.	**5.1.1.1**
	L-alanine = D-alanine.

Alanine Racemase; Origin: Bacillus stearothermophilus
Unitika: Alanine Racemase (AlaR)
Pilot

Aldose 1-epimerase.	**5.1.3.3**
Mutarotase. Aldose mutarotase.	Alpha-D-glucose = beta-D-glucose.

Mutarotase; Origin: hog kidney
Amano: Amano 2 [MUT-2]
Pilot

Mutarotase; Origin: pig kidney
Biozyme
Pilot

Triosephosphate isomerase.	**5.3.1.1**
Triosephosphate mutase. Phosphotriose isomerase.	D-glyceraldehyde 3-phosphate = glycerone phosphate.

Triosephosphate isomerase; Origin: rabbit muscle
Biozyme
Pilot

Xylose isomerase.	**5.3.1.5**
	D-xylose = D-xylulose.

Glucose isomerase; Origin: microorganisms
Novozymes: Sweetzyme®
Industrial

Glucose-6-phosphate isomerase.	**5.3.1.9**
Phosphoglucose isomerase. Phosphohexose isomerase. Phosphohexomutase. Oxoisomerase.	D-glucose 6-phosphate = D-fructose 6-phosphate.

Phosphoglucose Isomerase; Origin: Bacillus stearothermophilus
Unitika: Phosphoglucose Isomerase (PGI)
Pilot

Phosphoglucose Isomerase; Origin: baker's yeast
Fluka
Pilot

Protein disulfide isomerase.	**5.3.4.1**
S-S rearrangase.	Rearrangement of both intrachain and interchain disulfide bonds inproteins to form the native structures.

Protein disulfide Isomerase; Origin: bovine liver
Fluka
Lab

Protein disulfide Isomerase; Origin: E. coli
Fluka
Lab

Phosphoglucomutase.	**5.4.2.2**
Glucose phosphomutase. Phosphoglucose mutase.	Alpha-D-glucose 1-phosphate = alpha-D-glucose 6-phosphate.

Phosphoglucomutase; Origin: rabbit muscle
Fluka
Lab

Long-chain-fatty-acid--CoA ligase.	**6.2.1.3**
Acyl-activating enzyme. Acyl-CoA synthetase. Fatty acid thiokinase (long-chain). Lignoceroyl-CoA synthase.	ATP + a long-chain carboxylic acid + CoA = AMP + diphosphate + anacyl-CoA.

Acyl-CoA Synthetase; Origin: microorganisms
Asahi
Pilot

Table 20.5. (cont.).

Acyl-CoA Synthetase; Origin: Pseudomonas sp.
Amano: Amano 2 [ACS-2]

Pilot

Acyl-CoA Synthetase; Origin: Pseudomonas sp.
Amano: Amano 3 [ACS-3]

Pilot

Acyl-coenzyme A Synthetase; Origin: Pseudomonas sp.
Fluka

Lab

Glutamate--ammonia ligase.

6.3.1.2

Glutamine synthetase.

ATP + L-glutamate + NH(3) = ADP + phosphate + L-glutamine.

Glutamine Synthetase; Origin: Bacillus stearothermophilus
Unitika: Glutamine Synthetase (GS)

Lab

NAD(+) synthase.

6.3.1.5

NAD(+) synthetase.

ATP + deamido-NAD(+) + NH(3) = AMP + diphosphate + NAD(+).

NAD Synthetase
Asahi

Pilot

Urea carboxylase.

6.3.4.6

Urease (ATP-hydrolysing). Urea carboxylase (hydrolysing). ATP--urea amidolyase. Urea amido-lyase.

ATP + urea + CO(2) = ADP + phosphate + urea-1-carboxylate.

Urea Amidolyase; Origin: yeast
Toyobo

Pilot

Index